Frozen Soil Engineering in Cold Regions

Qingbai Wu • Changjiang Tong

Frozen Soil Engineering in Cold Regions

Springer

经典中国国际出版工程
China Classics International

Qingbai Wu
State Key Laboratory of Cryospheric Science and Frozen Soil Engineering, Northwest Institute of Eco-Environment and Resources
Chinese Academy of Sciences
Lanzhou, Gansu, China

Changjiang Tong
State Key Lab of Cryospheric Science and Frozen Soil Engineering, Northwest Institute of Eco-Environment and Resources
Chinese Academy of Sciences
Lanzhou, Gansu, China

Translated by
Hongxia Gao
School of Foreign Languages and Literatures
Lanzhou University
Lanzhou, China

ISBN 978-981-95-2526-3   ISBN 978-981-95-2527-0 (eBook)
https://doi.org/10.1007/978-981-95-2527-0

Jointly co-published with Lanzhou University Press Co., Ltd., Lanzhou, China
The print edition is not for sale in the mainland of China. Customers from the mainland of China please order the print book from: Lanzhou University Press Co., Ltd.

Translation from the Chinese language edition: "Hanqu Dongtu Gongcheng" by Qingbai Wu and Changjiang Tong, © Lanzhou University Press Co., Ltd 2020. Published by Lanzhou University Press Co., Ltd. All Rights Reserved.

This Springer imprint is published by the registered company Springer Nature Singapore Pte Ltd.
The registered company address is: 152 Beach Road, #21-01/04 Gateway East, Singapore 189721, Singapore

Preface

Permafrost engineering in cold regions is an important component of frozen soil science and a crucial part of engineering geocryology. This book, titled *Frozen Soil Engineering in Cold Regions*, does not address the interaction theory or research methodology between permafrost and engineering buildings; instead, it solely focuses on the concepts and applications of geotechnical engineering in engineering geocryology.

Over the past decades, numerous cases of frost damage and frozen soil engineering issues have arisen in engineering construction in frozen soil areas. Simultaneously, numerous accomplishments in the field of frozen soil research, governance experience, and engineering technology have been achieved. To obtain more comprehensive experience, all of these need to be examined and summarized.

This book briefly describes the distribution of permafrost and its engineering properties in China, which is the basic content that should be known in permafrost engineering in cold regions. More in-depth explanations of permafrost engineering in cold regions can be found in *Chinese Permafrost* and *Mechanics of Frozen Soil*. With the development of China's economy, more expertise is required in the design, construction, and operation of diverse engineering projects in permafrost regions (seasonally frozen soil areas and permafrost regions). The book summarizes the research accomplishments, experiences, and lessons of frozen soil scientific research and engineering technicians working in a variety of industries for decades in engineering investigation, design, construction, and engineering disease treatment, as well as a summary of the results and experiences of the authors participating in the research. What's more, the research achievements and experience of the construction of the Qinghai-Tibet highway and railway are summarized. The purpose is to provide scientific researchers and engineering technicians with reference materials for design, construction, and operation of engineering projects in permafrost regions.

In permafrost regions, the expansion of frozen soil in frozen soil foundations in the freezing process and the thaw settlement of ground ice in frozen soil foundation during the thawing phase are the primary causes of engineered construction diseases. The thermal and mechanical interaction between engineering buildings and permafrost causes the change of foundation soil temperature (positive and negative temperature) and moisture (water and ice), leading to the deformation and the instability of the buildings. When using frozen soil as a foundation, the design and construction of engineering buildings should prioritize the preservation of the permafrost environment, the temperature, and the ice content of frozen soil. In order to maintain the stability of buildings in cold regions, it is recommended to make full use of atmospheric cold energy, adopt physical and chemical methods of various effective engineering measures, and protect the surrounding frozen soil environment to minimize or completely eliminate changes in the frozen soil foundation state that may occur during the interaction between engineering and frozen soil.

For the purpose of practicality, the book discusses in detail the selection of lines, principles for the utilization of frozen soil foundations, foundation design, construction, foundation treatment, and frost prevention measures in construction projects. It also covers road (roadbed) engineering, bridge engineering, culvert engineering, water conservancy engineering, pipeline engineering, overhead transmission line engineering, and airport construction in permafrost regions.

I would like to express my heartfelt thanks to my colleagues in various units in terms of data and literature, and the State Key Laboratory of Frozen Soil Engineering of the Chinese Academy of Sciences for its assistance in authoring this book.

Lanzhou, Gansu, China
October 2019

Qingbai Wu
Changjiang Tong

Competing Interests The authors have no competing interests to declare that are relevant to the content of this manuscript.

Contents

Chapter 1
Permafrost and Engineering Property

1.1 Permafrost

Frozen soil refers to rock-soil or soil containing ice at temperatures of 0 °C or lower. Soil that does not contain ice is classified as cryotic ground. Cryotic ground can be subdivided into dry permafrost, which lacks ice and gravitational water, and cryopeg, which contains no ice but rather saline brine at sub-zero temperatures. In natural settings, frozen soil layers and entire frozen soil regions encompass both frozen soil and cryotic ground. Rock-soil at temperatures above 0 °C is designated as non-cryotic ground, whereas thawed ground refers to previously frozen rock-soil in which all ice has completely melted.

As a foundation material, the engineering geological properties of frozen soil essentially differ from that of cryotic ground. Due to the cement pore ice and ice, the change in temperature, moisture, and stress within frozen soil (rock) can directly change its physical and mechanical properties, including its bearing capacity and thaw settlement, as well as its engineering geological properties. Frozen soil has the concrete strength, as well as the properties of soft soil and sludge (after thawing), but it is not the case for cryotic ground.

China is one of the countries in the world where permafrost is mainly distributed, with the areas of 2.15 million km^2 (Zhou et al., 2000). According to *1:4,000,000 Map of the Glaciers, Frozen soil and Deserts in China* (Wang et al., 2006a, b, c), permafrost degradation causes the shrinking of its areas to 1.7539 million km^2, accounting for 22.4% of China's land areas. The permafrost in China is mainly distributed in the Da Xing'anling Mountains and Xiaoxing'anling Mountains, and the Northern Songnen Plain, Qinghai-Xizang Plateau and high-mountains in the west.

Q. Wu, C. Tong, *Frozen Soil Engineering in Cold Regions*,
https://doi.org/10.1007/978-981-95-2527-0_1

1.1.1 Permafrost Regions in Daxing'anling and Xiaoxing'anling Mountains

1.1.1.1 Permafrost Distributions

The permafrost regions in Daxing'anling and Xiaoxing'anling Mountains as well as the Songnen Plain, situated at high latitudes, constitute an integral component of the Eurasian permafrost system.

The southern boundary of permafrost extends to 46°30′N in the west and 47°48′N in the east, forming a "W"shape. The southern limit fluctuates around the mean annual temperature isoline between −1. 0 °C and 1.0 °C, specifically ranging from −1.0 °C to 0 °C in the western region and from 0 °C to 1.0 °C in the eastern region. In the central section (Songnen Plain and the southeastern and southwestern foothills of Daxing'anling and Xiaoxing'anling Mountains), this limit is approximately aligned with the 0 °C isotherm (Fig. 1.1) (Zhou et al., 2000). The distribution of permafrost is predominantly influenced by latitudinal zonality.

In recent years, several studies and meteorological data analyses have indicated a significant northward shift in the southern boundary of the permafrost region within Daxing'anling and Xiaoxing'anling Mountains. Investigations conducted between 2008 and 2013 revealed the presence of permafrost in various locations, including Yakeshi City, Manzhouli City, the Yichun-Wuying-Tangwang River area, and the Yichun to Wudalianchi section. Additionally, permafrost was found beneath certain wheat fields in the outskirts of Yakeshi, with a permafrost table depth of 8 m.

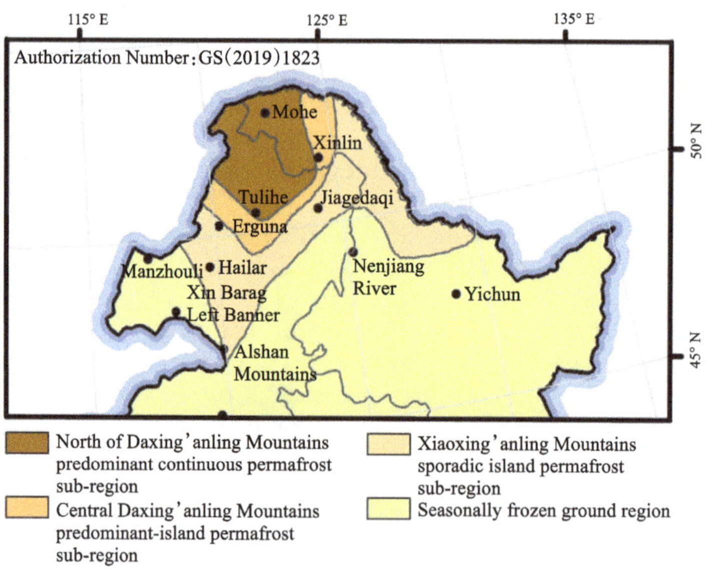

Fig. 1.1 Permafrost distributions in Daxing'anling and Xiaoxing'anling Mountains (Zhou et al., 2000)

These findings suggest that permafrost persists in regions such as wetlands, marshes, shaded slopes, low-lying areas, and peat layers along the mountains, even in areas significantly impacted by global warming and human activities. Notably, these locations still fall within the southern limit of permafrost as defined in the 1970s and 1980s.

Permafrost in Daxing'anling and Xiaoxing'anling mountains can be categorized into three distinct regions (Fig. 1.1). Northern Daxing'anling Mountian predominant continuous permafrost subregion are bounded by the Erguna River and Heilongjiang River to the west and north, extends along the Moerdaoga, Deerbuer, and Tuli River to the southwest, and reaches Mohe along the eastern ridgeline of the mountains. The mean annual air temperature here is generally below −5 °C, while the mean annual ground temperature of frozen soil typically ranges from −1 °C to −2.5 °C, with occasional readings as low as −4 °C. Permafrost thickness usually varies between 60 m and 100 m but can reach up to 130 m in certain areas. The continuity of permafrost distribution within this region ranges from 65% to 75%.

Daxing'anling Mountains predominance-island permafrost sub-region is predominantly situated on the peripheral margins of discontinuous permafrost zones. Specifically, this region encompasses areas such as the Sanhe River and Yituli River basins in the southwest, as well as along the Keyi River, in the Songling District (Xiaoyangqi Town), and Shibazhan Town in the northeast. The mean annual air temperature typically ranges from −3 °C to −5 °C, while the mean annual ground temperature of the frozen soil generally falls between −0.5 °C and −1.5 °C. Permafrost thickness typically varies between 20 m and 60 m, with a distribution continuity ranging from 30% to 65%.

Xiaoxing'anling Mountains sporadic island permafrost sub-region is located in the northern part of the southern limit of permafrost distribution. This region extends from the western bank of Hulun Lake in the west to Wuyiling in the east. Permafrost coverage ranges from approximately 5–20%. Island permafrost is commonly found around wetlands and limnetic zones, as well as in fine-grained soils on low-lying shady slopes. The mean annual air temperature typically varies between –0.5 °C and 2 °C, while the mean annual ground temperature of frozen soil generally ranges from 0 °C to −1 °C. Permafrost thickness predominantly measures between 5 m and 10 m, although it can reach depths of 10–20 m in certain areas. The continuity of permafrost distribution ranges from 5% to 30%.

1.1.1.2 Permafrost Thickness and Temperature

The upper boundary surface of the permafrost layer is referred to as the permafrost table, while the lower boundary surface is designated as the permafrost base. The vertical distance between these two boundary surfaces defines the permafrost thickness (Fig. 1.2).

The permafrost thickness in Daxing'anling and Xiaoxing'anling Mountains is primarily governed by latitudinal zonality. As latitude increases from south to north, the extent, thickness, and continuity of permafrost also increase, while the mean

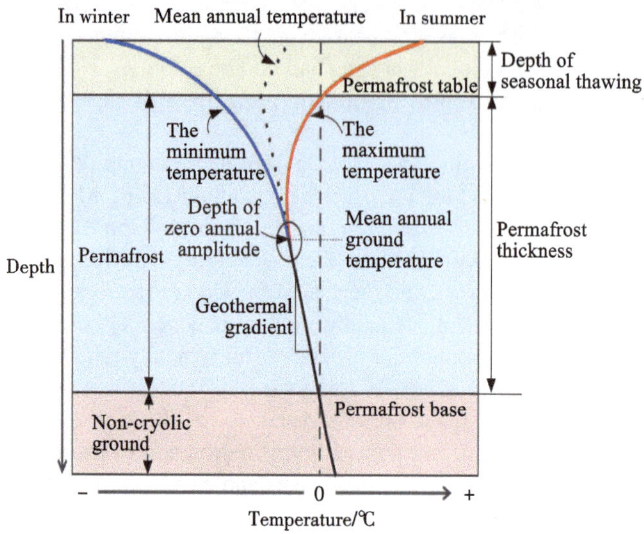

Fig. 1.2 Terms used to describe the ground temperature relative to 0 °C in a permafrost environment

annual ground temperature of permafrost decreases. Near the southern boundary of permafrost distribution, the thickness typically ranges from a few meters to approximately 10 m. In contrast, measurements indicate that maximum permafrost thickness can reach 120 m ~ 130 m in northern regions, exemplified by the Gulian coalmine area. Within the discontinuous permafrost zone between the northern and southern boundaries, permafrost thickness generally varies between 40 m and 70 m in discontinuous permafrost regions. However, local geological and geographical factors, including geomorphology, lithology, geological structure, surface water conditions, and slope aspect, can influence permafrost thickness, sometimes deviating from the general latitudinal zonality. The deepest permafrost layers are usually located in marshy areas within basin centers, characterized by dense vegetation cover and thick peat soil layers. Permafrost on shady slopes tends to be thicker, whereas it is absent or minimal on sunny slopes and mountain tops. Human activities have led to a decline in the permafrost table, and most permafrost appear in suburban areas while they only disappear in urban centers.

The mean annual ground temperature of the permafrost in Daxing'anling Mountains generally decreases with increasing latitude from south to north, ranging from −1.0 °C to −2.5 °C in the northern regions and reaching a minimum of −4 °C. In the southern boundary, temperatures range from approximately −0.1 °C to −0.8 °C. The mean annual ground temperature decreases by approximately 0.5 °C for every 100 km moving northward. However, due to local factors such as vegetation, lithology, surface water, topography, and slope aspect, some areas exhibit significant deviations from this latitudinal zonality. The mean annual ground temperature in wetlands is lower than that in mountain terrain. Altitude can directly affect the mean annual ground temperature, for instance, on the eastern slope of Yilehuli Mountain within Daxing'anling Mountains, the temperature changes from

−2.0 °C to −2.6 °C. Under the influence of human activities and rivers, the permafrost temperature has increased significantly, for example, from −0.4 °C to −0.7 °C in the Tahe area.

1.1.1.3 Permafrost Table

The depth of seasonally thawing (maximum seasonal thawing depth) in Daxing'anling Mountains exhibits a gradual increase from north to south. In ravine, lowland, and piedmont gentle slope areas within permafrost regions, vegetation flourishes. These areas are characterized by soil layers rich in fine-grained material with high moisture content, resulting in a shallow permafrost table typically ranging from 1 m to 2 m. The permafrost table varies generally from 0.4 m to 1.0 m in the marshes, wetlands and lakeside areas with high vegetation cover and thick peat layers, where massive thickly layered ground ice can be found at a certain depth below the natural permafrost table. (Fig. 1.3 and Table 1.1).

Fig. 1.3 Ground ice layer under the permafrost table (Provided by Tong Changjiang)

Table 1.1 Maximum seasonal thawing depth of different latitudes in the Daxing'anling Mountains

	Mohe	Amuer	Mangui	Niuerhe	Tulihe	Yakeshi
Latitude	53°05′	52°50′	52°02′	51°32′	50°30′	49°24′
Soil properties	Peat, silty clay				Humus silt clay	
Permafrost table (m)	0.8 to 1.1	1.3 to 1.5	1.1 to 1.9			

1.1.2 Permafrost Regions on Qinghai-Xizang Plateau

1.1.2.1 Permafrost Regions in Qilian Mountains

1) **Permafrost Distribution**

The distribution of permafrost is predominantly influenced by altitude, latitude, and geographic location. Specifically, on the southern Qilian Mountains, the lower limit of permafrost distribution is 3700 m ~ 3950 m on the southern slopes along Laji Mountains-Qinghai Nanshan-Chaidamu Mountains, coincident with the mean annual temperature isoline of −2 °C. On the northern Qilian Mountains, the lower boundary of permafrost distribution is 3450 m ~ 3650 m on the northern slopes along Lenglongling-Zoulangnanshan-Danghenanshan, coinciding with the mean annual temperature isoline of −2 °C. In contrast, on the northern Qilian Mountains, the lower boundary of permafrost distribution is 3450 m ~ 3650 m on the northern slopes along Lenglongling-Zoulangnanshan- Danghenanshan, which aligns approximately with the mean annual temperature isoline of −2.5 °C (Fig. 1.4) (Guo, 1983; Zhou et al., 2000).

Overall, the lower limit of permafrost distribution is lower in the east than that in the west. In the central section, continuous permafrost prevails due to the high altitude, covering approximately 80% to 90% of the area. Conversely, the eastern and western sections consist of discontinuous permafrost zones, comprising roughly 20% to 30% of the total area.

2) **Permafrost Thickness and Temperature**

The permafrost thickness increases with the altitude increasing in Qilian Mountains, with the maximum of 139.3 m while its mean annual ground temperature decreases correspondingly (Fig. 1.5) (Zhou et al., 2000). In the altitude range from 3550 m to 3700 m, permafrost changes slightly in its thickness and mean annual ground temperature; the ground temperature decreases by 0.2 °C at most for every 100 m elevation in altitude, with the increase of less than 8 m in permafrost thickness. In the altitude range from 3700 m ~ 4200 m, permafrost changes greatly in its thickness and mean annual ground temperature; the ground temperature decreases by 0.4 °C to 1.1 °C for every 100 m in altitude, with the increases of 10.9 m ~ 42.5 m in permafrost thickness. The permafrost thickness increases with its decrease in ground temperature, with the increase by 31.9 m for every decrease of 1 °C in ground temperature. The mean annual ground temperature changes from −1.5 °C to −2.4 °C and the permafrost thickness varies from 50 m to 139 m in the continuous permafrost regions in Qilian Mountains. The mean annual ground temperature is generally 0 °C to −1.5 °C and permafrost thickness varies from a few meters to tens of meters with the range from 3 m to 50 m due to the height of topography in the patchy permafrost regions.

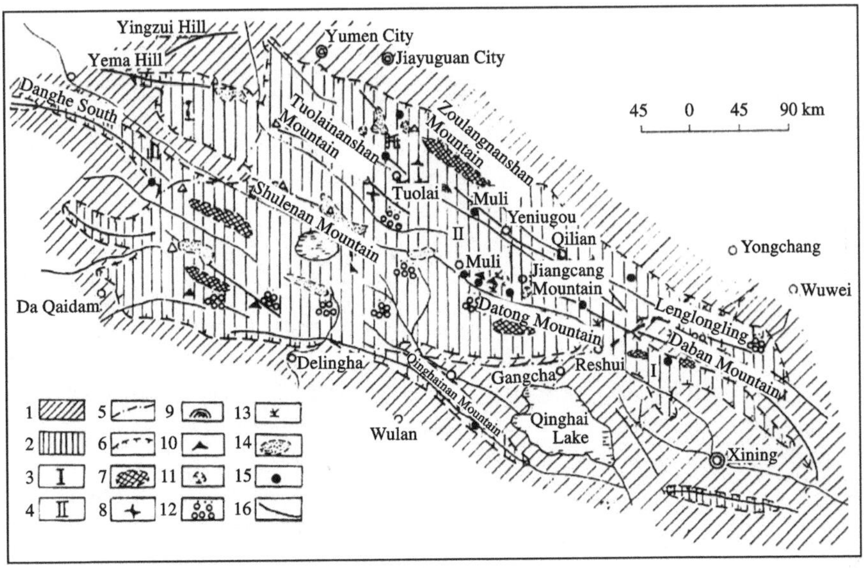

1. Seasonally frozen soil
2. Permafrost
3. Patchy permafrost regions
4. Predominantly continuous permafrost regions
5. The boundary line between 3 and 4
6. The boundary line of permafrost distribution
7. Cryoplanation surface
8. Frost mound
9. Debris flow
10. Icing
11. Thaw depression
12. Debris slope
13. Wetlands
14. Modern glacier
15. Borehole
16. Ridge line

Fig. 1.4 Permafrost distribution map in Qilian Mountains

3) **Permafrost Table**

Under natural conditions, the depth of permafrost table is 1.0 m ~ 4.5 m. The permafrost table generally declines 1 m ~ 3 m under the highway subgrade (with asphalt pavement), and 4 m ~ 6 m beneath the heating houses. As for the open-pit mining zones, the depth of permafrost table may be more than 10 m below the exposed surface or the whole permafrost layer thaws entirely.

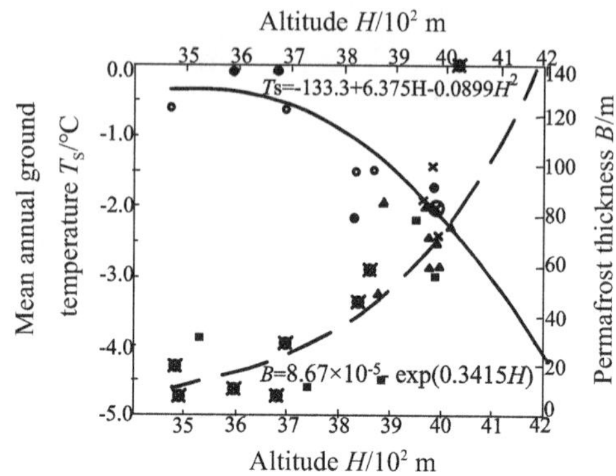

Fig. 1.5 Relationships among altitude and the mean annual ground temperature and permafrost thickness in Qilian Mountains

Regions with a mean annual temperature exceeding −2.5 °C predominantly feature deep seasonally frozen soil, and the maximum seasonal freezing depth can be 3 m to 4 m.

4) **Ground Ice**

In the central regions of marshes, there are more extensive ground ice layers and thicker permafrost compared to the peripheral areas. In the Qilian Mountains, the ground ice layer within permafrost regions is predominantly situated beneath the permafrost table, with a thickness ranging from 2 m to 3 m and can extend up to 7 m to 9 m in certain locations. Additionally, buried ice exists within glacial till at elevations above 4000 m.

1.1.2.2 Permafrost Regions in the East of Qinghai-Xizang Plateau

1) **Permafrost Distributions**

The permafrost distribution in the eastern Qinghai-Xizang Plateau is primarily influenced by altitude. As elevation increases, temperatures decrease, leading to more favorable conditions for permafrost development. The permafrost zones evolve from sporadic patches in the southeast to continuous sheets in the northwest (Fig. 1.6) (Luo et al., 2012). There are strip and belt-shape taliks in the predominant contribution permafrost regions, mostly near rivers, lakes and tectonic fault zones. Due to topographic and geological factors, the distribution of permafrost is highly complex. The lower limit of permafrost distribution ranges from 3840 m to 4380 m

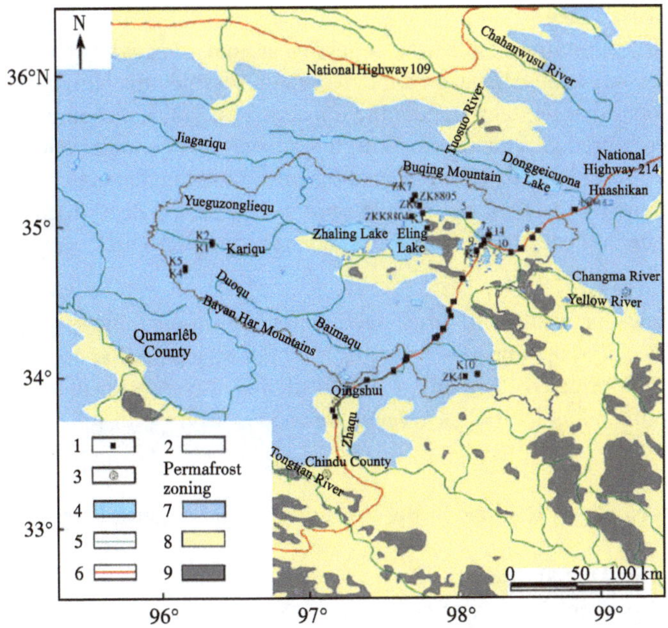

1.Borehole; 2.The source area of the Yellow River; 3.Towns; 4.Lakes; 5.Rivers; 6.Highways; 7.Predominant continuous permafrost; 8.Discontinuous permafrost; 9.Spordic island discontinuous permafrost

Fig. 1.6 Permafrost distribution map in the east of Qinghai-Xizang Plateau

where its mean annual air temperature is about −2.5 °C to −3.5 °C. Permafrost will degrade under global warming, and the lower limit of permafrost distribution will rise by 50 m to 100 m.

2) **Ground Temperature and Permafrost Thickness**

The relationship between altitude and permafrost development is such that higher altitudes promote better permafrost formation and increased thickness. Specifically, for every 100 m of elevation gain, the permafrost thickness increases by approximately 10–15 m. The mean annual ground temperature is also related to its altitude; the mean annual ground temperature of permafrost in mountains is lower, ranging from −1.0 °C to −2.0 °C, which belongs to the basically stable cold permafrost, while the mean annual ground temperature in plain is higher, ranging −0.5 °C to −1.0 °C, which belongs to the unstable warm permafrost zone. Permafrost has degraded obviously under the impact of climate warming. The patchy permafrost regions degraded into seasonally frozen soil regions with the sporadic island permafrost in the areas of Yeniugou and Huangheyan, and the deep buried permafrost and detachment of permafrost appear in the discontinuous permafrost regions.

3) **Permafrost Table**

In the 1980s, permafrost table was generally around 2 m, with the minimum of 1.3 m in marshes and wetlands and the maximum seasonal thawing depth of 2.5 m in sunny slopes in mountains. Under the impacting of global warming, the seasonal thawing depth gradually increases, with the greater change under the influence of human activities. Generally, the seasonal thawing depth increases to more than 2 m with maximum of 4–8 m at most while seasonally freezing depth reduces to about 2 m under the influence of climate warming and human activities.

4) **Ground Ice**

In the bedrock or weathering crust sections in high mountains, permafrost with gravel soil mostly contains low ice content, including ice-poor and icy frozen soil. In plain or platforms areas, permafrost with silt and silty clay mostly contains high ice content, including ice-rich, saturated-ice and ice layers with soil.

1.1.2.3 Permafrost Regions in the West of Qinghai-Xizang Plateau

1) **Permafrost Distribution**

The permafrost regions in western Qinghai-Xizang Plateau specifically encompass the Southern Qinghai-Northern Xizang Plateau, which includes southwestern Qinghai and northern Xizang. This area represents the largest permafrost region on the plateau in China, extending from 31°41′N to 35°45′N. The northern boundary of the permafrost is located at Xidatan Town at the northern foot of the Kunlun Mountains, with its lower boundary situated at an altitude of approximately 4150 m. The permafrost changes from patchy discontinuous into predominantly continuous permafrost regions, with a difference of about 200 m in elevation and a width of 550 km from north to south. The predominantly continuous permafrost regions turn into patchy discontinuous permafrost regions, with its distribution lower limit of about 4640 m ~ 4700 m from the south of Northern Anduo County to the valleys on the Zangbei Plateau where is the south limit of permafrost distribution on plateau, with a width of more than 200 km from south to north. The elevation at the lower limit of patchy discontinous permafrost is roughly equivalent to the mean annual air temperature isoline of −2 °C to −3 °C. There are different types of taliks in the predominantly continuous permafrost regions in Qinghai-Xizang Plateau, including river taliks, tectonic taliks, radiation taliks, and lake taliks.

2) **Permafrost Thickness and Mean Annual Ground Temperature**

In patchy discontinuous permafrost regions, the mean annual ground temperature is relatively higher, generally ranging from 0 °C to −0.5 °C and the permafrost thickness varies from several meters to 20 m. In the predominantly continuous permafrost, the mean annual ground temperature is relative lower, ranging from −1.0 °C to −4.0 °C in the high mountains and hilly areas and −0.5 °C to −2.0 °C in high plains, and −0.1 °C to −1.0 °C in valleys. The permafrost thickness varies from 30 m to 130 m in the high mountains and hilly areas, 20 m to 100 m in high-plain, and several meters to a dozen meters with its thickness of no more than 50 m in valleys.

Under the impact of climate warming, permafrost degraded significantly, with permafrost warming and seasonal thawing depth increasing but seasonal freezing depth decreasing while permafrost becomes thinner or even disappears. In patchy discontinuous permafrost regions, the mean annual air temperature is generally higher than −0.5 °C in the edges of taliks and sporadic patchy permafrost regions, and the degrading of permafrost both in horizontal and vertical directions is more significant. Degrading permafrost temperature curve are mainly found at the edges of continuous and discontinuous permafrost regions, such as valleys and basins on high plains, and in patchy permafrost regions, with the mean annual ground temperature higher than −1.5 °C and permafrost warming and thaw-sandwich occur in upper permafrost. Stable permafrost temperature curve can be found in continuous permafrost regions in high mountains and hilly areas, with the mean annual ground temperature lower than −1.5 °C, and with permafrost thickness thinning and permafrost warming within the depth of zero annual amplitude.

3) **Permafrost Table**

Permafrost table decreases with the increase of latitude and altitude and the higher the altitude is, the smaller permafrost table becomes. The maximum seasonal thawing depth for fine-grained soil varies generally from 0.8 m to 2.5 m and from 3 m to 5 m at the hilltops and slopes of exposed bedrock respectively. Moreover, permafrost table ranges from 2 m to 5 m in the high plain, basin and valley but 0.3 m to 0.8 m beneath the well-developed vegetation cover. The permafrost table in shady slope is smaller than that in sunny slopes, ranging from 0.1 m to 1.0 m due to soil properties and soil water content (Table 1.2). In the talik of flood plain, the maximum seasonal freezing depth varies from 4 m to 6 m, for instance, ranging from 4 m to 7 m in Xidatan and 3–4 m in high plains areas. Under the impact of global warming, seasonal freezing depth will decrease and seasonal thawing depth will increase. The thaw-sandwich will appear in the upper permafrost, especially at the edges of taliks.

4) **Ground Ice**

A substantial quantity of ground ice is embedded near the permafrost table, with the thickness of 3–5 m, especially within the grey-green silty clay with intensely weathered mudstones formed by lacustrine sediments, and often forming ground ice layer with medium and thickness layered, including ice-rich frozen soil, saturated-ice frozen soil and ice layer containing soil. Ice-poor and icy frozen soil can mostly found in the sections with sandy gravel. The genetic types of Quaternary soil, vegetation cover and slope gradient influence greatly on ground ice formation (Table 1.3).

Table 1.2 Permafrost table along the Qinghai-Xizang Highway (m)

Clay	Silty clay	Silt	Sand	Gravel	Gravel soil
0.8 to 2.5	1.0 to 3.6	1.2 to 3.2	1.5 to 3.5	2.0 to 4.3	2.0 to 4.5

Table 1.3 Proportion of frozen soil types for different Quaternary origin along the Qinghai-Xizang Highway

Frozen soil types*	Quaternary origin types						
	Lacustr. sediments	Slope washes	Resid. soil and slope washes	Pluvial sediments	Pluv. alluv. Sediments	Fluviological deposit	Alluvial sediments
H	52.2	34.7	30.7	21.9	8.9	3.9	5.1
B	23.5	20.5	22.7	23.9	25.8	12.0	0.5
F	24.3	12.7	2.3	20.0	7.6	15.5	–
D	–	7.6	44.3	7.0	9.9	12.2	0.4
S	–	24.5	–	16.9	17.9	51.4	–
TA	–	–	–	10.3	29.8	–	93.0

*H means ice layer containing soil, B means saturated-ice frozen soil, F means ice-rich frozen soil, D means icy frozen soil, S means ice-poor, and TA means thawed areas.

1.1.3 Permafrost Regions in High Mountains

The permafrost regions in high mountains primarily encompass the Altai and Tianshan mountain ranges in Xinjiang Province. Additionally, the Huanggangliang, Changbai Mountains in Northeast China, as well as the Taibai and Wutai Mountains in the Midwest of China are also classified as high mountain permafrost areas. However, due to a lack of comprehensive data, these latter regions are not discussed in this context.

1.1.3.1 The Altai Mountains Regions

The northern boundary of the Altai Mountains is located at 49°20′N, while the southern boundary is approximately at 46°N. The lower limit of patchy discontinuous permafrost is at the height of 2200 m ~ 2800 m where the mean annual temperature is −5.4 °C to −6.7 °C. The permafrost table ranges generally from 1 m to 2 m, the mean annual ground temperature is estimated to be 0 °C to −1 °C, and the permafrost thickness varies from about several meters to 19 m.

1.1.3.2 Tianshan Mountains Regions

Tianshan Mountains in China is 100 km ~ 400 km wide from north to south, about 40° ~ 45° N, 1700 km from east to west, spanning 21° longitude (74° ~ 95° E). The lower limit of permafrost distribution ranges from about 2700–2900 m on the north slope and 3100–3250 m on the south slope. Above this altitude, the permafrost distribution increases to be continuous permafrost. Due to the lithology, topography, slope aspect and ground coverage, patchy and strips of permafrost are greatly different in the mean annual ground temperature and permafrost thickness. The mean annual ground temperature ranges from −0.1 °C to −1.0 °C, and the permafrost table is about 2.5 m ~ 3.5 m and the permafrost thickness is about 16 m ~ 32 m.

Within the inter-bedded layer between humus soil and angular gravel soil, there is conglomeratic tectonic permafrost, with the volumetric ice content of 20–30% and some up to 50%, and they can be classified as ice-rich frozen soil.

With the increase of altitude, there are predominantly continuous permafrost regions, with the increase of permafrost areas the permafrost thickness. The mean annual ground temperature gradually decreases from −0.1 °C to −2.0 °C, and can reach −4.9 °C above 3900 m and its maximum permafrost thickness can reach about 230 m.

1.1.4 Engineering Problems of Frozen soil

The primary distinction between frozen and non-frozen foundations lies in the presence of ground ice and ice layers of varying thicknesses within frozen foundations. When in a frozen state, these soils typically exhibit enhanced strength and improved impermeability. Conversely, upon thawing, the soil's strength diminishes, potentially resulting in a weaker structure than its initial state. Moreover, repeated cycles of freezing and thawing (cryotic processes) can progressively degrade the foundation soil, leading to a range of engineering challenges.

1.1.4.1 Engineering Problems of Frozen Soil in Freezing Process

In both seasonally frozen soil areas and permafrost regions, the frost heave of foundation soils during the freezing process can cause significant damage to structures. Examples include frost-induced uplift of bridge piles (Fig. 1.7) (Chen et al., 2006), frost heave affecting channel lined boards (Fig. 1.8) (Wu & Liu, 2005), frost heave impacting oil pipelines (Fig. 1.9), and frost heave leading to cracks in buildings and walls.

Fig. 1.7 Uneven frost-heaving of bridges

Fig. 1.8 Frost-heaving damage of concrete channel line

Fig. 1.9 Frost-heaving of Golmud-Lhasa Oil Pipelines (Photo by Liu Yongzhi)

1.1.4.2 Frozen Soil Engineering Problems Caused by Thaw Settlement

The permafrost contains various forms of ground ice layers, which are classified as ice-poor, icy, ice-rich, saturated frozen soil and ice layers according to their ice contents, and their relative thaw settlements are less than 1%, 1–5%, 5–10%, 10–25% and more than 25% respectively. Engineering construction will change the thermodynamics balance of frozen foundation and cause thaw settlement and damages to buildings, such as subgrades (Fig. 1.10), houses (Fig. 1.11), culvert pipes (Fig. 1.12) and other common deformations.

Fig. 1.10 The settlement deformation of northeast Nenlin Railway (Photo by Li Qingwu)

Fig. 1.11 House settlements in Alaska (Provided by Jin Huijun)

1.1.4.3 Engineering Problems Caused by Repeated Freezing–Thawing

Annual freezing and thawing cycles induce frost heaving and thaw settlement, resulting in slope instability and building damages, such as road frost boiling (Fig. 1.13), gelifluction (Fig. 1.14), retrogressive thaw slumping.

Fig. 1.12 The settlement deformation of culverts in northeast Nenlin Railway (Provided by Li Qingwu)

Fig. 1.13 Road frost boiling in shallow layers (Photo by Tong Changjiang)

Fig. 1.14 Retrogressive thaw slump caused by slope excavation in Hongliang River Regions (Photo by Tong Changjiang)

1.2 Engineering Properties of Frozen Soil

Frozen soil constitutes a four-component system consisting of soil particle, ice, water and gas (vapor). In the process of freezing and thawing, complex physical and chemical reactions on the interface among the four components lead to change in composition and structure of frozen soil, resulting in its specific engineering properties.

1.2.1 Thermodynamic Equilibrium Between Unfrozen Water and Ice Within the Frozen Soil

The fundamental engineering properties of special frozen soil are intrinsically linked to the transformation between ice and unfrozen water. When water within the soil freezes, it significantly enhances the strength of cemented soil particles. Conversely, when ice within frozen soil thaws into water, it accumulates, leading to saturation and softening of the soil, thereby weakening or even eliminating the cohesion among soil particles and reducing its overall strength. The dynamic equilibrium between unfrozen water and ice in frozen soil indicates that the quantity, composition, and properties of unfrozen water and ice vary in response to external factors. The content of unfrozen water in frozen soil is notably influenced by negative temperatures and external pressure. With the temperature of frozen soil decreases, unfrozen water content decreases and the ice content increases (Fig. 1.15) (Tsytovich, 1985), so there are three-phase transition zones (Table 1.4). An increase in external pressure leads to a rise in unfrozen water content and a decrease in ice content.

Fig. 1.15 Relationships between the unfrozen water content and temperature in frozen soil

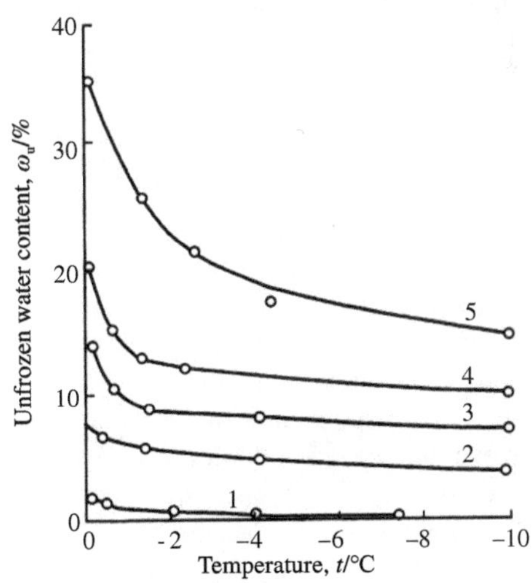

1. Quartz sand; 2. sand soil; 3. silty clay; 4. Clay; 5. Montmorillonite clay

Table 1.4 Temperature range of soil phase-change areas (°C)

Soil types	Extensive phase change areas	Passive phase change areas	Slowly phase change areas
Sand	0 to −0.2	−0.2 to −0.5	< −0.5
Silty Clay	0 to −2.0	−2.0 to −5.0	< −5.0
Clay	0 to −5.0	=5.0 to −10.0	< −10.0

The relationship between unfrozen water content and negative temperature in frozen soil can be described by the following equation (Xu et al., 2001):

$$\omega_u = at^{-b} \tag{1.1}$$

where ω_u represents the unfrozen water content in frozen soil (%); t denotes the absolute value of negative temperature (°C); a and b are empirical coefficients that are associated with soil properties.

It can be inferred that, for a given soil quality, the unfrozen water content in frozen soil decreases as the soil temperature drops, while the ice content increases as the negative temperature magnitude in frozen soil rises.

Experimental results indicate that as external pressure increases, the unfrozen water content in frozen soil rises while the ice content correspondingly decreases (Ding et al., 2011). This phenomenon fundamentally reflects the influence of external loading on the initial freezing temperature of soil.

It has been demonstrated that soil salinity significantly influences its freezing process and freezing temperature. Specifically, as soil salinity increases, the freezing temperature decreases while the unfrozen water content increases. Generally, the unfrozen water content in frozen soil varies with different types of salts in the following order: sulfate, carbonate, nitrate, and chloride. Moreover, a higher concentration of salt solution correlates with a greater unfrozen water content.

1.2.2 Soil Frost-Heaving

It has been demonstrated that soil salinity significantly influences its freezing process and freezing temperature. As the salinity increases, the freezing temperature decreases while the unfrozen water content rises. Generally, higher soil salinity leads to an increase in unfrozen water content during the freezing process. During this process, both inherent soil water and water migrating from external sources freeze into ice, transitioning from liquid to solid states, forming various ice structures such as ice layers, ice lenses, and polycrystalline ice. This transformation results in relative displacement among soil particles and varying degrees of expansion. And then after the segregated ice, the ice interlayer and ice lens will be precipitated on the interior, the soil layer on the subsurface will be heightened, hence resulting in soil frost heaving. It is usually expressed by the frost heaving amount (Δh, mm) and the frost heaving ratio (η, %).

Under natural conditions, the magnitude of frost heave in unidirectional freezing is:

$$\Delta h = h_2 - h_1 \tag{1.2}$$

The frost heaving ratio of the soil is as follows:

$$\eta = \frac{\Delta h}{h_f} \times 100\% \tag{1.3}$$

where Δh represents the increment of frost heave, measured in millimeters; h_1 and h_2 denote the soil heights before and after freezing, also measured in millimeters.

1.2.2.1 Influence Factors and Laws of Frost-Heaving of Soil

Water migration and ice formation during the freezing process can directly lead to soil frost heave. The severity of frost heave is influenced by multiple factors, including soil particle size distribution, mineral composition, water content and recharge sources within the soil, freezing conditions, external loading, and cation exchange capacity (Tong & Guan, 1985).

1) The granule composition of soils significantly influences frost heaving. The frost heaving of coarse-grained soil is generally smaller, but that of fine-grained soil is larger due to the water absorption increasing with the decrease of the soil particle size in the process of freezing. The silts (with the soil particles diameter of 0.005–0.2 mm) has the strongest frost heaving attributed to largest water absorption. The more silt clay in coarse-grained soil, the larger its frost heaving. When the coarse-grained soil with the content of silt clay exceeding 5% frost heaving will occur. Generally, the coarse-grained soil does not produce frost-heaving under the following two necessary conditions, one is that the content of silt clay is less than 5% (clay particle) to 12% (silt grain); two is that water can discharge freely in the frozen state.
2) The mineral compositions of soil particles. Montmorillonite exhibits a high ion exchange capacity and strong adsorption ability, enabling it to firmly retain a significant amount of water. Therefore, water migration reduced to form small frost heaving when freezing. Kaolinite has a strong connective force between crystal lattice, and the water molecules cannot be allowed to enter the unit cell, so it has small hydrophilicity and weak ion exchange capacity. As a result, kaolinite has a thick mobile water film on its surface, which contributes to pronounced frost heave. The properties of hydromica lie between those of montmorillonite and kaolinite.

3) Soil water content plays a critical role in frost heaving phenomena. Frost heaving initiates when the soil water content exceeds the threshold water content for initial frost heaving (ω_0). This phenomenon is primarily driven by the migration and accumulation of excess water during the freezing process. For fine-grained soils, under specific conditions, the threshold water content for initial frost heaving is approximately 0.8 times the plastic limit water content (ω_p). When the water content (ω) of fine-grained soils falls within the range of liquid limit water content (ω_L) plus 17%, it can be approximated using a linear equation:

$$\eta = 0.277\left(\omega - \omega_0\right) \tag{1.4}$$

In an open system with external water recharge, the rate of soil frost heave will increase significantly and will be influenced by the conditions and timing of the water recharge. The total frost heave amount (Δh) of the soil can be expressed as follows:

$$\Delta h = 0.09\omega \cdot \rho_d \cdot h_f \cdot i + 1.09\int_0^T U_\omega \mathrm{d}T \tag{1.5}$$

where ω is the water content of soil, %; ρ_d is the soil density, kg/ m^3; h_f is the freezing depth of soil, m; i is the freezing rate, %; U_ω is the rate of water migration, m/h; T is the time of water migration, h.

The groundwater level directly affects the frost heaving of soil. The shallower the groundwater table is, the stronger the frost heaving of soil is. The amount of precipitation in late autumn directly affects the frost heaving of soil. The influence of groundwater on soils frost heaving is often determined by the groundwater level and the capillary height of soil. The minimum height between the maximum freezing depth and groundwater table in various types of soils is approximate 2.0 m ~ 3.5 m for clayey soil, 1.0 m ~ 2.0 m for sandy soil and 0.5 m ~ 1.0 m for sand soil.

4) Soil Density. Based on the tests indicating a degree of saturation (Sr) exceeding 90% for cohesive soil, the relationship between soil frost heave (η) and dry density (ρ_d) can be described using the following empirical formula:

$$\eta = 9.666 - 4.928\rho_d \tag{1.6}$$

5) The relationship between soil temperature and frost heave. Generally, the initial frost heave temperature of soil is lower than its freezing point. Specifically, for cohesive soils, this temperature is approximately 0.5 °C to 0.8 °C lower, while for sandy soils, it is about 0.2 °C to 0.3 °C lower. In a closed system, due to the weak-binding film water migration stopped under a certain temperature condition, the temperature that the soil stops its frost heaving is −8 °C to −10 °C for clayey soil, −5 °C to −7 °C for silty clay, −3 °C to −5 °C for silt and about −2 °C for sand soil.

The faster the freezing rate, the less pronounced the ice segregation and water migration, resulting in weaker frost heaving during the formation of a cohesive frozen soil structure. Conversely, a slower freezing rate leads to more significant ice segregation and water migration, causing stronger frost heaving when forming a layered and reticulated frozen soil structure.

6) The external load exerts a restraining effect on soil frost heave. However, frost heave may still occur as soil temperature decreases. When the water content and the freezing conditions are determined, the frost heaving rate of clayey soil will decrease sharply with increasing external loads and ultimately reaches zero under specific conditions (soil quality, water content, and soil temperature), a phenomenon referred to as "interruption pressure". Coarser soil particles result in lower interruption pressure values. Additionally, as soil temperature decreases, the interruption pressure increases.
7) The influence of soil salinity on frost heaving: The extent of frost heaving in soil diminishes as the concentration of salt solution increases. The degree of frost heaving can be influenced by cation exchange capacity in the following order: Na+, K+ > Ca2+, Mg2+ > Fe3+, Al3+. When soil temperature remains above −12 °C, the minimum salt solution concentrations required to prevent frost heaving are as follows: 2% for sandy soil, 5% for silty soil, 8–10% for silty clay, and over 10% for clay. Chloride ions can temporarily reduce frost heaving, and typically, a desalination effect is observed in the soil after 2 ~ 4 winter seasons.

1.2.2.2 Estimation of Frost Heaving of Foundation Soil

Methods for determining the extent of soil frost heave:

1) The depth of seasonal freezing (h_f) and the groundwater level (Z_ω) must be accurately determined, typically through observational methods.
2) The extent of frost heaving on the ground surface can be quantified using the following methods:
   (1) Frost heaving can be minimal in coarse-grained soils and soils containing a high proportion of coarse grains.
   (2) The degree of frost heaving in low liquid limit clay can be determined using the formula 1.7 presented. When the groundwater table is buried at a depth exceeding 2.0 m, calculations should be performed under conditions simulating a closed system.

$$h = 2.437 h_f^{0.71} e^{-0.0013 Z_\omega} \tag{1.7}$$

where h represents the total frost heaving amount of the ground surface, defined as the elevation difference between the ground after and before the freezing period, measured in millimeters; h_f denotes the design freezing depth, which corresponds to the freezing depth of the natural ground surface or the designed ground elevation, also measured in millimeters. The design freezing depth Z_{df} of the foundation soil can be utilized to calculate its corresponding frost heaving amount h_d. Z_{ω}refers to the groundwater level at the natural ground surface or the designed ground elevation prior to the onset of freezing (during the initial freezing period).

The depth of groundwater level adopted from the bottom elevation of the floor is used to calculated the frost heave of foundation soil (h_d).

(3) The degree of frost heave for silt, high liquid limit clay, and fine soil sand (gravel) with particles smaller than 0.075 mm and comprising 20–50% of the total mass can be determined using Eq. 1.8. In cases where the groundwater depth exceeds 2.0 m, the calculation should follow the methodology applicable to closed system conditions.

$$h = 5.371 h_f^{0.56} e^{-0.0013 Z_w} \tag{1.8}$$

(4) The frost heave for sandy and gravel soils, where particles smaller than 0.075 mm account for 10–20% of the total mass, can be calculated using Eq. 1.9. If the groundwater depth exceeds 1.5 m, the calculation should follow the method applicable under closed system conditions.

$$h = 0.13 h_f e^{-0.002 Z_w} \tag{1.9}$$

(5) The ground surface frost heave under closed system conditions can be determined using Eq. 1.10.

$$h = 0.45 Z_d \left(\omega - 0.8\omega_p\right) \tag{1.10}$$

where h is the ground surface amount of frost heaving (mm); h_f is the design value of freezing depth (mm); ω is the average water content of frozen soil layer (%); ω_{p} is the plastic limit water content (%).

1.2.3 *Frost Heave Forces During the Freezing Process of Soil*

When the soil is frozen, the water in the soil (containing external supplementary water) will freeze (ice extraction), resulting in the expansion of soil volume and the displacement of soil particles. This expansion force is called frost-heave force. The influencing factors of soil frost heaving also restrict frost-heave force. According to the direction acting on the foundation surface, it can be divided into tangential frost-heave force, vertical normal frost-heave force and horizontal frost-heave force (horizontal normal frost-heave force).

Fig. 1.16 Relationship between tangential frost-heave force, adfreezing force and water content of silty clay

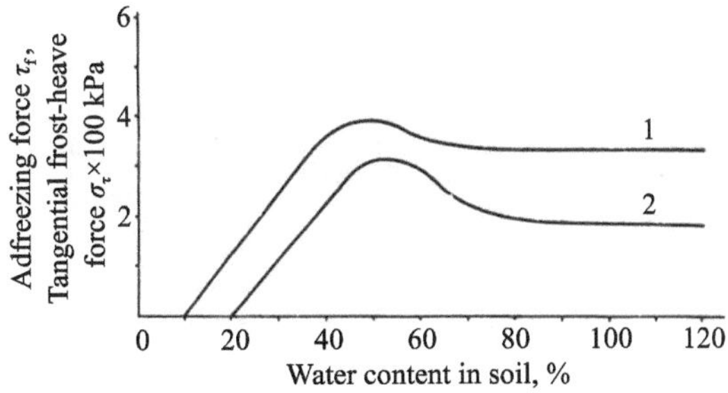

1—Adfreezing force (soil temperature is −3.1 ~ −3.2 °C);
2—Tangential frost heaving force (maximum value when soil temperature is −7 °C)

1.2.3.1 Tangential Frost-Heave Force σ_τ

The tangential frost-heave force is the expansion force acting on the side surface of the foundation, which must meet two conditions: ① the foundation surface is hydrophilic and can be frozen with the foundation soil, and there is adfreezing force (or freezing strength) between them (Fig. 1.16); ② The foundation soil produces frost heave in the freezing process. These two factors are indispensable.

The tangential frost-heave force increases with the increase of freezing depth of foundation soil, and the average and total tangential frost-heave force of clayey soil is greater than that of coarse-grained soil. The tangential frost-heave force develops at its greatest speed mainly at 1/3 ~ 2/3 freezing depth, that is, the main frost heaving zone. Each code provides the standard value of tangential frost heave force.

1.2.3.2 Normal Frost-Heave Force σ_{no}

The normal frost-heave force is defined as the force along the normal direction of the building foundation, that is, the vertical normal frost-heave force. The greater the frost heaving of foundation soil and the rigidity of foundation are, the greater the normal frost heaving force becomes. Under the condition that water cannot expand at all during freezing, it will produce huge expansion force, that is, crystallization pressure. When the temperature is −22 °C, the maximum expansion force reaches 2115 kPa (Tsytovich, 1985). According to Japanese scholars' research, the critical temperatures and ultimate frost heaving forces of standard sand, Nanao silt, Negishi silt and clay studied by Japanese scholars are: −0.36 °C/0.4 MPa, −2.7 °C/3 MPa, −11.4 °C/13 MPa and −30.7 °C/35 MPa, respectively.

1.2.3.3 Horizontal Frost-Heave Force σ_{ho}

The frost heaving force parallel to the ground and perpendicular to the foundation side surface along the frost heaving direction of the soil is called horizontal frost-heave force, which is the horizontal normal frost-heave force perpendicular to the foundation side surface. In the process of soil freezing, the horizontal frost-heave force on the wall back is much larger than the earth pressure at the same depth, from several times to dozens of times. The maximum horizontal frost heaving force measured at present (Ding et al., 2011): 261 kPa for cohesive soil and 188 kPa for coarse-grained soil. The horizontal frost-heave force is unevenly distributed along the wall back. In most cases, the retaining wall bears large horizontal frost heaving force within the height range of 0.3 ~ 0.8 h (h is the exposed height of the retaining wall), and the maximum value often appears at 0.6 h. Standard value of horizontal frost heave force given in the code.

1.2.4 Strength Characteristics of Frozen Soil

Soil in non-frozen soil area is a three-phase loose system composed of solid particles, water and gas. Its strength against external force mainly depends on the bonding force between solid particles: the molecular bond and structural bond between solid particles. In addition to inheriting the phase composition and binding force of soil, what is more important is the binding force of "ice" phase composition and "ice cementation" caused by water freezing, which is a factor affecting the properties of frozen soil, such as strength and deformation. Frozen soil has the instability of mechanical properties of freezing soil, frozen soil and thawing soil. The important factor determining this instability is the change of temperatures. The anisotropy of frozen soil strength is determined by the anisotropy of ice. Stress state and load action time are also the important factors affecting the instability of mechanical properties of frozen soil.

1.2.4.1 Freezing Strength of Soils

During the freezing process, the wet soil around the building foundation is frozen into ice closely cementing with the foundation. This cementation force is called the freezing strength between soil and foundation, commonly known as adfreezing force. The freezing strength can appear only under the action of external load, and its main direction is always opposite to the action direction of external load. Freezing strength can play the role of anchoring against frost expansion and supporting force against subsidence (thawing settlement). The important factors determining its value are frozen soil temperature, ice content, soil particle composition and density, external load action time, foundation surface properties and freezing conditions.

When the soil water content is constant and the soil temperature is lower than the freezing temperature, the freezing strength increases with the decrease of soil temperature. When the soil water content gets to a certain degree, the freezing strength

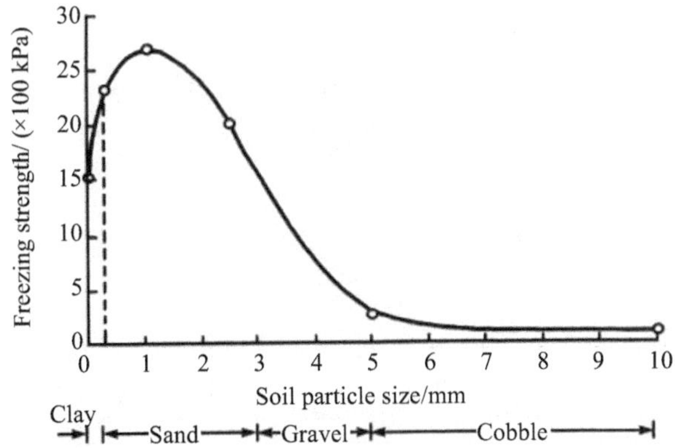

Fig. 1.17 Maximum freezing strength of frozen soil and wood of different soils

will increase slowly with the decrease of soil temperature. When the soil water content and soil temperature are the same, the freezing strength of coarse-grained soil is greater than that of fine-grained soil. Figure 1.17 (Tsytovich, 1985) shows the relationship between the maximum freezing strength between wood and soil particles when the saturation is 80% ~ 90% and the soil temperature is −10 °C.

The material and surface roughness of the foundation directly affect its freezing strength. The freezing strength of concrete pile is the largest, steel pile is the second, and wood pile is the smallest. Under the same conditions, the external load acting on the pile for a long time will rapidly reduce the freezing strength and finally reach the lasting freezing strength, which is only 1/3 of the transient freezing strength.

1.2.4.2 Compressive Strength of Frozen Soils

The compressive strength of frozen soils is essentially the result of the interaction of various factors such as the characteristics of frozen soil particles (particle size and composition, dry density, moisture content, salt content, etc.) and external conditions (temperature, pressure, loading rate, frost cycle, etc.).

1) The elastic modulus of frozen soil is more than ten to tens of times larger than that of thawed soil (melted soil, unfrozen soil), with a variation of 300 MPa ~ 30000 MPa.
2) At the standard loading rate, the compressive strength of frozen soil can reach several MPa to dozens of MPa. At a higher loading rate, the compressive strength of frozen sand can reach more than 15.4 MPa and that of frozen clay can reach 75 MPa when the soil temperature is −40 °C.
3) The compressive strength of frozen soil decreases with the duration of load. The long-term limit compressive strength is many times smaller than the transient compressive strength, sometimes 80% ~ 90% low.

1.2.4.3 Shear Strength of Frozen Soils

The cohesion (c) and internal friction angle(φ) of frozen soil are functions of soil temperature, water content and load duration. The shear strength of frozen soil are affected by the mineral composition, particle size, structural connection and arrangement, density, saturation, exchangeable ion composition and concentration in soil, as well as the temperature, ice content, structure, external pressure and load duration of frozen soil.

The strength of the soil at the freeze–thaw transition zone is between those of the thawed soil and the frozen soil. Compared with the thawed soil, the cohesion is larger and the internal friction angle is smaller. Compared with the frozen soil, the cohesion is smaller and the internal friction angle is larger. The long-term shear strength of the soil in the freeze–thaw transition zone is much smaller than the transient shear strength. The cohesion is about 1/5, and the internal friction angle is only 1/12. In the process of quick shear, the soil temperature in the shear zone will rise, and the underlying frozen soil will rise with the largest degree.

1.2.5 Deformability of Frozen Soil at Low Temperature

In fact, frozen soil is still compressible, and it can be considered as "incompressible body" generally under the condition of extremely low temperature of frozen soil. The compression deformation of warm frozen soil is a very complex physical process, which is controlled by the deformation and migration of its components—gas, liquid, viscoelasticity (ice) and solid (mineral particles).

In the process of frozen soil compaction, the recoverable deformation is small, which usually accounts for only 10% ~ 30% of the total compression deformation (Zhu & Zhang, 1982), and the plastic frozen soil with high ice content is less than 10%. Irreversible deformation of structure, which accounts for 70% ~ 90% of the total deformation, is mainly caused by irreversible shear of particles and particle aggregates.

Both temperature and moisture content have an effect on the compressibility of frozen soil, among which the effect of soil temperature plays the most significant role. With the same moisture content, the higher the temperature of frozen soil is, the greater the volumetric compressibility coefficient becomes. For example, for clay, when the soil temperature is −1.5 °C, the volume compressibility coefficient is less than 0.1 MPa^{-1}, and when the soil temperature is −0.3 °C, the volume compressibility coefficient is greater than 0.2 MPa^{-1}. With the same temperature, when the sample is supersaturated, the volume compressibility decreases with the increase of ice content, and finally reaches the volume compressibility of pure ice (Zhu & Zhang, 1982). What can be seen is that the volumetric compressibility coefficient of warm frozen soil (>−1.0 °C) with the high ice content (>40%) is generally greater than 0.067 MPa^{-1}, which belongs to medium compressibility soil and increases significantly with the increase of soil temperature.

1.2.6 *Thaw Settlement and Compressibility of Frozen Soil*

1.2.6.1 Characteristics of Frozen Soil During Thawing

When the frozen soil thaws, the ice in the frozen soil (cemented-ice, segregated-ice, etc.) gradually thaws, leading the volume of the soil reducing and the water discharging. The thaw of frozen soil of structural deformation caused by various ice during freezing, such as cracks, cavities or defects, will lead to a slight compaction under the action of dead weight, but compaction due to residual structural defects destructs under external load (Fig. 1.18).

When frozen soil thaws, its structure will change dramatically, which will affects not only its compressibility, but also its water permeability. The rethaw of silty clay after freezing leads to the compaction speed under the first level load (depending on the permeability and compressibility of the soil) increases 7 ~ 10 times. Therefore, frozen soil ice saturation is a very important index to evaluate the time-varying process of thawing soil settlement. The consolidation process of frozen soil rich in ice (such as frozen soil with high ice content) is completely different from that of frozen soil with low ice content. The frozen soil with layered and reticulate structure has its particularity when melting: ① the change of void ratio is much larger than that of unfrozen soil when compacted under load; ② The permeability of thawed frozen soil with high ice content is dozens or even hundreds of times higher than that of similar thawed soil, but decreases with time.

In the thawing process of frozen soil, the settlement caused by the weight of soil itself (10 kPa in the test) can be described by thaw-settlement coefficient (A_0):

$$A_0 = \frac{\Delta h}{h} \tag{1.11}$$

where $\triangle h$ (mm) represents the settlement of frozen soil after thawing under the stress of 10 kPa; h (mm) represents the height of frozen soil sample before the test.

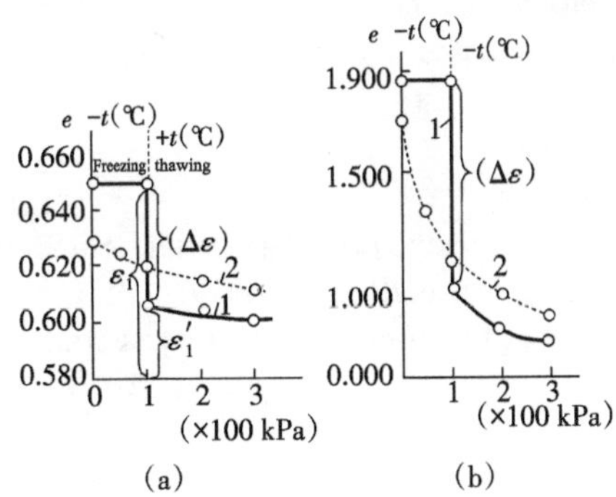

Fig. 1.18 Thawing compression curves of frozen sand (**a**) and clay (**b**) (Tsytovich, 1985). 1—In the process of frozen soils thawing; 2—Unfrozen soil (Under-positive temperature)

1.2.6.2 Compression Subsidence Characteristics of Frozen Soil After Thawing

After thawing, the frozen soil is compressed, and the soil particles move, reducing the pore volume in the soil. If the deformation of soil particles is ignored, the reduction of soil volume after compression is the reduction of soil pore volume. The compression deformation process of thawed frozen soil and non-frozen soil (thawed soil) is the same.

1.2.6.3 Influence of Moisture Content of Frozen Soil on Thaw-Settlement Coefficient

The research results show (Zhu et al., 1983) that the thaw settlement occurs when the frozen soil thaws and when the water content of frozen soil exceeds the plastic limit water content. When the total water content in frozen soil is less than ω_p+(30% ~ 40%), the thaw settlement coefficient can be described according to the following formula.

$$A_0 = k_1\left(\omega - \omega_0\right) \tag{1.12}$$

where k_1 represents the test coefficient; ω_0 represents the water content of initial thaw-settlement, that is, the water content of frozen soil with thaw-settlement coefficient in the range of 0 ~ 1%:

clayey soil: $\omega_0 = 5 + 0.8\ \omega_p$
coarse grained soil: $\omega_0 = 6 + k(\%)$

The value of k is determined by the test, generally 3%–6%. k tends to a large value when the silt clay content in coarse-grained soil is high and k tends to a small value when the silt clay content is low.

The thaw-settlement coefficient of ice layer with soil inclusions with water content greater than ω_p + (30% ~ 40%)belongs to a curve segment, and the thaw-settlement tends to parabola with the increase of water content in frozen soil. It can be calculated by superposition of straight-line segment and curve segment.

$$A_0 = k_2\sqrt{\omega - \omega_c} + A_0^{'} \tag{1.13}$$

where ω_c is the boundary water content between the straight segment and the curve segment, usually taken as ω_p + 35%; $A_0^{'}$ corresponds to the thaw-settlement coefficient of ω_c; $k_2$represents the test coefficient, 0.26 ~ 0.30.

1.2.6.4 Effect of Soil Dry Density on Thaw Settlement Coefficient of Frozen Soil

Within the range of dry density (ρ_{dc}) of ω_p + (30~40)%, relationship between thaw settlement coefficient (A_0) and soil dry density (ρ_d) changes in a curve, obeying the relationship of quadratic equation:

$$A_0 = k_3 \frac{\rho_{dc} - \rho_d}{\rho_d} \quad (1.14)$$

where k_3 represents the test coefficient; ρ_{dc} represents the initial dry density for initial thawing, that is, the soil density is the best one when the thaw-settlement coefficient is less than 1%.

The dry density of frozen soil with high ice content is small. The thaw settlement coefficient increases approximately linearly with the decrease of dry density when the dry density is less than ρ_{dc}. Therefore, the thaw-settlement coefficient of frozen soil with high ice content can also be calculated according to the superposition of two sections:

$$A_0 = k_4\left(\rho_{dc} - \rho_d\right) + A_0' \quad (1.15)$$

where k_4 is the test coefficient, of which ice layer with soil inclusions is 60 to 70, and numerical value 60 (k_4) means that A_0' is the frozen soil thaw-settlement coefficient corresponding to ρ_{dc}.

During in-situ field survey, it is usually required to measure the total moisture content of frozen soil (ω) and dry density (ρ_d). The thaw-settlement coefficient A_0 of frozen soil can be calculated respectively according to the above formula. If the two calculation results are inconsistent, the higher value shall be taken as the design value.

1.3 Engineering Classification of Frozen Soil and Foundation Design

The description and engineering classification of ice in frozen soil are paid high attention both at home and abroad. Generally, the engineering classification of frozen soil could be used to determine the design principles and foundation treatment measures and formulate their own technical code and frozen soil classification system accordingly. At present, the engineering classification of frozen soil is gradually considered from qualitative to quantitative description, and from single content to comprehensive index. At the same time, foundation design factors are also gradually considered towards the comprehensive consideration of main factors and multiple contents.

1.3.1 Engineering Classification of Frozen Soil

1.3.1.1 Engineering Classification by Ice Content of Frozen Soil

When using permafrost as foundation, the fabric, physical and mechanical properties of permafrost should be unified, and the difference of engineering properties of permafrost should be determined according to the proportion of ice in permafrost. The engineering properties of frozen soil are determined by the content of ice in frozen soil, that is, volume ice content or total water content of frozen soil, which can reflect the frost heaving and thawing settlement of frozen soil.

The engineering categories of frozen soil can usually be simplified into two categories: low-ice content and high ice content frozen soil.

The low-ice content frozen soil includes ice-poor and icy frozen soil, and the amount of thaw settlement after thawing is small. Engineering in frozen soil areas can usually be designed according to the design principle of melting soil or permitting thawing, but a little simple engineering measures are needed.

The high-ice content frozen soil includes ice-rich, ice saturated and ice layer with soil inclusions, and the amount of thaw settlement after thawing is large. Engineering in frozen soil areas should be designed according to the design principle of keeping frozen state, and some strong engineering measures are needed.

In practical application, the thickness of frozen soil foundation should be taken into consideration to judge the engineering category of frozen soil. For heating structure buildings, it is general to classify them according to the total water content of frozen soil layer within the maximum thawing depth, and engineering design should pay attention to the position of frozen soil with high ice content. For cold foundation buildings, the engineering category of frozen soil can be determined by the total water content within the range of 1.0–3.0 m deep below permafrost table under the conditions of cold permafrost, however, the engineering category of frozen soil can be determined by the total water content within the range of 3.0–5.0 m deep below the permafrost table under the conditions of warming permafrost.

1.3.1.2 Permafrost Classification by Ground Temperature

The mean annual ground temperature (MAGT) of permafrost can not only reflect the mean annual air temperature, permafrost thickness and distribution, but also reflect the thermal stability of permafrost. According to the sensitivity of permafrost to climate change and human activities and the analysis in the impact of ground temperature on the engineering properties and engineering stability, permafrost is divided into three ground temperate zones (Table 1.5).

Table 1.5 Ground temperature zoning of permafrost

Zone Name	MAGT (°C)	Permafrost thickness (m)		MAAT at the boundary of permafrost (°C)		Permafrost distribution regions	
		Northeast	Northwest	Northeast	Northwest	Da-Xiao Xing'anling Mts	Qinghai-Xizang Plateau
Extreme stable zone (I_1)	<−5.0		>150	<−6.0	<−8.5	Predominantly permafrost regions	High mountains zone
Stable zone (I_2)	−5.0 to −3.0	>100	100 to 150	−4.5	−6.5		High-middle Mountains zone
Basically stable zone (II_1)	−3.0 to −1.5	50 to 100	60 to 100	−3.5	−5.5	Permafrost regions with patchy talik	Low mountains and hill, marsh and peat areas
Transition zone (II_2)	−1.5 to −1.0	20 to 50	40 to 60	−2.5	−3.5		High-plain, hill and river valley
Unstable zone (III_1)	−1.0 to −0.5	10 to 20	20 to 40	−1.0	−2.5	Island and patchy permafrost regions	
Extremely unstable zone (III_2)	−0.5 to 0.5	0 to 10	0 to 20	±0.0	−2.0		River valley and island permafrost areas

MAGT = mean annual ground temperature, MAAT = mean annual air temperature

Zone I is a stable zone, where the change of permafrost temperature inducing by the warming of climate slightly influence engineering properties of permafrost. Zone II is a basically stable zone, where the warming of climate will have a great impact on these zones, resulting in the more complex engineering geological environment in these zones and weakening the permafrost thermal stability. Zone III is an unstable zone, where the permafrost thermal stability is extremely poor and permafrost may disappear with the increase of air temperature, however, the engineering geological conditions of some sections may become better and some sections may become worse, resulting in the largely uneven subsidence and deformation for permafrost foundation.

To facilitate engineering application, permafrost is usually simply classified into warm permafrost and cold permafrost based on the threshold of mean annual ground temperature (−1.5 °C for highway and −1.0 °C for railway). According to the engineering investigation and subgrade surface properties, the division of ground temperature zone in highway differs from that in railway in permafrost regions. First, the threshold of permafrost stability zone is −3.0 °C for highway and −2.0 °C for railway. Secondly, threshold between the warm and cold permafrost is divided by −1.5 °C for the highway and −1.0 °C for the railway.

1.3.2 Frozen Soil Foundation Design

The principle of building foundation design is to maintain the mechanical stability of building foundation system. In the design of foundation in permafrost area, the mechanical stability and thermal stability of frozen soil foundation system must be considered at the same time, which is a significant difference from the design of foundation in non-frozen soil area. "Thermal problem" has become the premise of maintaining the mechanical stability of frozen soil foundation system. Measures must be taken to maintain the design temperature of permafrost foundation in foundation design, or measures must be taken to maintain the structural stability of buildings with the temperature change of frozen soil foundation. Therefore, the following contents should be considered in the design and the calculation of building foundation in permafrost area:

(1) Thermotechnical calculation of frozen soil foundation mainly includes mean annual ground temperature of permafrost in construction site (or observational, statistical value), the highest soil temperature under thaw bulb, temperatures in ventilation basement and freezing depth and thaw depth of foundation soil.
(2) Mechanical design and calculation of frozen soil subgrade and foundation mainly includes bearing capacity of frozen soil foundation, thaw-settlement of frozen soil foundation, deformation, checking calculation of anti-frost-heaving stability of foundation soil with frost heaving, and foundation type and size selection.
(3) Calculation of thermal protection measures for frozen soil foundation mainly includes the design and calculation of thermal protection measures in permafrost area, such as thermosyphon foundation design, heat insulation layer design, ventilation (or elevated ventilation) foundation design, as well as their composite measures design, sunshade and artificial refrigerating structure design, etc.
(4) The design and calculation of the adaptive structural buildings in frozen soil areas mainly refers to elevated ventilated foundation, filled ventiduct foundation with filled soil, pile foundation, thermal pile foundation, deep foundation and extended foundation, heat insulation base plate, etc.
(5) Design of environmental protection engineering in frozen soil area. Permafrost foundation of many years shall be calculated according to two limit states at the same time (Former USSR State Construction Commission, 1990):

    It is calculated according to the bearing capacity (first limit state): the stress of permafrost foundation under the action of building foundation shall not exceed the allowable bearing capacity of foundation (Ultimate long-term strength).

It is calculated according to the settlement deformation (the second limit state): during the construction and operation of the building, the foundation settlement deformation speed and total deformation shall not exceed the allowable deformation of the building.

1.3.2.1 Selection of Foundation Design Principles

The building design principles need to fully consider engineering geological conditions of frozen soil in the construction site, the building features, and the changes of foundation soil properties. Engineering geological conditions of frozen soil in the construction site include geology and frozen soil fabric, mean annual ground temperature, ice content or total water content, permafrost table, thaw-settlement coefficient and compression coefficient of frozen soil foundation. The buildings features includes thermal state, building area, structure and sensitivity of uneven settlement.

When permafrost is used as building foundation, it can be designed in one of three states (Principles):

Principle I: keeping frozen state. During the construction and operation period of the project, the permafrost foundation always keeps frozen state.

This principle is applicable to all kinds of foundations in permafrost regions with mean annual ground temperature lower than −1.0 °C to −1.5 °C. The thermal protection measures adopted can effectively keep the permafrost foundation at the design temperature for a long time under the impact of climate warming and engineering thermal state. It is usually used in the regions of cold permafrost with high ice content and high-intensity earthquake region.

Principle II: gradually thawing state. During the period of engineering construction and operation, the permafrost foundation is in a state of gradually thawing.

This principle is applicable to the permafrost foundation with the mean annual ground temperature higher than −1.0 °C to −1.5 °C. Thawing of permafrost foundation does not lead to serious settlement, and the settlement amount is less than the limit deformation of this kind of buildings. Principle II is to limit the thawing of frozen soil foundation. During the period of building construction and operation, the permafrost foundation is only allowed to thaw to a certain design depth, and the foundation bearing layer still keeps frozen. Therefore, principle II is more suitable for permafrost with low ice content or frozen coarse-grained soil foundation.

Principle III: pre-thawing state. Before the engineering construction, the permafrost foundation can be permitted to thaw to the calculated depth or completely thaw. This principle is applicable to the foundation of warm permafrost with high ice content, but the premise is that the technology is feasible and the economy is reasonable.

According to the regulations of Russia (Former USSR) (State Construction Commission of the USSR, 1990), two principles are adopted in permafrost construction:

Principle I: keep frozen state in the permafrost foundation during the period of the building construction and operation.
Principle II: Permafrost foundation is permitted to thaw during the period of construction and operation or prethaw before construction.

It is noteworthy that in high-temperature permafrost areas, the whole building or the whole construction site must be designed according to one same architectural principle. For adjacent buildings in the same construction site, especially single buildings (even if the construction area is quite large), it is not allowed to mix two different building principles in the design. The linear infrastructure can adopt different design principles in different sections, but the adaptability of the linear infrastructure structure to the uneven deformation of the foundation should be considered in advance when the transition zone from one section to another is concerned.

1.3.2.2 Design Elements of Building Foundation in Permafrost Area

Main factors should be considered in the design of building foundation in permafrost areas: thermal influence, surface water and groundwater, frost heave and thaw settlement, foundation soil, load, building materials, etc.

1) **Thermal Effect**
Frozen soil temperature has a direct impact on the bearing capacity, freezing strength and creep speed of frozen soil. The lower is the mean annual ground temperature, the better the thermal stability of building foundation becomes. In the warm permafrost area where the mean annual ground temperature is higher than −0.5 °C ~ −1.0 °C, permafrost is easily affected by external thermal erosion due to worse thermal stability.

Due to the heat source of external thermal erosion from climate and buildings, the solar radiation and the heat of buildings cause the ground surface temperature to rise for a long time, resulting in an increasing in the mean annual ground temperature, and permafrost thawing, as well as drastic changes of permafrost table. In the ice-saturated permafrost zone, the thawing rate of permafrost decreases with the decrease of the mean annual ground temperature. The lower is the ground temperature, the smaller is the melting rate, and vice versa.

It is worth noting that thermal protection is one of the important measures in the design of building foundation in permafrost areas. The mean annual ground temperature and ice content of frozen soil determine the thermal stability of frozen soil foundation, and the heat source and the heat of the buildings determine the depth of thaw bulb under heated building. The combination of these two factors determines the design principle, foundation strength, foundation form and structure of the buildings, such as the ventilation form of elevated ventilated foundation, size, material, thermal insulation performance of building ground, etc. Simply using insulation layer can only delay the reduction of melting depth for a short time, and cannot prevent the heat flow from the atmosphere and the buildings. Natural ventilation or artificial refrigeration system can usually refreeze the subsoil and prevent the erosion of heat. Although the cold structure building itself cannot generate heat, its heat absorption property and its ability to change the original surface will cause the change of ground heat balance and lead to the melt of the frozen soil foundation.

2) **Frost and Thaw Action**

Whether it is in the seasonal frozen soil or permafrost areas, the foundation soil is subject seasonal to annual freezing and thawing process. During the period of freezing, the foundation soil will produce frost heave; and during the period of thawing, the frozen foundation soil will produce thaw settlement. Repeated freezing–thawing process reduces the strength of foundation soil, resulting in the buildings instability or even destruction.

In the design, the deformation and destruction of the building caused by the freezing and thawing process of the foundation soil shall be fully considered, and the foundation stability of the buildings in the frost heaving foundation soil shall be checked. According to the basic elements of frost heave and thaw settlement, effective measures should be taken to eliminate and weaken their effects. The factors affecting the frost heaving of foundation soil are properties of soil, moisture and negative temperature in soil. As long as one of the basic factors is eliminated, the frost heaving of soil can be eliminated or reduced in principle. Antifreeze measures include replacement method, physicochemical method, thermal insulation method, drainage and water-resisting method and isolation method. Anti-thaw settlement measures include cooling method (filling block stone, ventilation pipe, elevated ventilation, thermosyphon), thermal insulation method, pile foundation, drainage and water-resisting method.

3) **Permafrost Thaw Settlement**

Permafrost is a soil whose temperature below 0 °C and contains a variety of ice bodies. Thaw of ground ice leads to the foundation settlement due to discharge. If the ground ice thaw rapidly and the water cannot be discharged in time, the thawed frozen soil may become soft to mud and completely lose its bearing capacity. The higher is the mean annual ground temperature, the worse are the permafrost thermal stability and the ability to resist external thermal erosion. Under the impacting of global warming and engineering activities, the rapid thaw of warm permafrost can lead to moisture stay in the soil layer, accelerating permafrost thawing and soil layer softening.

When warm permafrost with high ice content is selected as building foundation soil, the calculated settlement should not exceed the limit settlement under the condition of gradually thawing of permafrost foundation during the period of building construction and operation. The foundation design can be carried out according to the "design principle of gradually thawing state". The deformation and the thaw settlement of permafrost depend not only on the permafrost properties, but also on the foundation form, size, and stiffness, foundation width, foundation load and the sensitivity of the superstructure to the uneven settlement of the foundation, etc. Therefore, special attention must be paid to the prediction of thaw settlement of permafrost foundation under the building. Its design and calculation work should include:

(1) Determining the thawing depth of permafrost under the building at different time and stability limit;
(2) Determining the strength (bearing capacity) of foundation soil when permafrost gradually thaws;
(3) Determining the reaction force of foundation by considering the uneven settlement due to the gradually thawing of permafrost;
(4) Determining the foundation size and structure;
(5) Checking the frost heave and thaw settlement of the foundation.

4) **Surface Water and Groundwater**

Water, be it the surface water or the groundwater, often produces immeasurable thermal erosion to the permafrost foundation. A large amount of free water within the freezing soil layer will form a huge of frost heaving and frost heaving force, even frost mounds and icing, causing great damage to buildings. A large amount of heat to the permafrost foundation from surface water and groundwater will lead to permafrost thawing, resulting in permafrost table deepening and the foundation softening to cause foundation subsidence and slope collapse. The thermal abrasion of surface water and groundwater greatly increases the thawing depth of permafrost foundation, resulting in serious thaw settlement. Therefore, the waterproof and drainage systems in permafrost areas have become an important measure to ensure the stability of the buildings. There shall not be soaks around the building, and there shall be good drainage systems on the slope top and bottom. The inlet and outlet of culvert and tunnel shall have good water-resisting and drainage facilities.

5) **Foundation Soil**

Foundation soil is the basis of the buildings, and the basic condition for judging frost heaving and thawing settlement is the particle size composition of foundation soil. Foundation soil prefers to select coarse grained soil layer and cold permafrost with low ice content. Therefore, it is necessary to conduct a detailed engineering geological survey of frozen soil on the construction site, obtain data in the ice content and mean annual ground temperature, divide the area in detail for permafrost, and evaluate the engineering geological conditions of frozen soil on the construction site.

The strength and thermal stability of frozen soil foundation are closely related to ground temperature. The lower is the ground temperature, the greater the strength and the better the thermal stability become, and vice versa. The design of the pile column foundation not only considers the frost heaving force of seasonal thawing layer, but also considers the thermal stability, strength change and creep property of permafrost.

6) **Load**

The building load can restrain and reduce the frost heaving of the foundation soil. Although the "pressure rupture" changes with the ground temperature, the frost heaving rate decreases with the increase of load on the foundation surface. As long as enough pressure is applied on the foundation surface, the frost heaving can be restrained. On the contrary, the building load is unfavorable to the permafrost foundation with thawing settlement. It is advisable to expand the foundation area, deep pile foundation, cooling foundation and use other methods to reduce the load and increase the bearing capacity of the foundation. Therefore, a comprehensive load and foundation treatment design should be adopted.

7) **Construction Material**

The properties and functions of the building materials shall be tested by long-term temperature expansion-contraction and seasonal freezing-melting under low temperature, so as to meet the requirements of building material strength and foundation thermal stability. There should be corresponding strict requirements and regulations for the mixing, pouring and curing of concrete in permafrost areas. It should be noted that: ① the early freezing of concrete seriously reduces its strength and durability; ② The freeze–thaw cycle continuously damages the firmness of concrete: ③ controlling the concrete mixing and molding temperature to reduce the adverse impact on the foundation and concrete; ④ The hydration heat of concrete will melt the frozen soil foundation and cause the foundation soil to sink, etc.

1.4 Frozen Soil Engineering Investigation

Frozen soil is a kind of special soil, which not only has the commonness of general soil, but also has the characteristics of multi-phase and complex frozen soil—ice and temperature. Under different permafrost temperatures, ice-water dynamic equilibrium directly affects the engineering properties of permafrost and the evaluation of engineering geological conditions of permafrost. The engineering geological investigation of different engineering buildings has its own key contents, but we must understand the ice content and ground temperature state of the frozen soil in the construction site. We should consider not only the impact of buildings on the permafrost environment and the interaction between them, but also the impact of climate change on the engineering properties of permafrost.

1.4.1 Basic Characteristics of Frozen Soil Engineering Investigation

At present, the engineering investigation of frozen soil still adopts the traditional exploration methods in geoscience, including drilling, pit prospection, costean, geophysical investigation, geochemical investigation, as well as indoor test and in-situ test. In addition to meeting the basic requirements of geotechnical engineering investigation, it also meets the requirements of special projects of permafrost, that is, the ice and temperature of frozen foundation soil.

Permafrost contains ice of different degrees —ground ice, which is its basic characteristic different from general ground. In the process of engineering investigation, it is necessary to clarify and describe in detail the buried depth, content, structure and distribution characteristics of ground ice, which involves the foundation treatment method required for engineering design.

Under different geological and environmental conditions, permafrost has different thermal regime. The mean annual ground temperature reflects not only the thermal stability of permafrost and permafrost thickness, but also the relationships among permafrost and climate, geological landform, and hydrogeological conditions. The most important thing is that the mean annual ground temperature can exhibit the thermal stability (inertia) of permafrost to external (climate and human activities) thermal disturbance. Engineering investigation needs to obtain the mean annual ground temperature and its relationship with the environment, which involves the design principles adopted in the engineering foundation and the safety and cost of engineering buildings. The equipment and instrument of ground temperature observed shall be provided for engineering investigation.

Under different environments (ground properties, geotechnical properties, hydrology and hydrogeology, geological structure, landform, solar radiant heat, etc.), the existence and development degree and burial depth of permafrost on its upper interface (the permafrost table) should have a clear conclusion in the engineering investigation of each construction site. The season for engineering investigation should be around the end of September to obtain maximum seasonal thawing depth by frozen state and ground ice in bore core or the process map of seasonal thawing depth in this area.

The basic physical parameters that must be measured in engineering investigation: the total water content of the frozen soil, the natural density of the frozen soil in the original state, the weight water content of the unfrozen water (measured in the laboratory under the natural temperature state) or the relative ice content of the frozen soil that replaces the ratio of the weight of ice in frozen soil to the weight of water.

The thermophysical and mechanical parameters of permafrost are important basis for engineering design. The thermophysical parameters (thermal conductivity, freezing temperature, etc.) can be measured by the laboratory, and the mechanical parameters can also be measured by the laboratory or selected from empirical data. However, the mechanical parameters required by some important projects must be measured by the laboratory studying the undisturbed soil.

The cryogenic phenomenon of unhealthy soil in frozen soil area has obvious seasonal characteristics. In case of frost heave and ice cone in freezing season (November to March), supplementary investigation and survey of frost mound and icing in March. During the thawing season (May to September), regressive thaw slumping, gelifluction, thaw settlement etc. shall be investigated (if necessary). The cryogenic phenomenon of frozen soil will affect the selection of engineering route, design scheme and foundation treatment, and even endangers the safe operation of the project.

1.4.2 Frozen Soil Engineering Investigation

The basic data of frozen soil must be obtained by engineering investigation, including permafrost type and distribution, permafrost table, mean annual ground temperature, frozen soil ice content and engineering type, basic physical and mechanical parameters, the cryogenic phenomenon of frozen soil impacting engineering safety. Frozen soil engineering investigation is suitable to be carried out simultaneously with hydrogeological investigation. These contents are of great significance to the determination of design principles of building foundation, foundation types and foundation treatment measures in permafrost areas.

1.4.2.1 Basic Tasks of Frozen Soil Engineering Investigation

Engineering investigation must obtain the most basic frozen soil data: such as permafrost type and distribution, upper limit depth, annual average ground temperature, frozen soil ice content and engineering type, basic physical and mechanical parameters, the cryogenic phenomenon of unhealthy frozen soil, etc. Frozen soil engineering investigation should be carried out simultaneously with hydrogeological investigation. These contents are of great significance to the determination of building foundation design principles, foundation types and foundation treatment measures in permafrost areas.

1.4.2.2 Identification of Frozen Soil Structure

Frozen soil structure refers to the mutual arrangement relationship between ice and soil layer (also known as cryostructure), which mainly includes the integral, layered and reticulated. More combination relationships can be seen in the actual engineering investigation. The structural diagrams and photos of seven types of frozen soil commonly and widely described in China are listed in Table 1.6. In addition, there are many intermediate or transitional types.

In the engineering investigation of frozen soil, the field discrimination of frozen soil structure can be carried out according to the Appendix of *the Code for Engineering Geological Investigation of Frozen Soil*. In the field survey, the frozen soil structure can be preliminarily determined by using the core name and referring to the origin of soil.

1.4.2.3 Determination of Permafrost Table

In the large continuous permafrost regions, the freezing layer, formed by the surface soil layer thawed in the warm seasons refreezing during the cold seasons, is connected to the underlying hidden permafrost layer. This freezing layer is called seasonally thawed layer, which belongs to the thawed layer of the permafrost layer in

Table 1.6 Brief illustration and photos of frozen soil structure type (photo taken by Yu Qihao)

Name of frozen soil structure	Diagram*	Lithology	Structure description
Wholeness	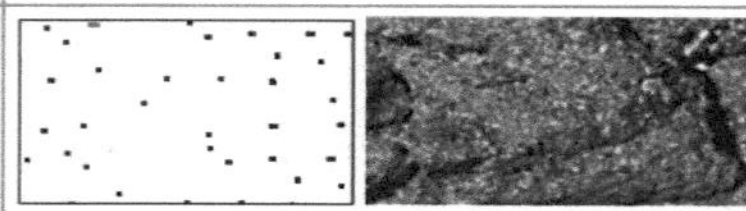	Clay, silty clay, homogeneous silty fine sand	Ice is evenly distributed in the voids and fissures of rock and soil. The water freezes in the pore. Fine ice crystals with uniform distribution can be seen by naked eye.
Patchy state	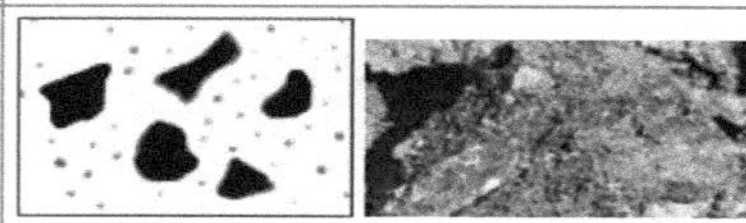	Gravel soil	There are "spot ice" lenses with irregular shapes, such as coarse ice particles and small ice blocks, distributed in the monolithic frozen soil. There are cementation cemented-segregated ice formations in cohesive soil and sandy gravel soil. The upper and lower parts of the formation with high ice content are the most common.
Wrapped (crustose) state		Gravel soil, sand gravel-cobble soil	Ice wraps gravel or small clods of earth in the middle. The gaps and fissures between coarse particles and small soil blocks are also basically filled with ice. It is common to form ice crust around gravel and cobble, with cemented ice, and occasionally with scattered small ice xenoliths.

Layers	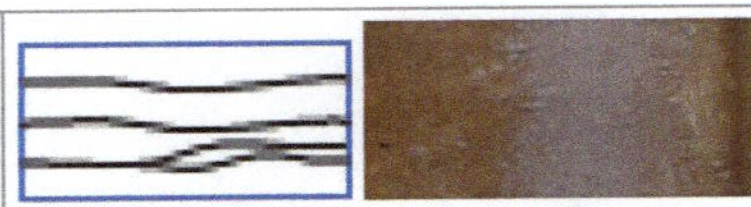 Micro-layer 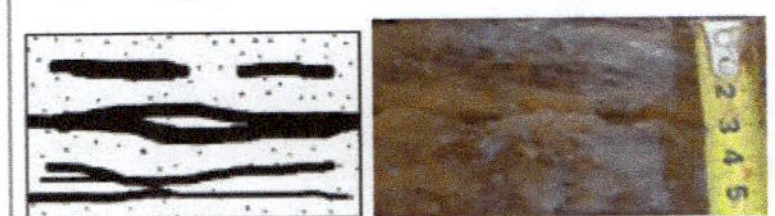 Thick layer	Clay, silty clay and silt interbedding	Ice is distributed in soil in layers or similar layers, which can be pure ice or soil containing ice. The thickness of the ice layer ranges from 1 mm to several meters, even more than 10 m. It is mainly composed of fractional ice and cemented ice. When the thickness of ice layer is less than 0.1 mm ~ 0.2 cm, it is called thin layer structure; When it is larger than 25 cm, it is generally called thick layered structure; The middle layer structure lies between the two. It is widely distributed in cohesive soil, sandy gravel soil and weathered argillaceous rock stratum, mainly located below the permafrost table to weathered bedrock.
Reticulated state	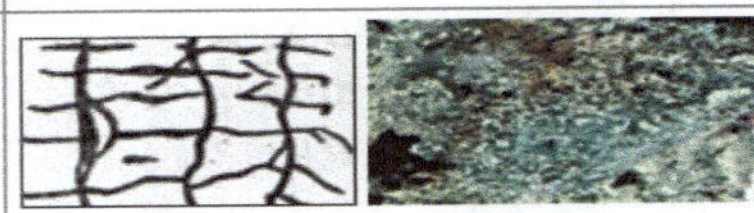	Silt, silty clay and silty fine sand	Ice is crisscrossed in the frozen soil layer in different directions, mainly horizontally, and vertically or obliquely, forming an ice net with different ice thickness. It is a mixed type of cementation and segregation, which is common in some cohesive soil strata with high density.
Basal state	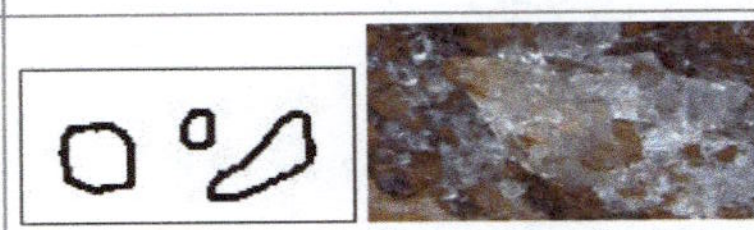	Soil block, crushed stone, pebble and gravel	Mineral aggregates and coarse clastic materials are distributed disorderly in the ice body, and the ice body is dominant.

the warm seasons. In vertical section, seasonally thawed layer is jointed to permafrost layer (Fig. 1.19a). The interface between the two is the permafrost table (the top layer of the permafrost). The minimum depth of seasonal thawing is about 0.05 m ~ 0.3 m (peat bogs) and the maximum is about 4.0 m (sand and gravel soil).

In the patchy permafrost regions, the freezing layer, formed by the surface thawed layer refreezing in the cold seasons, cannot be closely connected with the underlying hidden permafrost layer, and there exists an unfrozen thawed soil layer in the middle (Fig. 1.19b). This layer is called seasonal freezing layer, belonging to the freezing layer that freezes in the cold seasons. The thickness of the thawed soil layer between the seasonal freezing layer and the permafrost layer varies greatly, ranging from tens of centimeters to 7 m ~ 8 m. When the drilling depth is insufficient, it is often misjudged as a seasonally frozen soil area.

The surface freezing layer in seasonally frozen soil area is connected with the underlying non-frozen soil layer (Fig. 1.19c). When the thickness of seasonal freezing soil reaches 20 cm, it will have an impact on the stability of the engineering building. According to China's map of standard freezing depth distribution, the minimum freezing depth is 0.6 m and the maximum is 2.5 m. In the permafrost region, the maximum seasonal freezing depth in the talik can reach 3.5 m ~ 4.0 m.

The top depth of permafrost under natural condition is called "natural permafrost table", and the top depth of permafrost table under the impact of engineering is called "artificial permafrost table". The methods of determining permafrost table mainly include:

(1) Direct survey method. From the end of September to the beginning of October, permafrost table can be directly obtained by drilling, pit exploration, needle exploration (for peat bogs) and geophysical exploration. Because an ice-rich zone appears near the permafrost table, the initial depth of the ice-rich zone is

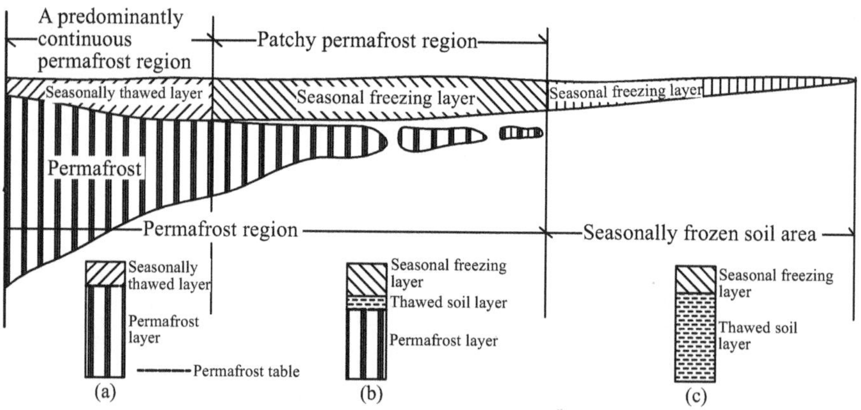

Fig. 1.19 The horizontal distribution area of frozen soil, and vertical section of seasonally thawed layer and seasonal freezing layer

generally designated as the permafrost table. Near the permafrost table, the water content of the frozen soil presents an "S" curve with the depth distribution, the permafrost table is at the peak zone depth of water content.

(2) Ground temperature observation method. The thermistor temperature sensor can be used to measure the ground temperature and help to draw the ground temperature thawing (freezing) process curve. The 0 °C (or freezing temperature) isotherm can be adopted to determine the maximum seasonal thawing depth. Permafrost table is estimated as the maximum thaw depth in the late autumn through linear interpolation of soil temperature profiles between two neighboring points above and below the 0 °C isotherm at all sites.

(3) Thaw Tubes observation method. An estimate of maximum thawing depth can be obtained using a liquid- filled or sand-filled tube. The outer tube of the device is anchored in permafrost and is not subject to heave. An inner tube is filled with water or sand containing a dye. The presence of ice in the tube or the boundary of the colorless sand records the approximate position of the thawed active layer. Each summer, measurements of current thaw penetration, ground level, and maximum heave or subsidence are made relative to the outer tube, which forms a stable reference level. Thaw tubes provide an inexpensive annual record of both maximum thaw penetration and active-layer thickness suitable for multi-year comparison. Thaw tubes do not, however, provide information on local variability, and individual thaw tube measurements therefore should be evaluated relative to a mean active-layer thickness field derived from more extensive sampling wherever possible.

(4) Calculation method. There are a variety of methods for calculating the seasonal thawing depth, such as mathematical analytical method, Stephen's analytical method, Kudriavtsev formula, empirical formula, etc. In the light of their boundary conditions and adaptation ranges, researchers can select the appropriate calculation method according to the data.

1.4.2.4 Determination of Mean Annual Ground Temperature of Permafrost

The main methods for determining the mean annual ground temperature of permafrost can be obtained by direct measurement of borehole temperature. Borehole drilling is taken in the permafrost layer, which the borehole depth should reach 15 m ~ 20 m (larger than depth of zero annual amplitude). Then, a steel tube or an aluminum-plastic tube (generally about 30–50 mm in diameter) should be inserted into the borehole, and the bottom of the pipe should be sealed. And a string of thermistors should be put into the tube (the wire should be vertical without bending), and a data acquisition instrument should be used (manual observation should be made at specific interval, using a 4.5-digit-high-precision digital multimeter) to measure the resistance value of the temperature and calculate the ground temperature value.

1.4.2.5 Investigation on the Cryogenic Phenomenon of Unfavorable Frozen soil

The survey method is generally used, and surveying and mapping are required when necessary. The visual method, semi-instrumental method or instrumental method can be adopted. In the frozen soil district with complex engineering geological conditions, observation points should be set up when it has a significant impact on engineering buildings. Survey and mapping reports (including exploration, observation, test and image data) should be in offering, and comprehensive frozen soil engineering geological maps and profiles should be provided if necessary. In March, investigation and survey on icing, pingo and frost heaving cracks (frost cracks), etc., will be carried out. From August to September, surveys such as gelifluction, thaw slumping, thermokarst lakes will be implemented.

1.5 Protection of Frozen Soil Environment

From the perspective of frozen soil engineering foundation and environmental protection, the existence, development or degradation of frozen soil is the main research subject involving the safety and security of frozen soil engineering. All the factors affecting changes of frozen soil can be considered as environmental factors of frozen soil. Both of the frozen soil engineering foundation and environmental protection are interdependent and mutually transformable research subjects. Hereby, only the effective engineering measures adopted to protect the survival environment of frozen soil and maintain the stable change of frozen soil are selected as the research object of frozen soil environmental protection.

1.5.1 The Particularity of Frozen Soil Environmental Protection

1.5.1.1 Special Frozen Soil Environment

Natural geographical-geological factors, such as climate, topography, geology, hydrology and hydrogeology, vegetation cover and so on, are involved in the heat exchange process of the lithosphere-soil-atmosphere system, influencing the formation and development of permafrost. Climate is one of the most important and most sensitive factors.

1) **High-Latitude Permafrost**
The characteristics of high-latitude permafrost are controlled by latitude and also affected by altitude. In other words, the higher the latitude is, the colder the climate becomes, and the more developed is the permafrost. The permafrost region in

Daxing'anling Mountains belongs to the cold temperate zone and the northern part of the mid-temperate in the severe cold zone of China. From south to north, southeast to northwest, the annual average temperature gradually decreases from 0 °C ~ 1 °C to about −5 °C, and the extreme minimum temperature value reached −52.3 °C (in 1969); the annual temperature range increases from 35 °C to 50 °C, the freezing time lasts for more than 6 months (about 150 ~ 200 days), and the amount of precipitation decreases from 500 mm ~ 600 mm to 200 mm ~ 300 mm. The percentage of the area occupied by the permafrost distribution has also gradually extended, from sporadic and patchy distribution to large-scale distribution. The thickness of frozen soil gradually increases, and the annual average ground temperature of the permafrost gradually decreases, and the thermal stability of the frozen soil changes from extremely unstable to basically stable. The average annual surface temperature of 0 °C is the basic condition for the formation and development of permafrost.

(1) The influence of terrain slope is more obvious. The permafrost on the west slope (shady slope) of the Daxiao Xing'anling Mountains is more developed than that on the east slope (sunny slope). Besides, the ground temperature of the west slope is almost a half lower than that of the east slope and the east slope is almost half as thick as the west slope.
(2) The permafrost is most developed in the marsh wetland of valley area, with low ground temperature and large thickness. Under the impacting of the latitudinal zonality, the altitude influences permafrost development, which can be found in many other mountains of Daxing'anling Mountains.
(3) The patchy, sparse, and sporadic permafrost in the Daxing'anling and Xiaoxing'anling Mountains is quite widespread, whose coverage area is larger than that of the predominantly continuous permafrost areas and widespread talik areas. In particular, most Xiaoxing'anling Mountains are such permafrost distribution areas, which are the mixed transition zones between permafrost and deep seasonally frozen soil, and especially is the sporadically distribution permafrost area.
(4) Vegetation often reduces surface temperature change by other environmental factors, such as reflecting solar radiation, shading, evapotranspiration, blocking wind and snow, moisturizing and absorbing water, etc. The relative thick permafrost develops well in the marsh wetlands and other lowlands where shrubs and birches grow and there is no permafrost in the pine forests on the hillsides. The marsh wetlands in the southern area have become patchy and sporadic permafrost residual zone.

2) **High-Altitude Permafrost**

High-altitude permafrost is also known as plateau permafrost. In China, it is mainly distributed on the mountains in Qinghai-Xizang Plateau.

Altitude restricts the formation and development of permafrost. Above a certain minimum altitude, the lowest altitude line is the lower limit of permafrost distribution, called the lower bound of natural geography (Fig. 1.20) (Cheng & Wang, 1982). With the increase of altitude, the patchy (discontinuous) permafrost gradually transits into predominantly continuous permafrost. Meanwhile, the mean

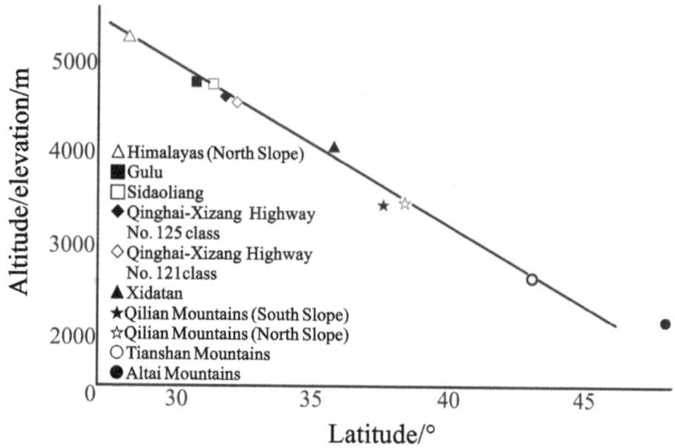

Fig. 1.20 Relationship between the height of the lower limit and latitude of high-altitude permafrost in Western China

annual ground temperature of the permafrost gradually decreases, the permafrost thickness increases, and the cryogenic phenomenon of permafrost also changes regularly with obvious vertical zonality. The lower limit of discontinuous permafrost roughly accords with the mean annual air temperature isotherm of −2 °C ~ −4 °C. With the increase of altitude, the patchy (discontinuous) permafrost gradually transits into predominantly continuous permafrost. Meanwhile, the mean annual ground temperature of the permafrost gradually decreases, the permafrost thickness becomes relative large, and the cryogenic phenomenon of permafrost also changes regularly, which has obvious vertical zonality. Glaciers and snow lines have an important impact on the development of permafrost.

The ingredients and properties of soil produce large influence on permafrost through its thermophysical properties and water content near the southern and lower limit of permafrost. Under the same condition, the permafrost thickness in hard-rock stratum is about 1.3 to 1.5 times of the loose soil layer, or even up to 2 times. Therefore, permafrost islands are often developed beneath the crushed rock near or below the lower limit of permafrost in high mountain areas, and permafrost can be found beneath crushed rock far from the permafrost region where the mean annual air temperature is positive.

1.5.1.2 The Dominant Influence of Temperature in Changing the Frozen Soil Engineering Properties

Under the natural environment, the mean annual ground temperature of permafrost is the result of the comprehensive effects of zonal and regional factors, which can, to a certain extent, reflect the continuity of permafrost distribution, the thickness of permafrost and talik on the vertical profile section, and also reflect the engineering thermal stability of permafrost and temperature-dependent changes in physic-mechanical properties.

The changes of permafrost temperature are mainly influenced by global warming and engineering activities. The global warming gradually changes the mean annual ground temperature of permafrost, which influences the development and degradation of permafrost extensively and regionally. Engineering activities changes the ground surface properties, solar radiation, and the heat exchange conditions of frozen soil resulting in permafrost warming and active layer thickness deepening.

1) **The Impact of Global Warming in China**

From 1905 to 2001, the annual average temperature in China had increased 0.79 °C, and the warming rate was about 0.08 °C/10a. From 1951 to 2001, the temperature change rate had reached 0.22 °C/10a, meaning that the temperature has increased about 1.1 °C in the last 51 years, the data of which is slightly higher than that of the global or the northern hemisphere average values in twentieth century. China and the Northern Hemisphere experienced two periods of high temperature in twentieth century. China's high-temperature period appeared in the 1930s and 1940s and after the mid-1980s (Fig. 1.21) (Tang & Ren, 2005), while the Northern Hemisphere it appeared in the 1940s and after the early 1980s. In twentieth century, China's temperature increased significantly in 1986, and the low temperature in the early stage was also more significant. Like the rest of the world, the warming in China in the past century mainly occurred in winter and spring, and there was a slight cooling trend in summer.

Under the background of global warming, the mean annual ground temperature of permafrost generally rises, the area and thickness of permafrost decrease, the permafrost table increases, the seasonal thawing depth increases, and the seasonal freezing depth decreases.

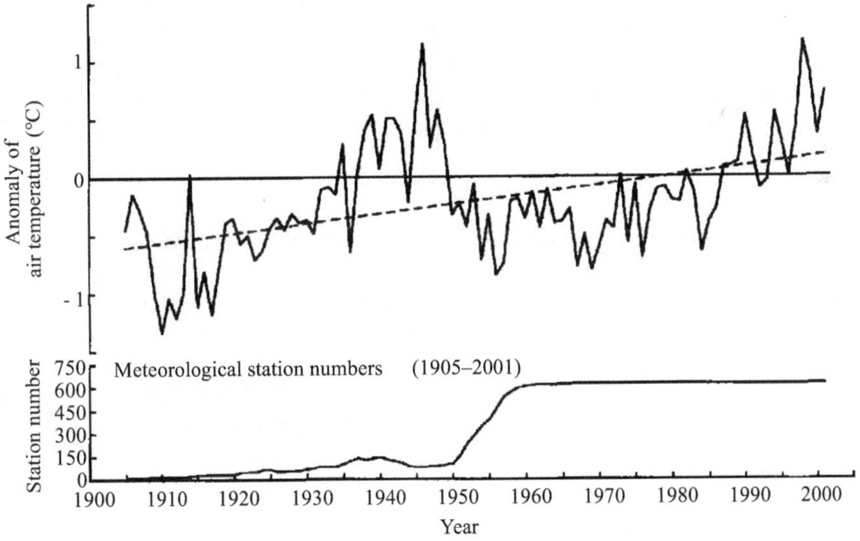

Fig. 1.21 Anomalies of annual average surface air temperature in China from 1901 to 2001

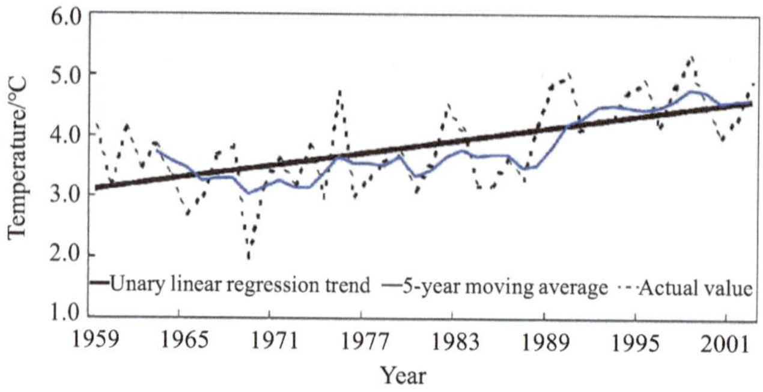

Fig. 1.22 Change in mean annual air temperature changes in North-Eastern China from 1959 to 2002

(1) **Changes of Air Temperature in the Permafrost Region of Daxing'anling Mountains**

Northeast China is one of the most significant regions affected by global warming. The mean annual air temperature in Northeast China from 1959 to 1988 was 3.05 °C and from 1989 to 2002 was 4.58 °C, with an increase of 1.08 °C from 1951 to 1988 (Fig. 1.22) (Sun et al., 2006a, b, c). In general, the air temperature shows a significantly warming trend, with the warming rate of 0.34 °C/10a. The annual warming trend in the coldest air temperature is almost twice as much as that of the highest temperature, indicating a relatively larger increase in the air temperature at nighttime. The air temperature increase in winter is the largest, which is three times as much as that in autumn. The 30-year annual average air thawing index and freezing index of the eight meteorological stations in the Da Xing'anling Mountains were 2606.2 °C·d and 2579.4 °C·d respectively. The annual average air thawing index shows an evident rising trend, while the freezing index shows an insignificant change trend, indicating a significant warming trend in warm seasons.

(2) **Temperature Changes in the Permafrost Region of the Qinghai-Xizang Plateau**

From 1950 to 2000, the annual average temperature change curve of 50 meteorological stations on the Qinghai-Xizang Plateau shows that (Fig. 1.23) (Zhou et al., 2005) the temperature is in an overall upward trend. The temperature had risen significantly in the 1950s, and there had been a temperature declining process in the 1960s; it began to rise in the 1970s, and continued to heat up after the mid-1980s, and the 1990s became the peak temperature period in the past 50 years. In the past 50 years, the average temperature in winter had increased 0.32 °C/10a, which was much higher than the warming trend in China (0.16 °C/10a).

Warming of the climate in the Plateau shows the linear rise of air temperature with different altitude and latitude (Table 1.7) (Li et al., 2005a, b). The air temperature in the part of northern and southern along the Qinghai-Xizang Railway increase greatly, especially in the southern part where the temperature rising rate is 0.52–0.58 °C/10a and the maximum rising rate in Lhasa is 0.58 °C/10a, whereas the data in the middle part is much lower, at 0.32 °C/10a ~ 0.39 °C/10a, and the

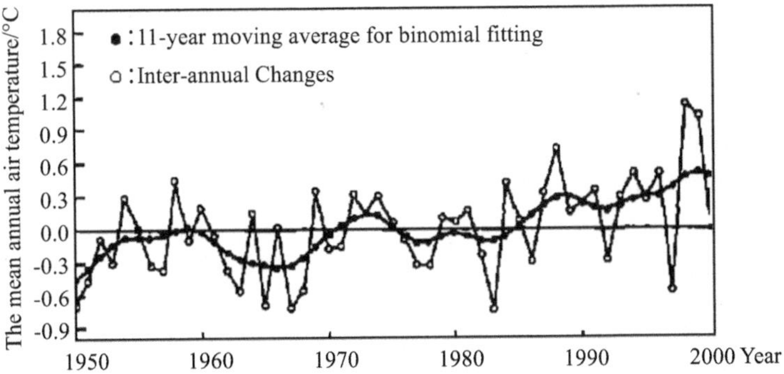

Fig. 1.23 The change of annual average air temperature in the Qinghai-Xizang Plateau

Table 1.7 Relationship between ground temperature and altitude and latitude along the Qinghai-Xizang Railway

Station Name	Golmud City	Wudaoliang Town	Tuotuo River	Amdo County	Nagqu City	Damxung County	Lhasa city
The Altitude of the Station/m	2807.6	4612.2	4533.1	4800.0	4507.0	4200.0	3648.7
Latitude/°N	36.25	35.13	34.13	32.21	31.29	30.29	29.40
Temperature Rising rate/°C·$10a^{-1}$	0.423	0.475	0.317	0.386	0.322	0.524	0.584

minimum rising rate in Tuotuo River is only 0.32 °C/10a. Generally, the rising rate of air temperature decreases with the increasing of the altitude for the meteorological station. The rising rate of air temperature in seasonally frozen soil regions and patchy permafrost regions is higher than that in permafrost region.

2) **Permafrost Degradation**

Due to the climate warming, the general trend of climate change in permafrost regions is warming, which started to rise significantly after 1986. Affected by climate warming and human production activities, the permafrost table has decreased, and the ground temperature has increased. Therefore, the area of permafrost has shrunk, and the southern limit (northern limit) of the permafrost has changed. The phenomenon of permafrost degradation under the influence of human activities such as urbanization, highway and railway construction, mining, pipeline building, hydraulic engineering construction, and land reclamation.

(1) Change of Permafrost Table

Under the influence of climate and human activities, the change of seasonal thawing depth of permafrost is the most obvious, especially in areas with intensive and frequent human activities, as shown in Fig. 1.24 (Jin et al., 2006). In the areas of southern limit of permafrost seasonal thawing depth deepens to 5 ~ 8 m (such as new district of Yakeshi eastern suburb, and abutment of Zhadunhe Reservoir, etc.).

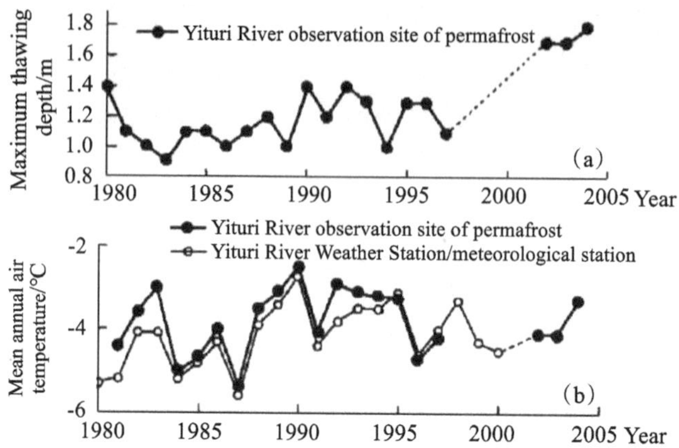

Fig. 1.24 Variation of the maximum seasonal thawing depth

Observation data in the permafrost regions of the Qinghai-Xizang Plateau show that the increase in air temperature has caused the changes of different extent in the maximum seasonal thawing depth over the past decade (Fig. 1.25). The change of active layer thickness in the cold permafrost regions is averagely 31 mm/a, such as No. 2, No. 7, No. 8, and No. 9 sites, while that in warm permafrost regions is averagely 81 mm/a, with the maximum 126 mm/a and the minimum 62 mm/a. The overall change trend of the average active layer thickness in the permafrost region of the Qinghai-Xizang Plateau is in concord with the change trend of the annual average temperature (Fig. 1.26) (Xu et al., 2017).

The permafrost table under the asphalt pavement of the Qinghai-Xizang Highway has changed greatly. In the cold permafrost area, the permafrost table under the asphalt pavement has a declining rate of 25 mm/a ~ 45 mm/a, which is slightly different from the natural site. In the warm permafrost region, the declining rate of the permafrost table is 167.5 mm/a ~ 205 mm/a (Wu et al., 2002), which is much higher than that of the natural state.

(2) Permafrost Warming and Thinning or Disappearing

In rivers and urban areas the permafrost temperature increases and the permafrost thickness becomes thin, or even disappears because of the urban heat island effect. The data of the observation borehole CK14 from the Yituli Railway Research Institute in Inner Mongolia (located on the first terrace on the north bank of the Yituli River) can be cited as example. Observation results show that the permafrost temperature at various depth of the CK14 borehole in the permafrost gradually increases, and the mean annual ground temperature at the depth of 13 m has increased by 0.2 °C in the past 13 years. Meanwhile the ground temperature from the YT-2 borehole, 30 m away from the CK14 borehole, has increased by about 0.4 °C from 1997 to 2010 (Fig. 1.27) (Chang et al., 2013). In the floodplain of the gold mining area Hanjiayuan, the lower reaches of the Huma River, the patchy permafrost has almost disappeared by 1995, and in the residential area of Amur Town,

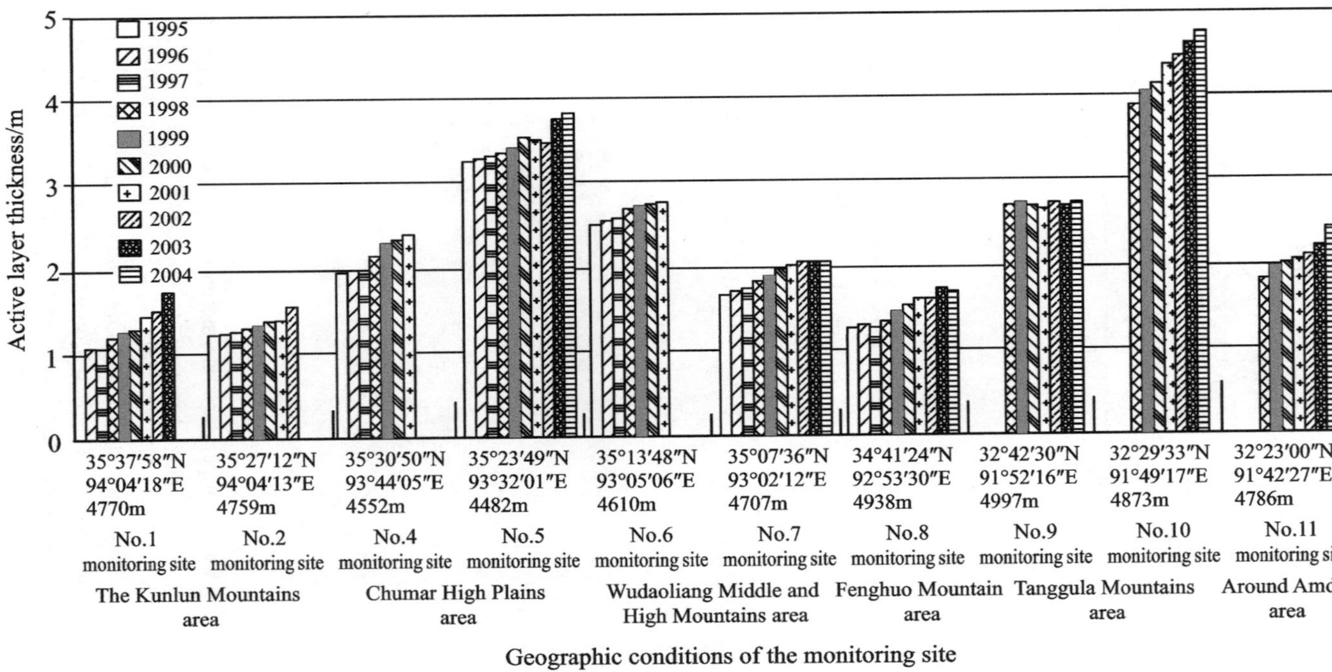

Fig. 1.25 The interannual variation of the active layer thickness in the permafrost region of the Qinghai-Xizang Plateau (Wu et al., 2005a, b, c)

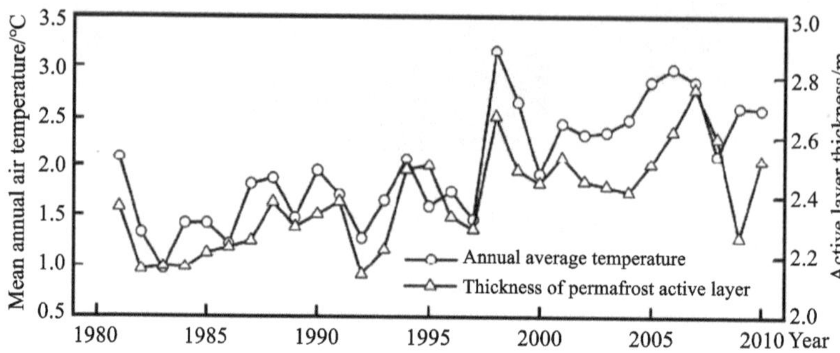

Fig. 1.26 Changes of active layer thickness and mean annual air temperature in Qinghai-Xizang Plateau

Fig. 1.27 Ground temperature changes of the No. 14 Borehole (1984–1997) and the adjacent YT-2 Borehole (2010) in Yituli River frozen soil station in Daxing'anling Mountains

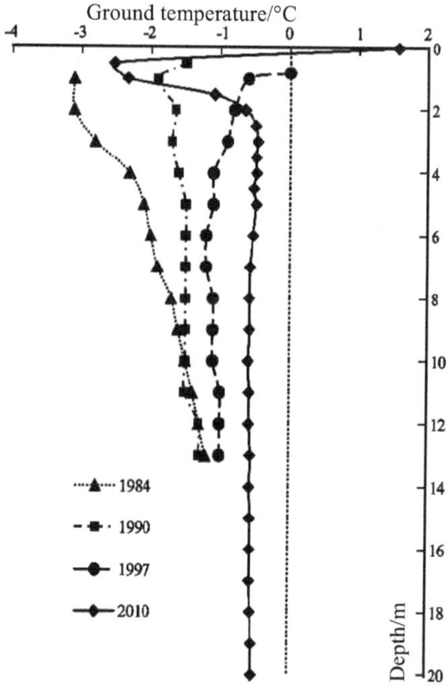

Mohe County the ground temperature has increased by 1.2 °C in 3 years. In the railway residential area of Mangui Town, Inner Mongolia, the mean annual ground temperature of the permafrost increases from −1.9 °C in 1973 to −0.6 °C by 2012.

Based upon the observation data of ground temperature from the borehole of 15 m depth in the sunny slope of Fenghuo Mountain on the Qinghai-Xizang Plateau, the mean annual ground temperature increased from −2.3 °C in 1978 to −1.43 °C in 2014, with an increasing rate of 0.26 °C/10a. From 2006 to 2011, the average increasing rate of mean annual permafrost temperature at the depth of 15 m from eight observation sites (4 sites for warm permafrost and 4 sites for cold permafrost) in permafrost region along the Qinghai-Xizang Railway was 0.18 °C/10a, with the rising rate of 0.32 °C/10a in cold permafrost regions (Cai et al., 2016).

(3) Changes in the Areas of Permafrost

The rising of air temperature leads to permafrost degradation, and especially in cities and towns, highways, railways and other areas with frequent human activities (such as deforestation, etc.) permafrost degrades more intensively. In the past 30 years, under the impact of the heat island effect, the area of permafrost has shrunk from continuous permafrost to discontinuous permafrost, and even disappears in some local areas.

Based on the relevant statistics relationships between the mean annual air temperature and the mean annual ground temperature of permafrost, the projection of the permafrost change under the scenario of climate warming shows the area of ground temperature zone of permafrost in the Qinghai-Xizang Plateau will change greatly (Fig. 1.28). The ground temperature zone of the cold permafrost (extremely stable zone, stable zone and basically stable zone) with the mean annual ground temperature at or below −1.5 °C changes little, and the distribution areas of the three zones decreased from 5.59%, 16.32%, and 25.5% in 2000, to 0.65%, 3.28%,

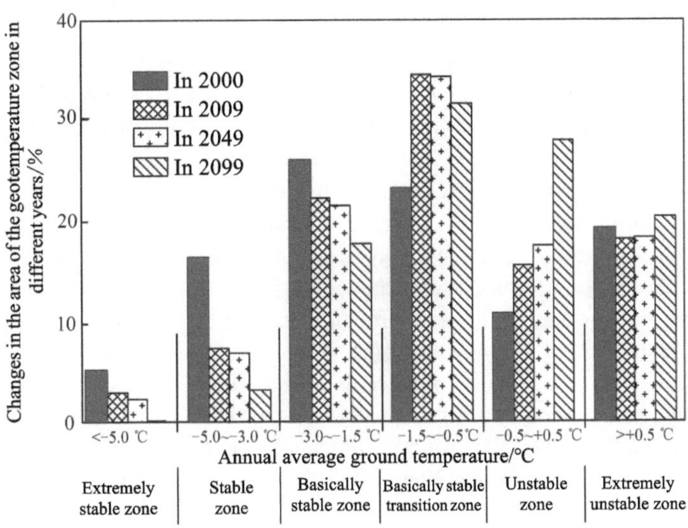

Fig. 1.28 Changes of permafrost geotemperature zone along the Qinghai-Xizang Highway (Wu et al., 2005a, b, c)

17.43% in 2099. The ground temperature zone of the warm permafrost (transition zone and unstable zone) with the mean annual ground temperature above −1.5 °C will gradually expand, and the distribution areas of the two zones increased from 22.85% and 10.8% in 2000, to 31.01% and 27.46% in 2099 respectively. With the gradually increasing of the mean annual ground temperature under the impact of climate warming, the warm permafrost will gradually be in the stage of degradation.

1.5.1.3 The Influence of Permafrost Degradation on Ecosystem and Engineering

The permafrost degradation causes permafrost table or active layer thickness to increase. The permafrost thawing releases a large amount of greenhouse gases, such as CH_4, CO_2 and N_2O, from the shallow permafrost layer. The research shows that the active layer thickness increases by 0.1 m, and the net emission of CH_4 in the wetland increases by 38% (Chu et al., 2017). According to the observation data, the CH_4 emission rate in the wetland of 13.3×10^4 km^2 in the Qinghai-Xizang Plateau is about 1 Tg a^{-1} (Jin et al., 1998).

Serving as the water-resisting layer for wetlands, the degradation of the permafrost will lead to the wetlands shrinking, drying up or even disappearing. The deepening of the active layer thickness will lead to a lowering groundwater table, the declining in vegetation coverage and an enhancing in aeolian action. This may further elevate the risk of adverse ecological successions, such as those from alpine meadow to alpine steppe, and further to alpine deserts.

As the climate warms up, alpine meadows, alpine steppe and swamp meadows are significantly degraded, and the vegetation coverage is reduced. The area of alpine meadows from Kunlun Mountains to Tanggula Mountains has decreased by 7.98% over the past of 15 years. The permafrost degradation in the Da Xing'anling Mountains leads to the northern coniferous forest belt gradually succeeding into the deciduous coniferous and broad-leaved mixed forest belt, and the Primordial Xingan larch degrading into the betula platyphylla secondary forest.

While the degradation and disappearance of permafrost leads to the suprapermafrost water transformation into an unfrozen water flow system with free water surface, and the decline of groundwater table causes the disappearance of spring water, and the water level of the well is reduced and even dries up. The surface water in the drainage basin is turned into groundwater, which increases the groundwater flow and seepage path, changes the groundwater circulation system, expands the reserves of the groundwater reservoir, and reduces the surface water flow and drying up the rivers as well.

Engineering construction is inevitable to excavate the surface, removing vegetation, build roads, and these activities will change the properties of the natural surface, destroy the thermal properties and heat transportation status of the frozen soil.

The permafrost warming and thawing will cause many frozen soil engineering problems. For example, thermal regressive slumping and thermokarst lake and lakeshore slumping will threaten embankment stability. Thermal erosion caused by the water depression in both sides of embankment results in thaw settlement of embankment. Human destruction and human intervention in ground surface environment cause the change of the groundwater runoff process, water flow and water distribution. The changes of hydrogeological conditions of frozen soil lead to freezing damages. The thermal thaw settlement and frost heave of frozen soil foundation endanger engineering stability.

1.5.2 Frozen Soil Environmental Protection Under Engineering Activities

1.5.2.1 Requirements and Measures for Permafrost Environmental Protection

1) **Environmental Protection**
The overarching objective of environmental protection in permafrost regions is to ensure the effective preservation of the permafrost environment, maintain the ecological functions of nature reserves, prevent pollution of rivers and water sources, ensure that wildlife migration remains unimpeded, preserve surrounding vegetation, protect trees and landscapes from damage caused by construction projects, safeguard wetlands, and control desertification. By achieving these aims, sustainable development can be realized.

2) **Measures to Protect Frozen Soil Environment**

(1) Establishing nature reserves;
(2) Strictly implementing national regulations of environmental protection;
(3) Legislating relevant regulations and rules.

The changes of natural environment will cause changes and imbalances of the permafrost environment. In the process of obtaining new thermal balance and stability, there will be a series of permafrost engineering problems, which will not only bring damages and hidden risks to engineering stability, but also cause permafrost environmental damage. Therefore, taking measures to minimize the damage of engineering activities to the permafrost environment can reduce the hidden dangers caused by the changes of permafrost and also give the maximum protection to the thermal stability of the permafrost. In the implementation of specific projects, corresponding regulations and rules should be formulated and implemented in order to achieve the overall goal of permafrost environmental protection.

1.5.2.2 Observation and Evaluation of Permafrost Environment

1) **Observation of Permafrost Environment**
The change of permafrost engineering properties mainly related to the change of permafrost temperature due to climate warming and the thermal disturbance from engineering. The changes of permafrost foundation are mainly reflected in the change of ground temperature, and the engineering stability is reflected in the building deformation. So, the long-term monitoring of permafrost engineering is the key factors. The changes of the surroundings of permafrost environment often affect the hydrothermal state and the engineering stability.

The implementation of long-term monitoring should take into account the climate, geology, geomorphology, and permafrost characteristics specific to these regions, with a database established accordingly. Building upon research into the interactions between climate, frozen soil, and engineering structures, a predictive model should be developed. Long-term monitoring data should then be utilized to issue early warnings regarding potential issues.

(1) Principles for Establishing Long-term Monitoring Systems. Monitoring sections should be established based on typical characteristics and a comprehensive evaluation of various factors, including climate, geological landforms, permafrost conditions (ground temperature and subsurface ice), engineering structures and mitigation measures, susceptible segments, complex topographical changes, and environmental requirements, among others.
(2) Monitoring items. The primary focus should be on monitoring the permafrost temperature and building deformation within the engineering construction area, as these factors can significantly impact or pose potential hazards to the project's permafrost environment and the unfavorable cryogenic phenomena.

2) **Evaluation of permafrost Environmental Changes**
Under the influence of climate and human engineering activities, the engineering geological conditions of permafrost are constantly changing and losing balance. The change of some factors will be accompanied by the change of other factors, which may be greater than the primary damage forms in quantity or quality, causing the change of permafrost environment and engineering damages. Especially, the coupling effect of water (ice) and heat (temperature) is the core factor to balance the changes of engineering geological conditions of permafrost (Fig. 1.29) (Tong et al., 1996). The thawing of underground ice is the essential factor that causes foundation deformation and engineering damages, and the heat exchange condition is the "source" of the change of engineering geological conditions of permafrost.

(1) Evaluation of Changes of Frozen Soil Engineering Geological Conditions
The magnitude of ground deformation and engineering damage resulting from changes of the permafrost environment depends on the ice content of the permafrost. The mean annual ground temperature of permafrost reflects the thermal stability of permafrost. The warm permafrost is more sensitive to the thermal effect caused by human engineering activities, resulting in the rapid response to

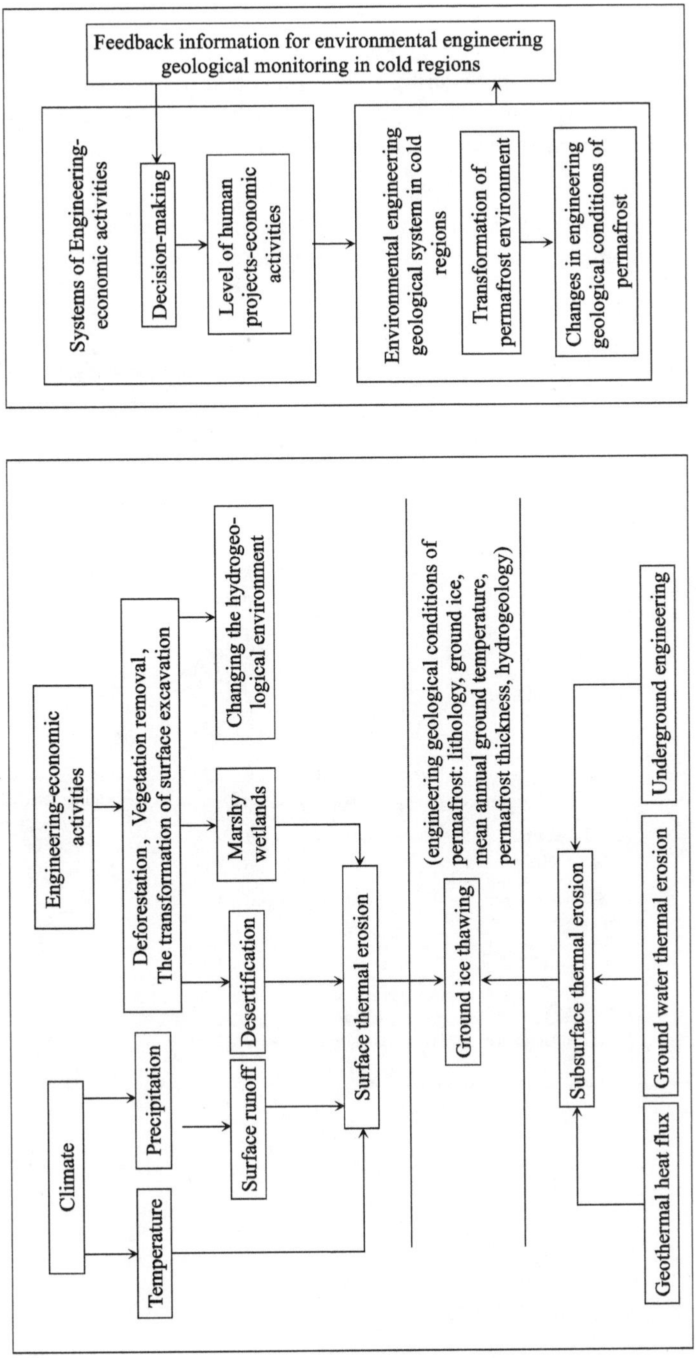

Fig. 1.29 Influencing factors of permafrost environmental protection

deformation. However, the cold permafrost region is more "inert" to respond to the thermal effect, resulting in correspondingly slower response to deformation. In the evaluation of permafrost engineering geological conditions under human engineering activities, the ice content of the permafrost layer is used as the preferred factor to evaluate the thaw settlement of the permafrost (Table 1.8).

(2) Evaluation of Changes in Permafrost Thermal Stability
Based on the relationship between the thermal stability of permafrost and the mean annual ground temperature, the thermal stability permafrost can be divided into four

Table 1.8 Evaluation in the change of permafrost engineering geological conditions under the Engineering Action

		Permafrost types and mean annual ground temperature/°C		
Lithology	Ice content i/%	Permafrost types	Mean annual ground temperature/°C	Human activities
Coarse-grained permafrost (sand, gravel, crushed stone)	Ice-poor frozen soil $i < 15$	Warm Permafrost	>−1.0 (−1.5)	Removing vegetation, filling the excavation, and changing the ground properties
		Cold Permafrost	<−1.0 (−1.5)	
	Icy frozen soil $i = 15 \sim 25$	Warm permafrost	>−1.0 (−1.5)	
		Cold permafrost	<−1.0 (−1.5)	
	Ice-rich frozen soil $i = 25 \sim 35$	Warm permafrost	> − -1.0 (−1.5)	
		Cold permafrost	<−1.0 (−1.5)	
	Ice-saturated frozen soil $i = 35 \sim 50$	Warm permafrost	>−1.0 (−1.5)	
		Cold permafrost	<−1.0 (−1.5)	
Fine-grained permafrost (silt, silty clay, and clay)	Ice-poor frozen soil $i < 10$	Warm permafrost	>−1.0 (−1.5)	Removing vegetation, filling the excavation, changing ground properties
		Cold permafrost	<−1.0 (−1.5)	
	Icy frozen soil $i = 10 \sim 20$	Warm permafrost	> −1.0 (−1.5)	
		Cold permafrost	<−1.0 (−1.5)	
	Ice-rich frozen soil $i = 20 \sim 30$	Warm permafrost	>−1.0 (−1.5)	
		Cold permafrost	<−1.0 (−1.5)	
	Ice-saturated frozen soil $i = 30 \sim 50$	Warm permafrost	>−1.0 (−1.5)	
		Cold permafrost	<−1.0 (−1.5)	
	Soil-containing ice layer $i > 50$	Warm permafrost	>−1.0 (−1.5)	
		Cold permafrost	<−1.0 (−1.5)	

Table 1.8 (continued)

Lithology	The chang of permafrost engineering geological conditions	Permafrost Deformation	Evaluation of permafrost engineering geological conditions
Coarse-grained permafrost (sand, gravel, crushed stone)	The active layer thickness increases rapidly, the permafrost temperature rapidly rising, and the extremely warm permafrost turns into a talik	Settlement deformation on the ground is relatively smaller, and engineering properties of soil do not change	Good
	The active layer thickness increases greatly, the permafrost temperature rises, and disconnected permafrost with thawed interlayer appears in most sections		
	The active layer thickness increases rapidly, the permafrost temperature increases rapidly, and disconnected permafrost with thawed interlayer appears in the extremely warm permafrost areas	Weak deformation can be observed on the ground, with weak thaw settlement; the thawed soil is saturated with water, and the strength decreases slightly	Better
	The active layer thickness increases greatly, the permafrost temperature rises, and disconnected permafrost with thawed interlayer appear in some sections		Good
	The active layer thickness increases greatly, the permafrost temperature rises quickly, and disconnected permafrost with thaw interlayer appear in some sections	The relatively larger deformation on the ground can be found in the warm permafrost area, some relatively smaller deformation on the ground can be found in the cold permafrost area, and the soil is saturated after permafrost thawed, showing thaw settlement with the thaw settlement coefficient of 3% ~ 10%, and the strength is reduced	Bad
	The active layer thickness increases, the permafrost temperature rises, and disconnected permafrost with thawed interlayer appear in some sections		Better
	The active layer thickness increases, the permafrost temperature rises, and disconnected permafrost with thawed interlayer appear in some sections	Uneven deformation on the ground can be found in the warm permafrost area, and obvious deformation on the ground can be found in the cold permafrost area. The soil is saturated after permafrost thawed, showing strong thaw settlement with the thaw settlement coefficient of 10% ~ 25%, accompanied with frost heaving and thermal thawing erosion on the slopes surface	Worse
	The active layer thickness increases, the permafrost temperature increases, and disconnected permafrost with thawed interlayer appears in some sections		Worst

(continued)

Table 1.8 (continued)

Lithology	The chang of permafrost engineering geological conditions	Permafrost Deformation	Evaluation of permafrost engineering geological conditions
Fine-grained permafrost (silt, silty clay, and clay)	The active layer thickness increases rapidly, the permafrost temperature rises rapidly, and the extremely warm permafrost degrades into taliks or disconnected permafrost with thawed interlayer	A slight settlement deformation on the ground can be found, and the engineering properties of the soil change slightly	Good
	The active layer thickness increases greatly, and the permafrost temperature rises more, and disconnected permafrost appears in most areas		
	The active layer thickness increases greatly, the permafrost temperature rises greatly, and disconnected permafrost with thawed interlayer occurs in the extremely warm permafrost area	The ground deformation is obvious, the permafrost engineering geological properties are obviously deteriorated, the thaw settlement is weak, the thawed soil is wet, the clay is hard plastic, and the strength is obviously reduced. Steep slopes may occur thermal slump; soil has weak frost heave	Better
	The active layer thickness increases, the permafrost temperature increases, and the disconnected permafrost with thawed interlayer appear in some sections		Good
	The active layer thickness increases more, the permafrost temperature rises, and the disconnected permafrost with thawed interlayer appears in some sections and there is water within thawed interlayer	Uneven deformation on the ground can be found in the warm permafrost area, and the ground in thecold permafrost area has obvious deformation. After thawing, the soil is in the state of saturated and soft plastic, showing the properties of thaw settlement with the thaw settlement coefficient of 3% ~ 10%, and soil has frost heave. And there is thaw slumping in the slope	Bad
	The active layer thickness increases, the permafrost temperature rises, and the disconnected permafrost with thawed interlayer appears in some sections		Better
	The active layer thickness increases, the permafrost temperature rises, and the disconnected permafrost with thawed interlayer appears in some sections and there is water in the thawed interlayer	Uneven deformation on the ground can be found in the warm permafrost area, and the ground in the cold permafrost area has obviously deformation. After thawing, soil is in the state of saturated and flow-plastic, showing strong thaw settlement properties with the thaw settlement coefficient of 10% ~ 25%; and showing strong frost heave, and there are thaw slumping, gelifluction and other phenomena in the slope	Worse
	The active layer thickness increases slightly, the permafrost temperature rises, and the disconnected permafrost with thawed interlayer may appear in some sections		Bad
	The active layer thickness increases slightly, the permafrost temperature rises slightly and the disconnected permafrost with thawed interlayer may appear in local sections	Uneven deformation on the ground can be found in the warm permafrost area, and the ground in the cold permafrost area has obviously deformation. After thawing, the soil is in the state of flow-plastic, showing thaw settlement with the thaw settlement coefficient of more than 25%, and showing extremely strong frost heave, and there are thaw slumping, gelifluction and other phenomena in the slope	Worst
	There are some small changes in active-layer thickness, the permafrost temperature slightly increased, and the disconnected permafrost with thawed interlayer may appear in local sections.		Bad

Note: The mean annual ground temperature in brackets is the division standard of highway system, and the ones without brackets is the division standard of the railway system categories (Table 1.9) (Wu et al., 2002): Iis thermally stable type; II is thermally stable transition type; IIIis thermally unstable type; IVis thermally extremely unstable type. That is to say, the higher the mean annual ground temperature is, the poorer the thermal stability of permafrost will be.

(3) Evaluation of Cryogenic Process in Permafrost Regions
Cryogenic phenomenon of permafrost results from thermal thawing effects in permafrost regions, and the core element of resulting in such phenomenon is hydrothermal (ice-heat) effect. Two conditions are required for their occurrence and development: ① the underground ice layer with a adequate thickness; ② the depth of seasonal thawing should exceed the buried depth of underground ice. There is no direct correlation between the development of the thawing process of the permafrost layer under the thermokarst lake and its ice content. Gelifluction and thaw slumping mainly occur in the slope areas, and thermokarst lake appears in flattish terrain. It can be evaluated by combining factors such as the ice content characteristics of permafrost, ground temperature dynamics, freeze–thaw activities, and human disturbance on the surface, etc., (Table 1.10), or the thermal sensitivity of permafrost can be used to evaluate the stability of the permafrost surface landscape (Wu et al., 2002).

Table 1.9 Evaluation of permafrost thermal stability

Permafrost Thermal Stability	MAGT/°C	Thermal stability Index	The impact of engineering activities on permafrost
I	<−3.0	>2.4	The engineering activities will not have a great influence on the permafrost thermal stability, the active layer thickness will increase, and the mean annual ground temperature will increase slightly.
II	−3.0 ~ −1.5	2.4 ~ 1.1	The engineering activities has a great influence on the permafrost thermal stability, the active layer thickness increases greatly, the mean annual ground temperature increases obviously, and the ground ice thaws affects the stability of engineering buildings.
III	−1.5 ~ −0.85	1.1 ~ 0.5	The permafrost thermal stability is greatly changed, the active layer thickness and the mean annual ground temperature are greatly increased. The permafrost degraded and the buildings are greatly deformed, which greatly affects the stability of engineering buildings.
IV	> − 0.85	<0.5	The engineering activities have a serious impact on the permafrost thermal stability. The permafrost is in serious degradation, and the engineering structures exhibit significant deformation, which significantly affects the stability of the engineering.

Notes: MAGT means mean annual air temperature; I means thermal stable permafrost; II means thermal stable transition permafrost; III means thermal unstable permafrost, and IV means thermal extremely unstable permafrost

Table 1.10 Cryogenic phenomena of frozen soil under the impact of thermal thawing action

Permafrost Thermal Stability	Engineering Types of Frozen Soil				
	S	D	F	B	H
I	None	None	Weak	Strong	Extremely strong
II	None	None	Weak	Strong	Extremely strong
III	None	Weaker	Weak	Extremely strong	Extremely strong
IV	None	Weaker	Strong	Extremely strong	Extremely strong

Notes: Cryogenic phenomena of frozen soil under the thermal thawing action include surface thaw settlement, thaw slumping, gelifluction, thermokarst lakes, thermal thawing erosion, and so on; S means Ice-poor frozen soil; D means Icy frozen soil; F means Ice-rich frozen soil; B means Ice-saturated frozen soil; H means soil-containing ice layer; I means thermal stable permafrost; II means thermal stable transition permafrost; III means thermal unstable permafrost; IV means extremely thermal unstable permafrost

Table 1.11 Classification index of hydraulic erosion intensity

<table>
<tr><td rowspan="2" colspan="2">Land types ground slope</td><td colspan="6">Ground slope</td></tr>
<tr><td><5°</td><td>5° ~ 8°</td><td>5° ~ 8°</td><td>5° ~ 8°</td><td>5° ~ 8°</td><td>>35°</td></tr>
<tr><td rowspan="5">Trees and grass coverage in non-arable lands(%)</td><td>>75</td><td>Slight</td><td colspan="3">Mild</td><td colspan="2" rowspan="2">Moderate</td></tr>
<tr><td>60 ~ 75</td><td colspan="4">Mild</td></tr>
<tr><td>45 ~ 60</td><td colspan="3">Mild</td><td rowspan="2">Moderate</td><td>Moderate</td><td>Strong</td></tr>
<tr><td>30 ~ 45</td><td colspan="2">Mild</td><td>Moderate</td><td>Strong</td><td>Extremely strong</td></tr>
<tr><td><30</td><td colspan="3">Moderate</td><td rowspan="2">Strong</td><td rowspan="2">Extremely strong</td><td rowspan="2">Severe</td></tr>
<tr><td colspan="2">Slope croplands</td><td>Slight</td><td>Mild</td><td>Moderate</td></tr>
</table>

Frost heave, frost mounds, icing, frost heaving cracks and other cryogenic phenomena caused by the frost action in permafrost are the result of the combined action of soil, water and temperature, which can be judged according to the frost heaving classification of soil in the *Code for Frozen Soil Engineering Geological Investigation.*

(4) Evaluation of Permafrost Environmental Changes

The evaluation of permafrost environment includes water and soil loss, desertification, freezing–thawing erosion and so on.

Hydraulic erosion refers to the type of soil erosion with surface water as the main erosion force. It is related to precipitation, surface runoff, subsurface runoff, topography, vegetation, and the material composition of soil, soil body and other ground components (Table 1.11).

Wind erosion refers to the type of soil erosion with wind as the main erosion force. The intensity of wind erosion is related to surface configuration and vegetation coverage (Table 1.12).

Table 1.12 Classification of wind erosion (Desertification) degree

Level	Standard of erosion (Desertification) intensity		Comprehensive landscape features	Degree of soil desertification
	Vegetation Coverage	Proportion of sand area		
Slight	>60%	<5%	There is no quicksand in most of the lands, the distribution of quicksand is spotted, and there is basically no wind erosion on the windward slope of the sunken sand dunes	Potential desertification
Mild	30% ~ 60%	5% ~ 25%	There are small pieces of quicksand and long wind-eroded gullies, wind-eroded breaches, dense clusters of sand mounds and wind-eroded pits appear on the windward slope of the sunken sand dunes, the ground is thinly covered with sand or exposed sand and gravel, and the wind erosion loss of soil humus layer is less than 50%	Mild desertification
Moderate	10% ~ 30%	25% ~ 50%	The area of quicksand is large, the sand mounds are dense, the wind-eroded breaches on the windward slope of the sunken dunes is up to 1/2, and wind-eroded depressions with medium depth appear, which cannot be covered by shrubs, and most of the wind-eroded pits are exposed. The wind erosion is strong, and the loss of wind erosion is greater than or equal to 50%	Moderate desertification
Strong	<10%	>50%	Dense moving dunes are dominant, and large wind-eroded pits appear. The breaches of the windward slope of the sunken sand dunes reach the top of the mounds. The ground is desertified or gravelified by wind erosion. The sand pile vegetation presents residual piers and residual columns, and the soil humus layer is almost completely eroded by wind.	Strong desertification
Severe	0	100%	Moving dunes, presenting a desert, Gobi or Yardang landform with the complete loss of soil humus layer	Severe desertification

Table 1.13 Classification index of freeze-thawing erosion intensity

Grade	Comprehensive landscape features
Slight	Polygonal cracks and water erosion are common in flat ground and lush vegetation areas. And there are some sorted circles in the exposed areas.
Mild	On the gentle slopes with vegetation developed and relatively developed, the intensity of freezing–thawing action is general, polygone cracks are developed, with water erosion, and there are often spotted vegetations on the ground. There are some gelifluction phenomena under man-made and natural damage. Icing and frost mounds develop well.
Moderate	On the middle and high mountain slopes or the steep slopes, freezing–thawing weathering is severe, and a large number of gravel slopes or rock glaciers are formed under the action of thawed water and gravity. On the slopes with vegetation generally-developed, freeze–thaw action and water runoff are relatively large, and there are fish-scale hillsides and small-scale thaw slumping and gelifluction, and upfrozen stones. On the general flat land and the land with vegetation developed, there are large and medium-sized thermokarst lakes, as well as depressions formed by thermal thaw erosion.
Strong	On the snow-covered mountains with the altitude of 4000 ~ 5000 m, there are exposed bedrocks, strong freezing–thawing weathering, and large runoff by thawed water. On the slopes with relatively developed vegetation, destroyed by man-made or nature damages, the ground ice is exposed, resulting in multiple or large thaw slumping and gelifluction.

Freezing–thawing erosion refers to the softening and weathering of rock and soil, due to the action of multigelation in cold regions.The ground ice thawing in permafrost causes soil creep, thaw slump, gelifluction and so on. Freezing–thawing erosion is related to altitude, landform location, vegetation, surface composition, temperature conditions, ground ice layer in permafrost, icemelt (snowmelt), and human activities (Table 1.13).

Chapter 2
Construction Engineering in Frozen Soil Regions

The "construction engineering" in this chapter focuses on industrial and civil building in permafrost regions. In permafrost regions, the foundation of construction engineering is a special foundation soil containing "ice" at negative temperature, and its engineering properties are restricted by "ice" and negative temperature. With low load and negative temperature, the frozen soil is a good foundation with high strength; with high load and negative temperature, it has high strength but great creep characteristics. Therefore, under the influence of load and thermal effect, the permafrost will thaw gradually, showing the properties of thaw settlement and compressive deformation (Figs. 2.1 and 2.2).

The seasonal freezing layer and thawing layer will refreeze every winter due to air temperature decreasing, and the water in the foundation soil and the migrated water also freeze, leading to the "frost heave" deformation for the construction engineering. The construction engineering on the frozen soil foundation constitutes a thermodynamic system of construction engineering-foundation-subgrade. There are two important methods dealing with the system engineering. One is the site selection and the layout of the construction, and the other is the adoption of different design principles and measures. Three measures, such as eliminating the thermos-osmosis of construction engineering, intercepting its thermos-osmosis and heat leak, and cooling the frozen soil foundation, should be considered to prevent frost heave and thaw settlement.

Q. Wu, C. Tong, *Frozen Soil Engineering in Cold Regions*,
https://doi.org/10.1007/978-981-95-2527-0_2

Fig. 2.2 The obsolete office building with box-type foundation in Deerbuer due to thaw settlement (by Li Yingwu)

Fig. 2.1 Thaw settlement deformation of a two-story building in Alaska (by Jin Huijun)

2.1 Site Selection of Construction Engineering

The engineering stability in permafrost regions depends on the thermal and mechanical stability of the foundation soils at the construction site, and the construction site should be built at sections with good geological conditions for frozen soil engineering. Firstly, the site should be selected in the exposed bedrock sections or with shallow buried depth. Secondly, the site is chosen in the sections of coarse-grain soils without frost heave or with slight frost heave, and without thaw settlement or with slight thaw settlement. The best site is in the sunny drying sections where there are coarse-grain soils with low ice content at the radiation talik. Thirdly, the sites should avoid the sections with pingo, icing, thermokarst lakes, massive ground ice (high ice content), transition zones between taliks and permafrost, regressive thaw slumping sections and other areas with worse geological conditions. At the same time, foundation soil within the thaw depth affected by building should comprehensively analyze its physical and mechanical properties. The sites should be selected at the areas that are conducive to discharge surface water and industrial and domestic sewage. For the sites with already-existing buildings, it is necessary to avoid the drainage direction of the buildings and pay full attention to the frozen soil environment and its change after the engineering construction. For the unavoidable frozen soil sections with high ice content, the relevant thermodynamic calculation and special foundation design and engineering measures are necessary.

Based on the frozen soil geological conditions, construction sites should be chosen as much as possible in the sections of taliks, the exposed bedrock areas or with shallow buried depth, coarse-grained soil layers, sunny slopes, ice-poor and icy frozen soil areas, and the lowest areas of mean annual ground temperature, no surface water erosion and good drainage and so on. Although it is difficult to find satisfying construction sites that can meet all the conditions, at least the sites should be selected on the coarse-grained soil layer with ice-poor and icy frozen soil and good drainage on the sunny slope. Otherwise, in order to ensure the long-term stability of the construction engineering, it is necessary to adopt different measures, foundation types and building structures according to various foundation conditions of frozen soil.

2.2 Plane Layouts of the Buildings

The general plane layout of urban buildings in permafrost regions should consider the following requirements. For example, construction may be combined with the natural landscape of frozen soil in order to protect the environment of frozen soil. The plane layout of the buildings needs to be as simple as possible. The basement loads and heat sources of adjacent buildings should not be greatly different, and with relatively uniform foundation soil. If there are no special requirements, the best orientation of buildings is to the south. Under the same conditions, the daytime room temperature in a sunny room is about 5 °C higher than that in a north-facing room. At the same time, plane layout of buildings should consider the design of traffic roads, site drainage and heating pipe trenches and maintain a safe distance from the thermal effect on buildings. The minimum distance between main and secondary roads and the adjacent buildings should be more than 10 m. Attention should be paid to the mutual thermal effects among buildings, and the building units should be uniformly arranged according to the singular design principle of frozen soil foundation, so as to build similar buildings. Traffic arteries and green belts can be used against thermal effects among building units with different design principles.

The design principle of buildings for the frozen soil foundation mainly depends on the conditions of frozen soil foundation, including the buried depth of the bedrock surface, the lithology and soil physical-mechanical properties (freezing and thawing property, etc.) on the vertical section, and permafrost types (such as continuous permafrost, patchy discontinuous permafrost, connection between seasonally thawed layers and permafrost, and ice content), permafrost table, and the mean annual ground temperature of permafrost.

These conditions determine the structure of the buildings, the design principle of the foundations, the buried depth of the foundations, the treatments of the frozen foundations and engineering measures. Therefore, when laying out and designing a project, suitable building sites for specific building conditions should be selected based on the technical and economic analysis.

Firstly, the design principle of keeping the frozen state (principle I) should be applied in cold permafrost regions where the seasonally thawed layer is connected with the permafrost. Building construction in an area with many existing houses should consider the thermal effects between adjacent buildings, and use the same design principles as those for the existing buildings.

Secondly, in the permafrost areas where there is talik in vertical profile, the design principles ought to be determined on the basis of the minimum housing cost to ensure its stability. Design principle of keeping the frozen state (principle I) or keeping thawed state (allowing the frozen soil foundation to thaw gradually, namely the principle II, or allowing the frozen soil foundation to pre-thaw completely, namely the principle III) are permitted to consider to ensure buildings stability. When building in an area with many existing buildings, the thermal effects among buildings should be factored in particularly and the design principles should be consistent with the best principle for the whole building.

Thirdly, the building design can use the principle of pre-thawing (namely principle III) in the sections with deep buried permafrost or the sections with shallow bedrock.

Probabilistic method of the thermal-mechanical interaction between building and foundation is used to design the plane layout of buildings in Russian (Khrustalev & Pustovoit, 1988). There are three design methods. The first method is the optimal design of security based on the economic criterion. The second method is the design consideration based on safety operation of analyzing the similar buildings. The third method is the design based on the code. In the "foundation-subgrade" system, the stability of the permafrost foundation and thermal regime is a dynamic stochastic process. If the dynamic process is not taken into account in the design, it is difficult to ensure the safety of buildings in frozen soil area. If the first and the second methods are combined and the standard security is regarded as a lower limit, the optimal security can be obtained in an economical way under any circumstances. The higher the security is, the lower is the risk of the system.

Obviously, the economic cost of buildings includes the structural cost, the cost of cooling (or elevated ventilated foundation) frozen soil foundation, foundation cost and foundation removal cost. The "foundation-subgrade" system mainly involves the buried depth of foundation (h_m) and the thawing depth of permafrost foundation (H_r) (the natural permafrost table or the calculated depth of thaw bulb under heated buildings). According to the probability method of the foundation design, the design of cooling permafrost foundation by the cost estimation based on the minimum risk and the maximum safety can obtain the estimation of most optimal cost by considering the buried depth and thawing depth of the foundation. By comparing the total cost converted from several plane layout schemes (considering its architectural and functional purpose), the optimal plan of the building layout can be obtained after drawing on the frozen soil map with frozen soil characteristics.

2.3 Elements of Engineering Design in Permafrost Regions

The physical properties of frozen soil foundations are influenced by temperature, such as the freezing process of the foundation soil, the thawing of the frozen soil foundations, and changes in strength due to seasonal or long-term temperature changes. In the frozen state, most of the frozen soil foundations have relative high strength and impermeability. Under the heating structures, the thaw bulb caused by the thermal effects has a relative strong thaw settlement and compressive deformation. These characteristics are very important for the design of construction engineering which should be considered carefully.

2.3.1 The Design Principles of Frozen Soil Foundation

The permafrost regions can be divided into continuous, discontinuous, and patchy permafrost areas, as well as seasonally frozen soil areas. The permafrost foundation and seasonally frozen soil foundation can be used as the foundation of the buildings, which have the properties of thaw settlement and frost heaving in addition to general functions, causing the deformation and instability of building foundations.

As the foundation of buildings, permafrost thickness ranges from several meters to tens of meters, and the mean annual ground temperature is higher than −1.0 °C or lower than −1.0 °C. Permafrost foundations may show different thermal stability when they are interfered by external heat. When the construction site is chosen in the talik within the permafrost regions, the factor that the freezing depth may be up to 2 m ~ 4 m should be taken into consideration.

Frozen soil foundation not only bears the load of buildings, but also experiences the thermal effects of heating buildings and climate warming. In foundation design, not only the bearing capacity and deformability of frozen soil foundation need to meet the requirements of building stability, but also the influence of annual freeze-thaw cycle needs to be considered.

When permafrost are used as building foundation, it is necessary to consider how to maintain the firmness and stability of buildings in the whole process of design, construction and operation, and any deformation cannot be allowed. It is essential to consider how to keep the frozen soil foundation in a frozen state under various conditions, or to control the thawing of the frozen soil foundation during the construction and the usage of the buildings. All of these relate to the choice of construction principle in the whole design, construction and operation. The same construction principle within a building area is required to avoid mutual conflicts and interference with the foundation of the buildings.

The selection of design principles depends on the evaluation of the frozen soil geological conditions in the construction site. It is of great significance for selecting the design principles to determine ground temperature zoning of the construction site based on the thermal stability zoning by the mean annual ground temperature. At the same time, According to the classification of frozen soil engineering based on the ice content, it is important to determine whether the construction site belongs to the frozen soil with low ice content or with high ice content. It is necessary to evaluate the material compositions of frozen soil foundation, and to determine whether it belongs to coarse-grained soil layer or fine-grained soil layer. In the same ground temperature zone, the former belongs to the hard frozen soil foundation, and the latter may belong to the plastic frozen state. It is essential to judge the thaw settlement and compressibility of frozen soil foundation.

When choosing design principles for buildings, it is necessary to consider the characteristics of the buildings, including heat sources, floor area and structural treatment. As for permafrost, the mean annual ground temperature of permafrost often reflects its thermal stability influenced by external heat disturbance. The warm permafrost with the mean annual ground temperature (t_{cp}) higher than −1.0 °C is easily warmed or thawed by external heat disturbance, such as climate warming, anthropogenic heat source, and the environmental change of frozen soil. And the foundation of warm permafrost with high ice content has high compressibility and thaw settlement.

When permafrost is used as the foundation, there are usually in the frozen or the thawing state. If the permafrost is relatively thick in construction site, it will cost greatly to maintain or pre-thaw the frozen soil foundation. Considering permafrost conditions, foundation types and structural measures, it is still feasible to allow the permafrost foundation to thaw gradually to ensure the buildings stability during their service life. Although the above two design principles are still applicable, two more conditions need to be considered in the design. One is that the permafrost foundation under the maximum thaw bulb should still keep the frozen state, and the other is that the thaw settlement and mechanical strength of the thawed frozen soil foundation should meet the design requirements, or the foundation lies on the thawed soil layer crossing a thin permafrost layer.

If the permafrost is used as the foundation, there are three states that can be used in the design.

Principle I: Keeping the permafrost foundation in a frozen state during the whole service life of the buildings

The principle I is applicable to two situations. One is that the building structures and the engineering measures must ensure the thaw bulb nonoccurrence in the permafrost foundation and the other is that the foundation base must lie on the permafrost layer under the maximum thaw bulb estimated. In either case, it must be ensured that the load stress on the permafrost foundation is less than its ultimate long-term bearing capacity, and that the deformation of the foundation in operation does not exceed the allowable deformation.

Principle II: Gradual thawing state

The principle II is only applied to the bedrock, coarse-grained soil foundation with the ice-poor and icy. Even if the permafrost foundation thawed, its thaw settlement and compression can still meet the design requirements. The pile foundation is economic and reasonable to pass though the permafrost layer lying on the underlying non-frozen soil layer if the permafrost is relatively thin. When the permafrost is adopted as the foundation, no matter whether it is the ice-poor or icy permafrost, the load stress on the permafrost foundation should not exceed the allowable bearing capacity of thawed frozen soil, and the thaw settlement and compressive deformation of permafrost foundation should not exceed the allowable deformation values.

Principle III: Pre-thawing state

The principle III is applicable to the following five conditions. The first case is suitable to relatively high temperature and relatively thick permafrost areas without talik and engineering measures difficult to avoid the thawing of the foundation. The second case is suitable to permafrost areas with talik, which is relatively in high ground temperature and deep in permafrost table. For the above mentioned two cases, permafrost can be pre-thawed to reach the maximum depth of thaw bulb during the service life of the houses, which enables the thawed permafrost foundation to meet the design requirements using the methods of the compaction, reinforcement and the anti-frost heaving. The third case is suitable to patchy discontinuous permafrost areas with relatively high ground temperature and thin permafrost thickness. The permafrost can be pre-thawed artificially or naturally. The fourth case is suitable to sparse and sporadic permafrost areas with extremely high ground temperature and extremely thin permafrost thickness. The permafrost can be excavated or thawed naturally.

When the principle III is used to design the permafrost foundation, the load stress on the thawed foundation soil should not exceed the allowable bearing capacity of the foundation soil, and the compressive settlement deformation should not exceed the allowable deformation values. Notably, no matter what design principles are adopted, the foundation soil under the buildings will refreeze in the cold season every year. Therefore, the buried depth of the foundation and the structural strength of buildings must meet the stability requirements against frost heaving, and anti-frost heaving measures should be taken for the foundation within the seasonally freezing depth.

2.3.2 *The Ground Temperature of Permafrost*

Under the impact of climate, the influence of the variations of the mean annual ground temperature of permafrost on its strength determines the selection of plane layout and design principles of construction engineering. Although the indoor temperature in heating buildings has limited thermal effect on frozen soil foundation, it will produce greater strength and deeper depth influence on the thermal regime of permafrost.

As for the buildings in permafrost regions, the formation of "thaw bulb" under the frozen soil foundation must be fully considered under the influence of the temperature field change and thermal effect under heating buildings. The thermal regime of permafrost at construction sites is not only affected by the original permafrost conditions and around constructions and other subsequent public facilities, but also affected by the total density of buildings (especially over 30–40%) and local centralized heat sources (such as, heat workshops, boilers, public bathhouses, underground thermal pipelines). The permafrost will degrade after construction due to various thermal impacts. In cold permafrost areas, the lateral influence range of the thaw bulb under heating buildings is relatively small and cannot produce large thermal impact of permafrost on the buildings (Fig. 2.3) (Tsytovich, 1985). In warm permafrost areas, the lateral influence range of the thaw bulb under heating buildings is relative large and can lead permafrost thaw around the buildings.

For the heating houses with their width much larger than the depth of the seasonally active layer (annual thawing and freezing depth) on permafrost, if the lateral

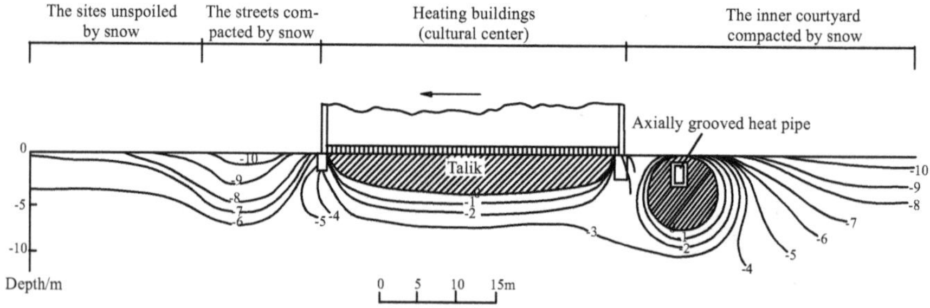

Fig. 2.3 The ground temperature field of permafrost layer under the heating buildings and streets

heat loss is omitted, here is the following equation according to the Fourier equation of steady heat transfer:

$$Q = \frac{\left(t_s - t_f\right)}{R_0} T \tag{2.1}$$

where Q is the heat transferred into the soil through the floor, kJ/h; t_s is the indoor temperature (positive temperature), °C; t_f is the mean ground temperature of permafrost, °C; R_0 is the thermal resistance of the indoor floor, °C/W; T is the time, hour.

During the time with positive temperature, there is a heat flow from the house into the soil layer (its strength depends on the temperature difference (t_s-t_f) and the thermal resistance of the floor (R_0)). When more heat continuously enters the permafrost layer, the result is the significant changes in the permafrost temperature and the formation of a thaw bulb under the floors of buildings.

When the permafrost is used as the foundation, that is, the foundation penetrates the thaw bulb into the permafrost (Design Principle I), the bearing capacity of the permafrost foundation is determined by its ground temperature. Hence, it is necessary to measure or calculate the highest ground temperature under the thaw bulb.

It is well known that the atmospheric temperature is transmitted downward from the ground to the soil, and its temperature amplitude changes with an exponential decay curve (Fig. 2.4). The temperature amplitude at y depth below the ground within its influence scope is:

$$T_y = T_0 e^{-y\sqrt{\frac{\pi}{Ta}}} \tag{2.2}$$

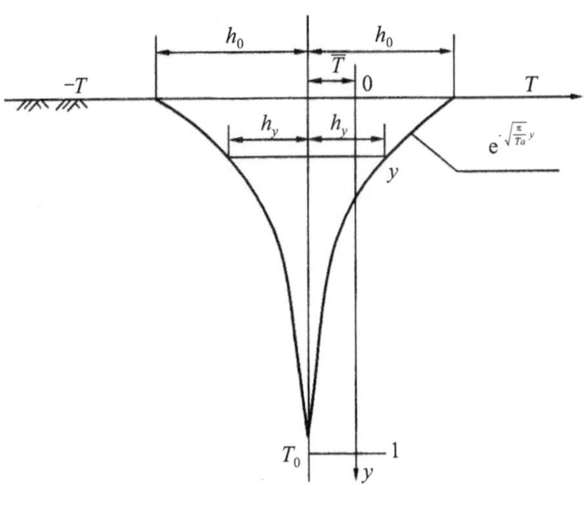

Fig. 2.4 The map of natural ground temperature transmission

where T_0 is ground temperature amplitude, °C; T is the climatic periodic variation, h; a is the thermal diffusivity of soil, m²/h; y is the depth from the ground, m.

The heating building means adding a new artificial heat source on the natural ground. Although its heat is minimal for the nature, it can interfere with the ground temperature under the buildings. Therefore, it is necessary to add an artificial heat source influence coefficient ξ to make the temperature amplitude increased (Fig. 2.5), and the temperature curve is described as:

$$T_y = T_0 e^{-y\sqrt{\frac{\pi}{Ta}}\cdot \xi y} \tag{2.3}$$

The envelope part in Fig. 2.5 shows the maximum and minimum value of permafrost temperature variation. Generally, the mean annual temperature $\overline{T}$ of frozen soil under the stable thaw bulb is basically equal to the mean annual ground temperature of permafrost, namely, $\overline{T} = t_{cp}$. The maximum mean monthly temperature of frozen soil at any depth of y under the stable thaw bulb is:

$$T_y = t_{cp}\left(1 - e^{-\sqrt{\frac{\pi}{Ta}}\xi y}\right) \tag{2.4}$$

where t_{cp} is the mean annual ground temperature of permafrost, °C; ξ is the influence coefficient of artificial heat sources, as shown in Fig. 2.6.

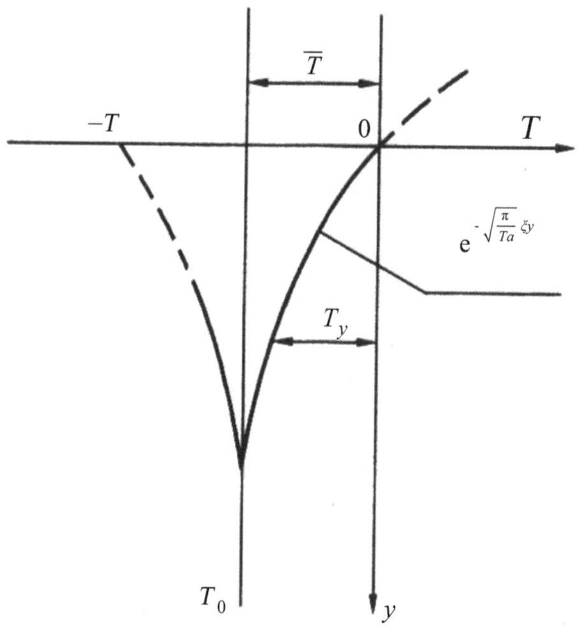

Fig. 2.5 The map of temperature wave downward transmission under the stable thaw bulb

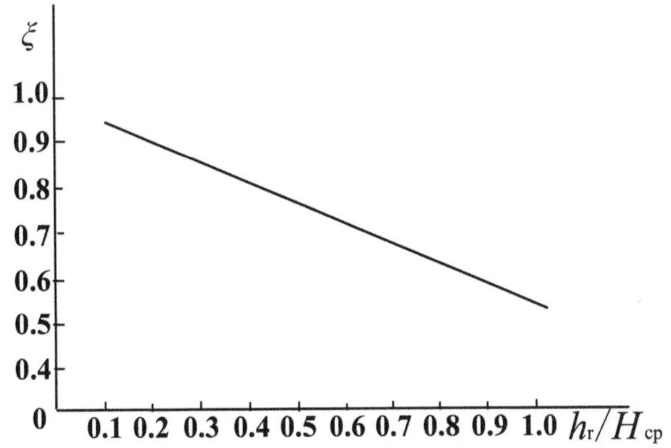

Fig. 2.6 The influence coefficient of artificial heat sources. h_r—The depth of the thaw bulb from the indoor ground, m; H_{cp}—The depth of zero annual amplitude of permafrost temperature, m

Figure 2.6 is determined by the observation data of temperature (The Third Railway Survey and Design Institute Group Corporation, 1994). Under the same conditions, the maximum mean monthly temperature under the thaw bulb increases with the increase of the thawing depth and it is related to the ratio of the thawing depth (h_r) and the depth of zero annual amplitude of ground temperature (H_{cp}). The larger the ratio is, the higher the ground temperature becomes. When h_r is higher than H_{cp}, it is not economical to make the frozen soil under the thaw bulb to keep the frozen state (Principle I).

2.3.3 *Thawing Depth*

The size of thaw bulb under heating buildings in permafrost regions depends on the mean annual ground temperature of the permafrost, the size of building basement, indoor heating temperature and thermal insulation of floors. In all cases, a thaw bulb will form below the heating buildings, except that the building is small, with its width slightly larger than its refreezing depth in winter.

It is a very complicated research to analyze and calculate the ground temperature of frozen soil during the period from its formation to its steady state for the thaw bulb under heating houses. The complexity lies in the precipitation of latent heat after pore ice thawing at the boundary between thawed zone and frozen zone of permafrost. However, thaw bulb depth is also an important index in the design of the heating houses for the frozen soil foundation, especially a primary design index in adopting the design principle of allowing the permafrost foundation to thaw.

The depth calculation of the thaw bulb of permafrost foundation under the heating houses involves three-dimensional temperature fields with unstable thermal conductivity. China has also put forward many calculation methods, some of which are complicated and others are quite different from the actual situation.

The formula for calculating the maximum thawing depth (H_{max}) of permafrost foundation under heating houses is:

$$H_{\max} = \psi_j \frac{\lambda_u T_B}{\lambda_u T_B - \lambda_f T_{cp}} B + \psi_c h_c - \psi_\Delta \Delta h \quad (2.5)$$

where λ_u is the weighted average thermal conductivity of foundation soil (including the structural materials of the indoor and outdoor height difference, such as the floor and insulation layer) in the thawing state, [W/(m °C)]; λ_f is the weighted average thermal conductivity of foundation soil in the freezing state, [W/(m °C)]; T_B is the mean surface temperature of the indoor, °C. If the indoor ground is provided with enough thick insulation layer, T_B approximately equals to the room temperature subtracting by 2.5 °C ~ 3.5 °C. T_{cp} is the mean annual ground temperature of permafrost (the ground temperature at zero annual amplitude depth), °C; B is the width of the house (the distance between the center of the front and rear exterior walls), m. Ψ_J is the comprehensive influence coefficient, as shown in Fig. 2.7. Ψ_c is the soil quality coefficient of coarse-grained soil, which is valued as shown in Fig. 2.8. Ψ_Δ is the influence coefficient of indoor-outdoor height difference, as shown in Fig. 2.9. h_c is the thickness of coarse-grained soil within the calculated thawing depth, m; Δh is the indoor-outdoor height difference, m.

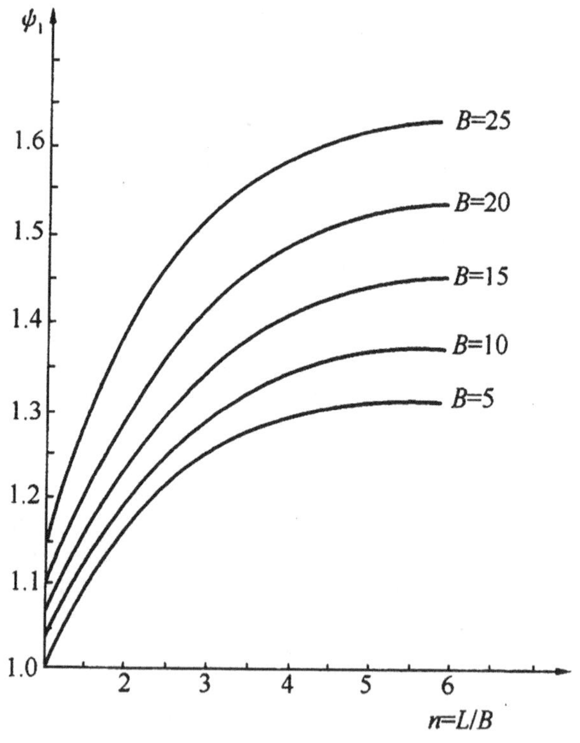

Fig. 2.7 Comprehensive influence coefficient Ψ_J. B—the width of houses, m; L—the length of houses, m

Fig. 2.8 Soil quality coefficient Ψ_c. 1—Sandy gravel; 2—Crushed stone; 3—Pebbles

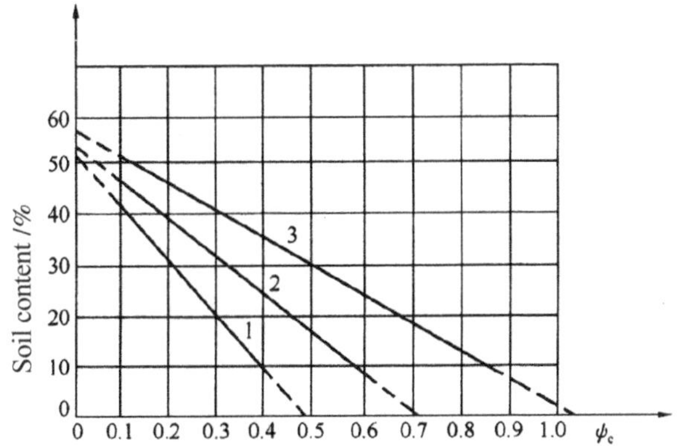

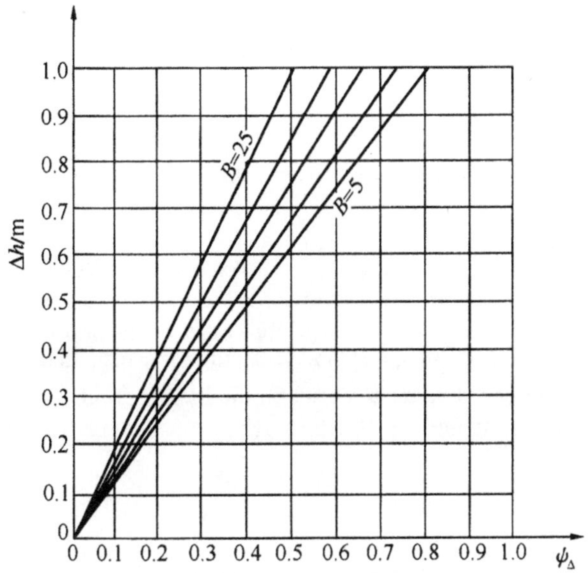

Fig. 2.9 The influence coefficient Ψ_Δ of indoor and outdoor height difference

Generally, the indoor-outdoor height difference will be no less than 0.45 m after the thawed foundation soil is compacted. Buildings in permafrost regions should be equipped with sufficient thermal insulation layer and thick plinth.

The drawing of the thaw bulb based on the temperature observation data shows that the shape the thaw bulb is reasonably expressed by the parabolic equation $y = \alpha x^2$ which takes the center line of the cross-section of the house as the coordinate y-axis. Due to indoor temperature and the house width, the focus position of the parabola of thaw bulb is also different, that is, the shape coefficient α is different. In addition, solar energy around the ground can cause the different heat amount of the houses with various orientations. Besides, the indoor heat sources

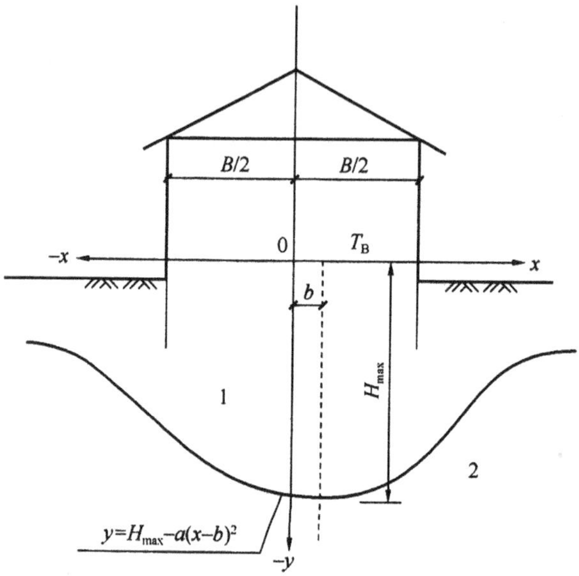

Fig. 2.10 The shape curve of the cross-section of the thaw bulb of houses. 1—Taliks; 2—Frozen soil area

(firewall, stoves, heated brick bed, etc.) are not distributed homogeneously, and the maximum thawing depth deviates towards the heat source, making the vertex of the parabola deviate from the y-axis of the center of the house by a certain distance b, which is also called b the shape coefficient. For the convenience of calculation, the origin of coordinate can be set on the indoor ground, taking the indoor ground as the x-axis, and moving H_{max} up (Fig. 2.10), and then the equation $y = \alpha x^2$ becomes:

$$y = H_{\max} - \alpha\left(x - b\right)^2 \tag{2.6}$$

where $H_{\max}$ is the maximum thawing depth of foundation soil of buildings, m; α is the shape coefficient of the thaw bulb, 1/m; b is the distance from the maximum thawing depth away from the center of the building, m; x is the distance between the desired point of thawing depth and the origin of coordinate, m.

Under the external wall on the south or east: $x = B/2$; under the external wall on the north or west: $X = -(B/2)$. α and b are referred to as shape coefficients, which can be determined by Table 2.1 obtained from statistics.

2.3.4 *Freezing Depth*

For heating houses of any foundation type in frozen soil area, the freezing process of the foundation soil also changes with that of the outdoor environment. And the outdoor freezing line still intrudes into the inside of the foundation of the heating

Table 2.1 Values of the shape coefficient a and b of the cross-section of the thaw bulb of houses

The types of houses		Dormitory and houses	Apartments and hotels	Infirmary and telephone offices	Stores	Offices	Station buildings or analogous houses
a/m^{-1}		0.06 ~ 0.16	0.04 ~ 0.10	0.05 ~ 0.11	0.05 ~ 0.14	0.05 ~ 0.12	0.04 ~ 0.09
b/m	North-south (eastward)	0.10 ~ 1.00	0.20 ~ 1.20	0.50 ~ 1.40	0.30 ~ 1.00	0.30 ~ 1.20	0.30 ~ 1.60
	East-west (westward)	0.00 ~ 0.30	0.00 ~ 0.60	0.00 ~ 0.40	0.00 ~ 0.40	0.00 ~ 0.50	0.00 ~ 0.70

houses, forming the smallest thaw bulb for the coldest seasons. Regardless of the permafrost regions or seasonally frozen soil areas, the foundation soil outside the heating houses will freeze as the ground freezes in the cold season, but the development process of the foundation soil freezing outside the walls is inconsistent. When the indoor floor of the heating houses is in direct contact with the ground, the maximum intrusion range of the freezing line is related to the indoor ground temperature of the heating houses. The higher the indoor temperature is, the smaller the freezing depth of the foundation soil is. When the temperature of the indoor ground is up to 13 °C ~ 18 °C, the intrusion range of the freezing line reaches about 0.5–2.0 m inside the exterior wall foundation. Normally, the intrusion range of the shady side is larger than that of the sunny side. The freezing depth outside the wall varies with the freezing strength of the ground, and its freezing depth is roughly 70% ~ 90% of the natural maximum freezing depth (80% ~ 90% for the north wall, 70% ~ 80% for the south wall, and 90% for the corners). The maximum freezing depth inside the wall is about 40–75% of the exterior wall (about 40% in the middle of the foundation and about 75% in corners). Whether the freezing line reaches the depth of the foundation base depends on the buried depth of the foundation.

The influence of height difference between indoor and outdoor floors on freezing depth cannot be ignored. When the height difference is 0 m, 0.3 m, 0.5 m and 0.75 m, the indoor freezing depth is 82%, 86%, 98% and 100% of the maximum depth of the ground. When a heating house was set with an elevated ventilated foundation, the freezing line under the house will be deeper than the maximum freezing depth of the outdoor ground, because the temperature of the elevated ventilated foundation is colder than the natural temperature.

Frost heaving will occur when the foundation soil refreezes under the heating house, causing frost heaving damage to the building. When the foundation base is below the freezing depth, the foundation soil will exert tangential frost heaving force on the side of the foundation of the house. If the foundation base is within the freezing depth, the combined actions of tangential and normal frost heaving force can be found on the side of the foundation and on the foundation base as well. Therefore, the maximum seasonally freezing depth of the construction site should be understood before designing.

2.3.5 The Stability of the Foundation under Frost Heaving Force

In frozen soil area (including seasonally frozen soil areas and permafrost regions), the foundation of buildings is buried under the seasonal freezing depth or in the permafrost layer. The active layer will seasonally freeze every year in the cold season and some of them can refreeze to the foundation base, and some only around the foundation. In either case, the foundation of the buildings will inevitably be affected by the frost heaving force generated by the refreezing of foundation soil. Buildings will subject to the combined impacting of the tangential and normal frost heaving force on the side of the foundation and on the foundation base. In the seasonally frozen soil areas the permafrost regions will only be impacted by the tangential frost heaving force on the side of the foundation, causing the foundation to be pulled up. When the uplift force of frost heaving exceeds the load stress of the buildings on the foundation, the foundation will lose its stability and cause the damage to the buildings. Therefore, when designing houses in frozen soil areas, the stability of the foundation under the action of frost heaving forces should be checked during foundation soil refreezing.

Under the action of tangential frost heaving force, the stability of all foundations must meet the requirements of the following formula:

$$G_0 + R_{ta} > nA_{\tau}\sigma_{\tau} \tag{2.7}$$

where G_0 is the standard value of permanent load acting on the foundation, kN, including the self-weight of the foundation (the masonry and the plain concrete foundation) and all the weight (the pile foundation with tensile reinforcement); if the foundation is in the groundwater, the buoyant unit weight should be taken. R_{ta} is the characteristic value (this value is not considered for the foundation of plain concrete and masonry structure) of the anchoring force (or frictional resistance, or the freezing strength, etc.) generated by the foundation soil when the foundation extends under the frost heaving soil layer or frozen soil, kN. σ_{τ} is the standard value of unit tangential frost heaving force when the foundation soil is frozen, kPa, which should be taken according to the actual measurement data and the larger value shall be taken for the same type of soil with higher water content. A_{τ} is the lateral surface area frozen together with the foundation soil, m^2; n is the safety factor, which is 1.1 for statically determinate structures and 1.3 for statically indeterminate structures.

2.3.5.1 Pile Foundation and Pier Foundation

The following formula should be met:

$$0.9G_0 + R_{ta} \geq \Sigma\sigma_{\tau i}A_{\tau i} \tag{2.8}$$

where $\sigma_{\tau i}$ is the standard value of the unit tangential frost heaving force in the *i* layer of soil, kPa; $A_{\tau i}$ is the lateral surface area of the pile and pier foundation frozen together with *i* layer, m^2; R_{ta} is the anchoring force, which is calculated according to the following conditions:

1. In the seasonally frozen soil areas, the anchoring force (the frictional resistance) between the lateral surface of the pile and pier foundation and the non-frozen soil, it can be calculated according to the formula:

$$R_{ta} = \Sigma\left(0.5 q_{sia} A_{qi}\right) \tag{2.9}$$

where q_{sia} is the characteristic value of frictional resistance of the *i* layer and the lateral surface between soil and pile and pier foundation, kPa; A_{qi} is the lateral surface area of the pile and pier foundation in the *i* layer, m^2.

2. In permafrost regions, the design principle of keeping the frozen state of the foundation is considered, it can be generally divided into the following situations:
   1. When the elevated ventilated foundation is adopted without thaw bulb under the ground of the buildings, the anchoring force (the freezing strength) between the lateral surface of the foundation and the frozen soil can be calculated as follows:

$$R_{ta} = \sum_{1}^{i}\left(\tau_{fi} \cdot A_{fi}\right) \tag{2.10}$$

where τ_{fi} is the characteristic value of the freezing strength between *i* layer of frozen soil and foundation, kPa, determined according to the measured value. A_{fi} is the lateral surface areas of the foundation in *i* layer of frozen soil, m^2.

   2. When there is a thaw bulb under the buildings, in addition to the anchoring force of freezing strength between the foundation under the thaw bulb and the frozen soil, the frozen soil foundation in the thaw bulb gradually thaws, the negative frictional resistance (f_n) in the process of thawing and thawing frictional resistance (a relatively long time needed) can be found when the frozen soil foundation is completely thawed and consolidated. Therefore, the negative frictional resistance must be factored in while calculating the bearing capacity of piles in thawed soil, which accounts for about 8% according to the experiments (Orlando & Branko, 2011). Then the calculation formula is turned into:

$$R_{ta} = \sum_{1}^{i}\left(\tau_{fi} \cdot A_{fi}\right) - f_n \tag{2.11}$$

where τ_{fi} stands for the freezing strength between the frozen soil foundation and the pile under the thaw bulb, and A_{fi} is the lateral surface areas of frozen soil foundation and the pile under the thaw bulb.

3. The design principles of gradually thawing (principle II) or pre-thawing (principle III) are adopted in the warm permafrost or patchy permafrost region. The anchoring force (the frictional resistance) between the lateral surface of pile and pier foundation and non-frozen soil could be generally determined according to the calculation formula of seasonally frozen soil areas.

2.3.5.2 The Foundation of Heating Buildings

The tangential frost heaving force of pile and pier foundation can be estimated in the following cases.

1. Under the heating condition, the frost heaving force P_h acting on the foundation is calculated according to the following formula:

$$P_h = \frac{\psi_t + 1}{2} \psi_h P_e \tag{2.12}$$

where P_h is the frost heaving force acting on the foundation under the heating condition, kN; Ψ_t is the influence coefficient of heating on freezing depth, determined by Table 2.2. Ψ_h is the influence coefficient of the distribution of frozen soil around the foundation on the frost heaving force due to the heating of the building, determined by Table 2.3, and its applicable locations are shown in Fig. 2.11. P_e is the frost heaving force acting on the foundation in the exposed building, kN.

Table 2.2 Influence coefficient Ψ_t of heating on the freezing depth

The height difference between indoor and outdoor ground/mm	The middle section of the exterior wall	The section of corners of the exterior wall
≤300	0.70	0.85
≥50	1.00	1.00

Notes: 1. The corner of the exterior wall refer to the exterior wall within the range of 1.5 times of the design freezing depth of the partition on both sides, counted from the apex of the sunny corner of the exterior wall, and the rest part is the middle section

2. As for the non-heating rooms in the heating buildings (enclosed entrance porches, aisles and staircases, etc.), its influence coefficient of heating is the same as that of the exterior wall

3. The influence coefficient of heating on the freezing depth is applicable to the indoor ground built directly on the soil, and the average indoor temperature during the heating period is not less than 10 °C. When the temperature is less than 10 °C, Ψ_h should be 1.00

Table 2.3 Influence coefficient Ψ_h of the distribution of frozen soil around heating buildings on frost heaving force

Locations	Ψ_h
The convex corner of the wall (the sunny corner of the wall)	0.75
Straight lines (the vertical corner of the wall)	0.50
The concave corner of the wall (the shady corner of the wall)	0.25

Note: The side length of the corner section of wall is 1.5 times of the design freezing depth

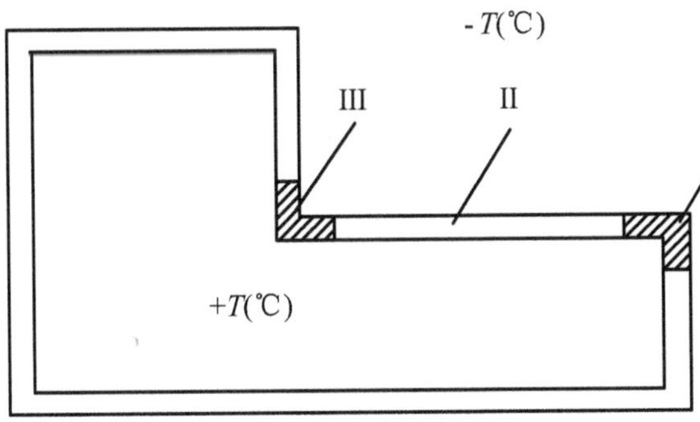

Fig. 2.11 Applicable locations of Ψ_h. I is the sunny corner of wall; II is the vertical corner of wall; III is the shady cirner of wall

2. The P_e can be calculated as follows:

$$P_e = \sum \tau_{fi} A_{fi} \tag{2.13}$$

3. The stability of the foundation should be calculated as the following formula:

$$0.9G_0 + R_{ta} \geq P_h \tag{2.14}$$

4. The influence coefficient of the freezing depth of the foundation for non-heating buildings shall comply with the following requirements:
   1. In non-heating buildings, the influence coefficient of the freezing depth for the interior and exterior wall foundation (Ψ_t) equals to 1.10. Non-heating buildings refer to those buildings whose indoor temperature is close to the natural air temperature and are rarely exposed to the sun.
   2. The influence coefficient of non-heating on the freezing depth shall not be used together with the shady slope coefficient of the influence coefficient of topography on freezing depth.

2.3.5.3 Self-Anchor Foundation

For self-anchor foundation, mainly including pedesdal pile, blasting belled pile and compacted expanded base pile foundation, its expanded part of the foundation must be placed below the maximum freezing depth, and the stability of the foundation under the action of the tangential frost heaving force on the foundation side must meet the following requirements:

$$0.9G_0 + A_{qi} R_{ta} \geq \sum \tau_{fi} A_{fi} \tag{2.15}$$

where R_{ta} is the reaction force generated by the overlying soil layer on the top of the expanded part of the foundation when the foundation is moved upward by the tangential frost heaving force, kPa, which it is taken according to the calculated value of the bearing capacity of the foundation under compression. When the overlying soil layer on the foundation is the backfill soil layer, the reaction force should be multiplied by a reduction coefficient of 0.6 ~ 0.8 based on the actual backfill soil quality. A_{qi} is the areas of top surface of the expanded part of the foundation, m^2.

2.3.6 *The Buried Depth of the Foundation*

The buried depth of the foundation of construction engineering shall be determined according to the following conditions:

1. Purpose of the building, the provision of a basement, the equipment and underground facilities, the type and structures of the foundation;
2. Load acting on the foundation and its nature;
3. Engineering geological and hydrogeological conditions;
4. buried depth of the foundation of adjacent buildings;
5. Effects of the frost heaving and thaw settlement on the foundation soil.

For the buried depth of the foundation of construction engineering in frozen soil area, special attention should be paid to the frost heaving and thaw settlement on the foundation soil. For buildings with more than two floors, the buried depth of the foundation should be increased, and pile and column foundation is mostly used for the foundation, and strip foundation are seldom used.

2.3.6.1 Seasonally Frozen Soil Areas

For buildings (structures) on the foundation soil with frost heaving susceptibility, the foundation usually should be buried below the maximum seasonal freezing depth, whether or not it is heated.

Because climate change will cause a 15–20% difference between mutli-year average maximum of seasonal freezing depth and maximum seasonal freezing depth, the seasonal freezing depth will probably exceed the standard freezing depth by extreme climate, which may result in adverse effects of the frost heaving soil on building foundation. Therefore, for the strong or extremely strong frost heaving soil, the buried depth of the foundation should be 0.25 m deeper than that of the design freezing depth. In the regions with shallow and medium seasonally freezing depths, the buried depth of the foundation should not be less than that of the design freezing depth. While in the regions with deep seasonally frozen soil, as for the non-frost heaving, weak frost heaving and frost heaving foundation soil, the foundation base can be buried within the design freezing depth, and the foundation could be buried shallowly, but it should be determined by the local engineering experiences and calculations.

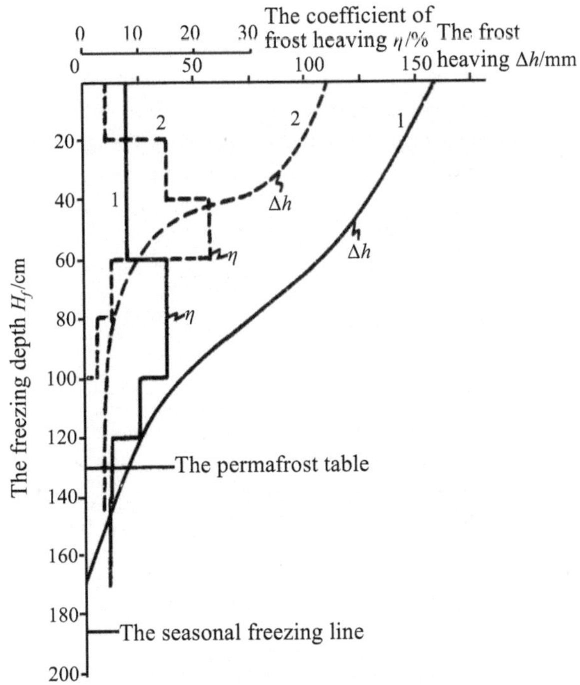

Fig. 2.12 Distribution of frost heaving and frost heaving rate along the freezing depth. 1—Seasonally frozen soil areas; 2—Permafrost regions

Generally, the main frost heaving zones of the foundation soil appears in the upper and middle parts of the seasonal freezing depth (Tong & Guan, 1985), that is, within 2/3 of the freezing depth, and more than 90% of the frost heaving occurs at this soil layer and the remaining 1/3 foundation soil has little frost heaving (Fig. 2.12). Therefore, as long as the bearing capacity of the foundation soil can meet the design requirements and anti-frost heaving measures are taken, it is possible to bury the foundation shallowly. Within the scope of the bearing layer of the foundation, the load of the buildings has a certain inhibitory effect on the frost heaving of the foundation soil.

When a certain thickness of frozen soil layer is allowed under the foundation base of the buildings, the following formula can be used to calculate the minimum buried depth of the foundation:

$$d_{\min} = z_d - h_{\max} \tag{2.16}$$

where d_{min} is the minimum buried depth of the foundation, m; z_d is the design freezing depth, m; h_{max} is the maximum thickness of the allowable residual frozen soil layer under the foundation base. In the regions with deep seasonally frozen soil, when the soil layer on the foundation base is non-frost heaving, weak frost heaving or frost heaving soil, the buried depth of the foundation may be less than the freezing depth of the site, and the allowable thickness of the frozen soil layer of the foundation base shall be determined according to the local experience.

What needs to be emphasized here is that anti-frost heaving measures must be taken on the lateral side of the foundation, whether the foundation base is buried below the seasonal freezing depth line, or there is a certain thickness of frozen soil layer under the foundation base. The simplest method is to backfill clean coarse sand and sandy gravel with a thickness of no less than 300 mm on both sides. If the foundation soil is silty clay or clay layer, a geotextile can be laid between the bottom of coarse-grained soil and the foundation soil to prevent the infiltration of fine-grained soil, and to avoid the anti-frost heaving layer having the frost heaving because of the infiltration of fine-grained soil.

2.3.6.2 Permafrost Regions

Before the 1980s, most of the heating houses in permafrost regions in China adopted strip foundations, and the foundation base was buried at the permafrost table, with a sand cushion of about 0.25 m under the base. However, a few years later, the houses were damaged to different degrees by thaw settlement and frost heaving of frozen soil foundation. Therefore, Russia, the United States, Canada and other countries basically adopted pile and column foundation in permafrost regions. At present, most of the construction engineering in permafrost regions in China use pile foundations and column foundations or elevated ventilated foundations, except for strip foundations or pier foundations in the talik (river talik) and the sections with the exposed bedrock or shallow buried depth.

The buried depth of the foundation of the buildings in permafrost regions varies according to the design principle:

(1) **Principle I: Maintaining the Frozen State and Using the Permafrost as the Foundation**

1) Elevated Ventilated Foundation

The elevated ventilated foundations in the United States, Canada and Russia generally have an elevated height of 1 m ~ 1.5 m, and some are about half a floor high [equipped with heating, water supply and drainage pipes [Fig. 2.13 (Chen et al., 2006) and Fig. 2.14 (Wu et al., 2005a, b, c)]. The elevated height of construction engineering in China is mostly 0.5 m ~ 1.0 m [Figs. 2.15 and 2.16 (Wu et al., 2005a, b, c)].

The depth of the thaw bulb under the elevated ventilated foundation is basically the same as the seasonally thawing depth. The maximum depth of thaw bulb is 2.1 m ~ 2.9 m about in the DaXiao Xing'anling Mountains and about 1.0 m ~ 2.0 m in the Qinghai-Xizang Plateau. Most of them are pile foundations in foreign countries, while most are column foundations in China. The buried depth of column foundations in China is 0.8 m ~ 2.5 m deep into the permafrost table. In this case, the buried depth of the foundation should not only meet the requirements of the bearing capacity, but also meet the requirements of stability against upfreezing (Figs. 2.17 and 2.18). The bearing capacity of the frozen soil foundation depends on the freezing strength between the permafrost and the lateral surface of pile

Fig. 2.13 Elevated houses in Alaska

Fig. 2.14 Elevated houses in Russia

foundation and the bearing capacity of the frozen soil foundation under the base of the pile and column foundation. As long as the freezing anchoring force of the foundation deep into the permafrost is greater than the frost heaving force, the depth is the buried depth of the pile foundation.

The minimum buried depth of the pile foundation is:

$$d_{\min} \geq \frac{n\sigma_{\tau} h_f}{\tau_f} - \frac{0.9G_0}{\pi D \tau_f} \tag{2.17}$$

$$d_{\min} = H_f + h_f$$

Fig. 2.15 Elevated buildings in Daxing'anling Mountains

Fig. 2.16 Elevated buildings in Wudaoliang on the Qinghai-Xizang Plateau

But, the following conditions could be met:

$$\sum_{i=1}^{n}\tau_{fi}A_{fi}+0.9G_0\geq\sum_{i=1}^{n}\sigma_{\tau i}A_{\tau i} \tag{2.18}$$

where σ_τ is the standard unit tangential frost heaving force during foundation soil refreezing, kPa; h_f is the seasonally freezing depth of refreezing foundation soil, m; τ_f is the freezing strength of permafrost, kPa; H_f is the depth of the pile foundation deep into the permafrost, m; D is the diameter of the pile foundation, m; G_0 is the permanent load acting on the pile foundation, kN; n is the safety factor.

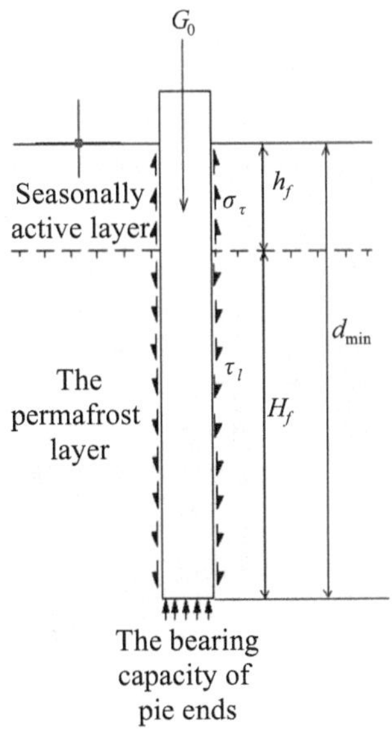

Fig. 2.17 Diagram of the system of forces acting on pile foundation in permafrost

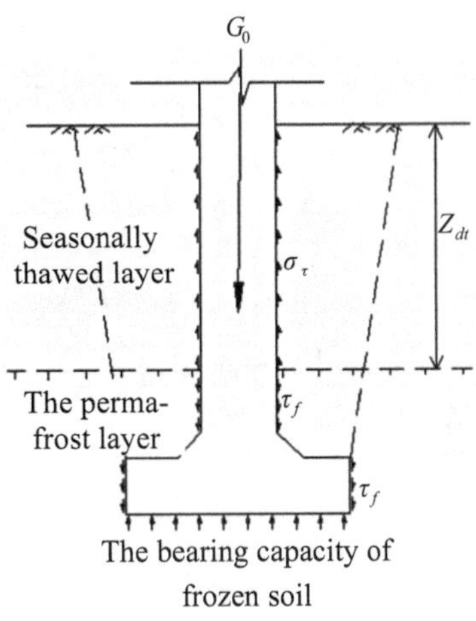

Fig. 2.18 Diagram of the system of forces acting on the expanded foundation in permafrost

For the pile and column shallow foundations, the minimum buried depth of the foundation base shall be greater than 0.5 m below the stable artificial permafrost table. When there is no stable artificial permafrost table for the buildings, the minimum buried depth of the foundation can be determined by the design thawing depth (Z_{dt}) of the permafrost.

The design thawing depth is equal to the standard thawing depth multiplied by various influence coefficients. The standard thawing depth can be estimated by the average maximum thawing depth measured in more than 10 years in an open and flat natural site in the permafrost region without talik, where soil layer within the seasonally thawed layer belongs to cohesive soil and cannot produced thaw settlement.

2) No Elevated Ventilated Foundation

The floors of the house are directly built on the ground (or elevating the indoor floor), and the thaw bulb will form under the floor. At this time, the foundation of houses still uses the permafrost under the thaw bulb as the foundation, and the foundation should be placed in the permafrost under the maximum depth of the thaw bulb. Therefore, the maximum depth of the thaw bulb for heating houses is about 7 m ~ 10 m, which is at least 0.8 ~ 1.4 times of the width of the house. And the heating house should use pile foundation instead of shallow foundation.

(2) **Principle II: Using the Permafrost that is gradually thawing as the Foundation**

The principle II is only applicable to the permafrost layer within the maximum thawing depth, with no thaw settlement or weak thaw settlement and it is only suitable for ice-poor and icy frozen soil, rather than the frozen soil with high ice content.

In addition to meeting the requirements of the bearing capacity, the buried depth of the foundation requires calculation of settlement deformation during gradual thawing of the permafrost foundation, as well as stability verification of the foundation under frost heaving forces during winter refreezing of the foundation soil.

The foundation can adopt pile and column foundation or enlarged shallow foundation. Regardless of the foundation type, the buried depth of the foundation base should be deeper than the maximum thawing depth, preferably 1.5 m to 3.0 m deeper.

(3) **Principle III: Using the Permafrost that is pre-thawed as the Foundation**

The principle III is mainly applied to the coarse-grained soil layer in warm permafrost regions with low ice content. Otherwise, the compaction and consolidation of the soil layer after pre-thawing permafrost will consume a huge cost.

The buried depth of the foundation can be handled according to the design of seasonally frozen soil. However, the foundation base should still be placed 0.5 m below the maximum freezing (refreezing) depth line of the foundation, and the calculation of stability against frost heaving is required. Particularly, no matter what design principle is adopted, anti-frost heaving measures should be provided within the maximum seasonal thawing depth to avoid deformation and damage of buildings caused by frost heaving of the foundation soil.

2.3.7 The Foundation in Patchy Permafrost Regions

The patchy permafrost regions are mainly distributed in the boundary zone of continuous permafrost regions, low-lying marshes, and fine-grained soil regions with better vegetation cover, belonging to permafrost with talik. The permafrost thickness varies greatly, most of which are 3 m ~ 10 m, and some sections can reach more than 15 m, and the mean annual ground temperature is 0 °C ~ −0.5 °C, which is the extremely unstable permafrost. The buried depth of the permafrost table in patchy permafrost regions varies greatly, generally 1.0 m ~ 2.0 m for marshes, 1.5 m ~ 3.0 m for piedmont slopes, and 3.0 m ~ 8.0 m for flood plain and the sites affected by human activities. The patchy permafrost is distributed in isolation from each other, covering an area of reaching more than 1000 m^2 to as small as 100 m^2.

In patchy permafrost regions, the air temperature is usually higher, with more heat absorption than heat release. Under the global warming, the patchy permafrost is in the process of degradation and even extinction, with poor and extremely weak thermal stability to resist external and human-induced thermal disturbances. In discontinuous and patchy permafrost regions, ground temperature is very high, and the talik with suprapermafrost ground water can be found.

The site selection of building should avoid the patchy permafrost areas as far as possible, and should select the site of the dry and sunny slope with high terrain and poor vegetation cover. Additionally, it is suitable to select coarse-grained soil (sandy gravel and crushed stone) as foundation soil, and the building site should lie in taliks sections with good drainage conditions. If it is unavoidable due to various factors, it also needs to avoid the low-lying marshes.

The foundation design for construction engineering in patchy permafrost regions should consider the following factors.

1. When the permafrost is thin and the mean annual ground temperature is relatively higher, the design principal III is supposed to be adopted, that is, the design of pre-thawing state. The foundation base is set on the thawed soil layer or bedrock under the maximum seasonal freezing depth.
2. When the permafrost layer is thick and the mean annual ground temperature is relatively lower, the design principle II can be adopted, that is, the permafrost foundation is allowed to thaw gradually during the construction and operation, and the foundation base should be passed through the talik layer between permafrost and seasonal freezing layer and set on the permafrost layer with weak thaw settlement.

2.3.7.1 Natural Thawing

The method of natural thawing is suitable for patchy permafrost regions with thin permafrost and high mean annual ground temperature, and the principle III should be used in the engineering design. One or two years before the method is used, the ground cover (vegetation and peat layer, etc.) of the proposed construction site will be pre-removed in May and the surface water needs to be drained, and the permafrost layer will gradually thaw naturally relying on the sun and high temperature in the warm season. When the foundation is designed by the method of natural thawing, the most reasonable foundation is pile and column foundation. The following work need to be considered in foundation design:

1. The pile and column foundation base shall be placed on the reliable bearing layer through the maximum seasonal freezing depth and talik layer between permafrost layer and seasonal freezing layer.

2. The amount of thaw settlement and compression deformation shall be calculated when the bearing layer under the foundation base thaws naturally. If it cannot meet the requirements, measures of compression and reinforcement should be carried out, such as mechanical method, compaction of deep soil layer; physical method, mainly sand drain, etc.; chemical method, mainly chemical reinforcement using chemical solvents; electrochemical method, mainly injecting water glass and other reinforcement agent at the same time in the electro-osmosis.
3. The bearing capacity of the foundation should be taken into account of the foundation reaction of the thawed soil layer at the pile end and the friction of the thawed soil layer.
4. The foundation design doesn't consider the friction within the maximum seasonal freezing depth, but needs to check the stability of the foundation under frost heaving force. If stability of the foundation cannot meet the design requirements, anti-frost heaving measures should be taken to reduce or eliminate the tangential frost heaving force on the base side.
5. When the bearing capacity of the foundation soil does not meet the design requirements, the expanded foundation can be used, and the foundation base should be placed below the maximum freezing depth.

2.3.7.2 Controlling Thawing

The method of controlling thawing is applicable in patchy permafrost regions with thick permafrost layer and low mean annual ground temperature. The foundation design can adopt the principle II, which is, allowing the permafrost foundation to thaw gradually during construction and operation. The foundation design needs to consider two cases:

1. When the soil quality of permafrost foundation is coarse-grained soil with poor-ice or icy conditions, the principle II can be adopted, which is suitable for pile and column foundation. The foundation design needs to calculate the following factors:

   The depth of the largest thaw bulb of building below the ground;
   The thaw settlement of permafrost and compression deformation of foundation soil after thawing within the expected thawing depth;
   The bearing capacity of the permafrost foundation;
   The buried depth of the foundation, and the foundation base should be placed 3 m below the permafrost table;
   The stability of the foundation against frost heaving.

2. When the soil quality of foundation is permafrost with high ice content, it is appropriate to adopt the principle II. Firstly, the foundation should pass through the permafrost layer below the depth of the maximum thaw bulb, suggesting to adjust the pile foundation penetrating into the permafrost layer to a depth not less than 3 m. The foundation design needs to calculate the following factors:

   The depth of the maximum thaw bulb under the ground;
   The bearing capacity of the frozen soil foundation;

The stability of the foundation against frost heaving;
The thermal resistance of the floor of houses to reduce or eliminate the heat into the permafrost foundation.

Secondly, the relatively heat stability of permafrost foundation is maintained by using the elevated ventilated foundation and eliminating thermal effects of heating houses. Applicable foundations need to consider building structure adapted deformation, such as the elevated ventilated foundations with strip foundation, column foundation, or raft foundation that foundation base should penetrate into the permafrost layer to a depth not less than 3 m. At the same time, building design should take the effective measures to protect the construction site and environment around the buildings. The foundation design needs to calculate the following factors:

The area of the ventilation hole and the elevated height of the elevated ventilated foundation;
The thermal resistance of the floor and to keep the house warm and reduce or eliminate the heat into the frozen soil foundation;
The bearing capacity of the frozen soil foundation;
The stability of the foundation against frost heaving.

2.4 The Foundation in Frozen Soil

No matter what principles are adopted in the foundation design of construction engineering in permafrost regions, two basic requirements should be met: first, the load on the foundation should be less than the bearing capacity (frozen soil is the ultimate long-term bearing capacity; thawed soil is the allowable bearing capacity) of the foundation soil (frozen soil or thawed soil); second, the deformation during the operation of the foundation (the thaw settlement and compression deformation of the frozen soil) should not exceed the allowable value. At the same time, the buried depth and structure of building foundation should meet the requirements of the stability against frost heaving.

2.4.1 Issues to Be Considered in the Design of the Foundation

2.4.1.1 The Foundation in Seasonally Frozen Soil

The design of foundations in seasonally frozen soil areas, in addition to meeting the requirements of buildings, needs to consider foundation types, the buried depth of foundation and the stability against frost heaving. According to the survey of building damages in seasonally frozen soil areas, building destruction is caused by the displacement of frost heaving that is caused by water migration during the freezing process of the foundation soil to form ice lenses.

1. **Soil Frost Heaving**

Soil quality, water content and negative temperature are the three main factors causing frost heaving of the foundation soil. In the design of the foundation, if one of these three elements is eliminated or weakened, the frost heaving of the foundation soil will be greatly weakened or disappears and reduce the tangential or horizontal frost heaving force on the side of the foundation and the normal frost heaving force on the foundation base. Many measures can prevent the foundation from frost heaving, such as improving soil quality, setting thermal insulation, coating the side of the foundation with adhering grease materials, installing casing isolation, lowering the level of ground water, and increasing the buried depth of the foundation (Fig. 2.19). Comprehensive measures of anti-frost heaving is often more effective than the single measure.

2. **The Problem of the Buried Depth of the Foundation**

Sufficient attention should be paid to the buried depth of the foundation. In principle, the foundation base should be placed 0.5 m below the maximum seasonally freezing depth, especial in the sections where ground water exists. The shallowest buried depth of the multi-story buildings must exceed (2/3)–(4/5) the maximum seasonal freezing depth, due to inhibitory effect on the frost heaving of the foundation soil from the load of the multi-story buildings.

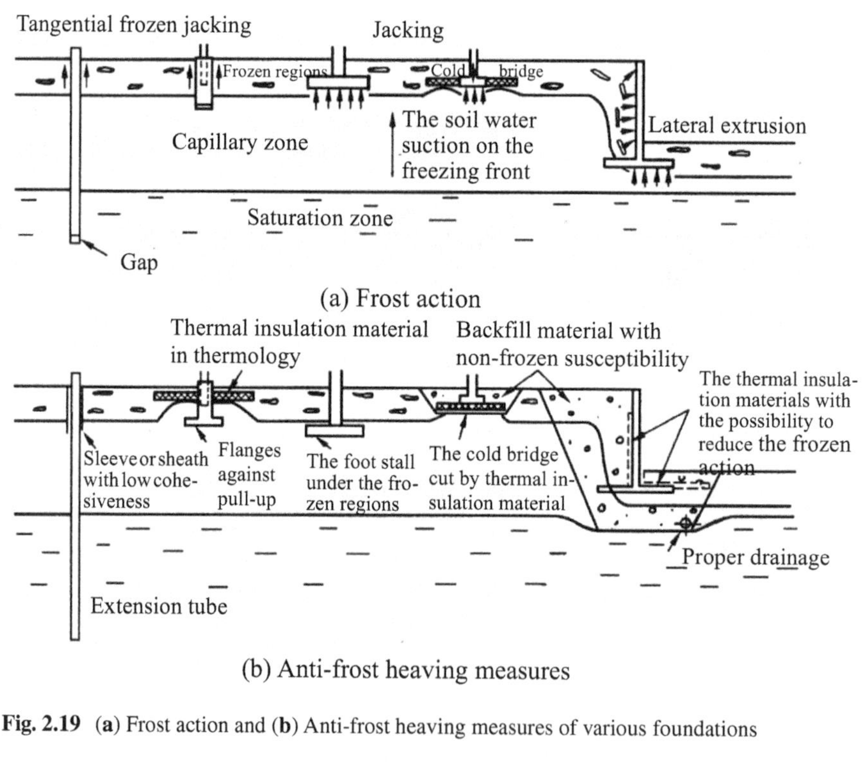

Fig. 2.19 (**a**) Frost action and (**b**) Anti-frost heaving measures of various foundations

3. **The Foundation Types**

The types of foundation suitable for buildings in seasonally frozen soil areas include column foundations, pier foundations, slope foundations (Fig. 2.20), and self-anchor foundations (Fig. 2.21). These foundations can use shallow foundations, but some special buildings can use pile foundations, mainly depending on the bearing capacity of the foundation soil, thaw settlement and the stability against frost heaving.

2.4.1.2 Foundations in Permafrost

The buildings destruction in permafrost regions is caused by the thaw settlement of the permafrost (Figs. 2.1 and 2.2), however, a few caused by the frost heaving (tangential and normal frost heaving forces) during the refreezing of the foundation soil.

1. **The Thaw Settlement**

The thaw settlement of permafrost mainly depends on the ice content, ground temperature and soil quality. The thawing of permafrost with high-ice content, such as massive ground ice, will produce larger thaw settlement, even thaw collapse (Fig. 2.1). The investigation of the thaw settlement and deformation of heating

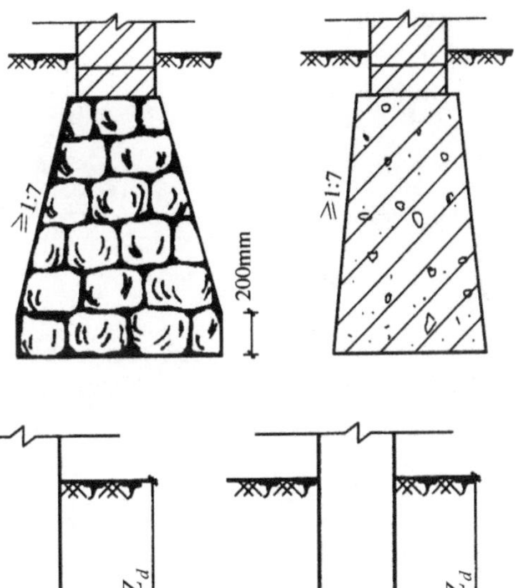

Fig. 2.20 Trapezoidal slope foundation

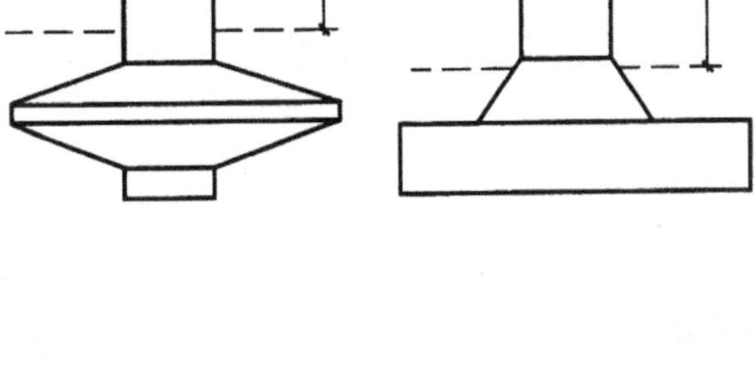

Fig. 2.21 Self-anchor foundation

houses in permafrost regions showed that most of which are caused by shallow foundations. For example, pile foundation with shallow buried depth (such explosive-expanded pile foundations) is found that the end of the pile foundation is set within the maximum thaw bulb.

2. **The Support Conditions of Foundations**

When the foundation end is set on the coarse-grained permafrost with low ice content, the foundation of permafrost thawed still has a relatively larger bearing capacity, and the foundation remains to be stable, which can be designed according to principle II. When the foundation end is set on the fine-grained permafrost with high ice content, under the effect of heating of building the foundation of permafrost thawed has a lower bearing capacity, resulting in thaw settlement and compress deformation, and the foundation stability become worse. Here the principle I and III can be adopted (Fig. 2.22), but anti-frost heaving measures should be taken in any case.

3. **Foundation Types**

The coarse-grained permafrost with low ice content can be used as the supporting foundation, adopting shallow foundations, column foundations and other foundations. In the fine-grained permafrost with high ice content, the supporting end of the foundation should be placed in the permafrost below the maximum thaw bulb through the seasonally active layer. For example, engineering design can use a pile foundation, an elevated ventilated foundation, or a cooling device to keep the foundation in frozen state, or thermal insulation layers reduce the infiltration of building heat, etc. (see Fig. 2.23).

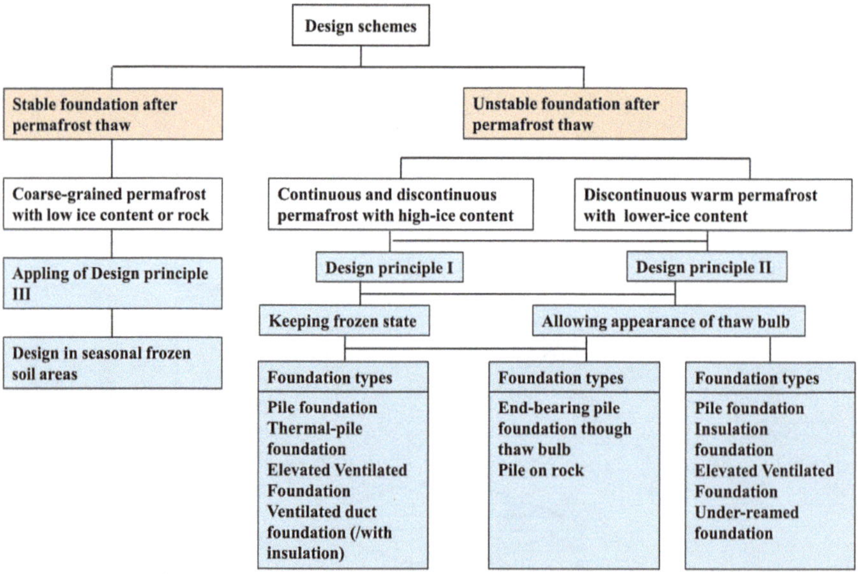

Fig. 2.22 The design schemes of foundations in permafrost regions

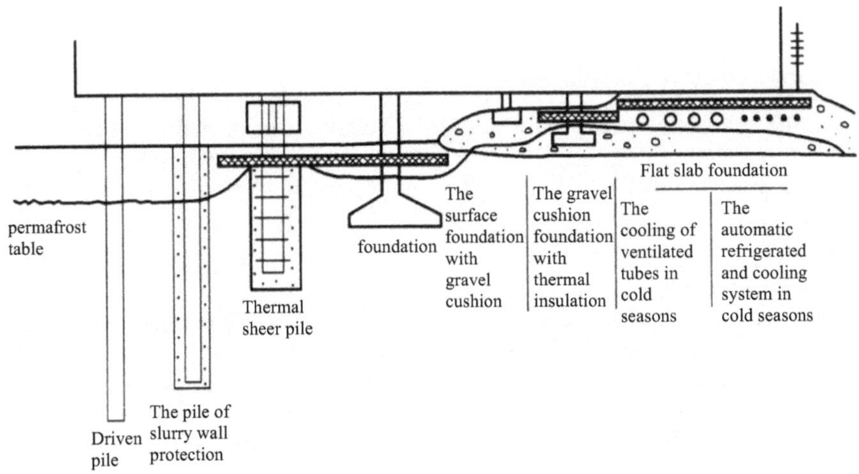

Fig. 2.23 Different types of foundation in permafrost

2.4.2 Shallow Foundations

2.4.2.1 Seasonally Frozen Soil Areas

The buried depth of the foundation base in seasonally frozen soil areas should be below the maximum seasonal freezing depth, or at least 2/3 below the maximum seasonal freezing depth, and shall be backfilled with a gravel cushion with a thickness of not less than 300 mm. The following methods can be used to reduce the buried depth of the foundation if necessary:

1. When the gravel cushion is used to reduce the buried depth of the foundation, two conditions must be met: first, the content of silty clay particles (the particle size <0.075) in the gravel cushion cannot be more than 15%; second, the groundwater in gravel cushion within the freezing depth should be able to be drained.
2. Expanding the width of apron on the outside of the foundation, and laying a thermal insulation layer under it to prevent the freezing depth from invading the foundation base and causing its normal frost heaving force.
3. The inner side of the foundation is equipped with heating trenches or other heating facilities, mainly on the shady side and sections with corners.

Regardless of the types of the foundations, the outside of the foundation should be backfilled with a clean sand gravel layer with a thickness of more than 300 mm to reduce the tangential frost heaving force.

2.4.2.2 Permafrost Regions

1. **The Methods of Foundation Treatment**

From the successful examples of shallow foundations in permafrost regions, the shallow foundations are located in (or on the top of) relatively higher fill gravel cushion layer, and then elevated ventilated or ventilated pipes are used to keep the permafrost foundations in a frozen state, and to preclude the erosion of surface water and groundwater. The types of the foundations include column, pier, expanded, raft foundations, etc. (Figs. 2.24, 2.25, 2.26 and 2.27).

For heating large structures with heavy load, such as oil depots, the sand-gravel cushion slab foundations with ventilated tubes can also be used to reduce the ground temperature of underlying permafrost (Fig. 2.28). It should be noted that the thickness of the cushion layer should be determined based on the maximum thawing depth in summer. When the air temperature rises to 0 °C, the vent should be closed in time to avoid warm air from entering the tube. In the freezing season, the vent should be open to let the cold air into the tube to accelerate the cooling of the cushion layer and protect the permafrost foundation; and the vent can be implemented by an automatic device. In the case of the huge heating structures, the combined foundation with thermosyphon and thermal insulation layer can be used to reduce the ground temperature of the permafrost foundation base.

2. **The Design of Shallow Foundations**

The design steps of shallow foundations:

1. Selecting the buried depth of the foundation based on the factors affecting the stability of the permafrost foundation;
2. Estimating the highest temperature under the foundation base to determine the influence on the permafrost foundation;

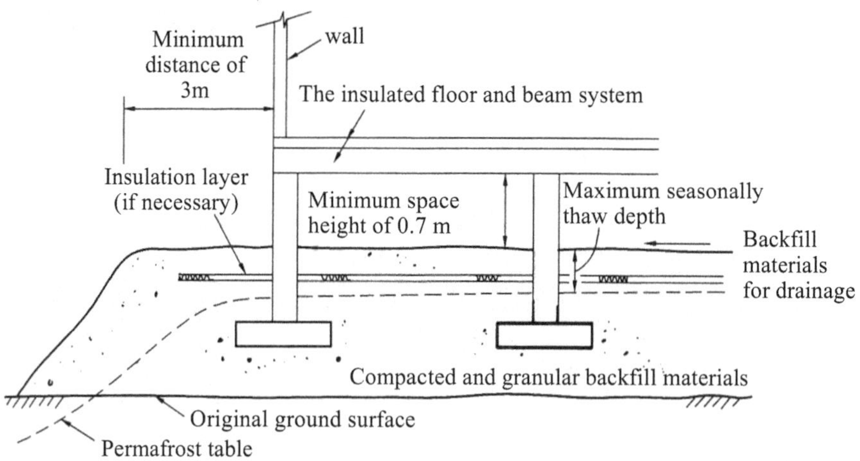

Fig. 2.24 The foundation placed on the ground covered with a thick insulated gravel cushion

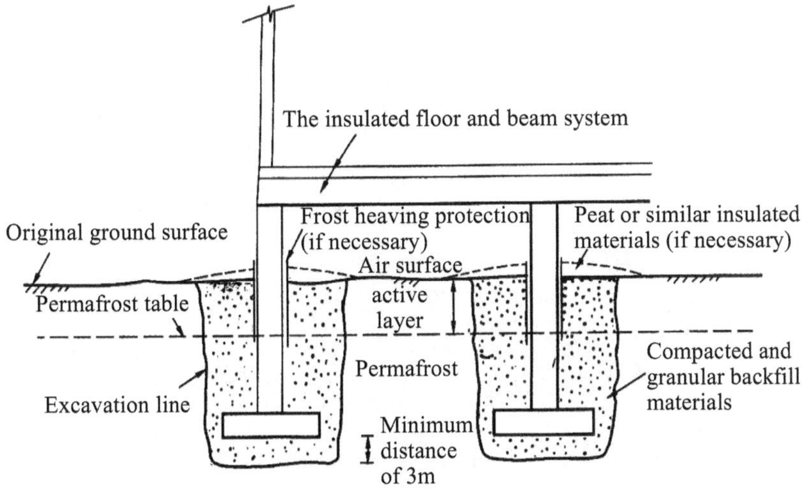

Fig. 2.25 The foundation placed under the original ground for excavation

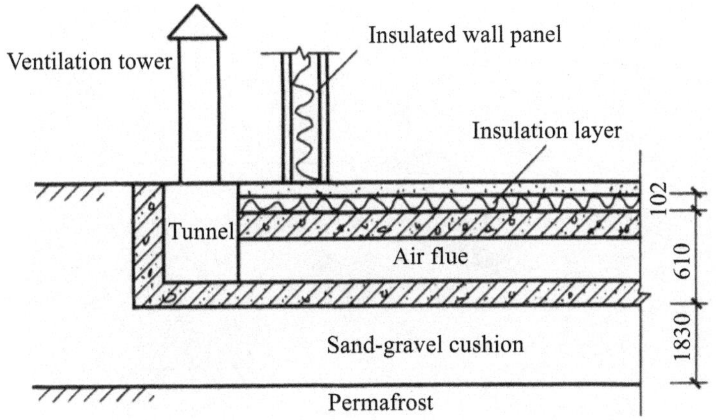

Fig. 2.26 Raft foundation in Fairbanks regions, Alaska (Both quoted from JGJ 118-2011)

3. Determining the foundation size under the condition of safety factor based on the bearing capacity theory;
4. Conducting the settlement analysis of the foundation based on the constitutive relation;
5. Correcting the foundation size or buried depth when the rate of settlement or deformation is beyond the required range.

The relationship between the mechanical properties of permafrost foundation and ground temperature needs to be considered in the design of shallow foundations. The distribution of ground temperature in the foundation is an especially important factor in the design, and the maximum temperature limit should be used for design.

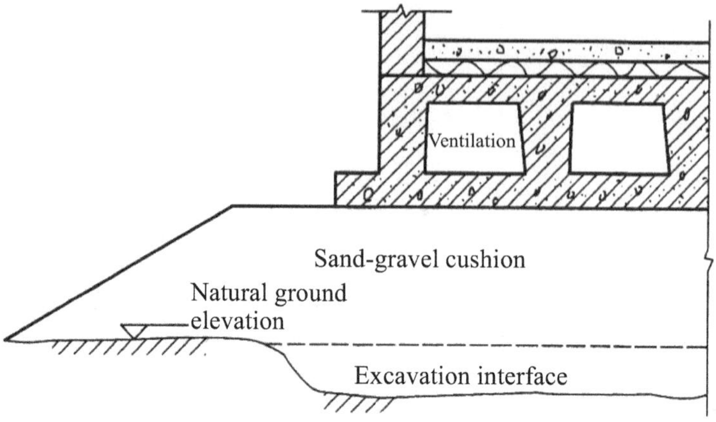

Fig. 2.27 Raft Foundation in Thule Regions, Greenland (both quoted from JGJ 118-2011)

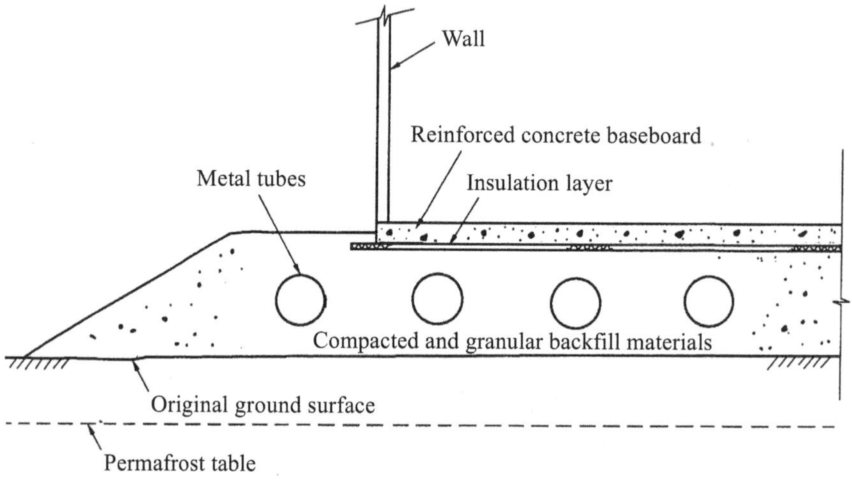

Fig. 2.28 Insulated concrete baseboard on the ventilated tubes on the compacted soil

3. **The Determination of Bearing Capacity and Deformation**

The allowable pressure of shallow foundations in non-frozen soil foundations is determined by the principle that the foundation soil resists the damage and settlement and deformation within the allowable range. Its standard is also applicable to the shallow foundations in permafrost foundations. Besides, the strength of permafrost foundation and its variation with temperature should be considered. The settlement deformation of foundation is mainly comes from the creep deformation rather than consolidation deformation of permafrost foundation. The bearing capacity and deformation of permafrost foundation shall be determined by the methods described in this chapter.

2.4.3 *Pile Foundations*

In deep seasonally frozen soil areas and the warm permafrost regions with high ice content, the use of pile foundations has many advantages. Its small thermal interference can prevent buildings from frost heaving and thaw settlement of the seasonally active layer, and can be supported on the deep permafrost layer with relatively stable strength. Furthermore, the pile foundation can be precast or cast-in-place to avoid the excavation process with greater thermal interference. At the same time, a slight change in the pile foundations can be changed into the elevated ventilated foundation and thermal pile foundation with greater cooling capacity.

2.4.3.1 Types of Pile Foundations

The most commonly used types of piles in permafrost usually include reinforced concrete piles, wood piles and steel piles. The type of pile depends on the soil type, ground temperature, bearing load, materials, construction equipment, transportation condition and the cost of installation, etc.

1. **Reinforced Concrete Piles**

Reinforced concrete piles are rarely used in engineering in North American permafrost regions, but are widely used in Russia and China permafrost regions. In North America and Russia, precast reinforced concrete piles are widely used, which are usually installed in pre-thawed or drilled holes, and occasionally cast-in-place; but in China, cast-in-place is more commonly used.

As for North American, concrete has a relatively lower tensile strength, and it tends to crack under the action of frost heaving force of the foundation soil. The steel bars in light load concrete piles with low reinforcement are elongated a lot, causing the corrosion of exposed steel bars. Under such a condition, precast pile is a better form, which can prevent concrete cracking, and prestressed steel bar have great advantages.

2. **Wood Piles**

In North America, the use of local wood is the most economical way. Commonly used wood is generally spruce, Douglas fir or pine, etc., with the length of 6 m ~ 15 m, a diameter of 150 mm ~ 250 mm at the top, and 300 mm ~ 350 mm at the bottom. Corrosion preventing treatment should be carried out in the seasonally active layer to prevent it from corrosion and reduce the frost heaving force on the side of piles.

The advantage of wood piles is that they have well insulated and integrated in saturated permafrost. However, it is relatively lower in structural bearing capacity and cannot be operated mechanically in permafrost layer.

3. **Steel Piles**

Steel pipe piles and I-shaped (H-shaped) steel pile are the most applicable forms in permafrost regions. Steel pipe piles with concrete or sand filled can improve their bearing capacity. Open steel pipe piles and H-shaped steel piles can be driven deep into the warm permafrost, can be lengthened or cut off and can withstand high loads. Steel pipe piles with closed or hooped ends are usually inserted into pre-drilled holes, and cement mortar or concrete is poured into the gap between the pile and the drilled hole. Corrosion has not basically occurred in the permafrost layer, primer can be first coated in the seasonally active layer and the lower 0.5 m range, and then preservatives can be applied on it. Corrosion preventing treatment cannot be adopted for steel pipe piles in the permafrost; otherwise there will be no freezing strength.

In design, steel pipe piles are usually selected in the case of the requirements of lateral load. The thickness of pipe shell should not be less than 2.5 mm ~ 3 mm; otherwise, it will have no effect on the bearing capacity. The average compressive stress of steel pipe piles and H-shaped steel piles should not exceed 63 MPa.

2.4.3.2 The Installation of Piles

The installation of piles depends on the ground temperature of the permafrost, soil type, pile type and the buried depth, and the construction equipment available on site. The common methods include pre-drilling method (rotary drill, twist drill, percussion drill, revolving drill, etc.), driven method (driven by thawing, driven by drilling), and artificial excavated method.

2.4.3.3 The Pile Size and the Buried Depth

The pile diameter and buried depth can be determined according to the load acting on the pile, the bearing capacity of the frozen soil foundation and the working conditions (the lithology of frozen soil, ground temperature, ice content, materials, construction technology and season, etc.). In order to ensure the structural stability of the pile, the bending strength of the pile should exceed the compressive strength of the soil by more than 15%. In designing the pile foundation, the thermal interference to the permafrost foundation should be reduced as much as possible, and the stability against frost heaving of foundation should be effectively increased, so as to reduce the number of piles and meet the requirements of building loads by increasing the buried depth of piles or their diameter.

During the installation of pile foundation, the formation of the bearing capacity for permafrost foundation is related to soil temperature around the pile, thermal disturbance, the heat amount of backfill (cast-in-place) material, the installation of pile foundation and refreezing time around the pile. In the initial stage, the pile foundation installed by drilling has a relatively large bearing capacity, and the resistance at the pile end and the freezing force on the pile side can be formed quickly. The bearing capacity of bored piles and bored cast-in-place piles is mainly borne by the resistance of pile end. If there is no freezing force on the pile side, the friction force can basically be ignored.

2.4.3.4 The Refreezing around the Piles

The refreezing time around piles in permafrost regions mainly depends on the ground temperature of permafrost, the amount of mud filled within unit length, the latent heat of melting mud, the temperature of casting concrete and the distance between piles (pile groups). Under the same site, the smaller is the volume and water content of backfill mud, the larger is the distance among piles and the shorter the refreezing time.

1. **Natural Refreezing**

The original ground temperature of permafrost plays an important role in the refreezing around the cast-in-place piles. If the pile end can refreeze faster, the process of refreezing is beneficial to improving the bearing capacity of the pile end. In designing pile foundations in permafrost regions, the ground temperature of permafrost, the diameter of pile, the temperature of casting concrete, and construction season must be fully considered. The construction season of natural refreezing should be in the period of the lowest ground temperature (from February to May), during which the frozen soil foundation has more "cool reserves" to accelerate the refreezing around the piles.

2. **Artificial Refreezing**

When the ground temperature of permafrost is too high or too much in heat amount introduced, the natural refreezing cannot be achieved during the planned construction period, the artificial refrigeration must be adopted to make the backfill materials in frozen state. The usual method is to circulate through the longitudinal or spiral steel pipes on the pile, or insert thermal probes (two-phase thermal pile) into the pile foundation. Artificial or natural cooling energy can be used to accelerate the refreezing of backfill around the pile.

2.4.4 Cooling the Foundation

Thaw bulbs are usually formed under heating houses in permafrost regions. In order to maintain the thermal stability and stability of permafrost foundation, special cooling facilities are adopted in making full use of the cold climate in permafrost regions. The houses design in permafrost regions, in addition to making the indoor ground warm and insulated, commonly used cooling systems, including the cooling system of elevated ventilation, ventilated pipes with filled soil, thermal probes and piles.

2.4.4.1 The Cooling System of Elevated Ventilation

The cooling system of elevated ventilation always uses pile foundations to elevate the building, with a certain height of ventilation space between the ground and the first floor of the building. The elevated ventilation is usually located directly on the ground (Figs. 2.13, 2.14, 2.15 and 2.16), or underground or semi-underground (Fig. 2.26). The simple ventilation space or ventilation stack, with the help of air convection, can intercept and take away the heat transmitted from the ground and floor of houses, and release it into the atmosphere, so as to prevent the heat from heating houses and from entering the frozen soil foundation, and maintain the heat stability of frozen soil. Such structure is called elevated ventilated foundation. Pile foundation is often used for elevated ventilated foundations, and ring beam and column foundations and strip foundations are mainly used for semi-underground elevated ventilated foundations.

The height of elevated ventilation is different. The elevated height of ground-based elevated ventilation cooling system in Russia and Alaska is mostly 1.0 m ~ 1.8 m (Figs. 2.13 and 2.14), and the elevated height of the workshop is 1.6 m ~ 1.8 m (Fig. 2.29) (Wu et al., 2005a, b, c). When used as an underground garage, the elevated height can reach 2.5 m. The elevated height of the ground-based elevated ventilation cooling system in the Qinghai-Xizang Plateau and Daxiao Xing'anling Mountains in Northeast China is mostly 0.5 m ~ 1.0 m [Figs. 2.15, 2.16 and 2.30 (Ding et al., 2011)]. The height of elevated ventilation of pile foundations should depend on the width of the building. Generally, when the width is not greater than 24 m, its height of elevated ventilation is about 1.0 m; and it should not be less than 1.5 m when the width is greater than 24 m.

The observation data of the elevated ventilated foundation houses in Qinghai-Xizang Railway show that the ground temperature at any depth under the foundation is lower than that of the natural state in both warm and cold seasons (Ding et al., 2011). The buried depth of the permafrost table is 1 m shallower than in natural conditions. The thermal insulation of floors should be emphasized in the design

Fig. 2.29 Elevated workshop in Yakutsk (1.6 m ~ 1.8 m)

Fig. 2.30 Elevated footing on pile foundations in Qinghai-Xizang Railway

because good thermal insulation of floors is not only conducive to protecting the thermal stability of permafrost foundation, but also providing a comfortable indoor living environment.

Some ventilation cooling systems are installed underground, or called elevated ventilated underground foundation (Fig. 2.26). The warm air in basements can be extracted (pumped) into the atmosphere through ventilation towers (chimney effect), and the cold air enters the basement from the ventilation stack, passing through the exchange of hot and cold air to maintain the frozen state and heat stability.

2.4.4.2 The Cooling System of Ventilated Pipes with Filled Soil

The cooling system of ventilated pipes with filled soil refers to filling a cushion of a certain height of coarse-grained soil (gravel soil) on the natural ground as the foundation of the building, burying ventilated pipes in the cushion, or laying ventilated pipes or raft foundation of ventilated pipes (Fig. 2.27) on the cushion. Through the air convection in the ventilated pipes, the heat transmitted from the indoor floor of houses to the fill cushion is brought out and released into the atmosphere, and the thawing depth is controlled within the fill foundation, so as not to be brought into the frozen soil foundation. In winter, the cold air can cool the fill cushion through the ventilated pipe, so as to maintain the frozen state and thermal stability of the frozen soil foundation.

The foundation of the cooling system of ventilated pipes with filled soil is mostly ring beam and column foundation, strip foundation, raft foundation and so on. The inner diameter D of the ventilated pipe should be selected according to the width of the house and the thickness of the filled soil, generally 0.3 m ~ 0.4 m. To ensure good ventilation in nature, the ratio of the length of ventilated pipes L to the diameter D should meet a certain relationship, that is, the length-diameter ratio of ventilated pipes L/D should be less than or equal to 30 in general (The Editorial Committee of *Qinghai-Tibet(Xizang) Railway*. 2016). The clear spacing of ventilated pipes should be determined by thermal engineering calculations. Generally, it should not exceed two times the pipe diameter, from 0.3 m to 0.6 m. It is advisable to set the buried depth of the ventilated pipe as 0.5 m (calculated from the indoor floor), and the area of the ventilated pipe mouth can be determined by the method used for the calculation of area of ventilated holes. Ventilated pipes should be arranged 2 m ~ 3 m beyond the longitudinal edge line, and the number of ventilated pipes outside the gable wall should not be less than 2.

2.4.4.3 The Cooling System of Thermal Probes and Piles

The cooling system of thermal probes and piles is to insert the thermal pipe (thermosyphon) into the side of the building or bury it in the fill foundation, transmitting the heat from the ground of the building to the atmosphere, through the device of heat exchange of convective circulation between the liquid and steam within thermosyphon pipe. Based on the cooling system of thermal probes and piles, the design of house foundation can effectively transfer the heat amount through the thermal piles from the ground, making the foundation soil refreeze to a certain thickness, intercepting the infiltrated heat amount and controlling the thermal cycle of the base-foundation, so as to maintain the thermal stability of the permafrost.

The following types of thermal probes and piles foundation are commonly used in permafrost regions:

1. The Foundation of Heat Dissipation of Straight Thermosyphon Pipe

The straight thermosyphon pipe is a conventional product, which is directly inserted around the building (Fig. 2.31) or near the foundation in the elevated ventilated foundation (Fig. 2.32). The cooling of the thermosyphon pipe makes the foundation soil around the foundation in refrozen state and maintains the frozen soil foundation in frozen state.

The freezing radius of the thermosyphon pipe is not only determined by its characteristic of heat transfer, but also closely related to the water content and the density of soil and the freezing index of air. In natural conditions, the freezing radius of thermosyphon pipe in the Daxing'anling and Xiaoxing'anling Mountains in Northeast China is 1.0 m ~ 2.0 m, and the experimental observation result of Qinghai-Xizang Railway is about 2.0 m. The freezing radius r of the thermosyphon pipe is the function of the freezing index of temperature (Fig. 2.33).

2. The Elevated Foundation of Thermal Piles of Steel Pipe Piles and Concrete Tube Pipes

The elevated ventilated foundation of steel pipe piles is filled with refrigerant liquid (kerosene or liquid ammonia) (Fig. 2.34) (Wu et al., 2005a, b, c) in the steel pipe pile foundation, or inserts thermosyphon pipe into the concrete tube piles (Fig. 2.35);

Fig. 2.31 Buildings with thermal probes for heat dissipation in Alaska (Photo by Chen Xiaobo)

Fig. 2.32 Pile foundation with thermal probes for heat dissipation in the elevated ventilated foundation in the cogeneration power plant in Yakutsk (Photo by Tong Changjiang)

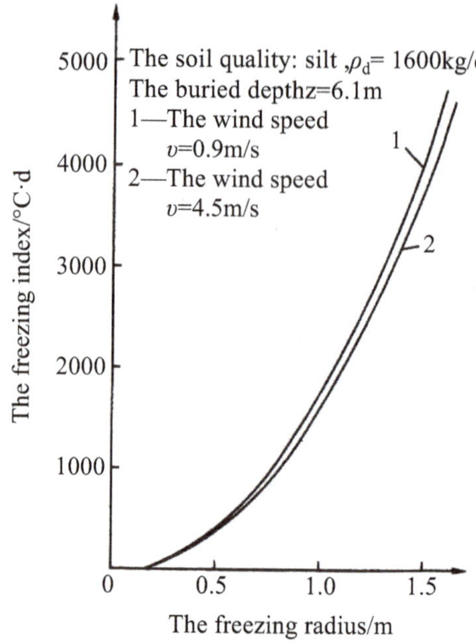

Fig. 2.33 The relationship between the freezing radius and the freezing index of thermal probes

Fig. 2.34 Refrigerated steel pipe pile foundation in Nadym Town, Siberia

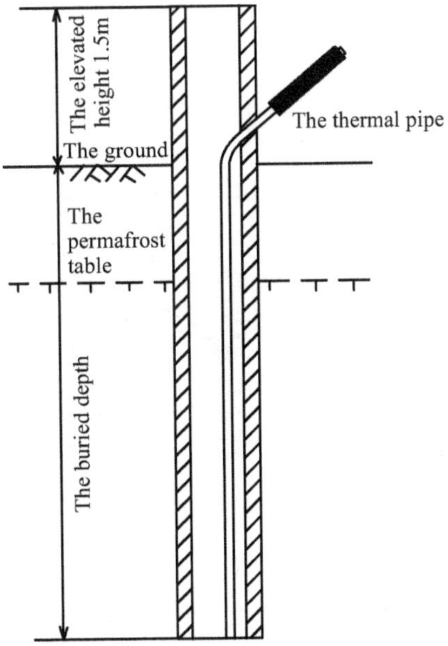

Fig. 2.35 The schematic diagram of refrigerated piles of thermal probes of reinforced Concrete Pipes

and then the hollow pipe pile is back-filled with wet sand, gravel sand and concrete. Thus, the ground temperature of permafrost around the pile is reduced by the cooling action of thermosyphon pipe, so as to keep the permafrost foundation of pile foundation in the frozen state.

3. The Ring Beam Foundation of Fill Thermosyphon Pipe

The ring beam foundation of fill thermosyphon pipe means that the fill foundation is filled with a certain thickness of non-frost heaving coarse-grained soil (gravel sand and gravel, etc.) on the ground, and the thermosyphon pipe (L-shaped) is buried in the fill foundation, which is combined with the ring beam and the filled soil to form a composite foundation (Fig. 2.36). During the warm seasons, in order to control and reduce the maximum thawing depth, the thermal insulation layer between the indoor floor and filled soil, as well as the thermal resistance of fill layer are used to prevent the heat of houses. In cold seasons, thermosyphon pipe can release the heat leakage from the floor and the heat in the foundation, making the fill layer and the foundation soil refreeze, and maintain the thermal stability of the permafrost foundation in the frozen state.

The periphery beam foundation with filled soil and thermosyphon pipe is essentially a kind of composite foundation. The fill soil layer can reduce or restrain the frost heaving susceptibility by the use of non-frost heaving materials, and the thermal insulation layer in the floor and fill soil layer can reduce the heat transmission and the thawing depth of the permafrost foundation. The thermosyphon pipe in the lower part of the buried depth can release the heat in the foundation and refreeze to keep the frozen state and thermal stability of frozen soil foundation.

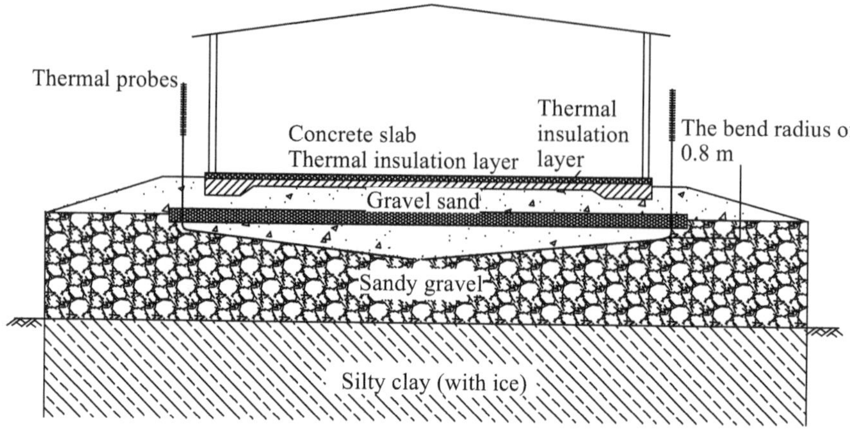

Fig. 2.36 The ring beam foundation with thermosyphon pipe

The content of silty clay particles in the fill soil layer should be controlled within 15%, and the thickness should be kept without decreasing the permafrost table which is generally not less than 1.0 times of the permafrost table at the construction site according to the thermal calculation. The filling boundary should exceed the outer contour line by 3 m ~ 5 m, and the width towards the sun should be properly widened. Before filling, the original ground should be set at a slope of 1%, making it incline to the lower side of the natural ground to constitute a good drainage condition and keep the fill foundation dry, without the runoff of surface water or groundwater. If there is no thermosyphon pipe buried in the fill soil layer, the fill soil foundation is only applicable to the continuous permafrost regions where the freezing index of air temperature is more than three times the thawing index.

The power, the buried space and buried depth of the thermosyphon pipe shall be determined by thermal calculations according to the heat leakage from the building floor, the thawing depth of the foundation in warm seasons and the mean annual ground temperature of the permafrost foundation.

2.4.4.4 The Cooling System of Rockfill

Crushed rock layer filled is a heat dissipation material. Due to high void space, the equivalent thermal conductivity of crushed rock layer is relatively larger, up to 10.55 kcal/(m · h · °C) in cold seasons; its thermal conductivity is relatively smaller, only 0.87 kcal/(m · h · °C) in warm seasons, and the ratio of the two is up to 12 or more. Therefore, the crushed rock layer has a good thermal effect: in the cold season, the cooling capacity can be passed directly to the crushed rock layer, depositing and storing in the lower part; in the warm season, the block stone layer can correspondingly prevent the transmission of heat.

Fig. 2.37 Buildings with block stone foundation on the Qinghai-Xizang Railway (the piece of stone layer of 1.5 m)

According to the thermophysical properties of crushed rock layer, the filled soil (sandy gravel) can be combined with the crushed rock layer to form a composite foundation of crushed rock and filled soil (Fig. 2.37) (Ding et al., 2011). The sandy gravel layer with a thickness of 0.3 m ~ 0.5 m should first be laid on the ground for drainage, and then a crushed rock layer of 1.2–1.5 m laid on it, and a layer of geotextile on the surface (to prevent the infiltration of fine-grained soil), and lastly, a certain thickness of non-frost heaving susceptibility sandy gravel layer on the top, so as to constitute a composite foundation with crushed rock and filled soil, which is an artificial foundation developed in recent years.

The technical requirements that the crushed rock layer should meet is that the compressive strength of crushed rock should not be less than 30 MPa, and the particle diameter should be 150–300 mm, and the filling thickness generally should be not less than 1.0 m ~ 1.5 m.

2.5 Design of Preventing Frost Heaving

The anti-frost heaving measures for foundation soil in cold regions usually include: the improvement method of the foundation soil (replacement method and physical and chemical modification method), the thermal insulation method (reducing the freezing depth) and the drainage and impermeable method (draining and reducing the water migration of the foundation soil).

The frost heaving force (upfreezing force or swelling force) of the foundation soil causes the deformation and damage to buildings. The frost heaving can be effectively prevented by bypassing isolating measures, increasing foundation pressure and anchoring force, and strengthening structural robustness or adaptability to deformation. From the perspective of the effect, it is advisable to adopt the comprehensive treatment methods to make an anti-frost heaving design, such as foundation treatment and structural measures.

2.5.1 Improvements of the Foundation Soil

The anti-frost heaving measures for the improvement foundation soil mainly adopt non-frost heaving soil to replace the frost heaving soil, and the usually used replacement filler is coarse-grained soil, such as sandy gravel, gravel, and coarse sand and so on.

The effect of the replacement method to prevent frost damage of buildings is related to the depth of replacement, the content of silt clay particles of replacement, drainage conditions, foundation soil texture, the groundwater level and the ability of buildings to adapt to uneven frost heaving and deformation. When using the replacement method, the material source, transportation conditions, the structural characteristics of the building, and the project costs need to be considered.

2.5.1.1 The Replacement Method

1. **The Determination of the Replacement Depth**

The replacement depth refers to use non-frost heaving material completely or partly to substitute the frost-heaving foundation soil within the maximum seasonal freezing depth on construction site, which is the replacement ratio. Two factors should be considered when determining the replacement depth of the foundation soil:

1. When the freezing depth of foundation soil is relatively larger, the replacement depth can be determined by the law that the frost heaving of the foundation soil decreases with the increase of the freezing depth. In the closed system, the frost heaving mainly occurs above the main frost heaving zone [the maximum freezing depth of (2/3)-(4/5)], below which the frost heaving of the foundation soil is very small. When the groundwater level is shallow, the influence factors of groundwater shall be considered. The frost-heaving and deformation of the foundation soil after replacement should be controlled within the allowable deformation of the building.
2. During the thawing period, the bearing capacity of the foundation after the thawing of the replacement and its lower foundation soil is likely to decrease. When determining the replacement depth, it should be considered that the settlement deformation can still meet the requirements of the allowable deformation of the building if the bearing capacity decreases.

The replacement method is often used in shallow foundations, such as pier foundations, column foundations, strip foundations, etc. When the groundwater level is

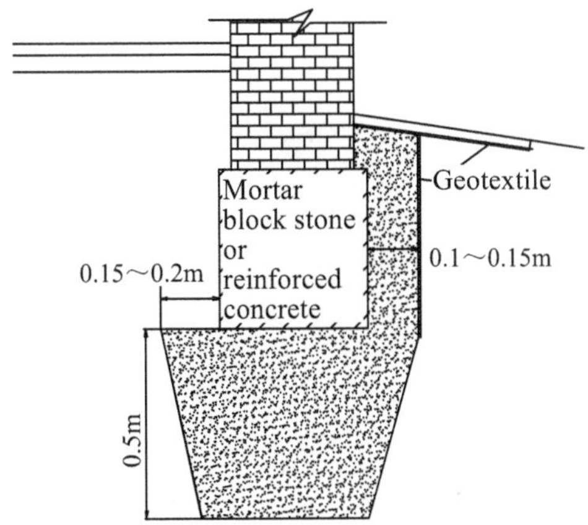

Fig. 2.38 The replacement of the foundation

deep, the main frost heaving zone trends to the upper part of the frozen layer. The total thickness of the foundation and replacement layer can be determined by 80% of the local maximum freezing depth. Heating buildings can be reduced by the coefficient of heating influence of different parts of the building (Table 2.2). When the groundwater level is high, the total thickness of the foundation and replacement layer should be equal to the local maximum freezing depth.

The replacement layer is mainly used as the cushion of the foundation to eliminate the normal frost heaving force at the foundation base. The replacement section is mostly inverted trapezoidal or rectangular, and the thickness is generally not less than 0.5 m (Fig. 2.38). The width of the replacement layer should be longer than that of the foundation, generally not less than 0.15 m ~ 0.2 m. The lateral side of the foundation will also be affected by the tangential frost heaving force and the outer side of the foundation should be replaced appropriately, with the thickness generally more than 0.1 m ~ 0.15 m.

2. **Control of Fine-grained Soil in Replacement Layer**

The frost heaving of coarse-grained soil increases with the increase of the content of silty clay particles (the particle size less than 0.075 mm). When the silty clay particle with a particle size of less than 0.02 mm reaches 3% of the total weight, water migration will occur during the freezing process, and the segregated ice will also be formed. When the content of silty clay particles is controlled within 5%, coarse-grained soil does not have frost heaving. The experimental data show that the content of silty clay particles is controlled within 12%, and its frost heaving rate is generally less than 2% (Tong & Guan, 1985). According to JGJ 118, *Code for Design of Building Foundations in Frozen soil areas*, the content of silty clay particles is controlled at 15%, and its frost heaving can still be considered as "non-frost heaving". This is the basic condition for classifying coarse-grained soil as non-frost heaving in construction engineering, but it is not applicable to the engineering of high-speed railway foundation.

Table 2.4 Frost heaving rates of replacement layers with different groundwater level and soil content

Groundwater level H_W/m	Frost heaving rate of soil particle size less than 0.05 mm, η/%					
	0 ~ 2.5	2.5 ~ 5	5 ~ 7.5	7.5 ~ 10	10 ~ 15	>15
>2	0.29	0.78	0.48	1		0.25
1.5 ~ 2.0	0.44	0.52	0.42	1	0.35	
1.0 ~ 1.5	0.67	0.85	0.82	0.43	1.7	1.3
0.5 ~ 1.0	0.24	0.17	0.1	0	0.23	0.25
<0.5	2	2.2	5.2	5	5.7	4.5

The frost heaving of coarse-grained soil is also controlled by water content (the water content of foundation soil and groundwater level). According to the experimental observation data, when the distance between the bottom of the frozen layer and the groundwater level (H_W) is less than 0.5 m, even if the soil content of the replacement layer (the particle size less than 0.05 mm) is less than 5%, the frost heaving rate η is more than 2%, even up to 4.5% ~ 5.7% (Table 2.4) (Zhu et al., 1988). When the groundwater level is larger than the requirements of JGJ 118-2011, *Code for Design of Building Foundations in Frozen soil areas,* the content of silty clay particles in the replacement layer should be reduced, or drainage measures should be strengthened. In other words, the sufficient condition for classifying the replacement of sandy gravel soil as non-frost heaving is the drainage condition during the freezing period.

3. **Preventing the "Infiltration" of the Replacement Layer**

The so-called "infiltration" of the replacement layer means that silty clay particles (smaller than 0.075 mm) is infiltrated into the replacement layer by the water seepage of the surface and the lateral foundation during the engineering operation, so that the content of silty clay particles in the replacement layer increases, and the frost heaving increases correspondingly, thus the anti-frost heaving measures finally fail. According to the data of engineering survey, this kind of "infiltration" mainly occurs within the range of 0 m ~ 0.4 m in the surface layer, some of which can reach 0.8 m, and the silty clay soil content of the replacement layer can increase to more than 30% (Zhu et al., 1988). Therefore, geotextiles against the infiltration of fine-grained soil need to be laid under the foundation of water dispersal and the outside of the replacement layer on the base side, so as to prevent the silty clay particles from infiltrating into the replacement layer with the water flow to maintain the original "purity" of the replacement layer.

4. **Engineering Evaluation of the Replacement Method**

1. The replacement method has great advantages for the areas that have plenty sand and gravel with low unit price.
2. The sandy soil and silty clay layer with a lower groundwater level require a shallow replace depth. However, there are heavy cohesive soil containing much clay under the replacement layer, the replacement depth should be increased. When the groundwater level is higher with thick heavy cohesive soil, it is often necessary to replace and fill the whole frozen depth of houses to achieve the effect of frost heaving.

3. The replacement depth can be reduced for buildings with strong ability to adapt to uneven frost heaving and deformation. For structures susceptible to uneven deformation, it is necessary to replace the whole depth, and the content of silty clay particles needs to be strictly controlled, and in such condition, the replacement method is not necessarily economical and reasonable.
4. Anti-frost heaving cannot achieve a great effect if the replacement is only applied on the foundation base, and the foundation side should also be replaced. Otherwise, the tangential frost heaving force on the foundation side is still an important cause of frost damage to buildings.

2.5.1.2 The Physical and Chemical Modification Method

The physical and chemical modification method is to use artificial materials to treat the foundation soil, based on the law of exchanges of cations and salinity on the frost heaving of soil, so as to change the interaction between soil particles and water, so that the intensity of water migration and its freezing point in the soil are reduced to weaken the frost heaving of the foundation soil.

The physical and chemical modification method to prevent the frost heaving has been used abroad for many years. There are various methods, but only three modification methods to improve the foundation soil are introduced in this book. They are artificial salinization modification method, hydrophobic materials modification method and the aggregation or dispersion of soil particles modification method.

1. **The Method of Artificial Salinization to Improve the Foundation Soil**

The artificial salinization method refers to adding a certain amount of soluble inorganic salts into the foundation soil to make it salinized, such as sodium chloride ($NaCl$), calcium chloride ($CaCl_2$), potassium chloride (KCl) and so on.

The addition of soluble inorganic salts into the soil can increase the electrolyte and the thickness of the water film on the surface of soil particles, and reduce the surface energy and capillary action, so as to reduce the intensity of water migration and the freezing temperature of soil. The freezing temperature of the solution decreased with the increase of its concentration (see Table 2.5).

The mixture amount is determined by the soil type and construction method. In general, according to the mass ratio, the sandy clay needs 2% ~ 4% sodium chloride and calcium chloride, and 1% ~ 2% for the sandy soil containing a small amount of silt and clay.

Three kinds of construction technologies are, ① salt is directly paved on the ground, permeating into the foundation soil through rain; ② replacement layer is

Table 2.5 Freezing temperature of soil with different salinity (t_f, °C) ($\omega = 25\%$)

	Salt concentration/%			
Soil type	0.4	0.55	0.85	1.2
Silty clay	−1.2	−1.5	−2.6	−3.0
Silty soil	−1.3	−1.7	−2.5	−3.4
Fine sand	−1.3	−1.6	−2.5	−3.5

first salinized and then back-filled into the foundation pit; ③ shallow holes are set around the foundation, filled with salt crystals, and then a saturated solution with the same kind of salt is injected into the holes.

2. **The Method of Hydrophobic Materials to Improve the Foundation Soil**

Using hydrophobic materials to improve the foundation soil means to add a small number of hydrophobic materials into the soil, making the surface of soil particles a good hydrophobicity, weakening or eliminating the infiltration of surface water or the ascent of groundwater, reducing the water content of soil, and then weakening the frost heaving of soil and the freezing strength between the foundation soil and buildings.

Hydrophobic soils are usually made by mixing petroleum products or by-products with other chemical surfactants into the soil. Petroleum products include heavy oil, diesel, liquid petroleum asphalt, liquid coal tar, etc. Chemical surfactants mainly include trimethyl stearyl ammonium chloride [$C_{18}H_{37}$-$N(CH_3)_3$·Cl] and dodecyl trimethyl ammonium chloride [$C_{18}H_{37}$-$N(CH_3)_3$·Cl] (Pei & Lu, 1982). Surfactants can make hydrophobic oily substances be firmly adsorbed by soil particles, playing a role in weakening the interaction between soil and water.

3. **Aggregation or Dispersion of Soil Particles to Improve the Foundation Soil**

According to the relationship between the frost heaving and granule composition of soil, the soil with the particle size larger than 0.1 mm or less than 0.002 mm has a weak frost heaving. Chemical additives can make soil particles aggregate (aggregation effect) or disperse to weaken the frost heaving of soil.

Polymerizing agents enable fine particles in the soil to aggregate into relative larger agglomerates, including maleic polymer (soil conditioner), polymer sodium acrylate, polyvinyl alcohol and high-valent cations such as Fe3^{+} and Al3^{+}, all of which have the effect of reducing frost heaving (Table 2.6). In addition to clay, the polymerizing agents have obvious anti-frost heaving effect on silt and silty clay.

Chemical dispersants increase the content of fine particles in the soil and the density of soil, although the natural agglomerates are dispersed. In this way, the permeability of soil and the intensity of internal water migration will decrease significantly, exerting an anti-frost heaving effect (Table 2.7).

Table 2.6 Anti-frost heaving effect of polymerizing agents (Pei & Lu, 1982)

Polymerizing agents		Relative frost heave ratio	
Types	Usage/%	Clay	Silt
CKD-197	0.5	2.27	2.31
Soil conditioner	1.0	1.87	2.91
Styrene and sulfuric acid	0.5	1.43	0.14
Methyl copolymer	1.0	1.79	0.30
Polyvinyl alcohol	0.5	1.17	1.76
(PVA)	1.0	0.20	0.78
$FeCL_3$		1.49	0.48

Note: Relative frost heave ratio = the average frost heaving ratio of treated soil/the average frost heaving ratio of untreated soil)

Table 2.7 Anti-frost heaving effect of phosphate (Pei & Lu, 1982)

Polymerizing agents		Relative frost heave ratio		
Types	Usage/%	Clay	Silt	Silty clay
Sodium tetra-metaphosphate	0.5	0.49, 0.21	0.31, 0.36	0.18, 0.22
	1.0	0.40	0.3, 0	0.06
Sodium hexametaphosphate	0.5	0.37	0.58	0.42
	1.0	0.25	0.29	0.06
Sodium tripolyphosphate	0.5	0.42	0.48	0.09
	1.0	0.32	0.46	0.00

Note: Relative frost heave ratio = the average frost heaving ratio of treated soil/the average frost heaving ratio of untreated soil

Phosphate has a significant effect on anti-frost heaving of soil, with an average heave ratio of 0.2. The higher is the natural frost heaving ratio of soil, the better is its anti-frost heaving and the stronger its durability. It is unnecessary for the soil to be treated and maintained in advance in case of using the dispersants, with its cheap price, little consumption, easy construction, and quick reaction.

4. **Engineering Evaluation of Physical and Chemical Modification Method**

The physical and chemical methods against frost heaving have advantage of simple and easy operation, a wide range of material sources and relatively economic cost. Its disadvantage is that the effective time of anti-frost heaving is relatively shorter, especially the modification method of salinization, whose effectiveness depends on the permeability of the foundation soil, drainage conditions and anti-permeability of the foundation. The modification method of salinization is the most effective for the soil with lower permeability. However, the method will generally lose efficacy after 5–6 years due to easy leach of salt in sandy soil. Therefore, it is necessary to adopt supplementary measures to prolong its effect against frost heaving, and the crystal salt usually needs to be added every 2–4 winter. The combination with artificial salinization and foundation temping in layers can prolong the effect against frost heaving. The experiment data show that the stability of anti-frost heaving can still be maintained after 16 freeze-thaw cycles in the open system, with the dry density of silty clay of 2000–2010 kg/cm^3 and the actual concentration of pore water solution is 150 kg/cm^3.

2.5.2 *The Method of Thermal Insulation*

The method of thermal insulation refers to setting a thermal insulation layer at the bottom or around the building foundation to increase thermal resistance. From the perspective of anti-frost heaving, the method of thermal insulation can delay the freezing of the foundation soil, increase the temperature of the foundation soil and reduce the freezing depth, so as to achieve the purpose of anti-frost heaving. From the perspective of anti-thaw settlement, this method can reduce the thawing depth of the frozen soil foundation to achieve the purpose of anti-thawing settlement.

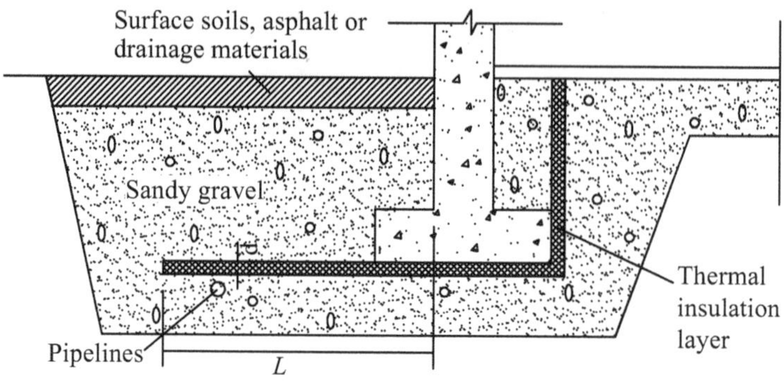

Fig. 2.39 Anti-frost heaving of thermal insulation layer in heating houses in seasonally frozen soil areas

Thermal insulation materials are mainly industrial products, such as polystyrene foam (EPS), extruded polystyrene foam (XPS) and so on. There are other materials including boiler slag, volcanic ash, foam concrete, glass fiber, etc. The technical performance indexes of industrial thermal insulation materials are that thermal conductivity should be less than 0.025 W/(m·k), water absorption rate should be less than 0.5%, and compression deformation under the pressure of 250 kPa should not exceed 5%.

As for anti-frost heaving measures for the thermal insulation layer of heating house foundation in the seasonally frozen soil areas (Fig. 2.39), the foundation of thermal insulation layer is laid on the inner side and bottom of the vertical wall to avoid freezing penetration by "cold bridge" causing by the wall foundation, resulting in freezing the foundation base and its inner side. At the same time, the temperature of the floor in non-thermal insulation room can resist freezing the foundation base and its inner side. However, in permafrost regions, some measures about the thermal insulation of indoor floor should be adopted to avoid indoor "heat leakage", resulting in deeper thaw bulb.

For non-thermal insulation houses, the combination of the thermal insulation layer and the drainage layer should be adopted (Fig. 2.40) and a permeable layer of sandy gravel is usually laid under each layer. Thermal insulation layer is continuously laid around the building and the foundation base, and local freezing penetration should be avoided to prevent "cold bridge".

The thickness and width of the polystyrene foam thermal insulation layer paved can be determined by thermal calculations or curve graphs (Tong & Guan, 1985). The thickness and width of the polystyrene foam insulation layer paved for heating houses and non-heating house can be determined by Figs. 2.41 and 2.42, respectively, according to indoor temperature and the local freezing index of air temperature.

Polystyrene foam and other industrial materials exhibit an advantage of good thermal insulation, low water absorption and high strength. The thermal insulation board that can be selected according to the technical requirements needs to

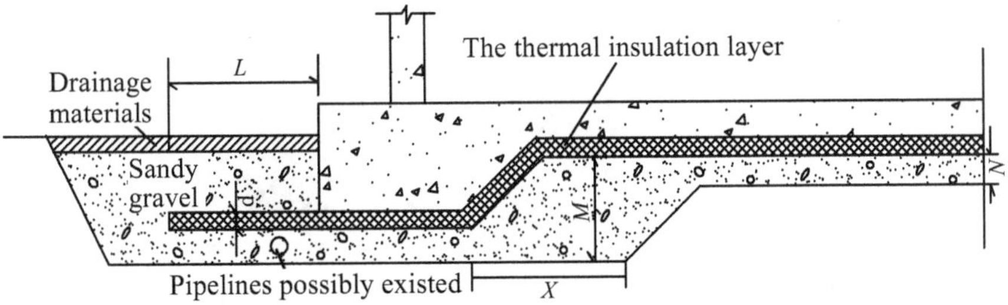

Fig. 2.40 Anti-frost heaving measures for thermal insulation layer of non-heating houses foundation

Fig. 2.41 Design curve of the minimum thickness of thermal insulation layer for heating houses

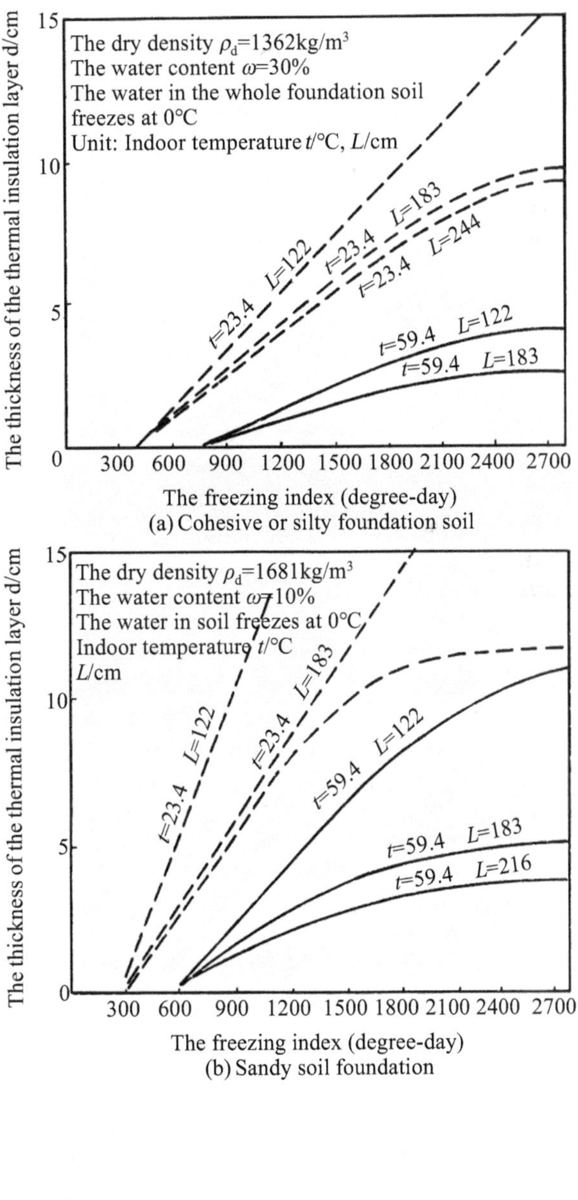

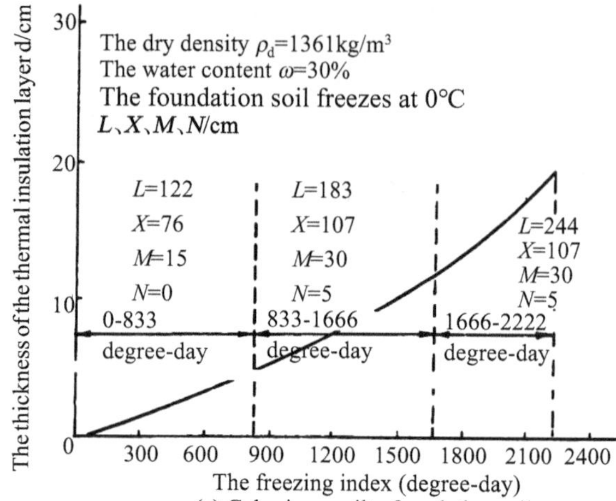

(a) Cohesive or silty foundation soil

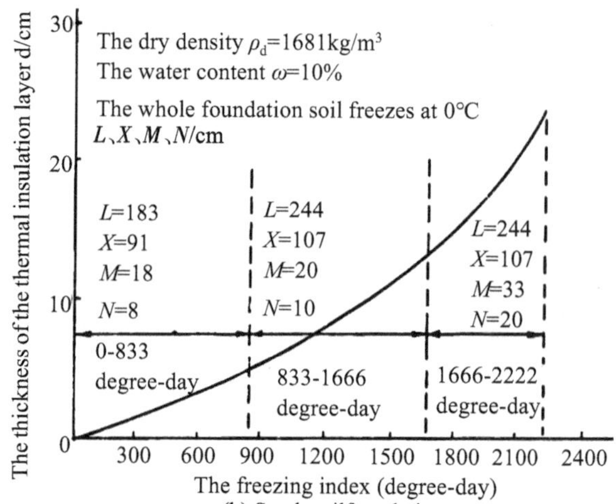

(b) Sandysoilfoundation

Fig. 2.42 Design curve of the minimum thickness of thermal insulation layer for non-heating houses

meet the requirements of the design strength to avoid compression deformation, especially for those laid on the foundation base. When the groundwater level is relatively higher, the engineering design should select the thermal insulation board with lower water absorption. Otherwise, the thermal insulation board after long-term immersion will increase its thermal conductivity and reduce its insulation.

2.5.3 *Waterproof and Drainage Method*

Water is one of the important factors causing frost heaving or thaw settlement of foundation soil. The waterproof and drainage method need to exclude and cut off the infiltration of external water such as the surface water, lower the groundwater level and reduce the water content in the foundation soil.

For surface waterproof and drainage, engineering design should treat the standing water in the apron slope outside the house and in foundation trench (Fig. 2.43), and set up drainage ditches to remove the standing water. When the underlying layer is sandy gravel with lower ground water level, it can be connected with the drainage pipe of the foundation trench and drain the water in the foundation trench (Fig. 2.44).

In permafrost regions, it is often difficult to drain the water in the foundation trench, so it is possible to set drainage wells on the slope away from the houses and connect the wells to the drainage pipes in the foundation trench with heating cables (Fig. 2.45). Otherwise, the water in foundation trench will be squeezed out into the drainage wells as it freezes and then be drained out to the surface. The soil around the drainage wells should be replaced with sand gravel, and the top of the wells can be covered first with block stones with a thickness of 0.3 m ~ 0.5 m, and then covered with a polystyrene foam insulation layer with a thickness of 0.1 m, and finally covered sodding and dry soil (Fig. 2.46). The heating cable is used as a backup to prevent drainage wells freezing.

The waterproof and drainage method is an important measure to prevent frost heaving by lowering the groundwater level and reducing water content. The engineering design should consider the topography, frost heaving and hydrogeological

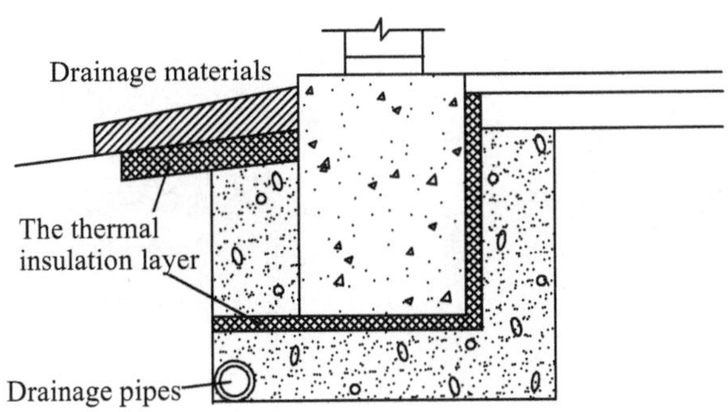

Fig. 2.43 The measures of water insulation and drainage of figure

Fig. 2.44 Foundation drainage to the underlying permeable layer in seasonally frozen soil areas

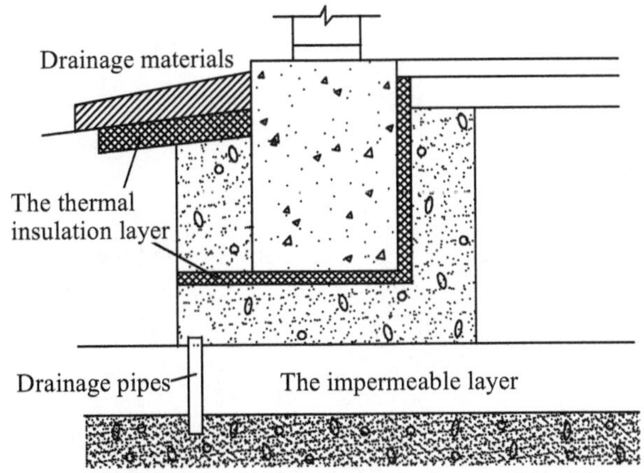

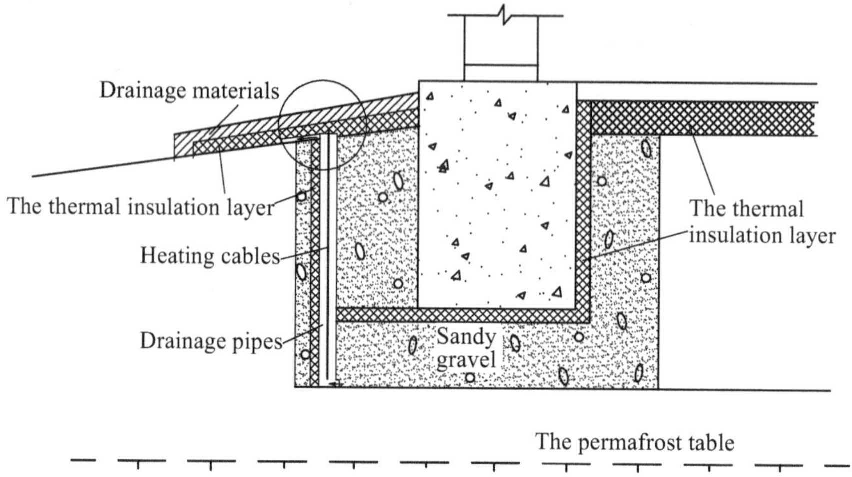

Fig. 2.45 Foundation drainage to the drainage wells in permafrost regions

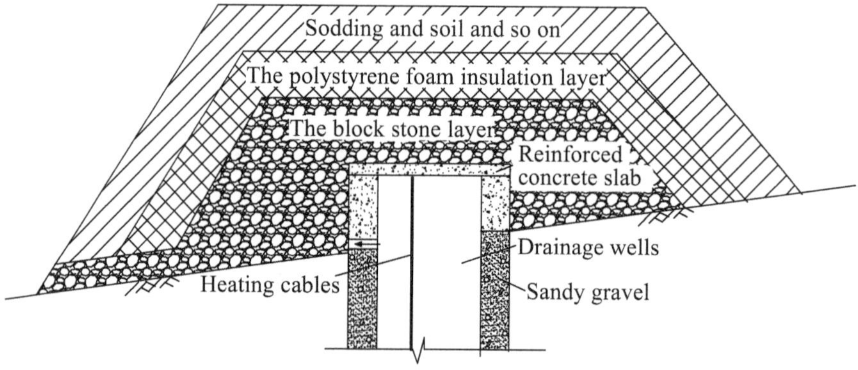

Fig. 2.46 Structural map of the outlet of drainage well

conditions of the engineering location, which is usually used in conjunction with the replacement method. In permafrost regions, it is necessary to make adequate anti-freezing measures at the outlet.

2.5.4 Isolation Method

Isolation method refers to adopting isolated measures between the foundation and its surrounding foundation soil, so that there is no freezing (no freezing strength) between the side surface of the foundation and the foundation soil, so as to eliminate

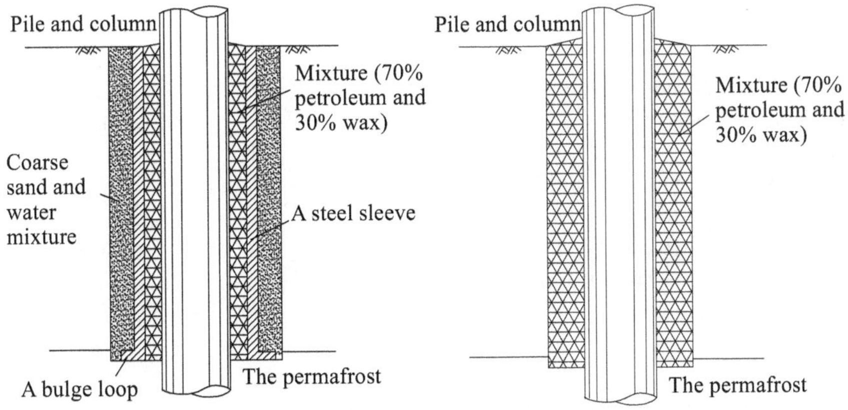

Fig. 2.47 The Measures against frost heaving by insulation method

the tangential frost heaving force of the foundation soil on the foundation side. This method can effectively prevent the frost heaving of buildings in pile foundation in either seasonally frozen soil or permafrost regions.

There are two kinds of isolation methods, including pile and column foundation with or without sleeves within the range of seasonally active layer. The isolation method against frost heaving with a sleeve is to cover the pile and column foundation with a steel sleeve with a bulge loop at the bottom, which is buried into the bottom of the maximum seasonally active layer (Fig. 2.47a). And the outer wall of the steel sleeve is coated first with a layer of hydrophobic material to avoid upfreezing when the foundation soil freezes. The gap between the sleeve and the pile and column foundation (usually 20 mm ~ 50 mm) is filled with a hydrophobic mixture, and the outside of the sleeve is back-filled with coarse sands and water. As for the isolation method without sleeves, the gap between the pile and column foundation and the foundation soil is directly filled with hydrophobic mixture (Fig. 2.47b).

For the foundation, pile and column foundation against frost heaving needs to be elevated and isolated between the ring beam on the pile and column foundation and the ground, so as to avoid the damage of buildings due to the direct contact between the frost heaving of ground and the ring beams of the foundation. The elevated isolation height of the foundation ring beam should be determined according to the maximum amount of frost heaving, and the external baffle (or bricks) can be set on both sides of the ring beams to prevent soil and other sundries from entering the elevated isolation space.

For pile and column foundations, it is ideal and easy to use the isolation method against frost heaving. In practice, the steel sleeve has upfreezing because the bottom bulge loop does not play the anchoring role, and the hydrophobic materials coated on the outer wall are too shallow to avoid the freezing of the foundation soil.

2.5.5 *Anchoring Method*

The anchoring method against frost heaving refers to anchoring the foundation in the foundation soil, but the strength of the foundation is supposed to resist the tangential frost heaving force of the foundation soil. And the stability against frost heaving of the foundation should carry out the checking calculation according to the foundation soil with frost heaving.

Foundations usually using the anchoring method include deep pile foundations, under-reamed foundations, explosive-enlarged pile foundations, squeezed cast-in-place pile foundations, expanded pile foundations, pile-bent foundations, etc.

2.5.5.1 Deep Pile Foundations

Deep pile foundation is always used in permafrost regions or other areas of low bearing capacity of the foundation, high groundwater level and some difficult construction conditions. The buried depth of pile foundation should not only meet the requirements of the bearing capacity, but also meet the requirements of stability against upfreezing, which are the controlling conditions of buried depth of pile foundation. The design of pile foundation in permafrost regions should refer to the Sect. 2.4.3.

2.5.5.2 Spread Foundations

The spread part of the spread foundation in cold regions should be buried under the seasonally active layer. First, it needs to meet the requirements of the bearing capacity of the foundation. Second, it can prevent the foundation from upfreezing by using its self-anchorage.

2.5.5.3 Explosive-Enlarged Pile Foundations

Holes are formed by blasting methods, and an enlarged head is formed below the seasonally active layer (Fig. 2.48). The enlarged diameter is usually 1.2 m ~ 1.5 m determined experimentally. The diameter and reinforcement of explosive-enlarged pile should meet the requirements of the strength under design load and the tensile strength under tangential frost heaving force. The method, with simple construction and fast progress, is suitable for cohesive soil foundation with shallow underground water, but sandy soil and gravel soil foundations are prone to hole collapses. Because it is difficult to accurately control the size of the hole-forming and the enlarging end, the construction quality should be controlled by experiment in application.

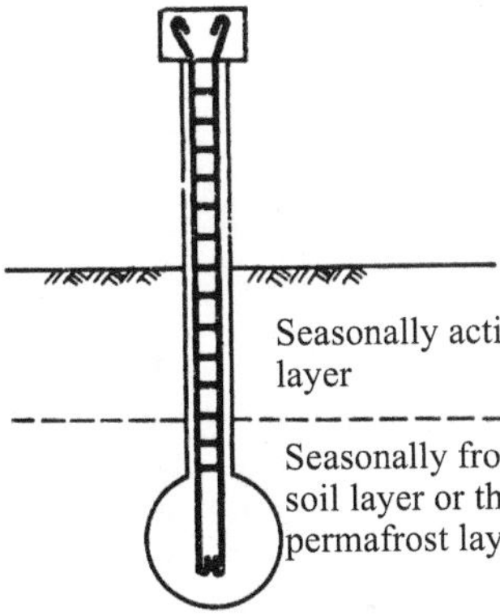

Fig. 2.48 Exploded pile foundations

2.5.5.4 Squeezed Cast-in-Place Pile Foundations

First, drilling is used to make a hole according to the design requirements, and then sandy gravel and cement mortar are filled in the bottom of the hole, and the sandy gravel is squeezed and expanded around the hole wall by impact method to form an expanded head, and then the construction procedures of cast-in-place piles is carried out to complete the construction of a squeezed and expanded pile. Squeezed cast-in-place piles are suitable for cohesive soil, or warm permafrost with cohesive soil.

2.5.5.5 Expanded Pile Foundations

First, drilling is used to form a hole, and then a drill bit with shrinkage and expansion is used to enlarge the bore. And the drill bit of expanded pile is suitable for cohesive soil foundations.

2.5.5.6 Pile-Bent Foundations

Like the spread slab of the spread foundation, pile-bent foundation is a "frame" formed by connecting two spread slabs of piles (Fig. 2.49). The pile-bent foundation has advantage to prevent frost heaving, but the excavation of the foundation trench is relatively larger. The engineering practice shows that the above anti-frost heaving methods can exert a certain effect, but each has its own advantages and disadvantages. In the process of application, the above methods should be used comprehensively according to the specific conditions of the foundation soil and engineering construction.

Fig. 2.49 Bent frame and pile blant foundations

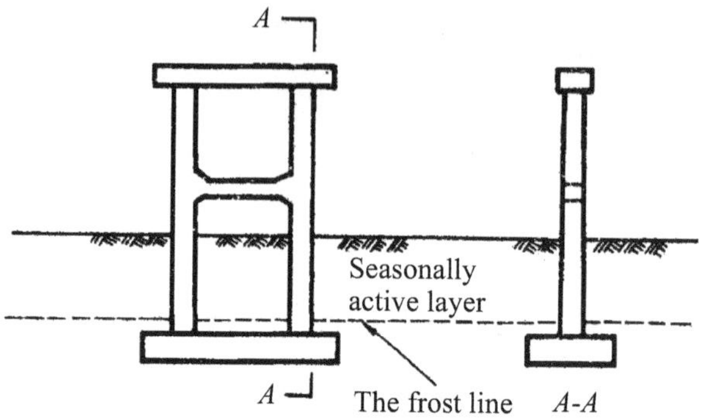

2.6 Design of Preventing Thaw Settlement

In addition to frost heaving damage in the construction engineering in permafrost regions, thaw settlement of permafrost foundation will cause building damage. Therefore, the design of building foundation in permafrost regions needs to consider the prevention of frost heaving as well as the thaw settlement of the foundation.

There are usually two methods to prevent thaw settlement of permafrost foundation. One is to cool permafrost foundation to decrease the ground temperature; the other is to passively maintain ground temperature of the permafrost foundation. The methods of cooling the permafrost foundation include elevated ventilated basement, ventilated pipes with filled soil, thermosyphon pipe (piles), and crushed rock foundation with filled soil, as well as the composite foundations, such as fill thermal insulation-ventilated pipes and thermal insulation-thermosyphon foundation.

The methods of passively maintaining the ground temperature of the permafrost foundation include thermal insulation foundation, and the high-fill foundation with sandy gravel, etc.

The engineering practice shows that the most effective way is to adopt elevated ventilated cooling system in decreasing the ground temperature of the permafrost foundation and preventing the formation of "thaw bulb" under building construction in permafrost regions.

The methods of preventing thaw settlement can be summarized as follows:

- The measures against thaw settlement
  - Physical method – the replacement method: using "single" coarse – grained material to replace the frozen soil foundation
  - The method of water insulation and drainage
    - The method of drainage: to drain the water around the foundation;
    - The method of water insulation: to block the flow of the groundwater into the foundation
  - The insulation method – thermal insulation method: to use industrial insulation materials to prevent atmospheric heat from invading the foundation soil and to avoid the thawing of frozen soil on the foundation base
  - The cooling method
    - Block stone paved: the foundation is covered with a layer of block stones, using its characteristic of "thermal switch" to cool down;
    - The method of ventilated pipes: the ventilated pipes buried under the foundation;
    - Thermal probes: to cool the foundation by the transmission of "temperature difference" and "latent heat" of the working medium in the sealed pipes;
    - Thermal insulation layer + thermal probes: thermal insulation on the upper part + the composite thermal probes to cool the lower part
  - Structural method
    - Conical column spread foundation: conical column + spread slab foundation at the bottom of column;
    - Thermal pile foundations: the composite pile and column foundations with thermal probes buried in the pile and column;
    - Spread foundation: the expanded head at the bottom of the pile foundation, manual or mechanical construction;
    - Conical + pile foundation + spread foundation: conical shape in the active layer + straight piles in the frozen soil layer + expanded head at the foundation base;
    - Deep pile foundations: pile foundations with reasonable growth
  - Comprehensive measures: composite measures

Chapter 3
Subgrade Engineering in Permafrost Regions

Highway and railway engineering in permafrost regions have the similarities in the forms and causes of subgrade damages (thaw settlement, longitudinal cracks and side slope cracks, et al) and prevention and control measures in frost damages. Numerous survey data show that the main damage of road engineering (or subgrade engineering) in permafrost regions is caused by thaw settlement of permafrost under embankment. It exhibits subgrade settlement (Figs. 3.1 and 3.2), longitudinal cracks in the pavement and at the road shoulder (Fig. 3.3 and 3.4) and side slope cracks in road shoulder (Figs. 3.5 and 3.6). The damage of part road engineering is caused by frost heave (not obvious), and ice damages (Figs. 3.7 and 3.8). Pavement fracturing and frost boiling in highway engineering are also common damages in permafrost regions (Fig. 3.9). Ice damage often occurs in areas with springs above the highway, such as in the sections from K3085 to K3087 on the south slope of Fenghuo Mountain along the Qinghai-Xizang Highway, K3129 to K3131 in Wuli, K140 of Ijia along the Yalin Railway Line, Mangui Riverside working areas, and Highway of Daxing'anling forest area in Inner Mongolia where there appear a lot of aufeis.

Embankment construction has changed the thermal state of permafrost thermal regime under embankment, resulting in thaw settlement due to the increase of permafrost temperature, the decline of permafrost table, and the melting of ground ice. The uneven heating of shady and sunny slopes for embankment causes the change of permafrost environment and water accumulation due to the poor drainage at the slope foot of the embankment, resulting in uneven thawing of the permafrost under embankment. So permafrost table on the sunny slope changes greatly, resulting in embankment thaw settlement, the instability of side slope, and the pavement cracking.

Embankment construction not only affects the original heat balance state between ground surface and atmosphere, but also changes the hydrogeological conditions and the environment of a certain range under and around the embankment. Therefore, the runoff condition of groundwater is changed, resulting in aufeis appearance.

Q. Wu, C. Tong, *Frozen Soil Engineering in Cold Regions*,
https://doi.org/10.1007/978-981-95-2527-0_3

Fig. 3.1 Thaw settlement of embankment for Qinghai-Xizang Highway subgrade (Photo by Tong Changjiang)

Fig. 3.2 Thaw settlement of embankment for the Yalin line railway in the northeast of China (Photo by Li Qingwu)

Pavement crack and frost boiling are mostly caused by water migration during the freezing process due to the high-content water of the base course, or the shallow groundwater level, poor drainage of the side ditch, water accumulation, or insufficient roadbed height, or poor roadbed filler. When the migration water is gathered in the basement layer under the pavement and freezes, coupled with the insufficient strength of the pavement structural layer, the pavement is softened during the spring thawing period to form crack and frost boiling.

There are many similarities between highway and railway embankment. However, the pavement structural layer that bears the external load action on the upper part of embankment is very different, resulting in the difference of thermal effect between the permafrost and the pavement. The surface layer of highway embankment is mainly black asphalt pavement, white or gray cement pavement,

Fig. 3.3 Longitudinal cracks of embankment for Qinghai-Xizang Highway (Photo by Tong Changjiang)

Fig. 3.4 Longitudinal cracks of embankment for the Qinghai-Xizang Railway (Photo by Bao Liming)

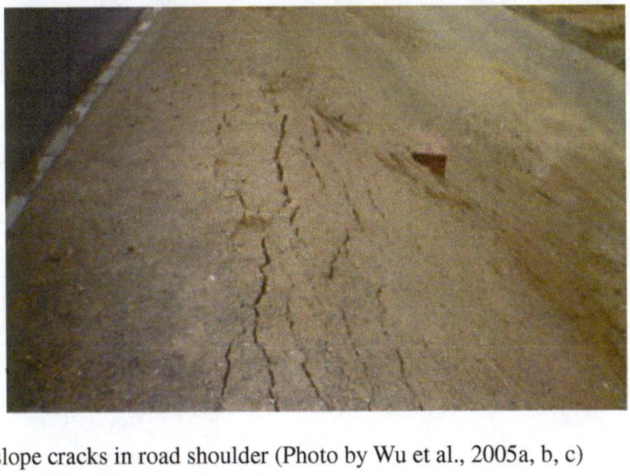

Fig. 3.5 Side slope cracks in road shoulder (Photo by Wu et al., 2005a, b, c)

Fig. 3.6 Cracks of embankment slope for the shoulder in railway (Photo by Li Yongqiang)

Fig. 3.7 Overflow ice on the Tian-Mu Highway in Qinghai province (Photo by Tong Changjiang)

which is a closed pavement structure; but the surface layer of railway embankment is mainly sand-gravel pavement with ballast and sleeper, which is an open pavement structure. The two pavement structures cause a great difference in the intensity of thermal effects between the permafrost under embankment and the atmosphere. The closed pavement structure has no surficial evaporative heat dissipation, and the heat flow direction is blocked and downward, which often has a "penetrating" heat storage effect. The heat direction of the opened pavement structure is upward and has a heat dissipation effect. The asphalt pavement with strong heat absorption and without evaporative heat dissipation increases the thawing depth of permafrost under embankment.

In permafrost regions, the surface heat mainly comes from solar radiation. Different surface characteristics have different reflectivity (or absorptivity) of solar

Fig. 3.8 Overflow ice on the northeast Yalin line railway (Photo by Tong Changjiang)

Fig. 3.9 Frost boiling and fracturing of asphalt pavement for National Highway 214

Table 3.1 Surface albedo of different surface types (Ding et al., 2011)

Surface properties	Asphalt pavement	New concrete pavement	Old concrete pavement	Gravel surface	Crushed rock pavement	Yellow quartz sand pavement	Sandy pebble pavement
Reflectivity α	0.06	0.20	0.27	0.12	0.14	0.34	0.23

shortwave, resulting in different surface temperatures. Table 3.1 shows that black asphalt pavement has the smallest reflectivity and the largest heat absorptivity. The n-factor of the ground surface (the ratio of the ground surface freezing or thawing index to the air freezing or thawing index) indicates the difference of heat between the thawing period and the freezing period (Table 3.2).

Table 3.2 Reference for surface *n* coefficient in permafrost regions of China (Ding et al., 2011)

Area	Qinghai-Xizang Plateau area		Northeast area of China	
Ground surface types	n_t factor	n_f factor	n_t factor	n_f factor
Mossy surface in Forest	–	–	0.4	0.3
Turf-covered surface	1.0	0.5	1.0	0.5
Crushed rock layer and pebble layer surface	1.65	1.04	–	–
Fine-grained soil surface	2.5	0.8	1.5	0.85
Gravel soil and gravelly soil surface	3.5	0.65	2.0	0.7
Cement concrete surface	3.5	0.7	2.0	0.7
Asphalt concrete surface	5.0	0.65	3.0	0.7

The characteristics of *n*-factor in Table 3.2 show that the surface temperature is always higher than the air temperature. The *n*-factor during the thawing period is greater than that during the freezing period. The *n*-factor of the asphalt pavement is the largest, and *n*-factor of the crushed rock layer surface is smallest. In Fenghuo Mountain area of the Qinghai-Xizang Plateau, the mean annual temperature of the silty clay surface is about 2 °C to 4 °C higher than the mean annual air temperature (Ding et al., 2011), and the mean annual surface temperature of asphalt pavement is 6 °C higher than the mean annual air temperature (Zang & Wu, 1999). Therefore, the influence of asphalt pavement on the permafrost layer under embankment is much greater than that under the sand-gravel pavement with ballasts.

As far as highway and railway embankment are concerned, within a certain range of the line passing through the permafrost zone, embankment construction will change the thermal state of permafrost under natural conditions and under embankment and re-establish a new thermal balance system between the atmosphere and the permafrost. During this process (may be 3–5 years, or a longer time), the permafrost table and permafrost thermal regime under embankment will be changed, resulting in the change of embankment stability. Therefore, the key point of embankment design in permafrost regions is to maintain or improve the thermal stability of the permafrost under embankment, thereby ensuring its mechanical stability.

3.1 Geological Route Selection for Frozen Soil Engineering

The road engineering has a long line and passes through various geomorphic units. The geological and hydrogeological conditions encountered are extremely complex. There are many engineering projects, including embankments, road cuttings, bridges, culverts, tunnels, communications, stationhouses, and environmental engineering, etc. From the perspective of engineering foundation, this chapter only discusses the route selection for subgrade engineering according to the engineering geological conditions of frozen soil.

The engineering properties of permafrost, in addition to geological, geomorphological and hydrogeological conditions, are closely related to permafrost temperature. When permafrost is used as the foundation, the relationship of heat balance among the atmosphere, embankment and permafrost environment must be considered. In the natural environment, under the geological period, the thermodynamic characteristics of permafrost environment composed of permafrost-geology-hydrogeology-environmental state are in a relatively stable state. The construction of subgrade engineering will disturb the original thermodynamic stability of the permafrost, and a new heat balance state will be established over a period of time, which may aggravate the variation and extinction of permafrost. Historical experience and lessons have shown that road project passing through a permafrost region is required to pay close attention to the heat balance in terms of line selection plan, design scheme, construction procedure and operation management, etc. The principle of "prevention first, priority to protection, equal importance of construction and protection" should be earnestly implemented. Based on the engineering geological conditions of permafrost, various measures and new technologies should be adopted to maintain or minimize thermal disturbance to permafrost, thereby protecting the permafrost environment and ensuring engineering safety.

3.1.1 Principle for Route Selection

In permafrost regions, the route selection of road (roadbed) engineering, in addition to complying with the basic principles for the non-frozen soil area, focuses on establishing the route position from engineering geological conditions and environmental protection of permafrost. At the same time, considering the rationality of technical economy, the feasibility of construction, operation and maintenance conditions, the fill-over-cut principle (prefer embankment to cutting) should be followed in the permafrost regions with high-ice content. The design for the route position should follow the following principles:

1. The route should be selected in the dry, sunny upper part of the gentle slope with ice-poor permafrost zone. Figure 3.10 shows the distribution of permafrost with high ice content in different topographical parts of the same geomorphic unit. From the view of slope upward, the sunny slope receives more solar radiation than the shady slope, the ground temperature is relatively high, the water evaporates more, and the ground ice is poorly developed. In the continuous permafrost and patchy permafrost zones, the shady slopes have permafrost with high ice content while the sunny slopes have not. Sunny slopes are mostly ice-poor or icy permafrost, with dry surface and sparse vegetation. From the perspective of slope, the sections of thin overburden layer, and good drainage conditions are not conducive to the growth of ground ice for steep slope; the section of thick overburden layer, sufficient moisture, and easy growth for vegetation are beneficial to the development of ground ice for gentle slope. The survey data along the

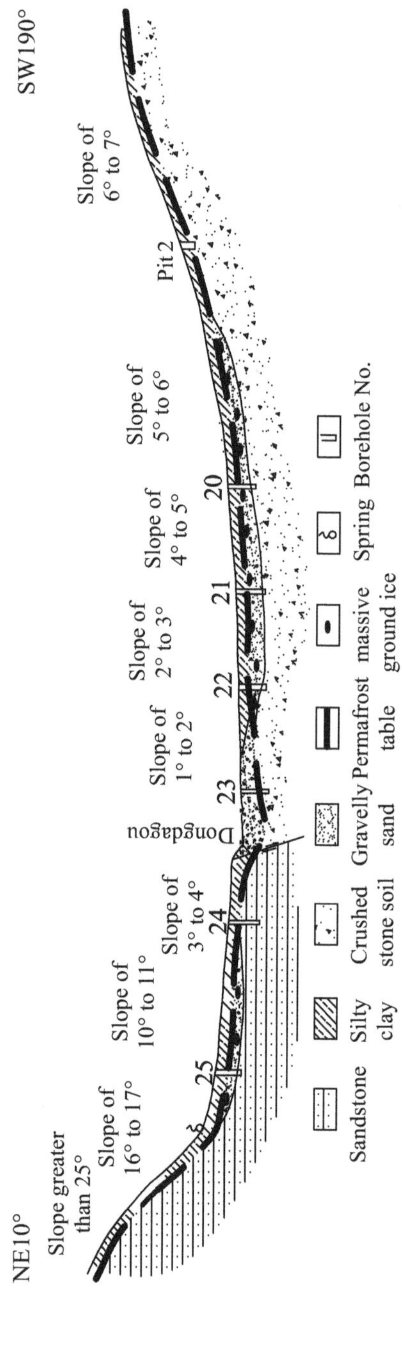

Fig. 3.10 Development of ground ice in different slopes and slope aspect of Dongdagou, Fenghuo Mountain (Qinghai- Highway research Group, 1983)

Qinghai-Xizang Highway show that on the slopes with the gradient of less than 10°, the ground ice generally develops, especially 4–8° is most conducive to the growth of ground ice; on the slope gradient of 10–16°, the conditions for the development of ground ice become worse; on the slopes greater than 16°, there is generally no thick ground ice, and when the slope gradient is greater than 25°, there only exists ice wedges. In terms of location, from the upper part of the hillside to the lower part, the lithology changes from coarse-grained to fine-grained, the loose material deposit changes from thin to thick, ground ice thickness gradually increases from top to bottom, and gradually transitions from ice-poor permafrost to ice with soil.

2. The route position in front of the mountain should choose a high line location, however, in the section of gelifluction and thermal thaw slumping, the route passes through in the form of embankment at its low outer edge. Ground ice is often distributed in the piedmont slopes, and there is often the possibility of unfavorable cryogenic phenomena such as gelifluction and thaw slumping, etc., and the route is usually below its outer edge and passes through in the form of embankment. Once the surface vegetation is destroyed, gelifluction and thaw slumping mayoccur, leading to a headward erosion. No timely restoration and protection during the excavation of the embankment side ditch results in gelifluction due to the thawing of ground ice.
3. When the route is laid along the river valley, it is advisable to choose high terraces or platform, and the high floodplains. The unstable zones such as the transition zone between the river talik and the permafrost zone, the interval distribution between the talik and the island permafrost should be detoured or avoided, and the line should pass through this area with the shortest distance.
4. The route should be selected on a good foundation soil as far as possible, that is, coarse-grained soil such as block stones, crushed stones, pebbles, and sand gravels, etc., and hard plastic to semi-hard dry cohesive soil with low moisture content. The surface of these areas is dry, the ice content of permafrost is small, and the foundation base is relatively stable.
5. The route should detour or avoid the permafrost zones with the development of various unfavorable cryogenic phenomena and with high ice content (ice-rich frozen soil, ice-saturated frozen soil and soil-bearing ice layer). If it is necessary to pass through, it is advisable to choose the place in the flat terrain, with a narrow distribution of cryogenic phenomenon, and a thin ice layer. The line should meet the following requirements:

    1. Route Selection in Alpine Swamps and Thermokarst Lake in Permafrost Areas
    For the route crossing stripe marshes in permafrost areas, special embankment design methods should be adopted to pass through with the shortest route, preferably crossing the swamp; but for the large swamp area with wide alluvial and proluvial terraces, the line should be selected at the edge of the swamp, because the peat layer is thin and the ice content is relatively small. When the route must pass in the middle of the swamp, it should be selected at the shallow swamp, and the route design should consider whether there is

an ancient riverbed and the size of transverse slope at the bottom of the swamp since thicker soft soil layers, such as peat and silt, may be found in ancient riverbed, with higher ice content and larger transverse slope easily sliding.

For the route crossing the gentle hillside swamp, it should be avoided to pass through the section with a steep transverse slope at the bottom of the swamp under the condition that the route direction should parallel to the trend of the hillside. Because the swamp and the bottom of steep transverse slope usually have the permafrost with high-ice content, in the warm season, the embankment on these sections is prone to slide, and the embankment often slides down suddenly under the action of vibration load. Special treatment and design should be made.

2. Route Selection in Massive Ground Ice Section
It is advisable that the routes through massive ground ice section or comparing with the road and bridge schemes choose the shortest distance or the section with thinner ground ice to pass. When the embankment is passed through, the zero section and the low fill with the height lower than 1.0 m should be minimized and avoid excavation, and the high fill embankment with the height higher than 3.0 m should be avoided. During the construction of the Qinghai-Xizang Railway, there was a cutting project in the permafrost section with high ice content. During the excavation, the ground ice was thawed and the aufies came into being (Fig. 3.11). After the cutting was formed and the skeleton slope protection was completed, thaw slumping (Fig. 3.12) occurred due to permafrost thaw at the foot of slop, while cracks and slips appeared on the side slope and the top of the cutting slope (Fig. 3.13).
3. Route Selection in Icing and Frost Mounds Sections
Suprapermafrost water can be found in the section of icing and frost mound, so hydrogeological surveys should be carried out to find out water sources, runoff directions, and estimate their flows.

Fig. 3.11 Drift ice in the ground ice section after digging

Fig. 3.12 Regressive thaw slumping of road cuttings in the ground ice section

Fig. 3.13 Slips and cracks on the top of the cutting slope in the ground ice section (Photoes by Liu Yongzhi)

Suprapermafrost water, causing icing and frost mounds, is usually seasonal phreatic water with limited flow in permafrost regions and generally deplete by the end of December. The route should be selected to pass under it in the form of embankment, and the distance from icing and frost mound should be outside the allowable range (tens of meters). It is worthy to note the embankment should be avoided to pass though above icing and frost mound. The foundation soil compaction of embankment or the increase of the permafrost table will hinder the channel of suprapermafrost water to cause icing and frost mounds to move up and appear on the embankment, so affecting the roadbed and driving safety.

The intrapermafrost or subpermafrost water, causing icing and frost mounds, has a certain confine water capacity, with high flow rates. The route should be selected above it since the foundation soil compaction of the embankment or the permafrost table will not block the groundwater channel.

In hillsides or watershed areas, when passing through, either the embankment or the road cutting may block groundwater channels, or expose the groundwater, causing ice damages. In this case, drainage measures should be carried properly. Insulating anti-seepage blind drains should be built above the embankment or on both sides of the road cutting. The depth of blind drains should be relatively larger, and the slope ratio at the bottom of the ditch should be increased as much as possible. The drains should connect with culverts or bridges in order to quickly discharge water. The outlet of the blind ditch should be insulated to avoid freezing and blocking water flow.

6. Route Selection in Ridge-crossing Sections. When the route crosses the watershed, it is mostly selected to pass at the mountain passes in order to reduce the height. According to the permafrost distribution characteristics and hydrogeological characteristics, mountain pass section is located in a saddle-shaped terrain where the stratum is often relatively fragmented, the loose overburden layer is thick, the groundwater is more enriched, or there is a tectonic zone distribution, which may become a section with relatively rich underground ice (Fig. 3.14). The survey in geology, frozen soil engineering geology and hydrogeology should be carried out during the route selection. The high vegetation coverage and more water accumulation in swamps indicate that the permafrost is more developed, with massive ground ice. Meanwhile, it is advisable to avoid the section with higher ground ice content and choose the section with lower ground ice content. These sections should avoid deep-excavated cutting and is advisable to pass through in the form of tunnel.

   Obviously, route selection for road (roadbed) engineering in permafrost regions should avoid these sections needing to excavate, light fill and shallow cut as far as possible and strictly prohibit taking soil on the spot from both sides of the embankment. The route should choose in the relatively higher position, and the foundation base should protect surface vegetation and peat layer. These sections should fill the crushed rock to increase the strength of embankment base

Fig. 3.14 Ground ice outside Fenghuo Mountain of Qinghai-Xizang Railway

with a reasonable embankment height. Engineering practice in the permafrost regions of the Da and Xiao Xing'anling Mountains in Northeast and the Qinghai-Xizang Plateau has proved that this basic principle cannot be ignored. And the only embankment construction by filling and cutting in order to save earthwork may be counterproductive.

3.1.2 *Frozen Soil Engineering Investigation*

The advantages and disadvantages of engineering geological conditions in permafrost regions, in addition to the assessment of the engineering geological conditions in non-frozen soil areas, should focus on ground-ice content and mean annual ground temperature. The extent of ice content of permafrost determines the thaw settlement and frost heave of foundation, and the mean annual ground temperature determines the thermal stability of permafrost and the thermal inertia affected by external thermal disturbances. Meanwhile the hydrogeological conditions in permafrost regions are often constrained by the extent of ice content and mean annual ground temperature. Their combined effects can reflect the development degree of unfavorable frozen soil phenomenon, especially the stability of slope zones.

The frozen soil engineering investigation should pay attention to the identification of the characteristics and extents of ice content and mean annual ground temperature in permafrost along the route.

Ice-poor and icy frozen soil are classified as low-ice content frozen soil while ice-rich frozen soil, ice-saturated frozen soil and soil-bearing ice layer are classified as high-ice content frozen soil, which determines the prevention and control measures and design for subgrade engineering. Based on mean annual ground temperature and thermal stability, permafrost can be divided into warm and cold permafrost, with the threshold of -1.5 °C for highway and -1.0 °C for railway, which related with the principle of engineering design.

3.2 Elements of Subgrade Design

3.2.1 *Thermal Stability of Permafrost and Ground Temperature Division of Subgrade*

3.2.1.1 Temperature Field of Permafrost Subgrade

The road project built on permafrost has changed the heat exchange conditions between the atmosphere and the ground surface, leading to the change of original heat balance state of ground surface in the permafrost region, and reconstructing the ground temperature field between the atmosphere and ground (engineering

Table 3.3 Comparison of changes in permafrost table under embankment with different pavement

Area	Location	pavement	Embankment Height/m	Permafrost table/m	Artificial permafrost table/m
	Qinghai-Xizang Highway				
Warm permafrost	Qingshuihe IIII	Asphalt		−2.2	−6.24
	Chumaerhe III	Asphalt		−2.3	−6.67
Cold permafrost	Fenghuo Mountain I	Asphalt		−2.1	−4.34
	Wudaoliang II	Asphalt		−2.2	−4.3
	Qinghai-Xizang Railway				
Warm permafrost	Qingshuihe III	Gravel	3.3	−2.0	−0.6
	Chumaerhe III	Gravel	2.9	−2.3	−0.6
Cold permafrost	Fenghuo Mountain I	Gravel	3.8	−2.6	+0.8
	Wudaoliang II	Gravel	2.7	−2.2	−0.4

Note: The permafrost table is estimated from the natural ground, especially the negative value is below the ground, and the positive value is above the ground; I is stable cold permafrost area, II is basically stable cold permafrost area, III is unstable warm permafrost area

environment). Meanwhile, the linear road engineering changes the properties of ground surface (such as vegetation coverage, soil properties and density, etc.), and surrounding hydrogeological environment (such as groundwater runoff and water content), and increases pavement materials (asphalt, concrete, gravelly crushed stone, etc.), resulting in the change of the surface reflectivity of the natural environment and the heat exchange capacity of the surface layer. Therefore, the coupled impact from the thermal disturbance and heat storage of embankment will increase permafrost temperature and lower permafrost table under embankment during the initial period of heat exchange process. After several years, under similar environmental conditions, the influence of asphalt pavement on the permafrost table under embankment remains to be still obvious (Table 3.3) (First Highway Survey and Design Research Institute of China Communications Construction, et al., 2006).

Figures 3.15 and 3.16 show the ground temperature envelope curves of permafrost under asphalt pavement in cold and warm permafrost areas respectively. It can be seen that permafrost table under the asphalt pavement decreases in both cold permafrost and warm permafrost areas, indicating a strong absorptive heat effect.

Figures 3.17 and 3.18 show ground temperature curves under embankment during the early stage (in the third year) of National Highway 214 being put into operation respectively (Zang & Wu, 1999). For the permafrost that the mean annual ground temperature is −0.9 °C to −1.0 °C (unstable warm permafrost), the ground temperature under the embankment changes slightly, and permafrost table rises slightly (by 0.2 m to 0.7 m). For the permafrost that the mean annual ground temperature is −0.3 °C to −0.5 °C (extremely unstable warm permafrost) the ground

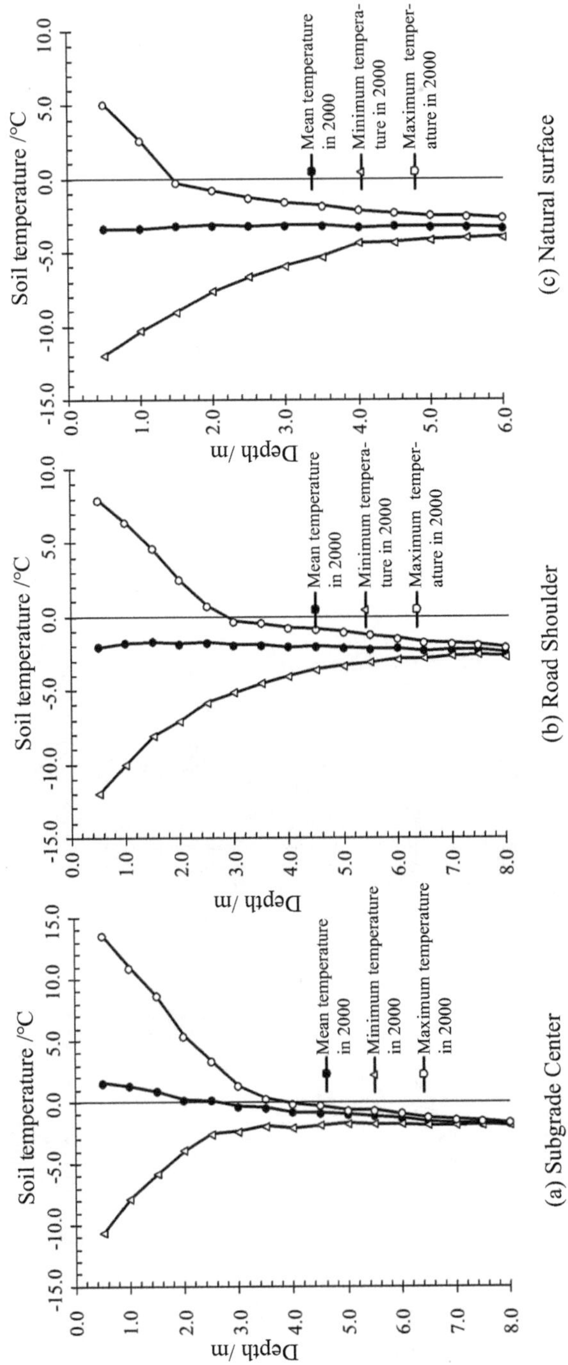

Fig. 3.15 Ground temperature envelope curve under embankment with asphalt pavement in cold permafrost areas

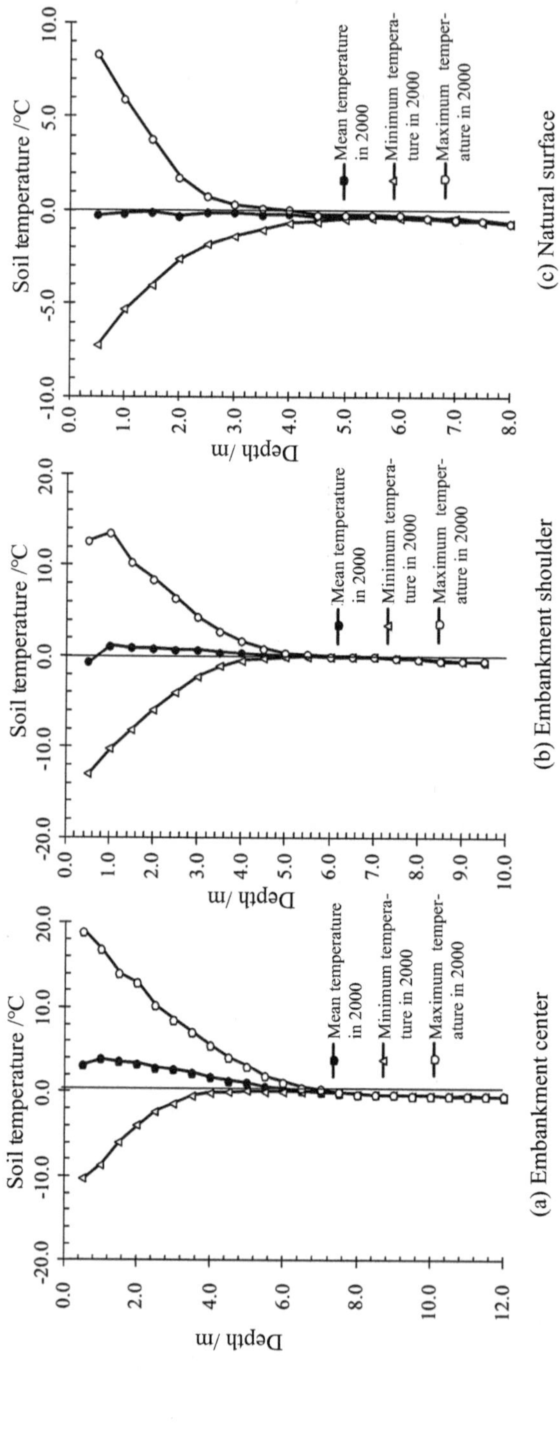

Fig. 3.16 Ground temperature envelope diagram of asphalt pavement material subgrade in warm permafrost area

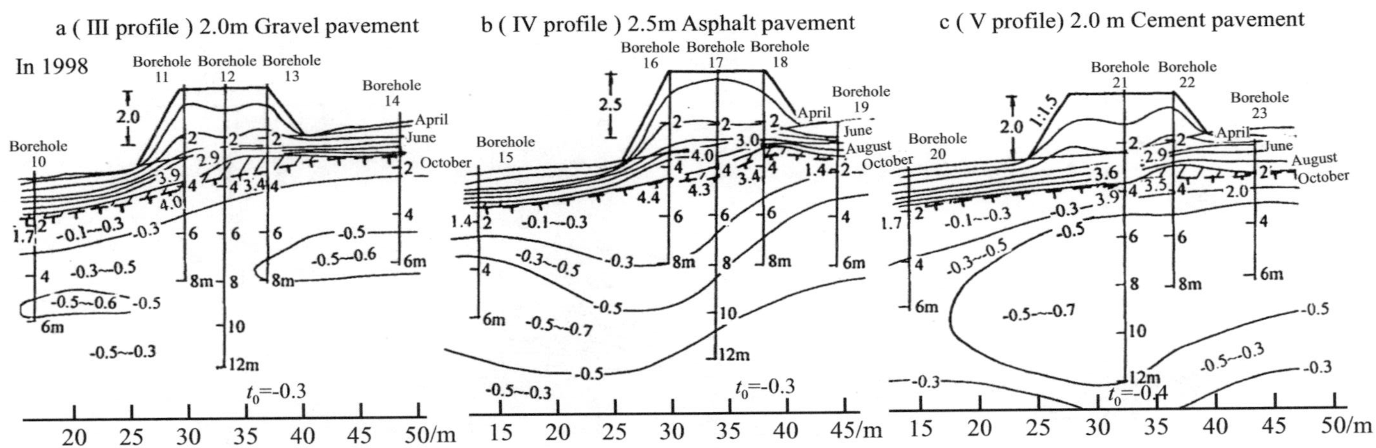

Fig. 3.17 Thawing process under embankment in the extremely unstable permafrost zone in continuous permafrost region (t_0 = mean annual ground temperature)

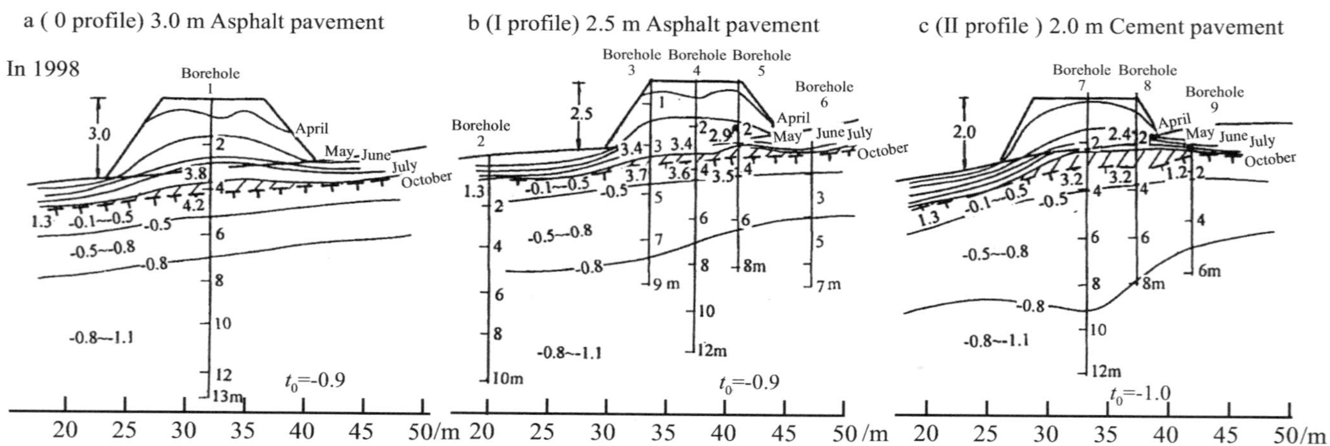

Fig. 3.18 Monthly thawing process of subgrade in the unstable permafrost zone in continuous permafrost region (t_0: Mean annual ground temperature)

temperature under the embankment changes greatly, and permafrost table declines (by 0.2 m). Compared with these three kinds of pavements (gravel, asphalt and cement), the permafrost table under the embankment with asphalt pavement is lowered, the ground temperature under cement pavement is increased, which is larger than that under gravel pavement for the ground temperature in the range from −0.1 to −0.3 °C. After the road is in operation, the ground temperature under embankment will change and the thawing depth will increase.

For the extremely unstable warm permafrost, the permafrost table under the embankment with gravel pavement along the Qinghai-Xizang Railway raises. The maximum rising extent of permafrost table at the center of embankment is up to 1.69 m, and permafrost table under the sunny slope of the embankment (on the left side) raises by 0.51 m and under the shady slope of embankment (on the right side) by 0.94 m. These results show that the gravel pavement has good heat dissipation performance.

3.2.1.2 Thermal Stability of Permafrost

Under the influence of engineering construction, changes in surface properties, hydrogeological conditions and vegetation eradication will lead to the increase in the thickness of active layer, melting of the ground ice, and the lowering in the permafrost table. However, due to the different mean annual ground temperature of the permafrost, the sensitivity of the permafrost to external thermal disturbances is different. In addition, the increase of ice content in permafrost is beneficial to increase its inertia against external thermal disturbance. Under the same environmental conditions, the lower is the mean annual ground temperature of the permafrost and the higher the ice content, the stronger is the inertia to the external thermal disturbance and the better the thermal stability.

Under the effect of engineering activities, the response extent of permafrost to the impacts of human activities can be used to show its sensitivity to external thermal disturbances. Here, the ratio of seasonal thawing depth to potential seasonal freezing depth is proposed to express the sensitivity. When the ratio of seasonal thawing depth to potential seasonal freezing depth increases, it indicates that the sensitivity to the impact of engineering activities on permafrost increases in the process of freeze-thaw heat balance; otherwise, the sensitivity to the impact of engineering activities decreases. The thermal thaw sensitivity (S_{Te}) of frozen soil is expressed as (Wu et al., 2002):

$$S_{Te} = \frac{h_t}{h_{qf}} \tag{3.1}$$

where h_t is the seasonally thawing depth (m); h_{qf} is the potential seasonally freezing depth (m).

Generally, heat amount entering ground is completely used to thaw seasonally freezing layer in summer season, that is, the seasonal thawing depth (h_t). Meanwhile, part of heat amount entering the ground is used to freeze the seasonally thawed layer, and other part is used to decrease permafrost temperature in the cold season. In this case, the impact depth of decreasing permafrost temperature is called the potential seasonally freezing depth (h_{qf}).

Obviously, under the impact of the external thermal erosion, the ability of permafrost to maintain the freeze-thaw process of the original permafrost and the mean annual ground temperature during the process of heat exchange is defined into the thermal stability of permafrost. The thermal stability of permafrost (S_T) is described by the ratio of the heat amount absorbed by the ground surface during the warm season to the total heat amount of thawing the sediments between the bottom of seasonally thawed layer and the potential seasonal freezing depth and of rising temperature at the bottom of seasonal freezing layer to 0 °C, that is:

$$S_T = \frac{Q_t}{Q_+} \tag{3.2}$$

where Q_+ is the total heat absorbed by ground surface during the warm season; Q_t the total heat amount of thawing the sediments between the bottom of seasonally thawed layer and the potential seasonal freezing depth and of rising temperature at the bottom of seasonal freezing layer to 0 °C, that is:

$$Q_t = \frac{\lambda_t \left| t_{dcp} \right| T}{h_t} + Q_\varphi \left(h_{qf} - h_t \right) \tag{3.3}$$

where λ_t is the thermal conductivity; t_{dcp} is the mean annual temperature at the top of the permafrost; T is the time period; h_t is the seasonal thawing depth; h_{qf} is the potential seasonal freezing depth; Q_φ is the heat of phase transition.

The annual heat flux of the soil (Q) is the sum of the total heat absorbed during the warm season (Q_+) and the heat released during the cold season (Q_-). Using thermal thaw sensitivity (S_{Te}) of permafrost to describe, the thermal stability of permafrost can be expressed by the following equation:

$$S_T = \frac{1 + S_{Te}}{S_{Te}} \times \frac{Q_t}{Q} \tag{3.4}$$

According to the in-situ measured data of ground temperature under natural surface along the Qinghai-Xizang Highway, the thermal stability of permafrost is closely related to the mean annual temperature at the top of permafrost and mean annual ground temperature, seasonal thawing depth and thermal erosion sensitivity.

1. **Relationship between the Thermal Stability of Permafrost and the Permafrost Temperature**

The thermal stability of permafrost is closely related to the mean annual temperature at the top of permafrost, mean annual ground temperature, surface temperature, and annual range of surface temperature. The mean annual ground temperature and

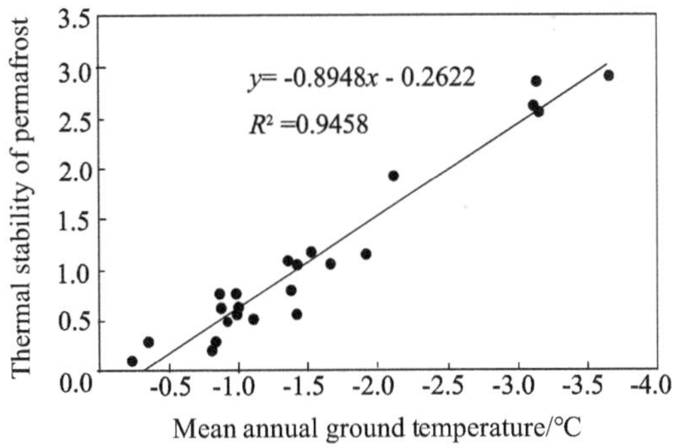

Fig. 3.19 Relationship between thermal stability of permafrost and mean annual ground temperature

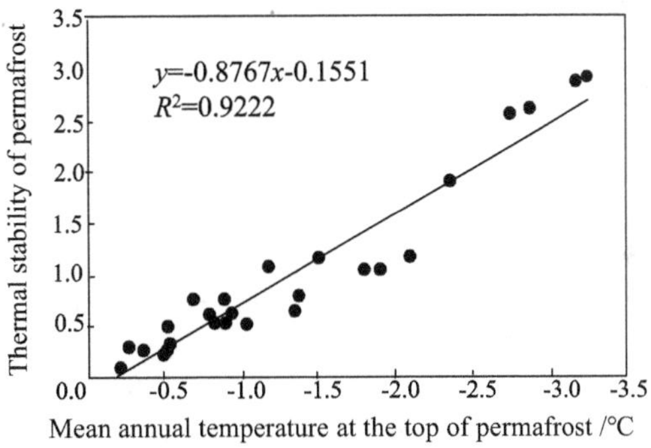

Fig. 3.20 Relationship between thermal stability of permafrost and mean annual temperature at the top of permafrost

the mean annual temperature at the top of permafrost have a good linear relationship (Figs. 3.19 and 3.20) (Wu et al., 2002). The lower the mean annual ground temperature of permafrost is, the more stable the effects of thermal disturbance of human activities on permafrost. When the thermal stability of permafrost is greater than 1, it indicates that the more heat (Q_t) in warm semi-year will be required to thaw the frozen soil layer between the bottom of seasonally thawed layer and potential seasonal freezing depth and no residual heat amount can be used to raise the temperature of seasonal freezing layer to 0 °C, and it is advantageous to decreasing the permafrost temperature. When the thermal stability of permafrost is less than 1, the heat absorbed in the warm season is greater than Q_t, the excess heat can be used to raise permafrost temperature, resulting in the permafrost degradation.

2. **Relationship between Thermal Stability of Permafrost and Seasonal Thawing Depth**

Figure 3.21 shows that the thermal stability of permafrost has a power function relationship with the seasonal thawing depth (Wu et al., 2002). Under the embankment with asphalt pavement along the Qinghai-Xizang Highway, when the thermal stability of permafrost (S_T) is greater than 1.5, S_T will decrease sharply and only lead to a slight change in the seasonal thawing depth. When S_T ranges from 0.5 to 1.5, S_T

Fig. 3.21 Relationship between thermal stability of permafrost and seasonal thawing depth

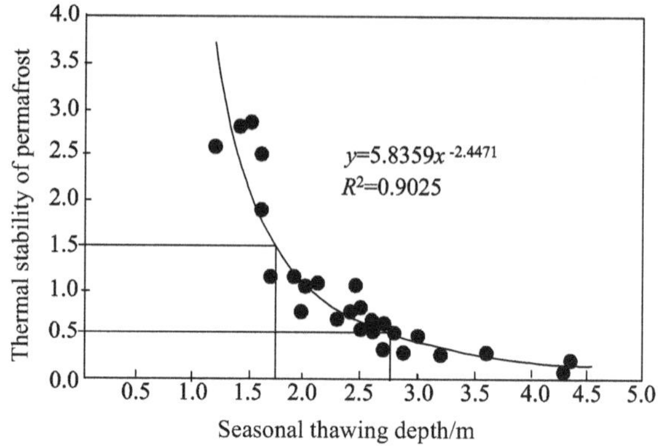

will vary with the seasonal thawing depth. When S_T is less than 0.5, the seasonal thawing depth will change drastically, and S_T will change slightly. These results show that when S_T is strong, the heat absorbed in the warm season will only cause a weak thawing of permafrost between the bottoms of seasonally thawed layer and the potential seasonal freezing depth, but it is not enough to raise the permafrost temperature at the bottom of the potential seasonal freezing layer to 0 °C. As S_T decreases, the heat absorbed in the warm season not only causes a large thawing, but also increases the permafrost temperature at the bottom of the potential seasonal freezing layer to 0 °C. When the S_T deteriorates, the permafrost temperature at the bottom of the potential seasonal freezing layer will be higher, and the heat absorption in the warm season will change slightly, which the permafrost temperature at the bottom of the potential seasonal freezing layer rise rapidly to 0 °C, making the permafrost between the bottom of the seasonally thawed layer and the potential seasonal freezing depth completely melt.

3. **Relationship between Thermal Stability and Thermal Thaw Sensitivity of Permafrost**

Under the embankment with asphalt pavement along Qinghai-Xizang Highway, if the ratio of Q_t/Q tends to be 1, for the thermal thaw sensitive permafrost, in order to keep the permafrost temperature under human activities, the minimum thermal stability (S_T) of permafrost should be greater than 2.4, and the mean annual ground temperature of permafrost should be lower than −3.0 °C. For the insensitive permafrost thermal thaw, S_T ranges from 2.1 to 2.4, which can keep the permafrost under human activities in a stable state, and the mean annual ground temperature is −3.0 °C to −1.5 °C. Therefore, when considering the stability of permafrost under human activities, the thermal thaw sensitivity of permafrost should also be taken into consideration.

Figures 3.22 and 3.23 show the relationship between thermal stability, thermal thaw sensitivity and seasonal thawing depth, mean annual ground temperature (Wu et al., 2002). For the weak thermal thaw sensitivity permafrost, seasonal thawing depth is relatively smaller while mean annual ground temperature is relatively

Fig. 3.22 Relationship between thermal stability, thermal thaw sensitivity of permafrost and seasonal thawing depth

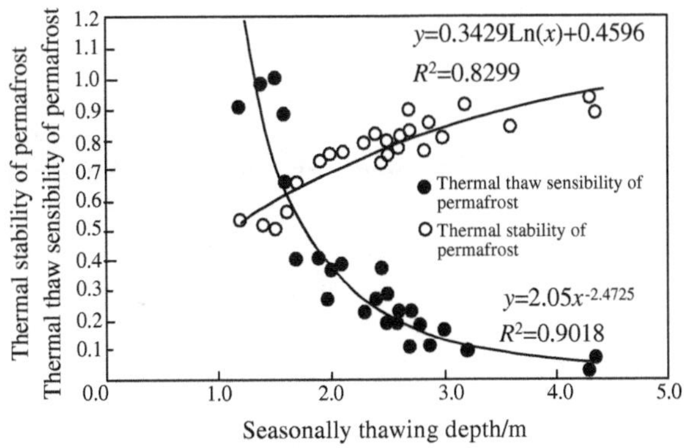

Fig. 3.23 Relationship between thermal stability, thermal thaw sensitivity and mean annual ground temperature

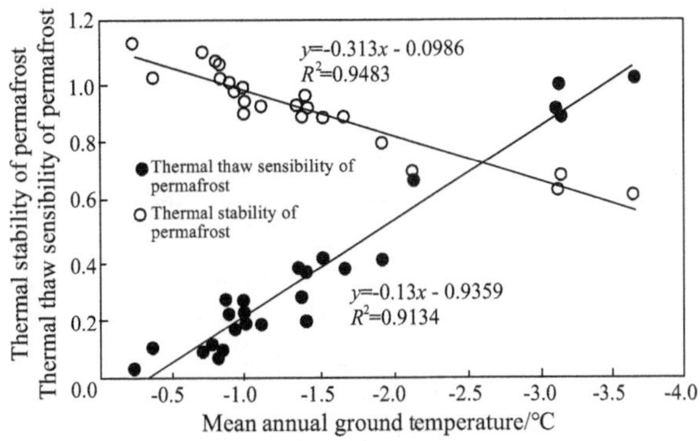

lower, permafrost is in a stable state. But For thermal thaw sensitivity permafrost, seasonal thawing depth is relatively larger while mean annual ground temperature is relatively higher, permafrost is in an unstable state. Therefore, the freezing-thawing process of cold permafrost is dominated by the freezing process, part of heat amount in the cold season is used to freeze seasonally thawed layer, and the residual heat is used to decrease permafrost temperature. Here the potential seasonal freezing depth is far greater than the seasonal thawing depth, which weakens the thermal thaw sensitivity of permafrost and strengthens the thermal stability of permafrost. However, the freezing-thawing process of warm permafrost is dominated by the thawing process. The heat amount in the warm season and cold season is mostly in the state of thermal balance, the seasonal thawing depth increases while the potential seasonal freezing depth decreases. The thermal thaw permafrost increases whereas the thermal stability decreases.

For permafrost with high ice content, although the mean annual ground temperature is relatively higher, thawed permafrost with high ice content requires a larger phase transition heat, which makes the change of seasonal thawing depth decrease and the thermal thaw sensitivity of permafrost weaken. Even so, as the deformation of permafrost with high ice content after thawing is still large, the influence of engineering buildings on underlying permafrost is larger than that of permafrost with low ice content.

3.2.1.3 Ground Temperature Division of Subgrade Engineering

For highway and railway engineering, there is difference in the width of roadbed section and pavement materials. In general, the pavement width of expressway and first grade highway is 23–28 m; the width of the 2-lane pavement of the second grade to the fourth grade highway is 7.5–12 m. And the pavement width for single line is 7.5–7.9 m and 11.2–12.1 m for double lines. At present, China's highway in permafrost regions is 2 lanes and the railway is single line, the roadbed width is basically 7.5–20 m. Highway pavement materials are asphalt and concrete which absorb heat and block evaporative heat dissipation, increasing the thaw rate of permafrost under the embankment. However, railway pavement materials are gravel with ballast with cold storage and evaporative heat dissipation.

Observations show that under the same permafrost and environmental conditions, whether it is in warm permafrost or cold permafrost, the permafrost table under the highway embankment often decreases (Wu et al., 2002). In the warm permafrost region, not only the artificial permafrost table is relatively deeper, but also a large amount of suprapermafrost water accumulates and forms a talik.

While the artificial permafrost table under the railway embankment is relatively shallower. According to the survey of the Qinghai-Xizang Highway, in areas where the mean annual ground temperature of permafrost is lower than or equal to −1.5 °C, even though the ice content of permafrost is high, there is fewer roadbed damage. In the permafrost areas with high ice content rate, where the mean annual average ground temperature is higher than −1.5 °C, the roadbed damage is relatively high, mainly distributed in high plains, hills and river valleys.

There are many reasons for roadbed damage in permafrost regions, and the ice content and thermal stability of permafrost are the main causes. According to the thermal stability index of frozen soil in Fig. 3.19, those with the thermal stability index of permafrost under the embankment with asphalt pavement greater than 1 belong to the stable permafrost and basically stable permafrost, and those less than 1 belong to the unstable permafrost and extremely unstable permafrost. Under the embankment with gravel pavement, if the thermal stability index of the permafrost is greater than 0.5, the permafrost table under the embankment can still maintain its original state, and such section still belongs to the basically stable permafrost. Engineering practice shows that under the condition that the mean annual ground temperature is −1.0 °C, the permafrost table under the embankment with asphalt pavement presents a descending trend, while the permafrost table under the embankment with gravel pavement and ballast rises or remains the same (see Figs. 3.16, 3.17 and 3.18). Therefore, under the two pavement conditions, the threshold between warm and cold permafrost is −1.0 °C for railway with gravel pavement and ballast and −1.5 °C for the highway with asphalt pavement. Therefore, the mean annual ground temperature of warm permafrost is higher than or equal to −1.0 °C for the railway and −1.5 °C for the highway (Fig. 3.24).

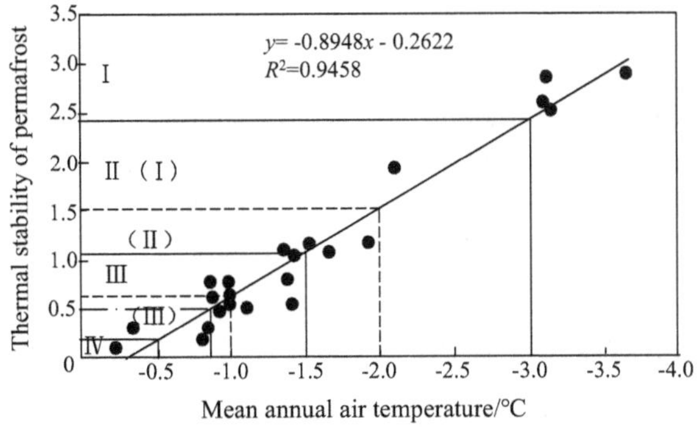

Fig. 3.24 Division of permafrost temperature zone

Solid line, the dotted line, and I etc., stand for the Qinghai-Xizang Highway; Dotted line and (I) are the Qinghai-Xizang Railway

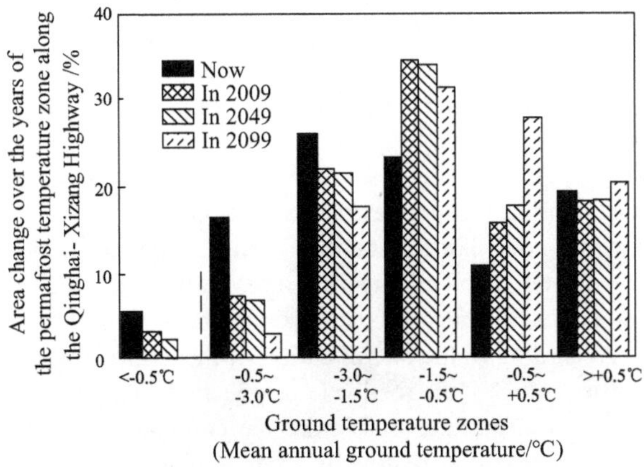

Fig. 3.25 Changes in permafrost temperature zones after temperature changes in different years

It should be noted that the classification for the warm and cold permafrost in highway and railway systems is based on the data of current mean annual ground temperature of permafrost. With a warming of climate, the ground temperature of permafrost also will change (Fig. 3.25) (Wu et al., 2001). At the same time, the zones of permafrost ground temperature will also change greatly, the area of cold permafrost will shrink, and part of it will be converted into warm permafrost; the area of warm permafrost will expand, and unstable permafrost will degrade. The design of subgrade engineering in permafrost regions should consider the factor of climate warming.

3.2.2 Frozen Soil Engineering Types

Ice content (or water content) of frozen soil is a factor that plays an important role in the various mechanical properties of frozen soil engineering characteristics. It not only reflects the difference of frozen soil cryostructure, but also reflects the characteristics in soil frost heave, thaw settlement of permafrost and frozen soil strength. As the foundation of buildings, frozen soil can be divided into five engineering types based on the ice content of frozen soil, ice-poor frozen soil, icy frozen soil, ice-rich frozen soil, ice-saturated frozen soil and soil-bearing ice (Table 3.4).

The thaw settlement deformation of permafrost depends on soil property, ice content and permafrost temperature. As the ground temperature rises and permafrost thaws, ice content of permafrost is the dominant factor causing the deformation of thaw settlement. The monitoring of embankment deformation shows that the deformation of embankment caused by thaw compression ranges from different permafrost engineering types. The volumetric ice content of ice-poor and icy permafrost is usually between 0.1 and 0.2, and the maximum thaw-settlement coefficient does not exceed 5%, and the deformation of thaw settlement is relatively smaller. The volumetric ice content of the ice-rich and saturated permafrost and the soil-bearing ice layer is more than 0.2, even more than 0.5, so that when the coefficient of thaw settlement exceeds 20–25%, the deformation of thaw settlement is very large, and the roadbed damage rate is very high. Figure 3.26 shows the relationship between different frozen soil engineering types corresponding to the roadbed damage rate of G214 in the eastern part of the Qinghai-Xizang Plateau (where the mean annual ground temperature of permafrost is relatively higher) (Xu, 2014).

As is shown, the thermal stability of permafrost is a comprehensive reflection of granulometric composition, ice content and the thermal regime of permafrost, and also the essential factor for embankment deformation. However, the thaw settlement characteristics of permafrost exhibit the embankment deformation and the control index of embankment stability in the permafrost regions. Therefore, the classification of frozen soil engineering must be based on the thermal stability and thaw settlement of permafrost to achieve the unity of essence and phenomenon (Table 3.4).

According to the classification of the frozen soil engineering for road engineering in Table 3.5, the roadbed design can be adopted corresponding principles.

For cold permafrost (I), the ice content of permafrost is relatively higher, its thermal capacity is relatively larger, with the better thermal stability, due to the relatively lower mean annual ground temperature. According to the monitoring data of permafrost temperature along the Qinghai-Xizang Highway, the permafrost table under the embankment with asphalt pavement has changed significantly (Fig. 3.27) (Liu et al., 2008a, b, c). For cold permafrost, permafrost table under the embankment evidently declines, but less than the change for the warm permafrost (Fig. 3.28). For cold permafrost, the thawing rate of permafrost under embankment with asphalt pavement is 3.0 cm/a to 5.0 cm/a under the case of high embankment (Liu et al., 2008a, b, c), lower than that under natural ground (Fig. 3.29). For warm permafrost, the thawing rate of permafrost under the embankment with asphalt pavement is up to 23.0–29.0 cm/a, much higher than that under natural ground (Fig. 3.29). Therefore, it is feasible to maintain the stability of the roadbed by increasing the height of the embankment in cold permafrost areas, while the stability of the roadbed cannot be maintained in warm permafrost area.

Obviously, for the cold permafrost (I), although its thermal stability is relatively better, permafrost will also change slightly under the natural surface and engineering influence. Engineering design should also take principle I to protect permafrost under embankment. However, engineering design should take some strengthening

Table 3.4 Comprehensive classification of frozen soil engineering (Supplemented by Wu, 1982)

Classification of frozen soil engineering		I	II	III	IV	V
Categories		Ice-poor soil	Icy soil	Ice-rich soil	Saturated frozen soil	Soil-bearing ice layer
		Frozen soil with low ice content		Frozen soil with high ice content		
Volumetric ice content	Cohesive soil	<0.10	0.10 to 0.20	0.20 to 0.30	0.30 to 0.50	>0.50
	Non-cohesive soil	0.15	0.15 to 0.25	0.25 to 0.35	0.35 to 0.50	>0.50
Thaw settlement	Grade	No	Weak thaw settlement	Thaw settlement	Strong thaw settlement	Thaw collapse
	Coefficient of thaw Settlement A_0/%	<1	1 to 5	5 to 10	10 to 25	>25
Frost heaving	Grade	No	Weak frost heaving	Frost heaving	Strong frost heaving	Extremely strong frost heaving
	Frost-heave coefficient η/%	<1	1 to 3.5	3.5 to 6	6 to 12	>12
Frozen soil strength	Grade	Ice-poor frozen soil	Icy and ice-rich frozen Soil		Ice-saturated frozen soil	Soil-bearing ice
	Relative strength	<1.0	1.0		0.8 to 0.4	<0.4
Cryostructure		Massive Cryostructure	Micro-layered or micro-reticulated Cryostructure	Layered Cryostructure	Ataxitic cryostructure	Ataxitic cryostructure
Atterberg limits ω/%		$<\omega_p$	ω_p to $\omega_p + 7$	$\omega_p + 7$ to $\omega_p + 15$	$\omega_p + 15$ to $\omega_p + 35$	$>\omega_p + 35$

Note: Taking cohesive soil as an example

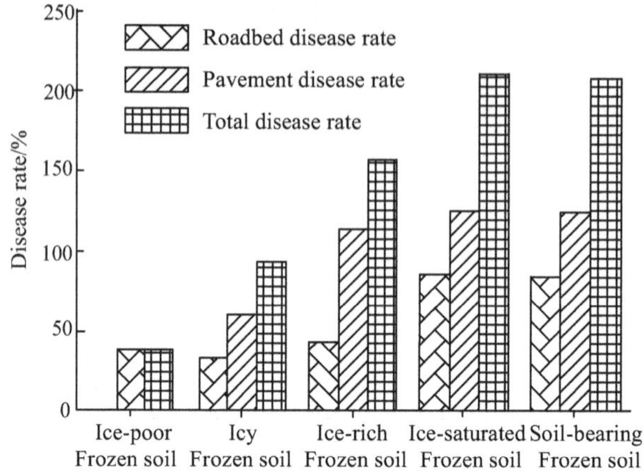

Fig. 3.26 Relationship between the types of frozen soil engineering along National Highway 214 and the roadbed disease rate

Table 3.5 Classification of frozen soil engineering types of road engineering

Frozen soil engineering types	Mean annual ground temperature(°C)			
	0 to − 0.5	−0.5 to − 1.5 (−0.5 to − 1.0)	−1.5 to − 3.0 (−1.5 to − 2.0)	< −3.0 (< −2.0)
Ice-poor frozen soil (S)	III	II, III	II	
Icy frozen soil (D)	III	II, III	II	
Ice-rich soil (F)	II, III	II	I	
Saturated permafrost (B)	II, III	II	I	
Soil-bearing ice layer (H)	II, III	II	I	

Note: The mean annual ground temperature of permafrost in brackets is for railway system

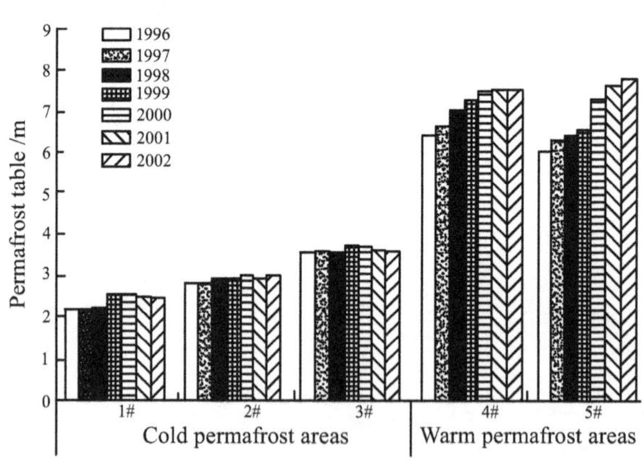

Fig. 3.27 Changes in permafrost table at the monitoring sites

Fig. 3.28 Comparison of permafrost thawing rates at monitoring sites

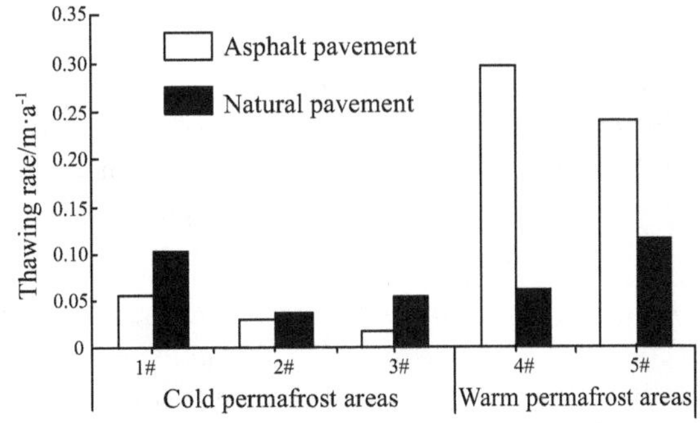

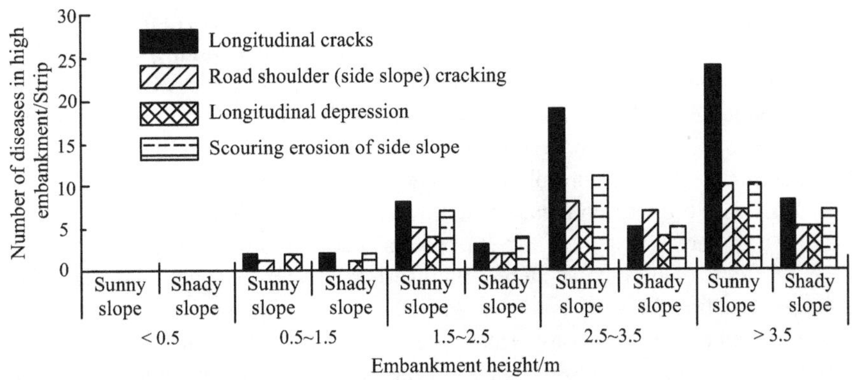

Fig. 3.29 Relationship between roadbed disease and embankment height in permafrost regions along Qinghai-Xizang Highway

measures to protect permafrost under embankment in the fringe of the ground temperature zone.

For the warm permafrost (II), its thermal stability is relatively worse. Under natural surface and engineering influence, permafrost will change greatly or fast and permafrost table will decrease greatly. Because the warm permafrost with high ice content will produce large thaw settlement, some effective engineering measures should be taken to mitigate the permafrost thaw. Generally, the engineering design can take some measures to control permafrost thaw by the principle II. There are two situations that should be considered.

1. For ice-poor frozen soil and icy frozen soil, because the deformation of permafrost thaw produced is relative smaller due to its low ice content, it will not cause embankment damage. When the permafrost thickness is relatively larger, engineering design should take the principle to control the permafrost thaw.
2. For warm permafrost with high ice content, such as ice-rich permafrost, saturated permafrost and soil-bearing ice layer, permafrost thaw causes large deformation of thaw settlement, resulting in large-scale or serious embankment

damages. Generally, the passive measures of protecting permafrost, such as increasing the height of embankment or paving the thermal insulation, are difficult to prevent permafrost thawing and decline permafrost table under embankment. Thus, cooling roadbed measures, such as crushed rock structure embankment, ventilation tube embankment, thermosyphon embankment, should be adopted to control or mitigate permafrost thaw, improving the permafrost thermal stability under the embankment.

For the warm permafrost (III), because the thermal stability of permafrost is relatively worse, permafrost under embankment will change under the natural surface or engineering influence, and permafrost will greatly degrade, with permafrost table decreasing greatly and permafrost warming. Two cases should be considered in the embankment design in permafrost regions.

1. For warm permafrost with high ice content, permafrost thaw will produce a relatively larger deformation of thaw settlement. When the permafrost thickness is more than 5 m, embankment design should take the measures of effectively cooling roadbed or replacing embankment with bridge. When the permafrost thickness is less than 5 m, embankment design should take the strengthening measures of cooling roadbed to control or mitigate permafrost thaw or remove warm permafrost with high ice content, through economic and technical comparison.
2. For warm permafrost with low ice content, permafrost thaw does not produce great thaw settlement. Here engineering design should usually take preconstruction thaw method (principle III) or destroying permafrost method, or take design principle in non-permafrost regions. But for the warm icy permafrost with the thickness of more than 5 m, the engineering design should take the method of controlling permafrost thaw.

3.2.3 *Deformation Characteristics of Permafrost Subgrade*

3.2.3.1 Main Forms of Deformation of Permafrost Subgrade

The results in field investigation, exploration and on-site observation along Qinghai-Xizang Highway show that the embankment deformations in permafrost regions are caused mainly by thaw settlement of underlying permafrost. The uneven subgrade subsidence caused by permafrost thaw accounts for more than 80% of embankment damage sections. The embankment damage exhibits longitudinal subsidence and wave-like subsidence, lateral inclination of roadbed, longitudinal cracks and roadbed cracking, etc. For the relatively slight section of uneven subsidence, embankment shows an overall subsidence of subgrade and basically smooth road pavement. For the road section with serious uneven subsidence deformation, embankment mainly shows the local subsidence of the roadbed, with the pavement waving.

The height difference between the wave crest and the wave trough within the range of more than 10 m is 0.3 m to 0.5 m, and the height difference of the frozen soil roadbed with high ice content can exceed 0.5 m within a few meters. Meanwhile, the longitudinal cracks of high embankments range from several millimeters to 0.4 m in width, and the length is usually tens of meters, and the longer ones can reach several kilometers. If the embankment height is more than 3.0 m, the cracks will increase sharply, with the number of cracks accounting for 66.2% of the total, and the length 70.6% of the total (Dou et al., 2002). Especially, the number on the sunny side of embankment shoulder is about five times the number on the shady side (Fig. 3.29) (*First Highway Survey and Design Research Institute of China Communications Construction*, 2006). The thawing rate of permafrost under the sunny shoulder of embankment is larger than that under the shaded shoulder, resulting in the embankment inclination. The amount of subsidence on the sunny side of embankment is larger than that on the shady side, with the maximum subsidence difference of 0.38 m and an average of 0.15 m, resulting in cracks appear in local sections (Fig. 3.30). With the change of permafrost temperature, frost heave and thaw settlement deformation appear in the subgrade. But the extent of frost-heaving deformation is slight, while thaw settlement deformation is severe.

The subgrade deformation of embankment along Qinghai-Xizang Railway includes thaw settlement, frost heave and cracks. During the period of railway construction, the deformation of permafrost subgrade is mainly subsidence and frost-heaving deformation. The resource of subsidence is contributed to the compression of seasonally thawed layer and the thaw settlement of permafrost caused by the lowering of permafrost table under the engineering disturbance (Li, 2008). During the middle period of railway construction, the freeze-thaw deformation mainly occurs within the seasonally thawed layer after the roadbed filling is basically completed. In the edge of permafrost regions and in warm permafrost area, subgrade deformation is relatively larger. Under the influence of the surrounding permafrost hydrothermal environment and lateral thermal erosion, subgrade deformation evidently increases, resulting in cracks and slumps of the embankment shoulder. The monitoring data of 40 sections show that the subgrade deformation was relatively

Fig. 3.30 Arc subsidence cracks in local sections of Qinghai-Xizang Highway (Photo by Yu Qihao)

Fig. 3.31 Left shoulder cracks (on the Left), longitudinal cracks (in the Middle) of the Qinghai-Xizang Railway and road shoulder crack developing towards the center of the road (on the Right) (Photo by Li Yongqiang)

stable, and the largest cumulative deformation of subgrade appeared near the southern limit of the permafrost until 2008.

The investigation of roadbed cracks shows that the number of new cracks is greatly reduced, but the extent of cracking is more serious than that during the middle stage of construction. Among them, the longitudinal cracks on the sunny shoulder of embankment in the Fenghuo Mountain slope have a maximum width of 0.3 m (Fig. 3.31), a length of 35 m, and a subsidence of 50–100 mm. Longitudinal cracks in the subgrade mostly occur in the fill subgrade, with a height of about 4–5 m, on the sunny side, and the slope is protected by skeleton slope protection (Fig. 3.31). Individual longitudinal cracks are connected with transverse cracks and develop toward the center of the road (Fig. 3.31).

The characteristics of railway subgrade deformation are mostly overall subsidence. In warm permafrost areas, the amount of subsidence is lower than that in cold permafrost areas, and is more serious in southern limit of permafrost and the fringe of the talik. Weak frost heave can be found. Longitudinal cracks mostly occur in the left shoulder (on the sunny side) and the skeleton slope protection section. Cracks are prone to occur in high embankment of cold permafrost areas. The embankment with longitudinal cracks is mostly related to the surrounding environment, such as pond on the ground, high groundwater level, etc.), and precipitation, etc.

3.2.3.2 Characteristics of Deformation of Permafrost Subgrade

The subgrade deformation in permafrost regions is thaw settlement of permafrost under embankment, which is mostly large wave-like settlement deformation. When the embankment height is less than the "critical height of embankment", most subgrade deformations tend to be the overall subsidences, with a talik under embankment. When the subgrade height exceeds the "critical height of embankment", under embankment with the asphalt pavement and cement pavement, there is still thaw settlement deformation of permafrost, mostly appearing on the sunny side of embankment. Most of permafrost table under the gravel pavement are on the rise, with a "permafrost core", and a large number of longitudinal cracks of embankment shoulder occurs in the sunny side of embankment.

1. **Embankment with Asphalt Pavement along Qinghai-Xizang Highway**

The observation data of eight monitoring sections along the embankment with asphalt pavement along Qinghai-Xizang Highway show that climate change and engineering activities have little thermal impact on the cold permafrost with the mean annual ground temperature of lower than −1.5 °C, thaw settlement deformation of permafrost under embankment is relatively smaller. So, raising the height of the embankment can ensure the basic stability of the permafrost under embankment. However, warm permafrost with a mean annual ground temperature higher than −1.5 °C is easily affected by climate change and engineering activities, and the stability of the embankment is extremely poor, with the serious deformation of thaw settlement. So, it is difficult to maintain the stability of the embankment only by raising embankment height. For the transition permafrost zone between the basically stable permafrost zone and the unstable permafrost zone with the mean annual ground temperature between −1.0 °C and −1.5 °C, thaw settlement deformation of embankment is much more, and the stability of embankment depends on the degree of thermal disturbance of engineering activities and the ice content of permafrost under embankment. Frost heave and thaw settlement of embankment change periodically with air temperature and ground temperature, and thaw settlement deformation is greater than that of frost heave. And permafrost subgrade mainly presents subsidence deformation, and longitudinal cracks caused by uneven thaw settlement.

For low embankment lower than the critical height of embankment, the deformation of embankment with asphalt pavement is dominated by overall settlement, permafrost table under embankment shows a declining trend, with a depression of the "pot". The permafrost table under the central of embankment is generally deeper 1–2 m, even deeper 4 m, than that under natural surface, resulting in a talik with water accumulation. So, in the cold season, the maximum freezing depth of soil above the talik under embankment is up to 3.5 m, leading to the maximum frost heaving amount of roadbed of 0.3 m. For the high embankment of much higher than the critical height of embankment, the damages of embankment with the asphalt pavement are mainly longitudinal cracks and side slope cracking, accounting for 58.9% of the total damages except the damages of thaw settlement (First Highway Survey and Design Research Institute of China Communications Construction, 2006). However, the "frozen core" can be found under the embankment for the embankment with gravel pavement. For zero-fill embankments and excavation embankments, the maximum frequency time of subgrade thaw settlement occurs from November to December, and the lag time is longer than that of the low embankment, which is a section with serious thaw settlement.

2. **Qinghai-Xizang Railway (with Ballast)**

During the beginning of construction (the first year after the embankment constructed), the engineering activities have thermal effects on permafrost under embankment and permafrost environment, and the heat of the fill itself cannot be dissipated in time. The overall subsidence deformations of the subgrade are relatively smaller, and deformation in subsidence and frost heave mainly appears within the soil layer between the original natural permafrost table and the surface ground. Subsidence deformation of embankment in warm permafrost areas (Chumar River, Tuotuo River, Buqu River, etc.) is larger than that in cold permafrost area, and the subsidence deformation on the sunny slope of embankment is greater than that on the shady slope, and the frost-heaving deformation of shady slope is relatively larger. The generation and width of longitudinal cracks appear with the development of settlement deformation.

During the middle stage of construction, with 1–2 years of freeze-thaw cycles after the completion of the roadbed, the ground temperature field tends to be stable, the permafrost table under embankment has different degrees of rising, and has been raised above the original natural ground in the local area. The permafrost table under embankment between the sunny slope and shady slope differs from each other. The total subgrade settlement deformation and the difference decrease; embankment cracks appear, but the number declines; and the hydrothermal environment changes significantly. The subgrade deformation is mainly the thaw compression deformation of soil in seasonally thawed layer above permafrost table under the influence of self-weight and ballast bed load, which is controlled by the thickness, density and water content of soil layer.

The roadbed crack is mainly longitudinal crack, with the crack width of more than 30 mm, accounting for 48.1%; the number on the sunny side of embankment accounts for 40% of the total of longitudinal cracks found, and on the shady side the number accounts for 60%, but the degree of cracks on the sunny side is more serious. The general subgrade with only soil berm has the highest probability of appearing longitudinal cracks and the most serious the cracks. And the embankment with thermal insulation layer and thermosyphon have a high probability of longitudinal cracks, especially in the thermosyphon section, there are more new cracks. However, the cracks of crushed rock structure embankment and reinforced embankment appeared on the embankment slope, with minimum probability of occurrence. The warm permafrost areas with saturated permafrost and the soil-bearing ice layer have a higher probability of longitudinal cracks than that in the sections with ice-poor and icy permafrost, while the fine-grained soil foundation is more likely to have cracks than the coarse-grained soil foundation.

The transverse cracks of embankment mainly occur in the transition section, including the transition section of permafrost types, the transition section of embankment and bridge or culvert, the transition section with different engineering measures, which has a higher probability of appearing transverse cracks. The probability of transverse cracks occurrence in the permafrost roadbed of fine-grained soil is larger than that of coarse-grained soil. And most of the cracks in the natural ground on both sides of embankment may extend into embankment. The probability of transverse cracks occurrence in the sections with saturated permafrost and soil-bearing ice layer is higher than that with ice-poor and icy permafrost.

The ground temperature field of the roadbed under various environments and permafrost conditions gradually tends to be stable. The artificial permafrost table under embankment rises, and its form tends to be stable. The artificial permafrost table will change slowly in the long climatic cycle (50 years).

For the areas where air temperature is cooler than −5.0 °C and mean annual ground temperature is cooler than −1.0 °C (such as Wudaoliang, Fenghuo Mountain, Kunlun Mountains), the artificial permafrost table shows an obviously rising trend, the embankment deformation is relatively smaller, and the cracks are fewer. For the areas where air temperature is warmer than −5.0 °C and mean annual ground temperature is warmer than −1.0 °C, (such as part of the areas of Beiluhe River, Chumaer River, Hongliang River), the permafrost table under embankment has been raised, even raised below the base of embankment by the cooling measures of crushed rock structure in the section of high ice content and soil-filled embankment with a berm in the section of ice-poor and icy permafrost (Table 3.6). Therefore, embankment deformation and cracks are relatively smaller for the mentioned sections while embankment deformation is relatively larger and cracks easily occur in the sections with fine-grained soil and high water content.

For the areas where air temperature is about −4.0 °Cand mean annual ground temperature is warmer than −1.0 °C, such as Wuli basin, Kaixinling, Tuotuo River, Tongtian River, Buqu River, and the southern piedmont of Tanggula Mountain, the permafrost table under embankment has a small rising trend, even arising to the bottom of embankment, by using the measures of "bridge replacing with embankment" in the section of warm permafrost with high ice content and the measures of crushed-based rock embankment and soil berm embankment in the section of warm permafrost with low ice content (Table 3.7). For the seasonally thawed layer under the original ground, embankment deformation is relatively larger, resulting in theoccurrence of cracks, due to the high water content and fine-grained soil.

Environmental engineering geological problems caused by embankment construction are mainly manifested in the changes in hydrothermal transport channels and lateral thermal erosion. Firstly, the heat exchange conditions between permafrost and the atmosphere are changed, leading to the rise of permafrost temperatures, change of permafrost table, and the thaw settlement of subgrade base. Secondly, the blocking of surface water runoff and waterlogging at slope toe of embankment cause thaw settlement of permafrost, resulting in embankment crack.

Table 3.6 Change of permafrost table and mean annual temperature of embankment under different engineering measures in Beiluhe (Li, 2008)

Embankment types	Crushed-based rock embankment		Rubble embankment		General embankment	
	Sunny shoulder	Shade shoulder	Sunny shoulder	Shade shoulder	Sunny shoulder	Shade shoulder
Shallow temperature/°C	0.99	−2.12	−0.39	−1.84	0.21	−2.28
Temperature at the bottom of embankment/°C	−0.08	−1.25	−0.09	−0.47	0.15	−0.42
Temperature at the original top of permafrost/°C	−0.49	−1.01	−0.09	−0.36	−0.32	−0.59
Extent of permafrost table raising/m	1.0	1.0	2.0	2.0	0.5	0.3

Table 3.7 Change of permafrost table and mean annual temperature of embankment under different engineering measures in Kaixinling Area (Li, 2008)

Embankment types	Crushed-based rock embankment		Rubble embankment		General embankment	
	Sunny shoulder	Shade shoulder	Sunny shoulder	Shade shoulder	Sunny shoulder	Shade shoulder
Shallow temperature/°C	0.04	−0.55	−0.37	−0.52	−0.08	−0.70
Temperature at the bottom of embankment/°C	−0.63	−2.10	−0.20	−0.31	0.20	−0.12
Temperature at the original top of permafrost/°C	−0.67	−1.43	−0.23	−0.41	−0.12	0.19
Extent of permafrost table rasing/m	0.70	1.10	0.20	0.70	0	0.20

Meanwhile embankment construction will change groundwater runoff conditions, resulting in icing and frost mounds on the upper side of the embankment and vegetation withered and desertification aggravated on the lower side of embankment. These are an obvious environmental engineering geological problem of permafrost in the middle stage of construction.

During the period of railway operation from the full completion to two or three freeze-thaw cycles, thermal regimes under the embankment are in a relatively stable heat exchange process and the embankment engineering is relatively stable, but factors harmful to engineering are becoming more and more obvious.

The monitoring data from the 40 sections during the two-year operation show that the roadbed deformation exhibits a general attenuation trend, and most of the total deformations are less than 50 mm. However, the settlement deformation is relatively larger near the southern limit of the permafrost. The number of new embankment cracks is significantly reduced, but the degree of embankment cracks is more serious than that in the middle period of construction. Most of the sections with severe embankment cracks occur in such areas as the slope of high embankment, embankment with both sides of the embankment with large difference in height, skeleton slope protection with large slope length on the sunny slope, the embankment shoulder and soil berm with seepage and other zones. The formation reason is caused by the difference of horizontal embankment deformation, thermal contraction cracking and the train dynamic load in the early stage of thawing. Once the cracks occur, lateral erosion leads to the development of cracking with the differences of vertical deformation.

The observation data of permafrost temperatures at 50 embankment sections show that the maximum thawing depth under the embankment exhibits large change. The artificial permafrost table under embankment is shallower than the original natural permafrost table which accounts for 78% of the sections, where the artificial permafrost table under the embankment accounts for 26% of the sections. The permafrost table has risen even more than the bottom of embankment. But there are two sections (near the southern limit of permafrost in Walibailitang and Anduo) that the maximum thawing depth under the embankment is deeper than the original permafrost table under natural surface. There are 18% sections in the observation sections whose maximum thawing depth under the sunny shoulder of embankment is deeper than the original permafrost table under natural surface and it is shallower under the shade shoulder of embankment. So the difference between the sunny and the shade should cause the sunny shoulder of embankment to sink or crack.

The survey shows that the water accumulated at the slope toe of the embankment cause the development of its cracks. For example, the amount of water accumulated at the foot of the embankment in the north of Budongquan is 2475 m^3, and the crack is as long as 300 m while in Quwu (Qushui River) area. The volume of accumulated water is 2889 m^3, and the crack length is 398 m. Therefore, the water accumulation at the slope toe of the embankment during operation will cause changes in the surrounding permafrost environment, resulting in embankment deformation and crack development.

3.2.3.3 Causes of Deformation of Frozen Soil Subgrade

In addition to the deformation under load, the change in heat exchange and radiation-heat balance between atmosphere and ground is the main reason for the variation of the thermal regime of permafrost under the embankment. And the thermal erosion effects caused by the change of permafrost environment accelerate the occurrence and development of embankment deformation.

1. **Change in Heat Exchange and Radiation-Heat Balance between Atmosphere and Ground**

After the construction of the embankment, the properties of the natural surface are greatly changed, altering the reflectivity (Table 3.1) and properties of the surface, the water content and the way of surface heat consumption, and then the radiation-heat balance between the atmosphere and the ground. Besides, the heat absorption on the surface of the embankment is greatly influenced, causing the changes in the total annual heat budget of permafrost under the embankment and the fluctuation of permafrost table under the embankment (rising or declining).

Obviously, both the seasonal freezing depth and the seasonal thawing depth in the frozen soil area depend on the freezing index and thawing index of the local surface. By the ground n factors (the ratio of the surface temperature index to the air temperature index), the freezing index or thawing index of the surface can be estimated into the freezing index or thawing index of air temperature.

The surface temperature depends on factors such as geographical location (latitude), solar radiation, the nature and area of the covering, the physical properties and reflectivity of the surface, humidity and climate environment, et al., and these factors also affect the n factors. According to relevant data, the n factors of different pavements vary greatly (Table 3.2). Asphalt pavement has powerful function of absorption of heat and water-resistant (without surface evaporation). In the warm season, the surface temperature of asphalt pavement can reach more than 50 °C. Even in the cold season, the surface temperature of asphalt pavement is higher than the air temperature. Therefore, the n_t factor of the asphalt pavement is very large, while the n_f is smaller.

During the thawing period, if the n_t factor of asphalt pavement is 3 times larger than that of the gravel pavement (Table 3.2), the thawing depth under the embankment with asphalt pavement will be much larger than that with gravel pavement. During the freezing period, the n_f factor of asphalt pavement is 1.6 times smaller than that of gravel pavement, and the freezing depth under the embankment with asphalt pavement is shallower than that with gravel pavement. As the observation data from Fenghuo Mountain Station (with the altitude of 4750 m), the freezing index of temperature is 4.3 times the thawing index, the freezing index of asphalt pavement is only 0.59 times the thawing index, and the thawing index of asphalt pavement is 1.68 times the freezing index, indicating that the thawing depth under asphalt pavement is 1.68 times the freezing depth. It can be known that under the same permafrost environment, the permafrost table under the embankment with asphalt pavement along the Qinghai-Xizang Highway shows mostly a declining trend, so it is easy to form a talik under the embankment; the permafrost table under the embankment with gravel pavement along the Qinghai-Xizang Railway shows a rising trend, so it is easy to form a frozen soil core under the embankment. According to the heat balance observation on the Qinghai-Xizang Plateau, the solar absorptivity increases by 20% due to construction of asphalt pavement (Dai et al., 2006), and the asphalt pavement hinders the evaporation process on the embankment surface and the heat consumption by evaporation cannot be released (Zhang et al., 2014), resulting in soil thaw 20–30 days earlier and freeze about 20 days later than the natural subsurface soil layer.

2. **Changes in the Properties of Ice-bearing in Permafrost**

The internal cause for the deformation of the embankment in permafrost regions is that there contains a large amount of ground ice layer in the permafrost, which produces a large amount of thaw settlement after thawing. A slight change in natural conditions will make the ice in permafrost change significantly, which determines the instability of ice properties and the instability of permafrost properties.

3. **Thaw Bulb Change of Permafrost under Embankment**

Under the condition of low embankment (with the height less than 1.5 m), because the absorptivity of solar radiation by asphalt pavement increases by 20%, the mean annual temperature of the embankment surface is about 4 °C higher than that of the natural surface, causing the thawing time within the embankment to be prolonged by 30–40 days. These processes make the permafrost table under the center of the embankment increase and move down to form a concave thaw bulb (Fig. 3.32). Meanwhile a large amount of water accumulation in embankment accelerates the development of the thaw bulb. Therefore, it can't be completely refrozen and a talik appears in the cold season, increasing the thaw settlement deformation of the

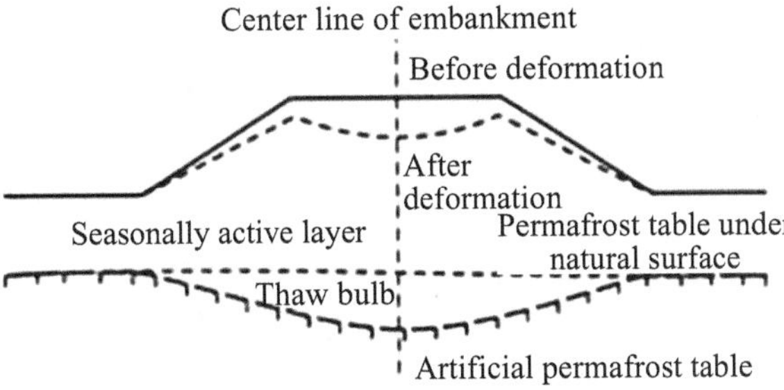

Fig. 3.32 Schematic diagram of the concave-shaped thaw bulb of the permafrost table for lower embankment

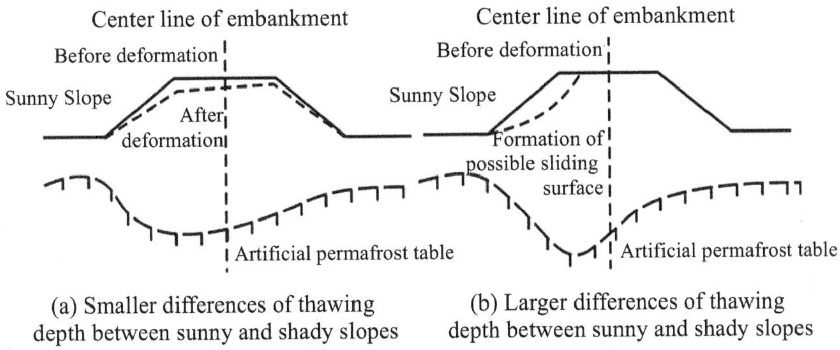

Fig. 3.33 Diseases caused by differences of thawing depth between sunny and shady slopes of high embankment

embankment. The drilling borehole data of highway show that talik with pressure-bearing water can be found in 47% of the boreholes.

The height of embankment in warm permafrost regions has been greatly improved after the renovation of Qinghai-Xizang highway. With the increase of embankment height, the difference between the sunny and shady slopes on the thermal regime of permafrost under embankment is highlighted (Fig. 3.33). The damage rate on the sunny slope of the high embankment (accounting for 66.5% of the total statistics) is significantly higher than that on the sunny slope (for 33.5% of the total statistics).

In the case where the height of the embankment approaches to the critical height, when the difference between thawing depth under the sunny shoulder and shady shoulder of the embankment is relatively smaller, the overall inclination to the deformation of the embankment appears (Fig. 3.33a); when the difference between thawing depth under the sunny shoulder and shady shoulder of the embankment is relatively larger, longitudinal cracks and slipping may appear on the shoulder side or side slope (Fig. 3.33b). In the warm permafrost areas, when the embankment height is much larger than the critical height of embankment, the depth of the thaw bulb under the embankment is relatively larger, a talik that does not freeze for 1 year or more is formed, so that the soil body in the thawed interlayer becomes a soft layer due to the thawed water of ground ice and the suprapermafrost water in talik. Under

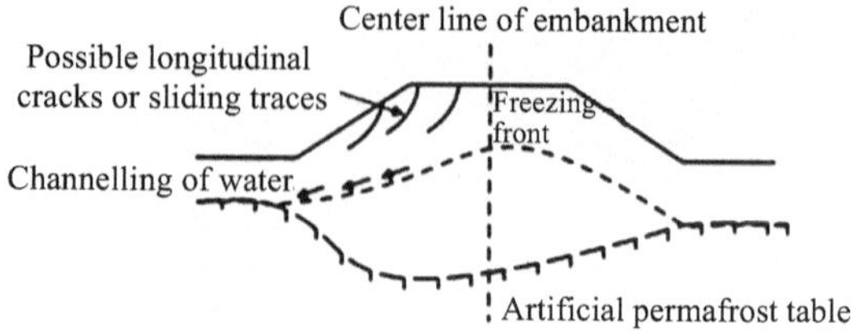

Fig. 3.34 Thawed interlayer under high embankment in warm permafrost regions

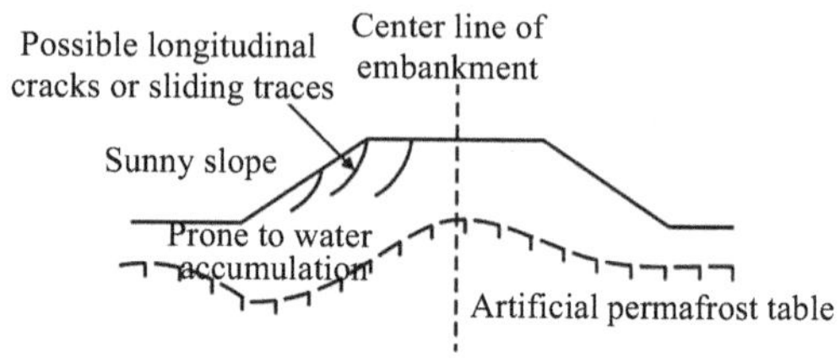

Fig. 3.35 Frozen soil core under high embankment in cold permafrost regions

the influence of embankment and traffic load, settlement deformation or slope slipping occurs (Fig. 3.34).

In the cold permafrost areas, under the high embankment with the asphalt pavement along the Qinghai-Xizang Highway, seasonally thawed layer can be completely frozen due to up freezing and down freezing and produce potential seasonally freezing, resulting in the rising of permafrost table under the embankment. However, under the influence of strong solar radiation, the thawing depth on the sunny shoulder of embankment is larger than that on the shady shoulder so that the frozen core under the embankment is deviated to the side of shady shoulder, and longitudinal cracks or sliding appear on the sunny shoulder (Fig. 3.35).

For the high embankment with gravel pavement along Qinghai-Xizang Railway, the permafrost table under the embankment in cold permafrost area is generally uplifted, even to the roadbed body above the original ground (Fig. 3.36a), and longitudinal cracks may appear on both sides of the embankment shoulders. When there is a sharp difference in permafrost table and permafrost temperature between sunny shoulder and shady shoulder of embankment, the frozen core may be deviated to the side of shady shoulder (Fig. 3.36b), and longitudinal cracks or sliding may appear on the sunny shoulder.

4. **Thermal Erosion in Permafrost Environment**

Removal of surface vegetation, surface water accumulation in the low-lying areas of the roadbed and the lateral erosion of thermokarst lake (Fig. 3.37) and water accumulation in side ditch (Fig. 3.38) will cause the increase of seasonal thawing depth and the decline of permafrost table, resulting in the thaw settlement of embankment and longitudinal cracks. Under the influence of asphalt pavement along Qinghai-Xizang Highway, the intrusion of suprapermafrost water into the talik under the embankment extends the lateral range of the artificial permafrost

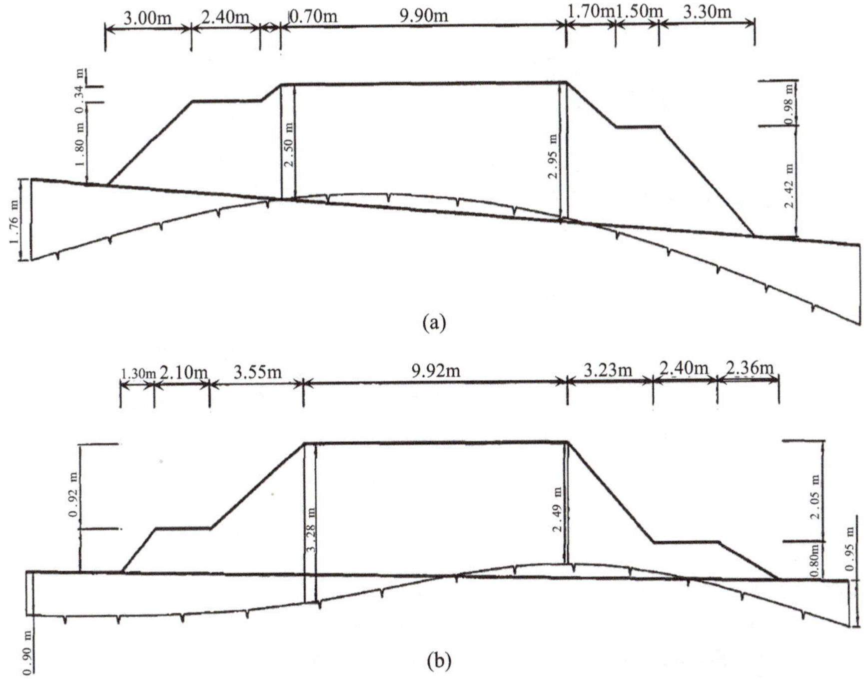

Fig. 3.36 Rising (Left) and shift rising (Right) of the permafrost table under embankment in Wudaoliang along the Qinghai-Xizang Railway (Li, 2008)

Fig. 3.37 Embankment settlement caused by the erosion from the roadside ponding (Left) and thermokarst lake (Middle) of Qinghai-Xizang Highway (Photo by Tong Changjiang) and roadside ponding of Qinghai-Xizang Railway (Right) (Photo by Li Yongqiang)

table under the embankment, resulting in increasing in thaw settlement deformation and longitudinal cracks. Under the influence of gravel pavement along Qinghai-Xizang Railway, the artificial permafrost table under the embankment rises and the water accumulation in the side ditch of the embankment causes the decrease of the permafrost table, which combines to form a water-logged depression on the side of the subgrade, resulting in longitudinal cracks and slipping on the subgrade slope.

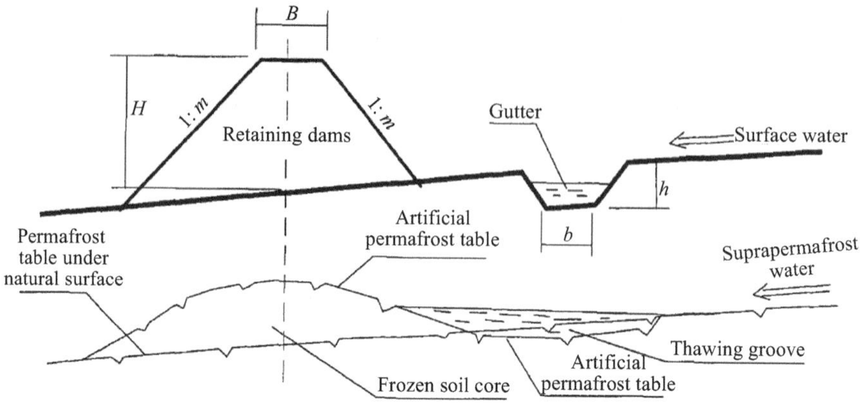

Fig. 3.38 Changes of artificial permafrost table under embankment caused by retaining dams and drainage ditch (Li, 2008)

3.2.3.4 Principles for Controlling Embankment Deformation

Embankment deformations in permafrost regions show that the permafrost change is determined by thermal stability and ice-bearing characteristics of permafrost. Meanwhile solar radiation and the thermal erosion from water intrusion also influence permafrost change. Therefore, the principle for preventing and controlling embankment deformation should meet the following requirements.

1. Engineering measures should be taken to prevent or reduce solar radiation heat entering into the subgrade, or to enhance the cold storage of permafrost under the embankment, so as to maintain or reduce the change of the permafrost table under the embankment, and to avoid the formation of thawed interlayer or frozen core in soil profile under the subgrade.
2. Thermally asymmetrical structure measures should be adopted to raise the permafrost table under the embankment and avoid the differences between permafrost table under the sunny and shady shoulder of embankment, and mitigate occurrence and development of roadbed cracks.
3. Different pavement properties have different heat absorptivity. The heat of embankment with asphalt pavement subgrade mainly comes from the embankment surface, and the heat of embankment with gravel pavement mainly comes from the embankment slope. Measures for embankment slope thermal protection can change according to the differences in conditions.
4. From the perspective of permafrost environment, engineering measures should be taken to protect the permafrost environment at the slope toe of embankment, avoiding waterlogging and protecting (maintaining) the ecological environment such as vegetation, etc., or add crushed rock or thermal insulation berms. Drainage ditch should not be set at the slope toe and the accumulated water should be drained out of the slope toe 5 m outside. The slope toe of embankment should be connected with the overall embankment drainage system and kept

unobstructed, so as to reduce the lateral thermal erosion of the subgrade by water, reduce the moisture content of the subgrade soil body, and improve the resistance of the subgrade to freeze-thaw cycle.
5. It's advisable to select good embankment fillers. The coarse-grained fillers may be the best, but fine-grained fillers need to meet the requirements for anti-frost heaving/frost heaving and anti-tensile fracture in moisture content, density, cohesion and other indicators.
6. Line location and embankment structure should meet the special requirements for permafrost areas. According to the engineering geological conditions of frozen soil, especially the annual average ground temperature and ice content of permafrost, different subgrade structures and engineering measures should be adopted to meet the thermal and mechanical stability requirements for the subgrade.

3.2.4 Utilization Principles for Permafrost Foundation

The design principles for permafrost foundation should be based on the engineering geological conditions of permafrost foundation, the temperature characteristics of buildings and the possible changes of foundation soil properties. The engineering geological conditions of permafrost mainly include frozen soil engineering types (Table 3.4), mean annual ground temperature (Table 3.7), permafrost table, permafrost thickness, thaw settlement coefficient and compressibility of permafrost foundation, material composition and material properties of foundation soil. The temperature characteristics of buildings refer to the absorptivity of solar radiation heat by the subgrade, which is determined by pavement properties. The possible change of foundation soil characteristics refers to whether the permafrost foundation can maintain the change of the frozen state or gradually thawing during the operation process, so as to avoid the formation of talik or frozen core.

Obviously, the design principles of permafrost foundation proposed by the *Code for Design of Building Foundations in Frozen Soil Areas* (JGJ 118-2011) are also applicable to the subgrade engineering in permafrost regions. In continuous / discontinuous / sporadic (patchy) permafrost regions, the design principles are mainly adopted in order to keep a frozen state (principle I) and gradually (controlled) thawing state (principle II). In the sporadic patchy permafrost regions in the southern limit of permafrost, the design principles mainly adopt the method of pre-thawing permafrost (principle III) or the design of excavating permafrost should follow the principle for non-frozen soil areas. Under the impacts of global warming, the thermal stability of permafrost should be taken as the most important indicator (Huang, 2001) while combining ice content of permafrost to select the design principle.

1. The design principle of keeping frozen state is applicable to cold permafrost with high ice content areas and frequent earthquake areas. The economy for using design principle I is more reasonable than for principle II.

2. The design principle for gradually thawing permafrost can be applied to warm permafrost with lower ice content areas. For the coarse-grained frozen soil foundation, the deformation of thaw settlement after thawing tends to be stable and can meet the design requirements.
3. The design principle for the pre-thawing or allowable thawing permafrost can be used in the permafrost areas with low ice content, and the amount of thaw settlement after thawing will not cause embankment damage. Meanwhile, this design principle is suitable to rock areas, extreme warm permafrost areas with relatively small thickness (less than 3 m). For the coarse-grain extremely deep-buried residual permafrost and permafrost areas in the southern limit of permafrost and sporadic (sparse) permafrost with lower ice content, this design principle is also suitable.
4. Embankment engineering in the talik of permafrost regions, such as river talik, radiation talik, and lake talik, could be designed by the design principle of embankment in non-permafrost regions. Embankment engineering in the small areal island permafrost with high ice content and larger thickness could be designed by the design principles of controlling the thawing state or maintaining the frozen state after comprehensive consideration of technology, economy and linear suitability. The residual permafrost island with small area, thin thickness, high ice content and shallow buried depth can be treated by replacement method.

When the design principle I is adopted, effective measures should be taken to prevent the warming and thawing of permafrost under embankment. When the design principle II is adopted, the measures taken should be able to control the thawing rate of the permafrost under the embankment to meet the design requirements in deformation rate and total deformation of embankment during the operation. When the design principle III is adopted, the measures should consider the detailed situation, either thawing or replacement. It is worth noting that different design states can be adopted for different sections, but embankment structures of the transition sections from one design principle sections to another design principle sections should be considered to adapt uneven deformation of embankment.

Engineering design usually adopts measures in lowering the surface temperature, increasing the thermal resistance of embankment, strengthening to cool the permafrost under embankment, preventing the thermal erosion of the surface water and groundwater, and protecting the permafrost environment around the embankment. However, engineering design in such areas should be specially treated according to the specific situation for the unfavorable cryogenic phenomenon in permafrost areas, such as the thick ground ice, thaw slumping, thermokarst lake and gelifluction, frost mounds and icing, frozen soil swamp, the transition sections between embankment and culvert and between embankment and bridge.

3.3 Permafrost Subgrade

The cause for subgrade engineering damage in the permafrost regions shows that under the impacting of permafrost environment change, the thermal regime of permafrost under embankment will change greatly, resulting in the breakdown of the hydrothermal balance of the permafrost. During the process of re-establishing a new balance state, the permafrost table under embankment changes greatly and the ground ice thaws or develops, and a talik or frozen core is possible to be formed under embankment. Such change will result in thaw settlement, cracks or soil frost heave. Therefore, for the embankment design in permafrost regions, in addition to complying with the requirements of relevant design code, it is more important to adopt engineering measures combining prevention and treatment according to the engineering geological conditions and damage characteristics of embankment to ensure the stability of permafrost engineering.

3.3.1 Soil-Filled Embankment in Permafrost Subgrade

Under the condition of permafrost, the permafrost engineering geological conditions under the embankment change with the change of temperature (thermal regime), which is the major difference between the design for embankment in permafrost regions and non-permafrost regions. Embankment construction changes the thermal regime of permafrost under embankment, resulting in the change of permafrost table and permafrost temperature. The general embankment design in permafrost regions involves the design principles, embankment height and waterproofing and drainage issues.

3.3.1.1 Design Principles for Permafrost Embankment

The design principle for permafrost embankment should abide by the principle of permafrost foundation utilization. Based on the permafrost engineering types and thermal states of permafrost under the embankment, the design principle and its application scope should be determined correspondingly.

3.3.1.2 Design Height of Permafrost Subgrade

In the design of embankment in permafrost regions, in addition to the strength and deformation of the embankment, the ice content (permafrost engineering types) and thermal stability (permafrost zone of ground temperature) of permafrost under embankment should be effectively controlled.

According to the survey on the permafrost under embankment along Qinghai-Xizang Highway and Railway, the changes of artificial permafrost table under the embankment in permafrost regions are related to the filling height of embankment, the properties of soil-filled and pavement, the embankment direction, the engineering geological conditions of permafrost (including mainly mean annual ground temperature and permafrost engineering types), climatic characteristics and construction (construction season and technology).

1. **Minimum (Lower) Height of Embankment**

In permafrost regions, the minimum embankment height that maintains the relative stable permafrost table under the embankment is defined into the critical height of embankment, also known as the lower critical height. In general, the critical height of embankment can basically control the hydrothermal balance between embankment-permafrost systems, prevent the permafrost from thawing and maintain the relatively stable permafrost table.

The critical height of embankment in permafrost regions has been noted in1983. Based on the study of Fenghuo Mountain area on the Qinghai-Xizang Plateau, in the condition that the permafrost table under the ground surface is 1.3–1.9 m, the minimum critical height of embankment with cohesive soil-filled is about 0.63 m (Huang, 1983). According to the relationship between the embankment height of gravel pavement and asphalt pavement, in the conditions that the permafrost table is about 1.3–1.5 m, the critical heights of embankment with gravel pavement is about 0.65 m and with asphalt pavement about 1.16 m respectively (Scientific *Research Group Rof Qinghai-Tibet(Xizang) Highway*, 1983). In the conditions that permafrost table under ground surface is about 1.5–2.0 m and 2.3–2.5 m, the critical heights of embankment is about 0.2 m (with gravel pavement) and 1.15 m (with asphalt pavement) respectively. Zang and Wu (1999) proposed a method of estimating the critical height of embankment (1.18 m) by the sum of average critical height and the variance of 1.5 times according to research results of the gravel pavement of National Highway 214. Dai (1983) summarized the critical height of railway embankment in the Daxing'anling and Xiaoxing'anling Mountains in Northeast China. For the predominantly continuous permafrost areas, the critical height of embankment is about 1.5 m, and for the permafrost areas with an island-like talik, the critical height of embankment is about 2.0 m.

As for the highway embankment in the permafrost regions, Canadian Arctic Circle, the soil-filled height of embankment should not be less than 1.4 m, and the width of the berms on both sides should be 6 m. For the embankment highway in permafrost regions, Alaska, USA, the minimum embankment height should be 0.9 m for the section with better permafrost engineering geological conditions (California bearing ratio (CBR) is larger than 3 after the permafrost thawed). And minimum embankment height should be 1.5–1.8 m for the section with worse permafrost engineering geological conditions (CBR is less than 3 after permafrost thawed). In fact, it is generally required that the embankment height be not less than 2.0 m and the embankment of soil-filled should use coarse-grained soil with weak frost heave.

Based on the investigation and research by the scientific research group established in 1979, the mean annual ground temperature that can assess the thermal stability of permafrost was applied to embankment design of the second renovation project in 1995. Embankment design recommended the minimum soil-filled height of embankment and the conversion factor of the soil-filled for the division of warm permafrost and cold permafrost, with the threshold of the mean annual ground temperature of −1.5 °C. The critical height of embankment under different ground temperature environment of Qinghai-Xizang Highway (Tables 3.8 and 3.9) and Qingkang Highway (National Highway 214) (Tables 3.10 and 3.11) are as follows.

In the design of the embankment in the permafrost region of the Qinghai-Xizang Railway, when the embankment is filled with cohesive soil in the permafrost regions with high ice content, the minimum design height of embankment should be determined according to the permafrost temperature under the division of ground temperature (Table 3.12). When the embankment is filled with non-cohesive soil, the influence of filler properties on the embankment height should be considered, and its thickness should be corrected (Table 3.13).

The surface temperature can be calculated on the basis of adherent layer theory. Average gradient of ground temperature can be taken into 0.03 °C/m, which is used as the lower boundary condition of the computational domain. The critical height of the embankment under different mean annual ground temperature conditions is calculated by finite element method (Table 3.14).

Table 3.8 Recommended Value for Minimum Fill Height (H_o) of Asphalt Concrete Pavement Subgrade along Qinghai-Xiang Highway in 1995 (First Highway Survey and Design Research Institute of China Communications Construction, et al, 2006)

Design principles	Protecting permafrost (Cold permafrost)			Controlling thaw rate (Warm permafrost)		
Frozen soil engineering types	ice-rich permafrost	Ice-saturated permafrost	Soil-bearing ice layer	ice-rich permafrost	Ice-saturated permafrost	Soil-bearing ice layer
H_o/m	1.6 to 2.0	1.8 to 2.6	2.4 to 3.2	1.8 to 2.4	2.2 to 3.2	2.6 to 3.4

Table 3.9 Conversion coefficient for the filler (*First Highway Survey and Design Research Institute of China Communications Construction*, 2006)

Soil	Silty clay	Sandy loam	Fine-grained sand soil	Sand, Sandy gravel	Dry and dense gravel
Conversion factor	0.6 to 0.65	0.7 to 0.75	0.8 to 0.85	1.0	1.1 to 1.2

Table 3.10 Critical height of embankment (H_o) with gravel pavement for National Highway 109 and 214 (Zang & Wu, 1999)

Permafrost	Unstable warm permafrost	Transition permafrost	Relative stable cold permafrost
H_o(m)	1.30*	1.17**	0.99**

Notes:*Envelop curve; **Sum of Statistic average and variance of 1.5 times

Table 3.11 Design height (H) of embankment (Zang & Wu, 1999)

Permafrost		Relative stable cold permafrost	Transitional permafrost	Unstable warm permafrost	Extremely stable warm permafrost
Mean annual ground temperature/°C		≤−1.8	−1.8 to −1.0	≥−1.0 to −0.3	≥ −0.3
Corresponding mean annual air temperature/°C		−6.0 ± 0.5	−5.0 ± 0.5	−4.0 ± 0.5	>−3.5
H/m	Asphalt concrete pavement	2.5 to 2.8*	3.0 to 3.4*	3.8 to 4.0**	>4.0**
	Cement concrete pavement	1.7 to 1.9	2.0 to 2.2	2.3 to 2.4	>2.5
	Gravel pavement	1.3 to 1.4	1.5 to 1.6	1.7 to 1.8	>2.0

Notes: * For the sections with swamp, peat soil, the surface water and groundwater with frequent activity and the section with developed vegetation growth, the larger values should be taken. ** The asphalt pavement is not recommended

Table 3.12 Minimum design height (H_o) of embankment filled with cohesive soil in permafrost regions of Qinghai-Xizang Railway*

Ground temperature zone	Stable zone of cold permafrost	Basically Stable Zone of cold permafrost	Unstable Zone of warm permafrost	Extremely Unstable Zone of warm permafrost
Mean annual ground temperature T_{cp}(°C)	$T_{cp} < -2.0$	$-2.0 \le T_{cp} < -1.0$	$-1.0 \le T_{cp} < -0.5$	$-0.5 \le T_{cp} < 0$
H_o(m)	1.50	1.90	2.30	2.50

Table 3.13 Conversion coefficient of the filled soil for minimum design height of embankment

Soils	General cohesive soil	Sandy soil	Mixed soil with sand and gravel	Block and pebble soil
Conversion coefficient	1.00	1.20	1.30	1.40

Based on the relationship between the critical height of the embankment and the thawing index of air temperature, the critical height of embankment can be estimated. It shows in Table 3.15 that the critical height of embankment is made of the mixed sand and gravel for the subgrade containing permafrost with soil-bearing ice layer and mean annual ground temperature of −2.0 °C, and the cohesive soil with a water content of 14% within the active layer. Table 3.16 shows the critical height of embankment for subgrade containing permafrost with ice-poor and icy permafrost and mean annual ground temperature of −0.8 °C, and coarse-grained soil with the water content of 10% within the active layer. The critical height of an embankment in the cohesive soil sections will be generally higher than that in the coarse-grained soil sections under the same thawing index. And the change of critical height of the

Table 3.14 Lower critical height of embankment under the mean annual average temperature conditions (Zhang, 2004)

Mean annual air temperature(°C)	−2.8	−3.0	−3.5	−4.0	−4.5	−5.0	−5.5	Design service life of embankment (Year)
Mean annual ground temperature(°C)	−0.3	−0.5	−1.0	−1.5	−2.0	−2.5	−3.0	
Asphalt pavement (highway)	–	6.07	4.40	2.57	1.25	0.86	0.66	20
Asphalt pavement (highway)	–	8.24	5.32	3.11	4.67	1.04	0.71	30
Gravel pavement (highway)	1.98	0.79	0.38	0.37	0.37	–	–	20
Gravel pavement (highway)	3.37	1.84	0.55	0.42	0.42	–	–	30
Gravel pavement (railway)	6.76	4.15	1.08	0.57	0.51	–	–	50

Table 3.15 Critical height (H_o) of cohesive soil road section (Qinghai-Xizang Highway Permafrost Subgrade Research Group, 1988)

Thaw Index (degree·day)	1400	1600	1800	2000	2200	2400	2600	2800	3000
Permafrost table(m)	1.00	1.05	1.10	1.15	1.20	1.25	1.30	1.35	1.40
(H_o)(m)	1.30	1.40	1.55	1.70	2.10	2.20	2.35	2.45	2.50
Soils	Embankment filled: Mixed gravel soil $\rho_d = 1850$ kg/m^3, $\omega = 8\%$; foundation soil: Gravel soil $\rho_d = 1700\rho_d = 1850$ kg/m^3, $\omega = 10\%$								

Table 3.16 Critical height (H_o) of gravel soil section (*Qinghai-Xizang Highway Frozen Subgrade Research Group*, 1988)

Thaw index (degree·day)	1600	1800	2000	2250	2400	2700	2950	3150
Permafrost table(m)	1.80	1.90	2.00	2.13	2.20	2.35	2.45	2.51
(H_o)(m)	1.40	1.45	1.55	1.65	1.70	1.75	1.80	1.90
Soils	Embankment filled: Mixed gravel soil $\rho_d = 1850$ kg/m^3, $\omega = 8\%$; foundation soil: Gravel soil $\rho_d = 1700\rho_d = 1850$ kg/m^3, $\omega = 10\%$							

embankment with the thawing index for the cohesive soil section change is larger than that for the coarse-grained soil section.

According to the simulation analysis of the moist road section of continuous permafrost in Changshitoushan of Qingkang Highway, the relationship between the changes of the natural and artificial permafrost table, under the center of embankment after 30-year operation of asphalt pavement, cement concrete pavement and gravel pavement, and the height of the embankment is shown in Figs. 3.39 and 3.40. The critical height for ensuring thermal stability of permafrost under embankment

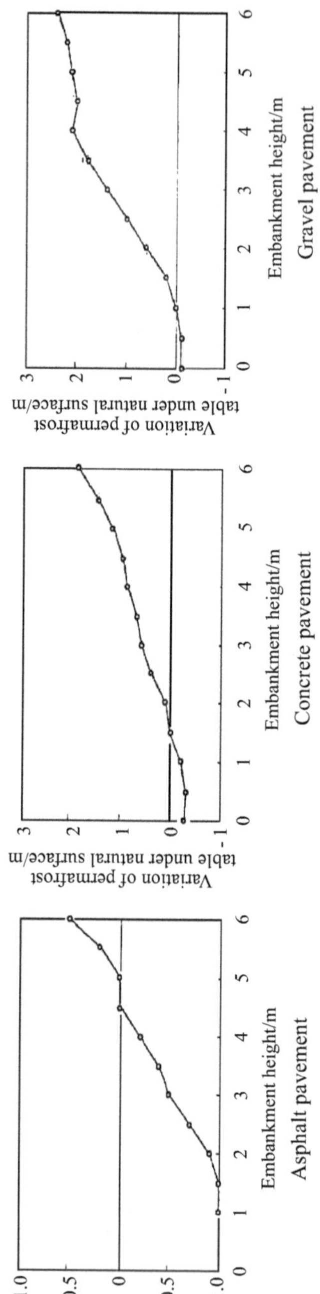

Fig. 3.39 Relationship between the change of permafrost table under natural surface and the embankment height (Li et al., 1999)

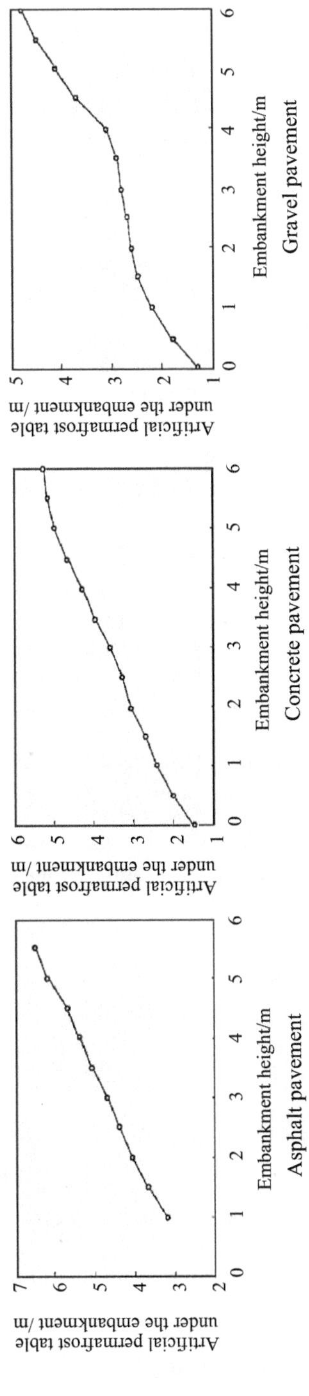

Fig. 3.40 Relationship between the change of artificial permafrost table under the embankment and the embankment height (Li et al., 1999)

with asphalt pavement, cement concrete pavement and gravel pavement are 4.50 m, 1.60 m, and 1.00 m, respectively.

It is noteworthy that increasing the height of embankment to critical height does not necessarily ensure the stability of permafrost table under embankment. Some sections exhibit declining trend of permafrost table under embankment, even appear a talik. So, determining the critical height of embankment in permafrost regions should consider the influence of many factors, such as the mean annual air temperature, relationship between air and ground freezing index and thawing index, mean annual ground temperature of permafrost, permafrost under ground surface, surface characteristics of embankment and soil-filled properties, the service life and the thermal effect of construction.

2. **Estimated Method for Critical Height of Embankment**

The basic requirement of adopting the design principle of protecting permafrost for the road engineering in permafrost regions is that the permafrost table under embankment remains stable state. The most basic method is to increase the soil-filled height of embankment that meet the minimum critical height by increasing the thermal resistance to reduce the heat amount entering embankment and to maintain the stability of permafrost table under the embankment. At present, there are two main methods for determining the critical height of the embankment. The first method is to obtain the empirical formula between the critical height of embankment and the permafrost table under ground surface for different conditions through a large number of field tests and observations. The second is to estimate theoretically by considering the soil parameters and climatic conditions.

1. Estimation method by the relationship between the critical height of embankment and permafrost table under ground surface.

   (a) Empirical formula for Qinghai-Xizang Highway (First Highway Survey and Design Institute of China Communications Construction, et al., 2006).
   Gravel pavement:

$$H_G = 1.0 - 0.15h_o \tag{3.6}$$

$$H_G = 0.933 - 0.088h_o \tag{3.7}$$

   Asphalt pavement:

$$H_A = 2.46 - 0.4h_o \tag{3.8}$$

where H_G and H_A is critical height of embankment with gravel pavement and with asphalt pavement, respectively (m). h_o is permafrost table under ground surface (m). For warm permafrost areas in the south of Tanggula Mountains, the calculation results should be multiplied by the coefficient of 1.1–1.2 in formula 3.6 and formula 3.8.

   (b) Empirical formula for Qingkang Highway (Zang & Wu, 1999).

Gravel pavement:

$$H_G = 0.07h_o + 0.73 \tag{3.9}$$

Asphalt pavement:

$$H_A = A - Bh_o \tag{3.10}$$

where H_G and H_A is critical height of embankment with gravel pavement and with asphalt pavement, respectively. h_o is permafrost table under ground surface(m); A and B are constants, and A represents the section parameter. For cold permafrost area in the mountain, A = 3.5, B = 0.70; For cold permafrost area at hill: A = 3.6, B = 0.72; For transition permafrost area in basin and flood plain: A = 5.0, B = 0.82; For transition permafrost area in high plain: A = 5.2, B = 0.90.

2. Estimation method by the thawing index

   The thawing depth under embankment in permafrost regions depends on the combined effects of mean annual ground temperature, duration of the warm season, the surface temperature, the thermophysical properties of embankment and subgrade soil, and the height of embankment.

   (a) The Empirical Formula for Qinghai-Xizang Highway

      The critical height (H_0) of embankment for any subgrade with any soil-filled under any foundation condition is this. (First Highway Survey and Design Institute of China Communications Construction et al., 2006)

$$H_0 = K_2\left(K_1 h_T - h_o\right) \tag{3.11}$$

$$K_1 = \sqrt{\frac{\lambda_1}{Q_1} \cdot \frac{Q}{\lambda}} \tag{3.12}$$

$$K_2 = \sqrt{\frac{\lambda_2}{Q_2} \cdot \frac{Q_1}{\lambda_1}} \tag{3.13}$$

When coarse-grained soil is used as the standard foundation, the thawing depth under asphalt pavement will be:

$$h_T = 1.39 + 0.239h_s \tag{3.14}$$

$$h_T = 46.8 - 0.394H_e - 0.699L_a \tag{3.15}$$

where h_T is the thawing depth under embankment with the asphalt pavement under the standard foundation (m); h_s is the thawing depth under embankment with the gravel pavement (m); h_o is the permafrost table under ground surface (m); K_1 is the conversion factor used for calculating the thawing depth under embankment at local areas by using the thawing depth under standard embankment; K_2 is the conversion factor of the soil-filled according to Table 3.17; λ and Q are the thermal conductivity and heat of phase transition of standard foundation soil, W/(m·°C), respectively; λ_1

Table 3.17 Conversion factor of the filler (K_2)

Soil types	Silty clay	Silt	Gravelly fine sand and sandy soil	Sand, Sandy gravel and gravel soil	Dry gravel
K_2	0.6 to 0.65	0.7 to 0.75	0.8 to 0.85	1.0	1.1 to 1.2

Note: The value can be taken by the type of the zone where the embankment is located, the low value will be taken for the wet area, and the high value for the dry area

and Q_1 are the average thermal conductivity and average heat of phase transition of the soil within the local active layer, respectively; λ_2 and Q_2 are the thermal conductivity and heat of phase transition of the replacement soil-filled, respectively. H_e is the elevation (100 m); L_a is the latitude (°). The heat of phase transition can be estimated by the following:

$$Q = 80(\omega - \omega_u)\rho_d \tag{3.16}$$

where ω is the total moisture content (%); ω_u is unfrozen water content (%); ρ_d is the dry density of soil (g/cm^3).

The thermal parameters of the standard foundation can be estimated by the weighted average of the whole section $\sqrt{Q/\lambda}$ = 149.892 to 138.508. If the drilling time of the road section is in the cold season, suggesting $\sqrt{Q/\lambda}$ take the low value; and in the maximum thawing season, take the high value. For the conversion factor of the soil-filled, $\sqrt{Q/\lambda}$ take the high value.

(b) The Empirical Formula for Qingkang Highway

The critical height of embankment can be estimated by the relationship between the height of embankment with gravel pavement and the change of permafrost table under the embankment (Zang & Wu, 1999):

Linear equation regression:

$$\Delta H = 1.27H - 1.06 \tag{3.17}$$

Logarithmic equation regression:

$$\Delta H = 1.48\lg H + 0.147 \tag{3.18}$$

where ΔH is the variation of the permafrost table under embankment (m); H is the height of soil-filled embankment (m).

(c) The Calculation Method for Critical Height of Embankment of Qinghai-Xizang Railway

For permafrost regions with mean annual ground temperature (T_{cp}) lower than −1.0 °C in the hinterland of Qinghai-Xizang Plateau, the height of low embankment filled with cohesive soil can be estimated by the following empirical statistical formula (Huang, 1983).

$$\Delta H = 1.10H - 0.61 \tag{3.19}$$

where $\triangle H$ is the variation of the permafrost table under low embankment (m); H is the height of soil-filled low embankment (m).

For basically stable cold permafrost areas in Qingshuihe experimental section, the empirical formula for low embankment filled with coarse-grained soil (China Railway Northwest Research Institute, 2000) is:

$$\Delta H = 0.70H - 0.38 \tag{3.20}$$

Considering the thermal effects of horizontal heat flux on the sunny shoulder, the thawing depth under the sunny shoulder is about 15% deeper than that under the center of embankment. Meanwhile, the thawing depth under the sunny shoulder multiplied by the correction coefficient of 1.15 takes the control index of critical height of the embankment.

3. Estimation Method by Conversion of Thermal Parameter
   According to the relationship between the thawing process of the embankment and the local natural stratum in the same climatic zone, the critical height of the embankment (H_0) adapted to the local climatic conditions is obtained by the conversion of the boundary conditions:

$$H_o = (m_t \cdot \alpha \cdot \beta - 1)h_o \tag{3.21}$$

where H_0 is the critical height of embankment (m), h_0 is the permafrost table under ground surface (m). m_t, α and β are the correction coefficients of horizontal additional heat flux, thermophysical properties and surface state, respectively, which are valued according to Table 3.18.

4. Approximate Solution by Using Stephen's Formula
   According to the definition of the critical height of embankment (H_0), then:

$$H_o = h_{\max} - h_o \tag{3.22}$$

$$h_{\max} = \sqrt{\frac{2\lambda \sum t\tau}{Q}} \tag{3.23}$$

Table 3.18 Calculation formula coefficient of thawing depth under embankment

Correction coefficient		Value
Coefficient of horizontal additional heat flux m_t		1.15
Physical properties coefficient α		$\sqrt{\lambda_H Q_T / \lambda_T Q_H}$
Correction coefficient of surface state β	Consistent surface	1.00
	Natural surface with turf and filled with cohesive soil	1.06
	Natural surface with turf and filled with gravel	1.12

Notes: λ_T and λ_H are the thermal conductivity of natural stratum and fill soil respectively (W/(m · °C)); Q_T and Q_H are the heat of phase transition of natural stratum and fill soil respectively (kJ/m^3)

where h_0 is the permafrost table under ground surface (m), and h_{max} is the maximum thawing depth under embankment. λ is the thermal conductivity of the soil-filled (W/(m·°C)); Q is the heat of phase transition (kJ/m^3): $\sum t\tau$ is the thawing index of the ground surface (°C·h).

3. **Upper Critical Height of Embankment**

The observation data of permafrost along highway and railway in permafrost regions show that the permafrost table increases with the increase in the soil-filled height of embankment. However, when the embankment reaches a certain height, the permafrost table will not continue to rise but will decline. When the height of embankment is relatively lower, the thermal impact mainly comes from the embankment surface. With the increase of embankment height, the thermal impact increases from the embankment slope except heat resource from embankment surface. So, thermal impact is influenced by embankment direction. As a result, the surface temperature on embankment slopes, the ground temperature under the slope and the permafrost table will change.

The observation data of permafrost at the experimental section in Beiluhe basin showed that the permafrost table under the embankment decreases with the increase of embankment height (Fig. 3.41). The embankment height in this area ranges from 2.5 m to 5.0 m (Zhang & Wang, 2004).

The upper critical height of the embankment is defined into the maximum soil-filled height at which seasonal freezing layer in the first winter after embankment is completely connected with the permafrost table under ground surface before embankment constructed (Huang, 1983).

The necessary condition for the rise of the permafrost table under the bottom of embankment is that the freezing capacity of the climate in the engineering areas must be greater than the thawing capacity. Only the residual (potential) freezing capacity after the freezing of the seasonally thawed layer can increase the permafrost table under the bottom of embankment. For areas with large potential freezing capacity (or freezing index), the height of the soil-filled embankment can be suitably increased.

In cold seasons, the embankment freezes in double directions from top to bottom and bottom to top. In the cold permafrost areas, the mean annual ground

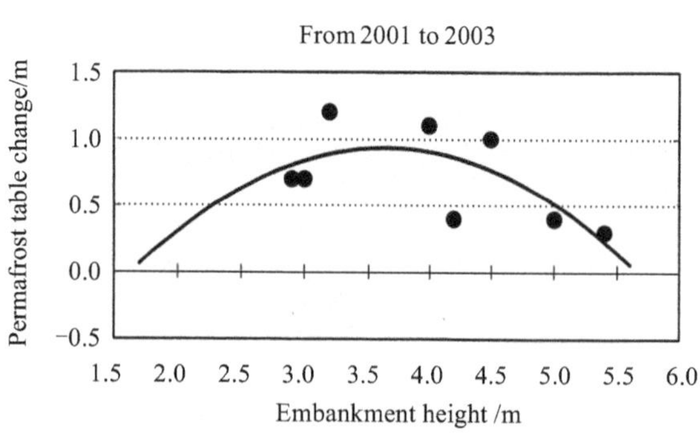

Fig. 3.41 Relationship between permafrost table change and embankment height

temperature is lower than −1.0 °C to −1.5 °C, which can make the seasonally thawed layer gradually refreeze. The lower the ground temperature is, the faster the refreezing speed and the higher the refreezing height will be. In the warm permafrost areas, the mean annual ground temperature is relatively higher, which can only maintain or be difficult to maintain the stability of permafrost table, so the height of refreezing is relatively lower. The cold permafrost area has a relatively cold climate, with a lower mean annual air temperature and a relatively larger freezing index. Therefore, after the embankment is frozen from top to bottom, there is still a large potential freezing capacity so that the seasonally thawed layer under the embankment can refreeze to connect with the permafrost table. The lower is the mean annual temperature, the greater the freezing index and the higher the upper critical height of the embankment will be. In the bidirectional freezing process of the embankment, the top-to-bottom freezing depth often exceeds 80%, while the soil layer near the bottom of embankment has the worst heat dissipation conditions and a lower refreezing height. Thus, under the slope feet of high embankment, cold permafrost layers are often formed, or even thawed interlayers.

Table 3.19 shows effects of the early and late construction on the thermal state of the embankment in the warm season. Early construction in the warm season reduces the effects of heat storage and freezing processes in the embankment.

According to the calculation data, when the height of the embankment on the Qinghai-Xizang Plateau is more than 4 m, the thawed interlayer still remains in the embankment after the winter season of that year, and increases with the embankment height. Under the condition of the absence of thawed interlayer, and considering the design service life of 30 years and the permafrost with different mean annual ground temperatures, the upper critical height of the embankment is obtained (Table 3.20).

4. **Conditions for the Existence of the Critical Height of Subgrade**

In a certain regional environment, the greater the freezing index is, the greater the freezing depth will be and the greater the possibility that the critical height of the embankment exists will be. Conversely, the greater the thawing index is and the greater the thawing depth will be. The critical height of the embankment may not exist. According to the observation data of Qinghai-Xizang Highway, the freezing index and thawing index are closely related to the annual average temperature (Fig. 3.42) (Ding & He, 2000).

Data on the Qinghai-Tibet Plateau show that the mean annual surface temperature in the same area is 2–5 °C higher than the mean annual air temperature. Regardless of the properties of the surface material, the n_t factor of the thawing period of ground surface in permafrost regions is always greater than the n_f factor of the freezing period (Table 3.9). The n_t factor of the thawing period is greater than 1, but the n_f factor of the freezing period is less than 1. In other words, the thawing index of the ground surface during the thawing period is greater than that of the air, and the freezing index of the ground surface during the freezing period is smaller than that of the air. The n_t factor of surface with cohesive soil, gravel and asphalt

Table 3.19 Comparison of the influence of the construction seasons on the permafrost table under embankment (China Railway Northwest Scientific Research Institute Company, Ltd., et al, 2006)

Section	Soil-filled height of embankment/m	Dates of filling and completion	Average temperature during filling period/°C	Average temperature of the soil-filled/°C	Average Freezing Rate of the filling year/cm·d^{-1}	Connecting duration of upper and lower freezing/d	Change of permafrost Table/m
0 + 280A	5.4	7 to 9月	4.4	6.5	3.0	150	−0.25
0 + 245	4.2	6 to 7月	2.0	4.2	5.1	130	+1.0

Table 3.20 Upper critical height of embankment under different mean annual temperature conditions (m) (Zhang, 2004)

Mean annual air temperature/°C	−2.8	−3.0	−3.5	−4.0	−4.5	−5.0	−5.5
Mean annual ground Temperature/°C	−0.3	−0.5	−1.0	−1.5	−2.0	−2.5	−3.0
Asphalt pavement (highway)	–	2.79	3.47	3.71	3.99	4.11	4.55
Gravel pavement (railway)	3.41	3.42	3.99	4.13	4.39		

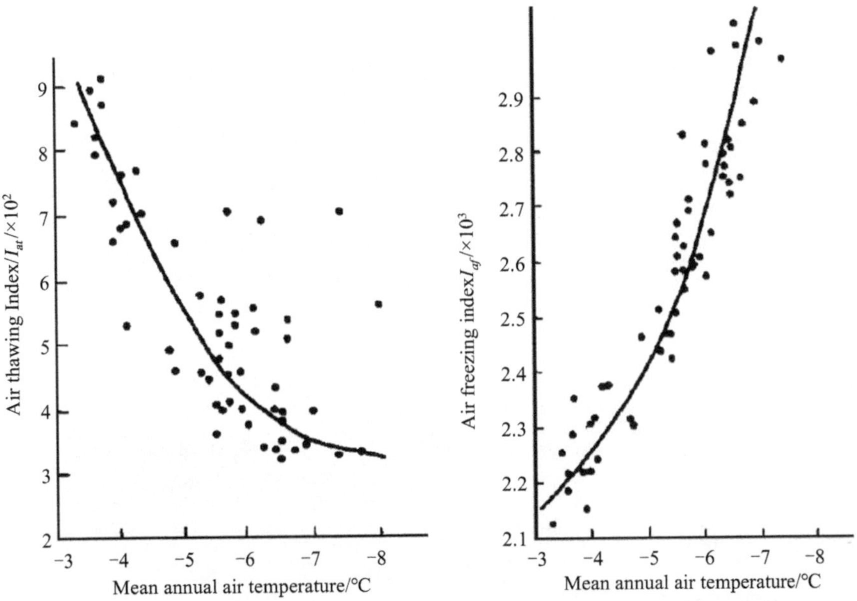

Fig. 3.42 Thawing index (**a**) and freezing index (**b**) vs. mean annual air temperature

pavement during the thawing period in permafrost regions of the Qinghai-Xizang Plateau are 2.5, 3.0 and 4.5, respectively, which means that the thawing index of the ground surface is 2 to 5 times larger than that of air temperature. The freezing index of the ground surface is smaller than the air freezing index, which is generally 0.6 to 0.9 times the air freezing index.

If ignoring the slight difference of thermal conductivity between frozen soil and thawed soil, the freezing depth and thawing depth estimated by Stephen's equation only change with the freezing index and thawing index. When $n_t I_{at} = n_f I_{af}$, the thawing depth under embankment can be considered equal to the freezing depth under the embankment. Obviously, the basic condition for keeping the permafrost table under the embankment unchanged is that the freezing depth is equal to or greater than the sum of the thawing depth and the permafrost table under natural surface. Otherwise, when n_t I_{at} is greater than n_f I_{af}, the thawing depth under embankment will be greater than the freezing depth, and the permafrost table under the embankment will decline, resulting in the occurrence of thawed interlayer. At that time, the mean annual air temperature corresponding to I_{at} and I_{af} is the critical value of the

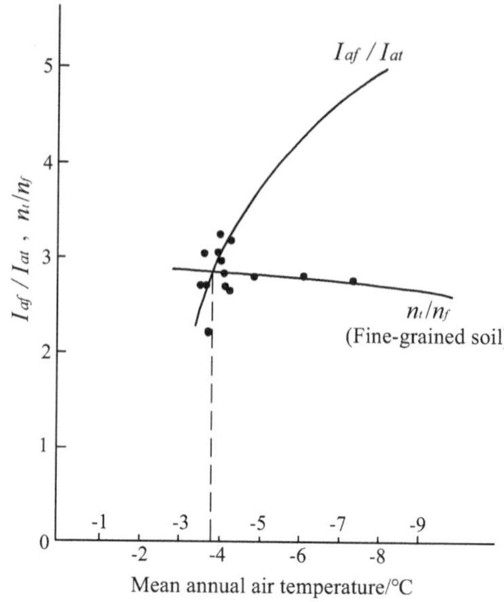

Fig. 3.43 Critical mean annualair temperature at critical embankment height

Table 3.21 Identification of the critical height of the asphalt pavement subgrade along the Qinghai-Xizang Highway (Ding et al., 2011)

Location	I_{af}/I_{at}	n_t/n_f	Critical Height of subgrade exists.
Qingshui River	5.04	5.56	No, with thaw interlayer
Kunlun Mountains	6.19	5.56	Yes
Fenghuo Mountains	7.11	5.56	Yes

critical height of embankment. The mean annual air temperature corresponding to the intersection of I_{af}/I_{at} and n_t/n_f is the critical value of the critical height of the embankment (Fig. 3.43). For the fine-grained soil embankment of the Qinghai-Xizang Highway, the critical value of the mean annual air temperature of the critical height is −3.8 °C which is equivalent to the corresponding mean annual air temperature is −4.0 °C when the mean annual ground surface temperature in permafrost regions on the plateau is 0 °C. Therefore, when the mean annual air temperature is higher than the critical value, there may be no critical height of embankment. When the mean annual air temperature is lower than the critical value, the critical height of embankment will exist. According to the corresponding upper boundary conditions and soil parameters, a theoretically estimated method is used to obtain the upper and lower critical heights of the embankment under different temperature conditions, and then obtain the critical value of the mean annual air temperature with or without the critical height of the embankment (Lai et al., 2009).

According to the drilling data under the embankment with asphalt pavement along the Qinghai-Xizang Highway, when the value of I_{af}/I_{at} is greater than n_t/n_f, critical height of the embankment will exist; otherwise, there may not exist the critical height of the embankment (Table 3.21). In the high plain area of Chumaer River, there are a large number of thawed interlayers containing plenty of water under the embankment. Within the range of 63 km from Budongquan to Chumaer River where the embankment height is 1.0–3.5 m, 39 boreholes among 59 boreholes under the center of embankment have thawed interlayers, accounting for 66.1% of the total. Only 12 boreholes (average embankment height is 1.1–1.8 m) do not have thawed interlayers, accounting for 20.4%.

The results of numerical modeling show that the conditions for the existence of the critical height of the embankment are closely related to the local mean annual air temperature and the material properties of the embankment surface. Taking the change of upper and lower critical heights of the embankment with the mean annual average temperature as the limiting condition, the air temperature corresponding to their intersection is the critical value existing at the critical height of the embankment. If the service life is 30 years for the highway design, the critical mean annual air temperature of embankment with asphalt pavement is −3.9 °C(Fig. 3.44), and the critical mean annual temperature of embankment with gravel pavement is −2.8 °C (Fig. 3.44). If the service life is 50 years for the railway design, the critical mean annual air temperature of the embankment with gravel pavement is −3.1 °C.

Fig. 3.44 The critical mean annual air temperature for the critical height of embankment with asphalt pavement (on the Left) and gravel (on the Right) pavement (Zhang, 2004)

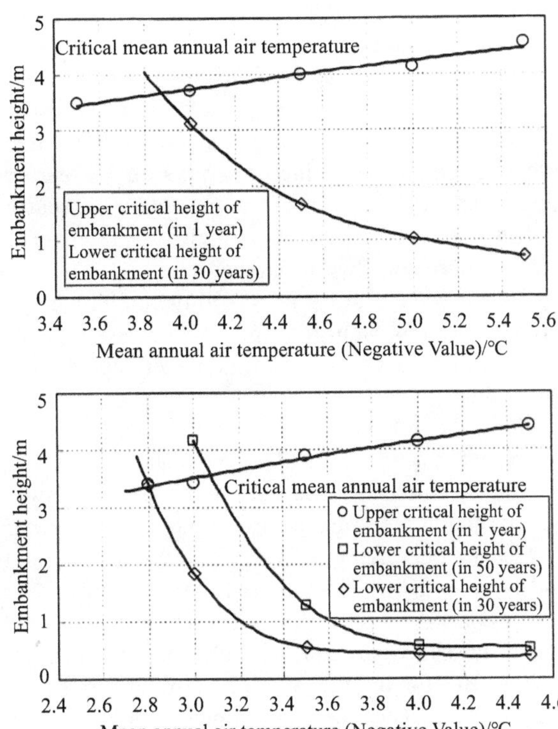

5. **Design Height of Embankment**

In the design of subgrade in permafrost regions, in addition to the design based on the deformation and strength, regulation temperature method should be adopted by controlling heat transfer path and influence scope in the embankment, aiming at the distribution characteristics of the temperature field under the embankment. Besides, considering the climate change and the service life of infrastructures, together with appropriate adjustments, the reasonable height of embankment in permafrost regions is finally determined.

The main factors that should be considered in the design of the embankment in permafrost regions include engineering properties of permafrost, mean annual ground temperature, compressibility of the seasonally thawed layer under the embankment, and a warming of climate.

The characteristics of ground ice in the range of 2–3 m below the permafrost table are closely related to the types and properties of seasonally thawed layer. For example, in the distribution area of seasonally thawed layer such as swamp, vegetation and peat, the underlying permafrost has a relatively thicker ground ice layer, and the engineering types of permafrost are mostly ice-saturated frozen soil or soil-bearing ice layer. In the distribution area of seasonally thawed layer of cohesive soil with developed vegetation, underlying permafrost engineering types are mainly ice-saturated frozen soil and ice-rich frozen soil. Engineering properties of the permafrost with high ice content will greatly affect embankment stability. Meanwhile, thawing depth in cold permafrost areas is relatively smaller, but it is relatively larger in warm permafrost areas. So, critical height of embankment should consider different mean annual ground temperature. Under the fluctuation effects of climate warming, in terms of a 30-year operation time, the increase of global temperature will have an impact on the seasonal thawing depth of the permafrost. The impact on cold permafrost is relatively smaller, but the warm permafrost is more sensitive to temperature rise. The results of numerical simulation show that the permafrost table will increase by 6 cm under the condition of mean annual air temperature of −6.0 °C, however, the permafrost table increases by 40 cm under the condition of the mean annual temperature of −3.0 °C (Zang & Wu, 1999).

Under the influence of repeated freezing-thawing cycles, the compactness of the seasonally thawed layer under the embankment is relatively smaller than that of the non-frozen soil area. Especially in the area with swamp, peat, and peat layers, the peat layers on the original surface has not be allowed to be eradicated in order to protect the underlying permafrost. However, the drainage consolidation on these soil layers is slow, generally only 30–40% of the compactness can be completed during construction. For the old line reconstruction of the roadbed, the compactness of the seasonally thawed layer under the embankment will not be taken into consideration. But for the new line of the roadbed, the compactness of the seasonally thawed layer under the embankment should consider the compacting settlement. According to the field and test data, the reserved soil-filled height of embankment can refer to Table 3.22.

Table 3.22 Reserved soil-filled height of subgrade in plateau permafrost regions (m) (Zang & Wu, 1999)

General Road section	Special road section		
	General soil conditions with swamp	Swamp sections with rich humus and turfy soil	Swamp sections with rich in peat soil
0.05 to 0.075	0.075 to 0.125	0.125 to 0.15	0.15 to 0.20

Table 3.23 Allowable extent of artificial permafrost table declining (Δh) for different types of frozen soil engineering (First Highway Survey and Design Research Institute of China Communications Construction et al., 2006)

Permafrost engineering types	Δh(m)
Soil-bearing ice layer	0.15 to 0.20
Ice-rich frozen soil	0.50 to 0.75
Ice-saturated frozen soil	1.00 to 1.50

As for the design height of embankment in permafrost regions, there are different design principles for different engineering geological conditions of permafrost. According to the design principle for protecting permafrost, the design height of embankment can be determined by the method of the critical height of embankment plus safety height. According to the principle for controlling the thawing rate of permafrost (without removing the turf and peat layer on the surface), the design height of the embankment should be carried out by the design method that the road-bed deformation of the road surface within the design service life should not be greater than the allowable deformation.

According to the data in survey, drilling and georadar of the Qinghai-Xizang Highway, and, under different frozen soil engineering geological conditions, the allowable extent of the artificial permafrost table decline can be determined by the allowable deformation index of the embankment pavement (Table 3.23).

Design height (H_d) of new highway embankment:

$$H_d = MH_0 + K \cdot P \cdot \phi \cdot t \cdot m + S \tag{3.24}$$

When reconstructing and renovating the original embankment with asphalt pavement section, the increased height of embankment can be determined:

$$\Delta h = K \cdot P \cdot \phi \cdot t \cdot m \tag{3.25}$$

where H_0 is critical height of embankment (m), M is comprehensive correct factor. K is the correction coefficient of air temperature, which the Qinghai-Xizang Highway can take 1.075. P is the average thawing rate ($P = \triangle h/\triangle T$, $\triangle h$ is the extent of artificial permafrost table declining (m) under the embankment with asphalt pavement in the exploration year. $\triangle h = h_1$-h_t-h_r, h_1 is the embankment height (m) in the exploration year; h_r is the artificial permafrost table (m) under the embankment in

Table 3.24 Values of empirical coefficient (m) for conversion of soil-filled equivalents

Permafrost engineering types	Applicable conditions		m^*
	Embankment height (h)/m	Extent of permafrost table declining (Δh)/m	
Soil-bearing ice layer	3.8 to 3.0	0.4 to 1.2	1.0 to 5.0
Ice-saturated permafrost	2.4 to 2.8	0.8 to 1.6	1.0 to 2.5
Ice-rich permafrost	1.8 to 2.0	0.8 to 1.8	1.0 to 2.0

the exploration year; h_t is the permafrost table (m) under natural surface in the calculated section; ΔT is the time from the completion of embankment with asphalt pavement to the exploration, calculated by year). ϕ is the attenuation coefficient of thawing rate, $\phi = 1/\ln t$, t is the design service life of pavement, calculated by year; m is the empirical coefficient for conversion of soil-filled equivalents (Table 3.24).

In addition to bedrock road section, the minimum height of the embankment should not be lower than 1.5 m, and the height of the embankment in the non-longitudinal slope or controlled infrastructure section should not exceed 3.5 m.

In any case, drainage measures should be taken on both sides of embankment in the permafrost regions to avoid water accumulation. The embankment base must be well-drained and cut off the capillary water connection between the soil and embankment. If the ground surface drainage is better and when the lower part of the embankment is filled with fine-grained soil, the cushion thickness of coarse-grained soil (the content of silt and clay should be controlled within 15%) not more than 0.5 m should be set at the bottom of the embankment. If the ground surface drainage is poor, the embankment should be filled with coarse-grained soil and drainage measures should be taken at the bottom of the embankment to cut off the capillary water connection of the base layer of embankment to avoid frost heaving and frost boiling.

3.3.2 The Embankment of Low Soil-Filled, Shallow Cutting, and Zero-Section

The embankment of lower soil-filled, cutting-shallow, and zero-section in the permafrost region is often the road section with the most serious damages such as thaw settlement and frost heave. Here the road section of lower soil-filled, cutting-shallow, and zero section refer to the road section where the soil-filled height of embankment and excavation depth of embankment are less than the minimum critical height of embankment. Considering the principle for protecting permafrost, it is not appropriate to use the lower soil-filled embankment and shallow-cutting embankment with the soil-filled height less than 0.5 m. Therefore, when selecting road routes and designing embankment, the engineering should try to minimize or shorten the length of such sections and segments.

Fig. 3.45 Relationship between thawing depth under embankment and embankment height after 30 years of highway operation for different areas of mean annual air temperatures

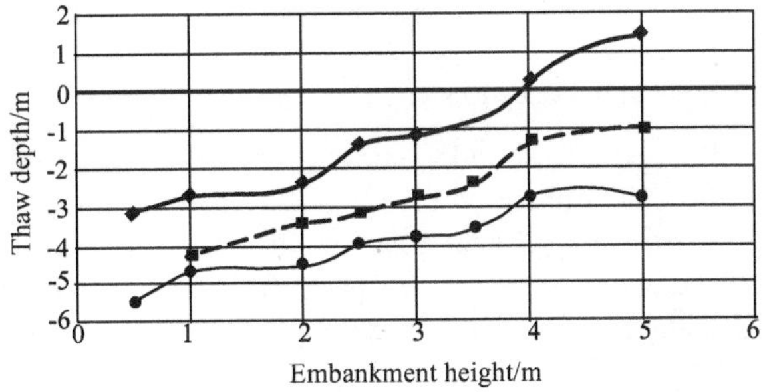

The low embankment responds rapidly to the changes in air temperature. With the continuous increase of external positive accumulated temperature, the thermal resistance of low embankment is small, and the thickness of thawed core under the center of low embankment will increase rapidly. The results of the numerical simulation under different temperature conditions (Wang, 2004) show that there is a good linear relationship between the maximum thawing depth under the embankment and the embankment height (the correlation coefficient reaches 90%) (Fig. 3.45). As the highway operation time is longer, the influence of the embankment height on the thawing depth under the embankment will gradually increase.

The design of the embankment of low soil-filled and shallow cutting and zero-section should pay close attention to the thaw settlement of permafrost under the embankment and the replacement and thermal insulation measures for the embankment base and the waterproofing and drainage measures, especially the prevention and control of suprapermafrost water. At present, the embankments of low soil-filled, shallow cutting, and zero-section in permafrost regions produce damages of thaw settlement. Firstly, the thickness or missing of the insulation layer at the bottom of embankment is relatively smaller. Secondly, the particle size of crushed stone and coarse-grained replacement layer is relatively smaller, it can only play a role in reducing frost heaving, and its cold storage effect is insufficient. Thirdly, measures in waterproofing and drainage fail to meet the design requirements, especially not completely block the infiltration of suprapermafrost water. Although the suprapermafrost water quantity is very small, it can also partly thaw permafrost layer under the embankment. Fourthly, the unsuitable constructed season will produce a large amount of heat storage entering embankment and replacement layer, resulting in the continuous thawing of the permafrost under embankment after construction. Finally, conventional methods of embankment have not been correctly constructed.

In the warm permafrost areas with high ice content (ice-rich permafrost, ice-saturated permafrost, and soil-bearing ice layer), engineering damages usually occur in the embankment of low soil-filled, shallow cutting, and zero-section, and there are often thawed interlayers under the embankment. So, commonly-used engineering design principles and permafrost subgrade treatment methods are as follows:

3.3.2.1 Design Principles of Destroying Permafrost

Its applicable conditions are:

1. The permafrost under embankment belongs to ice-poor and icy permafrost, and the water content in the permafrost layer is small and the conditions of the permafrost subgrade are relatively better.
2. The thickness of ice-rich permafrost, ice-saturated permafrost and soil-bearing ice layer under the embankment is not large, and the buried depth is shallow or less than 3.5 m.
3. The permafrost subgrade is bedrock, or coarse-grained soil with the water content less than 10%.

The method of foundation treatment of removing the permafrost with high ice content is used and embankment design can use the conventional design method for the embankment in non-permafrost areas.

3.3.2.2 Design Principles of Protecting Permafrost

Its applicable conditions are:

1. The thickness of permafrost with high ice content, such as ice-rich permafrost, ice-saturated permafrost and soil-bearing ice layer, under embankment, is relatively larger, which is not suitable to use replacement methods to prevent the permafrost from thaw settlement.
2. The ground temperature of permafrost belongs to stable or basically stable zone.
3. The foundation soil is hard frozen cohesive soil.

The method of foundation treatment of replacement and thermal insulation is used, and makes well longitudinal drainage and slope protection.

3.3.2.3 The Design Principle of Gradually Thawing Permafrost

Its applicable conditions are:

1. The ice content of permafrost under embankment is relatively smaller, and its buried depth is relatively larger.
2. The ground temperature of permafrost belongs to unstable or extremely unstable zone.
3. The thickness of coarse-grained permafrost with high ice content is more than 2 times the permafrost table under the natural surface.

The methods of foundation treatment are:

1. Permafrost is broken in advance and exposed in the warm season, then replacing with coarse-grained soil and then laying asphalt pavement, with primarily operation for 1–2 years.

2. The design height of the embankment should increase the reserved amount of thaw settlement from pre-thawing depth and the depth of extra-dredging.

3.3.2.4 Design Methods for Low Soil-Filled, Shallow Cut, and Zero-Section Embankment

In flat areas, when the fill height of subgrade is less than the minimum critical height of the subgrade and the zero section, frozen soil areas with high ice content can be treated as follows:

1. When the thickness of the permafrost layer with high ice content is thin and its buried depth is shallow, embankment design can choose the treatment of complete excavation and replacement (Figs. 3.46 and 3.47). The replacement filling material should use cohesive soil with better thermal insulation and water resistance, or set up XPS (extruded polystyrene foam board) thermal insulation layer, or adopt crushed rock. The sum of replacement depth and embankment height should not be less than the sum of minimum critical height of embankment and the permafrost table under natural surface.
2. When the thickness of the permafrost layer with high ice content is thick, it can be partially excavated and replaced (Fig. 3.48).
3. If the permafrost with high ice content is thick, it can be partially replaced by using cohesive soil with better thermal insulation and water-resistance proper-

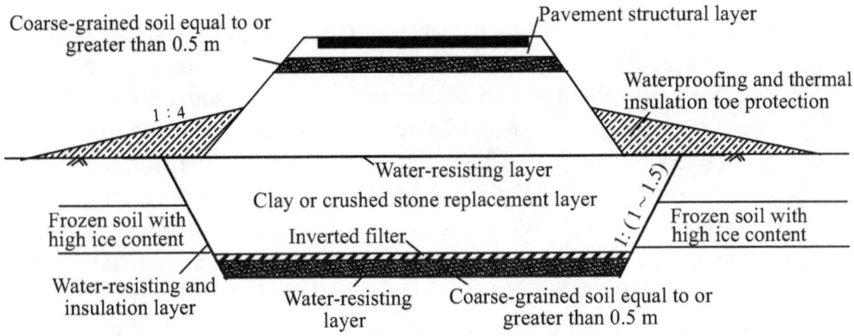

Fig. 3.46 Complete replacement of frozen soil with high ice content

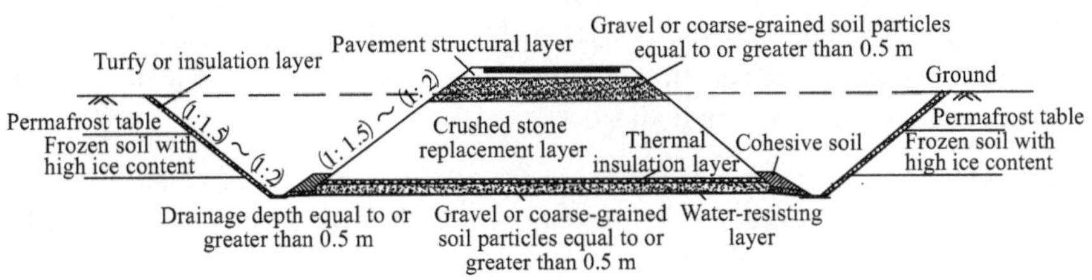

Fig. 3.47 Complete replacement sections

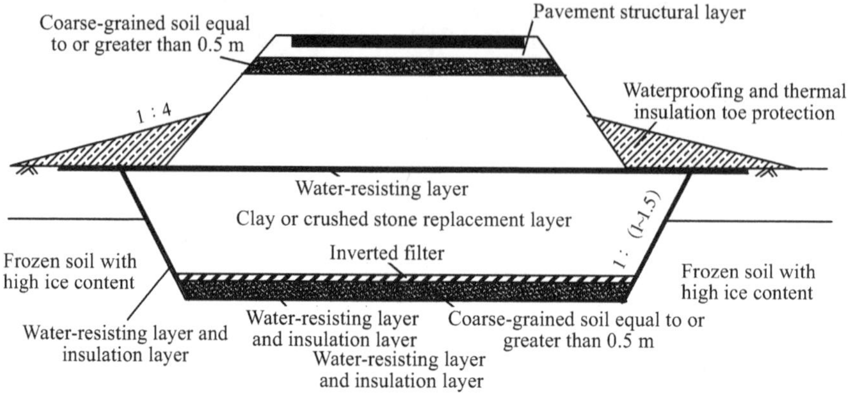

Fig. 3.48 Partial replacement of soil layers with high ice content

ties, and crushed rock. The sum of the replacement depth and embankment height should not be less than the sum of minimum critical height of the embankment design and permafrost table under natural surface.

4. The design for low embankment and zero-section embankment should be carried out according to the specific requirements. In order to protect the subgrade from thaw settlement caused by the declining permafrost table after the construction of embankment, the base of embankment should be filled with coarse-grained soil or cohesive soil, and its thickness can be determined by thermal calculation. If there is no calculation data, it can be replaced by 1.4–1.6 times the permafrost table under natural surface. When coarse-grained soil is used for replacement, in order to prevent the infiltration of surface water, a composite geomembrane impermeable layer should be laid on the ground, outward into a 4% lateral drainage slope. If using partial or complete replacement, when the embankment height obtained by thermal calculation is greater than the minimum design height of the embankment, the replacement should be based on the calculated thickness of replacement. When the embankment height obtained by thermal calculation is less than the minimum design height of the embankment, the replacement should be carried out based upon the minimum design height and the bottom of replacement layer should be outward into a 4% lateral drainage slope. When industrial thermal insulation materials are used as the thermal insulation layer of embankment, the thermal insulation layer should be set at 0.8 m below the road shoulder, and outward into a 4% lateral drainage slope. And a layer of 0.2 m medium-coarse sand cushion should be laid on the upper and lower surfaces of the insulation layer, with the width 1.2 m wider than the roadbed surface (with 0.6 m on both sides).

In the gentle slope section, when designing the embankment in the stable zone and the basically stable zone of cold permafrost, the following situations should be considered.

1. For the permafrost with high-ice content in the gentle slope sections that the ground transverse slope is less than 1:5 (including the soil-bearing ice layer with a cumulative thickness of more than 0.15 m in the range of 2 times the permafrost table under the embankment, or the ice-saturated permafrost with the thickness of 0.4 m or the ice-rich permafrost with the thickness of 0.6 m), the embankment structure can refer to the design of embankment in the flat section. Steps should not be dug in the base of embankment and water-stop ridge or intercepting ditch should be set on the top side of the embankment while a back pressure insulation berm should be set at the foot of the slope, with the width of 2.0–3.0 m and the height of 1.0–2.0 m.
2. For the permafrost with high-ice content in the slope section where the ground transverse slope is steeper than 1:5, water-stop ridge should be set on the upper side of the embankment, and back pressure insulation berm should be set on the lower side at the foot of the slope, with the width of 2.0–2.5 m, the height of 1.0–2.0 m, and a slope ratio of 1:1.5.

Whether smooth or not, the drainage system on the embankment side often has a great impact on the permafrost under the embankment. In the warm permafrost areas, if the original permafrost environment remains completion, it is usually not suitable to build thermal insulation berms, but on the surface of the waterproofing and thermal insulation berm (Fig. 3.49) and the thermal insulation of slope toe protection (Fig. 3.50), strong water-resistant cohesive soil, peat and turf or gravel layer should be laid. If gravel, coarse-grained soil or other materials with strong water-resistance are used as thermal insulation berms, the surface should be covered with 0.2 m thick cohesive soil as a waterproofing protective layer. The dimensions of the thermal insulation berms and the thermal insulation toe protection are determined according to Table 3.25.

For the design of controlling thawing rate, the stepped embankment design should be carried by different permafrost conditions, and the specific road sections should be determined based on the allowable permafrost table declining (Table 3.26). The embankment height should be determined according to the design critical height of the embankment reconstructed in the cold permafrost areas.

Whether the low soil-filled and shallow cutting and replacement section can achieve the expected effect, the season and method of embankment construction

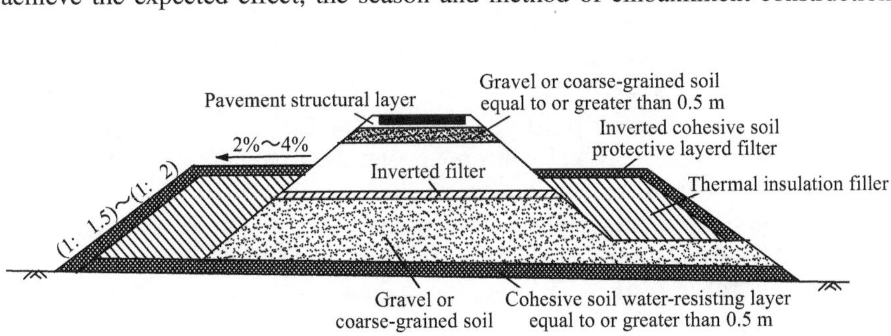

Fig. 3.49 Thermal insulation berm

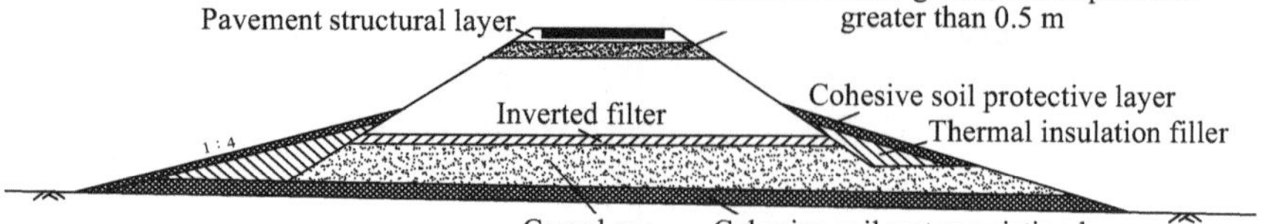

Fig. 3.50 Thermal insulation toe protection

Table 3.25 Dimensions of thermal insulation berms and thermal insulation toe of embankment

Embankment height (m)	Berm or toe pad	Height (m)	Width (m)
≤3	Toe pad	0.8 to 1.2	2.0 to 2.5
>3	Berm*	1.5 to 2.5	2.0 to 3.0

Note: According to *Interim Regulations on Engineering Design of Qinghai-Xizang Railway in Plateau Permafrost Regions*, the general height of the berm is 0.8 to 1.6 m and the width is 1.5 m to 2.5 m; When the embankment height is greater than 6.0 m, it will be the same as the data in the table

Table 3.26 Allowable artificial permafrost table declining (Δh) for ice-rich permafrost

Frozen soil engineering types	Soil-bearing Ice Layer	Ice-saturated permafrost	Ice-rich permafrost
Δh (m)	0.15 to 0.20	0.50 to 0.75	1.00 to 1.50

play an important role. That is, whether it can ensure suprapermafrost water from entering the foundation pit during the construction period; otherwise, it may cause the permafrost thawing. Therefore, the following works should be done:

1. Prepare enough gravel cushion materials, replacement filler, water-resistant materials, etc. in advance.
2. The construction season should be at the end of the cold season (from mid-March to early May) when suprapermafrost water is still in a frozen state, which is favorable for foundation trench construction, and water erosion should be cautioned.
3. Waterproofing and anti-seepage measures for the seasonally thawed layers on both sides of the foundation trench should be taken, embedding the water-resisting layer or water-stop board at the depth of 0.25 m below the permafrost table, and laying a water-resisting layer between the replacement layer and the permafrost layer, to prevent suprapermafrost water from infiltrating the replacement layer (referring to Figs. 3.48 and 3.50).
4. When the excavation reaches the design elevation, the protective layer with 0.5 m thick should be reserved, and then the rapid stepped construction should be carried out in sections, digging, laying impermeable layer, gravel cushion, insulation layer and replacing, between the side wall and the embankment base

according to the design requirements, and all the process should be done in one in order to avoid the occurrence of thawing in permafrost layer.

5. On both sides of the embankment, waterproofing and thermal insulation toe should be built, with the height not less than 1.0 m and the width not less than 2–3 m. The fill material can adopt cohesive soil or crushed rock (but the impermeable layer should be made).

3.3.3 *Road Cuttings*

Road cutting engineering in permafrost regions should consider the thermal and mechanical stability. The change of mechanical stability will lead to the change of thermal stability. The thawing of ground ice will lead to gelifluction and thaw slump, resulting in the slope instability (Fig. 3.12). Therefore, road cutting engineering in permafrost regions, under the condition of meeting the requirements for mechanical stability, should pay more attention to the thermal stability of slope and foundation base.

The main problems of the road cutting engineering in permafrost regions include section forms and treatment measures and calculation in the thickness of insulation replacement, the slope stability, and the base strength (Huang, 1983).

3.3.3.1 Section Forms and Treatment Measures

Reasonable section forms (Fig. 3.51) and treatment measures should minimize the influence to permafrost, so as to facilitate the restoration of the heat balance state of permafrost and reduce the infiltration of atmospheric precipitation, wetting and the harm of suprapermafrost water. At the same time, the amount of engineering should be minimized, which is beneficial to maintenance and conservation. Specific requirements for sections are as follows:

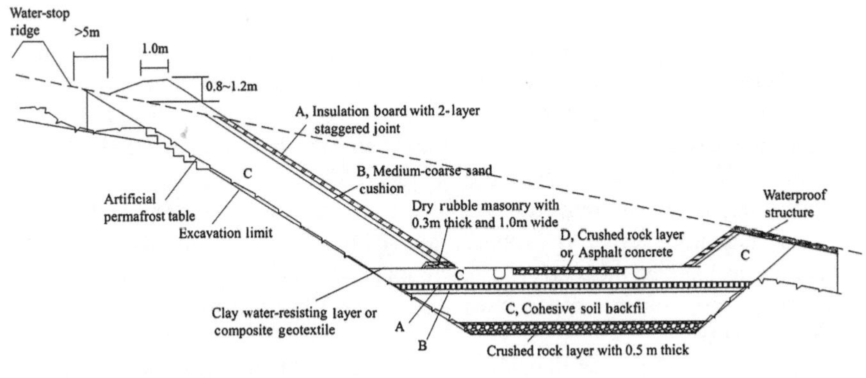

Fig. 3.51 Rational section of road cutting

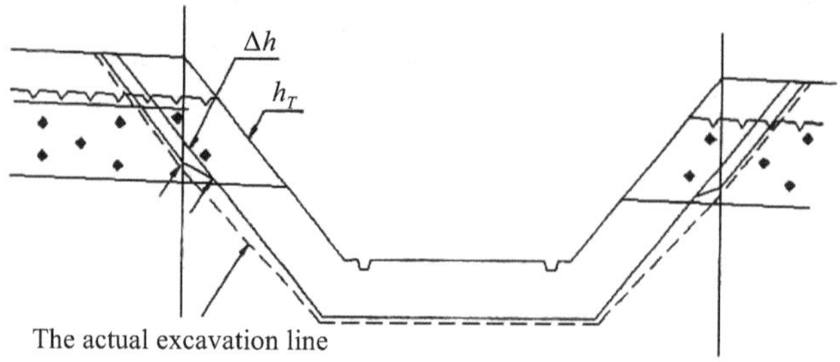

h_T - Calculating the thickness of the thermal insulation layer;

Δh - The increased value of the thermal insulation layer at the break angle of the slope top;

$\Delta h = (0.06 - 0.10)\ h_T$.

Fig. 3.52 Schematic diagram of road cutting excavation section

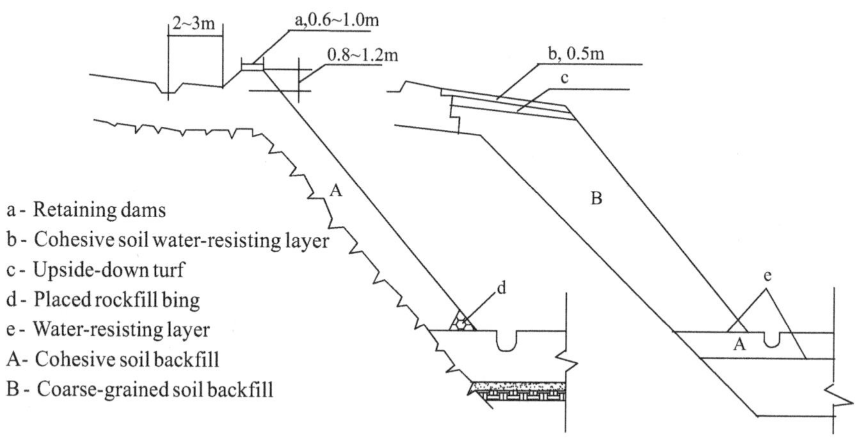

Fig. 3.53 Schematic diagram of facilities in water interception, drainage and resistance on the foundation base of the slope

1. The excavation of general road cutting section need to abide by the principle of "Rather exceeding than owing", the excavation section should, in terms of the determined thickness of the insulation layer, reserve an appropriate amount of overbreak (especially the ground ice part) so that the seasonal thawing depth remains above the replacement interface (Fig. 3.52).
2. According to the requirement of "Drainage system first, project later", the drainage system at the top of the slope, comprised by the water-stop ridge and gutter outside the bank on the road cut top, should be well built to prevent the suprapermafrost water of the upper transverse slope from harming the side slope (Fig. 3.53).
3. A shallow and wide side ditch section (preferably a U-shaped section) should be used, and the bottom of the ditch should be paved with flexible water-resistant

material, or cohesive soil, or reinforced composite waterproof geomembrane, which is convenient for maintenance and keeps the drainage of side ditch smooth.
4. By setting the platform above the side ditch, the stability of the slope will be enhanced, or by setting the platform and configuration of the dry rubble buttress at the foot of the slope, the drainage of the slope will be strengthened and the stability of the side slope will be improved. (Fig. 3.53).

Treatment measures should take the following factors into account:

1. The replacement materials should be mainly local materials (such as crushed rock, pebble gravel, cohesive soil, turf, etc.). The form of slope protection, with cohesive soil replacement and turf laid on the surface layer, is beneficial to reduce the heat absorption of the slope and reduce the thawing depth. The foundation base of road cut should be replaced with crushed gravel with a certain particle size range and sealed with impermeable cohesive soil. The foundation base should be replaced with coarse-grained soil at the depth of 1–1.5 m.
2. In order to reduce the excavation and replacement amount, industrial thermal insulation materials should be laid on the side slope and the foundation base. When laying a certain thickness of coarse sand should be set at the bottom of the thermal insulation board. When the thermal insulation board is laid on the slope surface, drainage openings should be left properly (Fig. 3.54).
3. In order to prevent water from immersing into the foundation base, the impermeable insulating layer should be set up at the appropriate parts of the side slope and the foundation base, and the water content and the density of compaction of the filling material should be controlled (Fig. 3.53). Besides, drainage measures for side ditch should be strengthened.

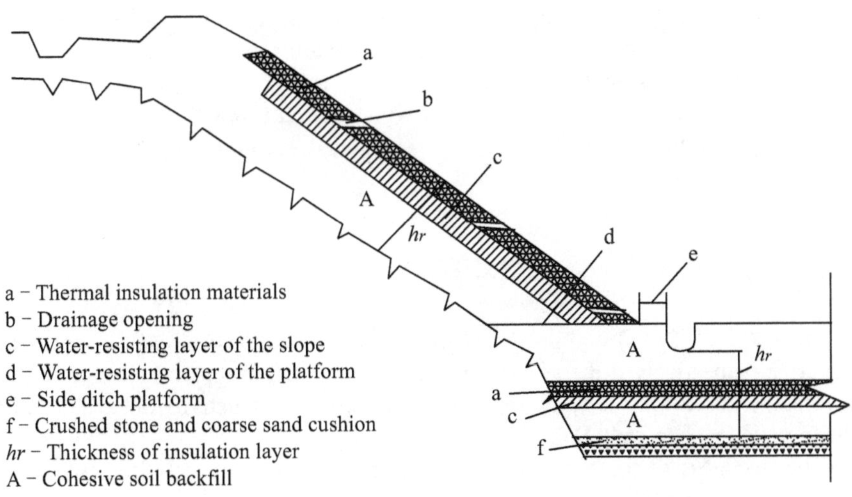

Fig. 3.54 Schematic diagram of laying thermal insulation materials on the ground

4. When there is no water-stop ridge set on the top of the road cut, due to the influence of the bidirectional heat source on the slope of the road cut top, the thawing depth is about 6 to 10% larger than that in the middle of the slope. Therefore, according to the thickness of the ground ice, the thickness of the insulation layer on the upper half of the slope should be appropriately increased (Fig. 3.54).
5. The construction time should be arranged reasonably. It is advisable to choose the early cold period (from September to November) and the early warm period (from March to May). In the process of excavation, rapid construction and continuous operation should be adopted to reduce the exposure duration of permafrost and strengthen thermal insulation.

3.3.3.2 Determination of the Thickness of Insulation Replacement

The design thickness of the thermal insulation replacement layer can be determined by the empirical formula or the method of equivalent thermal resistance, and the correction factor should be considered in the calculation. When there is no data, generally, the thickness of the replacement insulation layer can be 1.4–1.6 times the permafrost table under natural surface, and the slope ratio will be 1:1.75–1:2.0. The calculation formula for the thickness of the thermal insulation layer is determined by the formula:

$$h_T = k \cdot \frac{\lambda_0}{\lambda_t} \cdot h_t \tag{3.26}$$

where h_T is the design thickness of slope protection or the thickness of foundation replacement (m); k is the safety factor, when the slope insulation layer is designed, the value k will take 1.2–1.5. When the foundation replacement filling is designed, it will take 1.5–2.0; h_t is the local permafrost table under natural surface (m); λ_t is the average thermal conductivity of the local seasonally thawed layer in the thawing state, W/(m·K); λ_0 is the thermal conductivity, W/(m·K) of the selected insulation material or replacement material in the thawing state. If it is a multi-layer material, the average thermal conductivity will be calculated according to the principle of equivalent thermal resistance, that is:

$$\lambda_0 = h_T \Big/ \left(\frac{h_1}{\lambda_1} + \frac{h_2}{\lambda_2} + \ldots + \frac{h_n}{\lambda_n} \right)$$

where h_n (n = 1, 2, …) will be corresponding to the thickness of each layer of the material, respectively if the slope protection or foundation replacement are multi-layer materials; λ_n (n = 1, 2, …) represents the thermal conductivity of each layer of the material, respectively [W/(m·K)].

The thickness of the cut insulation layer can be determined by the following methods (Huang, 2006).

The thickness of the insulation (or replacement) layer can be determined by adding the artificial permafrost table under the cutting slope or the foundation base, and the safety reserve required by the construction conditions and filling conditions, namely:

$$Z = K_{\Phi} \cdot H_{\max} \tag{3.27}$$

where Z is the thickness of the thermal insulation (or replacement) layer (m); H_{max} is the artificial permafrost table under the slope or the foundation base (m); K_{Φ} is the engineering safety factor, depending on the construction and filling conditions.

For slopes:

$$H_{smax} = K \cdot K_{s}^{'} \cdot K'' h_{\max} \left(\text{Sunny slope}\right)$$

$$H_{Nmax} = K \cdot K_{N}^{'} \cdot K'' h_{\max} \left(\text{Shaded slope}\right)$$

For the foundation base:

$$H_{Dmax} = K \cdot K'' \cdot K_{D} h_{\max}.$$

When there is no water-stop ridge set on the road cut top, due to the maximum thawing depth on the road cut top slope, after determining the thickness of the thermal insulation (or replacement) layer according to the above formula, $\triangle Z$ should be increased above the middle of the slope:

$$\Delta Z = \left(0.06 \text{ to } 0.10\right) Z \tag{3.28}$$

where K_D is the structural correction coefficient, and the value K_D is 1.10; K is the correction coefficient of orientation, when the slope ratio is 1:1.5 to 1:1.75, the sunny slope will be 1.15 while the shady slope will be 0.95; K is the correction coefficient of surface state: the cohesive soil surface is 1.05; the turf surface is 1.00; K is the correction coefficient of filler. When the road cut side slope is replaced with cohesive soil:

$$K = \sqrt{\frac{\lambda_1}{W_1 \cdot \rho_1} \times \frac{W_0 \cdot \rho_0}{\lambda_0}} \tag{3.29}$$

When the slope is replaced with coarse-grained soil:

$$K = \sqrt{0.8 \times \frac{\lambda_1}{W_1 \cdot \rho_1} \times \frac{W_0 \cdot \rho_0}{\lambda_0}} \tag{3.30}$$

where λ_0 and λ_1 are the weighted average values of the thermal conductivity of soil layer above the permafrost table under natural surface and the slope replacement material, respectively, W/(mK); W_0 and W_1 are the weighted average values of soil moisture content corresponding to λ_0 and λ_1, %; ρ_0 and ρ_1 are the weighted average values of soil dry density corresponding to λ_0 and λ_1, g/cm^3; h_{max} is the local maximum permafrost table under natural surface (m).

When thermal insulation material is laid on the foundation base, a certain thickness of sand cushion should be set on the foundation base. When thermal insulation material is laid on the side slope, drainage openings should be reserved.

3.3.3.3 Design of Protective Measures for Supporting and Blocking Structures

If the site allows, in order to reduce the amount of road cut excavation and replacement earthwork volume, the supporting and blocking structure of "L"-type retaining wall or sheet-pile retaining wall (Fig. 3.55) can be used. Usually, it is suitable for low road cut or deep road cut combined with temperature measures.

The following factors should be considered in the design:

1. Road cut should be selected in sections with small surface transverse slopes.
2. The retaining wall foundation should be buried at the depth of 0.3–0.5 m below the artificial permafrost table, or located on the bedrock.
3. According to the ice content in soil body, the location of permafrost table under natural surface and the slope ratio of the stable slope, the range and the height of the slump mass behind the wall can be estimated, and the section of the retaining wall will be designed according to the principle of the earth pressure behind the wall.
4. Drainage openings should be set at different heights of the wall, and the longitudinal drainage of the cut and the side ditches in the form of shallow width should be strengthened.

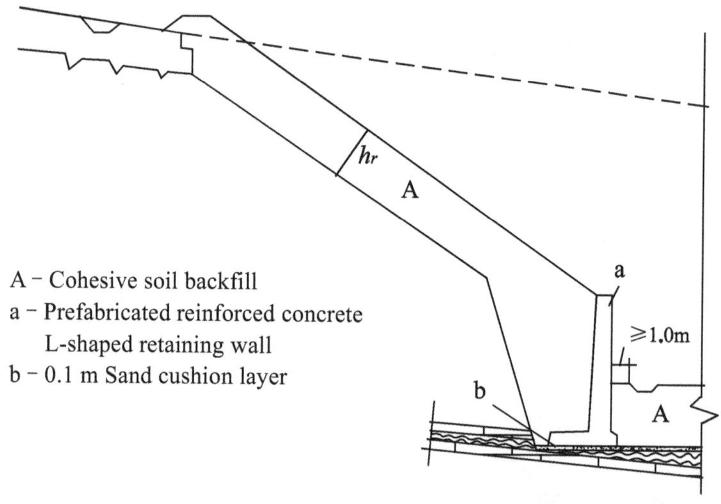

Fig. 3.55 Schematic diagram of the lower retaining and the upper protecting measures on the cutting slope

For deep road cut, the use of a section with the lower retaining and the upper protecting can reduce the amount of excavation and replacement. But it is required to reset a sufficient thickness of filling behind the wall. In the design, structural protection measures such as the stability of the retaining wall under the influence of the horizontal frost-heave force and the thickness of the thermal insulation layer, etc., should be taken into consideration. The use of reinforced concrete "L"-type retaining wall or anchor plate structure or sheet-pile retaining wall has its advantages.

3.3.4 *Transition Sections*

The road transition sections mainly include fill/filling (roadbed), cut/excavation (cut), roadbed and bridge (or culvert), the permafrost region and the talik; transition sections in different permafrost engineering geological conditions (or different roadbed structures).

The investigation and research results show that the settlement deformation in the transition section of the roadbed and bridge is the typical and the commonest type of roadbed damage. Based on the settlement investigation and related factors analysis in the roadbed-bridge transition section of 164 bridges on the Qinghai-Xizang Railway from Xidatan to Chiqu Valley (Niu et al., 2011), the embankment settlement in the transition section is related to the northern and southern ends of the bridge strike, embankment slope direction, embankment height, permafrost engineering type (ice content), ground temperature, embankment structure and geological conditions, etc. The average settlement of the northern end of the road-bridge transition section exceeds that of the southern end, and the total amount on sunny slopes is greater than that on shady slopes (Fig. 3.56); the settlement increases with the embankment height in a logarithmic trend (Fig. 3.57). The thaw settlement in permafrost areas with high ice content, such as ice-rich permafrost and ice-saturated permafrost, is significantly higher than that in icy or ice-poor permafrost area, and the thaw settlement in warm permafrost regions is larger than that in cold

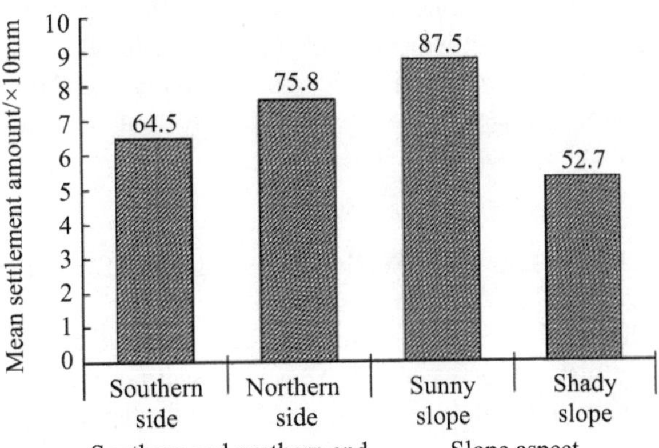

Fig. 3.56 Settlement of northern and southern bridge ends and slope aspect of embankment at transition section of road-bridge (Fig. 3.56 ~ 60, Niu et al., 2011)

Fig. 3.57 Relationship between settlement at the transition section of road-bridge and embankment height

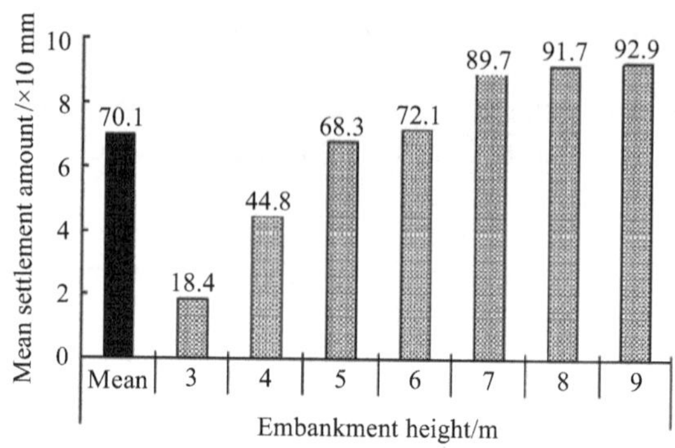

Fig. 3.58 Relationship between settlement and permafrost type and ground temperature

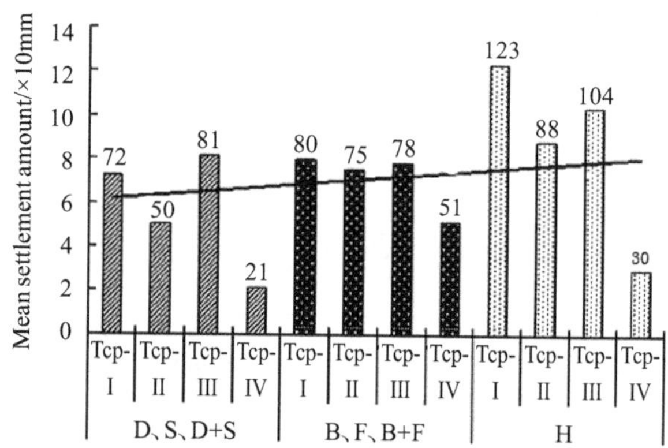

S − Ice-poor frozen soil D − Icy frozen soil
F − Ice-rich frozen soil B − Ice-saturated frozen soil
H − Soil-bearing ice layer

permafrost regions (Figs. 3.58 and 3.59). The embankment structure also has an impact on thaw settlement in the transitional section, with relatively smaller thaw settlement of special embankment structures. The thaw settlement of transition section with fine-grained sections such as silt and silty clay is larger than that with other lithological sections such as gravel soil (Fig. 3.60). The correlation analysis shows that the thaw settlement in the transition section is mainly affected by warm permafrost, warm permafrost with high ice content, and thermal impact of sunny slop. The treatment measures for the transition section are mainly to protect permafrost and prevent thermal erosion.

3.3.4.1 Transition Section of Filling and Excavation

The shallow excavation sections, zero sections, and low filling sections with a certain distance between the excavation section and the filling section belong to the design of longitudinal transition section for the filling-excavation transition section,

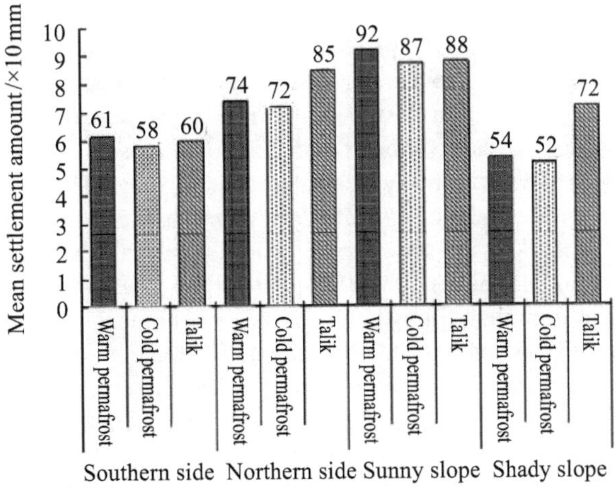

Fig. 3.59 Relationship between settlement in different embankment orientation, slope aspect, and ground temperature

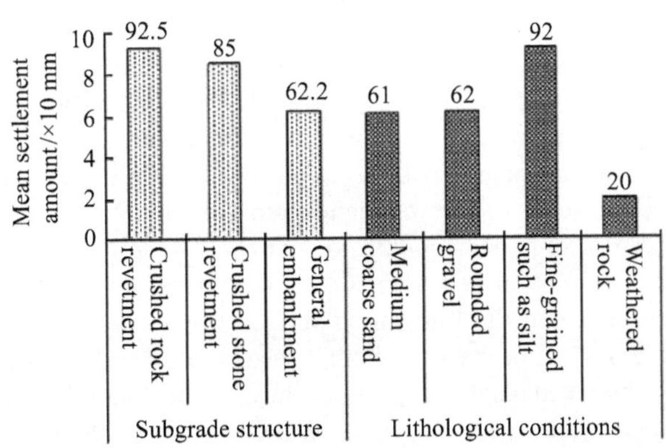

Fig. 3.60 Relationship between settlement at the transition section of road-bridge and subgrade structure and lithology

and the continuity of subgrade should be considered. When the minimum filling height of the embankment is equal to the critical height of the embankment (or equal to 1.5 m based on embankment height in the cold permafrost area) it can be considered as the starting point (or end point) of the filling and excavation transition section of the embankment. When the excavation depth of subgrade is equal to the thickness of subgrade bed (or considered as 0.8 m), it can be considered as the end point (or starting point) in subgrade excavation transition section of filling and excavation transition section. The design of the transition section should be based on the design of the excavation section and extends to the filling section. When the section whose filling and excavation height equal to "zero" is the starting point (or the end point), thermal insulation materials should be used as the replacement layer in the foundation. The excavation section should be subject to base replacement, and the replacement thickness can be determined by referring to formula 3.21 or 3.22. The excavation section should be extended to the filling section, and thermal insulation layer should be set in the filling embankment. In addition to the replacement and filling of the subgrade in the excavation section, the embankment of the filling section should also be provided with a drainage slope along the longitudinal direction of the line to drain water toward the filling direction of the road section.

Cross-section design of the transitional subgrade. When the cross slope of the ground surface is greater than 1:3, the cross-section of the embankment base should be excavated step by step to prevent slippage. The longitudinal step length should be greater than or equal to 2.0 m, and the horizontal width should not be less than 1.0 m. The cross-section direction should not be less than 1.0 m, the step depth should not be less than 0.3 m, and a 2% inward inclined cross slope should be set. The minimum filling height of the embankment should meet the requirements of the critical height of the embankment. When the minimum filling height of the embankment cannot meet the requirements of the critical height of the embankment, the XPS thermal insulation layer can be set. Its thickness should be determined by thermal calculation, but the minimum thickness should not be less than 60 mm. The embedded depth of the XPS thermal insulation layer should be set within the inner structural layer, or other measures regulating and controlling temperature should be adopted to keep the stability of permafrost subgrade.

Figure. 3.61 shows the treatment scheme for filling and excavation transition section of Qinghai-Xizang Railway. For the excavation section, the unstable permafrost layer should be treated by excavating the ice-rich and saturated permafrost below the subgrade surface, and then backfilling the coarse-grained soil, with the 3.5 m depth of the foundation replacement. Meanwhile, the composite geomembrane should be laid at 0.2 m depth below the foundation surface and the XPS thermal insulation board should be laid at the depth of 0.7 m below the foundation surface, and the foundation surface should be widened by 0.6 m. U-shaped side ditches should be set at both sides of transition section, and a 2 m platform should be reserved outside the side ditches. The U-shaped side ditch on the composite geomembrane should be poured with C15 concrete, and a drain hole should be reserved every 2 m, the water-stop ridge outside the cutting top should be filled with the soil layer above the natural table of the original layer. In the filling section, XPS thermal insulation board should be laid 0.7 m below the embankment surface, and the slope

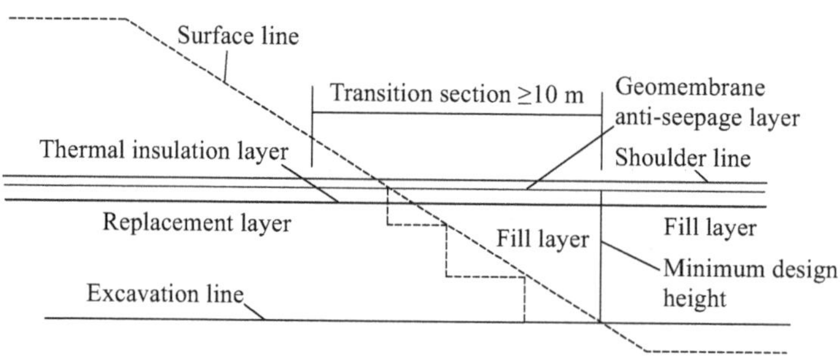

Fig. 3.61 Transition section of embankment-cut

should not be less than 1:10 from replacement to non-replacement within a certain range of zero section. When the embankment height reaches a certain height, no thermal insulation board should be laid.

3.3.4.2 Design of Transition Section Between Embankment and Bridge (Culvert)

The embankment design at the contact section between embankment and bridge (culvert) is the design of transition section of road and bridge (culvert) (Figs. 3.62 and 3.63). The length of transition section of road and bridge (culvert) should not be less than 20 m and not more than 50 m. If the embankment height cannot meet the critical height of embankment, engineering measures of thermal insulation or XPS thermal insulation layer should be adopted. When the foundation depth of the bridge (culvert) is large, the transition section between the embankment and the bridge (culvert) should be replaced or provided with thermal insulation layer to protect the permafrost subgrade. In addition to the thermal insulation layer, the transition section between the embankment and the bridge (culvert) should be replaced with gravel soil, with the content of silt and clay not more than 5% and meeting the requirements of compactness.

In the Qingshuihe River experimental section, the bottom layer of the embankment of the road and bridge transition section is stacked with a certain thickness of crushed rock to form a Crushed rock structure embankment. An inverted trapezoidal transition section is used for filling above the crushed rock layer, and thermal insulation berms are added on both sides of the embankment (Liu et al., 2004) (Fig. 3.62).

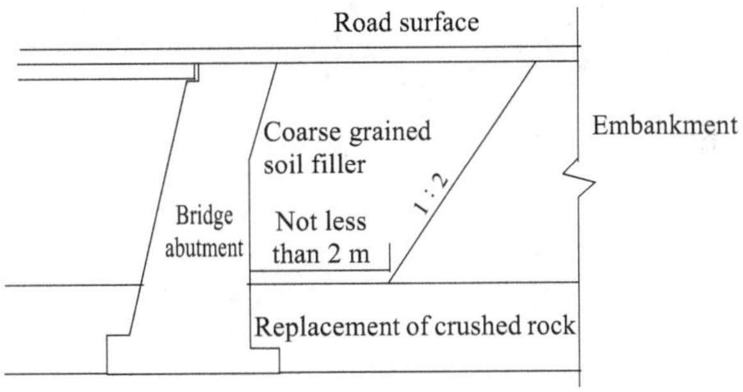

Fig. 3.62 Road and bridge transition section

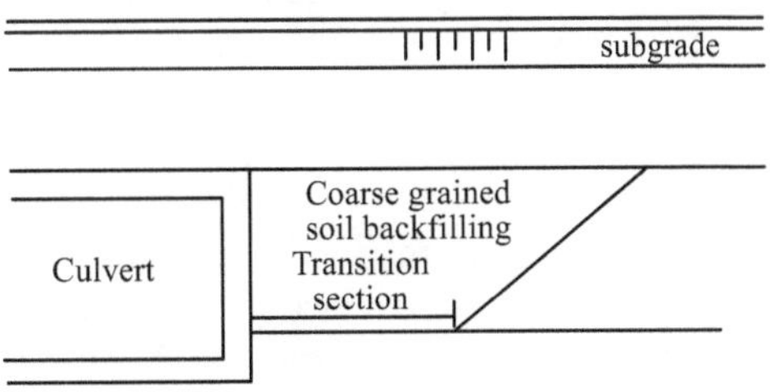

Fig. 3.63 Transition section between road and culvert

3.3.4.3 Embankment Design of Transition Section Between Talik and Permafrost Zone

The embankment design of the transition section between talik and permafrost zone need to prevent soil frost heave (in talik) and thaw settlement (in permafrost areas). Based on this principle, the minimum filling height of embankment should not be less than 1.5 m. It is advisable to use crushed rock structure embankment, or lay waterproof and thermal insulation layer in the filled embankment. The treatment of the transition section should be based on the permafrost section and extended to the talik.

Other transition sections with different engineering geological conditions can also refer to this scheme, which should mainly focus on permafrost with high ice content and extended to the permafrost sections with low ice content.

3.3.5 Subgrade Drainage

The water that erodes the subgrade mainly includes surface water and supra-permafrost water. It is one of the important measures to maintain the stability of permafrost subgrade to take good waterproof and drainage measures in permafrost region.

The subgrade drainage in permafrost regions should be closely combined with the local topography, permafrost characteristics, hydrology and hydrogeological conditions to achieve the combination of drainage and the protection of permafrost environment. The water-stop ridge, drainage side ditch, ground water seepage prevention, bridge (culvert), river and embankment lateral protection form a complete and unimpeded subgrade waterproof and drainage system. No depression should be formed near the embankment, and no water should be accumulated at the toe of the embankment slope.

The drainage side ditch of embankment in permafrost regions should take the combining types of wide-shallow ditches (Fig. 3.64). The size of the drainage side ditch should be determined according to the terrain, permafrost characteristics, catchment area, slope height and other conditions. The distance between the drainage side ditch and the foot of the embankment slope is generally not less than 5 m, and their distance in the permafrost section with high ice content should not be less than 10 m. The longitudinal slope of the drainage side ditch should be consistent with that of the road route, generally not less than 0.3%. The drainage side ditch can be dry masonry side ditch or U-shaped (or trapezoidal) prefabricated concrete assembled side ditch. The structural form of the drainage ditch is consisted by geotextile (with 0.2–0.3 mm thick), coarse-grain filling (with 0.2 m thick), and semi flexible and semi-rigid structural form of prefabricated parts (Fig. 3.64). The drainage ditch should be designed with large longitudinal slope and reinforced to facilitate smooth drainage. The section of drainage ditch should be designed according to the surface runoff conditions, with the bottom width no less than 0.6 m, the depth

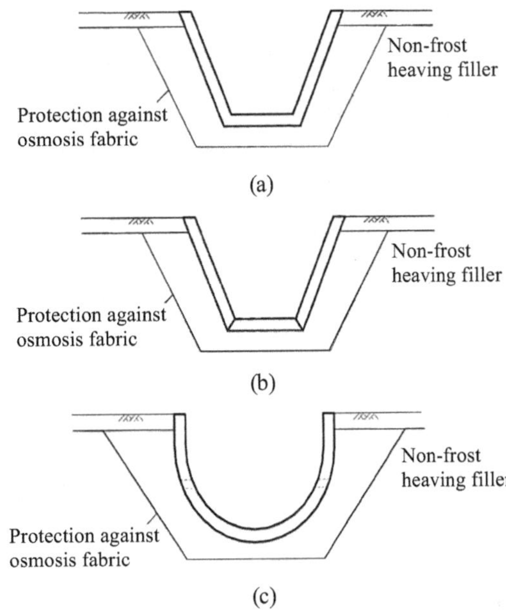

Fig. 3.64 Drainage ditch with precast concrete structure for seepage prevention and replacement. (**a**) Precast concrete trapezoidal drainage ditch; (**b**) Assembled prefabricated concrete trapezoidal drainage ditch; (**c**) Precast concrete U-shaped drainage ditch

no more than 0.4 m, and the slope ratio of 1:1. When the soil is soft plastic and flow plastic cohesive soil and coarse-grained soil containing a certain amount of cohesive soil, the ratio of drainage ditch slope should be 1:1.5 ~ 1:2. And when the soil is peat, the slope ratio should be 1:0.5 ~ 1:1. The bottom of the trench should be paved with 0.2 m thick fine sand gravel layer or coarse sand layer, and a composite geotextile of "two geotextiles and one film" should be added within the fine sand gravel layer, or impermeable film should be laid on the top and bottom, or a thermal insulation layer should be added in the middle to keep heat and prevent water infiltration.

When the side of foundation topography is relatively high or the catchment area on the side of excavation slope is relatively large, the water-stop ridge and intercepting ditch should be set 10 m away from the side above the foundation. The top width of the water retaining dam should not be less than 1.0 m and the height should not be less than 0.8 m. The internal slope rate should be 1:0.5 ~ 1:1 and the external slope rate should be 1:1.5 ~ 1:2. For hillside sections with loose soil and containing more crushed gravels, the water-stop ridge is prone to leakage, so anti-seepage measures should be strengthened, or the size of the water retaining dam should be increased for paving and reinforcement, or anti-seepage geomembranes should be laid outside the water retaining dam and vertical waterproof plates should be embedded. Drainage ditch should be set at the base side to eliminate the erosion of foundation caused by water retaining and ponding at the base side.

When the terrain on both sides of the foundation is relatively flat, the longitudinal and transverse slopes of the line are not large, and the line position is low, a large arc continuous water-stop ridge can be set, and the water-stop ridge can be connected with the culvert and drainage ditch. Waterproof and thermal insulation berms (not less than 5 m) should be set on both sides of the foundation. The height of the general height section is 1.5 m ~ 2.0 m. The ponding section should be 0.5 m higher than the maximum water level. A 4% outward cross slope should be set on the surface to prevent the surface water outside the foundation from approaching and eroding the frozen soil foundation. When the frozen soil environment in the vicinity of the foundation is damaged and there is accumulated water on the roadside, a waterproof and thermal insulation berm with a certain height can be set (0.8 m below the surface is buried with thermal insulation layer) to extrude the accumulated water 5 m ~ 10 m away. The purpose of this method is to reduce the replenishment of the foundation lateral ponding to the basement ground water, and to have a back pressure effect on the foundation slope, which is conducive to the mechanical stability of the foundation.

To prevent the side ditch from collapsing and cracking caused by repeated multigelation and frost heaving of the side ditch soil, non-frost heaving soil should be used as cushion around and at the bottom of the rigid side ditch such as wet masonry.

In the permafrost section with soil-bearing ice layer, it is appropriate to set up water-stop ridge, and try to avoid setting catchwater and drainage ditches. The natural cross-section drainage system should be maintained as much as possible to avoid changing rivers and ditches. When the embankment passes through ravines, it is advisable to "set culverts and bridges in every ditch". In principle, it is not allowed to set culverts and bridges in two ditches, nor is it allowed to let the water from one ditch flow into another ditch along the embankment, causing damage to the permafrost environment and affecting the stability of the foundation.

3.3.6 *Retaining Wall*

The retaining wall is set according to the needs of the project. In case of the following conditions, the retaining wall can be considered.

1. Excavation of slope and road cut slope with cross slope steeper than 1:5, or strengthening the stability of embankment slope.
2. Avoiding or reducing a large number of excavated road cut and high slope sections.
3. Road sections with unfavorable cryogenic phenomena in permafrost regions, such as thick ground ice, thaw slumping and other sections.
4. Road sections with severe water erosion.
5. Protecting permafrost environment or adjacent existing buildings.

The selection of retaining wall types in permafrost regions should comprehensively consider the factors such as engineering geology, hydrogeology, environment, construction conditions and engineering cost, etc. Light and flexible structures with prefabricated assembly construction technology, such as anchor rod retaining wall, anchor plate retaining wall, reinforced soil retaining wall and the cantilever retaining wall containing reinforced concrete. Gravity mortar rubble retaining wall should not be adopted. Light and flexible structure retaining wall can speed up the construction progress, reduce the exposure time of permafrost in pit and reduce the thermal erosion of permafrost foundation. This retaining wall has a strong adaptability to deformation and can reduce the effect of horizontal frost heave force. The gravity retaining wall has the worst adaptability to deformation and has great restraint on frost heave. Under the horizontal frost heave force of the wall back soil, it is quite easy to damage the retaining wall.

Generally, the retaining wall needs a good foundation, such as ice-poor and icy permafrost. Otherwise, the foundation bottom should be buried at least 0.25 m below the stable artificial permafrost table. In general, the buried depth of the foundation should not be less than 1.3 times of the permafrost table under natural surface at the building site. If the foundation is on ice-rich and saturated permafrost, the bottom of the foundation should be buried at least 0.5 m below the artificial permafrost table, and a sand gravel cushion with a thickness of not less than 0.3 m should be laid under the bottom, with 0.5 m wider than the bottom of the foundation in order to make the foundation stress uniform and prevent the permafrost from thawing due to local stress concentration (Fig. 3.65). The long-term strength of the soil-containing ice layer is very small, decay creep occurs, even large settlement deformation occurs under the action of external load. Therefore, replacement treatment should be carried out to reduce the stress acting on the soil- containing ice layer. The depth of replacement soil should not be less than the foundation width.

The back of the retaining wall needs an insulation layer with sufficient thickness. Industrial insulation materials such as EPS or XPS can be used, and other insulation materials can also be used. The thermal insulation material laid on the wall back is

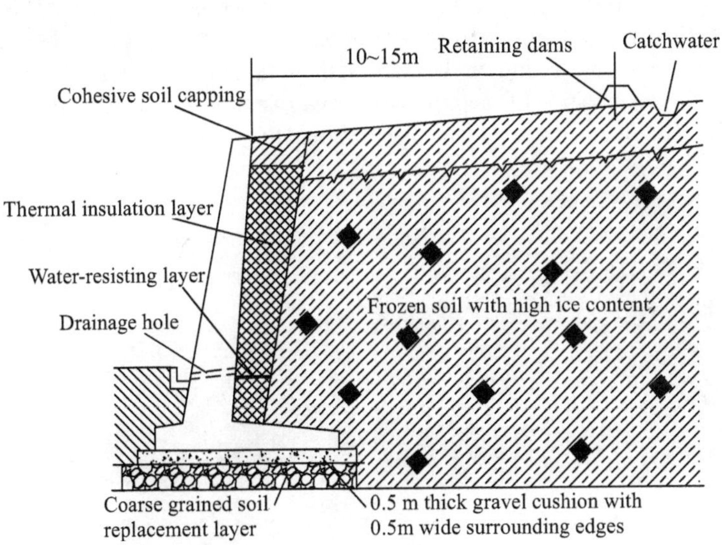

Fig. 3.65 Structure of thermal insulation retaining wall

mainly to protect the permafrost with high ice content on the wall back from melting. The thickness of the insulation layer can be determined by the calculation of the insulation layer. The total thickness of the insulation layer and the retaining wall should be greater than the maximum seasonal thawing depth. If cohesive soil is used as the insulation layer, the wall back should be filled with 0.8 m thick gravel layer to avoid frost heaving force of the retaining wall caused by freezing of cohesive soil. The retaining wall in the road cut section can be set according to the relevant measures of the road cut design (Fig. 3.66).

The construction of supporting and retaining structures in permafrost region often has a large free face due to slope excavation, which on the one hand makes the slope lose mechanical stability and on the other hand makes the permafrost exposed to losing thermal stability. The retaining wall structure needs to address thermal and mechanical problems directly. However, the soil at the back of the retaining wall in permafrost regions is easy to form a new freezing front during the freezing process, with the earth pressure absence, resulting in frost-heave stress of soil (Fig. 3.67), causing the forward displacement or overturning of the wall. During the thawing period, the frozen layer thaws, the frost heaving force gradually disappears and the earth pressure gradually increases. The freezing and thawing process of the soil at the back of the retaining wall is an alternating cycle of earth pressure and frost heaving force. The test results show that the horizontal frost heaving force acting on the retaining wall is several times to ten times larger than the earth pressure. Therefore, earth pressure and frost heaving force cannot be considered simultaneously. The magnitude of horizontal frost heaving force is not only related to the nature of the wall back fill, but also related to the height of the wall and the degree of constraint on the frost heaving of the fill. The higher is the wall and the greater the constraint and the greater the frost heaving force become. So, there is the variation characteristic of the acting stress on the back of the retaining wall in permafrost regions. This is also the reason why the gravity retaining wall cannot be adapted and is easy to be damaged.

The earth pressure acting on the back of the retaining wall should be calculated according to the position of the artificial permafrost table on the back of the retaining wall. When the permafrost table is gentle and the sliding plane can be formed in

Fig. 3.66 L-shaped retaining wall of Beiluhe River of Qinghai-Xizang Railway (Provided by Zhang Jian)

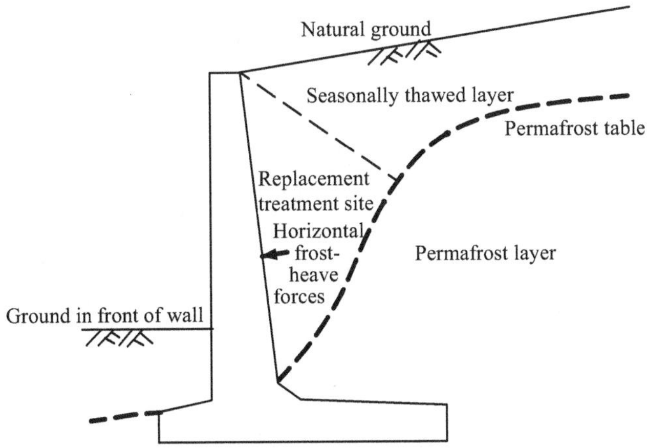

Fig. 3.67 Frost-heavingforce caused by retaining wall back

Table 3.27 *Φ*, *C* Standard values of soils within freeze-thaw transition zone soil value

Soil types	Internal friction angle *Φ*(°)	Cohesion *C*(kPa)
Fine-grained soil	20 ~ 25	10 ~ 15
Sandy soil	25	
Crushed and gravelly soil	30	

the wall back thawed soil layer, it can be calculated according to the Coulomb Theory or the Rankine Theory. When the permafrost table is steep, the thickness of thawed soil layer at the wall back is small, and the sliding plane cannot be formed in the thawed soil layer, the earth pressure should be calculated according to the filling within a limited range. At this time, the permafrost table should be taken as the sliding plane, and the internal friction angle and cohesion of the soil in the freeze-thaw transition zone should be taken to calculate the active earth pressure. When the freeze-thaw interface is difficult to be determined, it can also be calculated according to Coulomb theory.

The internal friction angle and cohesion of the soil within the freeze-thaw transition zone should be determined by the test. If there is no test data, the value should be taken according to Table 3.27. The distribution of horizontal frost-heave stress acting on the wall back should be determined by the field test. In general, the distribution of horizontal frost-heave stress on the wall back is shown in Fig. 3.68.

For coarse-grained fill, it can be assumed that the horizontal frost-heave stress is a right triangle distribution as in Fig. 3.68a. For cohesive soil and silty soil, when the wall height is less than or equal to 3 times of the permafrost table (Z_a), the distribution diagram shows in Fig. 3.68b. When the wall height is 3 times higher than the permafrost table (Z_a), the distribution diagram is showed in Fig. 3.68c. For various

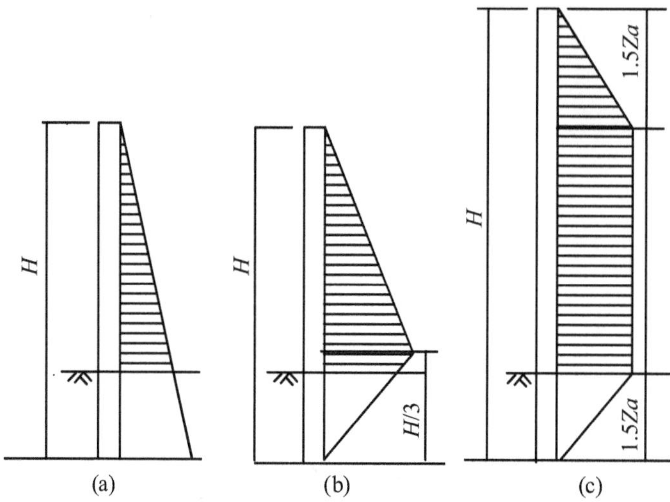

Fig. 3.68 Distribution diagram of horizontal frost-heave force along the wall back. Z_a—Permafrost table in the middle of wall back; H—Height of retaining wall

diagrams, the horizontal frost heave force of the buried part of the foundation may not be considered in the calculation. When the calculated section of the retaining wall is too large, the insulation layer should be laid on the wall back and the wall top ground, and the frost heaving soil in the wall back slope should be replaced with the coarse-grained soil with non-frost-heaving to reduce the horizontal frost heaving force.

According to the observation of some reservoir retaining walls in Heilongjiang Province, there are cracks between the upper part of the retaining wall and the fine-grained filling (Fig. 3.69). Because the water content of the upper part is small, the fine-grained soil at the back of the wall will freeze and shrink at the initial stage of freezing, and it will be separated from the retaining wall, with a depth of about 1.0 m ~ 1.3 m. In this case, the horizontal frost heaving force generated when the wall back filling frozen only appears in the middle and lower part of the retaining wall, and there is a "non-frozen area" in the upper part, which does not generate the horizontal frost heaving force on the wall.

Therefore, the design value of horizontal frost heaving force at the back of the retaining wall should be obtained according to the soil quality of the construction site, surface frost heaving amount, freezing depth, ground water level during freezing period, and maximum horizontal frost heaving force, and then amended to obtain the design value by taking into account the horizontal allowable deflection of the retaining wall structure and other factors. In the engineering design, the standard value of the maximum horizontal frost heaving force in the fully constrained state cannot be directly used for the horizontal frost heaving force of the retaining wall. Because any wall will have some deformation after being stressed, the design value of horizontal frost heave force of retaining wall will change with the wall deformation.

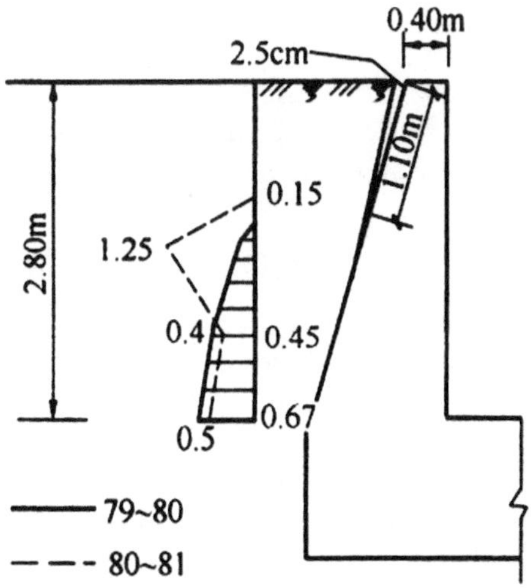

Fig. 3.69 Distribution of horizontal frost-heave force along the wall back (JGJ 118-2011)

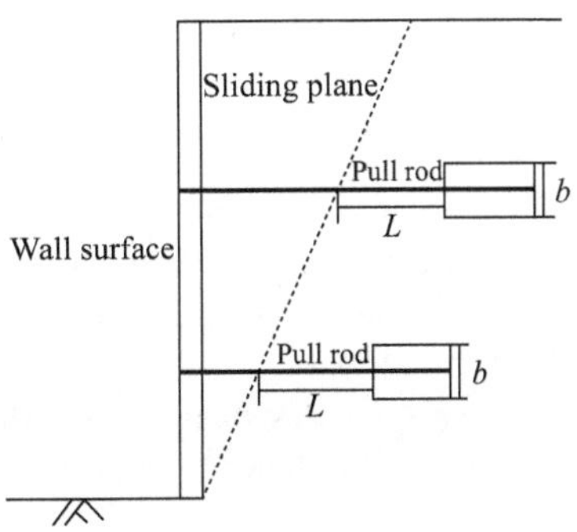

Fig. 3.70 Schematic diagram of anchor plate retaining wall

3.3.6.1 Anchor Plate Retaining Wall

There are few studies and examples of anchor plate retaining wall in permafrost regions. The current experimental research data are also used in hydraulic engineering in seasonally frozen soil. Anchor plate retaining wall is a composite structure composed of wall surface, pull rod, anchor plate and soil between wall surface and anchor plate (Fig. 3.70). In permafrost region, the use of anchor plate retaining structure has its advantages, that is, the high strength of permafrost is used to balance the horizontal frost heave force. However, under the action of small external loads, stress relaxation and creep deformation of permafrost often occur. Therefore, the calculation theory and design method of anchor plate in permafrost region are determined by the rheological characteristics of permafrost. It is necessary to determine the pull-out capacity of anchor rod in permafrost and the horizontal frost heave force acting on the wall back.

The design of anchor plate in permafrost regions is mainly to determine the area and embedded depth of the anchor plate. The area of the anchor plate should be determined according to the design load and the allowable bearing capacity of the permafrost in front of the anchor plate. The depth of the anchor plate embedded under the permafrost table and the length of the pull rod above the permafrost table should be determined according to the position of artificial permafrost table. The design methods of anchor plate in permafrost regions (Ding et al., 2011) are the following:

1. **Determination of Embedded Depth of Anchor Plate**

When the strength of permafrost changes with the depth and the area of anchor plate is fixed, the embedded depth of anchor plate can be changed to meet the requirements of design load. When the strength of permafrost changes, the area of anchor plate can be changed to meet the requirements of design load. Considering the overall stability of the anchor plate, the embedded depth should not be less than a certain limit value, that is, the minimum embedded depth of the anchor plate.

If the overall stability of the anchor plate is destructed, the permafrost and active layer thawed soil in front of the anchor plate will undergo shear failure along the conical surface in Fig. 3.71. At this time, the external load should be balanced with the shear strength of the shear plane, that is:

$$F_1\tau_1 + F_2C = P \tag{3.31}$$

where F_1 is the area of the shear cone surface of the active layer thawed soil, cm^2; F_2 is the area of shear cone of permafrost, cm^2; τ_1 is the shear strength of the thawed soil in the active layer, kg/cm^2; C is the shear strength of permafrost, kg/cm^2; P is the external load, kg.

If the shear resistance of the thawed soil is omitted, the shear resistance of the permafrost is balanced with the external load. For the circular anchor plate:

$$\pi\left(R_1 + R_2\right)L \cdot C \cdot \cos\alpha = P \tag{3.32}$$

$L = \dfrac{H}{\cos\alpha}$ is substituted into:

$$\pi\left(R_1 + R_2\right)L \cdot C = P \tag{3.33}$$

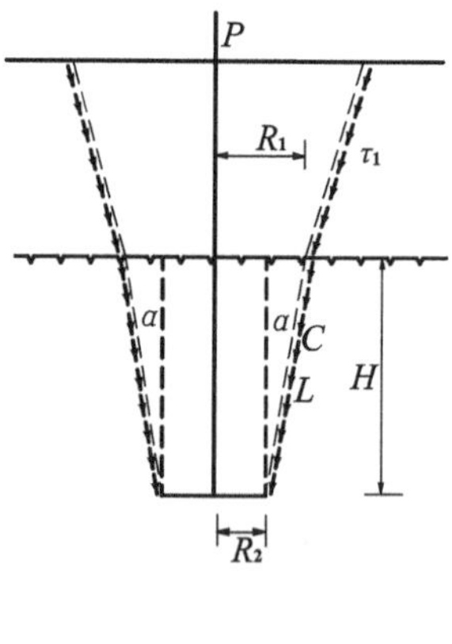

Fig. 3.71 Calculation diagram of minimum depth of anchor plate

where R_1 is the radius of the top surface of the permafrost shear frustum of a cone; R_2 is the radius of the anchor plate; H is the length of the busbar of the shear frustum of a cone; α is the stress diffusion angle of permafrost.

Substituting $R_1 = R_2 + H \times \tan\alpha$ into the formula (3.33) to obtain:

$$H^2 \cdot \tan\alpha + 2R_1 H - \frac{P}{\pi C} = 0 \tag{3.34}$$

The formula (3.34) is the limit state equation of anchor plate. The minimum embedment depth of anchor plate in permafrost can be obtained by solving this equation.

In addition to the minimum burial depth, the design embedment depth of the anchor plate should also consider the interannual fluctuation of the permafrost table and the reliability requirements of the retaining structure.

2. **Allowable Bearing Capacity of permafrost in front of Anchor Plate**

The allowable bearing capacity of permafrost in front of the anchor plate can be determined according to the ultimate load P_K determined by the field test of the anchor plate, namely:

$$[P] = \frac{P_K}{f_K} \tag{3.35}$$

where f_K is the all-coefficient, generally 2.5.

In the absence of field test data, P_K can also be calculated by Prandtl formula according to the ultimate long-term cohesion C_{gi} determined by spherical compression test.

For strip anchor plates: $P_K = (\pi + 2)\ C_{gi}$
For square anchor plates: $P_K = 5.71\ C_{gi}$
For circular anchor plates: $P_K = 5.68\ C_{gi}$

Table 3.28 shows the cohesion and allowable bearing capacity of some frozen soil for reference in the design of anchor plate.

3. **Design and Calculation of Anchor Plate in Permafrost Regions**

The design of anchor plate in permafrost regions is carried according to the first limit state (allowable stress) and second limit state (allowable deformation). When the anchor plate is designed according to the allowable stress, the cohesion and allowable bearing capacity of the permafrost in front of the anchor plate should be determined by the test. If there is no test data, refer to Table 3.28 for values. When the anchor plate in permafrost regions is designed according to the allowable displacement, the compressive stress of the permafrost in front of the anchor plate should be determined according to the design load P, and then it is calculated that the creep displacement S (=2/3 times allowable deformation of the building) within the service life of the building under the stress. Meanwhile the embedded depth of anchor plate should meet the requirements of the minimum embedded depth. In

Table 3.28 Cohesion and allowable bearing capacity of frozen soil (Ding et al., 2011)

Soil name	Rate of water content	Temperature (°C)	Dry density(g/cm³)	Frozen soil structure	Ultimate long-term cohesion (kg/cm²)	Ultimate load (kg/cm²)	Allowable bearing capacity(kg/cm²)	Remarks
The Fenghuo Mountain silty sand clay	24.2	−2.5	1.45	Microlayer	1.33	7.55	3.78	According to the calculation of circular anchor plate, the safety coefficient is 2
	28.6	−2.3	1.37	Microlayer	1.34	7.61	3.81	
	28.4	−1.5	1.39	Microlayer	0.82	4.66	2.33	
	31.7	−2.7	1.34	Microlayer	1.33	7.55	3.78	
	36.2	−2.5	1.32	Microlayer	1.30	7.38	3.69	
	33.8	−2.0	1.33	Microlayer	1.55	8.80	4.40	
The Fenghuo Mountain clay interlayer	21.6	−1.6	1.69	Reticular	1.39	7.90	3.95	
	19.6	−1.5	1.53	Reticular	1.15	6.53	3.27	
The Fenghuo Mountain contains a large amount of ice layer sand clay	78.2	−2.0	1.03	Stratiform	0.98	5.57	2.78	
	119.4	−2.0	0.58	Stratiform	1.16	6.59	3.29	
	118.5	−2.0	0.76	Stratiform	0.80	4.54	2.27	
Beiluhe River clay and sandy clay	33.8	−1.0	1.37	Stratiform	0.61	3.46	1.73	
	33.1	−1.7	1.41	Stratiform	1.08	6.13	3.07	
	37.5	−1.5	1.32	Stratiform	0.72	4.09	2.04	
	44.7	−1.1	1.32	Stratiform	0.73	4.15	2.07	
	38.9	−2.0	1.30	Stratiform	1.45	8.24	4.12	
Qingshuihe River sand clay and clay sand	62.8	−1.5	1.00	Stratiform	1.08	6.13	3.07	
	54.3	−1.5	1.02	Stratiform	1.22	6.93	3.46	
	70.4	−1.8	0.93	Stratiform	1.48	8.41	4.20	
	50.9	−1.8	1.02	Stratiform	1.74	9.88	4.94	

addition, reinforced soil retaining wall or pile plate retaining wall can also be used as the retaining wall in permafrost regions.

3.3.6.2 Measures to Prevent Freezing and Thawing Action of Retaining Wall

The following measures should be taken to prevent freezing and thawing action of retaining wall:

1. Both ends of the retaining wall should be embedded in the undisturbed soil layer. The embedded depth should not be less than 1.5 m for the soil slope, 1.0 m for the strongly weathered rock slope, and 0.5 m for the slightly weathered rock slope.
2. The foundation of retaining wall should be buried 0.5 m below the artificial permafrost table or 1.3–1.5 times of the permafrost table under natural surface on the construction site. In case of permafrost layer with high ice content, 0.3 m sand cushion should be laid under the base.
3. When the accumulated thickness of ice layer in permafrost on the back of the wall is greater than 0.3 m, or the frost heave level of the soil belongs to grade III, IV and V, coarse-grained soil should be used for replacement, and the replacement range should not be less than the range shown in Fig. 3.72, or the insulation layer should be laid.
4. Drainage holes should be set along the wall height and wall length, alternately arranged every 2 m ~ 3 m. A water-resisting layer should be laid at the bottom of the bottom drain hole.
5. When thermal insulation materials are used to prevent the soil behind the retaining wall from freezing, the physical properties of the materials should meet the design requirements, and the laying thickness should be determined by thermal calculation. One way or two-way laying methods should be adopted.

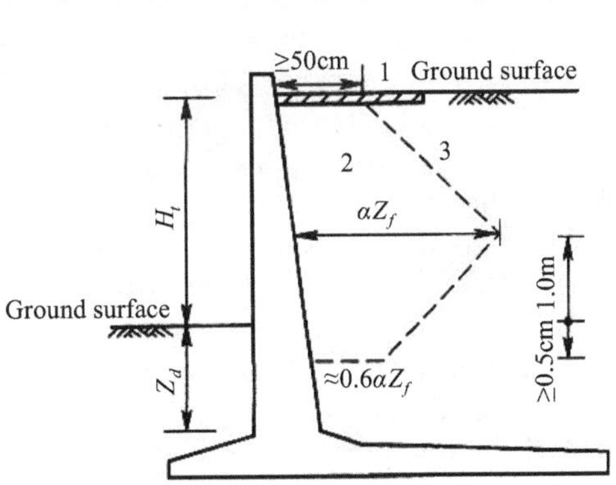

Fig. 3.72 Schematic diagram of replacement range of retaining wall back. 1—Confining bed; 2—Non-frost heaving material; 3—Replacement range line; Z_d—Design freezing depth of soil in front of wall (m); Z_f—Design freezing depth of replacement soil (m); H_t—Filling height behind the wall calculated from the ground in front of the wall (m); α—Coefficient (Table 3.29)

Table 3.29 Coefficient α

Frost heave level of the calculated point soil behind the retaining wall	I, II	III	IV
α	≤0.3	0.3 ~ 0.6	0.6 ~ 0.9

3.4 Subgrade with Special Structure

A large number of engineering damages in permafrost regions are caused by the imbalance of thermal and mechanical stability due to the rise of ground temperature. Therefore, in the engineering design of permafrost regions, regardless of site selection, structure, measures, construction and environmental protection, various methods must be adopted to regulate the thermal balance among the atmosphere, engineering and permafrost, so as to cool permafrost under embankment as much as possible and ensure the thermal stability of permafrost. In addition to the method of replacing roads with bridges, the effect of some special structure embankment in decreasing permafrost temperature has been verified by railway and highway in the world. These measures can better protect the permafrost under embankment, increasing the stability of permafrost and engineering.

3.4.1 Crushed Rock Structure Embankment

There are numerous reports about using crushed rock structure embankment as cooling permafrost methods. The Soviet Union widely uses large crushed rock (average particle size 20–40 cm) to decrease permafrost temperature and raises the permafrost table under embankment in the Baikal-Amur Mainline, especially in the swamp peat area. The road cut slopes in permafrost regions in Canada are also covered with crushed rock layer, with a thickness of 0.3 m ~ 1.0 m. The railway base is covered with crushed rock layer of 0.8 m ~ 1.2 m thick to improve the construction conditions and protect the permafrost stability. Crushed rock embankment and crushed rock revetment, called "air convection embankment", are widely used for highways in Alaska, with a thickness of 2 m ~ 3 m. The experimental results show that the air density inside the crushed rock layer is uneven, forming convection between warm air and cold air within crushed rock layer. In 1979, the experimental section of crushed rock embankment in the permafrost region was established in Reshui Town, Qinghai Province. The experimental results showed that the average ground temperature at the bottom of crushed rock embankment was much lower than that of the ordinary embankment, and the permafrost table under the embankment was raised significantly. A 3.2 m-thick crushed rock embankment is set in the moss peat swamp area in the northwest of Heilongjiang Province, and the ground temperature under embankment is 0.9 °C ~ 1.7 °C lower than the ground temperature under natural surface in the same section. At the same time, many areas such as the Alps, the East Tianshan Mountains, the Urumqi River Valley, Kuandian County of Liaoning Province, the Daqianling tunnel entrance of the Tianheng railway, Pingquan County of Hebei Province, Hokkaido and the extinct volcano vent of Hawaii have reported that in the environment where the mean annual air temperature is positive, there are permafrost or "cold holes" under the crushed rock layer. In the Kunlun mountain Pass of the Qinghai Tibet Plateau, the mean ground temperature under the 0.15 m thick crushed rock layer is 7.1 °C lower than that under the adjacent fine-grained soil layer. These reports show that the crushed rock layer has an effect of thermal insulation in the warm season and a cooling effect in the cold season.

3.4.1.1 Characteristics and Application of Crushed Rock Embankment

In the cold season in the permafrost region, the air temperature is extremely low, the cold air with large density flows downward and falls through the porous medium of the crushed rock layer and sinks to the embankment base. Under the effect of the density difference and temperature difference of the cold and hot air, the air in the crushed rock layer is driven to generate natural convection, so as to achieve the purpose of protecting the permafrost. On the contrary, during the warm season, the external temperature is high, and the heat is transferred into the embankment only through the embankment surface in the form of heat conduction, heating the air in the pores of the upper layer of the embankment. The hot air with light density is only in the upper layer of the pore, and the cold air with heavy density is in the lower layer of the pore. It cannot form convection, and only conducts heat exchange in the opposite direction of heat flow.

The thermal conductivity of the air is very small, only 0.025 $W \cdot (m \cdot K)^{-1}$, which greatly reduces the heat exchange rate and can only slightly increase the temperature of the upper layer in the embankment. Therefore, the crushed rock layer can enhance the heat release in cold season and reduce the heat absorption in warm season. It plays the role of "thermal diode effect" formed by the characteristics of "heat resistance and cold transmission" (Fig. 3.73). Finally, the mean annual temperature in the crushed rock embankment will drop, and the artificial permafrost table rises. According to Goering's (1996) numerical analysis, compared with the ordinary foundation, the natural convection of crushed rock embankment reduces the mean annual temperature under the embankment by 5 °C.

The crushed rock layer has variable thermophysical characteristics. Under the positive temperature condition, its thermal conductivity is very small, and under the negative temperature condition, it has a large thermal conductivity, which is more than 12 times the thermal conductivity at positive temperatures. According to the calculation of the test data in Fenghuo Mountain area of the Qinghai Xizang Plateau (He et al., 2006), after the crushed rock embankment is filled with 1.3 m thickness, the "cold influx" entering to the embankment in the cold season is 3.7 times of the heat amount in the warm season. Compared with the coarse-grained soil with the

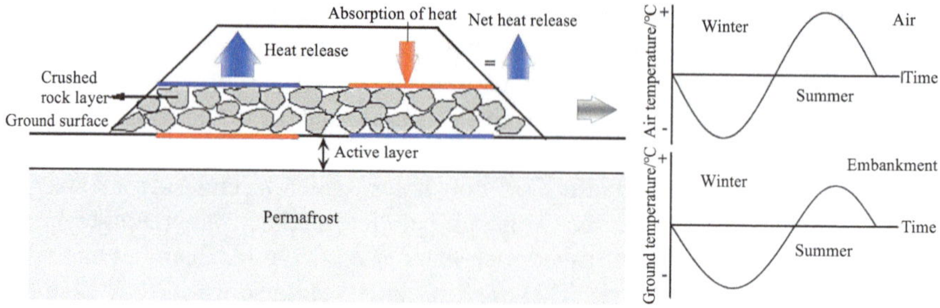

Fig. 3.73 Schematic diagram of cooling mechanism of crushed rock embankment (Wu et al., 2007)

Fig. 3.74 Construction of crushed rock embankment of Qinghai Xizang Railway (left) and National Highway 214 (right) (Left photo by Ge Jianjun, right photo by Fang Jianhong)

same thickness, the heat input from the crushed rock layer in the warm season is only 46% of that from the coarse-grained soil layer, while the cold input from the cold season is 1.12 times of that from the coarse-grained soil layer. So, the crushed rock layer has a certain "heat shielding effect", indicating a "heat insulation but not cold insulation" bulk heat insulation material (Fig. 3.74).

The influence of embankment construction on permafrost is to change the heat balance between ground and atmosphere. By changing the structural characteristics of the embankment engineering and increasing the temperature difference regulation in the embankment, the permafrost under the embankment base is changed from heat absorption to heat dissipation to achieve the purpose of regulating and cooling the permafrost under embankment.

The main factors affecting the thermal stability of crushed rock embankment include climate factor, such as air temperature, wind speed and direction, and dryness, permafrost characteristics such as mean annual ground temperature, frozen soil engineering type, and depth of seasonal thawing, and other environmental factors, such as lithology, terrain (slope direction and gradient) and ground surface characteristics (vegetation). The characteristics of subgrade engineering are the human activity factors that change the thermal state of permafrost, such as

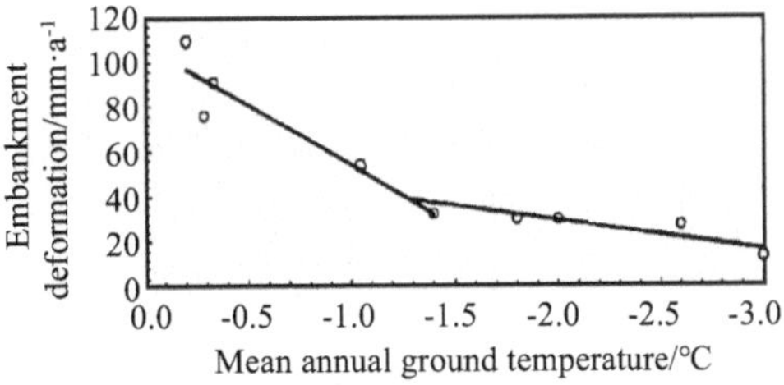

Fig. 3.75 Relationship between mean annual ground temperature and embankment deformation

embankment height, direction and structure (pavement material, particle size and thickness of crushed rock layer).

The mean annual ground temperature of permafrost and its engineering type (ice content) are the essential factors of embankment deformation, while the climatic conditions and embankment engineering characteristics are external factors. The investigation in embankment damage of Qinghai-Xizang Highway shows that most of the serious embankment damages in sections of different frozen soil engineering types occur in warm permafrost with high ice content areas with mean annual ground temperature higher than −1.5 °C, especially in soil-bearing ice layers, ice-saturated permafrost and thick ground ice regions. In areas where the mean annual ground temperature is lower than −1.5 °C, the embankment is relatively stable even in permafrost regions with high ice content, and the embankment damages are relatively fewer (Fig. 3.75) (Liu et al., 2002).

A large number of practices such as Qinghai-Xizang Railway and Qinghai-Xizang Highway have proved that crushed rock embankment can decrease the permafrost temperature and maintain the thermal stability of permafrost under embankment. Table 3.30 shows the engineering effect of crushed rock embankment measures under different permafrost environments. Although the heat storage effect of the embankment construction has not been eliminated, it has produced obvious cooling effects in only 1 year. The annual mean temperature at the bottom of crushed rock embankment is 0.5 °C ~ 1.0 °C lower than that of the comparison section. The observation of 8 sections along Qinghai-Xizang Railway also shows (Table 3.31) that after 2–3 freezing and thawing cycles, the permafrost table under the crushed rook structure embankment has generally raised by 1.4 m ~ 5.3 m. The lower is the mean annual ground temperature of permafrost, the more obvious the cooling trend becomes.

In 2003, comparative observation of crushed rock embankment test section (K3006 + 300) and general soil-filled embankment (K3006 + 600) was also conducted in Wudaoliang area of Qinghai-Xizang Highway. The observation shows that the mean annual temperature of the embankment base (5.7 m below the road surface) is lower than that of the general filled embankment, especially in the cold seasons, both at the center of embankment and at the shoulder of embankment. The mean annual ground temperature at 15 m under embankment is −0.70, −0.73, −0.70 °C for the sunny shoulder, the center and the shaded shoulder of the crushed rock embankment respectively, and −0.49, – 0.46 and −0.48 °C for the general soil-filled foundation of the comparison section. The numerical simulation results show

Table 3.30 Comparison of ground temperature (°C) change at the original permafrost table under crushed rock embankment (Editing Committee of Qinghai-Xizang Railway, 2016)

Location	Permafrost features	Measures	Experimental miles	2004			2005			Extent of decreasing temperature		
				LR	RR	CE	LR	RR	CE	LR	RR	CE
WDL	Soil-bearing ice layer, $T_{cp} = -2.3$ to-2.4 °C	Embankment with crushed rock berm	DK1082 + 625	−0.16	−1.17		−0.68	−1.99		0.52	0.82	
			DK1082 + 675	−0.36	−1.40		−0.66	−1.91		0.50	0.51	
			DK1082 + 725	−0.37	−1.25		−0.73	−1.75		0.36	0.50	
			DK1082 + 825	−0.06	−0.85		−0.16	−1.10		0.04	0.25	
			DK1082 + 775	−0.21	−1.00		−0.36	−1.85		0.15	0.85	
BLH	Soil-bearing ice layer, $T_{cp} = -1.4 \sim -1.7$ °C	Crushed rcok embankment with crushed rock berm	DK1142 + 660	−0.41	−0.89	−0.59						
			DK1142 + 700	−0.49	−1.01	−0.61						
			DK1142 + 530*	−0.32	−0.59	−0.49						
KXL	Soil-bearing ice layer,$T_{cp} = -0.65$ to −0.78 °C		DK1262 + 390	−0.54	−0.91	−0.55	−0.67	−1.43	−0.98	0.13	0.52	0.43
			DK1262 + 430	−0.20	−0.32	−0.30	−0.28	−0.85	−0.50	0.08	0.53	0.20
			DK1262 + 530*	−0.08	−0.17	−0.07	−0.12	−0.19	−0.10	0.04	0.02	0.03

Notes: * stands for comparison test section. LR means sunny shoulder, RR means shaded shoulder, and CE means center of embankment. WDL is Wudaoliang, BLH is Beiluhe, and KXL is Kaixinling

Table 3.31 Change of permafrost table under construction of crushed rock embankment of Qinghai-Xizang Railway (2002 to 2004) (Editing Committee of Qinghai-Xizang Railway, 2016)

Design miles	Location	GT zone	T_{cp}/°C	PET	EH/m	ES	PT/m Nature	PT/m Artificial	PT change/m
DK1102 + 000	Kekexil	IV	−2.4	D,F,H	3.6	UCRE	1.6	1.51	+3.7
DK1160 + 592	Fenghuo Mountain	IV	−2.1	F,B,D	6.55	CRBE	1.5	2.75	+5.3
DK1053 + 600	Chumaerhe high-plain	III	−1.5	D,B,F	3.3	UCRE	2.6	3.9	+2.0
DK1141 + 374	Beiluhe Basin	III	−1.36	H	4.84	CRRE	2.0	3.64	+3.2
DK1191 + 770	Wuli Basin	II	−0.54	F,B	3.7	CRBE	2.8	4.3	+2.2
DK1272 + 120	Kaixinling	II	−0.8	F,H	3.6	CRRE	2.9	5.1	+1.4
DK1273 + 455	Kaixinling	I	−0.5	F,H	4.0	CRRE	3.0	5.0	+2.0
DK1297 + 930	Buquhe	I	−0.34	B,D	3.0	UCRE	2.4	3.4	+2.0

Notes: D, F, B, H is icy permafrost, Ice-rich Permafrost, Ice-saturated permafrost, Soil-bearing ice layer, respectively.UCRE means U-shaped Crushed cock embankment; CRBE means Crushed Rock-based Embankment, CRRE means Crushed rock revetment embankment. Tcp means mean annual ground temperature. I means extremely unstable warm permafrost, $T_{cp} \geq -0.5$ °C,II means Unstable warm permafrost, $-1.0 \leq T_{cp} < -0.5$ °C, III means basically stable cold permafrost, $-2.0 \leq T_{cp} < -1.0$ °C and IV means stable cold permafrost, $T_{cp} < -2.0$ °C

that the ground temperature under crushed rock embankment is obviously lower than that under general soil-filled embankment.

3.4.1.2 Selection of Particle Size of Crushed Rock

Crushed rock layer is a porous medium with high permeability. According to the porous media seepage theory, the discriminating parameter of natural convection in porous media is Rayleigh value R_a, which is expressed as (Kong, 1999):

$$R_a = \frac{\rho g C \beta K H \Delta T}{\mu \lambda} \tag{3.36}$$

where C, β, μ are the volumetric thermal capacity, volumetric thermal expansion coefficient and kinematic viscosity of the fluid in the pores; g is the gravitational acceleration; K is the medium permeability; H is the thickness of the medium layer; $\triangle T$ is the temperature difference between the bottom and the top of the medium (applicable to the temperature of the bottom plate higher than that of the top plate); λ is the equivalent thermal conductivity of the medium.

The study shows that there is a critical Rayleigh number R_{ac}. When $R_a < R_{ac}$, the crushed rock layer is in a stable simple heat conduction state, while when $R_a \geq R_{ac}$, natural convection will occur in the crushed rock layer. Therefore, R_{ac} becomes the criterion of whether natural convection can occur, and its size depends on the geometric status and boundary conditions of the crushed rock layer. The critical Rayleigh number R_{ac} for a three-dimensional rectangular porous medium region with a length to height ratio of h_1, a width to height ratio of h_2, six sides closed (impermeable), upper and lower boundaries at defined temperature (the lower boundary is higher than the upper boundary), and side insulation is:

$$R_{ac} = \left(b + \frac{l^2}{b}\right)^2 \pi^2 \tag{3.37a}$$

$$b = \left[\left(\frac{m}{h_1}\right)^2 + \left(\frac{n}{h_2}\right)^2\right]^{1/2} \tag{3.37b}$$

where l, m, and n are the number of cells in the vertical height, horizontal length and horizontal width directions of the porous medium respectively.

Obviously, from formula (3.41a), the minimum critical Rayleigh number R_{ac} is $4\pi^2$. With the increase of the length to height ratio h_1 or the width to height ratio h_2, the minimum R_a value rapidly tends to be $4\pi^2$. Except for the thin and high cuboid, that is, h_1 or h_2 is far less than 1, the side wall has little influence on the critical Rayleigh number. When the width to height ratio h_2 is greater than 1, the minimum R_a is close to $4\pi^2$.

The modeling experimental results show that the temperature change under the center of the bottom of the crushed rock layer with different particle sizes decreases with time under the top sealing condition (Lai et al., 2006). The temperature fluctuation range is all below 0 °C, and the temperature with the particle size of 221 mm is the lowest. Compared with the average temperature of the 8th cycle when the temperature tends to be stable, the cooling effect of the crushed rock layer with the mean particle size of less than 100 mm is significantly worse than that of the crushed rock layer with the mean particle size of more than 100 mm. The bottom temperature of the crushed rock layer with the mean particle size of 221 mm is the lowest, indicating that its cooling effect is the best (Figs. 3.76 and 3.77).

Table 3.32 shows the average periodic temperature of the bottom center of the 8th period of each particle size crushed rock. It can be seen from the table that the temperature at the bottom of the crushed rock with an average particle size of 221 mm is the lowest, which is −1.57 °C, 0.52 °C and 0.19 °C lower than the temperature at the bottom of the crushed rock layer with an mean particle size of

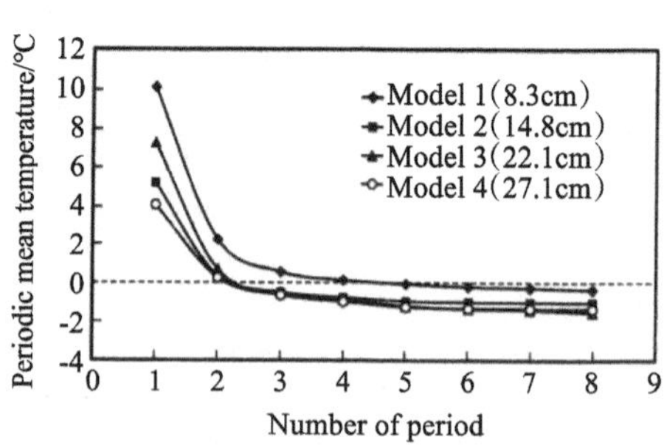

Fig. 3.76 Change of mean soil temperature in the center bottom of the closed crushed rock layer with time

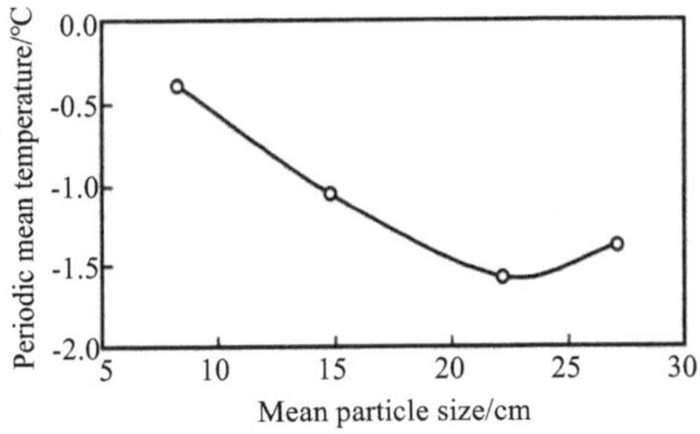

Fig. 3.77 Mean soil temperature in the center bottom of closed crushed rock layer in the 8th cycle

Table 3.32 Mean periodic temperature at the center bottom of the closed crushed rock layer in the 8th cycle

Mean particle size/mm	83	148	221	271
Mean temperature/°C	−0.39	−1.05	−1.57	−1.38

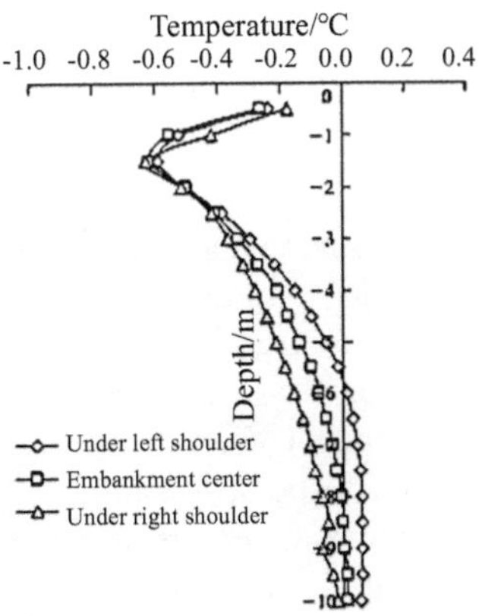

Fig. 3.78 Change of mean soil temperature at the center bottom of open crushed rock layer with time

148 mm and 271 mm. This indicates that the cooling effect of the crushed rock layer with a particle size of about 220 mm is the best. In other words, the crushed rock layer with the particle size of 200 mm to 300 mm can achieve good cooling effect.

Under the open condition at the top of crushed rock layer, the temperature change at the bottom of the crushed rock layer with different particle sizes is relatively stable, but the periodic temperature range is large, fluctuating above and below the 0 °C. The larger is the particle size, the greater the periodic temperature range becomes. The fluctuation for the particle size of 83 mm is the smallest, but basically above the 0 °C. From the mean periodic temperature at the bottom center of the crushed rock layer with time (Fig. 3.78), when the average particle size of the crushed rock is 83 mm, the mean temperature at the bottom center always decreases very little (Table 3.33), which is obviously inferior to the state under the top sealing condition. Under the condition of open top, the crushed rock layer is easily affected

Table 3.33 Temperature periodic range at the center bottom of open crushed rock layer

Mean particle size/mm	83	148	221	271
Mean temperature/°C	8.85	16.26	17.78	19.08

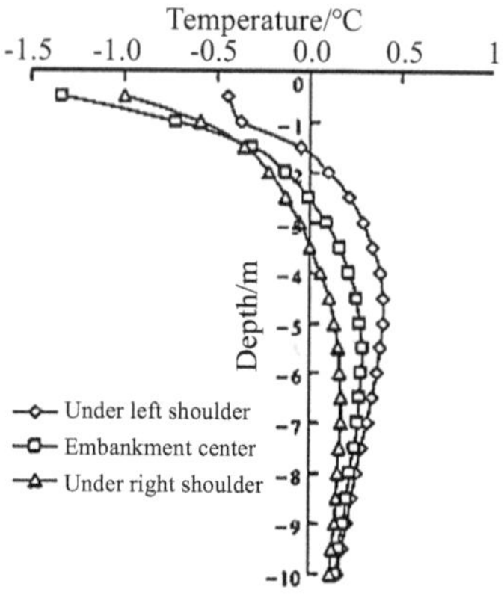

Fig. 3.79 Cooling of soil under crushed rock embankment under open and closed conditions

by external factors. Forced convection is the main internal part of the crushed rock layer, and natural convection is in a secondary position or difficult to occur.

The cooling roadbed effect of crushed rock embankment in permafrost sections with soil-bearing ice layer was experimentally carried out in the Beiluhe River of the Qinghai-Xizang railway. The top width of the crushed rock embankment is 7.6 m, the bottom width is 18.6 m, with the slope ratio of 1:1.5. The lower part of the embankment is filled with a crushed rock layer with a particle size of 200 mm ~ 300 mm (containing 10% of the crushed rock with a particle size of 100 mm and 400 mm) and a thickness of 1.2 m. The upper part is filled with 2.5 m coarse-grained soil after laying a geotextile. The mean annual ground temperature in the non-railway test section is −1.4 °C, the mean annual temperature is −5.2 °C, the average wind speed is 4.1 m/s, and the open and closed crushed rock structure embankment and soil-filled embankment are set for 30 m long each. The closed state is that both sides of the shady and sunny slopes of the crushed rock are covered with 0.2 m thick soil-filled. The observation data show that the influence depth of the crushed rock embankment on the soil thermal regime under embankment can reach 6.0 m ~ 10.0 m in the open state and only 1.5 m ~ 3.0 m in the closed state (Fig. 3.79) (Wu et al., 2006). These results shows that the forced convection within the open crushed rock layer was found, however, forced convection has not been found but only natural convection within the close crushed rock layer. It is difficult for the laboratory simulation to achieve the field effect. Therefore, the crushed rock embankment along the Qinghai-Xizang Railway is covered by aeolian sand, which will have a certain impact on the cooling effect of the crushed rock embankment.

Table 3.34 Permafrost table rising for crushed rock embankment along Qinghai-Xizang Railway (Wu et al., 2005a, b, c)

Location	MAGT/°C	PT/m	Embankment parameters/m		APT/m	Change of PT/m
			Soil-filled height	Embankment height		
Chumaerhe	−1.48	2.60	2.1	3.1	3.90	+1.80
Xieshuihe	−0.72	2.45	5.1	6.3	6.15	+2.60
Kekexili	−2.64	1.60	2.4	4.0	3.40	+1.60
Beiluhe basin	−1.48	1.90	2.3	4.5	1.50	+1.90
Fenghuo Mountain	−2.23	1.50	3.5	5.7	6.45	+1.70
Wuli Basin	−0.50	2.80	2.5	3.7	4.60	+1.90
Buquhe Basin	−0.34	2.40	1.8	3.0	3.60	+1.80

Many railway sections in the permafrost region with high ice content adopt the crushed rock structure embankment along Qinghai-Xizang Railway. The particle size of the crushed rock is 200 mm ~ 400 mm and the thickness is 1.2 m ~ 1.5 m. The upper part is filled with a certain thickness of gravel soil and other A-level filling layers, and the thickness varies with the embankment height. The minimum embankment soil-filled layer is not less than 0.8 m. Through soil temperature observation of 2 ~ 3 years at 6 monitoring sites of crushed rock embankment, preliminary results have been obtained on the ground temperature state, permafrost table change and "cold capacity" accumulation under crushed rock embankment in permafrost region along Qinghai Xizang Railway (Table 3.34). After the construction of the crushed rock embankment, the permafrost table has been greatly raised, with a range of 1.7 m ~ 2.6 m. Obviously, the crushed rock structure embankment exerts cooling effect and suppresses the temperature rise of the soil under the embankment.

The experimental results in the Beiluhe River show that the crushed rock with the particle size of 200–300 mm is superior to the crushed rock with the particle size of 100 mm in terms of the ground temperature under the embankment and the rising extent of permafrost table under embankment.

After the optimal particle size is selected for the crushed rock embankment, the particle size, pore size and porosity within the crushed rock will change through long-term use, resulting in change of the long-term cooling effect. Additionally, the long-term effect of convection heat transfer of crushed rock embankment is affected by the crushed rock and its weathering degree, permafrost environment, accumulation of aeolian sand and other factors.

3.4.1.3 Thickness of Crushed Rock Layer

The main parameter affecting the natural convection generated by the crushed rock embankment is the Rayleigh number (R_a). In formula 3.40, H is the actual filling height of the crushed rock layer, ΔT is the temperature difference between the lower boundary and the upper boundary of the crushed rock layer (applicable to the bottom plate temperature higher than the top plate temperature). Only when $R_a \geq R_{ac}$, the air in the voids of the crushed rock layer can start to generate natural convection, which makes the heat transfer effect from the lower boundary of the crushed rock layer to the upper boundary greater than the simple heat transfer. R_{ac} becomes an important control parameter for the generation of natural convection, and its extent of change depends on the shape and boundary conditions of the crushed rock layer.

For the crushed rock embankment, under the given reasonable design height of the embankment, the basic requirements in the structure layer of the highway pavement should be met, and the cooling effect of the crushed rock layer in winter should also be ensured. The crushed rock layer in the embankment should meet a certain thickness. At this time, whether natural convection can occur in the crushed rock layer depends entirely on the temperature difference between the upper and lower boundaries (ΔT). For example, height (h) and temperature difference of crushed rock layer ΔT must reach a certain value so that the Rayleigh number R_a can be greater than the critical R_{ac} for generating natural convection, which can be seen from formula (3.38).

$$H\Delta T \geq \frac{\mu\lambda}{\rho g\beta KC} R_{ac} \tag{3.38}$$

If the height (H) of the crushed rock layer is given, the critical temperature difference (ΔT_c) of natural convection occurs can be expressed as:

$$\Delta T_c = \frac{\mu\lambda R_{ac}}{\rho g\beta KC}\frac{1}{H} = \frac{M_p}{H} \tag{3.39}$$

Only if $\Delta T \geq \Delta T_c$, the natural convection effect will occur in the crushed rock layer.

Figure 3.81 shows the specific changes of temperature difference between the lower and upper boundary of the crushed rock layer embankment in permafrost regions (Sun et al., 2006a, b, c). Only when temperature difference (ΔT) is greater than the critical temperature difference (ΔT_c) has an impact on the strength of natural convection effect of the crushed rock layer and determines the duration of natural convection effect (vertical line in the Fig. 3.81). Where, t_1 is the start time of natural convection in the crushed rock layer in a period, and t_2 is the end time of natural convection. Obviously, t_1 and t_2 are the two real roots of the following equation in the negative temperature half period.

$$\frac{M_p}{H} = T(0, -H, t) - T_b \tag{3.40}$$

In order to reflect the characteristics that the larger the surface temperature drop and the longer the duration, the stronger natural convection effect can be defined as accumulated temperature:

$$A_t = \int_1^{t_2} \left[T(0, -H, t) - T_b - \Delta T_c \right] dt \tag{3.41}$$

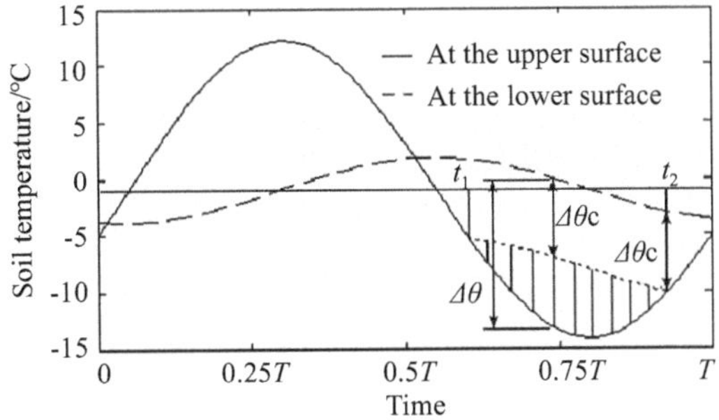

Fig. 3.80 Soil temperature difference caused by natural convection in crushed rock layer

where A_t is the accumulated temperature index to measure the intensity of natural convection cooling effect of crushed rock layer, which is called natural convection index, °C·d.

The index of natural convection (A_t) generated by the crushed rock embankment in winter should be greater than zero. In order to make the convection effect as large as possible, its value should be as large as possible. Figure 3.80 shows the variation law of natural convection index at the crushed rock layer with the height under given specific boundary conditions (Sun et al., 2006a, b, c). When H is small, the accumulated temperature index remains zero, and the crushed rock layer has a minimum height H_{min}. When $H>H_{min}$, the natural convection index will be greater than zero, and then, the index will also increase sharply with the increase of H. When H reaches a certain value, it enters a slow change stage. When H increases to the lower boundary $T(0,-H,t)$ of the crushed rock layer and the fluctuation approaches 0, it reaches the maximum value, that is $0 \leq A_t \leq A_{max}$. The natural convection index of the crushed rock embankment in permafrost regions varies with the height of the crushed rock layer in three areas: zero area, rapid increase area and slow change area (Fig. 3.81). Zero area reflect that natural convection within the crushed rock layer does not occur, rapid area shows that natural convection is rapidly formed, and slow change means that natural convection is slowly increased.

It can be seen that only the area with large temperature gradient within the crushed rock layer has significant natural convection effect. In view of the attenuation of temperature in the process of propagation, the temperature gradient at the lower part of the crushed rock layer has been very small, which does not play a great role in enhancing the natural convection effect. Therefore, as long as the height of the crushed rock layer meets 0.8 times of the natural convection index, it can be positioned as the maximum height value H_{max} of the crushed rock layer.

$$H_{\max} = H\Big|_{A_t = 0.8A_{t\max}} \tag{3.42}$$

Under the environmental conditions of permafrost regions, based on the natural convection cooling effect in the crushed rock layer, the filling height of the crushed rock layer only needs to take an appropriate value, that is:

$$H_{\min}\ ''\ H\ ''\ H_{\max} \tag{3.43}$$

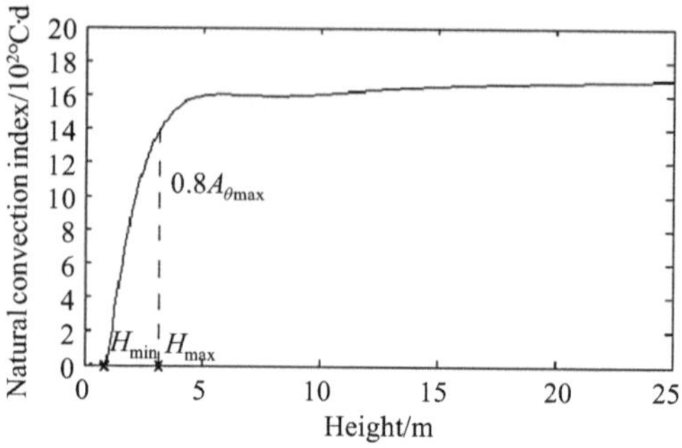

Fig. 3.81 Change of natural convection index with the height of crushed rock layer

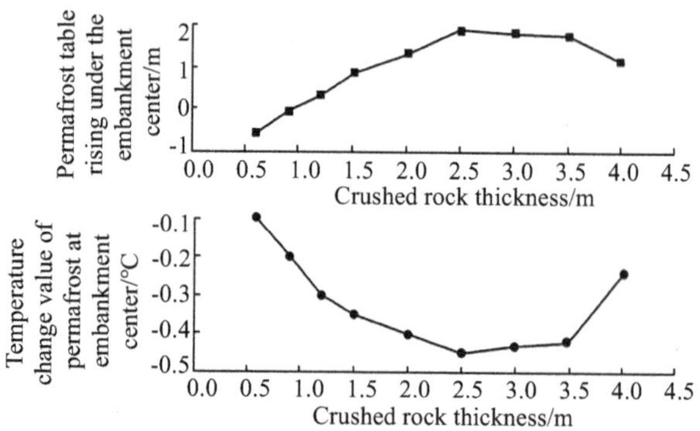

Fig. 3.82 Change of permafrost table and permafrost temperature at the depth of 15 m with the thickness of crushed rock embankment

According to the results of numerical simulation (Wang et al., 2006a, b, c), when the height of embankment is 3.5 m and the thickness of crushed rock layer is 0.6 m, the permafrost table under the center of embankment will decline, but the permafrost temperature at the deep of 15 m will decrease. The thickness of the crushed rock layer is 0.6–1.5 m, with the maximum thickness of the crushed rock layer, the high-temperature area within the crushed rock layer is gradually divided into upper and lower parts by the crushed rock layer. When the high-temperature area in the crushed rock layer is reduced, its range is also gradually reduced, and its temperature is gradually decreased. It shows that the convective cooling capacity of the crushed rock layer gradually increases with thickening the crushed rock layer. This trend can be continued until the thickness of the crushed rock layer is 2.5 m. If the thickness of crushed rock layer is more than 2.5 m, the permafrost table will decrease, and the deep permafrost temperature will also increase (Fig. 3.82). Therefore, under the simulated embankment structure, the optimal thickness of the crushed rock layer is 2.0 m ~ 3.5 m. When the thickness is 2.5 m, the permafrost table under the embankment center rises to the highest (Fig. 3.83). According to the experience of Qinghai-Xizang railway and Qinghai-Xizang Highway, the thickness of the crushed rock layer should range from 1.2 m to 1.5 m.

Fig. 3.83 Permafrost table with different thickness of crushed rock layer

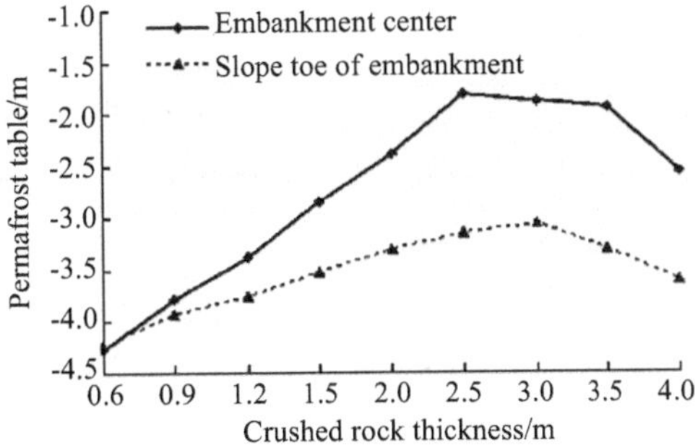

3.4.1.4 Paving of Crushed Rock Layer

According to formula (3.40), under certain engineering environmental conditions, the temperature difference (ΔT) between the upper and lower surfaces of the crushed rock layer determines the internal air condition. According to the exponential attenuation law of the ground temperature decreasing with the depth, it is gradually transmitted to the surface of the crushed rock layer in the embankment. Under the condition of meeting the requirements of the embankment structural layer and mechanical stability, the setting position of the crushed rock layer in the embankment should be upward as much as possible to reduce the thickness and to enhance the cooling effect. In this case, the thickness of overlying soil-filled layer and the local climate environment often affects the cooling effect of the crushed rock layer. The lower is the mean annual temperature, the larger is the freezing index, and the more favorable the cooling effect of the crushed rock layer becomes.

Under the condition that the mean annual surface temperature is −4 °C, and the thickness of the crushed rock layer is fixed at 1.5 m, the change in the permafrost table and permafrost temperature at the depth of 15 m under the center of the embankment with different thickness soil-filled layer is numerically simulated (Wang et al., 2011). The permafrost table under the embankment center rises with the increase of soil-filling thickness on the crushed rock layer, however, the permafrost temperature at the depth of 15 m shows obvious changes (Fig. 3.84b). When the thickness of the soil-filling on the crushed rock layer increases from 1.0 m to 2.5 m, the decreasing extent of the permafrost temperature at the depth of 15 m is increasing. When the thickness of the overlying soil-filling is greater than 3.5 m, the permafrost temperature at the depth of 15 m is decreasing. These results indicate the crushed rock embankment with the overlying soil-filling of 2.5 m ~ 3.5 m thick is the best cooling effect. So, the thickness of the overlying soil-filling increases can reduce the temperature difference on the top surface of the crushed rock layer, reducing the air convection velocity in the crushed rock layer and weakening the cooling capacity of the crushed rock layer. According to the research results, for the 1.5 m-thick crushed rock layer, the thickness of the overlying soil-filling should not exceed 6 m. Only in this way can the overall cooling capacity of the crushed rock

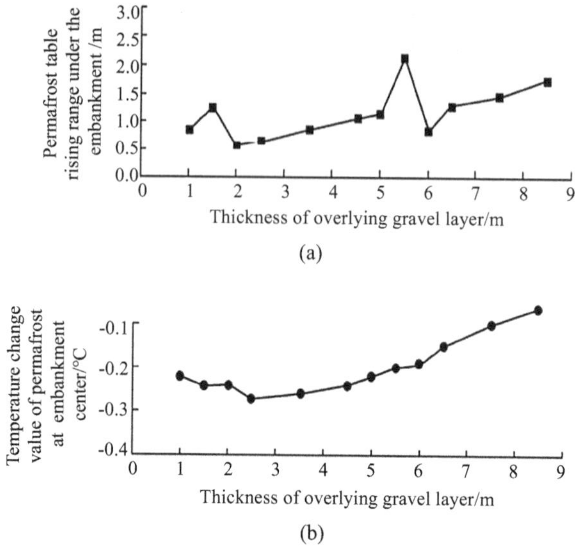

Fig. 3.84 Variation of the permafrost table under the embankment center (**a**) The permafrost temperature at the depth of 15 m (**b**) With the thickness of overlying gravel layer

embankment be ensured so that the permafrost under the embankment will be cooled, and the maximum elevation of the permafrost table can be achieved.

According to the observation data of Qinghai-Xizang Railway in Kaixinling (Table 3.35) (Compilation Committee of *Qinghai Tibet(Xizang) Railway*, 2016), the crushed rock layer should be set at 0.3 m above the natural ground. With different embankment and crushed rock berm heights, the cooling effect of the crushed rock embankment is better than that of the ordinary soil embankment. The ground temperature under the crushed rock structure embankment is 0.5 °C ~ 1.0 °C lower than that under the ordinary embankment. This indicates that increasing the berm height is more conducive to the cooling permafrost under embankment than that increasing the berm width.

Therefore, under the condition that the mean annual temperature of −4.0 °C and the crushed rock layer of 1.5 m thick, when the embankment height is 3–5 m, the crushed rock layer should be set at 0.3–0.5 m above the ground; When the embankment height is higher than 5 m, the thickness of the soil-filling layer above the crushed rock layer should not exceed 4–5 m. If the embankment is higher, the paving height of the crushed rock layer should be raised to keep the thickness of the soil-filling layer 2.5 m ~ 3.5 m. If the mean annual temperature is higher than −4.0 °C, the top surface of the crushed rock layer should be set at 0.3 m ~ 0.5 m below the top surface of the roadbed.

Table 3.35 Mean annual temperature at the embankment base (EB) and original permafrost table under natural surface (PT) for crushed rock structure embankment (CRSE) and ordinary embankment (OE) (°C)

	DK1262 + 390, CRSE				DK1262 + 430, CRSE				DK1262 + 530, OE			
	2004		2005		2004		2005		2004		2005	
	EB	PT	EB	PT	EB	PT	EB	PT	EB	PT	EB	PT
Sunny berm	−2.41	−1.01	−2.13	−1.23	−1.6	−0.43	−1.41	−0.68				
Shaded berm	−0.40	−0.54	−0.63	−0.67	−0.21	−0.20	−0.28	−0.28	0.14	−0.08	0.20	−0.12
Center of embankment	−0.89	−0.55	−1.47	−0.98	−0.46	−0.30	−0.74	−0.50	−0.22	−0.07	−0.13	−0.10
Sunny shoulder	−1.46	−0.91	−2.10	−1.43	−0.70	−0.32	−1.33	−0.85	−0.12	−0.17	−0.12	−0.19
Shaded shoulder	−2.07	−0.98	−2.41	−1.75								

Notes: DK 1262 + 390, height of CRSE is 5.3 m, with the crushed rock sunny and shaded berm of 6 m and 4 m width, respectively, and the soil-filling above crushed rock layer of 3.5 m thick. DK1262 + 430, height of CRSE is 3.9 m, with the crushed rock sunny and shaded berm of 5 m and 3 m width, respectively, and the soil-filling above crushed rock layer of 2.1 m thick

3.4.1.5 Auxiliary Protection Structure

The auxiliary protection structure refers to some auxiliary measures required to ensure the cooling effect of the crushed rock embankment. For example, a layer of geotechnical cloth and gravel layer at the top and bottom of the crushed rock layer is required to prevent the fine-grained soil from entering into the crushed rock layer, blocking the pores, and affecting the air convection in the pores. Meanwhile, gravel berms are laid on both sides of the embankment to reduce the thermal erosion of the slope toe of embankment. In order to reduce the influence of the temperature difference between the shady and sunny slopes, gravel slope protection is required. Additionally, the transition section between the crushed rock embankment and the adjacent embankment should be treated to reduce the uneven deformation. With the construction method of dumping and filling crushed rock, it is difficult to maintain a certain slope rate on both sides of the crushed rock layer. The method of crushed rock suitably stacked can be adopted to maintain the slope rate. However, attention should be paid to maintaining the voids and connectivity of the rock slope.

The experimental results show that the temperature at the bottom of the embankment with widened crushed rock berm is slightly lower than that of the embankment widened with gravel. However, the temperature rise at the bottom of crushed rock berm in warm season is slightly higher than that under the center of embankment in the test of block stone berms with the same particle size and different widths. To reduce the temperature at the bottom of crushed rock berm, it is not advisable to increase the width of crushed rock berm, but to increase the thickness of crushed rock berm. The general trend is that the cooling effect of crushed rock is better than that of ordinary soil-filling embankment, and the crushed rock embankment widened by berm has obvious effect on cooling permafrost and rising permafrost table.

The crushed rock embankment and crushed rock revetment can be combined into a U-shaped crushed rock embankment to enhance the cooling effect of the crushed rock embankment. The observation data of the experimental embankment in the Beiluhe basin on the Qinghai Xizang Plateau show that the U-shaped crushed rock embankment has a mutually reinforcing cooling effect. Soil temperature at the depth of 0.5 m, the depth of original permafrost table, and 5 m depth under the embankment reduced to −3.67 °C, −3.31 °C and −2.36 °C, with the extent of decrease of 1.13 °C, 1.28 °C and 0.77 °C, respectively. The change of permafrost table is also very obvious. From 2004 to 2006, the permafrost table under the U-shaped crushed rock embankment raises 0.9 m ~ 1.4 m, and even enters the embankment body. Its rising extent is 0.86 m larger than that under the ordinary embankment (Table 3.36). Not only the permafrost table under the U-shaped crushed rock embankment raises year by year, but also the soil temperature decreases year by year (Fig. 3.85). The permafrost temperature at the depth of 5 m under embankment obviously decreases, and even the permafrost temperature at the depth of about 10 m still shows a cooling trend. This fully shows that the U-shaped crushed rock embankment has a very significant cooling effect.

Table 3.36 Permafrost table under the ordinary embankment and U-shaped crushed rock embankment (m) (Wu et al., 2010)

Year		2004	2005	2006	2007
Ordinary embankment	Sunny shoulder	2.26	2.08	2.28	2.12
	The center of embankment	2.27	1.40	1.56	
	Shaded shoulder	2.00	0.70	0.84	0.68
U-shaped crushed rock embankment	Sunny shoulder	1.50	0.86	0.64	
	The center of embankment	1.42	0.64	0.20	0
	Sunny shoulder	1.20	0.62	0.10	*

Notes: * means permafrost table rising to the crushed rock layer

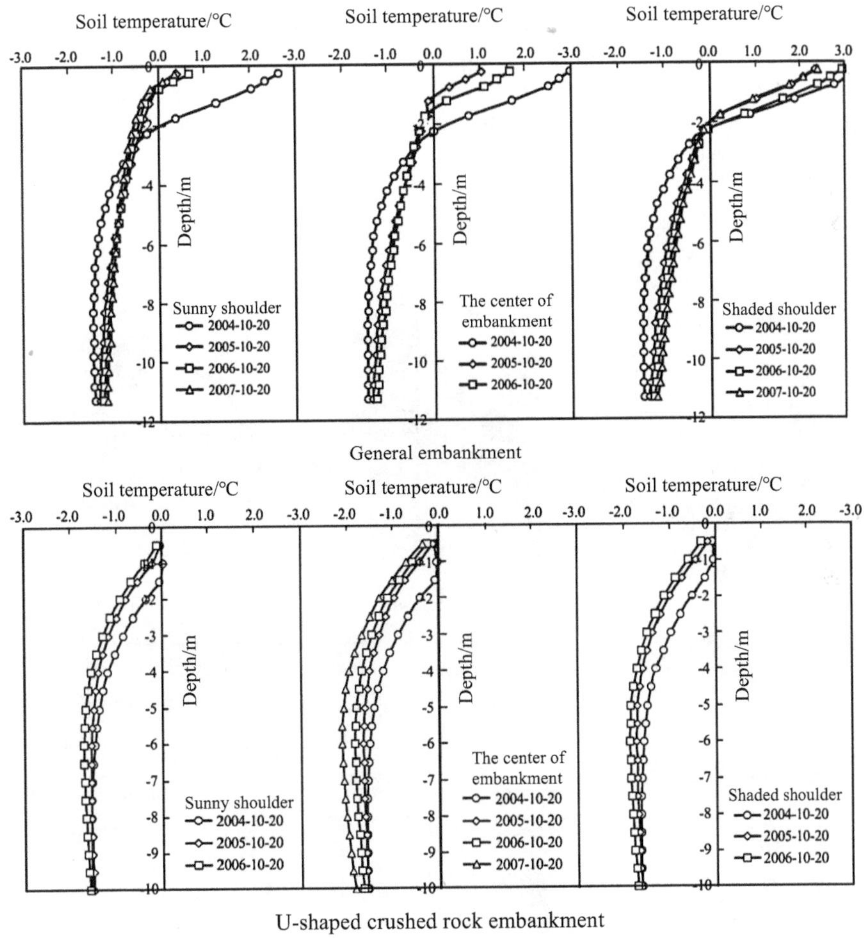

Fig. 3.85 Change of soil thermal regime under general embankment and U-shaped crushed rock embankment from 2004 to 2007

3.4.1.6 Adaptability of Crushed Rock Embankment

The overall cooling effect of crushed rock embankment is closely related to the mean annual ground temperature. The lower the mean annual ground temperature, the better the cooling effect, and the more conducive it is to the stability of deep permafrost. In the extremely unstable permafrost zone, the cooling effect of the crushed rock embankment is relatively worse, resulting in deep permafrost temperatures rising. It is necessary to take reinforcement measures to maintain the thermal stability of the permafrost under the embankment. At the same time, it also shows that in areas with strong wind speed, the cooling effect of crushed rock embankment is better than that in areas with weak wind speed, which is more conducive to the cooling of permafrost under crushed rock embankment.

U-shaped crushed rock embankment can significantly reduce the temperature of shallow soil under the embankment and maintain the stability of deep permafrost temperature (Wu et al., 2005a, b, c) (Fig. 3.86). Crushed rock revetment embankment can raise the permafrost table under the embankment and decrease the permafrost temperature. At the same time, in the cold permafrost area, the crushed rock

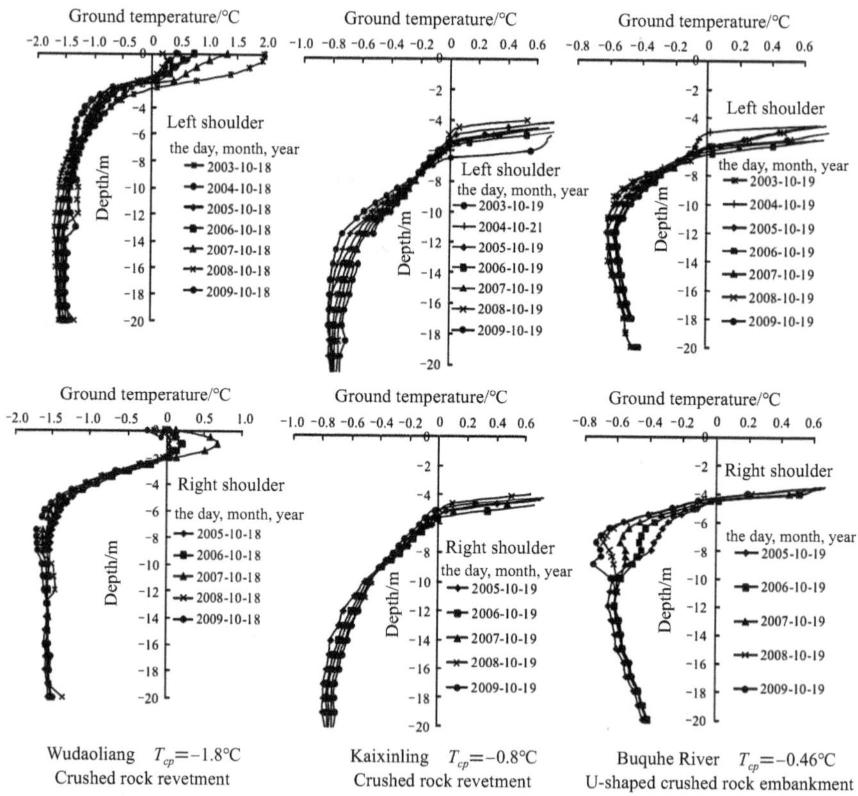

Fig. 3.86 Ground temperature under left and right shoulders at different monitoring sections (Mu et al., 2010)

embankment can effectively decrease the permafrost temperature under the embankment, and deep permafrost temperature is relatively stable. In the extremely unstable warm permafrost areas, the cooling effect under the crushed rock embankment and crushed rock revetment is relatively slight, and the permafrost temperature under the embankment raises, accompanied by slow degradation from the bottom. These phenomena indicate that the cooling effect of crushed rock embankment is also affected by the mean annual ground temperature. It should be noted that the thickness of the overlying soil-filling will weaken the intensity of negative temperature at the top of the crushed rock layer in winter and affect the cooling effect of the crushed rock layer.

3.4.2 Ventilation Duct Embankment

Ventilation foundation is widely used in buildings engineering, but rarely used in road engineering. In 1974, a field test on the 40 km highway at the west of Fairbanks, Alaska was conducted. The metal corrugated pipe with an inner diameter of 0.2 m ~ 0.5 m was buried in parallel near the slope toe of the embankment. There were vertical air inlets and outlets at both ends, forming a "chimney effect" (Fig. 3.87) to promote air flow within duct. The test site in Beaver River, Yukon River, Canada (Fig. 3.88) and the national highway 214 test site in China had also carried out experimental study in the ventilation tubes buried in parallel at the foot of the foundation slope (Fig. 3.89) (Wang et al., 2004a, b, c). A ventilation duct embankment was conducted in Russia and was buried at the bottom of the soil-filling embankment. The air inlet and outlet had a certain height difference. All experiment results show that ventilation duct has a better role of cooling underlying permafrost.

In 2001, the ventilation duct was applied to the subgrade engineering for the first time in Qinghai-Xizang Railway, and the test of ventilation duct embankment was carried out in Qingshuihe River and Beiluhe River. In the warm season, the air convection in the ventilation duct can bring out the heat from the soil-filling

Fig. 3.87 Air inlet and outlet of ventilation duct at the Beaver River test site in Yukon, Canada

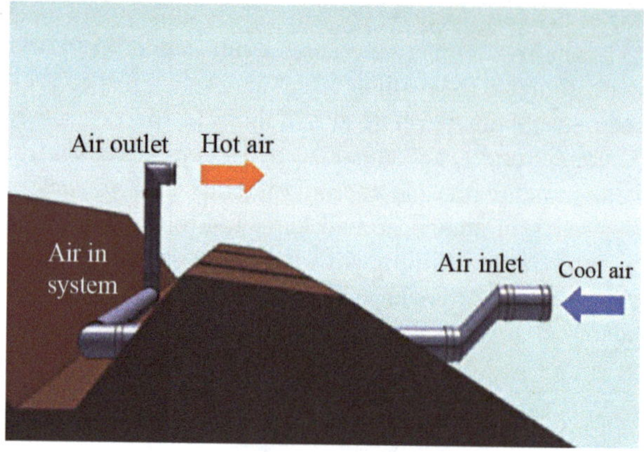

Fig. 3.88 Ventilation duct test site in Beaver River, Yukon, Canada [materials of the international seminar on road operation technology in cold regions of China and Canada in 2012 (Paul Murchison)]

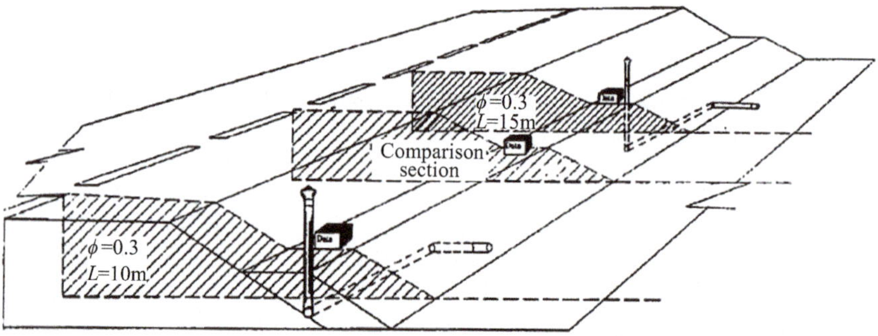

Fig. 3.89 Embankment test section of the National Highway 214

construction and the upper part of the embankment. In the cold season, the soil temperature within the embankment can be decreased, the cold storage of the permafrost under the embankment can be increased, enhancing the thermal stability of the permafrost under the embankment. When the freezing index of air temperature within the ventilation duct is more than twice of the thawing index, the potential freezing capacity can keep the thermal stability of permafrost under embankment.

3.4.2.1 Working Characteristics of Ventilation Duct Foundation

When the fluid enters the circular pipe from a large space, the flow boundary layer has a process from zero to convergence at the center of the pipe. Similarly, when there is heat exchange between the fluid and the pipe wall, the heat exchange on the pipe wall surface also has a process from zero to convergence with the pipe centerline. When the flow boundary layer and the heat boundary layer meet at the center line of the pipe, the flow and heat transfer are fully developed, and the heat transfer intensity will remain unchanged in the future. The area from the inlet to the fully developed section is defined as the inlet section. The thermal boundary layer in the inlet section is thin, and the local surface heat transfer coefficient is higher than that in the fully developed section, and gradually decreases along the mainstream direction. If turbulence occurs in the boundary layer, the local surface heat transfer coefficient will be increased due to turbulence disturbance and mixing, and then gradually tends to a fixed value (Fig. 3.90) (Huang et al., 2004).

The experimental study shows that the inlet length L in laminar flow is determined by the following formula:

$$\frac{l}{d} \oplus 0.05 R_e P_r \tag{3.44}$$

where d is the inner diameter of the pipe; R_e is the Reynolds number; P_r is the P_r and t_l number.

In the case of turbulence, as long as $L/d > 60$, it indicates that the heat transfer coefficient is not affected by the inlet section, that is, $d/L > 0.017$. The d/L of the test section of Qinghai-Xizang Railway is substantially greater than 0.02. Therefore, it is necessary to make full use of the heat exchange effect of the inlet section to strengthen the heat exchange, so that the measures of ventilation duct can effectively protect the permafrost under embankment.

There are two kinds of air flow in the ventilation duct. One is that when the fluid is forced to pass through the ventilation duct by external force, the air flow has a certain pressure, which is forced convection. The other is caused by the buoyancy force generated by the density difference between the cold and hot air fluids, which is natural convection. When air flows through the wall of the ventilation duct, the heat exchange process between the air and the soil outside the pipe is called convection heat exchange. This process includes the convection caused by the fluid

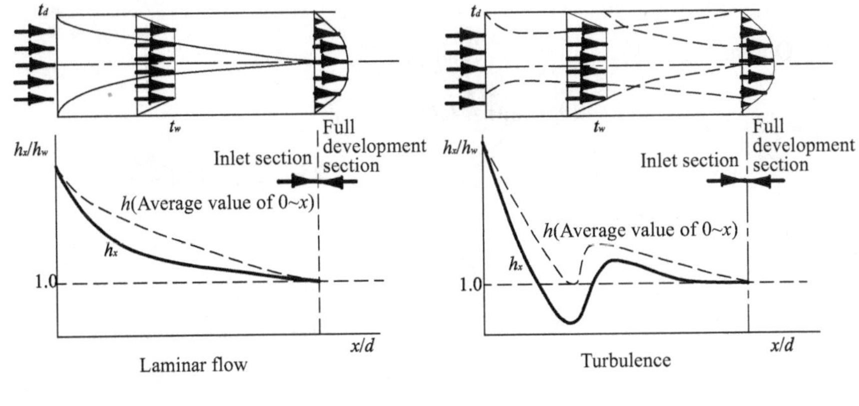

Fig. 3.90 Change of local surface heat transfer coefficient of flow in pipe

displacement and the heat conduction between the fluid molecules, and has a combined effect of heat conduction and convection.

Forced convection means that the whole fluid has a regular macro movement, and the flow velocity of the fluid has a great impact on the size of the heat transfer coefficient. However, natural convection means that there is no regular macro movement in the fluid, and the buoyancy force is the acting factor affecting the size of the heat transfer coefficient. When the flow of air through the pipe is conducted by natural convection, the movement state of the fluid is mainly laminar flow. When the flow of air through the pipe is conducted by forced convection, the motion state of the fluid is mainly turbulent. The heat release coefficient of forced convection is 20 W/(m^2·K) to 100 W/(m^2·K), and that of natural convection is 1 W/(m^2·K) ~ 10 W/(m^2·K). So, the heat release coefficient of forced convection is more than twice that of natural convection (Jiang & Yang, 2008).

The forced convection is different in the case of different ratio between the length and the diameter of ventilation duct. As the air is affected by the roughness, Reynolds number and pipe curvature in the flow process, its wind pressure will decrease along the way. In order to ensure that the air passes through the ventilation duct at a certain initial velocity, the initial velocity of the air flow must be greater than the wind pressure lost when the air flow passes through the ventilation duct. The flow state (laminar flow or turbulent flow) in the pipe is based on its critical velocity. The air below the critical velocity flows through the ventilation duct in a laminar flow, and the air above the critical velocity flows through the ventilation duct in a turbulent flow. The critical velocity depends on the physical properties of the fluid and the shape and size of the flow channel. For laminar flow, the heat transfer along the normal direction of the wall depends on heat conduction. For turbulent flow, the thin layer closest to the wall has laminar flow property. Outside the thin layer, the heat transfer depends on the convection mechanism of turbulent disturbance in addition to the heat conduction mechanism.

In general, the critical Reynolds number that can maintain laminar flow in the pipe is:

$$R_e = 2300.$$

In the Qinghai-Tibet Plateau, the viscosity coefficient of air movement is.

$$\nu = (2.30 \sim 2.73) \times 10^{-5}\ \text{m}^2/\text{s}.$$

For a ventilation duct with an inner diameter d of 0.4 m, its critical speed is:

$$V_{lj} = \frac{\nu R_e}{d} = 0.132\ \text{m/s} \tag{3.45}$$

For the ventilation tube with inner diameter d of 0.3 m, its critical speed:

$$V_{lj} = 0.176\ \text{m/s}$$

Table 3.37 Air temperature (T_0), freezing index (Ω_f) and thawing index (Ω_t) in permafrost region of Qinghai-Xizang Plateau

Area	T_o (°C)	Ω_f(°C·d)	Ω_t(°C·d)	Ω_f/Ω_t	Remarks
Chumaerhe high-plain	−4.9	−2703	539	5	Ding et al. (2011)
Wudaoliang	−4.7	−2576	476	4.4	
Fenghuo Mountain	−5.6	−2770	354	6.8	
Tuotuohe River	−3.4	−2323	717	2.2	
Wenquan	−3.1	−2043	589	2.5	
Tumengela	−4.2	−2406	529	3.5	
Maduo	−3.69	−2079	756	2.75	Luo and Jin (2014)
Huashixia	−4.2				

According to the calculation results, in the current design of Qinghai-Xizang Railway, the critical velocity of the boundary between laminar flow and turbulent flow in the ventilation duct is very small, and it is almost impossible for the air flow to flow in laminar flow. Therefore, the heat exchange of forced convection in the ventilation duct is turbulent.

The ventilation duct is used to cool the permafrost under embankment. The cold input into the embankment in the cold season must be greater than the heat input into the embankment in the warm season. The data of freezing index and thawing index of the Qinghai-Xizang Plateau (Table 3.37) shows that the freezing index is more than twice of the thawing index. Except for the weak freezing capacity in the Tuotuohe River and Wenquan area, other areas have a large potential freezing capacity, which can cool the permafrost under the embankment.

3.4.2.2 Test Results of Qinghai-Xizang Railway

In 2001, the experimental section of ventilation duct embankment was established in Qingshuihe River and Beiluhe River for Qinghai-Xizang Railway construction project. The three-year monitoring data shows that no matter whether the ventilation duct is buried in the middle or lower part of the embankment, the air temperature in the ventilation duct in the center of the embankment for the two experimental sections is higher than the ambient temperature, showing consistency with the ambient air temperature outside the pipe. The difference of temperature gradually increases along the wind direction in the ventilation duct (from right to left). In the warm season, the difference between the temperature in the pipe and the ambient temperature is small, the temperature gradient is small, and the heat exchange is relatively weaker. In the cold season, the ambient temperature is lower than the temperature in the pipe, and the temperature difference is relatively larger, being conducive to cooling the permafrost under the ventilation duct embankment. When the temperature is lower than −2.1 °C, the ventilation duct embankment can play a role of cooling permafrost under embankment. However, from the perspective of engineering safety, suggesting that the mean annual temperature applicable to the ventilation duct embankment should be lower than −3.5 °C.

Regardless of the buried position of the ventilation duct, the soil temperature around the ventilation duct embankment will be reduced to below 0 °C within 2 years after the construction, and the extent of cooling soil will increase with the increase of the buried height and pipe diameter (Tables 3.38 and 3.39) (Editing committee of *Qinghai Tibet(Xizang) Railway*, 2016). The thawing index in the ventilation duct is only 50% of the thawing index of ordinary embankment at the same depth. The freezing index is 2 ~ 4 times of the thawing index and 4 ~ 6 times of the ordinary embankment. It shows that the ventilation duct embedded in the embankment has obvious "heat resistance and cold conduction" effect, which is conducive to the cooling permafrost under the embankment and plays a role in protecting the permafrost stability.

Figure 3.91 shows the temperature isoline of ventilation duct embankment section in the time of thawing depth in Beiluhe basin. In the third year after the construction, the maximum thawing depth under the embankment becomes shallow, basically rises to the original ground, and the −1.0 °C isotherm rises as a whole, indicating that the cold storage of permafrost under the embankment gradually increases. The 0 °C isotherm of ordinary embankment only rises on the right side of the road center, while the −1.0 °C isotherm just moves down, indicating that the cold storage of underlying permafrost is gradually decreasing.

3.4.2.3 Ventilation Duct Material and Pipe Diameter

At present, the ventilation ducts used in the embankment engineering are mainly reinforced concrete pipes and PVC pipes. The ventilation ducts of different materials have different thermal conductivity, which will affect the thermal conducting properties of the ventilation ducts. The ventilation ducts with large thermal conductivity have greater heat transfer effect. PVC pipes and metal corrugated pipes have good continuity and ductility, but the thermal conductivity of PVC pipes is small, with the poor durability and the high vulnerability rate, but the price of PVC pipes is relatively lower. The biggest advantage of metal corrugated pipes is better thermal conductivity, strength and durability, but its cost is relatively higher. The thermal conductivity of reinforced concrete pipes is moderate, but the continuity and ductility are not as good as PVC pipes and metal corrugated pipes. The prefabrication process is usually adopted, and the length of each reinforced concrete pipes is generally 2 m, which has high strength and durability under the premise of quality assurance.

According to the test results of ventilation duct embankment in the Beiluhe River, there is little difference between the heat transfer effects of reinforced concrete pipes and PVC pipes (Table 3.40) (China Railway First Survey and Design Institute Group Co., Ltd. 2007), while the damage rate of PVC pipes is large, especially in the air inlets and air outlets at both ends of the duct, and its serious damage rate and poor durability may be related to the construction, ultraviolet radiation, freeze-thaw cycle and other factors. Therefore, the inlet and outlet of duct should be replaced by other materials with high strength.

Table 3.38 Temperature and accumulated temperature on the outer wall of ventilation duct and 0.5 m below the pipe in Beiluhe basin along Qinghai-Xizang Railway

Sections	Particular year	Mean annual temperature (°C)		Accumulated temperature (°C·d)					
				Duct wall (DW)			0.5 m Depth below duct wall ($DW_{0.5}$)		
		DW	$DW_{0.5}$	Positive	Negative	Difference	Positive	Negative	Difference
DK1140 + 015 (ZBH300)	2003	−0.93	−0.24	876.25	−1216.9	−340.67	633.15	−721.52	−88.37
	2004	−2.08	−1.51	537.56	−1295.46	−757.90	252.54	−802.71	−550.18
DK1140 + 075 (ZBH400)	2003	−1.01	−1.13	956.82	−1326.07	−369.25	579.20	−990.53	−411.33
	2004	−2.48	−1.91	+517.03	−1422.52	−905.49	242.31	−941.06	−698.75
DK1140 + 882 (DB-M)	2003	−0.24	−0.34	−380.31	−46.85	−87.54	233.57	−358.92	−1125.36
	2004	−0.46	−0.35	−101.24	−269.05	−167.81	57.61	−185.29	−127.68
DK1141 + 237 (XBH300)	2003	−1.20	−1.17	579.60	−1018.42	−438.82	210.89	−638.95	−428.06
	2004	−2.23	−1.89	380.27	−1195.60	−815.33	112.12	−803.48	−691.36
DK1141 + 284 (XBH400)	2003	−1.38	−1.34	618.51	−1121.63	−503.11	190.02	−679.58	−489.56
	2004	−2.09	−1.82	545.64	−1309.78	−764.14	131.95	−794.50	−622.55
DK1141 + 882 (DB-B)	2003	−0.17	−0.10	35.31	−95.62	−60.31	112.71	−48.84	−36.14
	2004	−0.11	−0.07	0.19	−38.59	−38.40	0.00	−25.75	−25.75

Note: Z means the ventilation duct buried in the middle of the embankment; X is the lower part of the embankment of the ventilation duct buried; P is PVC ventilation duct; H is reinforced concrete pipe; The number is the pipe diameter (mm)

Table 3.39 Comparison of thawing index (°C•d) and freezing index (°C•d) of ventilation duct embankment and ordinary embankment (DB) in Qingshuihe River

Sections	Location	2002		2003		2004	
		Thawing Index	Freezing Index	Thawing index	Freezing Index	Thawing Index	Freezing Index
DK1026 + 37(ZBH300)	Embankment surface	4332.86	−59.19	4130.06	−6.08	4223.45	−55.31
	Inner wall of ventilation duct	679.96	−1611.36	424.36	−1207.59	318.20	−1151.15
	Outer wall of ventilation duct	875.03	−1677.90	626.24	−1231.36	431.83	−1172.06
DK1026 + 252 (DB)	Embankment surface	3641.20	−202.12	3312.61	−83.25	3606.32	−163.17
	EIP	754.95	−458.14	844.78	−301.26	607.77	−181.61
	EOP	968.80	−606.86	670.23	−187.44	786.14	−287.64

Notes: EIP means Equivalent to the position of the inner wall of the pipe, EOP means Equivalent to the position of the outer wall of the pipe

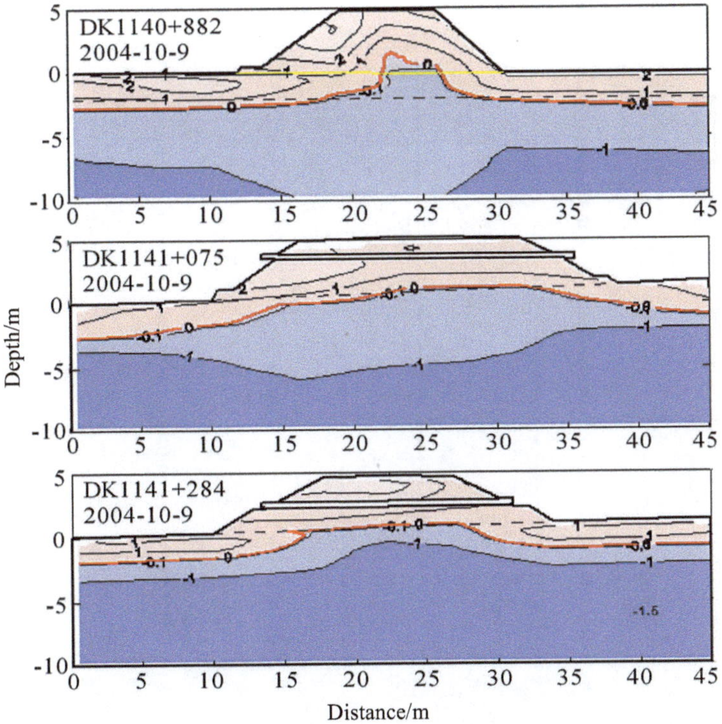

Fig. 3.91 Distribution of ground temperature in the time maximum thawing depth of ventilation duct embankment in Beiluhe basin of Qinghai-Xizang Railway

The diameter of the ventilation duct will affect the operation mode of the fluid in the pipe, and then affect the heat exchange effect between the ventilation pipe and the soil.

For the design of ventilation duct embankment, the heat exchange area of ventilation duct should be considered under the given embankment width. The heat exchange area of the ventilation duct is related to the diameter and spacing of ventilation duct. Under the assumption that other conditions are unchanged, the temperature changes only along the radius direction, and the temperature difference of the cylinder coordinate changes only according to the radius of the ventilation tube. According to the basic law of heat conduction:

$$Q = -\lambda F \frac{dt}{dr} = -\lambda \pi r l \frac{dt}{dr} \tag{3.46}$$

After the integration:

$$Q = \frac{2\pi\lambda l}{\ln \dfrac{d_2}{d_1}} \left(t_1 - t_2\right) \tag{3.47}$$

Table 3.40 Temperature and accumulated temperature in the outer wall of reinforced concrete pipe, PVC pipe outer wall and soil at the 0.5 m depth below the pipe in the test section in Beiluhe River

Sections	Particular year	Mean annual temperature/°C		Temperature accumulation/(°C·d)					
				Duct Wall (DW)			0.5 m depth below duct wall ($DW_{0.5}$)		
		DW	$DW_{0.5}$	Positive	Negative	Total	Positive	Negative	Total
ZBP300 (DK11410 + 927)	2003	−0.95	−0.64	594.4	−942.3	−348.0	444.0	−677.5	−233.5
	2004	−2.08	−1.40	310.9	−1069.7	−758.8	249.6	−761.4	−511.9
ZBH300 (DK1141 + 015)	2003	−0.93	−0.24	876.3	−1216.9	−340.7	633.2	−721.5	−88.4
	2004	−2.08	−1.51	537.6	−1295.5	−757.9	252.5	−802.7	−550.2
XBP400 (DK1141 + 175)	2003	−1.90	−1.48	296.5	−989.0	−692.5	123.5	−664.5	−541.0
XBH400 (DK1141 + 284)	2003	−1.38	−1.34	618.5	−1121.6	−503.1	190.0	−679.6	−489.6

Notes: Z means the ventilation duct buried in the middle of the embankment; X is the lower part of the embankment of the ventilation duct buried; P is PVC ventilation duct; H is reinforced concrete pipe; The number is the pipe diameter (mm)

It can be seen that the amount of heat through the ventilation duct every hour is directly proportional to the thermal conductivity λ, the length L of the pipe and the temperature difference $(t_1 - t_2)$, and inversely proportional to the logarithm of the ratio of the inner and outer diameters of the pipe. Table 3.41 also roughly reflects this situation.

In the condition that the pipe spacing is twice the pipe diameter, the ventilation area per unit of the ventilation duct embankment with a diameter of 400 mm accounts for 10.47% and the ventilation area with a diameter of 300 mm accounts for 7.85% for the ventilation duct embankment in Beiluhe along Qinghai-Xizang Railway. Table 3.42 shows that increasing the diameter of ventilation duct can increase the cooling capacity of ventilation duct embankment. It should be noted that in warm season, large pipe diameter brings more heat than small pipe diameter, which will increase the thawing depth under the embankment.

In the case of given embankment width, the comprehensive analysis of simulation test shows that the pipe diameter of 0.7–1.0 m is appropriate (Table 3.47) (Li et al., 2005a, b). Based on the test road in Wudaoliang along Qinghai-Xizang Highway, the results of the calculation model show that the cooling effect of ventilation duct embankment with large-diameter is better than that of ventilation duct embankment with small-diameter.

The field test observation and calculation show that the duct diameter of the ventilation duct embankment should be controlled by the cooling effect, and the engineering cost and construction technology should be considered. If the pipe diameter is too small (e.g. $D < 0.4$ m), the cooling effect is relatively worse, its cost and compaction of soil-filling are not good. If the pipe diameter is too large (e.g. $D > 1.2$ m), the cooling effect is relatively better, but the duct strength is highly required, the compaction of soil-filling is relatively worse. The relationship between D and the influence radius (R) of cooling effect (Li et al., 2005a, b) is roughly as follows:

$$R = k\left[\frac{D}{D_0}\right]^{\alpha} \tag{3.48}$$

where $D_0 = 1.2$ m; $k \approx 3.5 \sim 4.5$ m; $\alpha \approx 0.3 \sim 0.5$.

Table 3.41 Total accumulation temperature of ventilation duct embankment with different diameters in Beiluhe Basin

Sections	Particular year	Total accumulation temperature (°C·d)	
		Below Duct Wall	0.5 m depth below Duct Wall
DK1140 + 015(ZBH300)	2003		−272.56
DK1140 + 075(ZBH400)			−326.29
DK1141 + 015(ZBH300)	2004		−492.35
DK1140 + 075(ZBH400)			−538.16
DK1141 + 237(XBH300)	2003	−567.63	−413.64
DK1141 + 284(XBH400)		−601.72	−456.95
DK1141 + 237(XBH300)	2004-01-01 and 2004-03-09	−718.23	−449.41
DK1141 + 284(XBH400)		−750.53	−456.82

Notes: Z means the ventilation duct buried in the middle of the embankment; X is the lower part of the embankment of the ventilation duct buried; P is PVC ventilation duct; H is reinforced concrete pipe. The number is the pipe diameter (mm)

Table 3.42 Maximum depth of temperature (MDT) (−4 °C) and maximum thawing depth (TD) in each year under different pipe diameters (D) and duct spacing (L) (m)

Program		The second year		The 4th year		The 6th year		The 8th year		The 10th year	
		MDT	TD	MDT	TD	MDT	TD	MDT	TD	MDT	TD
D = 0.4 m	L = 0.8 m/15	−1.68	−1.31	−3.48	−1.12	−5.94	−0.95	−7.01	−0.94	−8.11	−0.94
D = 0.7 m	*L* = 1.4 m/9	−1.39	−1.23	−3.22	−0.96	−5.78	−0.89	−6.80	−0.86	−7.83	−0.86
D = 1.0 m	*L* = 2.0 m/6	−1.35	−1.14	−3.05	−0.92	−5.40	−0.87	−6.31	−0.85	−7.17	−0.83

According to the overall analysis of the ventilation duct test, design and construction in the Beiluhe Rive, the duct diameter of the ventilation duct embankment should range from 0.4 m to 0.7 m, and the ratio of duct diameter (*D*) to length (*L*) should be greater than 0.02, or the ratio of length to diameter should be $L/D \leq 30$.

3.4.2.4 Determination of Ventilation Duct Spacing

The observation data of ventilation duct embankment in the Qingshuihe area along Qinghai-Xizang Railway shows that the influence depth of ventilation duct on the underlying soil temperature is about 3.5 m (Fig. 3.92) (China Railway First Survey and Design Institute Group Co., Ltd. etc, 2007). The temperature of the ventilation duct (XBP40) buried in the lower part of the embankment is 0.4 °C lower than the temperature of the ordinary embankment (DB) at the depth of 6.5 m. The range that the temperature decrease is greater than 0.4 °Cis the influence range of the ventilation duct. Therefore, the effective influence radius of the PVC ventilation duct is 2.5 m. Similarly, at the depth of 6.5 m, XBH30 is 0.5 °C lower than the ordinary embankment, so the effective influence radius of concrete ventilation duct is 2.5 m. From the perspective of safety, the spacing of ventilation duct with a diameter of 0.3 m ~ 0.4 m is about 1.0 m to 1.5 m.

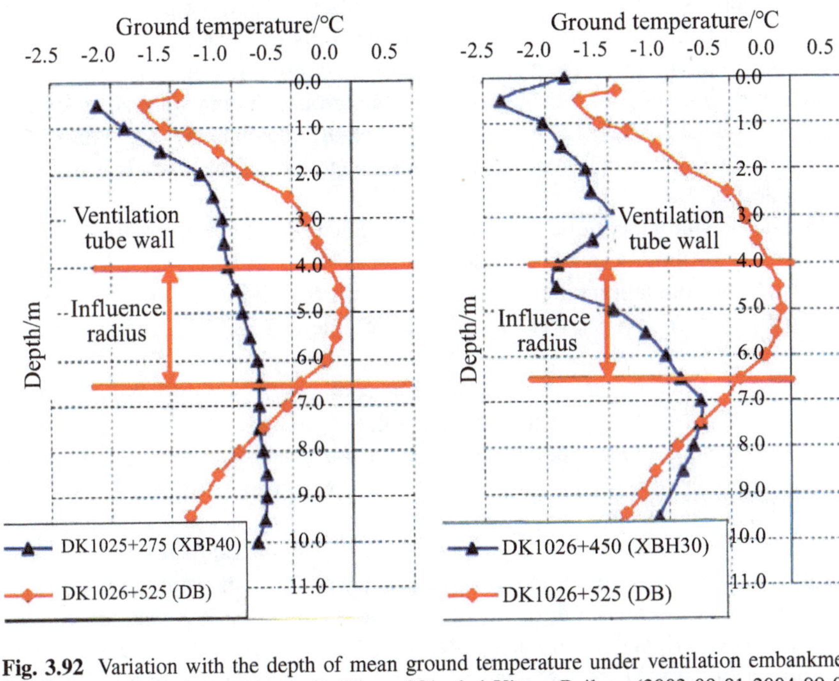

Fig. 3.92 Variation with the depth of mean ground temperature under ventilation embankment and general embankment at Qingshuihe River of Qinghai-Xizang Railway (2003-09-01-2004-09-01)

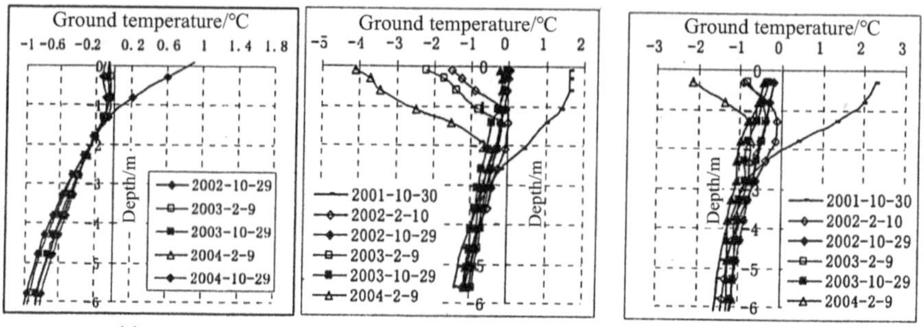

Fig. 3.93 Ground temperature with depth change under the ventilation duct at Beiluhe along Qinghai-Xizang Railway (calculated from the original ground surface) (**a**) General embankment; (**b**) Reinforced concrete ventilation duct embankment, (XBH300, pipe diameter 0.3 m,); (**c**) Reinforced concrete ventilation duct embankment, (XBH400, pipe diameter 0.4 m)

The observation data of concrete ventilation duct embankment (XBH300, XBH400) in Beiluhe along Qinghai-Xizang Railway show that the influence depth of ventilation duct on the underlying soil temperature can exceed 2.0 m under the original ground (Fig. 3.93) (Compilation Committee of *Qinghai-Xizang Railway (Xizang)*, 2016). According to the fact that the ground temperature under the ventilation duct is about 0.5 °C lower than that under the ordinary embankment, the influence depth of the ventilation duct on the underlying soil temperature is 2.6 m. If the ground temperature under the ventilation duct is 1.0 °C lower than that under the ordinary embankment, the maximum influence depth is 2.1 m. The buried height of the ventilation duct is 0.7 m above the original ground, and the influence depth of ventilation duct can reach 2.8 m. Therefore, when the duct diameter is 0.3 m ~ 0.4 m the space between ventilation ducts is far more than twice of the duct diameter in the current design.

In order to study the influence of interval, diameter, and buried depth of ventilation duct on cooling permafrost, a simulation experiment is conducted (Li et al., 2005a, b). The simulation scheme designed the pipe spacing L as 0.8 m ($2D$), 1.2 m ($3D$) and 1.6 m ($4D$) respectively, and the pipe diameter D as 0.4 m, the pipe wall as 60 mm thick, the embankment height as 3.0 m, the buried depth of the ventilation duct as 2.5 m under embankment surface. The cooling effect of different pipe intervals shows that the smaller is the pipe spacing, the greater is the negative temperature wave transmission speed, the better is the cooling effect, and the smaller is the thawing depth in warm season (Table 3.43). The pipe spacing ($S/D \leq 4$) set in the simulation test can achieve good cooling effect.

To sum up, the effective cooling influence radius (R) should be taken as the optimization basis for the determination of the spacing between the ventilation ducts. In general, the spacing between the ventilation ducts should be less than the cooling influence radius (R) and should be greater than the compaction distance of the construction machinery. Therefore, it is appropriate to select 3–4 times of pipe diameter as the pipe spacing, that is, the net spacing of ventilation duct is 1.0–1.5 m.

Table 3.43 Annual maximum transmission depth (H_{max}) of temperature (−4 °C) wave and annual maximum thawing depth (H_t) under different pipe spacing (L) (m)

	The second year		The 4th year		The 6th year		The 8th year		The 10th year	
Scheme	H_{max}	H_t	H_{max}	H_t	H_{max}	H_t	H_{max}	H_t	H_{max}	H_t
$L1 = 2D$	−1.60	−1.37	−4.07	−1.12	−5.58	−1.04	−6.48	−0.96	−7.30	−0.96
$L2 = 3D$	−1.40	−1.23	−4.00	−1.03	−5.50	−0.95	−6.32	−0.95	−7.13	−0.87
$L3 = 4D$	−1.35	−1.20	−3.90	−0.96	−5.40	−0.86	−6.20	−0.85	−7.00	−0.82

Note: D stands for pipe diameter, D = 0.4 m

3.4.2.5 Burial Depth of Ventilation Duct

The buried depth of the ventilation duct is usually determined by the soil-filling embankment under the duct and the seasonally thawed layer rapidly freezing in the cold season under the embankment. When the thermal resistance of soil-air interface is neglected, the approximate calculation of heat transfer coefficient between buried duct and soil is as follows (Wu et al., 2003a, b, c):

$$\beta = \frac{2\lambda}{D \ln \frac{4h}{D}} \tag{3.49}$$

where λ Is the thermal conductivity of soil, W/(m·K); D and h is the diameter and burial depth of the ventilation duct respectively.

Under certain simplified conditions, the deeper is the embedded depth of the ventilation duct above embankment, the lower is the heat transfer coefficient, and the worse is the cooling effect on the surrounding soil. Considering the influence of surface water flow, sandy wind, and snow cover, the distance between the buried height of the ventilation duct above embankment and the ground should not be less than 0.5 m (Fig. 3.94). It can be seen from formula (3.50) that the pipe diameter and burial depth of the ventilation duct affect the heat exchange coefficient between the ventilation duct and the underlying soil. According to the extreme value theorem:

$$\frac{h}{D} = \frac{e}{4} \tag{3.50}$$

It can be seen that the heat transfer coefficient should reach a certain value. The buried depth of the ventilation duct is directly proportional to the pipe diameter. The deeper is the burial, the larger the pipe diameter becomes.

According to the test data (Table 3.43) of Ventilation duct embankment in Beiluhe along the Qinghai-Xizang Railway, the cooling effect of the embankment with ventilation duct buried at low level (0.5 m ~ 0.7 m above the ground, code XB) is better than that of the embankment with ventilation duct buried at high level (middle of the embankment, code ZB). Simulation research shows that when the embankment height is 3.0 m, the buried depth of the ventilation duct is 2.5 m, 2.0 m

Fig. 3.94 Ventilation duct embankment of National Highway 214 (photo by Fang Jianhong)

and 1.5 m respectively from the embankment surface (Li et al., 2005a, b). The calculation results show that the greater is the buried depth of the ventilation duct (the closer to the original ground), the better the cooling effect becomes. Considering the strength of the ventilation duct and the requirements of subgarde stability, the ventilation duct should be buried in a relatively lower position, and the suggested buried depth h should be greater than 3–5 times of the pipe diameter D (Li et al., 2005a, b).

3.4.2.6 Other Ventilation Duct Embankment

1. **Automatic Control Damper Ventilation Duct**

In order to prevent the hot air from passing through the ventilation duct in the warm season, a barrier plate can be set at the inlet and outlet ends (or the inlet end) of the ventilation duct to reduce the permafrost temperature rising under the ventilation duct embankment. For example, an automatic temperature control switch damper device was designed (Yu et al. 2004) (Fig. 3.95). During the period of warm season, the damper at the ventilation duct orifice is closed when air temperature is higher than 5 °C. During the period of clod season, the damper is opened, effectively blocking hot air from entering the vent duct, and improving the cooling effect of the ventilation duct embankment.

Comparative test observation (DK1141 + 237 ~ +280) was carried out on the ventilation duct embankment with and without dampers in the Beiluhe along the Qinghai-Xizang Railway. The height of two ventilation duct embankments is 3.5 m, and the height between the center line of the ventilation ducts and the natural ground is 1.5 m. The observation results show that during the cold season, the automatic control damper is in the open state, and the temperature change under the embankment with and without the automatic control damper ventilation duct is basically the

Fig. 3.95 Working status of ventilation duct embankment with automatic temperature control damper device (Photo by Yu Qihao) (**a**) Opening state in the morning (**b**) Closed state at noon

Table 3.44 Mean temperature of inner wall of ventilation ducts for different embankment structures in warm season

Embankment	Mean temperature of inner wall for different location of ventilation ducts/°C		
	Left side of the road	Center	Right side of the road
Without automatic damper	4.95	3.06	3.14
With automatic damper	3.46	1.93	2.99
Temperature difference	1.49	1.13	0.15

same. During the warm season, automatic control damper is closed, and the soil temperature under the ventilation duct embankment is lower than that without automatic control damper, the soil temperature under the center of embankment is the lowest, and the soil temperature under the sunny slope is higher (Table 3.44). This phenomenon shows that the heating effect of air convection on embankment during the warm season is almost eliminated by the automatic damper device. The mean temperature at the center of the inner wall of the ventilation duct is 1.13 °C lower than the air temperature, and the temperature of duct inner wall near the slope is 1.49 °C.

Observation data of the ventilation duct embankment show that the soil temperature at the depth of 3.0 m under embankment with automatic dampers can be more rapidly decreased than that without automatic dampers and the difference of the lowest soil temperature between the embankment with and without automatic dampers gradually expand. Meanwhile, ventilation duct embankment with automatic dampers can make the exothermic time of the underlying soil prolong 40 days. According to the data of temperature measurement, the heat release time of the ventilation duct embankment with an automatic damper is twice as long, indicating that the ventilation duct embankment with an automatic damper can more rapidly decrease the soil temperature around the pipe. During the period of warm season, the thawing rate and thawing depth of permafrost under the ventilation duct embankment with an automatic damper are relatively smaller (Fig. 3.96), and the time to reach the maximum thawing depth is 20 days earlier.

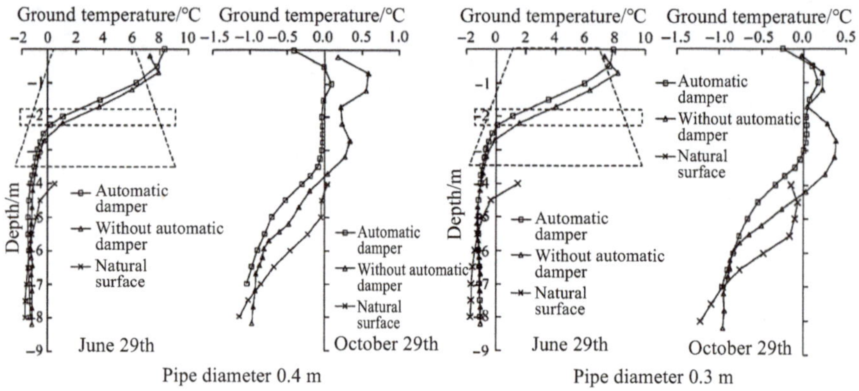

Fig. 3.96 Comparison of ground temperature under the ventilation duct embankment with and without automatic damper during the warm season (Yu et al., 2004)

Fig. 3.97 Perforated ventilation pipe buried on field

2. **Perforated Ventilation Duct Embankment**

At present, the ventilation duct embankment used in the Qinghai-Xizang Railway, Qinghai-Xizang Highway and National Highway 214 are all non-perforated reinforced concrete pipes and PVC corrugated pipes. The perforated ventilation duct installs many holes on the pipe wall to enhance heat exchange between ventilation duct and the surrounding soil (Fig. 3.97) (Hu et al., 2004). So, the cold air entering the ventilation duct not only transfers heat through the ventilation pipe, but also can accelerate the cooling effect of the embankment through the direct convection heat exchange between the holes in the pipe wall and the soil outside the pipe. However, it is necessary to prevent fine-grained soil and ground water from entering the pipe and affecting the soil-filling effect.

The in-situ experiment results show that under the same conditions, perforated ventilation duct has a good cooling capacity for the permafrost, and the thaw bulb under the embankment is very small. When the air temperature is higher than the ground temperature, the automatic damper at the duct should be added to reduce the influence of hot air. Compared with ordinary ventilation duct embankment, many procedure and protective measures of perforated ventilation duct will increase the project cost.

3. **Vertical Ventilation Duct Embankment**

Under a certain pipe diameter, the effective area of the air inlet of the ventilation duct depends on the relationship between the axial direction of the ventilation duct and the wind direction. When they are in the same direction, the most effective air intake can be achieved. When it is at oblique intersection, only the wind speed component in the axial direction of the parallel ventilation tube can enter the ventilation duct.

When the wind speed is constant, the larger is the projected area of the cross-sectional area of the air inlet of the embankment ventilation tube in the wind speed direction, the larger the air inlet volume of the ventilation tube will be, and the air inlet volume in cold season will be more conducive to the cooling permafrost. Therefore, the ventilation duct is equipped with a device of automatic windward air intake, or automatic damper. That is, a vertical air intake pipe with a certain height higher than the ventilation duct is installed at the air intake of the ventilation duct with a fixed or movable "hood" with the wind (Fig. 3.98). The air collection outlet can be increased as much as possible to enhance the air volume entering the ventilation duct and avoid pipe blockage. Ventilation pipes that are vertically higher than the air inlet or the road surface are also set at the air outlet. By using the air density difference and the "chimney" effect, the cold air that enters the ventilation pipe is quickly brought out the heat air in the embankment.

Under the existing subgrade conditions, the ventilation duct can also be used as an engineering measure to control the embankment settlement. The ventilation duct can be buried on the embankment slope (one side or both sides) to decease the ground temperature under the embankment slope, for example, the test section of

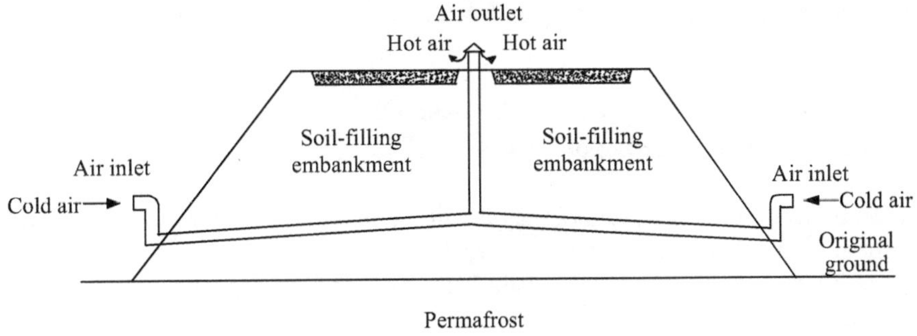

Fig. 3.98 Air inlet of ventilation duct embankment (Yang et al., 2011)

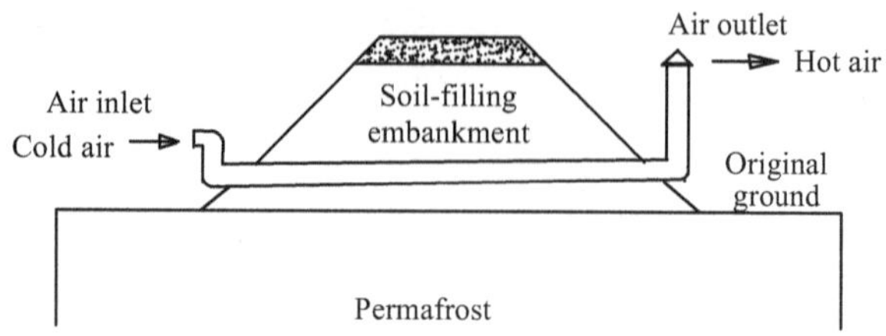

Fig. 3.99 Air inlet and air Intake of wide ventilation duct embankment with asphalt pavement

ventilation duct for national highway 214 (Fig. 3.89) and the Beaver River test site in Yukon, Canada (Figs. 3.87 and 3.88).

When the embankment is a wide pavement, the air outlet can be installed in the middle isolation belt, and the air inlet of the ventilation tube is still provided with a movable air intake (Fig. 3.99) to increase the air collection volume of the ventilation duct. The ventilation duct is laid in herringbone shape from the center of the foundation, with a slope rate of about 4%.

3.4.3 *Thermosyphon Embankment*

The thermosyphon (called as thermal probe in China) is a heat transfer device for circulating heat exchange between steam and liquid. Thermal probes are widely used in engineering in cold regions to solve many engineering problems such as foundation frost heave and thaw settlement. It is used to improve the foundation stability during the construction and operation of industrial and civil buildings in permafrost regions, such as foundations, coal yards, low-temperature storehouses, pile foundations and foundations, above-ground pipelines, water tanks, electricity transmission lines, bridge piers, culverts, etc. Meanwhile, it is used to prevent frost heave and thaw settlement of roads, airport runways and dams during construction and operation. It is used in the construction of "frozen soil wall" and impervious curtain, ice island, ice road and ice ferry. It has its advantages to form freezing wall, which becomes the freezing wall of underground engineering and mining engineering, and the natural container wall for storing hazardous wastes in cold regions.

3.4.3.1 Principle of Thermal Probes

It can be seen from the structural diagram of thermal probe (Fig. 3.100) that for the thermal probes inserted into the permafrost foundation in the cold season, when the permafrost temperature (t_2) at the evaporation section of the thermal probe is higher

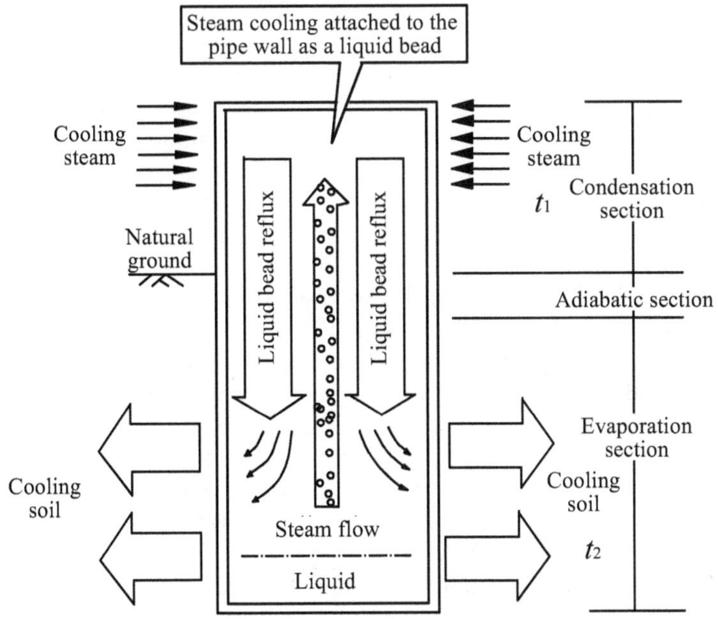

Fig. 3.100 Schematic diagram of working principle of thermal probe

than the ambient temperature (t_1) at the condensation section of the thermal probe, the working medium liquid within the thermal probe evaporation section is heated and evaporated into stream, forming a stream flow and rising to the condensation section. When the stream in the condensing section is cooled by the cold air outside the pipe to form liquid beads, it flows back to the evaporation section along the pipe wall under the action of gravity. Through the evaporation and cooling cycle process of working medium in the thermosyphon pipe, the heat of the permafrost foundation is continuously carried to the atmosphere for emission, so that the permafrost temperature is cooled. Until the permafrost temperature at the evaporation section is equal to or lower than the ambient temperature (t_1) at the condensation section, the process stops. In the warm season, when $t_1 > t_2$, the stream in the condensation section of the thermosyphon pipe cannot be cooled into liquid beads, and the two-phase flow circulation between stream and liquid cannot be generated, and the cold capacity cannot be automatically exchanged with the stream in the evaporation section, and the heat outside the foundation cannot be carried to the permafrost under the foundation. Therefore, the thermal probe is a kind of unidirectional and efficient heat transfer device. When the air temperature is lower than the ground temperature, it can produce a cooling effect to decrease permafrost temperature. But when the air temperature is higher than the ground temperature, it cannot transfer heat to the foundation to increase the temperature.

3.4.3.2 Design Parameters of Thermal Probe

The structural parameters of the thermal probe refer to the design parameters of the thermal probe itself, including the length of the condensation section and the evaporation section, the number of fins, and the working medium. The heat release of the condensing section depends on the heat release coefficient between the condenser surface and the air. The heat transfer capacity of the evaporation section depends on the heat release coefficient between the evaporator surface and the liquid working medium (Fig. 3.101).

The thermal probes buried in the permafrost foundation to form a heat exchange system of atmosphere—thermal probe—foundation. The transmission power of the thermal probe in this system mainly depends on the geometric size of the thermal probe, type and filling amount of thermal probe working medium, thermophysical properties of the material constituting the thermal probe and the soil around the thermal probe, the heat release coefficient between the inner wall of the thermal probe and the working medium and the heat release condition from the condenser wall to the air, soil and temperature etc. These involve the thermal resistance of each part (between the thermal probe and the soil and the atmosphere), the control of the heat transfer limit (carrying limit, drying limit and boiling limit), the optimization of the length of the evaporation section and the condensation section and the liquid filling volume that need to be considered in the production process of the thermal probe. The liquid filling amount is the decisive factor that affects the heat transfer limit of the thermal probe and is determined by the total length of the thermal probe.

The cooling capacity of the thermal probe, that is, the effective power of the thermal probe in the actual work, depends not only on the rated power of the thermal

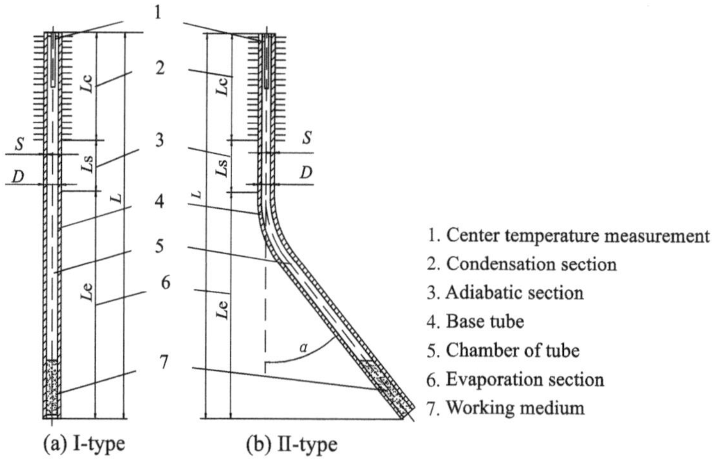

Fig. 3.101 Composition and shape of thermal probe (GB/T 27880). *D*—Outer diameter of base tube, mm; L_c—Length of condensation section, m; L_s—Length of insulation section, m; L_e—Length of evaporation section, m; *L*—Thermal probe height, m; *S*—Nominal wall thickness of base tube, mm; α—bend angle, °

probe itself, but also on the actual working time and the soil temperature, air temperature, wind speed and thermal resistance of the surrounding soil. Under different geological conditions of permafrost, the cooling capacity of thermal probes is different. Therefore, relevant data should be collected during the design and calculation of thermal probe, including air temperature, wind speed, surface temperature, geometric dimension of foundation and physical and thermal physical parameters of each layer of soil, lithology of foundation, permafrost type, natural and artificial permafrost table, depth of zero annual amplitude and mean annual ground temperature. In fact, in the thermal probe foundation system, the effective power of the thermal probe is much smaller than its rated power. Therefore, the heat transfer (cooling capacity) of the thermal probe cannot be directly estimated by simply using the working time and rated power. The cooling capacity (Q_y) during the cold season can be obtained by estimating the effective power of the thermal probe.

3.4.3.3 Applicable Conditions of Thermal Probe

The thermal probe is one of the best engineering measures for preventing the thawing settlement of permafrost foundation in cold regions. It can decrease the permafrost temperature and reduce and inhibit the thawing settlement deformation. The in-situ experiment of a single thermal probe was carried out in Qingshuihe River (warm permafrost) and the Fenghuo Mountain (cold permafrost) areas of the Qinghai Tibet Plateau. The thermal probe is 7 m long (the evaporation section is 3 m long and the condensation section is 1.5 m long). The mean ground temperature at the depth of 6–8 m under the ground in March is −1.4 °C. Observation data show the mean annual temperature of the soil around 3–8 m from the evaporation section of the thermal probe at the initial stage of embedding (March 10 to April 20 of the first year) (Table 3.45).

According to the simulation results of the finite element method, in the cold permafrost regions (for example, Fenghuo Mountain), the ground temperature curve moves toward the relatively lower temperature direction from the first year to the 30th year. Within the 30 years of its life, the thermal probe is mitigating the influence of a warming climate, and the soil is in a "cooling-down" state of releasing heat (Fig. 3.102a. In the 30th year, the thermal probe stops working. In the 50th year, the ground temperature curve of foundation soil is similar to that of foundation without thermal probe. In the warm permafrost regions (Qingshuihe is an example), since the first year, the ground temperature change of the thermal probe is similar to that of the cold permafrost regions. However, when the thermal probe stops working in the 30th year, the ground temperature has returned to the state of warm permafrost regions in the first year (Fig. 3.102b). This shows that after the thermal probe fails to work, it is not enough to mitigate the impact of a warming climate, and the cold storage in the early stage begins to be consumed during the work period, this state can only be changed by selecting an appropriate area of the evaporation section of the thermal probe and the density of the thermal probe arrangement.

Table 3.45 Mean annual temperature (°C) of soil at different distances (D) from thermal probe (3 m ~ 8 m) in 2004 (Guo et al., 2009)

Location	MAAT/°C	MAGT/°C	TWTP/°C	D = 0.3 m	D = 0.8 m	D = 1.3 m	D = 1.8 m	D = 5.0 m	Mean
Qingshuihe	−4.0	−0.7 to −0.9	−4.71	−4.08	−1.42	−1.30	−1.21	−1.08	−2.54
Fenghuo Mountain	−5.6	−1.8 to −2.6	−6.33	−5.08	−3.73	−3.55	−3.31	−3.15	−4.40

Note: MAAT stands for mean annual air temperature, MAGT stands for mean annual ground temperature

3.4.3.4 Spacing of Thermal Probes

The maximum spread range of the thermal probe in cooling soils under the ground is the influence radius of the thermal probe. The temperature gradient, that is, the horizontal distance from the wall of the thermal probe increases and then decreases to zero. According to the observation data of the swampy permafrost regions along the Chaimu Railway in Qinghai, if 5 °C decrease is used as the criterion, the effective influence radius based on the lowest temperature is 4.5 m, and the effective cooling radius based on the mean annual temperature is 2.4 m (Fig. 3.103). The effective influence radius of thermal probe in the Qingshuihe test section along the

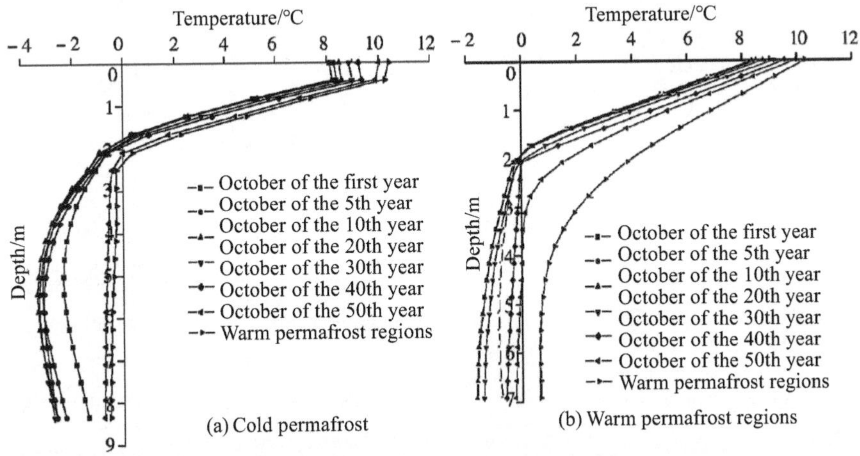

Fig. 3.102 Ground temperature change at the foot of the sunny slope of thermal probe embankment in warm and cold permafrost regions (Guo et al., 2009)

Fig. 3.103 Horizontal ground temperature curve of thermal probe at the 3.5 m depth for Chaimu Railway, Qinghai (Chen et al., 2011)

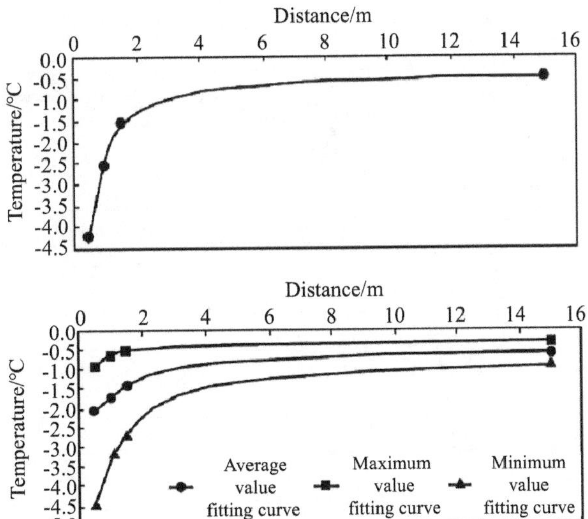

Qinghai-Xizang Railway is 1.45 m (China Railway First Survey and Design Institute Group Co., Ltd. et al., 2007). The observation data in Anduo test section show (Fig. 3.104) that the effective influence radius of thermal probes with different diameters is also different. The effective spread range of ϕ60 mm, ϕ89 mm and ϕ108 mm are 1.8 m, 2.3 m and 2.55 m, respectively (Table 3.46). According to the observation results of Yakeshi-Linhai Railway and Nenlin Railway in permafrost regions (Qiqihar Railway Research Institute of Harbin Railway Bureau, 2007), the influence radius of thermal probe is 2.0 m (Fig. 3.105).

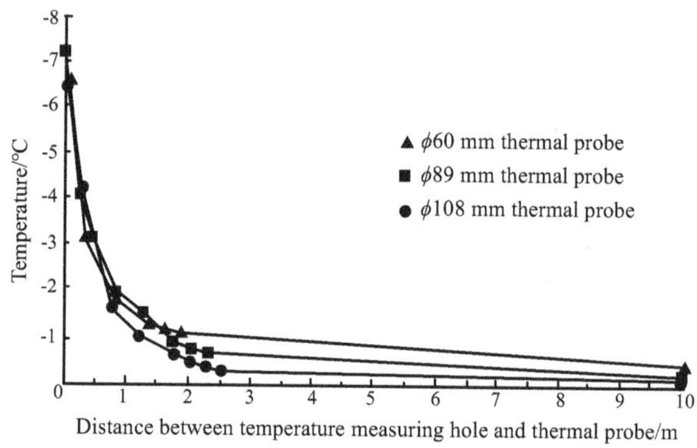

Fig. 3.104 Soil temperature change curve of horizontal distance of thermal probe at Anduo (Li et al., 2006)

Table 3.46 Mean ground temperature around thermal probe (ϕ89) under natural ground in Anduo

Time	Distance from outside the thermal probe wall (m)							
	0	0.3	0.8	1.3	1.8	2.05	2.3	10
2004-11-10 to 2005-4-20	−6.47		−2.10	−1.73	−1.11	−0.95	−0.87	−0.30
2005-11-9 to 2006-4-20	−6.06	−3.38	−1.94	−1.58	−1.05	−0.91	−0.83	−0.32

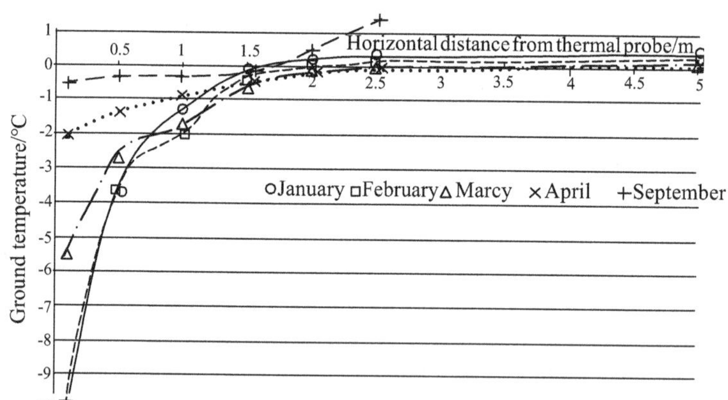

Fig. 3.105 Soil temperature change curve of horizontal distance of Thermal probe in the Da Xing'anling Mountains

The observation data in test sections of thermal probe in permafrost regions show that the temperature of the evaporation section gradually decreases and the soil temperature around the pipe also start to decrease after the thermal probe starts working (about October). After reaching the minimum temperature (March to April), with the air temperature rising, the cooling role of the thermal probe weakens and the soil temperature around the pipe rises slowly. The early stage is the cold storage stage of the thermal probe which mainly collects "cold energy" and a large amount of cold energy can rapidly decrease the soil temperature around the pipe and gradually transfer and diffuse to the soil around the pipe. In the later stage, the temperature of the evaporation section decreases and it continues to spread out under the influence of the temperature difference of the soil around the pipe, increasing the heat transfer range. Obviously, the lower the atmospheric temperature is, the longer the duration of cooling role is. That is, the larger is the freezing index, the more the cold storage of the thermal probe is, and the farther the diffusion distance to the soil around the pipe is. The total cooling capacity of the thermal probe is in direct proportion to the temperature difference between the side wall of thermal probe and ground soils.

If there are no experimental conditions, the working period of the thermal probe can be analyzed and determined according to the local meteorological data and the thermophysical properties of permafrost (Ding et al., 2011). The surface temperature difference of the evaporation sections (the difference between the lowest temperature on the surface of the evaporation section and the initial temperature), and the temperature fluctuation period (the time from the beginning of work to the lowest temperature on the surface of the thermal probe) can be estimated by the modified Fourier equation:

$$L = k\sqrt{\frac{\lambda T}{\pi C}}\ln\frac{A_0}{A_1} \tag{3.51}$$

where L is the heat transfer influence range of the thermal probe, m; k is the correction coefficient, which can be 0.20 ~ 0.25 in the permafrost region of the Qinghai Xizang Plateau; λ and C is the average thermal conductivity kcal/(m·h·°C) and heat capacity kcal/(m^3·°C) of permafrost around the evaporation section of the thermal probe respectively; T is the temperature fluctuation period of the thermal probe evaporation section, hour; A_0 is the temperature range in the evaporation section of the thermal probe during the calculation period,°C; A_1 is the temperature range at the boundary of the heat transfer influence range of the thermal probe in the calculation period, which can be 0.1 °C to 0.2 °C.

In Qingshuihe areas, the heat transfer influence radius of the thermal probe is 1.57 m from October to January of the second year, however, the test data of K1025 in Qingshuihe section show that the influence radius of thermal probe in natural site is 1.45 m.

Numerical simulation results show that the smaller is the longitudinal distance of thermal probes, the more the ground temperature under the center of embankment shifts to the negative temperature side along the depth distribution curve (Fig. 3.106)

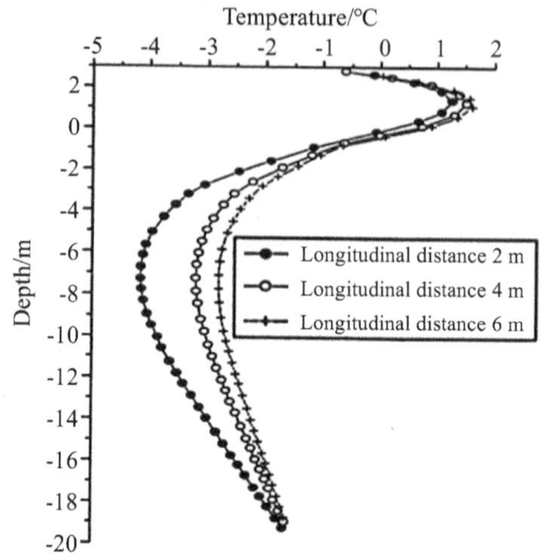

Fig. 3.106 Ground temperature curve under the center of embankment with different spacing of thermal probes (The 5th year)

(Qinghai provincial local railway administration, et al., 2010), which is more conducive to decrease the ground temperature under the embankment and rise the artificial permafrost table.

According to the results of in-situ experiments and numerical simulations in Qinghai Tibet Plateau, Qilian Mountain and Daxing'anling & Xiaoxing'anling Mountains in Northeast China, the effective cooling radius of the thermal probe range from 1.5 m to 2.0 m, exhibiting the effective cooling capacity in maintaining the thermal stability of permafrost. Therefore, it is reasonable to arrange 3.0–4.0 m longitudinal intervals for thermal probes. When the embankment width is relatively larger, the burial method of the thermal probe should be changed, that is, the method of directly inserting into embankment should be changed to the method of obliquely inserting into embankment. Meanwhile, the bent thermal probe (II type thermal probe) should be used to achieve the effect of cooling permafrost under the center of embankment. Of course, the cooling capacity of the thermal probe is related to the local freezing index and wind speed. The larger is the freezing index, the higher is the cooling capacity of the thermal probe, and the larger is the effective cooling radius.

3.4.3.5 Reasonable Burial Depth of Thermal Probe

The burial depth of the thermal probe depends on the intention and purpose of utilization. For the subgrade engineering in cold regions, thermal probes can be used to eliminate thermal disturbance during construction and operation of the engineering and prevent permafrost thawing and cool the permafrost under embankment to improve the thermal stability of the permafrost. For buildings in cold regions, thermal probes can be used to adjust the artificial permafrost table under the building and reduce the thickness of the active layer to prevent the thaw settlement and soil frost heave and to maintain the stability of engineering in cold regions. Additionally, thermal probe have significant advantages in artificial ground freezing, for example constructing the underground frozen soil wall to prevent water and seepage, and to stabilize underground working face.

In terms of subgrade and building engineering, the use of thermal probes is one of the important engineering measures to prevent and control the temperature rise of the permafrost under the foundation and thaw settlement. The thermal probes should be buried in the bearing layer of permafrost foundation, so as to cool the permafrost under foundation, increase the bearing capacity of the permafrost, raise the artificial permafrost table, and enhance the thermal stability of permafrost. As a treatment measure to prevent the freeze-thaw damage of engineering buildings, the thermal probe should be buried at the place so that the engineering damage can be restored to its original thermal regime and its bearing capacity can be improved.

1. If the evaporation section of thermal probe is buried in seasonally thawed layer under the original ground, a thermal probe can cause the refreezing area of the seasonally thawed layer to greatly increase (Fig. 3.107) (Qinghai provincial local railway administration, etc., 2010). In the middle and early stage of the cold season, cooling role of thermal probe can cause the seasonally thawed layer under the embankment to refreeze more rapidly, but the permafrost is in a relatively higher thermal regime. During the warm season, the seasonally thawed layer gradually turns to the positive temperature, however, the local permafrost core can remain or disappear near the thermal probe.

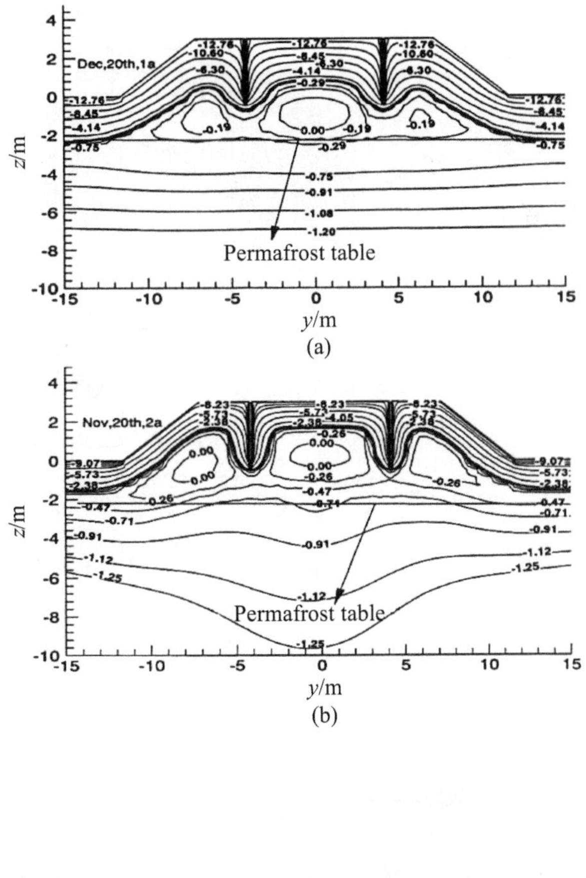

Fig. 3.107 Ground temperature field in the evaporation section of thermal probe buried in the thawed layer of the embankment on December 15 in the 1st (**a**) and 2nd (**b**) year of operation

2. If evaporation section of thermal probe is buried into permafrost layer under embankment (Fig. 3.108), the refreezing area of the seasonally thawed layer is small, but the permafrost temperature will be greatly decreased and will still kept at a low temperature in the second year. During the warm season, the seasonally thawed layer gradually turns to positive temperature, while the permafrost temperature can still be maintained in a relatively lower temperature.
3. For the new project, the evaporation section of the thermal probe is generally buried in the seasonally thawed layer and permafrost layer. Figure 3.109 is the variation curve of maximum thawing depth in the test section of thermal probe in Qingshuihe along Qinghai-Xizang Railway. In the first year after the completion of the embankment, the permafrost table under the thermal probe embankment is obviously raised, with an average increase of 1.29 m, while the ordinary embankment only raised by 0.3 m. The average rising of permafrost table in 2003 was 0.55 m compared with that in 2002, and the average increase in 2004 was 0.31 m compared with that in 2003. Figure 3.110 is the variation of the maximum thawing depth under the thermal probe embankment in the Beiluhe along the Qinghai-Xizang Railway. Compared with ordinary embankment, the permafrost table under the thermal probe embankment is raised by 0.34–1.64 m, however, the permafrost table under the ordinary foundation exhibited a declining trend of about 0.2 m.

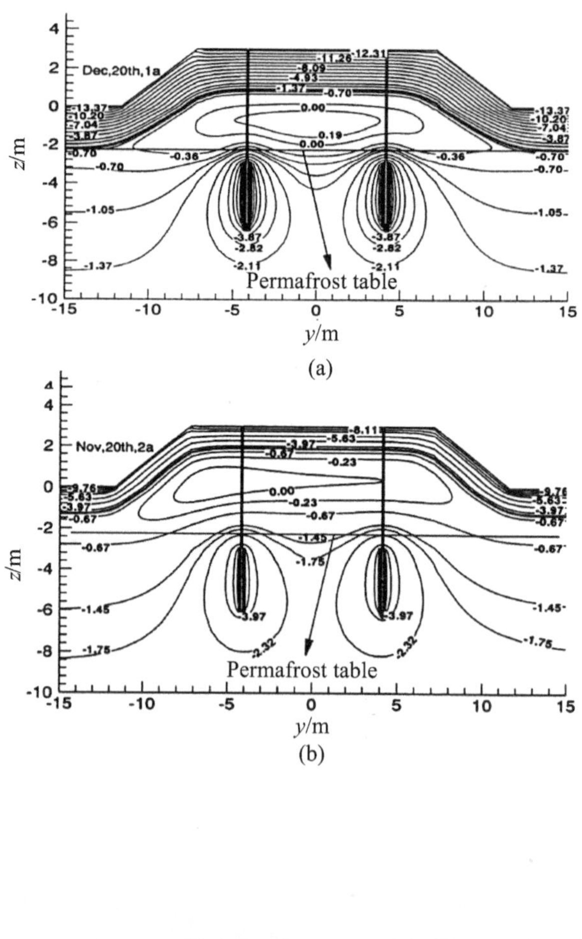

Fig. 3.108 Ground temperature field in the evaporation section of thermal probe buried in permafrost layer on December 15 of the 1st (**a**) and 2nd (**b**) year of operation

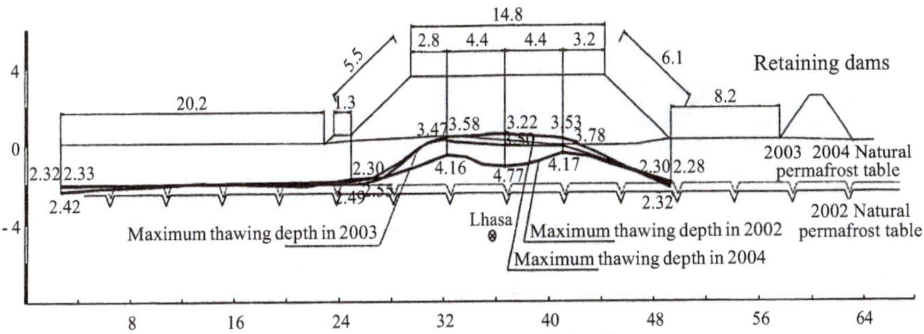

Fig. 3.109 Variation of maximum thawing depth under embankment for DK1024 + 425 section along Qinghai-Xizang Railway from 2002 to 2004 (Edit Committee of *Qinghai-Tibet(Xizang) Railway*, 2016)

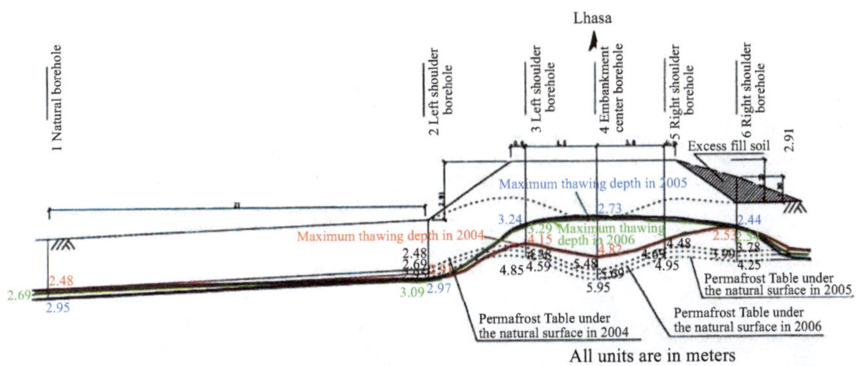

Fig. 3.110 Variation of maximum thawing depth under the embankment at DK1505 + 275 along Qinghai-Xizang Railway from 2004 to 2006 (China Railway First Survey and Design Institute Group Co., Ltd., 2007)

4. For the remediation section with engineering damages, the evaporation segment of thermal probe is buried in the thaw bulb and permafrost layer under the embankment. The survey and research aim to control the freezing and thawing disease of Yakeshi-Linhai Railway and Nenlin railway in the permafrost region of the Daxing'anling & Xiaoxing'anling Mountains in Northeast China, which has been operated for nearly 50 years (Qiqihar Railway Research Institute of Harbin Railway Bureau, et al. 2007). The thermal probe is inserted obliquely at an angle of 75 into the foundation. The length of the thermal probe is 9 m, the condensation segment is exposed for 4 m, and the longitudinal distance of the thermal probe is 3 m ~ 3.5 m. The evaporation segment is placed below the seasonally thawed layer and the permafrost table. In the second year (2005) after the construction, the permafrost table under the railway embankment generally rises, reaching 1.5 m ~ 3.0 m (Figs. 3.111 and 3.112), and the maximum is 4.0 m. Frozen cores in soil profiles are formed in 25% ~ 40% of the railway sections.

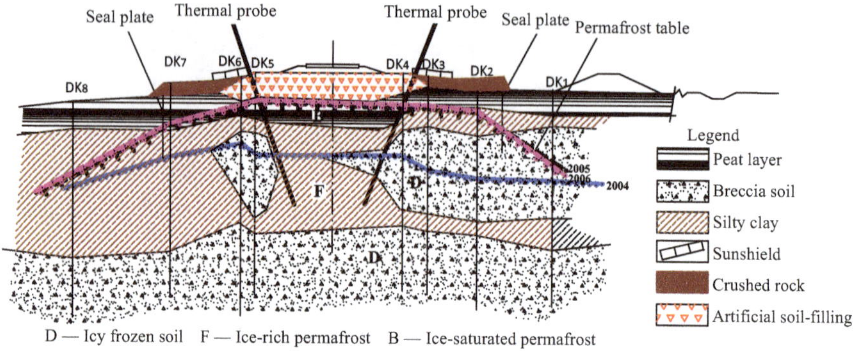

Fig. 3.111 Geological section and permafrost table at the 100 K100 section along Yakeshi-Linhai Railway

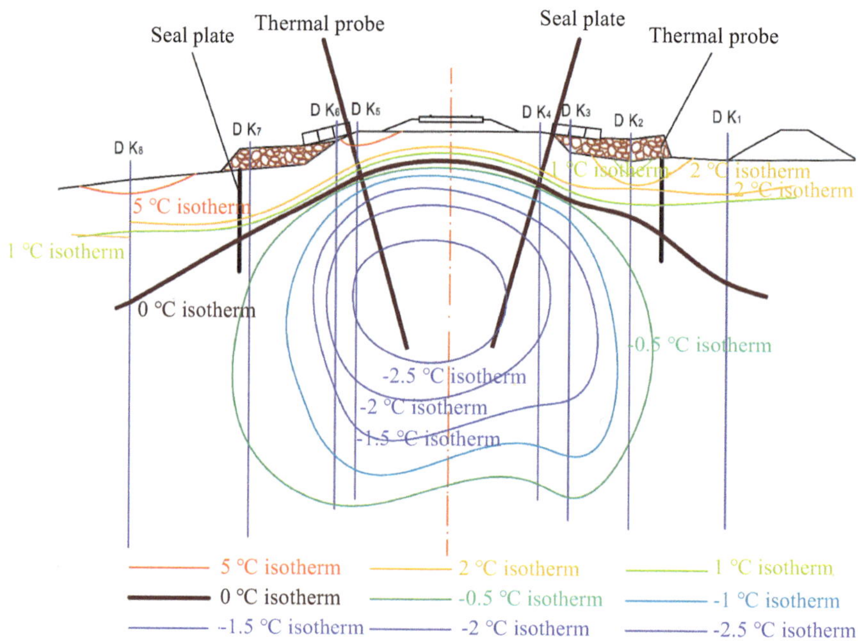

Fig. 3.112 Isothermal curve in October 2006 at the 101 K100 section along Yakeshi-Linhai Railway

Generally, a thermal probe should be directly placed in the thaw bulb under the embankment, and the bottom end of the thermal probe should reach the bottom of the thaw bulb. If possible, it should be inserted into a certain depth below the permafrost table, so that the thaw bulb under the embankment can be connected with the underlying permafrost, and no talik will occur.

3.4.3.6 The Installation Method of Thermal Probes

The installation methods of thermal probes include the vertical and oblique insertion of single pipe or of double pipes (Figs. 3.113 and 3.114). At present, the thermal probes used in embankment engineering in China are mainly “straight form”, namely, the bending angle α is 0° (90° in the product standard). “Curved” thermal probes, commonly known as “L” shaped thermal probes with its bending angle of 45° are used in experimental sections along the Qinghai-Xizang Highway. Some “L” shaped thermal probes have a larger bending angle, or are presented in “hairpin-shaped”.

Fig. 3.113 Measures against thaw settlement of thermal probes in existing embankment in the Nenlin Railway in northeast China (Oblique Insertion) (Photo by Tong Changjiang)

Fig. 3.114 The embankment with (straight insertion) thermal probes in Qinghai-Xizang Highway

"Hairpin-shaped" thermal probes, also known as flexible (or rigid) thermal probes, have the same working principles as the rigid ones. However, when "hairpin-shaped" thermal probes are used in the subgrade engineering, the whole thermal probe needs to be buried in the embankment, so it is required that the local area has a large freezing index, so that the condensation segment can have more "cool energy" and longer working periods, so as to cool soil in evaporation segment and maintain the thermal stability of subgrades.

The installation method of thermal probes should be determined according to the engineering requirements and environment, as well as the construction conditions. The vertical insertion of thermal probes can be adopted in cold region engineering. For road engineering or building engineering with small cross-sections, thermal probes are often used to meet the cooling range and area of the foundation base. In flat and open warm permafrost regions for road engineering with a wide pavement, it is suitable to use the oblique insertion installation method for thermal probe, with an angle of 75° (the included angle between the thermal probe and the horizontal ground). For road engineering with wide pavements (more than 10 m), it is advisable to use L- shaped thermal probes (Fig. 3.115), and the bending angle should be determined by the engineering requirements. For expressways, L-shaped thermal probes can be used on both sides of embankments, and one or two rows of straight or L-shaped thermal probes can be used in the middle "isolation belt" of expressway (Fig. 3.116), so as to avoid the concave artificial permafrost table under embankment. In warm permafrost regions with high ice content and the areas of permafrost degradation, when the terrain is relatively flat and a height of embankment is less than 2.5 m, it is advisable to set two rows of thermal probes, multiple rows of thermal probes (Fig. 3.117), as well as L-shaped thermal probes.

Fig. 3.115 Construction of L-shaped thermal probes in Qiannen Highway (Photo by Guo Zhirong)

Fig. 3.116 Multiple rows of thermal probes on both sides of the high-grade highway (214) and the isolation belt (Photo by Fang Jianhong)

Fig. 3.117 Double rows and two-way arrangements of thermal probes in Qinghai-Xizang Railway

When the permafrost table under the sunny shoulder of embankment is deeper than that under the shady shoulder due to sunny-shady slope of high embankment, it is appropriate to set thermal probes on the sunny slope, but the evaporation segment should be buried in the thaw bulb. If the evaporation segment of thermal probes is buried below the permafrost table, it will have little effect on the distribution of thaw bulb under embankment.

The numerical calculation results for different installation methods of thermal probes show that the mean annual soil temperature under thermal probes can obviously decrease soil temperature under embankment. The soil temperature for double-rows thermal probes is lower than that of single-row thermal probes, soil temperature for the oblique insertion lower than that of the vertical insertion (Fig. 3.118). There will be no thawing core under the center of the embankment when choosing a suitable thermal probe and a reasonable spacing. (Fig. 3.119).

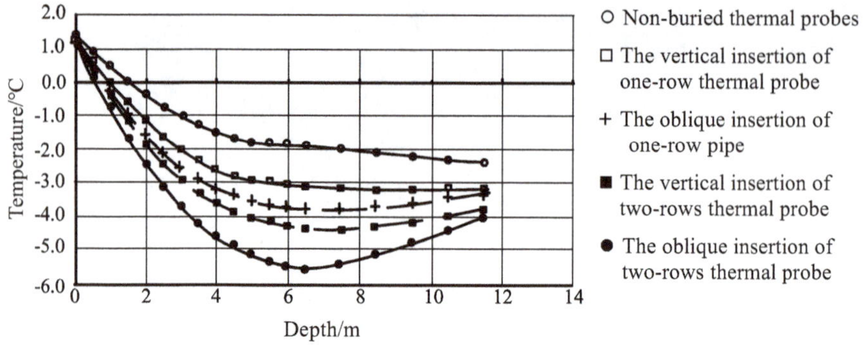

Fig. 3.118 Mean annual soil temperature under the center of embankment with different installation methods of thermal probes

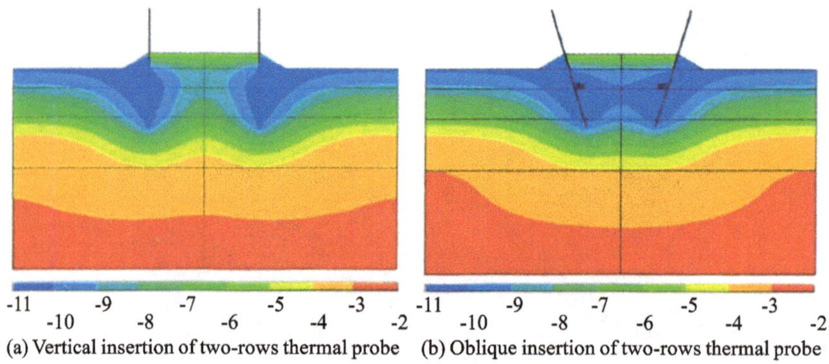

Fig. 3.119 Temperature field of different installation methods of thermal probes (March, 15) (China Communications Construction First Highway Consultants, etc. 2006). (**a**) The vertical insertion of two-rows thermal probe (**b**) The oblique insertion of tow-rows thermal probe

3.4.4 Thermal Insulation Subgrades

For subgrade construction in permafrost regions, especially in low embankments, zero section or road cuttings, the atmospheric heat can easily cause the rising of the permafrost under embankment and affect the thermal stability of the subgrade. The thermal conductivity of new industrial insulation material is smaller than that of the soil. It can prevent the atmospheric heat from entering the subgrade in warm seasons if new industrial insulation material is buried within the embankment, to meet the minimum height of embankment in permafrost regions and maintain the thermal stability of the permafrost under embankment. But the insulation layer can also prevent the atmospheric "cold energy" from entering the subgrade. Under the conditions of global warming, the embankment with thermal insulation layer can only mitigate the degradation of the permafrost.

Table 3.47 Physical property of thermal insulation materials after freezing-thawing cycles

Materials	Times of freezing-thawing cycles	Thermal conductivity/ $W \cdot (m \cdot k)^{-1}$	Volume water absorption rate/%	compressive strength/kPa
EPS	5	0.0253	2.6	401*(347)**
	10	0.0249	2.5	372*(335)**
	20	0.0253	2.8	383*(352)**
	30	0.0242	2.5	337*(326)**
	Average	0.0249	2.6	373*(340)**
PU	5	0.0193	0.5	308
	10	0.0184	1.1	306
	20	0.0188	1.0	263
	30	0.0181	1.1	282
	Average	0.0187	0.9	290
XPS	5	0.022*(0.024)**	0.422*(0.469)**	646*(560)**
	10	0.023*(0.025)**	0.362*(0.402)**	637*(552)**
	20	0.021*(0.023)**	0.39*(0.433)**	628*(544)**
	30	0.019*(0.021)**	0.38*(0.423)**	633*(574)**
	Average	0.021*(0.023)**	0.369*(0.432)**	636*(557)**

Note: * The apparent density of the EPS board is 42 kg/cm^3; ** the apparent density is 30 kg/cm^3; the apparent density of the PU board is 59 kg/cm^3. * Theapparent density of the XPS board is 44.9 kg/cm^3; ** The apparent density of X350 is 45 kg/cm^3

At present, industrial thermal insulation materials used in subgrade engineering in China include polystyrene foam (EPS), polyurethane foam (PU) and extruded polystyrene foam (XPS). Extruded polystyrene foam (XPS) is widely used in subgrade engineering. The test results show that the thermal conductivity of the EPS board is larger than that of the XPS board, and the PU board has the lowest thermal conductivity. The volume water absorption rate of the EPS board is more than 6 times that of the XPS board, but its compressive strength is only half of that of the XPS board (Table 3.47) (China Communications Construction First Highway Consultants and China Railway First Survey and Design Institute, 2006).

3.4.4.1 Working Principles of Thermal Insulation Subgrades

The Thermal insulation embankment refers to laying a thermal insulation layer in the embankment, using its low thermal conductivity (thermal resistance) to prevent the upper heat into the lower part, delay the thawing of the permafrost, and reduce the downward shift of the permafrost table, and thus maintain the relative stability of the permafrost.

Under the yearly atmospheric temperature boundary condition approximate to sinusoidal fluctuations, the temperature of heat input into the soil layer also presents periodic fluctuations in which the amplitude decreases gradually with depth and the phase lags. The ground temperature shows an envelope line, whose maximum and

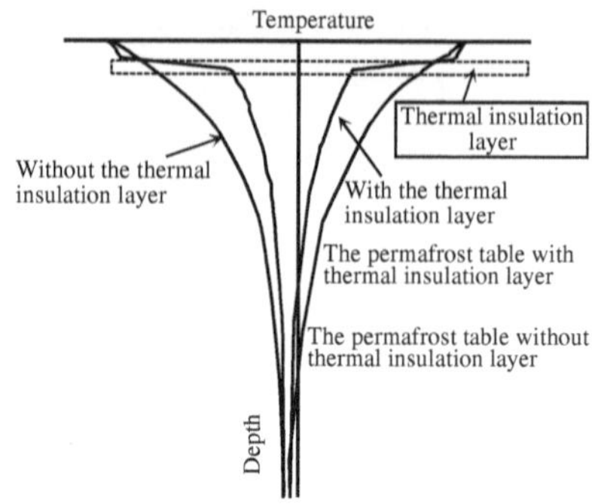

Fig. 3.120 Temperature variation curve with or without thermal insulation layer within embankment

minimum temperature changes dynamically with the variation of depth. When the maximum temperature envelope line varies with time and depth, the depth at which the permafrost reaches the frozen state (usually judged by 0 °C, but actually lower than 0 °C) is the position of the permafrost table. When there is no thermal insulation layer within the embankment, the envelope line of the maximum and minimum temperature in the soil is usually a relatively smooth attenuation curve (the thick line in Fig. 3.120). After laying the thermal insulation layer, because of the huge difference of thermal conductivity between the thermal insulation layer and the soil layer, the heat on the upper part of the thermal insulation layer is intercepted, so there is a large soil temperature difference between the lower and the upper part, so that the annual amplitude of the soil temperature reduces, and the range between the maximum and minimum temperature envelope lines is decreased (the thin line in Fig. 3.120). Sometimes, the depth at which the maximum temperature envelope line reaches 0 °C decreases, and it is higher than that without thermal insulation layer. This means that the permafrost temperature will not be affected by heat due to the effect of thermal insulation, no matter how much the position of the original permafrost table moves or just keep in place (namely, the temperature of the original permafrost can be maintained).

Based on this principle, a thermal insulation layer laid within a certain depth of the embankment can increase the thermal resistance of the embankment, reduce the heat under the thermal insulation layer into the permafrost under embankment (Fig. 3.121), delay the permafrost thawing and decrease the extent of permafrost warming, so as to maintain the relative stability of the permafrost.

3.4.4.2 Applicable Conditions for Thermal Insulation Embankment

The combination of different thicknesses of thermal insulation layer and the soil-filling embankment height can achieve the effect of thermal insulation in collaboration and ensure the thermal stability of subgrade. The calculation results show that after the thickness of the thermal insulation layer is increased to a certain extent (100 mm), the corresponding effect of thermal insulation will not be improved obviously. Considering the economy and strength, 100 mm is an appropriate thickness

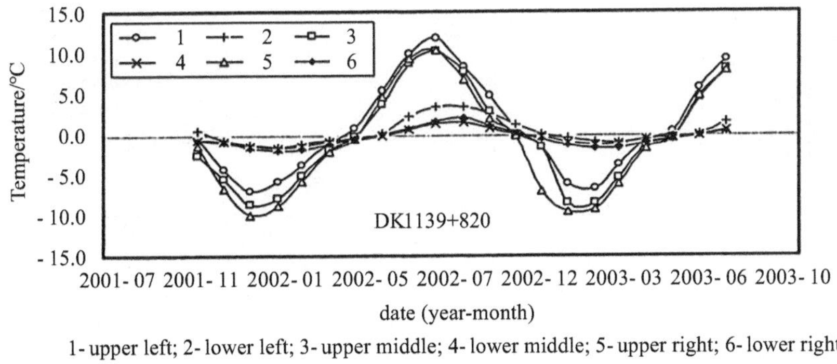

Fig. 3.121 Soil temperature curve above and below the thermal insulation layer (The Compilation Committee of *Qinghai-Tibet(Xizang) Railway*, 2016)

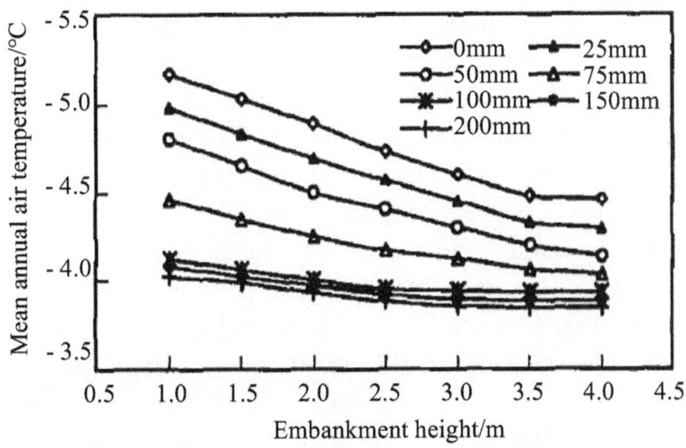

Fig. 3.122 Applicable scope of thermal insulation method for embankment in Qinghai-Tibet Plateau (Wen et al., 2005)

of the thermal insulation layer. When the mean annual air temperature is higher than −3.84 °C the thermal stability of the subgrade during the operation period cannot be guaranteed by the thermal insulation method. When the mean annual temperature is lower than a certain value (−5.20 °C), there is no need to lay a thermal insulation layer within the embankment height of 1 m, and the thermal stability still can be ensured. However, a thermal insulation layer placed in the embankment can reduce the deformation due to the periodic freezing-thawing process, which is beneficial to the stability of the subgrade. Taking the economy and engineering into consideration, the thermal insulation method can be used to maintain the relatively thermal stability of the subgrade in permafrost regions on the Qinghai-Tibet Plateau where the mean annual temperature is between −3.8 °C and −5.2 °C.

For the thermal insulation with the same thickness within embankment that is 0.5 m height above ground surface, the simulation calculation shows that the soil-filling embankment height can also be correspondingly reduced with the decreasing of mean annual ground temperature of the permafrost under the same air temperature (Fig. 3.122). At the same height of subgrade, if the mean annual ground

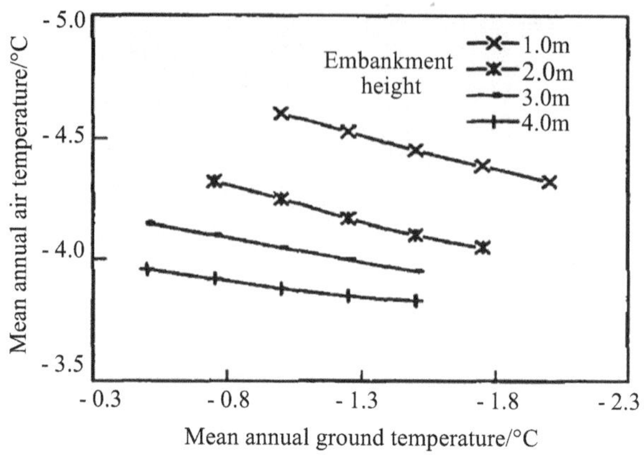

Fig. 3.123 Influence of permafrost thermal regime under the different embankment heights on applicable scope of thermal insulation layer (Wenet al., 2005)

temperature is relatively lower, the embankment with thermal insulation can be applied to the areas with higher mean annual air temperature. The mean annual ground temperature has different influences on the different height soil-filling embankment with thermal insulation, having a great influence on low embankment and a slight influence on high embankment (Fig. 3.123).

According to the test section (K2896) of embankment with thermal insulation in Kunlun Mountains along the Qinghai-Xizang Highway, the permafrost under ground surface degrade seriously, with a large increase rate of permafrost table, however, increase rate of permafrost table under embankment with thermal insulation is relatively smaller. Compared with permafrost table under ground surface in 1996, the permafrost table under the embankment with thermal insulation was increased about 0.41 m in 2002. In cold permafrost regions, the thermal insulation layer within embankment can delay the degradation of permafrost under embankment, maintain the thermal stability of the subgrade, and prolong the service life of the subgrade.

The experimental observation data of permafrost table under the embankment in Beiluhe of the Qinghai-Xizang Railway show the effect of thermal insulation on permafrost table (Table 3.48).

Three years later (2004), the permafrost table under the center of embankment with thermal insulation reached 1.29 m to1.50 m, but the permafrost table under the center of soil-filling embankment without thermal insulation layer reached 2.90 m. It shows that the thermal insulation layer plays an obvious role in thermal insulation, elevating the permafrost table under the embankment by 1.40 m. Because the thermal insulation layer is not buried under embankment slope, the effect of lateral thermal erosion on the embankment is relatively stronger. So, the permafrost table under the shaded and sunny shoulder of embankment is relatively approaching to the permafrost table under the embankment without thermal insulation. At the same time, the difference of permafrost table between the sunny and shaded shoulders of embankment is relatively larger. The permafrost table under the sunny shoulder of embankment is by 1.0 m greater than that under shady slopes. The change of the permafrost table under the embankment with thermal insulation can be gradually stabilized, while that under the soil-filling embankment is still in change.

Table 3.48 Changes of the permafrost table under the embankment in the test sections in Beiluhe along the Qinghai-Xizang Highway (m) (China Railway First Survey and Design Institute, 2006)

Mileage	Thermal insulation		S (m)	EH (m)	2002			2003			2004			PTs (m)
	M	T(mm)			L	C	R	L	C	R	L	C	R	
DK1139 + 900	PU	40	0.5	3.34	2.50	2.00	1.50	2.85	1.50	1.00	3.00	1.25	0.90	2.30
DK1139 + 780			0.8*	2.94	2.30	1.40	2.00	2.50	1.40	1.70	2.75	1.35	–	2.70
DK1139 + 820	EPS	80	0.5	2.69	2.00		2.00	–	–	1.10	2.50	1.50	1.00	2.60
DK1139 + 740			0.8*	3.03	2.25	2.00	2.30	2.50	1.60	1.85	2.65	1.50	1.85	2.60
DK1139 + 670	No		–	4.12	2.50	1.80	1.75	2.65	1.75	1.50	2.75	1.50	0.80	2.50
DK1139 + 940			–	4.34	2.52	2.10	1.50	2.85	3.10	0.90	3.05	2.90	0.50	2.52

Notes: S means site of Thermal insulation layer within embankment above ground (m); *means site of Thermal insulation layer under the embankment shoulder. EH means embankment height. PTs mean permafrost table under ground surface. M means materials of thermal insulation, T means thermal insulation thickness. L means under the sunny shoulder of embankment; C means under the centre of embankment; and R means under the shaded shoulder of embankment

When the mean annual air temperature is lower than −4.0 °C, thermal insulation measures can be adopted in the following cases to prevent the atmospheric heat entering into the subgrade, elevating the permafrost table under the embankment, and reducing the embankment settlement caused by the decrease of the permafrost table.

1. The design height of embankment does not meet the minimum critical height of subgrade due to the longitudinal slope.
2. There are low embankments of soil replacement at sections of cuttings or passes.
3. There are sections of low embankment or shady slopes (or low embankment sections in sunny slopes).
4. There damage sections of embankment where the thaw bulb deviates under the embankment.

It is worth noting that even if the embankment with thermal insulation can elevate the underlying permafrost table, the underlying permafrost temperature is generally increased, and there is still heat accumulation under the thermal insulation layer. Therefore, within a certain period, when the mean annual air temperature is higher than −3.8 °C, the permafrost table under the embankment cannot be guaranteed not to decline by the thermal insulation. Of course, the permafrost table cannot be lowered even if the ordinary embankment without thermal insulation boards at low mean annual air temperatures (for example, the air temperature lower than −6.0 °C).

3.4.4.3 Selection of Thermal Insulation Materials

The aging problems of EPS, PU and XPS thermal insulation materials used in cold regions engineering, including the thermal conductivity, water absorption and strength changes, need to be tested under the repeatedly freezing-thawing cycles and different loads. The freezing-thawing cycles of these three kinds of thermal insulation materials can be tested in laboratory, that is, after immersing in water for 7 h, the materials are placed in refrigerator with the temperature of −15 °C for 17 h, and then are placed in water for 7 h. Under such repetitive freezing-thawing cycles, the variations of thermal conductivity, water absorption and compressive strength can be measured. Multiple parallel samples have been tested under repeatedly freezing-thawing cycles for 5, 10, 20 and 30 times and the test results are shown in Table 3.52.

The experiment results show that the change in thermal conductivity and volumetric water absorption of thermal insulation materials is relatively smaller after repeatedly freezing-thawing cycles, while their compressive strength fluctuates slightly, exhibiting a decrease trend with the increase of the freezing-thawing cycles. XPS board has the best durability under a humid condition, and its insulation performance varies slightly with time. After 2 years of use, the thermal resistance still maintains more than 80% in a stable state; the thermal resistance still keeps in 98.4% even after 200 times of freezing-thawing cycles from −45°C to 20 °C. After being immersed in water for 24 h, the compressive strength of XPS boards basically has not decreased, while that of PU board, EPS board and phenolic foam board all have decreased. XPS board has the advantages of small thermal conductivity, low water absorption, high compressive strength and good durability.

The strength selection of thermal insulation boards should not only consider the vehicular load and stress dispersal beneath the pavement, but also consider the compressive stress transmitted to the thermal insulation board by the road roller during the period of the compaction of construction. It is difficult to control the compressive stress transmitted from the road roller to the thermal insulation board to be no more than 0.3 MPa if the soil layer on the thermal insulation board is compacted with high quality.

3.4.4.4 Determination of the Thickness of the Insulation Layer

The thickness of the thermal insulation layer directly affects the thermal insulation effect of the embankment with thermal insulation. The effect of thermal insulation is not directly proportional to the increase of its thickness, but gradually decays with the increase of thickness after reaching a certain thickness.

When the thickness of the thermal insulation layer increases from 10 mm to 50 mm the thawing depth under the embankment decreases rapidly. When it increases from 50 mm to 100 mm, the thawing depth decreases slowly (Fig. 3.124). The test data of thermal insulation embankment in Fenghuo Mountain show that when the thickness of the thermal insulation layer increases from 50 mm to 100 mm, the thawing depth under the embankment only reduces from 32.7% to 42.3%, and the thickness of the thermal insulation layer doubles, and the thawing depth under the embankment only increases by 9.6%. When the thickness of the thermal insulation layer in embankment in Kunlun Mountains increases from 10 mm to 20 mm, the thawing depth under the embankment only decreases by 4.8%; when it increases from 90 mm to 100 mm, the thawing depth decreases by only 1.9%. When the thickness of the thermal insulation layer in embankment doubles in Wudaoliang to 50 mm, the thawing depth under the embankment decreases from 46% to 14.4%.

The thickness of XPS boards should be 80 mm ~ 100 mm in the case of a general embankment height (2.5 m). When the embankment height exceeds 3.0 m, the

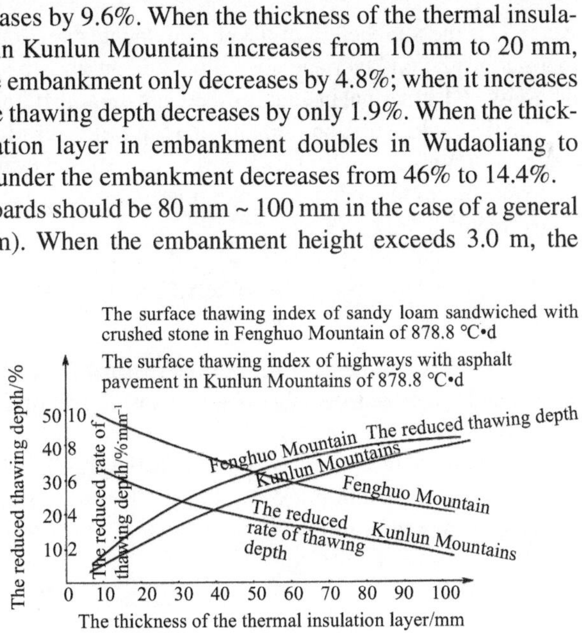

Fig. 3.124 Relationship between the effect of thermal insulation and the thickness of thermal insulation layer (Ding et al., 2011)

lateral heat transfer of the side slope of the embankment should be considered, and the buried depth of the thermal insulation layer should be reduced.

The relationship between the thawing depth of permafrost under the embankment with asphalt pavement and the thickness of thermal insulation layer can be estimated by the Stephen's equation for multilayer media:

$$I_s = \frac{L_n \cdot h_n}{24K^2}\left(\Sigma R_{n-1} + \frac{R_n}{2}\right) \tag{3.56}$$

where I_s is the thawing index of asphalt pavements, °C•d; L_n is the latent heat of volume thawing of the nth layer, kcal/m^3. h_n is the thawing depth of the nth layer, m; R_n is the thermal resistance of the thawing layer in the nth layer, (h•°C•m^2)/kcal; ΣR_{n-1} is the sum of the thermal resistance from the first layer to the nth layer, (h·°C·m^2)/kcal; K is the regional correction coefficient, $K = 1.20$–1.95.

Thermal resistance R_b of the thermal insulation layer:

$$R_b = \delta_b / \lambda_b \tag{3.57}$$

where δ_b is the thickness of the thermal insulation layer; λ_b is the thermal conductivity of the thermal insulation layer.

After the thermal insulation layer is laid in the embankment, the total thermal conductivity (λ) is calculated as:

$$\lambda = \frac{H}{\frac{\delta_b}{\lambda_b} + \frac{\delta_s}{\lambda_s}} \tag{3.58}$$

where H is the total height of the embankment (m); δ_b and δ_s are the thickness of the thermal insulation layer and soil-filling embankment; λ_b and λ_s are the thermal conductivity of the thermal insulation layer and the soil-filling embankment respectively.

The thermal insulation embankment and the elevation of embankment are both designed to reduce the heat transfer by regulating thermal resistance to protect permafrost. From the perspective of thermal resistance, they are treated equivalently, then:

$$\frac{\delta_x}{\lambda_x} = \frac{\delta_s}{\lambda_s} \text{Namely}, \delta_s = \frac{\delta_x \cdot \lambda_s}{\lambda_x} \delta_x = \frac{\delta_s \cdot \lambda_x}{\lambda_s} \tag{3.59}$$

where δ_x and δ_s are the thickness of the thermal insulation layer and the equivalent soil; λ_x and λ_s are the thermal conductivity of the thermal insulation layer and equivalent soil.

According to the concept of reasonable height (H_o) of embankment in permafrost regions, the reasonable thickness (δ_o) of thermal insulation layer in embankment can be determined as follows:

$$\mathrm{H_o} = \frac{(\mathrm{H_o} - \mathrm{h_u} - \mathrm{h_d})\lambda_b}{\lambda_s} \tag{3.60}$$

where: h_u is the thickness of the soil-filling on the thermal insulation layer; h_d is the thickness of the cushion soil layer below the thermal insulation layer; λ_b and λ_s are the thermal conductivity of the thermal insulation layer and soil-filling embankment. The calculated thickness should be multiplied by the safety factor k, which is 1.5–2.0 for the embankment and 1.2–1.5 for the side slope of embankment.

According to the observation of the test section with thermal insulation embankment of the Qinghai-Xizang Railway (China Railway First Survey and Design Institute, 2006), the temperature difference between the upper and lower sides of the thermal insulation board increases correspondingly with the increases in thickness of the thermal insulation layer (Table 3.49), and the elevated height of the permafrost table under the embankment also increases (Table 3.50).The ability to block the downward transmission of external heat also increases with the increase in thickness of the thermal insulation board, so that the temperature difference between the upper and lower sides of the board increases, which is good for the insulation effect of the subgrade.

3.4.4.5 The Buried Depth of the Thermal Insulation Layer

As for the reasonable buried depth of the thermal insulation layer not only the construction quality but also the damage of the thermal insulation layer, its thermal insulation effect under the influence of vehicular load and the self-weight of the subgrade structure after completion should be taken into consideration. As a result, the principle of the reasonable buried depth of the thermal insulation layer should be determined by the main protection object. Based on meeting the strength of thermal insulation layer and the thickness of pavement structural layer and maintaining the stability of permafrost table, the thermal insulation layer should be buried as shallow as possible to exert its function of thermal insulation and reduce the heat absorption of the embankment. The elevation of the permafrost table in the early stage is relatively apparent but its stable development stage is relatively shorter. Considering the protection of the permafrost to prevent the heat from the slope of embankment into the subgrade, the permafrost table beneath the embankment with a thermally insulated layer buried at a greater depth is the shallowest and persists for the longest duration.

1. Buried Depth of Thermal Insulation Layer Based on Stress Dispersal Theory

According to the characteristics of the vehicular load and the stress dispersal theory under the ground (Fig. 3.125), taking the allowable bearing capacity of the thermal insulation layer as the discriminant index, the following formula can be deduced to calculate the reasonable buried depth of the thermal insulation layer.

$$\frac{2Pd}{d+2htg\phi}+h\rho\leq\sigma \tag{3.67}$$

where P is the tire pressure, MPa; d is the diameter of equivalent circle of the pressure surface of a single tire, m; ρ is the weighted average of the density of each

Table 3.49 Comparison of temperature difference (°C) between the upper and lower sides of EPS boards with different thicknesses of thermal insulation embankment of Qinghai-Xizang Railway

The mileage of sections	The thickness of thermal insulation boards	The buried location	The height of embankments (m)	In 2002			In 2003			In 2004		
				At the upper side of the board	At the lower side of the board	Temperature difference	At the upper side of the board	At the lowerside of the board	Temperature difference	At the upper side of the board	At the lower side of the board	Temperature difference
DK1027 + 045	60 mm	0.8 m under the shoulder	3.1	6.1	5.74	0.36	6.82	5.12	1.7	7.36	5.49	1.87
DK1027 + 160		0.5 m above the ground	3.8	−0.21	−0.4	0.19	−0.16	−0.32	0.16	−0.16	−0.4	0.24
DK1139 + 820	80 mm	0.5 m above the ground	2.69	9.4	1.8	7.6	–	–	–	5.3	0.7	4.6
DK1024 + 775		0.8 m under the shoulder	3	14.25	4.72	9.53	11.89	3.85	8.04	7.34	1.71	5.63
DK1139 + 740			3.03	9.8	3.8	6	8.7	–	–	4.8	–	–
DK1024 + 625	100 mm	0.5 m above the ground	3	8.1	1.02	7.08	6.15	0.03	6.12	–	–	–
DK1139 + 670			4.12	8.4	1.1	7.3	–	–	–	14.5	11.2	3.3
DK1024 + 725		0.8 m under the shoulder	3	14.43	4.84	9.59	11.52	3.4	8.12	9.02	1.9	7.12

Table 3.50 Statistics of permafrost table (m) under the EPS thermal insulation embankment with different thicknesses of Qinghai-Xizang Railway (m)

The mileage of section	The thickness of thermal insulation boards	The buried location	The height of the embankment (m)	In 2002			In 2003			In 2004		
				The left shoulder	The center	The right shoulder	The left shoulder	The center	The right shoulder	The left shoulder	The center	The right shoulder
DK1027 + 045	60 mm	0.5 m above the ground	3.1	4.2	3.9	2.9	5.2	3.9	2.9	4.2	3.4	3
DK1027 + 160		0.8 m below the shoulder	3.8	–	1.6	–	–	1.6	–	–	1.6	–
DK1024 + 775	80 mm	0.8 m below the shoulder	3	4.92	3.85	3.85	5.03	4.35	4.95	5.01	2.34	3.27
DK1024 + 625	100 mm	0.5 m above the ground	3	3.29	3.29	2.79	3.22	3.09	2.74	3.57	2.99	3.04
1024 + 987 The off-line			1	2.1	1.5	1.9	1.8	1.2	1.68	1.8	1.22	1.68
1024 + 962 The off-line			1.5	1.3	1.2	1.2	1.2	0.95	1.1	1.21	1	1.1
DK1024 + 725		0.8 m below the shoulder	3	4.98	3.53	3.56	4.93	4.53	3.77	4.3	2.03	3.13

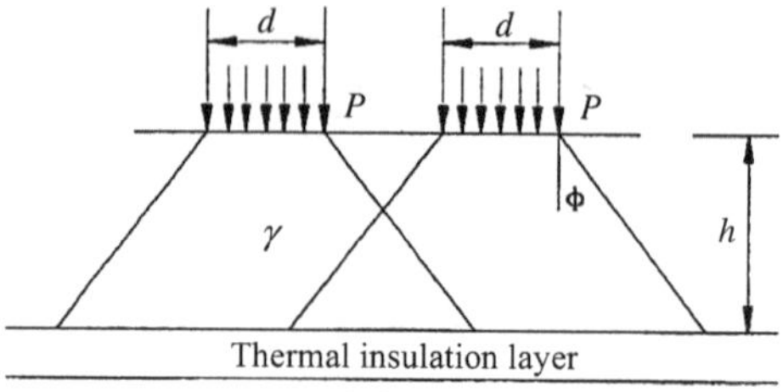

Fig. 3.125 Schematic diagram of vehicular load dispersal

Table 3.51 Pressure and tension above and below the XPS board caused by subgrade settlement (Wang, 2011)

XPS	Settlement/mm	The vertical pressure at different depths below the subgrade top/kPa				The transverse tension at different depths below the subgrade top/kPa			
		0	0.3 m	0.5 m	0.8 m	0	0.3 m	0.5 m	0.8 m
Above the board	30	250.30	243.69	239.04	231.86	7.06	5.98	5.29	4.26
	50	417.16	406.15	398.41	386.43	11.68	9.97	8.82	7.11
	65	542.31	527.99	517.93	502.36	15.30	12.96	11.47	9.24
Below the board	30	248.99	242.46	237.86	230.74	7.77	6.77	6.11	5.13
	50	414.99	404.10	396.43	384.57	12.95	11.28	10.18	8.55
	65	539.48	525.33	515.36	499.94	16.84	14.66	14.66	11.12

structural layer above the thermal insulation layer, MN/m^3; ϕ is the weighted average value of stress dispersal angle of structural layers above the heat insulation layer, °; h is the reasonable buried depth of thermal insulation layer, m; σ is the allowable compressive stress of the thermal insulation layer, MPa.

Different thermal insulation layers have different allowable compressive stresses (σ). The weighted average value of different stress dispersal angles (ϕ) and the weighted average value of the structural layer density (ρ) with different soil-filling above the thermal insulation layer should be calculated correspondingly. By substituting different parameters, the reasonable buried depth of the corresponding thermal insulation layer can be calculated. Since the material load may have a large load, the calculated value should be multiplied by the safety factor, usually 1.3–1.5.

2. Determination of the Buried Depth Based on the Tensile Stress of Thermal Insulation Layer Caused by Subgrade Settlement

Under the same conditions, assuming that the radius R of subgrade settlement is 13 m and the maximum settlement is 30 mm, 50 mm and 65 mm respectively, the pressure calculation results above and below the XPS board are shown in Table 3.51.

Under the effect of subgrade settlement, the pressure (tension) of XPS boards at the centerline of embankment reaches the most, and the stress at the edge fluctuates severely. The vertical pressure is far greater than the transverse tension, and the pressure above the board is 1–2 kPa larger than that below the board. Therefore, the maximum pressure above the XPS board should be taken as the control index to

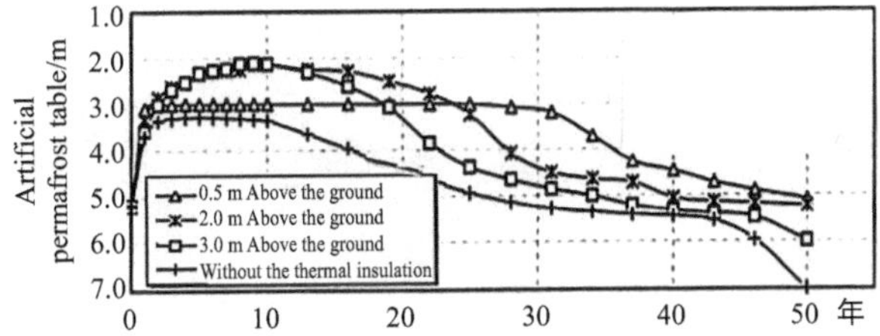

Fig. 3.126 Variation in permafrost table under embankment with thermal insulation layer of differently buried depths

determine the buried depth of thermal insulation layer. When the settlement reaches 65 mm, the pressure above the thermal insulation layer is 500 kPa ~ 540 kPa, and the difference between the maximum and the minimum pressure is 40 kPa, which is less than the compressive strength of the selected thermal insulation board, but there is an extremely large compression deformation (Table 3.51). Under the condition of large settlement deformation (the settlement greater than 65 mm), the XPS board will not be crushed if it is buried 0.5 m below the top of embankment.

3. The Analysis of Buried Depth of Thermal Insulation Layer Based on Thermal Effect

When the railway embankment height is 3.5 m, with the EPS thermal insulation layer of 0.1 m thick fully laid, the deeper is the thermal insulation layer buried, the better the insulation effect will be. (Fig. 3.126) (The Compilation Committee of *Qinghai-Tibet(Xizang) Railway*, 2016). For the embankment with thermal insulation buried 2.0 m and 3.0 m above the ground, the elevation of permafrost table under the centerline of embankment reaches nearly 3.0 m, but permafrost table under embankment with thermal insulation exhibits a decline trend after 9 years and the trend appears under the embankment without thermal insulation after 10 years. As for the embankment with the thermal insulation layer buried at 0.5 m above the ground, the underlying permafrost table will be maintained beneath the thermal insulation board, but exhibits a decline trend until 30 year later. At that time, the permafrost table under the embankment with thermal insulation is still 0.6 m and 1.0 m higher than that under the embankment with thermal insulation buried 2.0 m and 3.0 m above the ground, and almost 2.0 m higher than that under the embankment without thermal insulation layer. The deeper is the thermal insulation layer buried, the better it can prevent the heat from the road and the side slope of the embankment, and reduce the soil thawing depth under the embankment.

According to the numerical simulation results, the is higher the thermal insulation layer buried, the more obvious is the permafrost table rising under the embankment in the early stage, but the shorter is the time for the stable periods of the permafrost table (Fig. 3.127).

The temperature difference between the above and below thermal insulation layer shows that for the embankment of 1 m high, the soil temperature at the top of thermal insulation layer is about 47% higher than the soil temperature below the thermal insulation layer. For the embankment of 1.5 m high, the soil temperature at the top of thermal insulation layer is about 37% higher than the soil temperature

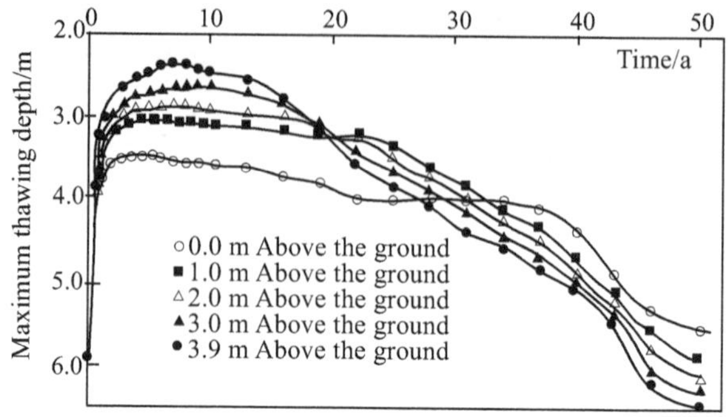

Fig. 3.127 Changes of the permafrost table under the centerline of embankment for thermal insulation layer of differently buried depths (with the embankment height of 4.0 m and mean annual air temperature of −4.0 °C) (Wen et al., 2005)

below the thermal insulation layer. This indicates that the effect of thermal insulation for low embankment is better than that relatively higher embankment. If the soil is in the frozen state under the embankment height of 1 m, there is a strong negative accumulated temperature at the depth of original permafrost table under ground surface, which can increase the cold reserves of the base stratum.

From the perspective of protecting the permafrost stability under the embankment, the thermal insulation layer within embankment should be buried 0.5 m above ground surface, but it is not suitable to be directly buried on the original ground, especially for the relatively higher embankments (more than 1.0 m high). The lateral heat erosion of the embankment slope can be reduced by a deeper buried depth in thermal insulation. For the low embankments or cuttings, under the premise of ensuring that the thermal insulation board will not be crushed, the thermal insulation layer should be buried as shallow as possible under the embankment surface. Meanwhile, some replacement can be carried out appropriately to meet the strength requirements of the embankment.

3.4.4.6 The Soil Minimum Compaction Height above Insulation Layer

At present, the strength of the thermal insulation layer used cannot meet the vehicular load directly placed on it. Therefore, when laying the thermal insulation layer, it should meet the requirements of the structural layer of thermal insulation embankment, and a layer of soil-filling should be laid on it. During construction, if the soil cushion layer is too thin, the thermal insulation layer may be squeezed and crushed, affecting its thermal insulation. If the the soil cushion layer is too thick, the degree of compaction conforming to the requirements of design cannot be met. After selecting a reasonable contact stress of road roller, the key is to determine the minimum thickness of structural cushion layer after compaction on the thermal insulation layer to satisfy the degree of compaction of the structural cushion layer and the condition that the thermal insulation layer is not crushed.

3.4.4.7 Settlement Deformation of Thermal Insulation Embankment

Generally, although there are differences in test sites, the total deformation of the embankment with the thermal insulation is generally smaller than that of ordinary embankment after adopting the thermal insulation layer (Table 3.52). Even in the case of the embankment of 3 m height, the total settlement of the left shoulder of the thermal insulation embankment (average 117 mm) is smaller than that of the ordinary embankment (199 mm), which can reduce the settlement of the embankment effectively. According to the comparison of the annual variation of the total settlement from 2002 to 2004, the average settlement of the left shoulder of the thermal insulation (PU, EPS) embankment with different buried depths (60 mm, 80 mm and 100 mm) is 37.9 mm, 26.4 mm and 30.3 m respectively, while the settlement of ordinary embankment is 55.4 mm, 36.0 mm and 44.7 mm respectively. On the whole, it can achieve good effects if the thermal insulation layer within embankment is buried 0.5 m above the original ground. The use of thermal insulation layer for low embankments has a positive effect on reducing the settlement of embankment. The observation data of stratified deformation shows that embankment deformation comes from soil layer settlement of the embankment basement, and the settlement of embankment itself is small, indicating that the thermal insulation layer plays a role in reducing the settlement. Generally, there is a large settlement difference between the sunny and shady shoulder of embankment, and the settlement on the sunny shoulder is relatively larger, with a larger amplitude of variation. Although the thermal insulation layer can reduce the settlement difference between the sunny and shady shoulder of embankment to a certain extent, it is still necessary to adopt reinforcement measures to reduce the settlement difference on both sides, such as thermal insulation berms or crushed rock revetment.

3.4.5 Composite Embankment with Crushed Rock and Ventilated Ducts

Composite embankment with crushed rock and ventilated duct is used to change the heat transfer mode of the embankment through laying ventilated pipes under large pores of crushed rock and the embankment, to reduces the temperature under the embankment and protect the permafrost.

Regardless of the warm seasons or cold seasons, laying ventilated duct within the embankment is adding a cooling surface in the embankment, which can not only intercept and take away the heat from the embankment surface, but also cool the soil around the ventilated pipes and increase the cold reserves of the subgrade.

According to the working characteristics of crushed rock and ventilated duct embankment, their advantages are concentrated in the same embankment to form a composite embankment structure. The special composite embankment cools the underlying permafrost through the convective heat transfer between the ventilated

Table 3.52 Settlement deformation of thermal insulation embankment with different thicknesses and buried depths (China Railway First Survey and Design Institute, 2006)

The mileage of section	The thickness of EPS thermal insulation boards/mm	The buried location	The height of the embankment/m	Total settlement/mm			
				The left shoulder		The right shoulder	
				The amount of settlement	The average value	The amount of settlement	The average value
DK1027 + 160	60	0.5 m above the ground	3.8	−129	−129	−38	−38
DK1139 + 820	80		2.69	−22.6	−22.6	−62.1	−62.1
DK1024 + 625	100		3.0	−58	−72.7	−32	−42
1024 + 987(the off-line)			1.0	−61		−16	
1024 + 962(the off-line)			1.5	−99		−78	
DK1027 + 045	60	0.8 m below the shoulder	3.1	−45	−45	−137	−137
DK1024 + 775	80		3.0	−185	−99.4	−113	−66.5
DK1139 + 740			3.03	−13.8		−19.9	
DK1024 + 725	100		3.0	−248	−248	−133	−133
DK1024 + 825	Contrastive section		3.0	−199	−199	−103	−103

pipe and the atmosphere, as well as the convection and conduction of the hot and cold air in the crushed rock, which increases the cold reserves of the embankment.

3.4.5.1 Principles of Composite Embankment with Crushes Rock and Ventilated Duct

Crushed rock layers have the thermal physical properties characterized by phase variation. Under positive temperature the thermal conductivity is very small, but under negative temperature the thermal conductivity is very large. The ratio of thermal conductivity of negative temperature to positive temperature is above 12, which can reduce the heat into the subgrade in warm seasons, and increase the cold reserves into the subgrade in cold seasons.

The ventilated ducts are buried in the embankment in permafrost regions, which not only expand the contact surface between the embankment and the air effectively, but also increase the way for air to transmit energy to the embankment and the subgrade. Influenced by forced convection in ventilated duct, the heat in the embankment soil is consumed, so the heat radiation absorbed by the surface of embankment is effectively prevented from being transmitted downwards. The cold reserves introduced into the embankment through ventilated pipes is larger than the heat transmitted into, which can increase the cold reserves of the permafrost subgrade and raise the permafrost table under the embankment.

The adding of ventilated ducts in the crushed rock embankment is equivalent to adding a cooling surface on the top surface of the crushed rock layer, which increases the temperature difference between the top and bottom of crushed rock layer and enhances the intensity of convective heat transfer of the air in the crushed rock layer and the cooling effect of the embankment.

3.4.5.2 Applicable Conditions of Composite Subgrade with Block Stone and Ventilated Pipes

According to the above mentioned, both the crushed rock and ventilated ducts embankment have the function of "heat shield" and "high cold transfer". In warm seasons, the heat into the embankment can be reduced; and in cold seasons, the cold reserves can be collected and transmitted to the embankment base to cool the underlying permafrost and maintain or raise the permafrost table. Therefore, in permafrost regions with warm temperature and high ice content where the mean annual air temperature is lower than −3.5 °C, the composite embankment with crushed rock and ventilated duct can be adopted to combine the advantages of the two kinds of embankments to increase the cold reserves of the cooling subgrade and the ability to cool the subgrade, thus protecting the underlying permafrost stability.

Figure 3.128 shows that the variations of the permafrost table under various embankment in the scenario of air temperature rising by 1.0 °C in the next 50 years. The cooling role of composite embankment is better than the single crushed rock or

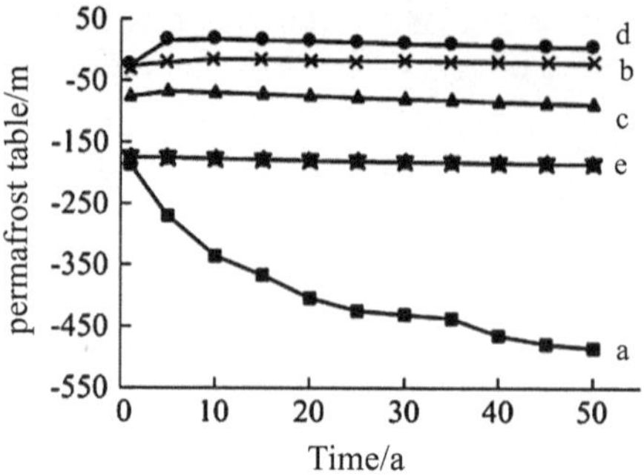

Fig. 3.128 Variations of natural permafrost table under natural surface and artificial permafrost table under various embankment (Zhang et al., 2011). a—General embankment; b—Closed crushed rock embankment; c—ventilated duct embankment; d—Composite embankment with crushed rock (closed) and ventilated ducts; e—Permafrost table under ground surface

ventilated duct embankment, which can greatly improve the artificial permafrost table and protect the stability of underlying permafrost.

In the permafrost degradation regions along the National Highway 214, if the mean annual temperature increases by 2.6 °C in the next 50 years, the composite embankment of crushed rock and ventilated duct with asphalt pavement can decrease the temperature of underlying permafrost. And the artificial permafrost table is always higher than that of the single ventilated duct embankment, which will increase from 0.55 m at the beginning to 0.90 m in the 50th year, with a total increase of 0.35 m after 50 years. It can be seen that the composite embankment with closed crushed rock layer and ventilated ducts can protect the permafrost in a certain period under the impact of long-term warming of climate.

3.4.5.3 The Thickness of Block Stone Layer

At present, it is still difficult to determine the optimal thickness of the crushed rock layer in the embankment with crushed rock and ventilated duct. Compared with the crushed rock embankment, the thickness of the crushed rock layer should be adjusted appropriately because of the cooling effect of ventilated duct. However, in order to give full play to the cooling effect of the embankment with crushed rock and ventilated duct and to adapt to the strong heat absorption of the wide asphalt pavement, it is suggested to design it according to the crushed rock embankment. Generally, the thickness of the crushed rock layer should not be less than 0.6 m, otherwise the artificial permafrost table at the center of the embankment will decrease, and the artificial permafrost table will gradually elevate when the thickness exceeds 0.6 m. When the thickness of the crushed rock layer is 0.6 m ~ 2.5 m, the artificial permafrost table under the embankment is approximately in the stage of rising (Wang et al., 2011). The thickness of the crushed rock layer should not be less than 0.8 m for the Qinghai-Xizang Railway, and 1.0 m ~ 1.5 m for the

Qinghai-Xizang Highway, 0.2 m ~ 0.5 m for the upper layer above the crushed rock layer, and 0.8 m ~ 1.0 m for the lower layer. The thickness of the crushed rock layer for the National Highway 214 in permafrost regions should be 1.0 m ~ 1.5 m, 1.2 m for the ice-rich permafrost regions and 1.5 m for saturated permafrost regions. The design of highways should meet the requirements of the pavement structural layer. Considering the design requirements of the subgrade and engineering cost, and to ensure the elevation of the artificial permafrost table under the embankment and the stability of permafrost, the crushed rock layer of 1.0 m ~ 1.5 m should be laid under the structural layer of embankment.

The observation data of the ventilation duct embankment shows that the ventilated duct can decrease the mean ground temperature of the soil at the depth of 2.5 m under embankment by 0.4 °C ~ 0.5 °C. Considering a certain safety factor, the prefabricated reinforced concrete pipe with a diameter of 0.4 m ~ 0.5 m (outer diameter) is adopted, and the net spacing of pipe laying should be 1.0 m ~ 1.5 m. According to the results of the ventilated pipe embankment of the Qinghai-Xizang Railway, when the pipe diameter is 0.3 m ~ 0.4 m, the spacing of the ventilated pipe should be 1.6 m ~ 2.0 m between the center of the pipe (the net spacing of pipe is twice the pipe diameter).

3.4.5.4 Location of Crushed Rock Layer and Ventilated Duct

For the composite embankment with crushed rock and ventilated ducts, the crushed rock layer should be 0.3 m ~ 0.5 m below the structural layer of embankment, or 0.5 m above the original ground, filled by the method of dumping.

The composite embankment with crushed rock and ventilated pipes can play the role of cooling subgrade by directly laying ventilated pipes on the top of the crushed rock layer (Fig. 3.129), so as to increase the temperature difference

Fig. 3.129 Composite embankment with crushed rock and ventilated duct

between the top and the bottom of crushed rock layer, to exchange heat and air convection and accelerate the cooling underlying permafrost. At the same time, the heat in the crushed rock layer is discharged through ventilated pipes. If the ventilated pipe is buried too deep, it cannot rapidly exert the cooling effect, especially in the advantages of forced convection and natural convection of the crushed rock layer. If the crushed rock layer is paved in double layers, the ventilated pipe should be buried in the upper part between the upper and lower layers of crushed rock layer, that is, on the top surface of the lower crushed rock layer (Fig. 3.130). Both ends of the ventilated pipe should protrude 0.3 m from both ends of the side slope, and an "automatic air door" can also be set to strengthen the cooling effect of the ventilated pipe.

3.4.6 Composite Embankment with Thermal Probes and Thermal Insulation Layer

In cold seasons, thermal probes can bring the atmospheric cold reserve into the permafrost subgrade, cooling underlying permafrost. In warm seasons, the external heat is gradually introduced into the permafrost through the embankment, making the cooling permafrost in cold seasons heat up gradually and thaw because the permafrost cannot be maintained permanently. In warm seasons, the thermal insulation layer prevents the heat transferring to the permafrost subgrade, and effectively reduces the soil thawing index under its lower interface (a reduction of 86%), and reduces the thawing depth of the underlying permafrost subgrade. In cold seasons, the soil freezing index between the upper and lower sides of the thermal insulation layer is also reduced by 83%, so that the permafrost at the base of embankment cannot be refrozen.

It can be seen that the thermal probes can refrigerate actively, but cannot avoid thermal erosion during warm seasons. The thermal insulation layer has strong thermal resistance, which can block the thermal erosion in warm seasons and cold reserves in cold seasons into the permafrost subgrade, reduce the thawing depth, and inhibit the elevation of the artificial permafrost table. The composite embankment with thermal probes and thermal insulation layer takes their own advantages. In warm seasons, the thermal insulation layer can block the thermal erosion and maintain the cold reserves of thermal probes of the previous year and the artificial permafrost table. In cold seasons, thermal probes can play the role of cooling that can continually decrease underlying permafrost temperature, elevate the artificial permafrost table further, and maintain the stability of the permafrost under embankment by repeated cycles.

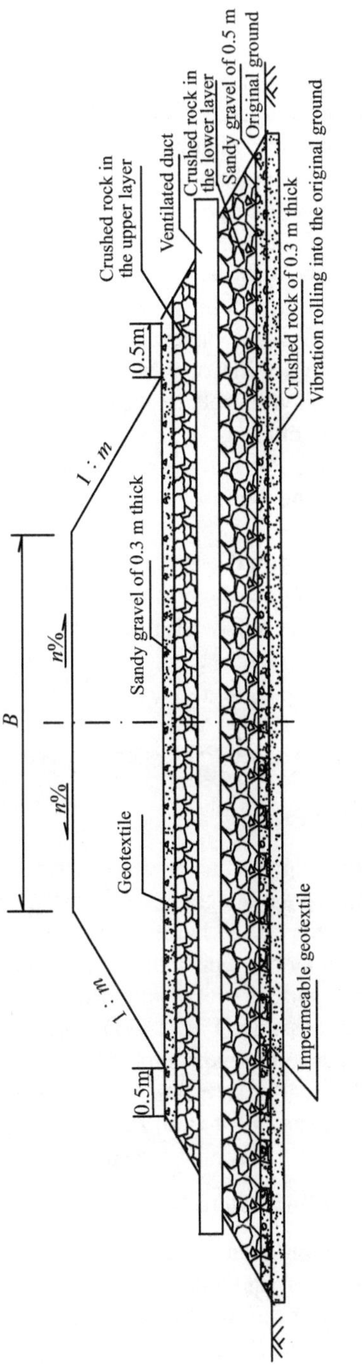

Fig. 3.130 Schematic diagram of composite embankment with crushed rock and ventilated duct

3.4.6.1 Application of Composite Embankment with Thermal Probes and Thermal Insulation Layer

In permafrost regions, the freezing index of air temperature is higher than the thawing index, and long periods of low temperature can be utilized to decrease the soil temperature around the evaporation section of the thermal probes. Thermal insulation materials can block most of the heat into the subgrade in warm seasons, so as to maintain the thermal state of the permafrost subgrade and mitigate permafrost change under the embankment.

The composite embankment with thermal probes and thermal insulation layer fully utlizes the advantages in the cooling role of thermal probes and the role of thermal insulation. The thermal insulation layer is set within the embankment above the ground, near the adiabatic section of thermal probes. During warm seasons, thermal insulation layer can play the role of thermal resistance to reduce the heat into permafrost layer. During cold seasons, the thermal probe can play its cooling role to continue the cooling of the permafrost and to elevate the artificial permafrost table, avoiding the defect that the thermal insulation layer blocks the cold reserves into the subgrade.

The composite embankment with thermal probes and thermal insulation layer is suitable for the embankment engineering in the warm permafrost regions, so as to reduce the permafrost temperature, raise the artificial permafrost table, and maintain the embankment stability.

3.4.6.2 Buried Position of Thermal Probes and Thermal Insulation Layer

1. **Buried Depth of Thermal Probes**

The buried depth of the thermal probe can be determined by the buried depth of the infrastructure and the depth of the subgrade to be strengthened. The evaporation section of the thermal probe should be buried in the permafrost layer to cool the permafrost layer under embankment. For both straight and curved thermal probes, the adiabatic section of the thermal probe should be placed beneath the thermal insulation layer as much as possible (Fig. 3.131), so that it is placed within the maximum seasonally thawed depth under embankment. When the height of the embankment is relatively higher, the adiabatic section can be appropriately higher than the thermal insulation layer, but it is still advisable to bury 1/2 ~ 2/3 of the evaporation section of the thermal probe into the permafrost layer.

2. **Buried Depth of Thermal Insulation Layer**

Generally, the thermal insulation layer should be buried 0.5 m above the original ground with full width, instead of being directly buried on the ground. The experience of the Qinghai-Xizang Highway shows that the thermal insulation layer with a thickness of 80 mm ~ 100 mm can block more than 80% of the heat intruded into

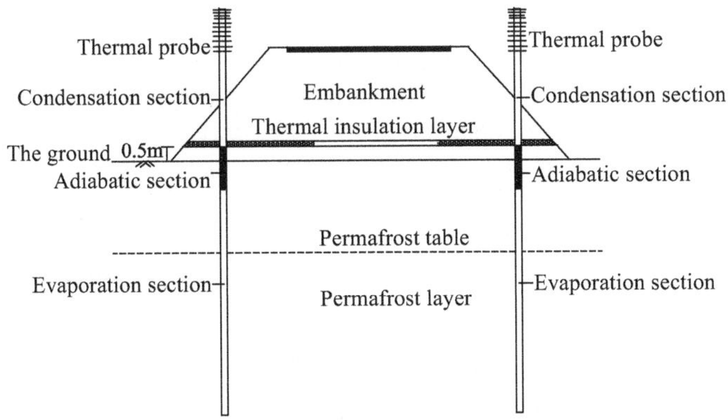

Fig. 3.131 Buried position of thermal probes (thermal probe of 9 m long)

the embankment. In the case of low embankments, under the condition that the thermal insulation layer is not crushed, the buried depth can be appropriately higher than the ground. If the thermal insulation layer is buried below the ground, the thermal insulation layer needs to be treated with some moisture-proof and waterproof measures.

3.4.6.3 Buried Method of Thermal Probes and Thermal Insulation Layer

The construction of the composite subgrade with thermal probes and thermal insulation layer should be divided into two stages: the thermal insulation embankment should firstly be constructed as usual to reach the design height of the embankment; and then the thermal probes can be installed and constructed according to the design requirements.

1. **Thermal Insulation Layer Construction**

For the composite roadbed with thermal probes and thermal insulation layer, the thermal insulation layer should be buried in the position as near the ground surface as possible. The heat from the side slope of the embankment with asphalt into the subgrade accounts for about 30% of the total heat amount, and the heat amount from the side slope of embankment is related to the embankment height, the radiation of the side slope, and embankment direction. The thermal insulation layer should be placed on the adiabatic section of the thermal probe, so that most of the evaporation sections of the thermal probe can be inserted into the permafrost. The thermal insulation layer should be laid with full width in the cross-section of the embankment and the surrounding of the thermal probe should be covered as tightly as possible. The connection among thermal insulation boards should use the lap joint or tongue-and-groove joint, bonded with special adhesive.

2. **Installation of Thermal Probes**

The thermal probe should be buried after the construction of the thermal insulation embankment is completed. The thermal probe can be buried by borehole and insertion according to the designed spacing. During the construction process, strict measures should be taken to ensure that the thermal insulation layer is not damaged.

Construction methods can be determined by the types of thermal probes: drilling and insertion for straight thermal probes, drilling on inclined surface, and oblique insertion for straight but oblique-inserted thermal probes according to its design angle. The above forms of thermal probes are suitable for the drilling method after the completion of the thermal insulation subgrade.

The curved thermal probes can also be carried out simultaneously with the construction of the embankment. The thermal probes can be placed in advance at the designed buried depth, and then the embankment can be filled. When the embankment height is slightly higher than the ground (0.3 m ~ 0.5 m) the thermal insulation layer shall be laid according to the construction requirements of the thermal insulation layer. Under the condition of integral embankment on wide pavement of expressway, the composite embankment with the combination of curved thermal probes and thermal insulation layer should be adopted as far as possible. In addition to the thermal probes buried on both sides of the subgrade, one or two rows of curved thermal probes should be buried in the middle isolation belt of expressway (Figs. 3.132 and 3.117).

3.4.7 *Revetment and Berm Embankment of Crushed Rock*

Under the same embankment height, more than 70% of the total amount can enter into embankment with sandy gravel pavement through embankment slope, and about 30% into embankment with asphalt pavement (the heat mainly comes from the asphalt pavement). With the increase of the embankment height, the thermal influence from the embankment slope becomes more and more obvious, therefore,

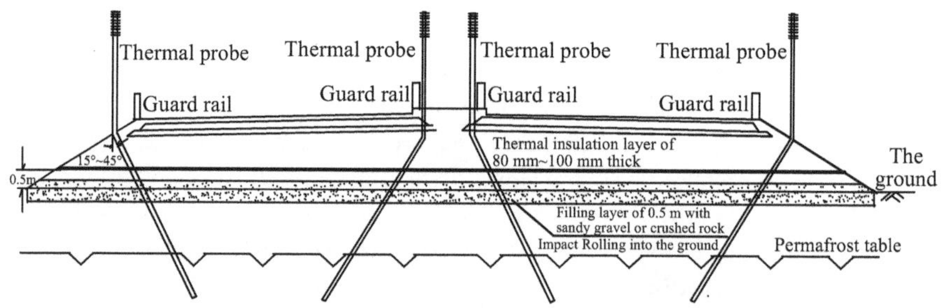

Fig. 3.132 Schematic diagram of composite embankment with curved thermal probes and thermal insulation layer on wide asphalt pavement

as the height of the subgrade increases, it is necessary to make full preparation for the heat flow protection on the slope.

Crushed rock layer laid on the slope of the subgrade can regulate the convection and conduction to reduce the heat into embankment. In warm seasons, it can reduce the heat from the atmosphere into the embankment. In cold seasons, it is conducive to the heat dissipation of the embankment to the outside. The heat transfer mechanism of the crushed rock revetment embankment is a mixed heat transfer mechanism of thermal insulation and air convection, which can effectively reduce the absorption of the solar radiant and play an effective role in cooling underlying permafrost.

In the 1970s and 1980s, soil berms were extensively built on both sides of the embankment along the Qinghai-Xizang Highway to prevent the lateral standing water from infiltrating into the embankment base, which can cause slumping of road shoulders and embankment side slopes. However, from the perspective of thermal stability for some warm permafrost areas, soil berm of embankment has not played its role in thermal insulation. The observation data of the ground temperature under soil berms along the Qinghai-Xizang Highway show that the ground temperature at the depth of 0.5 m below the surface is about 10 °C higher than that under the natural ground during warm seasons, indicating that the ground temperature has an obvious rising trend after laying the soil berm, which is not conducive to the thermal stability of the underlying permafrost.

Soil thermal regime under the soil berm shows that the soil temperature at different depths under the left berm (the sunny side) is higher than that under the natural surface, with a downward heat flow (Fig. 3.133). This indicates that the underlying permafrost is in a state of heat absorption, which will accelerate the permafrost degradation. The right berm (the shady side) is exactly the opposite, with an upward heat flow. This indicates that the permafrost is in a state of heat release, which will decrease permafrost temperature. The heat difference between the shady and sunny slopes on both sides of the embankment leads to the variations of the artificial permafrost table. The artificial permafrost table under the left berm declines, forming a

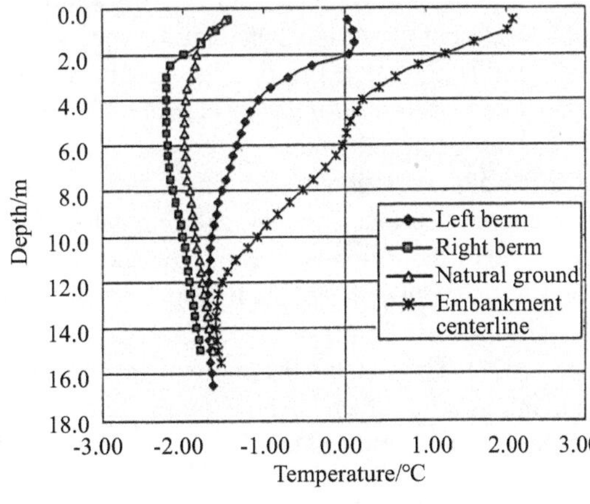

Fig. 3.133 Variations of the mean annual ground temperature with depth

residual thawed layer, while the artificial permafrost table under the right berm rises, forming a frozen soil core, which eventually causes the deformation of the subgrade and longitudinal cracks.

From this point of view, although the total amount of solar radiation on the soil berm and the natural ground is approximately the same, there is a difference in surface reflectivity. On the one hand, the vegetation on the natural ground can increase the surface reflectivity, and on the other hand, the root system has the characteristics of water retaining, and the water evaporation will consume a lot of heat, reducing the ground temperature and protecting the permafrost. Therefore, it is counterproductive to simply use soil berms to achieve the thermal insulation on the embankment in the warm permafrost areas because it can easily cause the appearance of the residual thawed layer under the embankment.

Therefore, it is necessary to change the surface characteristics and internal material composition of the soil berm to achieve the thermal insulation of the berm. The experimental results show that the embankment can be cooled by using the ventilation and convection effects of crushed rock layer. So, crushed rock berms replacing with soil berms play the role of cooling underlying permafrost. Meanwhile, slope protection and soil berms with vegetation can be utilized to decrease permafrost temperature by reducing the heat absorption and increasing the heat dissipation of slope protection and berms.

3.4.7.1 Working Principles of the Embankment with Crushed Rock Revetment and Berms

Crushed rock revetment is to dump a certain thickness of crushed rock layer on the slope of the embankment by its effects of thermal switch and sunshade to improve the permafrost stability under the embankment. In warm seasons, there are no convection within the crushed rock layer, but the crushed rock layer can play the role of thermal insulation. And the heat is only transmitted by the point contact of the crushed rock revetment to the embankment slowly. The hot air heated by the air in the crushed rock revetment returns to the atmosphere along the pores of the crushed rock revetment due to the density difference, reducing the heat into the embankment base, and achieving a strong "thermal shielding" effect. In cold seasons, chimney effect due to density difference between the cold air and the heat air within crushed rock layer is formed to cool underlying permafrost.

The solar radiation is the direct heat source for the temperature rise of the embankment surface (the embankment surface and the slope surface), resulting in the rising of the ground temperature. The observed data show that there is a strong correlation between net radiation and the surface temperature (Fig. 3.134) (Northwest Research Institute Co., Ltd. of China Railway Engineering Corporation, et al. 2006). The sunshield can effectively block solar radiation, reducing the radiation intensity by 82% (The Compilation Committee of *Qinghai-Tibet(Xizang) Railway*, 2016), and reduce the surface temperature (with the surface temperature

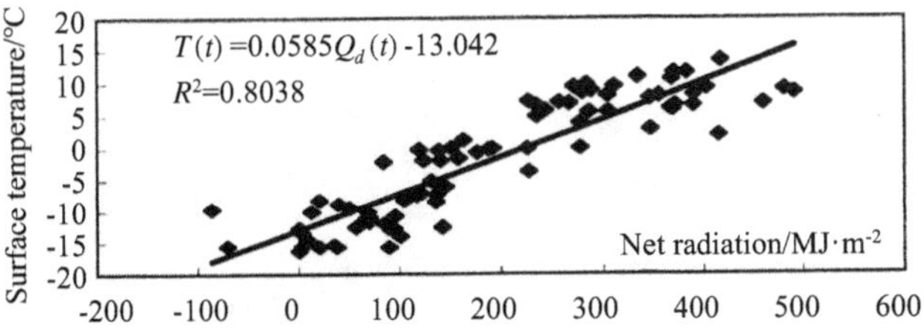

Fig. 3.134 Relationship between annual net radiation and surface temperature at Fenghuo Mountain meteorological station (2000–2005)

difference of 8 °C between the inside and outside of the sunshade) or the temperature of slope surface.

The crushed rock berm has the same function of a crushed rock embankment. It should be emphasized that, in addition to thermal insulation and convection, the more important function of the crushed rock berm is to drain the slope toe of the embankment to avoid standing water, preventing the soil temperature under the embankment from rising and the subgrade strength from weakening. Therefore, when crushed rock berms constructed, the standing water at the slope toe of the embankment should be drained in the original ground.

3.4.7.2 Applicable Conditions for the Embankment with Crushed Rock Revetment and Berms

Factors that the crushed rock revetment and berms can exert full play to its engineering effect are as follows.

1. **Climatic Conditions**

The mean annual air temperature is the basic requirement for determining the effective potential freezing capacity and meeting the occurrence and development of permafrost. According to the research data in the Qinghai-Xizang Plateau, if the corresponding mean annual air temperature is −2.74 °C, the baseline I can be drawn (Fig. 3.135), taking the basic conditions for the formation of permafrost in the cohesive soil zone in this region (lower temperature for the sandy gravel regions) as the standard. The excess freezing index (the difference of accumulated temperature) Ω_G is −1000 °C·d, and the freezing number (the ratio of temperature freezing index to thawing index) F is 2. So, there is more effective potential freezing capacity in the northern areas of 32°15′ to 35°30′, where the potential freezing capacity connecting permafrost (without talik) and continuing freezing can be formed in cold seasons. The lower part in Fig. 3.138, which is already at the edge of the northern and southern boundary of permafrost, has no surplus freezing capacity. The baseline II can be obtained, taking the corresponding mean annual temperature $T_0 \approx 4.0$ °C when the mean annual surface temperature in plateau is 0 °C as the standard. The excess freezing index Ω_G is −1500 °C·d, and the freezing number F is 3 (Fig. 3.138). The

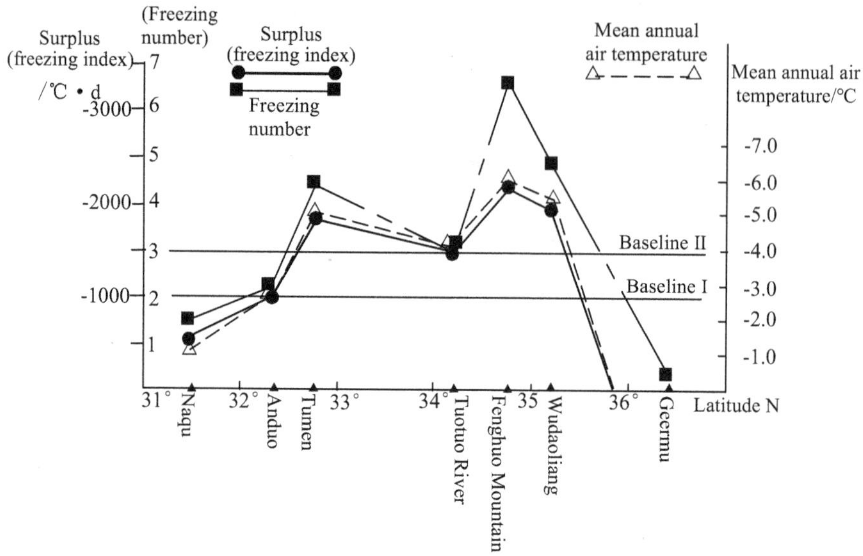

Fig. 3.135 Diagram of the potential freezing capacity in permafrost regions of the Qinghai-Xizang Plateau (According to Northwest Research Institute Co., Ltd. of China Railway Engineering Corporation, etc. 2006)

absolute value of the surplus freezing index is more than −1500 °C·d, approximately to the hinterland of the Qinghai-Xizang Plateau from 32°30′ to 35°23′, where there is a strong effective potential freezing capacity, so it is suitable for adopting the engineering structural measures of cooling the atmosphere.

It is reasonable to use the crushed rock revetment to maintain the permafrost stability under embankment in regions where the absolute value of the excess freezing index Ω_G corresponding to the baseline of I is more than or equivalent to 1000 °C·d and the freezing number F is more than or equivalent to 2.

2. **The Conditions of Surface and Stratum**

The conditions of surface and stratum refer to the thermophysical characteristics of the rock and soil subgrade, its thermal state, lithology, water (ice) content and hydrogeological conditions, all of which can maintain a stable geothermal field by crushed rock layer.

In the same exposed region, the mean annual surface temperature is higher than its air temperature, 2 °C ~ 4 °C higher on gravel ground, 5 °C ~ 6 °C higher on asphalt pavement, 10 °C ~ 15 °C higher in the warm permafrost, and 5 °C ~ 10 °C higher in cold permafrost. The surplus freezing index of the surface is only about 50% of the undex of the air temperature. Therefore, the corresponding relationship between the accumulated air temperature and the accumulated ground temperature should be fully considered when the air temperature is used to measure the potential freezing capacity of the ground. According to the boundary layer theory (Frozen Soil Subgrade Research Group of Qinghai-Xizang Highway, 1988), the same temperature fluctuations cause different temperature changes on different material surfaces, and different surface characteristics (material composition, topography, water state, vegetation and coverage, et al) have different absorption of solar radiation heat.

Under certain conditions, the influence of local factors will exceed the influence of the atmosphere. For example, the embankment construction often causes the change in orientation, resulting in the thawing index at the sunny slope being half larger than that of the shady slope. The average temperature at the exposed ground (gravel surface) is 8 °C higher than that of the ground with sun-shading measures (such as sunshade) (Northwest Research Institute Co., Ltd. of China Railway Engineering Corporation, et al. 2006). The mean annual soil temperature in the crushed rock layer is 4–7 °C lower than that of the adjacent soil. The mean annual soil temperature under the crushed rock layer (with the thickness of 0.15 m) in the Kunlun Mountain Pass measured by Harris is 7.1 °C lower than that of the adjacent fine-grained soil (Cheng, 2003).

When there is standing water at the slope toe of the embankment, the construction of soil berms can be used for drainage. However, the construction of the soil berms must be closely connected with the drainage system of the embankment to avoid losing the drainage function due to soil berms construction.

3.4.7.3 Particle Size of Crushed Rock Revetment and Berms

1. **Particle Size of Crushed Rock Revetment**

The intensity of natural convection within the crushed rock layer depends on the Rayleigh number R_a. Under the action of negative temperature gradient, the larger is R_a, the greater the intensity of natural convection will be. The experiment results show that R_a increases with the increase of particle size of crushed rock, mainly due to the increase of air permeability (*K*) of the crushed rock layer. The air permeability varies significantly with the particle size of crushed rock, while the thermal conductivity decreases with the increase of the particle size. (Table 3.53).

Table 3.53 Physical parameters of materials with different particle sizes (Sun et al., 2004a, b)

Material	Particle size/mm	Density/ Kg·m^3	Porosity/%	Permeability/ m^2	Thermal conductivity/ W·m^{-1}·K^{-1}	Volume heat capacity/J·m^{-3}·K^{-1}
Crushed rock	20 ~ 40	1520	44.4	7.80×10^{-7}	0.424	1.277×10^6
	40 ~ 60	1500	45.2	2.28×10^{-6}	0.407	1.260×10^6
	60 ~ 80	1400	45.7	4.38×10^{-6}	0.396	1.250×10^6
	100 ~ 150	1410	46.5	6.67×10^{-6}	0.385	1.184×10^6
Cobble & gravel	≤40	1790	36.1	3.00×10^{-8}	0.522	1.343×10^6
Sandy gravel	≤5	1800	–	3.00×10^{-10}	0.766(0.857))	$1.581(1.393) \times 10^6$
Air	–	0.900	–	–	0.024	0.909×10^3

Fig. 3.136 Block (left) and crushed (right) stone slope protection (Photo by Ge Jianjun)

According to the indoor model test and calculation results, the cooling effect of the crushed rock layer with a particle size of 60 mm ~ 80 mm is better (Sun et al., 2004a, b). Under the opened condition at the top of crushed rock layer and a certain wind speed, the cooling effect of the crushed rock layer with the average particle size of 70 mm is better than that of the crushed rock layer with the average particle size of about 220 mm (Yu et al., 2003). However, the observation results on the Qinghai-Xizang Railway in Beiluhe River show that the crushed rock layer with the average particle size of less than 70 mm has a strong thermal shielding effect, and its cooling effect is inferior to that of the crushed rock layer (Sun et al., 2004a, b).

According to the comparative experimental research of the Qinghai-Xizang Railway at the Beiluhe River test sections (Fig. 3.136), the particle size of crushed rock is 400 mm ~ 500 mm with the thickness of 0.8 m; and the particle size of broken rock is 50–80 mm with the thickness of 0.8 m. There are difference in permafrost temperature and permafrost table based on the comparison of observation data under the embankment of curshed rock revetment and broken rock revetment.

Comparative analysis of observation data in soil temperature during the period of 3 years show that the soil temperature difference between crushed rock revetment and broken rock revetment at the 0.3 m-depth under embankment surface is the largest, that is, about 0.2 °C ~ 0.4 °C lower; otherwise, the difference at the embankment base is middling, about 0.1 °C ~ 0.2 °C lower; the difference near the original natural permafrost under the embankment is the smallest, about 0.1 °C lower. Comparative analysis of accumulated temperature shows that there is more negative temperature accumulated under the center of crushed rock and broken rock revetment embankment than that under ordinary embankment, and the difference also decreases with the increase of depth (Table 3.54).

By comparing the ground temperature isotherm under the crushed rock revetment embankment and broken rock revetment embankment observed in Beiluhe River test section (Fig. 3.137) (Compilation Committee of *Qinghai-Xizang Railway*, 2016), the ground temperature under ordinary subgrades is generally higher than that of broken and crushed rock revetment embankment. The ground temperature difference between the sunny slope and the shady slope of broken rock revetment

Table 3.54 Comparison of mean annual soil temperature (MAST, °C) and accumulated temperature (AT, °C• d) at different depth under the centerline of embankment with crushed rok revetment (CRRE), embankment with broken rock revetment (BRRE) and ordinary embankment (OE) (The Compilation Committee of *Qinghai-Tibet(Xizang) Railway*, 2016)

Year	Sections	0.3 m-depth under embankment		Embankment base		Near the original permafrost Table	
		MAST	AT	MAST	AT	MAST	AT
2002	DK1141 + 374 CRRE	−0.93	6.96	−0.02	−2.40	−0.26	−36.26
	DK1141 + 324 BRRE	−0.68	41.36	−0.01	−0.55	−0.32	−45.28
	DK1140 + 882 OE	2.21	418.32	0.28	42.50	−0.21	−29.66
2003	DK1141 + 374 CRRE	0.33	55.80	−0.22	−81.67	−0.33	−120.85
	DK1141 + 324 BRRE	0.06	−40.11	−0.27	−98.18	−0.44	−160.91
	DK1140 + 882 OE	0.47	126.09	−0.06	−19.12	−0.33	−119.46
2004	DK1141 + 374 CRRE	−1.30	−477.93	−0.25	−90.44	−0.44	−159.44
	DK1141 + 324 BRRE	−2.28	−489.65	−0.20	−42.50	−0.52	−110.70
	DK1140 + 882 OE	−0.82	−294.35	−0.08	−30.16	−0.42	−152.56

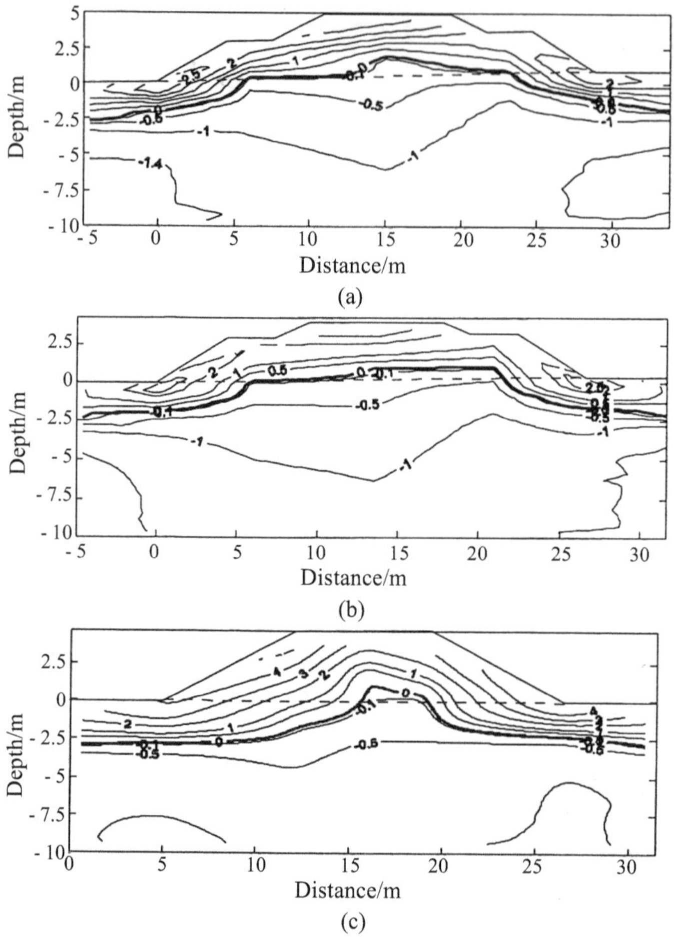

Fig. 3.137 Comparison of ground temperature Isotherm under broken rock revetment embankment. (**a**) Crushed rock revetment embankment (**b**) Ordinary embankment (**c**) In Beiluhe River Test Section (September 29, 2004)

embankment and crushed rock revetment embankment is small, but the difference of the ordinary subgrade is relatively large.

The observation data show that the permafrost table under the crushed rock revetment embankment rises by 2.5 m, 2.2 m for that under the broken rock revetment embankment, and 1.8 m for that under ordinary embankment. In comparison, the permafrost tables under the crushed rock revetment embankment and the broken rock revetment embankment rise by 0.7 m and 0.4 m respectively, compared with that under the ordinary embankment.

According to the calculation of the heat budget of the soil layer at the depth of 0.2 m ~ 0.7 m under the crushed and broken rock layer (Ma, 2006), the total heat budget of the soil under the broken rock layer was 3878.8 kJ m^{-2} during the

Fig. 3.138 Crushed rock berms (Left) and broken rock berms (Right) (Photo by Ge Jianjun)

observation period, which was in the state of heat absorption; while the total heat budget of the soil under the crushed rock layer was −10,940 kJ m^{-2}, which was in the state of heat dissipation. This shows that with the same thickness, the crushed rock layer with a particle size of 400 mm ~ 500 mm has better cooling effect on the permafrost than that of the broken rock layer with a particle size of 50 mm ~ 80 mm.

Within a certain range of particle size, the cooling effect of the crushed rock layer increases with the increase of the particle size, because the permeability of the crushed rock increases with the increase of the particle size. However, when the particle size exceeds 400 mm its permeability decreases with the increase of the particle size due to the decrease of surface area and the change in pore characteristics. The indoor experimental results in the optimal cooling particle size in closed crushed rock layer show that the cooling effect of the particle size of 221 mm and 271 mm is better than that of 83 mm and 148 mm.

According to the above indoor and field experimental test, as well as the situation in engineering practice, the particle size of the crushed rock should be 150 mm ~ 250 mm, and 80 mm ~ 100 mm for that of the broken rock.

2. **Particle Size of Broken Rock and Crushed Rock Berms**

The experimental research on broken and crushed rock berms is only conducted in the Wudaoliang section along the Qinghai-Xizang Railway (Fig. 3.138). Here are the experimental observation results (Table 3.55).

1. **Variation of Freezing-Thawing Process of Embankment with Berms**

Figure 3.139 shows that the surface of the left shoulder starts to thaw from late April to early May. The thawing time of section A3 is about 5 days later than that of the section A1 and A2. On the right shoulder, the surface of the section A1 and A3 begins to thaw in the middle of May, and the surface of the section A2 begins to thaw in late May, 5–10 days later. The thawing time of the left shoulder is 10 days to 1 month earlier than that of the right shoulder by comparison, and the refreezing is also delayed by the same time. The maximum thawing depth is also delayed

Table 3.55 Structural dimensions of the test section of crushed rock berms in Wudaoliang along Qinghai-Xizang Railway

Sections				Berms Width/m		
Mileage	Code	Particle Size/mm	Thickness/m	Sunny (Left)	Shade (Right)	Remark
DK1082 + 625	A1	100	1.5	6.0	4.0	The mean annual air temperature, 5.6 °C, mean annual ground temperature: −2.4 °C
DK1082 + 675	A2	200	1.5	6.0	4.0	
DK1082 + 725	A3	300	1.5	6.0	4.0	

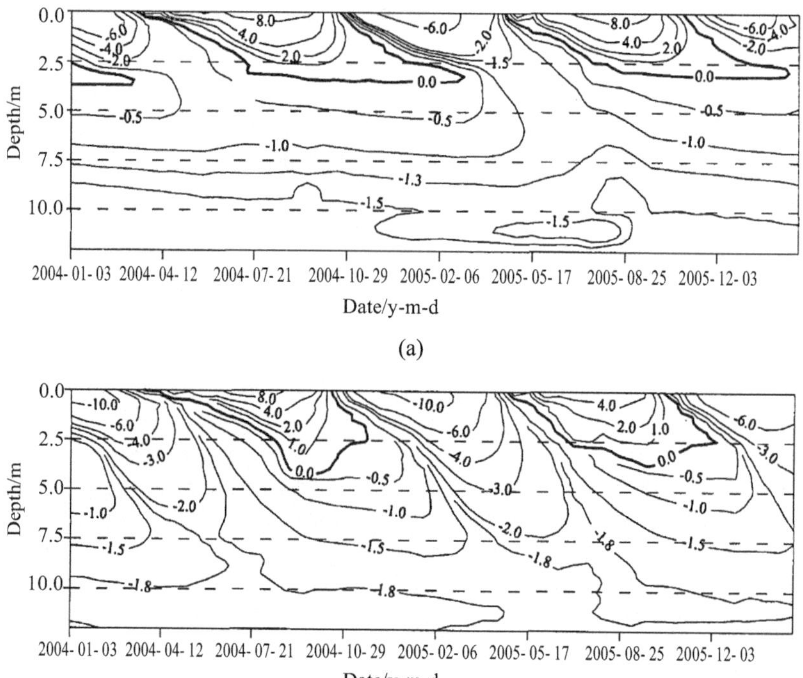

Fig. 3.139 Freezing-thawing process under the left (**a**) and right (**b**) road shoulders of the crushed rock berm DK1082 + 725 (A3) in Wudaoliang (Xu, 2006)

20 days compared with the right shoulder, and the complete refreezing is delayed by 2 ~ 3 months.

2. **Spatial-Temporal Variation of Ground Temperature of Berm Embankment**

The mean annual soil temperature at different depths on the left shoulder (the sunny slope) of the section A1 in 2005 is basically higher than that in 2004, but there is little change below the 4.5 m depth, while the ground temperature of the right shoulder decreases compared with the previous year. The mean annual soil temperature at different depths of the left and right road shoulders of the section A2 and A3 are lower than those of the previous year.

The positive and negative temperature period of the left shoulder is 5 ~ 7 months, that is, 7 ~ 9 months for the negative temperature period of the right shoulder, 3 ~ 5 months for the positive temperature period of the right shoulder, 9 ~ 11 months for the negative temperature period of the toe of the right slope, and 1 ~ 3 months for the positive temperature period of the toe of the right slope. Except that the mean annual values of the maximum and the minimum ground temperature at the original natural ground in the left shoulder (the sunny slope) for the section A1 and A2 are positive, the mean annual values of the section A3 as well as the right shoulder, the left and the right slope toe are negative. It indicates that crushed rock with a large particle size has a better cooling function. The broken rock in the section A1 has the smallest cooling function.

3. **Ground Temperature Field under Berm Embankment**

The ground temperature of each observation section in 2005 was lower than that in 2004, indicating that the cold reserves in cold seasons increase, so that there is no large talik under the left slope.The ground temperature field of the maximum seasonal thawing depth for each observation section shows that, on the same level, the ground temperature of the slope toe at the left and right berms and the left shoulder is higher, followed by the ground temperature of the slope toe at the left and right subgrade, and the ground temperature of the right shoulder is the lowest. The permafrost on the left and right shoulders of embankment, the slope toe of the embankment and the slope toe of the right berm are above the original permafrost table under natural surface, and the permafrost table under embankment has been raised. But the permafrost table at the slope toe of the left berm of each section is below the original permafrost table under natural surface.

The ground temperature at the permafrost table under the embankment is decreasing year by year. On the sunny or shady slope of embankment, the thermal stability of permafrost under the shoulder of each section, slope toes of berms, and slope toes of the embankment increases in turn, but the thermal stability of permafrost under the shady slope of the embankment is higher than that of the sunny slope. Obviously, it is related to the effect of solar radiation and different transmission distances of the ground temperature waves at the same depth. The crushed rock berm has a strong cooling effect on permafrost under the slope toes of the embankment, but has a weak cooling effect on permafrost under the slope toes of the berm.

The permafrost table under the embankment with crushed rock berm is generally raised. But the original permafrost table under the slope toes of the left and right shoulder of embankment shows a trend of rising and the mean annual soil temperature shows a trend of decreasing year by year. This indicates that the thermal stability of permafrost under the slope toes of the left and right embankment is higher than that under the road shoulder, which is because the slope toes of the embankment is affected by the cooling effect of the berm.

3.4.7.4 Thickness of Crushed Rock Revetment and Berm

According to the seepage theory of porous medium, in addition to using its variable equivalent thermal conductivity, it is also necessary to lay a minimum thickness to achieve the ideal cooling effect for the crushed rock layer with a certain particle size. This thickness needs to meet the following conditions: during the negative temperature period, the cold energy can be fully transferred to the bottom of crushed rock layer, while the heat cannot be transferred to the bottom during the positive temperature period. The minimum critical Rayleigh number R_{ac} for natural convection in the crushed rock layer is $4\pi^2$, and the Rayleigh number is directly proportional to the thickness of the crushed rock medium, so that the critical thickness for natural convection within the broken and crushed rock layer revetment can be obtained. Based on the convective cooling effect, when the ground temperature amplitude (θ) is a certain value, the minimum thickness (H) of the crushed rock layer corresponding to the critical state where the natural convection effect of the embankment that can start to "cool the subgrade" in winter can be given by the following formula (Sun et al., 2005):

$$\text{For airtight boundary } H_{\min} = -43.61 \operatorname{lin} \theta + 180.3 \tag{3.72a}$$

$$\text{For breathable boundary } H_{\min} = -34.94 \operatorname{lin} \theta + 129.3 \tag{3.72b}$$

For the embankment with ballast and crushed rock revetment, its thickness is 1 m, and the crushed rock shows that when the temperature amplitude is 13 °C, the minimum thickness of the crushed rock revetment is 0.7 m (in airtight boundary) and 0.4 m (in breathable boundary) according to the formula 3.72.

According to the crushed rock revetment test embankment of Fenghuo Mountain observation station of China Railway Northwest Research Institute, the particle size of crushed rock is 50 mm ~ 150 mm, and the thickness of the upper part is 0.7 m, 0.55 m to 0.6 m for the middle and lower part. When the surface temperature amplitude of the crushed rock layer is 14.2 °C, the minimum thickness of the crushed rock revetment is 0.66 m (in airtight boundary) and 0.37 m (in breathable boundary) according to the formula 3.72.

The natural convection effect of crushed rock revetment in winter has the following characteristics:

1. For the crushed rock revetment with a certain thickness, its natural convection effect in winter increases with the increase of the temperature amplitude of the pavement.
2. Under the same conditions, the air in the crushed rock layer in breathable boundary can form a convective circulation with the outside atmosphere, so that the natural convection effect in the crushed rock revetment in breathable boundary in winter is significantly stronger than that in airtight boundary.
3. Given a certain temperature amplitude of the pavement, the natural convection effect of crushed rock revetment in winter increases with the increase of the thickness of the crushed rock layer;

4. The natural convection effect of the crushed rock layer revetment in winter usually forms from the the slope toe of embankment. Therefore, in comparison, the natural convection in the lower part of the crushed rock revetment has relative stronger cooling effect.
5. The crushed rock revetment can not only have the cooling effect of natural convection in winter, but also play the role of thermal resistance in summer to prevent more heat from transmission.
6. When the thickness of the crushed rock revetment is small, its natural convection effect in winter is relatively weaker, and it mainly plays the role of thermal resistance materials.

In conclusion, the occurrence and evolution of natural convection cooling effects of crushed rock revetment in winter are mainly related to the thickness of crushed rock revetment, the particle size of gravel, the form of boundary and the conditions of permafrost in permafrost regions.

According to test results of crushed and broken rock revetment in Wudaoliang, Kaixinling, Beiluhe River and Qingshui River along the Qinghai-Xizang Railway and the structural dimensions of the observed sections (Table 3.56), main conclusions are as follows:

1. For the crushed rock layer with a thickness of 1.2 m, the ground temperature on the sunny slope of the embankment decreases by 1.9 °C and 1.65 °C respectively in warm and cold seasons, and that on the shady slope decreases by 1.26°Cand 0.47 °C. As for the crushed rock layer with a thickness of 0.8 m, the ground temperature on the sunny slope of the embankment decreases by 1.65 °C and 1.06 °C in warm and cold seasons, and that on the shady slope decreases by 1.22 °C and 0.07 °C.

Table 3.56 Structural dimensions of the crushed rock revetment of test section along the Qinghai-Xizang Railway (The Compilation Committee of *Qinghai-Tibet(Xizang) Railway*, 2016)

Location	Sections	Particle size/mm	Thickness/m		Remarks
			Sunny Slope	Shady Slop	
Qingshui River	DK1026 + 155	150–220	1.2	1.2	MAGT, −0.46 °C, PT_o, 2.8 m
	DK1026 + 190		0.8	0.8	MAGT, −1.2 °C, PT_o, 1.5 m
Wudaoliang	DK1082 + 375	100	1.0	0.6	MAAT, −5.6 °C; MAGT, −2.3 ~ −2.4 °C
	DK1082 + 425		1.3	0.8	
	DK1082 + 475		1.0	1.2	
Beiluhe River	DK1142 + 945	100	1.3	0.6	MAAT, -5.2 °C; MAGT, −1.4 ~ −1.6 °C; PT_o1.9 ~ 2.1 m
	DK1142 + 990		1.6	0.8	
Kaixinling	DK1262 + 575	70–120	1.2	0.6	年平均气温 −3.3 °C; 年平均地温−0.6 ~ −0.7 °C
	DK1262 + 625		1.6	1.0	
	DK1262 + 675		1.4	0.8	

Notes: MAGT means mean annual ground temperature, PT_o means permafrost table under natural surface. MAAT means mean annual air temperature

2. Compared with the ordinary embankment, the accumulated temperature of the crushed rock layer with a thickness of 1.2 m and 0.8 m at the 2.6 m depth under embankment shoulder on the sunny slope were 777 °C·d and 582°C·d lower in 2004; and the accumulated temperature at the 2.2 m depth under the embankment shoulder on the sunny slope were 273 °C·d and 220 °C·d lower. The crushed rock revetment with a thickness of 1.2 m has a better cooling effect than that of 0.8 m.
3. The mean annual temperature of slope surface under the crushed rock layer is lower than that of the slope surface of ordinary embankment. The crushed rock layer on the sunny slope of embankment can decrease the surface temperature of embankment slope by about 3.0 °C, and 1.6 °C on the shady slope of embankment. The annual range of slope surface temperature for ordinary embankment is significantly greater than that for the crushed rock revetment, and the difference is 13 °C and 20 °C respectively on the sunny and the shady slope.
4. The temperature difference between the the sunny and shady slope of ordinary embankment is about 2.2 °C, and it is 0.7 °C for that of the crushed rock revetment, indicating that the crushed rock revetment can reduce the temperature difference between the sunny and shady slope of embankment, which can weaken the asymmetry of the temperature difference of the embankment.
5. The elevation of the permafrost table under the embankment with crushed rock revetment is greater than that under ordinary embankment. The elevation of the permafrost table under the crushed rock revetment embankment with the thickness of 0.8 m and 1.6 m for shady and sunny slopes is greater than that of 0.6 m and 1.3 m, indicating that thick crushed rock revetment has a better engineering effect. The observation data for the ground temperature under crushed rock revetment embankment with three different thicknesses (0.6 m/1.2 m, 1.0 m/1.6 m, and 0.8 m/1.4 m) for shady and sunny slopes show that the thinner is the thickness of crushed rock revetment, the worse is the cooling effect. The thicker is the crushed rock revetment, the lower is the ground temperature at the original natural permafrost table under the embankment.

Comparative analysis of cooling effect on embankment between the broken rock revetment and crushed rock berms is the following (Xu, 2006):

1. The broken rock revetment and crushed rock berms have obvious cooling effects. The mean annual ground temperature at different depths under the sunny and shady road shoulders is decreasing year by year, and the ground temperature under the sunny road shoulder is lower than that under the shady road shoulder, and the cooling amplitude under the sunny road shoulder is slightly larger.
2. The permafrost table under the sunny and shady road shoulders is higher than the permafrost table under the natural surface. The permafrost table under the sunny and shady road shoulders with the crushed rock berm, under the slope toes of the embankment and under the slope toes of the shady berm are all smaller the original natural permafrost table, while that under the left berm is larger than the original natural permafrost table under embankment.

3. The broken rock revetment with the larger thickness and crushed rock berms with the larger thickness and the larger width have a good cooling effect, suggesting that design parameters as follows:

As for the thickness of block (crushed) rock revetment, the thickness on the sunny slope should not be less than 1.2 m and that on the shady slope should not be less than 0.8 m. For the width of crushed rock berm, under the condition of maintaining the thickness of revetment, the thickness on the sunny slope should not be less than 3 m ~ 4 m and that on the shady slope should not be less than 2 m ~ 3 m.

3.4.7.5 Structure of Block (Crushed) Revetment and Berm

The basic structural form of crushed (broken) rock revetment and crushed rock berm is shown in Fig. 3.140, which can be used as reinforcement measures for embankment engineering, such as crushed rock embankment, ventilated duct embankment, thermal insulation embankment, thermal probe embankment and other composite embankment. It is conducive to improving the cooling effect and the ability to collect and store cold energy of various engineering measures.

In order to make the berm drainage, the soil berm must be firstly made to prevent water from infiltrating into the slope toes of the embankment and the embankment base. The minimum height of the soil berm against drainage must be 0.5 m above the height of water overflow for many years, and then the crushed rock layer shall be laid on it.

3.4.8 Embankment Revetment with Sunshade (Sunshield)

The solar radiation in permafrost regions is the main factor to increase the ground temperature, especially in the Qinghai-Xizang Plateau at low latitude and high altitude, where the solar radiation is so strong that the exposed surface temperature is generally about 5 °C higher than the air temperature. Therefore, lowering the surface temperature of embankment slope can control the rise of the permafrost temperature under the embankment. The sun-shading methods can effectively reduce

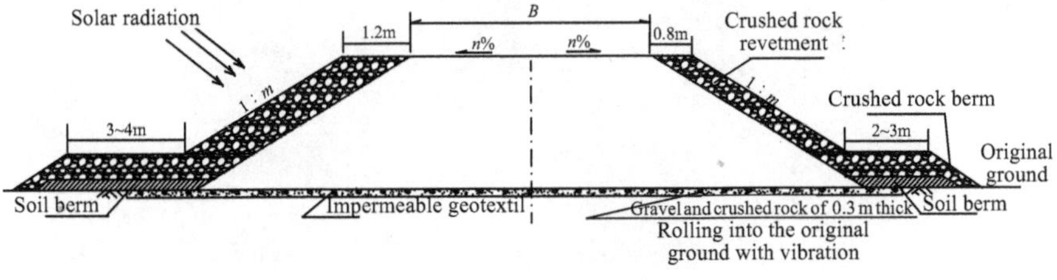

Fig. 3.140 Schematic diagram of embankment with crushed rock revetment and berm

the influence of solar radiation, especially for the protection of cutting slope facing the sun. In fact, the sunny slope of road cuttings and embankments are turned into shady slope, thereby reducing the permafrost temperature under the embankment and the decreasing amplitude of the permafrost table under embankment, which is conducive to the thermal stability of permafrost under embankment. Therefore, grass planting can be used in Northeast China to reduce the thermal effect of solar radiation on the subgrade and to reduce the surface temperature of its side slope, which is beneficial to the thermal stability of permafrost under embankment.

The sunshade (Fig. 3.141) and sunshield on the embankment slope (Fig. 3.142) have been built in the Fenghuo Mountain section for testing. According to the test data, the average surface temperature of the sunshade is 5 °C ~ 8 °C lower than that outside the sunshade, and the maximum temperature can be 15 °C ~ 20 °C lower than that outside the sunshade (Fig. 3.143) (The Compilation Committee of *Qinghai-Tibet(Xizang) Railway*, 2016). They differs slightly in winter but greatly in summer, but the mean annual temperature inside the sunshade can still be 0 °C. The results of indoor test show that the sunshield can reduce the soil temperature by about 91%.

Fig. 3.141 Tests of sunshade on Qinghai-Xizang Railway (Ding et al., 2011)

Fig. 3.142 Tests of sunshield on the embankment side slope on Qinghai-Xizang Highway (Photo by Li Yongqiang)

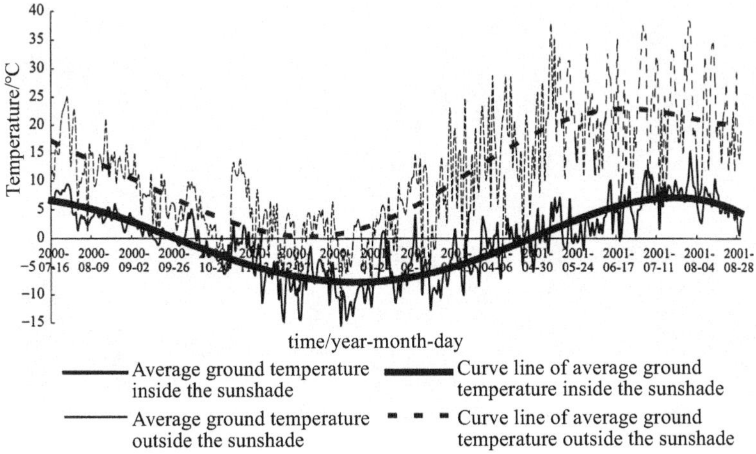

Fig. 3.143 Temperature comparison of ground temperature inside and outside the sunshade at 14 o'clock in 1 year

From the observation data of ground temperature inside and outside the sunshade 1 year later, it reveals that the elevation of the permafrost table under the embankment is 1.1 m (the sunny slope), 1.4 m (the center) and 1.27 m (the shady slope) respectively. Influenced by the sunshade, the elevation of the permafrost table under the top of cutting slope near the sunny and shady sides of the sunshade is 0.1 m (the sunny side) and 0.36 m (the shady side) respectively. As for the natural permafrost table far from the shady side of the sunshade, the permafrost table decreases by 0.11 m due to the atmospheric temperature. It can be seen that the sunshade has a significant cooling effect on the permafrost under embankment and can protect the permafrost actively.

On both sides of the Beiluhe embankment of Qinghai-Xizang Railway, color steel sandwich panels are used to pave the sunshield test section. The observation data show that the embankment surface temperature of the sunshield test section is 3.0 °C ~ 5.0 °C lower than that of the ordinary embankment. During the maximum freezing period, soil under the sunny road shoulders of the ordinary embankment is in the state of thaw, but the soil temperature under embankment slope with the sunshield remains at −2.0 °C ~ −3.0 °C or even lower. During the maximum thawing period, the soil temperature under the sunny shoulder of the ordinary subgrade is up to 4.0 °C ~ 5.0 °C, and the soil temperature under the embankment with sunshield is only 2.0 °C ~ 2.5 °C. By comparison, the soil temperature under the ordinary embankment is 2.0 °C ~ 3.0 °C higher than that under embankment with the sunshield. According to variations of the permafrost table under embankment in different sections (Table 3.57) (The Compilation Committee of *Qinghai-Tibet(Xizang) Railway*, 2016), the permafrost table under the centerline of embankment, and the sunny and shady slope toes and berms toes, has decreased (small decrease or slight elevation under the shady slope). The permafrost table under embankment with the sunshield basically rises, and the maximum elevation can be up to 2 m. Based on

Table 3.57 Maximum thawing depth (MTD) and the elevation of permafrost table (EPT) under embankment by temperature measurement (Relative to Natural Ground Surface)

Location	MTD in 2003		MTD in 2004		MTD in 2005		EPT between 2003 to 2005	
	OE	ES	OE	ES	OE	ES	OE	ES
SST	−1.74	−2.68	−1.65	−2.36	−2.07	−1.73	−0.33	0.95
SB	−2.24	−2.8	−2.31	−1.37	−2.47	−0.77	−0.23	2.03
SS	−1.62	−1.41	−1.26	0.19	−1.13	0.95	0.49	2.36
EC	−0.42	−1.14	0.16	−0.4	0.53	1.0	0.95	2.14
SSE	−0.32	−0.88	−0.02	0	0.08	0.47	0.4	1.35
SBE	−0.95	−1.92	−0.86	−0.86	−0.86	−0.5	0.09	1.42
SSTE	−1.72	−2.1	−1.7	−2.0	−1.79	−1.33	−0.07	0.77

Notes: SST means sunny slope toe of embankment, SB means Sunny Berm of Embankment, SS means sunny shoulder of embankment, EC means Centerline of Embankment, SSE means shady should of embankment, SBE means shady berm of embankment, SSTE mean shady slope toe of embankment. ES means Embankment with sunshield, OE means ordinary embankment

instructions for the use of sunshield on the Qinghai-Xizang Highway (China Communications Construction First Highway Consultant, et al., 2006), the mean annual surface temperature for sunshield used permafrost table can be reduced by 4 °C ~ 6 °C, a decrease of 0.5 °C ~ 1.0 °C in the permafrost under the subgrade, and the permafrost table under embankment is raised by 0.5 m ~ 1.0 m, effectively reducing the thaw bulb under the embankment and controlling the thawing and deformation of the embankment.

The cooling mechanism of sunshade (sunshield): ① the sunshade (sunshield) can effectively block the solar radiation and reduce the heating effect on the slope surface of embankment; ② The interspace between the sunshield and the slope surface (0.15 m ~ 0.2 m is the best) can increase the air flow and bring the heat out from the ground; ③ Shading plays a cooling role, which can reduce the temperature on the slope. Obviously the important function of sunshade (sunshield) is to reduce or avoid solar radiation on the slope of embankment and to increase air flow on the ground, thus reducing the ground temperature and the heat absorption of the permafrost under embankment.

3.5 Slope Embankment (Embankment Revetment)

The frost mounds, icing (ice flood or icefall), and collapse during the freezing process directly affect the safety of bridges, culverts, tunnels, and highways (Figs. 3.144 and 3.145). The engineering construction will change the hydrogeological conditions, including the ground temperature of surrounding permafrost layer, the permeability of the soil layer, the flow path of surface water and the seepage path of the groundwater. Meanwhile, thermal slumping, gelifluction and thermokarst lake during the thawing process will directly affect the stability of slope embankment and the embankment near lakeside.

Fig. 3.144 Ice floods endanger Mobei Highway in the north of Daxing'anling Mountains

Fig. 3.145 Regressive thaw slump in low and gentle slopes

3.5.1 The Slope Stability in Permafrost Regions

The main causes of slope instability in permafrost and non-permafrost regions are mainly related to the immersion of water (surface water and groundwater) and the thawing of ground ice, resulting in a decrease in the shear strength of the soil layer, even losing, and the appearance of sliding under the action of gravity or external force. Another important cause is the existence of ground ice. Influenced by thermal effect, the ground ice exposed naturally or artificially will cause thaw and water erosion of the soil layer, and then the slope is gradually thawing, cracking and collapsing along the slip surface, thus leading to slope instability with headward erosion till the thick ground ice in the permafrost layer disappears. Thermal slumping

Fig. 3.146 Qinghai-Xizang Highway detouring due to pingo

occurs even on low and gentle slopes (see Fig. 3.146). The slope instability occurs in annual warm seasons and is concentrated in the soil layer above the permafrost table.

According to the distribution of permafrost with high ice content along the Qinghai-Xizang Highway, thick ground ice is mainly distributed in the permafrost table. The ground ice is developed on slopes with a slope gradient less than 10° in low mountains and hills, especially 4° ~ 8° (Fig. 3.10). On the slope of 10° ~ 16°, the ground ice develops poorly. When the slope gradient is larger than 16°, there is few thick ground ice; while when the slope gradient is larger than 25° denudation can usually be found and there is crevasse ice (Fig. 3.147). The lithology of the soil layer hosting the ground ice is mostly silty clay, fine silt, silty soil, argillaceous glutenite and mudstone formed by glaciofluvial deposit. The overlying soil layer above the ground ice is mainly composed of silty clay, silt and fine silty sand.

In the permafrost regions of the Qinghai-Xizang Highway, the Qinghai-Xizang Railway, the source region of the Yellow River and the Daxing'anling Mountains in Northeast China, there are many engineering geological problems during the thawing process, including thaw slumping, thaw collapse and gelifluction. Thaw slumping on the Qinghai-Xizang Highway directly affects the stability of the embankment, for example K3035 (between Hongliang River and Qushui River), K3057 (between Beiluhe River and Fenghuo Mountain) and Qinghai-Xizang Railway DK 1500 (Yanggaer Burong Valley). In K3035 along Qinghai-Xizang Highway, the thaw slumping is attributed to the thick ground ice with the thickness of 2.0 m ~ 4.0 m below the ground (Fig. 3.148). In 1990, when the embankment drainage ditch was excavated, the exposed ground ice was covered in time, resulting in thaw slumping, which had developed to 50 m ~ 60 m by 1996 (see Fig. 1.14), and 90 m ~ 100 m by 2000. The width is limited by the range of ground ice, about 30 m.

In 1960, China Railway Northwest Research Institute carried out experimental researches on railway engineering at the Fenghuo Mountain station on the

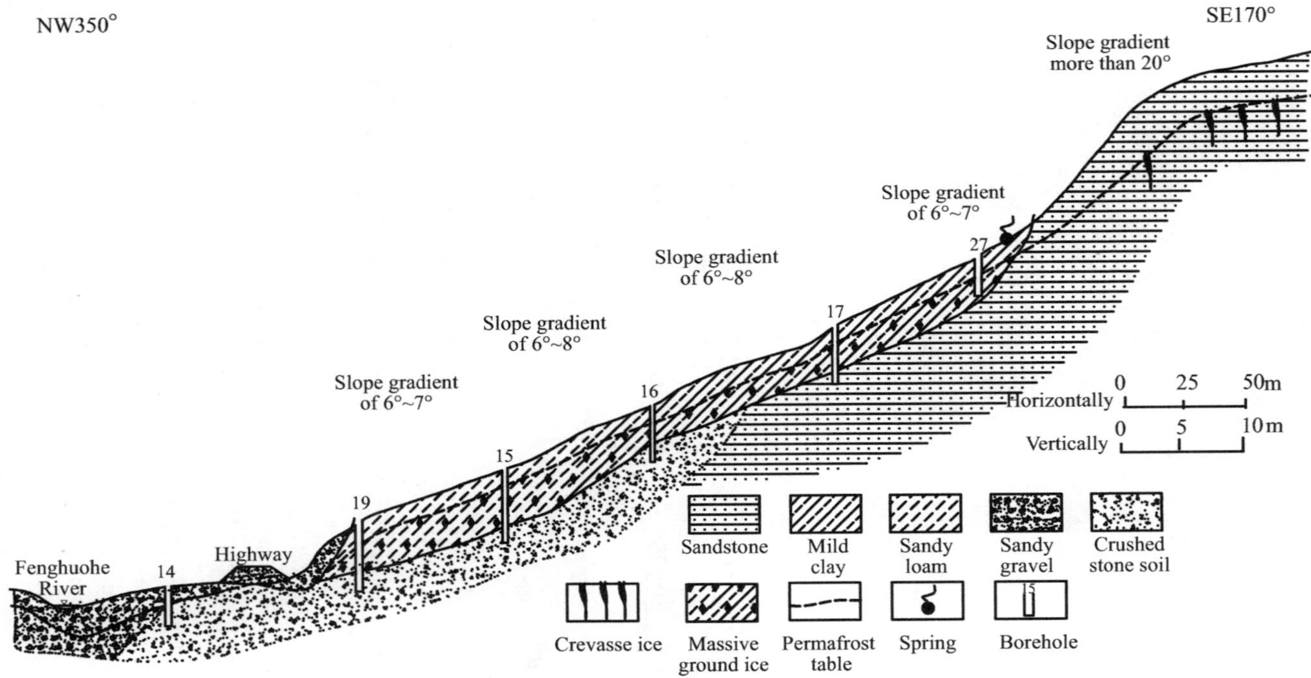

Fig. 3.147 Geological section of the hillside at the opposite of the Fenghuo Mountain meteorological station in the Qinghai-Xizang Plateau (the Scientific *Research Group of Qinghai-Tibet(Xizang) Highway*, 1983)

Fig. 3.148 Section of regressive thaw slump at K3035 along Qinghai-Xizang Highway

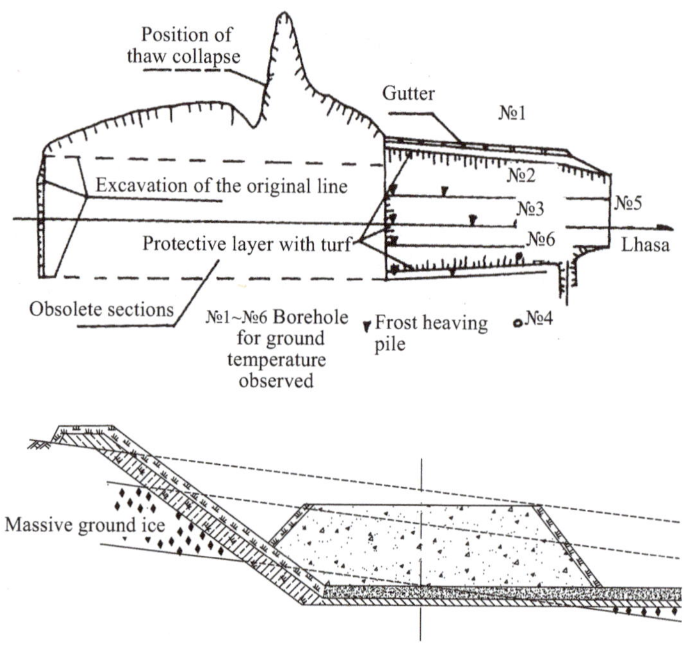

Fig. 3.149 Schematic diagram of experimental engineering of cutting on the section of massive ground ice in Fenghuo Mountain (Northwest Research Institute Co., Ltd. of China Railway Engineering Corporation, etc. 2006)

Qinghai-Tibet Plateau, including an experimental research on cutting engineering in the sections of thick ground ice (Fig. 3.149). When the road cutting (with 100 m in length) was excavated, the exposed ground ice (up to 5 m) was not covered in time, resulting in thaw slumping. Subsequently, the side slope of cutting was backfilled and compacted in time with the mixture of turf and cohesive soil with a thickness of 0.6–0.8 m and a slope ratio of 1:1.8, and the slope surface was laid with a turf of 0.2 m as the thermal insulation layer. After backfilling with silty clay of 0.3 m on the embankment base, the replacement thermal insulation layer was filled with local

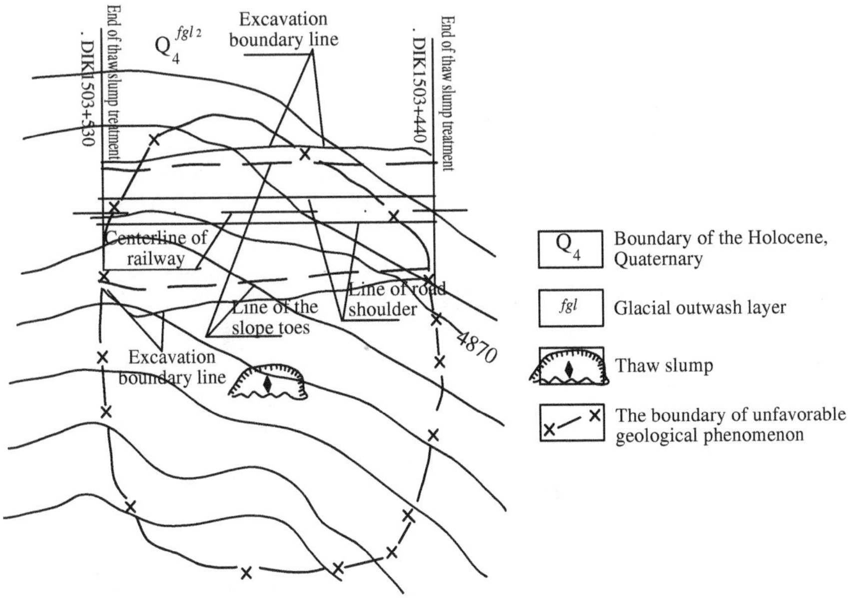

Fig. 3.150 Regressive thaw slump areas of experimental section in Ando along Qinghai-Xizang Railway (Zhao, 2007)

fine sandstone gravel with a particle size of 50 mm ~ 80 mm and a thickness of 1.1 m ~ 1.6 m. The road cutting test engineering is still in good condition.

In the Yanggaer Burong Valley (around DK 1500), the north of Amdo on Qinghai-Xizang Railway, there is a fault valley with extremely complex structural faults, where the suprapermafrost water is well developed, and there are frost mounds, icing and thick ground ice, as well as thaw slumping on the valley and slope (Fig. 3.150). The permafrost table in this section is 1.5 m ~ 2.8 m, and the mean annual ground temperature is −0.3 °C. There is saturated permafrost and ice layer containing soil below permafrost table. The total settlement has reached 254 mm 3 years after the embankment construction, and the total settlement of the embankment base is 200 mm. The maximum horizontal displacement at the sunny slope toes is 117 mm. The embankment has gradually been stabilized 3 years later after adopting special measures for crushed rock embankment.

For the stability of the slope embankment, the position of slope slip surface influenced by human activities should be considered from the stability of the natural slope. The slip surface of the slope embankment in permafrost regions occurs in three situations:

1. Slope of Embankment Base

The construction of the embankment will result in enriching the surface water and the suprapermafrost water at the slope toes, which promotes the permafrost thawing and reduces the soil strength of the embankment base. Soil layer will slide along the slope due to decreasing of soil strength, especially for the swampy wetland slope, where the soil strength is low and is prone to slip.

2. Permafrost Table

In the slope sections in permafrost regions, which are often rich in suprapermafrost water, the permafrost table will become a weak layer. And the strength of the freezing-thawing interface is reduced to form a slip surface, resulting in slope sliding. The embankment construction changes the thermal state of the permafrost under the embankment, and forms a new thermal balance and a permafrost table. Due to the heat difference between the sunny and shady slopes of embankment, there appears a slope which is higher in shady and lower in sunny side for the permafrost table under the embankment, resulting in a slope slip surface (Fig. 3.151).

The essence of slope stability in permafrost regions is that the permafrost containing ice controls the mechanical properties of rock and soil on slope. With the change of thermal regime of permafrost, the water content of the rock and soil also changes, resulting in changes in mechanical properties of the rock and soil. There are differences in mechanical properties of the same rock and soil in frozen or thawed state. The soil strength on freezing-thawing interface is the weakest (Table 3.58), prone to deformation and slip. In addition, during the annual variation period in air temperature, a maximum freezing-thawing interface can be found, that is, there is an ice-accumulated layer (belt) on the permafrost table. During the maximum seasonally thawed period, a large amount of water accumulates at the permafrost table, which becomes the weak surface of the rock and soil. When the ice-accumulated layer (belt) at the freezing-thawing interface will thaw in a large area, the soil layer will be supersaturated or be in the flowing state, and the safety factor will be the smallest (Fig. 3.152). In comparison, the safety factor of the slope embankment sliding along the permafrost table is the least, and it becomes the most dangerous slip surface. Moreover, in warm permafrost regions, the higher is the permafrost temperature at the freezing-thawing interface, the greater is the creep property, and the permafrost strength decays rapidly and tends to its long-term strength (generally only about 20% of the instantaneous strength).

The deformation and damage of slope embankments in permafrost regions is mainly caused by the soil layer sliding along the weak zone of the freezing-thawing

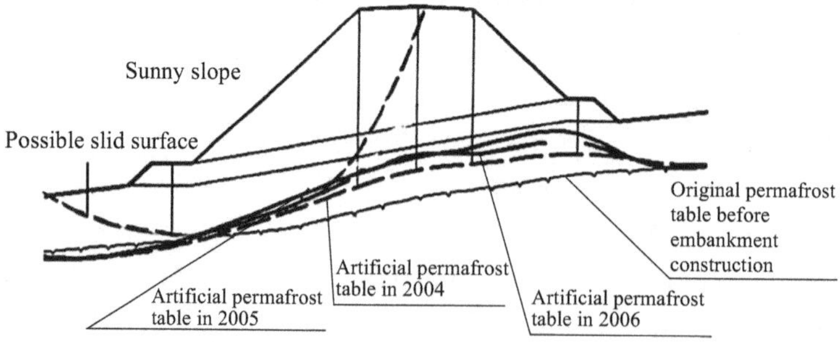

Fig. 3.151 Possible slip surface caused by permafrost table (Zhao, 2010)

Table 3.58 Quick shear test data of silty clay (Zhao, 2007)

State	Shear strengths at different vertical pressures/kPa				Internal friction angle (φ/°)	Cohesion (C/kPa)
	50	100	125	150		
Thawed	21.41	32.79	39.34	43.23	12.57	10.596
	22.53	34.75	41.58	45.12	13.07	11.338
	20.18	30.86	39.77	42.65	13.21	8.426
Freezing-thawing interface	19.37	28.45	37.59	41.76	13.02	7.226
	18.56	29.55	38.27	40.55	13.00	7.195
	17.97	27.46	36.87	40.11	13.01	6.501
Frozen	42.196	56.238	75.768	86.506	24.39	17.000
	43.792	61.236	80.444	87.724	24.57	19.723
	43.358	57.540	77.266	88.830	24.88	17.467

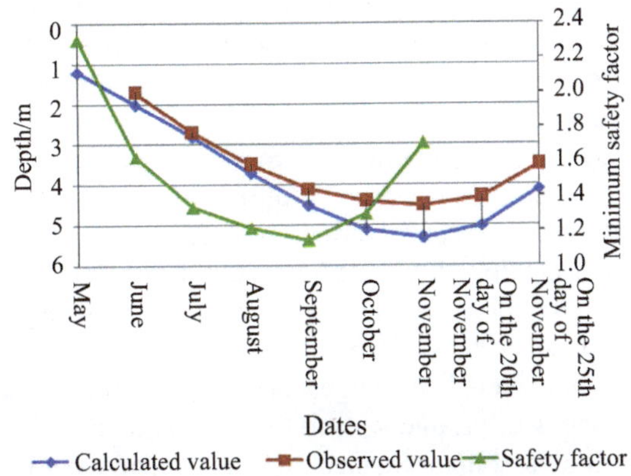

Fig. 3.152 Variation of freeze-thaw process and stability (Safety) Factor (Zhao, 2007)

interface under different temperature and humidity. In fact, the most dangerous slip surface for the slope embankment is slightly above the freezing-thawing interface.

According to the characteristics and change of slope embankment stability in permafrost regions, the position of slip surface is determined by the magnitude of ice content in permafrost layer and its variation with temperature. For the warm permafrost with high ice content, small changes in ground temperature near the permafrost table will cause a large compression or compaction deformation due to worse thermal stability. So, the increase in accumulative deformation will reduce the overall mechanical stability of the slope embankment, resulting in lateral slip deformation. For the warm permafrost with low ice content, even if it belongs to the extremely unstable ground temperature zone, the change of ground temperature and the permafrost table has little influence on the mechanical stability of the soil due to the low ice content, but it is still necessary to analyze the mechanical stability to determine the overall stability of embankment.

In slope subgrades, the thermal stability of permafrost is mainly manifested in the changes of the mean annual ground temperature and the permafrost table. The former reflects the stability of permafrost, which is characterized by the ground temperature zone, while the latter is represented by the shape of the permafrost table under the embankment:

1. Stable permafrost: the permafrost table under the embankment rises and enters into embankment with the core of frozen soil.
2. Basic stable permafrost: permafrost table under the embankment rises and is only positioned in the seasonally thawed layer under embankment base.
3. Slightly stable permafrost: the rising magnitude of permafrost table under the embankment is small and remains near the original permafrost table before embankment construction.
4. Unstable permafrost: the permafrost table under the embankment is lowered and it is deeper than the original permafrost table before embankment construction.

The mechanical stability of frozen soil is mainly manifested in the shear strength. The stability evaluation of geotechnical engineering mechanics is mainly judged by the analysis and calculation of the slope stability coefficient, and is evaluated by the stability coefficient of the four stable states of the slope. When the slope stability coefficient is less than its safety coefficient, some measures should be taken on the slope.

For slopes in permafrost regions, in addition to the natural factors that need to be considered in non-frozen soil regions, the key is to consider the characteristics of ground temperature variations and the permafrost table, as well as ground ice. It can be seen from Fig. 3.155 that the slope stability coefficient varies with freezing-thawing process. Generally, until the thawing depth reaches the maximum in September the soil temperature in the freezing-thawing interface is the highest and water content is the largest due to the unfrozen water content of the frozen soil. The thawing of ground ice and the infiltration of the suprapermafrost water at this time result in soil mechanical strength reduction to the minimum. The properties of embankment soil-filling should also be considered when evaluating the stability of slope subgrades in permafrost regions.

The engineering practice shows that the settlement deformation of the slope embankment is mainly caused by the decline of the seasonally thawed layer and the original permafrost table under embankment. The settlement deformation on the sunny slope is larger than that on the shady side, and there is a horizontal displacement. Therefore, the permafrost table under the embankment should be raised in the engineering of slope embankment, so that it can reach a stable or basic stable state, and measures should also be taken to reduce the difference between the permafrost table on both sides of the sunny and shady slopes, including cooling roadbed and waterproofing and drainage.

1. According to the design requirements of "crushed rock embankment", crushed rock should be dumped in the embankment base with the thickness of 1.2 m ~ 1.5 m.
2. The sunny slope is added with crushed rock revetment, or crushed rock berm, or thermal probes (combined with thermal insulation layer).

3. The retaining dams and drainage ditches are set on the upper reach of the embankment, 10 m away from the slope toe of the embankment. The retaining dam is 1.5 m height, 1.0 m for the top width, 1:1.5 for the slope gradient, and a drainage ditch is set at its outside. Thermal probes are longitudinally arranged along the center line of the retaining dam, with a spacing of 3.0 m; or PVC seal plates are buried (the bottom of board should be inserted into 0.5 m below the natural permafrost table), with the thickness of 1.2 mm. Drainage measures should be connected with the whole drainage system of subgrade.
4. The waterproof layer of cohesive soil should be installed at the slope toe of embankment (especially the upper reach) to squeeze out the standing water to 3 ~ 5 m away from the toe of the slope, and it is best to set it together with crushed rock berms.

Observations data on the Amdo test section along the Qinghai-Xizang Railway show that the above engineering measures can raise the permafrost table under the embankment by a maximum of 2.4 m (China Railway First Survey and Design Institute Group Co., Ltd., et al. 2007). The area of heating surface at the sunny slope toe is relatively large and the ground temperature is relatively higher, so thermal probes and other thermal protection measures need to be enhanced.

3.5.2 *The Stability of Cutting Slope*

In permafrost regions, the slope sections often contain different degrees of ground ice layer. The thermal balance of the permafrost is destroyed and disturbed by large-scale excavation of road cuttings, which is more serious than that of embankments, exposing large areas of permafrost with high ice content. It also increases the thawing depth of permafrost and the thawing of ground ice, which will cause slope instability, including frost heaving and thaw slumping, if thermal protection measures are insufficient.

Seriously damaged cutting slopes in cut and half-cut sections along the Qinghai-Xizang Railway account for 19.2% of the total cutting slopes (Hu & Du, 2015), and four of them were overall instability. Main damages of cutting slopes include longitudinal cracks at the top of cutting, damage to the slope protection structure, local slumping, the bulge of the slope toe, etc. (Figs. 3.11, 3.12, 3.13, 3.153 and 3.154).

The main reasons for the instability of the cutting slope are as follows: ① Permafrost is eradicated artificially due to large-scale excavation of the cutting, destructing the thermal balance of the permafrost layers; ② A lot of thick layers of ground ice in the permafrost is exposed to thaws during the process of cutting excavation, leading to a large amount of thawed water infiltration into the slope toe; ③ The shallow soil layer containing ground ice in the low mountains and hills in Qinghai-Xizang Plateau is mainly composed of silty clay (containing gravel), silt, and weathered and semi-weathered materials of marl, mudstone and shaly sand

Fig. 3.153 Slumping of cutting slope (Photo by Zhang Zhao)

Fig. 3.154 Bulge at the toe of cutting slope (Photo by Hu & Du, 2015)

whose strength decreases sharply after saturation; ④ During the construction, the slope absorbs heat strongly, and the permafrost with high ice content is affected by heat, which not only causes the ground ice to thaw, but also increases the soil temperature of the permafrost layer; ⑤ Although thermal insulation measures are actively adopted in engineering design and construction, the thermal balance of the permafrost in the cutting slope is destroyed by the heat storage effects of backfill, engineering measures and structures, leading to the thawing of permafrost under the cutting slope in the early stage and the slow recovery of the thermal balance of permafrost in the later stage.

According to the field surveys (Xiong et al., 2010), with the same ice content, the instability of cutting slopes occurs mostly on the higher side, which is not closely related to the direction of slope, but is obviously related to its area of heat absorption. Although remediation and reinforcement measures (addition with thermal probes) are adopted to cope with the instability of slope during the construction, the

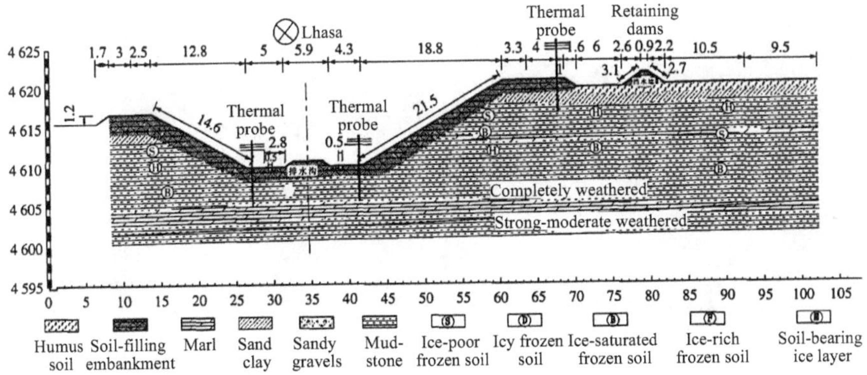

Fig. 3.155 Engineering and geologic conditions for the cutting slope instability cause by thermal thawing in the permafrost regions (Unit: m) (Xiong et al., 2010)

survey results after 5 years show that there are still deformation and slip cases in some road cutting sections, such as fluctuation and deformation on the original slope and wide cracks at the top of slope. The water content of soil below the depth of 1 m at the slope toe is still in a saturated state, indicating that the instability of the cutting slope is associated with the activity of thawing water in permafrost with high ice content (Fig. 3.155).

There are two kinds of cutting slope diseases in permafrost regions. One is for the soil slope in the permafrost regions with ground ice, and the slope diseases are mainly caused by the actions of thermal thawing. So, thermal protection measures should be taken in priority, as well as the method of drainage and interception (resistance) water, to avoid the loss of strength due to the thawing of ground ice and the soil-saturated at the slope toe. The other is for the rocky slopes, and slope diseases are caused by frost heave due to crevasse ice in the rock mass. In such a condition, the measures against frost heaving on drainage and interception (resistance) water as well as some other reinforcement and protection measures should be taken.

1. Thermal protection measures for cutting include thermal insulation, heat dissipation, grass planting, cooling permafrost. At the same time, some measures in water drainage and water interception should be strengthened.
2. As for slope top of road cutting, the composite measure of thermal insulation layer and thermal probes as well as grass planting can be adopted on the slope top. Retaining dam and seal plate and drainage ditches can be set up 20 m outside the road cutting.
3. As for slope surface, composite measures of assembled skeleton (anchor) revetment and artificial grass planting or native turf transplantation are adopted on the slope surface (Fig. 3.156). The tests in reinforced soil revetment, foam glass board revetment and improved soil revetment have been taken along the Qinghai-Xizang Railway, but it is found that they are not suitable for the plateau due to the complex prefabrication technology and poor anti-ultraviolet aging ability.

Fig. 3.156 Combination of assembled skeleton revetment (the Left) and grass planting revetment (the Right) (or Turf Transplantation) (Photo by Da Zhihua)

4. As for slope toe, it is suitable to take the composite measures of bind drain for thermal insulation and drainage, thermal probes and thermal insulation layer.
5. Cutting against frost heaving measures include shotcrete revetment which is used to seal cracks on rocks and reinforce the rock mass, or anchor cable which is used to strengthen rock mass. The slope should set up drainage holes, and the slope top is provided with retaining dam and drainage ditches should set up in the slope top of cutting, and blind drains for thermal insulation and drainage are set at the toe of the slope.

3.5.3 Embankment Stability Under the Effect of Vibration

The vibration response and deformation of embankment in permafrost regions caused by the vibration load from two sources, including the vibration load of moving vehicle on the embankment and the seismic load under the embankment.

The traffic vibration is the dynamic response of the subgrade section caused by the real-time acceleration, velocity variations and the vibration frequency of locomotive imposed on embankment. In different seasons and periods, traffic vibration repeatedly and normally occurs with its vibration frequency in a certain range. The embankment deformation is mainly settlement and slope slip, and the deformation will gradually become stabilized.

An earthquake is an instantaneous acceleration and velocity variation, which has a very strong destructive power. The earthquake will cause structure destruction of transportation infrastructure, such as settlement of embankment and pavement, bridge fracture, collapse of tunnels, slope slip and collapse and so on.

The soil layer structure in permafrost regions includes seasonally thawed soil and permafrost during warm seasons and seasonally frozen soil and permafrost layer during cold seasons. Relevant studies have shown that the thickness and hardness of permafrost layers have great influence on the acceleration response spectrum of the ground surface. Permafrost has the phenomenon of high-frequency amplification under different seismic intensities, but the peak seismic acceleration in winter maybe not necessarily greater than that in summer. The influence of permafrost layer on the seismic stress properties of the subgrade is much larger than that on the acceleration response spectrum, and the dynamic stress amplitude and frequency changes are much larger than those without permafrost layer. The maximum vertical acceleration of the permafrost is larger than that of non-frozen soil, but the horizontal acceleration of the permafrost is smaller than that of non-permafrost. The maximum dynamic shear stress fluctuates violently at the freezing-thawing interface layer. The thickness of the permafrost layer has the greatest influence on the vibration displacement of the subgrade; especially the slope subgrade with thawed soil layer has the largest displacement under the vibration load. From the perspective of seismic damage, whether there is permafrost layer in the sites will directly affect the ground motion and the damage type of the upper construction.

1. **Embankment Stability under the Effect of Vehicle Vibration**

The vehicle dynamic load of locomotives may have more influence on the embankment deformation than that of automobiles, and the load and speed of locomotives of railway are higher than those on highways. In the sections of Beilu River of the Qinghai-Xizang Railway, the research in real-time observations of strong vibrations in cold and warm seasons for five types of embankment was carried out. The five types of embankments include the combination of soil-filling embankment with crushed rock revetment on the shady slope, the crushed rock embankment with crushed rock revetment, the crushed stone subgrade with block stone slope protection, the ventilated duct embankment with broken rock revetment on the sunny slope, the thermal insulation layer embankment with thermal probes. The results of observation data showed the transfer law of locomotive vibration acceleration on different embankments. The following conclusions are obtained: (Chen, 2011)

1. In permafrost regions, when the locomotive passes, the vibration of ordinary soil-filling embankment is strong, and it is not conducive to maintaining the dynamic stability of the embankment.
2. The dynamic load of locomotive is transmitted through the embankment structure. The same components as the embankment structure show their excellence, while the components that are significantly different from the embankment structure are absorbed. The natural frequency of crushed (broken) rock embankment and ordinary soil-filling embankment structure is 30 Hz ~ 40 Hz, and that of ventilated duct embankment is 40 Hz ~ 50 Hz.
3. In transmitting locomotive dynamic loads through different embankment structures, the results show that the attenuation effect in warm seasons is greater than that in cold seasons. In cold seasons, the soil is frozen, so the attenuation effect is small, and with large structure stiffness and rapid transmission speed. In warm seasons, the acceleration energy spectrum from the road shoulder to the slope toe decreases to less than 5%, and a decrease of 10% ~ 20% in cold seasons.

4. The vibration load of the train transmits from the top of the soil-filling embankment to the interior, due to the reflection of the side slope of embankment, so the frequency of energy spectrum inside the embankment has no obvious attenuation. The vibration of the original natural ground in the centerline of the embankment is larger than that at the slope toe, and the soil under the original natural ground has obvious attenuation. There are structural characteristics of different material layers under the crushed rock embankment, and the interior vibration of the soil has an increasing effect when the dynamic load of the train is transmitted to the upper part of the crushed rock layer. After passing through the crushed layer, the acceleration amplitude of the vibration load decreases rapidly (Fig. 3.157) (Chen, 2011). It shows that the vibration transmission of the train is related to the embankment structure. In the centerline of embankment, the attenuation effect of the crushed rock embankment is greater than that of the ordinary soil-filling embankment when it reaches the natural ground.
5. The maximum settlement of ordinary soil-filling embankment and crushed rock embankment occurs in the soil at upper part of the embankment, and the maximum displacement is 0.018 mm and 0.020 mm respectively. However, the effect of crushed rock embankment in controlling vibration settlement is better than that of ordinary soil-filling embankment (Fig. 3.158).

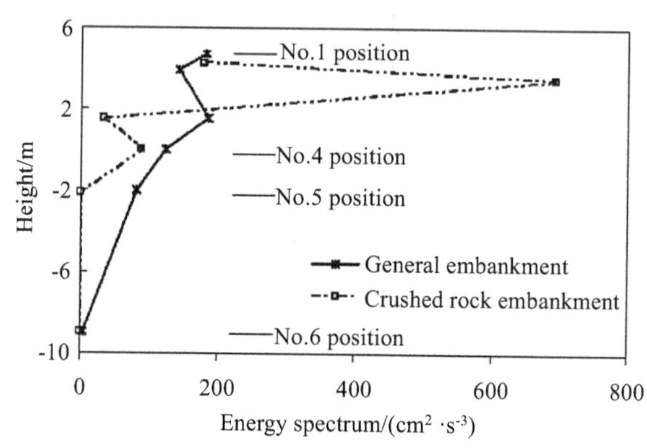

Fig. 3.157 Distribution of acceleration energy spectrum of different embankment structures with embankment height

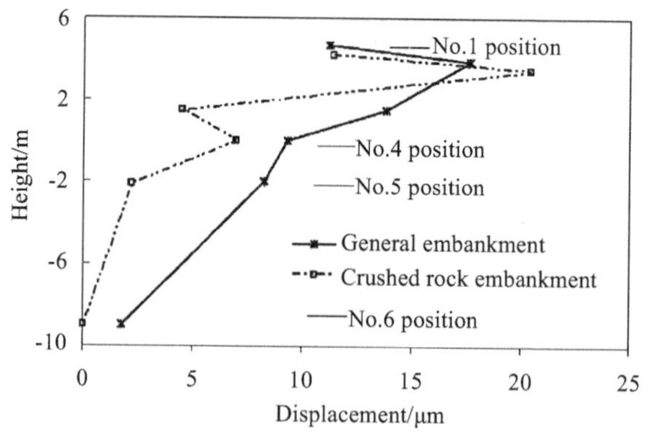

Fig. 3.158 Distribution of vibration settlement of different embankment structures (Chen, 2011)

6. According to the numerical calculation, the creep effect during the operation period of the embankment increases with the passage of time and the displacement and creep strain of each position of the ordinary soil-filling embankment increase continuously and finally tend to be stable. The maximum vertical deformation of the embankment gradually decreases from the top centerline of the embankment to the interior and sides of the embankment, and the maximum vertical deformation reaches 14 mm at the top centerline of the embankment. The creep strain under the ballast increases and the smallest strain lies at the original natural ground.

According to the observation and numerical calculation, the dynamic load cycle of the train occurs mainly in embankment in the permafrost regions. For the special embankment such as crushed rock embankment, the attenuation effect of the acceleration energy spectrum transfer is almost in the embankment, and it is seldom to reach the original natural ground. However, under the action of long-term locomotive load, the slope toe on both sides produces large shear strains with a maximum strain amplitude of 0.186, and gradually expands upward from the slope toe to form a connecting plane. The maximum creep strain appears at the slope toes on both sides, and the strain amplitude can reach 0.08. The creep characteristics have a trend of upward connection. Therefore, the reinforcement of the slope toes on both sides of the embankment should be considered in the practical engineering.

2. **Subgrade Stability Influenced by Earthquake**

Earthquake is formed by the sudden displacement of rocks on the fault and the subsequent release of energy, in the form of longitudinal waves and shear waves to all directions to form seismic waves. The vibrations on the ground reaches the strongest when shear and surface waves arrive, which is regarded as the main cause of earthquake damage. In the dynamic analysis of seismic design of road engineering, the influence of vertical earthquakes is generally not considered, but the effect of shear waves upward along the soil layer is the only factor to be considered. In the characteristics of seismic ground motion, the factors that control the reaction of structures include earthquake intensity, spectrum characteristics and duration of strong earthquake, i.e., the peak value of acceleration, waveform and duration of strong earthquake. There are three types of subgrade damages caused by earthquake: the damage caused by the deformation of the foundation base, mainly settlement; damage caused by the deformation of the embankment, mostly slope cracks, slips and collapse, and damage, collapse and slips caused by the deformation of hillside (cutting slope) and so on (Figs. 3.159 and 3.160).

Fig. 3.159 Damage to the subgrade of National Road 109 in permafrost regions caused by 8.1-magnitude earthquake in Kunlun Mountains (Photo by Tong Changjiang)

Fig. 3.160 Cracks on National Highway 109 (the Left) and the toe of slope (the Right) in permafrost regions caused by 8.1-magnitude earthquake in Kunlun Mountains (Photo by Yu Qihao)

3.6 Embankment in Frozen Soil Wetland

3.6.1 Environment of Frozen Soil Wetland

In permafrost regions, the special geography and environment is conductive to the coexistence of permafrost and wetlands. The swampy wetland develops well in continuous permafrost regions, moderately in discontinuous permafrost regions, and weakly in patchy or sporadic permafrost regions, indicating that swampy wetland closely relates to permafrost. In permafrost regions in the Daxing'anling Mountains in Northeast China, Qinghai-Xizang Plateau and Tianshan Mountains, the more developed the swampy wetlands, the more developed the permafrost, the thicker the permafrost, the more and thicker the ground ice layer, the shallower the permafrost table, and the thicker the clay layer, and vice versa.

The swampy wetlands in permafrost regions are mainly distributed in the following areas: ① Ancient river channels and beaches; ② Gentle terraces and alluvial plains in valleys; ③ Intermountain depressions; ④ Gentle slopes of Piedmont proluvial fans; ⑤ Gentle depressions in watersheds. The more developed permafrost regions, the weaker the vertical erosion and cutting of the river, the stronger the horizontal erosion expands, and the more developed the swampy wetland, which is conducive to the growth of vegetation. According to data (Sun et al., 2008), the heat flux of swampy wetlands accounts for a small proportion in their radiation balance, about 10%, while the latent heat flux accounts for a large proportion of about 70%. This suggests that the vegetation and standing water in swampy wetlands play a role in shielding heat, and the heat is mainly used for the evaporation of the surface of swampy wetlands, and little heat can enter the soil. In degenerative permafrost regions, patchy or sporadic permafrost persists in swampy wetlands.

3.6.2 Embankment Deformation in Swampy Wetlands in Permafrost Regions

The embankment deformation in swampy wetlands in permafrost regions mainly occurs in the subgrade soil under the embankment base which contains peat soil, silty clay, sandy clay and ground ice layer, including the seasonally thawed layer and permafrost thawing layer. The main forms of deformation are settlement deformation and frost heaving deformation.

The main factors affecting the embankment deformation in swampy wetlands in permafrost regions are the composition and structure of soil, as well as external environment.

1. Soil composition: under the action of external load, the particle size and mineral composition of the soil determine its deformation characteristics and deformation amount. Coarse-grained soil is a kind of single-grained structure, and the deformation of soil is determined by the rolling and sliding of soil particles and the change of soil density. The fine-grained soil is mostly flat scalelike, flocculence-like and dispersed structure, and its compression deformation is formed when the water film among the particles is extruded, and the flat soil particles slide and cause deflection deformation. The presence of organic matter (cellulose and humus) in the soil will increase the compressibility and contractility of the soil and affect its strength.

2. Organic matter: the main soil components in swampy wetlands are peat and peaty soil, with high water content and porosity, small specific gravity and high liquid-plastic limit. When the content of organic matter is high, it is easy to form an elevated structure with large pores, rich in pore water.
3. Pore water: most of the soil pores in swampy wetlands are filled with water. In the process of compression, the pore water is squeezed out and the water content is reduced. Coarse-grained soil, with strong water permeability, fast drainage under pressure, can complete the compression quickly, but the amount of compression is not necessarily large. As for fine-grained soil, with small pores and large pore volume, in addition to the existence of water film to make its water permeability weak, needs a long time to complete the compression under pressure because the pore water discharges slowly, and the compression amount is relatively large. The mineral composition of clay affects the thickness and compressibility of bound water film. In the compression of unsaturated soil, the gas escapes first, and then the pore water in a saturated state is discharged.
4. Stress history: the stress changes caused by previous environmental changes have different consolidations on wetlands, such as glacier cover, water level changes, surface exposure and dry shrinkage, etc., all of which will result in different degrees of consolidation on wetlands.
5. Temperature: in the non-frozen state, temperature has a great influence on the high compressibility of organic matter, because the change of temperature will cause the change in the volume of saturated soil pore water and the corresponding change in effective stress. In the frozen state, the decrease of soil temperature results in water migration and water accumulation, with the decrease in unfrozen water and compressibility and the increase in freezing rate and strength.

No matter what kind of engineering measures are taken, the soft soil subgrade in swampy wetlands in permafrost regions will gradually undergo consolidation, drainage, compression, settlement and lateral displacement in the process of embankment loading. Figure 3.161 shows the observation data of the highway construction process of soft soil subgrade in swampy wetlands, from Fuyu County to Nahe City of National Highway 111 in Northeast China. The surface layer is cultivated soil with a thickness of 0.3 m. The silty clay with a thickness of 2.1 m is below the surface. The lower layer is fine sand of 0.5 m, and the bottom layer is rounded gravel. The gravel sand drainage blanket was filled in the early stage, but ceased construction in cold seasons, so the settlement of the frozen soft soil did not increase significantly, but the settlement obviously increased when it thawed in the next year. The soil of a thickness of 2.4 m was concentratedly filled in May, June and July, with the embankment height of 4.0 m (Fig. 3.161a), the settlement of the soft soil subgrade increased significantly, and the closer it was to the embankment surface, the larger the amount of settlement (Fig. 3.161b). The settlement of soft soil accounted for 80% of the total settlement, and the pore water pressure increased accordingly (Fig. 3.161d). From mid-May to early July, the lateral displacement of the soft soil subgrade increased significantly, with a maximum displacement of 30.5 mm (Fig. 3.161c). Similar process was also found in the subgrade construction of swampy wetlands in permafrost regions.

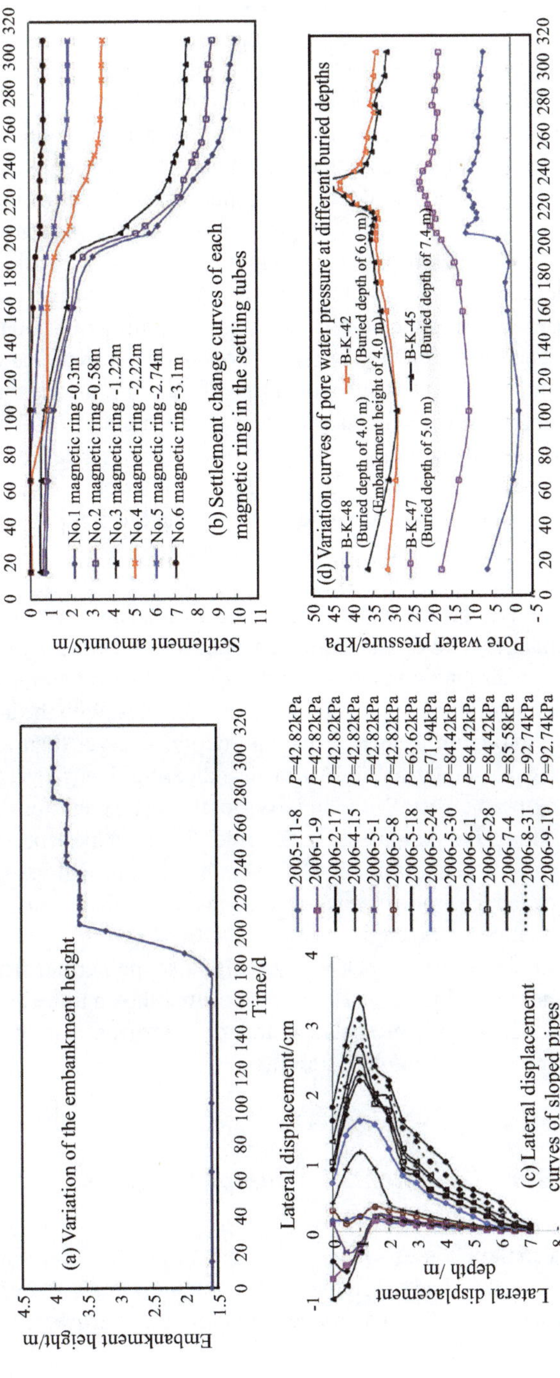

Fig. 3.161 Curves of compressive settlement of embankment in swampy wetland under loading (Liu, 2007)

The seasonally thawed layer in swampy wetlands in permafrost regions presents a freezing-thawing cycle state with the change of temperature, and the embankment deformation appears frost heaving-thaw settlement. In warm seasons, when the soil-filling embankment is loaded, the soft soil in swampy wetland above the artificial permafrost table will appear consolidation and compression settlement, and the closer it is to the surface of the embankment, the larger the amount of settlement becomes. During the loading process, the settlement rate of the embankment center is higher than that of the road shoulder. The permafrost layer below the permafrost table, with good integrality and high strength, can transfer the load in the middle to the edge areas on both sides of the embankments, and the soft soil subgrade has lateral displacement. The increase of lateral displacement and the maximum displacement depth are mainly affected by load size, the loading rate and the thickness of soft soil layer. The settlement on the sunny slope is larger than that on the shady slope due to the effect of shady-sunny slope. In cold seasons, the seasonally thawed layer under the embankment is gradually frozen, and then the soft soil subgrade will appear frost heaving, and the amount of frost heaving is affected by the residual water content and load stress of the soft soil subgrade after consolidation and compression. The settlement of soft soil under the embankment in swampy wetland treated by geogrids is large in warm seasons and small in cold seasons, and there is the existence of frost heaving (Fig. 3.162).

The freeze-thaw state of soil under embankment in the swampy wetland in permafrost regions changes with time. In the early stage after soil-filling embankment, the thawing depth under embankment increases due to the heat storage of the soil-filling, and the permafrost table under embankment declines, with the appearance of thawing core. The maximum seasonal thawing depth is larger than the measured natural permafrost table under natural surface, indicating that geogrids have poor effect on protecting permafrost. With the passage of time, except for the degenerative permafrost regions, the permafrost table under the embankment will gradually recover and rise, and the heat dissipation of the embankment will gradually exceed the heat absorption, and the area of the whole thawing core will decrease (Fig. 3.163). There is similar situation in the engineering measures of geocells, but the engineering measure of dumping crushed rock will greatly raise the permafrost table under embankment, without thawing core. This indicates that the engineering measure of dumping crushed rock not only protects the underlying permafrost, but also plays a role in strengthening the embankment stability.

3.6.3 Foundation Treatment of Swampy Wetlands

Swampy wetlands contain humus soil, peat soil, and ground ice layers with different thicknesses. When it is used as the foundation of a building (structure), it is necessary to treat the foundation, because its deformation and thermodynamic stability cannot meet the requirements.

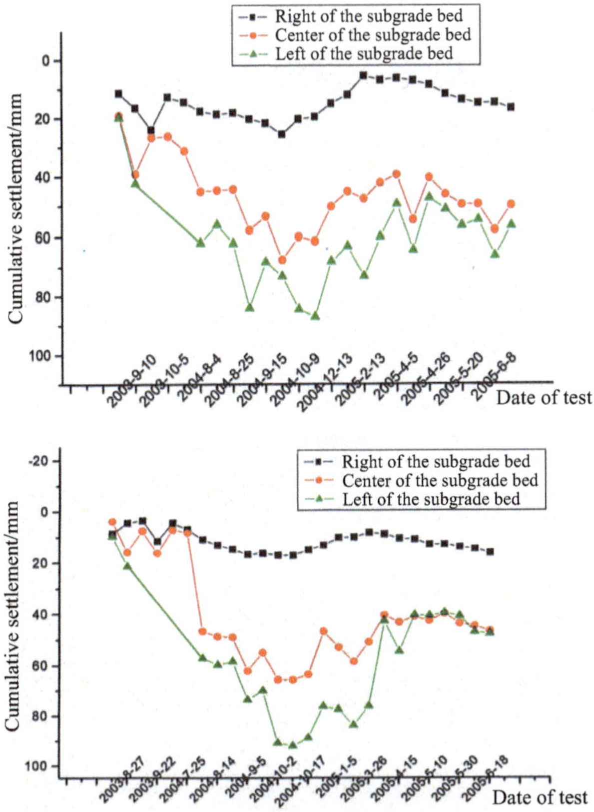

Fig. 3.162 Settlement curves with time at the bottom of the subgrade bed (the Left) and under the interface between revetment and embankment (the Right) (The Compilation Committee of *Qinghai-Xizang Railway*, 2016)

The foundation treatment of frozen wetlands should meet the following requirements:

1. Strength requirements: Foundation strength is required to meet the requirements in strength (compression, shear, etc.) of the superstructure (structure) under the action of its self-weight and external loads, and local or overall damage can not occur;
2. Deformation requirements: Foundation deformation is required to meet the requirements in deformation (settlement and lateral displacement) of the superstructure (structure) under its self-weight and external load, not exceeding the allowable differential settlement of the building (structure);
3. Thermal requirements: foundation is required to meet the requirements in thermal stability of the superstructure (structure) under its self-weight and external heat. The variation of the underlying permafrost table should not exceed the deformation that causes the thaw settlement of permafrost;

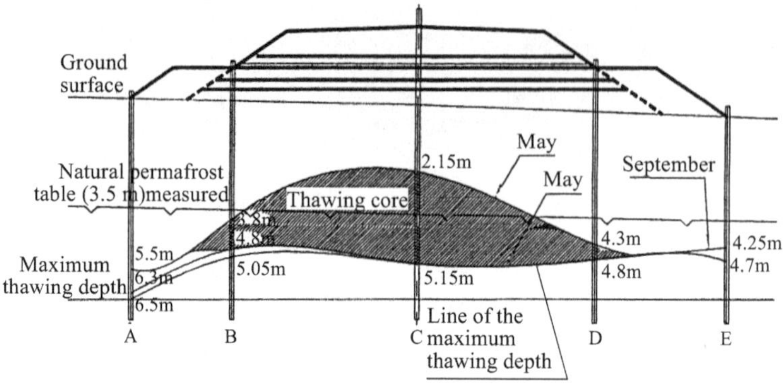

(a) 2004-2005

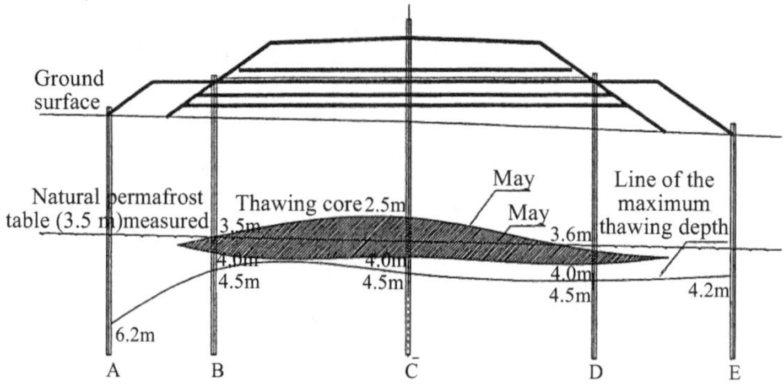

(b) 2005-2006

Fig. 3.163 Distribution of the maximum seasonally thawing depth under the embankment treated by geogrids at wetland areas of Za'gya Zangbo along Qinghai-Xizang Railway with seasonal variation (The Compilation Committee of *Qinghai-Tibet(Xizang) Railway*, 2016)

4. Dynamic stability requirements: Foundation is required to meet the dynamic load (earthquake, locomotive cycle) without liquefaction, instability, subsidence and other disasters

Foundation treatment in swampy wetlands in permafrost regions should be adopted in a comprehensive way: thermal and mechanical requirements should be met at the same time. In order to make good surveys on the engineering geological conditions of frozen soil, it is necessary not only to understand the basic properties of the foundation soil, but also to understand its characteristics, environment, hydrological (geological) characteristics, combined with the building structure type, requirements, building materials, construction and other aspects, and then to determine the method of foundation treatment comprehensively.

The purpose of wetland treatment in permafrost regions is: ① To improve the strength of soft foundation and ensure the stability of subgrade; ② To reduce the compressibility of the soft foundation and settlement deformation of foundations; ③ To decrease the temperature of permafrost under embankment and raise permafrost table to maintain the thermal stability of permafrost in swampy wetlands; ④ To prevent vibration liquefaction of foundation ground caused by earthquakes, etc.

The engineering diseases in swampy wetlands mainly include thaw settlement, frost heaving and frost boiling. The basic methods of foundation treatment in swampy wetlands are drainage and dewatering, water isolation, foundation reinforcement, foundation soil improvement and thermal protection of permafrost foundation. At present, there are many ways to deal with weak foundation of highway, including replacement, drainage consolidation, composite foundation and dynamic consolidation, etc. ① drainage consolidation method: the preloading consolidation method of plastic sheet drainage, the preloading consolidation method of sand well or sand bag well, and the drainage consolidation method of vacuum preloading; ② composite foundation method: the gravel compaction pile method, the method of cement mixing pile, the method of dry jet mixing pile, the pipe pile method of thin-walled concrete, CFG pile method, the composite foundation method of dynamic compaction of crushed (block) stone pier, and the reinforcement method of geopolymer, etc.; ③ replacement method: dredging for fill and throwing stones to packing sedimentation, etc.; ④ avoidance method: replacing embankment by bridge.

All kinds of treatment methods have their advantages and disadvantages, and all have their own applicable conditions. According to the properties of the soft soil foundation and the actual engineering conditions, one or several suitable methods should be adopted to treat the soft soil foundation in the frozen soil areas.

In the deep seasonally frozen soil areas, the maximum seasonal freezing depth is above 2.0 m ~ 2.5 m. In the talik within the permafrost region, the maximum seasonal freezing depth can reach 4.0 m, adopting the method of drainage consolidation, replacement, the gravel compaction pile, the combination of geogrids with EPS, and the combination of geocells with EPS.

In permafrost regions, the maximum seasonal thawing depth (permafrost table) is generally 0.8 m ~ 3.5 m, and the lower part of permafrost table often contains thick ground ice. Basically, the principle of protecting permafrost is adopted in foundation treatment, and combined with drainage and water isolation. Three methods are taken on the Qinghai-Xizang Railway: the combination of geocells with permeable soil subgrade, the combination of geogrids with permeable soil subgrade, and the dumping fill with crushed rock (with broken rock revetment) (Duan, 2010). And thermal insulation berms should be built at the slope toe of the embankment. According to the test results, the dumping fill with crushed rock can effectively protect the thermal stability of permafrost under the embankment. The thawing core appears, shrinks and disappears gradually under the embankment with time if using the other two methods. Therefore, if the combination of geogrids (cells) with permeable soil subgrade is used, the thermal insulation layer (EPS or XPS) also should be added, and thermal probes need to be installed if necessary.

The key measures to prevent and control frost heaving and frost boiling in swampy wetlands are: the isolation of the wetlands from water supply, drainage and dewatering, thermal insulation (keeping the foundation soil in thawed or frozen state), and foundation soil replacement.

Chapter 4
Bridge, Culvert and Tunnel Engineering in Frozen Soil Regions

Bridges and tunnels are the key and controlling engineering in the railway and highway constructions and operations. In frozen soil areas, the design and construction of the engineering of bridge, tunnel and culvert have different characteristics and requirements from those in non-frozen soil areas. The engineering problems, such as frost heave, thaw settlement and ice jam, etc., are not encountered in the construction in non-frozen soil areas. In addition to the mechanical stability, the thermal stability should also be considered in the design and construction of structures in frozen soil areas.

4.1 Bridge Engineering in Permafrost Regions

The construction of bridge engineering in permafrost regions has lasted for a long time. Many successful experiences at home and abroad show that the design of bridge engineering should be based on these principles including choosing the simply supported beam superstructure for different foundation deformations in permafrost regions, choosing the pile foundation with high thermal sensitivity to permafrost engineering geology and choosing the rapid construction methods that can adapt to permafrost environmental conditions, etc. In addition to the engineering measures of "bridge instead of embankment", the environment of bridge engineering in permafrost regions has the following characteristics:

1. In the permafrost regions of the plateau, the floodplain and terrace are wide, the water is shallow and oscillating, and the slope gradient of riverbed is small, wide and bare. Apart from the large rivers with large runoff, there are permafrost layers under the riverbed in the permafrost regions of the Qinghai-Xizang Plateau or the Daxing'anling Mountains, which belong to the non-penetrating taliks. The mean annual ground temperature of the permafrost is close to 0 °C, which is

Q. Wu, C. Tong, *Frozen Soil Engineering in Cold Regions*,
https://doi.org/10.1007/978-981-95-2527-0_4

warm and unstable permafrost. There are also permafrost layers in river flood lands and terraces, and its permafrost table is generally 3 m–4 m, or even deeper. And there are discontinuous permafrost layers in some sections.

2. In the permafrost regions, there are often river benches of various widths, and some of which are inundated, while others are not. In sections where the river water often flows, the thawing depth of permafrost is deeper. The thawing depth of non-inundated section is thinner, which affects the distribution of permafrost under the riverbed with frequent river diversion.
3. The groundwater level in the riverbed and floodplain is relatively shallow, and has a close hydraulic connection with the water flow of the mainstream. In winter, the freezing depth of the riverbed area is uneven, and it's easy to form the pingos and icings. Because the strata of the riverbed, floodplain and terrace vary greatly, there are silty soil, silty clay, sand, and gravel, etc., and the groundwater is abundant. Therefore, they are the areas with frequent pingos occurrence.
4. In bridge projects across rivers, the location, stratum, hydrology and other conditions of each bridge pier are different. As there are both permafrost and talik, for the great variations in permafrost parameters, such as the permafrost table, ground temperature, bearing capacity of foundation soil, adfreezing force, frost-heave force and thaw settlement, etc., they need to be treated differently.

In permafrost regions, the early bridge pier foundations were mainly pier foundations and column foundations that were excavated and poured. In the later period, the small-diameter group pile foundations were used. At present, the deep and large-diameter pile foundations are used. This kind of pile foundation has the advantages of adapting to the engineering geological conditions of permafrost, simple construction, large bearing capacity and minimum thermal disturbance to permafrost foundation. The pile foundations used in the United States, Canada and other countries mainly include the common steel pipe piles, the H-section steel piles, the spiral steel pipe piles, the steel thermal pipe piles, and the steel pipe piles with thermosyphon inserted. In Russia, it mostly uses factory precast reinforced concrete piles with circular and rectangular cross-sections. The H-shaped and cross-shaped piles can not only reduce the weight of pile foundation, but also increase the frozen area of pile foundation and improve the adfreezing force. In China, the cast-in-place reinforced concrete circular piles are mainly used, for the advantages of simple structure, convenient construction, in-situ aggregate and random pile foundation size, and large adfreezing force in the frozen section, which are more suitable for conditions in China. In recent years, the new technology of deep and large-diameter rotary drilling machinery for pore-forming has been adopted, which shortens the construction time of cast-in-place piles and greatly reduces the thermal disturbance to the permafrost foundation. The bridges in the permafrost area of the Qinghai-Xizang Railway were all made of the reinforced concrete cast-in-place piles.

4.1.1 Bridge Deformation

The characteristics of permafrost in river zones in permafrost regions are very complex, since there are penetrating talik and non-penetrating talik, continuous permafrost and discontinuous permafrost. There are complex changes in ice content of permafrost, variations of ground temperature of permafrost in the high-temperature range, strong thermal erosion of surface water and groundwater, climate and engineering thermal disturbance combined with each other, leading to frequent frost damage to bridges in permafrost regions. The main forms of common damage are:

4.1.1.1 Bridge Damage Caused by Thawing of Permafrost in Permafrost Foundations

1. Uneven sinking of bridge piers (Fig. 4.1);
2. The pier sinks and tilts (Fig. 4.2);
3. "Vehicle jolting on Bridge" occurrence when the bridge end sinks (Fig. 4.3) (Wu & Liu, 2005);
4. Thaw settlements of abutments and conical slopes (Fig. 4.4) (Wu & Liu, 2005).

Special attention should be paid to thaw settlement in road-bridge transition sections.

The reason for the thaw settlement of bridge piers is that the heat from the construction and operation of the bridge piers is transferred to the permafrost foundation, which causes the permafrost to thaw, and the warming climate causes the degradation of the permafrost. The buried depth of the pier foundation is insufficient, and when the permafrost thaws, the settlement deformation of the foundation is too large, resulting in the destruction of the bridge.

Fig. 4.1 Settlement deformation of Chumar River Bridge pier

Fig. 4.2 Sinking and tilting of the Yama River pier

Fig. 4.3 Sinking of bridge head

4.1.1.2 Bridge Diseases Caused by the Frost-Heaving Effects of the Active Layer in the Permafrost Regions

1. The overall deformation caused by differential frost heave of the bridge pier (Fig. 4.5);
2. The frost jacked bridge pier (Fig. 4.6);
3. Frost heaving of bridge pier and cracking of bridge (Fig. 4.7);
4. The bridge abutment pulled apart due to frost heave (Fig. 4.8);
5. Frost-heaving cracks in the bridge abutment cone (Fig. 4.9).

The reasons for the frost-heaving damage of the bridge can be due to the insufficient buried depth of the pier foundation or the improper construction method, as well as

Fig. 4.4 Sinking of abutment

Fig. 4.5 Frost-heaving hunched up bridge in Qing'an in Heilongjiang Province

the adverse texture of foundation soil and excessive moisture of the foundation soil. When the tangential frost-heave force of the bridge pile and the normal frost-heave force of the abutment base exceed the upper load the frost jacking occurs.

4.1.1.3 Freeze-Thaw Damages Caused by Changes of Water Level and Diurnal Variation of Temperature

1. Changes of water level causing the freeze-thaw denudation of pier concrete, resulting in exposure of the reinforced steel bars (Fig. 4.10).
2. Frost cracking of bridge piers or bridge concrete (Fig. 4.11).

Fig. 4.6 Frost-heaved and abandoned Zhangwanguan Bridge pier in Jilin Province

Fig. 4.7 Frost-heaving of bridge pier and the cracking of bridge

During the beginning and the end of the cold season in permafrost regions, the diurnal temperature difference is large, and the changes in river water level or the seepage of bridge deck often wets the micro-cracks of the bridge pier concrete and causes freeze-thaw cycle damage.

Under the low bridges of village roads, ice jams often occur during winter: frost heave under the bridge lifts the bridge deck, or ice covers the bridge deck. When in spring, the ice in the river thaws and cracks, the river ice will burst into the bridge piers and cause damages (Fig. 4.12).

Fig. 4.8 Frost-heaving of Daxing'anling Bridge abutment (Provided by Qiqihar Railway Science Research Institute)

Fig. 4.9 Frost-heaving cracks in the cone of the Daxing'anling abutment (Provided by Qiqihar Railway Science Research Institute)

4.1.1.4 Causes for Frost Damages on Bridges

Bridge foundations in permafrost regions are all built on permafrost foundations. The variety of permafrost conditions in river areas are very complex; the thermal stability of permafrost under the impact of water flow is poor; the external thermal disturbance has a strong influence on the thermal stability of permafrost foundation; frozen swamps or wetlands, upon which the bridges are built instead of roads, belong to the area with rich and thick ground ice; the seasonal thawing depth (the permafrost table) is usually deep, with higher moisture content and greater frost heave. The foundation soil of the river section in the seasonally frozen soil area often has strong-frost heaving properties.

Fig. 4.10 Freeze-thaw denudation of bridge pier concrete (Provided by Qiqihar Railway Science Research Institute)

Fig. 4.11 Frost cracking of bridge pier concrete (Provided by Qiqihar Railway Science Research Institute)

Bridge construction will disturb and destroy the thermal balance state of the permafrost in bridge construction area. The heat brought into permafrost foundation by human will make the temperature of permafrost foundation rise sharply, change the original environment of permafrost foundation, and cause the thawing of permafrost foundation and the change of engineering properties. After the completion of the bridge project, the permafrost foundation usually needs 2–3 years (by climate and permafrost itself) to adjust thermally and will establish a new thermal balance state and form a new permafrost table. In this process, the thaw settlement of

Fig. 4.12 Bridge pier of Daxing'anling Mountains damaged by the hit of drift ice (Photo by Tong Changjiang)

permafrost foundation has a certain influence on the stability of bridge foundation, or causes the thaw settlement deformation of bridge foundation and upper structure.

During the cold season, the foundation soil above the permafrost table (seasonally thawed layer in permafrost regions and seasonal freezing layer in seasonally frozen soil areas) freezes. In the freezing process, the action of the tangential frost-heave force of the foundation soil on the side of the bridge foundation or the normal frost-heave force under the bridge foundation will cause the stability problem in the balance of the force system between the bridge and the foundation soil system. When the frost-heave force of the local foundation soil exceeds the bridge load, the bridge pier will be pulled up or lifted by the frost heave, causing the bridge frost-heaving deformation.

The peeling, cracking and crushing damage of bridge concrete components are mainly caused by the micro-cracks or voids in the water-infiltrated concrete components. During the beginning of the cold season (in September and October) and the end of the cold season (in April and May), under the repeating action of positive and negative temperature in day and at night, congelifraction occurs, causing denudation and cracks of concrete components. Due to frost heave (or settlement) of the bridge foundation, the upper components are dislocated, squeezed and broken.

4.1.2 Bridge Foundation and Base

4.1.2.1 Foundation

The frozen soil foundation bearing the load of the structures belongs to a four-phase system: solid particles (soil), air, water (unfrozen water) and ice.

Soil In frozen state, the foundation soil is extremely hard, especially the coarse-grained soil, which has high strength characteristics, but when the soil temperature is high (slightly lower than the freezing temperature of soil), the cohesive soil (especially the clay) has the characteristics of plastic permafrost, with low strength and high compressibility and rheology. In some sections of the riverbed, there may be non-penetrating talik and groundwater enrichment regions. The permafrost table is often deep, and there is a discontinuous permafrost layer. When determining the strength of permafrost foundation, not only the grain size gradation, density, and moisture content of the foundation soil, but also the soil temperature, ice content, and the effects of the freeze-thaw process on the changes of foundation soil properties and strength should be considered.

Ground Ice The change of ice content of ground ice in foundation soil of riverbed in permafrost regions is complex. Usually, there is a layer of ground ice at the permafrost table. Thick ground ice is rarely found in coarse-grained soil, but there is lenticular ground ice with a thickness of 10 mm–300 mm in most cases. The fine-grained soil has a high ice content, which is cryptocrystalline-textured, or micro-layered, layered. The maximum water content can reach more than 40%–45%. While in a frozen state, the ice itself has a certain bearing capacity. As a transformable carrier, the greater significance of ground ice lies in that it cements the soil particles to form a hard frozen soil body, which has a higher bearing capacity than the original material composition (soil and ice). Once the ice is thawed, not only the strength of the ice itself is completely lost, but also the strength of the permafrost is also lost, and a large amount of settlement occurs, and its amount depends on the ice content and soil density.

Water The water that is existing in frozen soil is unfrozen water. When soil temperature drops and the amount of unfrozen water decreases, the frozen soil will have higher strength; when the soil temperature rises and the unfrozen water content increases, the frozen soil will be in a plastic state, and the bearing capacity will decrease and lead to a large plastic deformation. When the frozen soil thaws the thawed water will be drained, causing the sinking of foundation. And when the water in the cold season freezes, frost heave will occur again.

Bedrocks, Cracks or interlayers of frozen hard rock under riverbeds are often filled with ice (Fig. 4.13), and the weathered weak rock zones often contain layered ice and crack ice (Fig. 4.14).

It can be concluded that the strength of frozen soil foundation in bridge engineering is not only affected by particle size gradation of soil, density and water content, but also by soil temperature and ice content. The permafrost in the river area is mostly warm permafrost. Under the thermal impacts of humans and climate, the permafrost table is deep, and the thaw settlement and creep of frozen soil foundation are large. During the freezing period, the effects of frost-heave force on the pier foundation can't be ignored. Maintaining or reducing thermal disturbance to frozen soil foundations is not only important for bridge foundation construction and

Fig. 4.13 Layered ice in the interlayer of the bedrock

Fig. 4.14 Crack ice in the bedrock (Photos by Niu Fujun)

thawing control of frozen soil foundation, but also for the stability of bridge engineering operation.

4.1.2.2 Pile Foundation

The types of permafrost foundation should be selected according to the characteristics of frozen soil and foundation design principles. The commonly-used foundation types include the expanded foundations, such as the open-cut expanded foundations, the drilled-in foundation and expanded pile foundation, and the pile foundations, such as the driven pile in bore hole, the insert pile in bore hole, the cast-in-place pile, the excavated and cast-in-place pile. At present, most bridge engineering and construction projects in permafrost regions in China are using the cast-in-place piles. In the early stage, the open caisson foundation was used for the bridge foundation of large and medium-sized rivers in the talik of permafrost regions in Daxing'anling Mountains.

1. Driven pile in borehole. It is suitable for all kinds of frozen soil. It is difficult to form boreholes in all kinds of frozen soil and hard frozen rock layers containing large rocks and boulders. The pile foundation is a precast reinforced concrete pile with steel cap and steel tip. Drilling equipment is used to form a borehole in advance, the diameter of which is slightly smaller than the diameter of the pile, and then the pile hoisting is introduced into the borehole, and the pile is slowly driven to the design depth with a hanging hammer. In the plastic frozen cohesive soil and sandy soil, the direct driving method can be used for construction. The Wall protection measures should be taken in the active layer to prevent boreholes from collapsing.
2. Inserted pile in borehole. It is suitable for all kinds of frozen soil. The pile foundation is a precast reinforced concrete pile with steel tips. Drilling equipment is used to form boreholes in advance, and the borehole diameter is slightly larger than the pile diameter. First pour a small amount of concrete mortar into the borehole (just enough to fill the gap between the pile and the borehole wall), then insert the pile hoisting into the borehole, and use a hanging hammer to slowly drive the pile to the design depth. The wall protection measures should be taken in the seasonally thawed layer to prevent boreholes from collapsing.
3. Excavated and cast-in-place pile. It is suitable for all kinds of frozen soil at low temperature (below −1.0 °C). When digging boreholes, the borehole wall should be protected to avoid collapsing, and the construction should be carried out in the cold season, that is, the period when the air temperature is negative. The low-temperature concrete should be used, and the molding temperature should be as low as possible, and generally should not exceed 10 °C.
4. Cast-in-place Pile. It is suitable for all kinds of frozen soil. The Qinghai-Xizang Railway adopts high-power rotary drilling rig, which has the advantages of fast drilling speed and neat borehole walls. The wall inside the seasonally thawed layer should be protected to prevent the hole from collapsing.

In the 1970s, the Third Survey and Design Research Institute of the Ministry of Railways, Harbin Railway Bureau and other research institutes carried out pile foundation tests in the high-latitude permafrost region of the Yalin Line in the Da xing'anling Mountains.The Northwest Research Institute of the Ministry of Railways, the First Survey and Design Institute of the Ministry of Railways and other institutes have carried out pile foundation test research in Wudaoliang, Fenghuo Mountain and Qingshui River on the Qinghai-Xizang Plateau. The size of these test piles was generally not more than 0.65 m, and the pile length was less than 8.5 m. At the beginning of the twenty-first century, pile foundation tests of large pile diameters (greater than 1.0 m) and deep piles (pile length of 12–24 m) were carried out during the construction of the Qinghai-Xizang Railway.

Based on the fact that the cast-in-place piles are normally used in China's engineering construction in permafrost regions, the thermal effects on frozen soil foundation should be considered according to the design principle of maintaining the frozen state of frozen soil foundation. Based on the current experimental research on the Qinghai-Xizang Highway, the Qinghai-Xizang Railway and Bridge Engineering in the frozen soil area of the Daxing'anling Mountains, the following issues need special attention in the design of cast-in-place piles.

1. **The Influence Range of Hydration Heat of Cast-in-place Piles**

The thermal effects of cast-in-place piles on frozen soil foundation in permafrost regions is derived from a large amount of hydration heat released during the hardening process of concrete after pouring, which causes the rise of temperature and thawing of permafrost foundation around the piles.

The adiabatic temperature rises inside the concrete due to hydration heat:

$$T_{(f)} = \frac{G \times q}{C \times \rho}\left(1 - e^{-mt}\right) \tag{4.1}$$

The internal temperature of the concrete at different stages:

$$T_{u(f)} = T_m + T_{(f)} \cdot \alpha \tag{4.2}$$

where $T_{(f)}$ is the adiabatic temperature rise of the concrete after pouring for a period of time (°C); Q is the hydration heat per kilogram of cement (kJ); G is the amount of cement consumption per cubic meter of concrete (kg); C is the specific heat of concrete (kJ/kg °C); P is the mass density of the concrete (kg/m^3); $m$is the empirical coefficient related to the casting time of cement categories; t is the duration after pouring (day); T_m is the molding temperature of the concrete (°C); a is the heat release coefficient of the concrete.

After concrete is poured, the concrete temperature of the pile rises sharply with time under the impact of concrete hydration heat. With the dissipation of hydration heat and the heat absorption of surrounding frozen soil, the pile concrete gradually cools down (Fig. 4.15). The time for the temperature of soil to rise to the maximum value is different at different distances: the closer to the pile, the greater the temperature will rise and the shorter the time will be; the farther away from the pile, the smaller the temperature will rise and the longer the time will be. As time goes, the hydration heat gradually dissipates and the soil temperature gradually declines. Within a certain distance around the pile, the permafrost layer will be thawed. According to the calculation based on Fig. 4.15, under the mix proportion and molding temperature of concrete, the thawing depth of the permafrost layer around the cast-in-place pile is generally about 0.5–1.0 m.

The maximum temperature rising and decreasing of the pile foundation concrete and the thawing thickness of the permafrost layer around the pile foundation mainly depend on the amount of concrete that used (related to the pile diameter and length), the type, the molding temperature and the mean annual ground temperature, the ice content and the lithology of the permafrost layer.

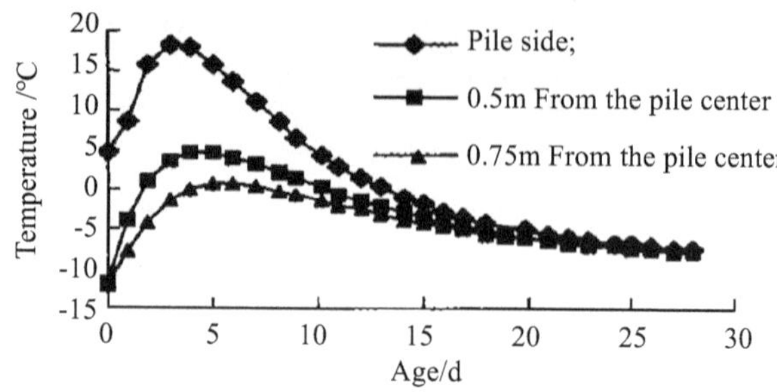

Fig. 4.15 Variation curve of temperature of soilat different distances around the pile (Jia et al., 2003)

In fact, the influence of cast-in-place piles on the permafrost layer around the pile will increase the ground temperature of the permafrost in a certain range around the pile in addition to the thawing of the permafrost layer around the pile to a certain thickness.

The observation data on ground temperature of the bridge piers of the Jiamo Highway in the southern part of Tahe County in the Daxing'anling Area, Heilongjiang Province, show that the area belong to isolated permafrost, the mean annual ground temperature is −2.0 °C, and the permafrost table is 1.6 m. From the ground surface to a depth of 3.6 m is the fill soil, the peat soil and the round gravel, which are ice-rich and ice-saturated permafrost. Within the depths between 3.6 m and 4.5 m is the soil-bearing round gravel and the gravel soil, which belong to icy permafrost; and beneath 4.5 m, it is the strongly weathered tuff. The diameter of the bridge pile is 1.2 m and the length of the pile is 15 m. Immediate observation of the completion of the pouring at the end of July (Fig. 4.16) shows the temperature at 1 m from the pile side: on the first day after the pouring, the temperature of the permafrost from the surface to the depth of 12 m increases successively, and above 2 m has the largest increase; on the fifth day, the temperature of the permafrost reaches the highest, with an increase of 0.1–1.0 °C, and the lateral transmission speed of hydration heat is 0.2 m/d; then, the temperature of the permafrost begins to decrease as a whole, and the temperature of the permafrost at the depths of 14–16 m recovers slowly. The temperature always maintains at −1.9 °C. On the 25th day, the temperature of the permafrost at depth of 4 m rises back to 0 °C; on the 90th day, the temperature of the permafrost below 2 m is stable at −1.9 °C (at the end of October). And it shows the temperature in the section 2 m away from the pile side: on the first day after pouring, the temperature of the permafrost increases, but the increase is not in large degree, and the temperature of the permafrost is below 0 °C at the depth of 4 m. Over time, the overall temperature of permafrost doesn't show a recovery, and the temperature of permafrost at the depths of 14–16 m remains at −1.9 °C. It can be speculated that the influence range of concrete hydration heat in the test site is between 1 and 2 times of the pile diameter.

The ground temperature of cast-in-place piles for bridges in different permafrost regions along Qinghai-Xizang Railway is observed (Wang et al., 2004a, b, c). The ground temperature of the soil adjacent to the pile (SB) increases sharply (Fig. 4.17) after the test pile is poured in the cold permafrost regions (T_{cp}-IV). The ground

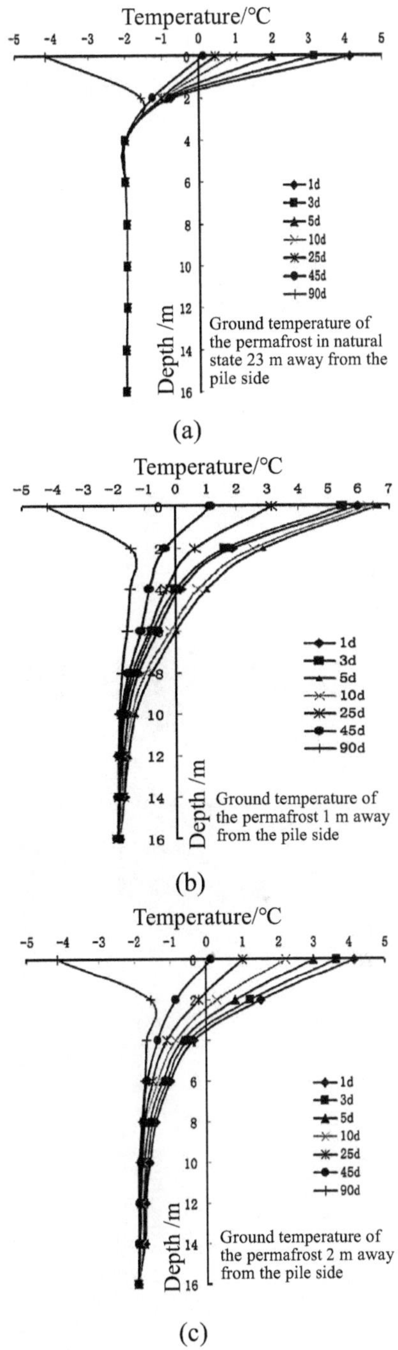

Fig. 4.16 Variations of permafrost temperature on the side of cast-in-place piles in Tahe isolated permafrost region of Daxing'anling Mountains (Ji, 2015). The diameter of the pile is 1.2 m, the length of the pile is 15 m, and the mean annual ground temperature of permafrost is −2.0 °C, with peat soil and round gravel, ice-saturated and ice-rich frozen soil. (**a**) Ground temperature of the natural borehole 20 m away; (**b**) Temperature of the permafrost 1 m away from the pile side; (**c**) Temperature of the permafrost 2 m away from the pile side

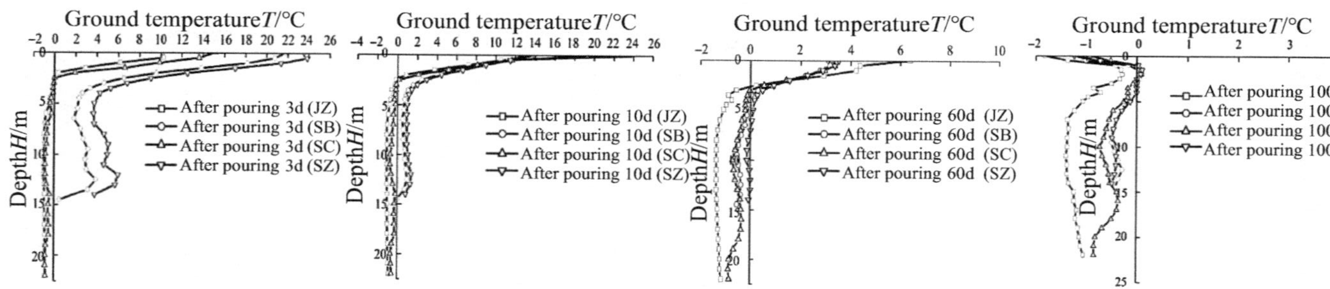

Fig. 4.17 Variation curve of ground temperature around cast-in-place piles in cold permafrost of Qinghai-Xizang Railway. (Wang et al., 2004a, b, c) JZ stands for the benchmark borehole for ground temperature measurement located 20 m away from the pile; SB stands for the borehole for ground temperature measurement adjacent to the pile; SC stands for the borehole for ground temperature measurement located 0.3 m away from the pile, SZ stands for the borehole for ground temperature measurement located at the center of the pile

temperature at the permafrost table (at 2 m) reaches 4.31 °C after pouring for 3 days, and the ground temperature at the lower end of the pile also increases to 1.59 °C, and then the ground temperature at each depth of the pile gradually decreases. On the 8th day, 3.5 m–7.5 m below the ground surface drops to −0.027 to −0.32 °C, and the negative temperature (at −0.35 °C) has restored at the lower pile end. On the 30th day, the ground temperature of the soil adjacent to the pile at a depth of 2.0 m below the ground is −0.43 °C, and the ground temperature at the lower end of the pile is −1.63 °C. On the 50th day, the ground temperature of the soil adjacent to the pile at depth of 2.0 m is −0.49 °C, and the ground temperature at the lower end of the pile is −1.85 °C. At this time, the ground temperatures of the reference pile are −2.98 °C and −3.43 °C respectively. It shows that the original ground temperatures of the test pile have not risen to the natural ground temperatures in this area in 50 days after pouring.

After the test pile concrete is poured in the warm (T_{cp}-II) permafrost region, the ground temperature of the permafrost layer around the pile increases sharply (Fig. 4.18). After 3 days, the ground temperature near the original permafrost table (2.5 m) reaches 6.51 °C, and the ground temperature at the lower end of the pile also rises up to 0.47 °C. On the 18th day, a negative temperature of −0.18 °C appears at the lower end of the pile. On the 30th day, the ground temperature adjacent to the pile at a depth of 8.5 m below the ground is −0.10 °C, and the ground temperature at the lower end of the pile is −0.57 °C. On the 60th day, the ground temperature adjacent to the pile at 4.5 m deep below the ground is −0.026 °C, and the ground temperature at the lower end of the pile is −0.57 °C. On the 100th day, the ground temperature adjacent to the pile at the permafrost table (2.5 m) is −0.03 °C, and the ground temperature at the lower end of the pile is −0.62 °C. At this time, the ground temperature at 2.5 m of the benchmark pile is −0.37 °C, and the ground temperature at the lower end of the pile is −1.26 °C. It indicates that in 100 days after pouring, the test pile has not yet refrozen to the original ground temperature in the area.

Based on the test piles at the Kunlun Mountain pass of Qinghai-Xizang Highway as an analysis prototype, considering the local geology and the conditions of the cast-in-place piles (with the pile diameter of 1.2 m, pile length of 15 m, and molding temperature of 5 °C), the calculation results (Fig. 4.19) are as follows: ① In 80 days after the construction, the temperature, including the concrete pile itself, has dropped below 0 °C; ② The time for the temperatures of the permafrost at different distances from the pile rising to the highest value is different: the closer to the pile, the shorter the time; the farther away, the longer the time; the temperature at 1.2 m away from the pile does not exceed 0.5 °C, and the highest temperature at 1.8 m away from the pile does not exceed 0 °C; ③ In 200 days after the construction, the soil temperature near the pile has not restored to the initial state; ④ The rise of ground temperature at the lower end of the pile is significantly smaller than that adjacent to the middle of the pile.With the increase of the distance from the pile, the difference is becoming smaller, and the ground temperature beyond 3.6 m away from the pile basically does not change (Fig. 4.20), indicating that the thawing thickness of the concrete heat release is about 0.6 m, and the influence range is 3 to

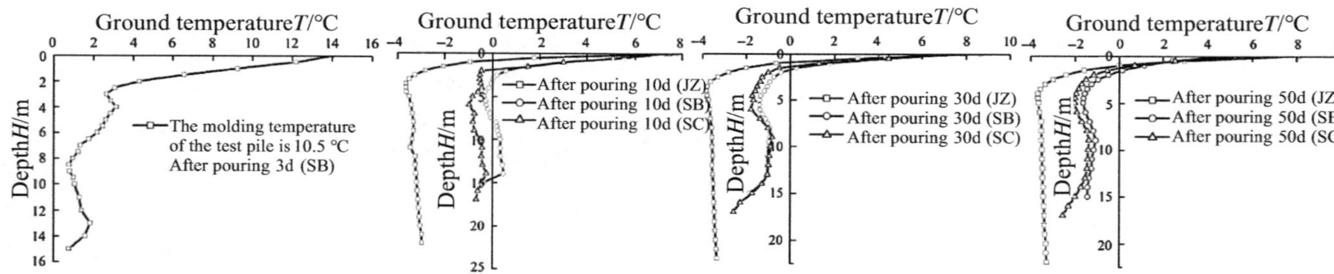

Fig. 4.18 Variations of ground temperature around cast-in-place piles in warm permafrost area along Qinghai-Xizang Railway (According to Wang et al., 2004a, b, c). JZ stands for the benchmark borehole for ground temperature measurement located 20 m away from the pile, SB stands for the borehole for ground temperature measurement adjacent to the pile, SC stands for the bore hole for ground temperature measurement located 0.3 m away from the pile, SZ stands for the borehole for ground temperature measurement located at the center of the pile

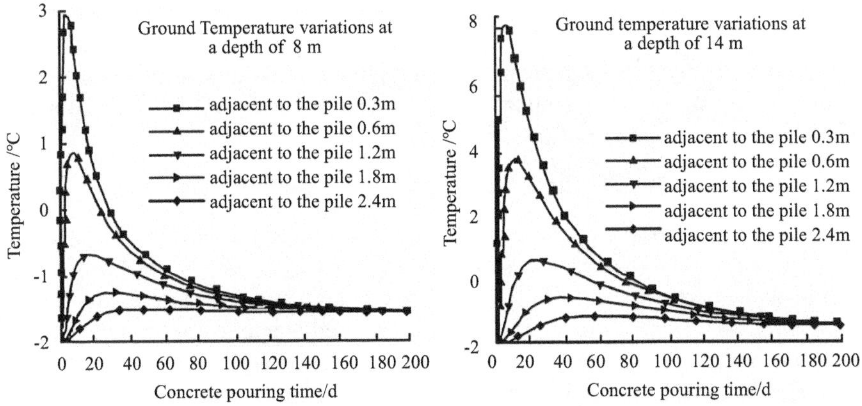

Fig. 4.19 Variations of the ground temperatures with time at different depths in the cold permafrost regions at Kunlun Mountain pass of the Qinghai-Xizang Highway (Xiong et al., 2009)

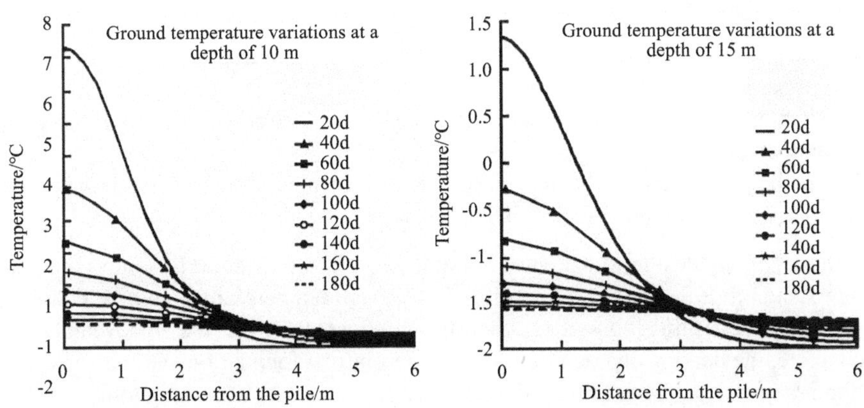

Fig. 4.20 Variations of ground temperature with time at different depths in the cold permafrost regions at Kunlun Mountain pass of the Qinghai-Xizang Highway (Xiong et al., 2009)

4 times of the pile diameter (larger than the influence range measured in Daxing'anling Mountains).

It indicates that the influence of the hydration heat of cast-in-place pile concrete on the permafrost around the pile dwindles with the increase of the distance from the pile. According to the experimental observation and model calculation analysis: ① In the radial direction, the hydration heat of concrete has the greatest influence on the permafrost within the range of 0.5 m, and the permafrost foundation is in a thawing state. Within the range of 1 m, it has a great influence on the ground temperature of the permafrost. The ground temperatures of the permafrost 2–3 times beyond the pile diameter in the radial direction from the pile are not affected by the hydration heat of concrete; ② In the vertical direction of the pile, the influence of concrete hydration heat on the permafrost foundation dwindles with the increase of depth, and the influence on the lower end of the pile is minor, and the influence is the largest at the permafrost table; ③ The higher is the molding temperature of the concrete, the more beneficial it is to the concrete strength, but it has a great influence on the ground temperature of the permafrost around the pile, and 5–10 °C is the reasonable molding temperature for cast-in-place piles.

2. **Refreezing Time of Permafrost Temperature around Cast-in-place Pile**

The thermal disturbance of the pouring of cast-in-place pile will inevitably cause the variations in the temperature regime of the permafrost around the piles, which will thaw the permafrost in a certain range around the pile and increase the permafrost temperature in a certain range. Before the load is applied to the pile foundation, it is better to ensure that the permafrost foundation around the pile has frozen back and met the design requirements. Therefore, the time and temperature of the permafrost refreezing around the pile have become the issues that are closely concerned in the design and construction of bridge engineering.

The refreezing time and temperature of the permafrost around the pile have different meanings for different requirements:

1. The time required for the thawed soil around the pile to freeze back to the "initial temperature when starting freezing".
2. The time required for the thawed soil around the pile to freeze back to the "design temperature".
3. The time required for the thawed soil around the pile to freeze and restore to the "original ground temperature in natural state".

Generally, the "design temperature" should be selected as the refreezing temperature standard of the cast-in-place piles. The design temperature should be comprehensively determined based on the strength increasing of the pile foundation concrete, the bearing characteristics of the permafrost foundation and the construction requirements. Determining the refreezing temperature according to the "initial temperature when starting freezing" cannot meet the requirements for the bearing characteristics of the permafrost foundation.And when according to the "natural state ground temperature" to determine the refreezing temperature, the refreezing time would be too long, and in the warm or degraded permafrost regions, the permafrost is often unable to freeze back to their original natural state.

Figure 4.21 shows the observation data of the refreezing process of concrete of the cast-in-place pile at Kunlun Mountain (the diameter of the test pile was 1.2 m, the pile length was 15 m, and it was a digging pile) (Institute of Transportation Science of Heilongjiang, 2006). Due to the diffusion of heat to the permafrost and the thawing of the permafrost around the pile, the actual concrete temperature of the pile is only 11–14 °C. Subsequently, the temperature decreases rapidly, and on the 7th day, the temperature drops to 3–4 °C. After 10 days, the cooling rate slows down, and on the 20th day, the temperature drops to 0 °C. On the 60th day, the temperature reaches −1.0 °C. It takes 110 days for the temperature of the concrete pile to drop to −2.0 °C, that is, the temperature of the permafrost in a natural state adjacent to the test pile.

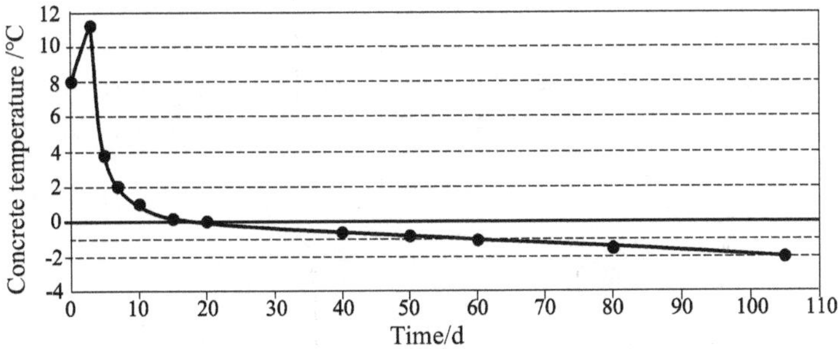

Fig. 4.21 Temperature variations of the pile center 5 m deep of the cast-in-place concrete pile with time. The pile, which is a digging pile, is with a diameter of 1.2 m and a length of 15 m, and the mean annual ground temperature is −2 °C, with silty clay and gravel soil in icy and ice-rich permafrost,

The variations of the ground temperature in the representative section of the test piles along the Qinghai-Xizang Railway are shown in Table 4.1 (Wang et al., 2004a, b, c). In the cold permafrost region of the Kunlun Mountain pass, the pouring is completed at the end of June. After 50 days, the ground temperature adjacent to the pile surface has not yet restored to the original ground temperature in natural state. In the warm permafrost area of the Qingshui River, the pouring is completed in late July. After the concrete has been poured for 100 days, it still doesn't reach the original ground temperature in natural state. According to the test data from another bridge pile (Editorial Committee of *Qinghai-Tibet(Xizang) Railway*, 2016), which is poured in November, the ground temperature adjacent to the pile surface at a depth of 3.5 m has risen to 3.6–4.6 °C in 4 days, and the ground temperature at the lower end of the pile has reached 1.26 °C. For the whole length of the pile, it has reached to the negative temperature (−1.03 to −0.24 °C) in 32 days, and in 53 days, the ground temperature adjacent to the pile surface is between −1.35 to −0.5 °C. When the cast-in-place piles are poured in the cold season, the temperature of the test piles in Qingshui River and Beilu River would drop to 0 °C in 30–60 days, and the piles would be basically re-frozen in 50–75 days.

In the transition zone from permafrost to talik in Tuotuo River, in the warm and extremely unstable permafrost regions, the lithology from top to bottom is the coarse sand, the silty clay, the strongly weathered marlstone (ice-bearing layer), the weathered marlstone, and the sandstone in order. Confined water is found at a depth of 16 m, and the water pressure is less than 160 kPa. The design pile diameter is 1.0 m, with a pile length of 20.3 m and a molding temperature of 8 °C. The pouring has finished in the early October. In 4 days after the concrete pouring, the temperature of the borehole wall 3.3 m under the ground surface is 11.4–12.5 °C, the ground temperature 0.3 m away from the borehole rises to 4.96–6 °C, and the temperature at the lower end of the pile is 6.75 °C (see Fig. 2.46) (Wang et al., 2005a, b, c), and the ground temperature 3.0 m below the lower end of the pile is 0.7 °C. Except that the ground temperatures adjacent to the pile within the depths from 4.8 m to 6.8 m

Table 4.1 Ground temperature variations in representative sections of test piles (Wang et al., 2005a, b, c)

Location	Section features (Below the natural ground)	Kunlun Mountain pass			Qingshui River*			
		3d	30d	50d	3d	30d	50d	100d
Pile surface	2.0(2.5*) m (natural permafrost table)	4.31	−0.43	−0.49	6.51	1.44	0.82	−0.03
	14.5 m (section at the lower end of the pile)	1.59	−1.63	−1.83	0.47	−0.57	−0.59	−0.62
0.3 m away from the pile (or pile center*)	2.0(2.5*)m (natural permafrost table)		−1.01	−0.65	9.1	1.61	0.92	0.0
	14.5 m (section at the lower end of the pile)		−1.91	−2.10	3.8	0.40	−0.05	−0.48
Natural borehole	2.0(2.5*)m (natural permafrost table)	−2.82	−2.79	−2.98	−0.21	−0.37	−0.56	−0.37
	14.5 m (section at the lower end of the pile)	−3.21	−3.48	−3.43	1.07	−0.97	−1.34	−1.26

Notes: At Kunlun Mountain pass, the permafrost table was 1.0–2.1 m, and the mean annual ground temperature was lower than −2.8 °C. The design pile diameter was 1.0 m and the pile length was 15 m. The molding temperature of the concrete was 11.0 °C, and the pouring was completed on June 27, 2002; * in the Qingshui River, the permafrost table was 2.0–2.5 m, and the mean annual ground temperature was 0.5–1.0 °C. The design pile diameter was 1.25 m and the pile length was 14.5 m. The molding temperature of the concrete was11.5 °C and the pouring was completed on July 21, 2002 below the ground surface are negative, the ground temperatures of the other observation points are positive. It shows that the permafrost layer in the transition zone between the permafrost region and the talik is relatively fragile, and the drilling in construction and the concrete pouring has a great influence on the ground temperature of the permafrost layer in this area so that the permafrost layer in some areas of the foundation of the pile group of the bridge can not re-freeze and almost does not exist any more.

For the cast-in-place piles with the pile diameter of 1.0 m–1.3 m, the time required for freezing back to the "initial ground temperature when starting freezing"can be estimated by the empirical formula:

$$t = k\sqrt{\frac{Q}{\lambda_f\left(T_{cp} - T_f\right)}} \tag{4.3}$$

where t is the time for the thawed soil around the pile to freeze back to its initial ground temperature when starting freezing, *(d)*; Q is the latent heat released when the thawed soil around the pile is freezing (the quantity can be considered to be equal to the hydration heat released by the concrete in 15 days) (kcal); λ_f is the thermal conductivity of the permafrost (kcal/(m·h·°C)); T_{cp} is the mean annual ground temperature of the permafrost (°C); T_f is the initial ground temperature of the thawed soil around the pile when starting freezing (°C); the value of (T_{cp}—T_f) takes the absolute value; k is the correction coefficient, and k takes 0.13–0.15.

If the mean annual ground temperatures at Qingshui River, Beilu River and Kunlun Mountains are −0.5 °C, −0.7 °C and −2.0 °C respectively, the thermal conductivity of permafrost is λ_f = 1.7 kcal/(m·h·°C), the initial ground temperature of the thawed soil around piles when starting freezing T_f, is −0.15 °C, and the molding temperature of the concrete is 15 °C, according to Formula 4.3, the time required for freezing back to the initial ground temperature is about 27 days in Qingshui River, about 23 days in Beilu River and about 15 days in Kunlun Mountain, which is basically consistent with the actual observation results.

It indicates that the re-freezing time of the cast-in-place pile depends on the factors such as the mean annual ground temperature of permafrost foundation, the lithology, the ice content, the permafrost environment, the total hydration heat of pouring concrete, the molding temperature and the construction characteristics, etc.

1. The lower is the mean annual ground temperature of permafrost, the shorter the refreezing time becomes. In the cold permafrost area, the pile will start to refreeze to negative temperature in 7 days, and it will basically refreeze to the initial ground temperature in 30 days. On the contrary, the higher is the ground temperature, the longer the refreezing time becomes. It takes 50–60 days for the warm permafrost area to basically freeze back to the initial ground temperature. For extremely unstable permafrost areas, it may be difficult to refreeze.
2. The higher is the ice content in the permafrost, the longer is the thawing time and the more slowly the refreezing becomes.

3. Regarding the construction method, if it is employing the rotary drilling to form the borehole, with the merits of a smaller thermal disturbance, it refreezes faster. If it is employing manual digging to build the pile, it takes the longest time for refreezing.The percussion drilling piles also take long time for refreezing. The time for insert pile in borehole to refreeze to 0 °C is approximately 20 days less than that of the cast-in-place pile.
4. The larger is the pile diameter, the greater is the thermal disturbance, and the thermal affecting range is about 2–3 times of the pile diameter. The thermal disturbance of the group piles is larger than that of a single pile. The smaller is the pile spacing, the greater are the combined effects of thermal disturbance. The distance between the cast-in-place piles should not be less than 3 times of the pile diameter.
5. The greater is the hydration heat of cement in concrete, the higher is the molding temperature. And the more is the heat brought in, the slower is the refreezing.
6. The refreezing time of pile foundation during construction in winter is shorter than that in summer.

3. **Refreezing Process of Thawed Soil around the Pile and Vertical Bearing Capacity of the Cast-in-place Pile in Early Stage**

During the refreezing process around the pile, there are two interfaces in the system of the pile foundation and the base: the interface between the pile and the thawed soil, and the interface between the thawed soil and the permafrost. The connection between the pile and thawed soil is the glue connection of water film between soil particles and the surface of pile foundation. The thickness of the round-shaped thawed soil around the pile caused by the hydration heat of concrete is small. In this case, there has generally no frictional resistance. When the permafrost around the pile is cohesive soil, and only if the thickness of thawed soil exceeds the critical value can there have frictional resistance.With the loss of hydration heat of concrete and the "cooling" effect of the permafrost, this kind of thawed soil gradually cools down and refreezes so that the range of the interface between the pile and the thawed soil gradually shrinks, the range of the interface between the thawed soil and the permafrost gradually expands, transforming to the interface between the pile and the thawed soil, and eventually turns into the interface between the pile and the permafrost, that is, from the water-glue bond between the pile and the thawed soil to the ice-glue bond between the pile and the permafrost. After this transforming, the freezing force between the pile and the permafrost has formed, which increases with the decreasing of the ground temperature during refreezing, and only then can the permafrost have the bearing capacity. Therefore, in permafrost regions, within a period of time after the cast-in-place piles are poured (before refreezing to the temperature below the freezing temperature), the frictional resistance of the thawed soil adjacent to the pile surface is very small, and the permafrost under the lower end of the pile also begins to thaw, and the bearing capacity at the lower end of the pile is also smaller.

1. In the warm and extremely unstable permafrost area (T_{cp} greater than −0.5 °C), it is difficult for the permafrost around the cast-in-place pile to refreeze when the concrete strength is forming. At this time, the bearing capacity of the concrete pile is composed of the lateral frictional resistance of the pile and the resistance of the permafrost at the lower end of the pile. For example, the cast-in-place piles in the transition zone between the permafrost region and the talik of the Tuotuo River cannot refreeze. In this case, the bearing capacity of cast-in-place piles (the lateral frictional resistance and the resistance at the lower end of the pile) can be calculated according to the calculation formula for the allowable bearing capacity of the bored or excavated and cast-in-place pile, which is from the design standard for the bridge and culvert foundation of the railway or highway in thawed soil area. However, it is necessary to consider that when the permafrost around the pile thaws and sinks, the presence of the pile prevents the development of the thawed soil deformation, and a certain lateral frictional resistance of the pile is generated on the interface between the pile and the thawed soil, namely, the negative friction, which causes the bearing capacity to decay and its value is about 8% (Sun et al., 2011).

The fundamental formula for calculating the allowable bearing capacity [P] of the bored or excavated and cast-in-place pile under the condition of thawed soil is:

$$[P] = \frac{1}{2} U \Sigma f_i l_i + m_0 A [\sigma] \tag{4.4}$$

where U is the circumference of the pile (m); l_i is the thickness of each soil layer (m); f_i is the lateral frictional resistance between each soil layer and the pile surface (kPa); A is the area of the lower end of the pile (m^2); $[\sigma]$ is the allowable value of bearing capacity of the soil at the lower end of the pile (kPa); m_0 is the reduction coefficient of the bearing capacity at the lower end of the pile.

The analysis of a bored and cast-in-place pile with a diameter of 1 m and a length of 20 m in a typically humid permafrost region on the Qinghai-Xizang Plateau shows that the average surface temperature of the pile surface only reaches 57.07% of the natural average ground temperature after 2 months of refreezing. According to the calculation results of the Formula 4.4, as long as the concrete strength meets the design standard, the bearing capacity of a single pile can meet the construction requirements of the pile cap and abutment (or bridge abutment) in front of the beam, whether it is refrozen or not (Wu et al., 2004).

The curve tracing the refreezing of the cast-in-place piles in a warm permafrost region of Qingshui River (the mean annual ground temperature was −0.5 to −1.0 °C) shows (see Fig. 2.32) that in 12 days after pouring, the ground temperature at the lower end of the pile has refrozen to 0 °C, and the ground temperature on the lateral pile surface at the permafrost table has not refrozen. During the first load test (Fig. 4.22), most of the soil around the pile is at positive temperatures and has not refrozen, which is carried out according to the slow loading method of pile foundation test in regular area. When loaded to 2400 kN, the maximum settlement is

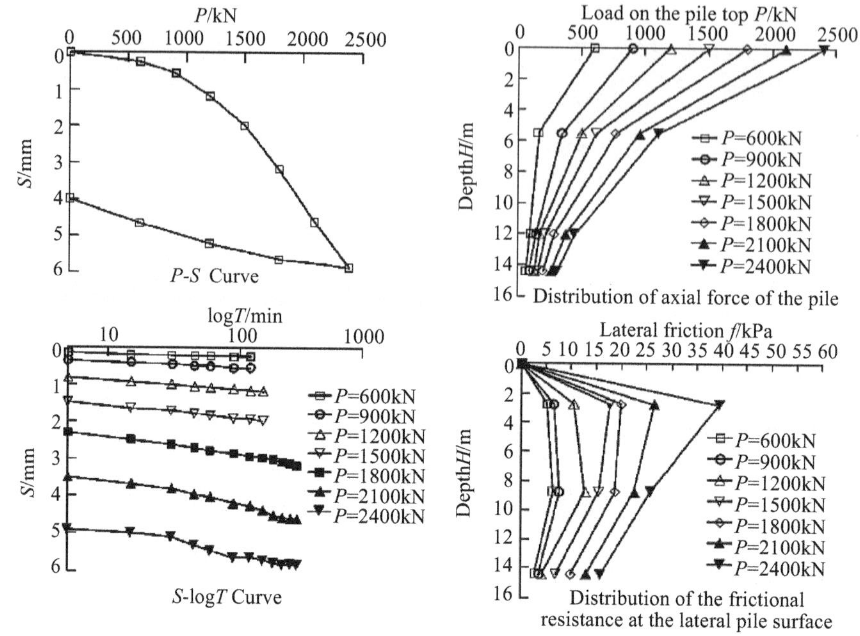

Fig. 4.22 Distributions of *P-S*, *S-log T*, axial force of the pile and frictional resistance at the lateral pile surface at Qingshui River in the first loading for the cast-in-place bored piles (Wang et al., 2005b)

5.85 mm, the residual settlement is 3.99 mm, the rate of rebound is 31.8%, and most of them are plastic settlement. At this time, the average frictional resistance of pile side sections is:

0.0–5.5 m: the friction resistance at the lateral pile surface of the medium sand (with 0.2 m hard clay on bottom) is 59.8 kPa.

5.5–11.5 m: the friction resistance at the lateral pile surface of the medium sand (with 0.5 m hard clay on top) is 25.8 kPa.

11.5–14.4 m: the friction resistance at the lateral pile surface of the silty clay (with 1.5 m medium sand on top) is 15.8 kPa.

The average of the friction resistance along the longitudinal direction of the pile is 36.9 kPa. The resistance at the lower end of the pile is 255.8 kPa. It shows that the bearing capacity of the pile-soil system is relatively small before it is refrozen.

During the second loading (Fig. 4.23), the ground temperature of the soil around the pile decreases, and the refreezing range increases. Considering the rheological property of permafrost and other factors, the test method of rapid loading of the pile foundation in permafrost region is adopted. When loaded to 4800 kN, the maximum settlement is13.92 mm, the residual settlement is 9.90 mm, and the rebound rate is 28.9%, resulting in a large plastic settlement.

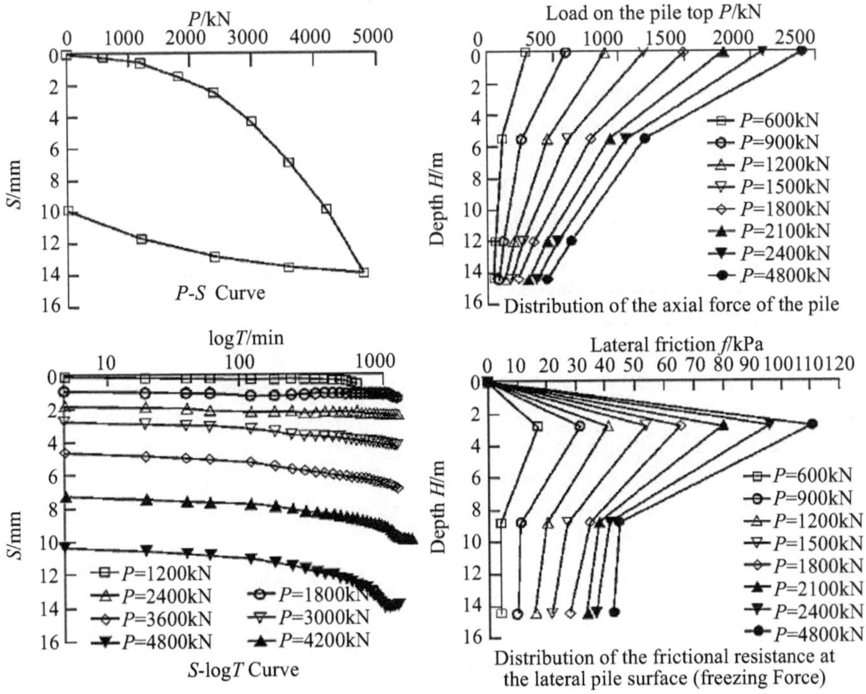

Fig. 4.23 Distributions of *P-S*, *S-log T,* axial force of the pile, and the frictional resistance at the lateral pile surface at Qingshui River for the cast-in-place bored piles in the second loading (Wang et al., 2005b)

When the load is up to 2400 kN, the average frictional resistances at the lateral pile surface are:

0.0–5.5 m: the freezing force at the lateral pile surface of the medium sand (with 0.2 m hard clay on the bottom) is 54.0 kPa.
5.5–11.5 m: the freezing force at the lateral pile surface of the medium sand (with 0.5 m hard clay on the top) is 27.01 kPa.
11.5–14.4 m: the freezing force at the lateral pile surface of the silty clay (with 1.5 m medium sand on the top) is 21.5 kPa.

The average frictional resistance along the longitudinal direction of the pile is 27.01 kPa. The resistance at the lower end of the pile is 292.0 kPa and the corresponding displacement is 2.42 mm. The displacement is smaller than that under the same load at the first loading. When loaded to 4800 kN, the average frictional resistance at the lateral pile surface is:

0.0–5.5 m: the freezing force at the lateral pile surface of the medium sand (with 0.2 m hard clay on the bottom) is 110.8 kPa.
5.5–11.5 m: the freezing force at the lateral pile surface of the medium sand (with 0.5 m hard clay on the top) is 44.5 kPa.

11.5–14.4 m: the freezing force at the lateral pile surface of the silty clay (with 1.5 m medium sand on the top) is 42.7 kPa.

The average value along the longitudinal direction of the pile is 68.9 kPa. The resistance at the lower end of the pile is 735.3 kPa, the corresponding settlement is 13.92 mm, and the residual settlement is 9.90 mm. It shows that the resistance at the lateral pile surface and at the the lower end of the pile increases with the refreezing of soil around the pile and at the lower end of the pile.

The refreezing of the bored and cast-in-place piles in the cold permafrost region (with a mean annual ground temperature lower than −2.8 °C) of the Kunlun Mountain Pass shows that (see Fig. 2.33 and Table 4.1), in 10 days after pouring, the ground temperatures in the middle part (3.5–7.5 m) and at the bottom of the test pile became negative. In 30 days, negative temperature will occur in the parts from the lateral pile surface at the permafrost table to the lower end of the pile of the permafrost. After 50 days, the ground temperature of the pile-soil system has not refrozen to the original ground temperature in natural state. When the time of the test pile after pouring reaches 30 days the load test of pile foundation is carried out by means of the rapid loading method (Fig. 4.24).

When the test is loaded to 7600 kN (the maximum load) the vertical displacement of the pile top is 4.93 mm, the unrecovered deformation after unloading is 1.01 mm, and the rebound rate is 79.5%. The deformation of the pile is mainly elastic deformation, indicating that the vertical bearing capacity of the pile is very high. When the time of the test pile after pouring reaches 30 days, the pile-soil system has not yet fully refrozen to the design temperature.The vertical bearing capacity of the pile foundation could meet the construction requirements of the pile cap, the bridge pier (or abutment) and the erecting beam.

When the test load was 1800 kN (the design load for the construction of pile cap, piers (or abutments) and erecting beams), the average freezing force along the pile was 31.1 kPa. When the load reaches 7600 kN, the average freezing forces in varying depths are:

0.0–5 m: the gravel soil (2 m thick, unfrozen) and silty clay (3 m thick), and the freezing force at the lateral pile surface is 222.2 kPa.
5–10 m: the silty clay, and the freezing force at the lateral pile surface is 48.9 kPa.
10–15 m: the silty clay (4 m thick) and silt, and the freezing force at the lateral pile surface is 44.6 kPa.

When the load reaches1800 kN, the resistance at the lower end of the pile is 335 kN, and the rate of the resistance there is 19%. When the load reaches 7600 kN, the resistance at the lower end of the pile is 2684 kN, and the rate of the resistance there is 35%.

4. **Vertical Bearing Capacity of the Bored and Cast-in-place Pile Foundation after Refreezing**

For the bored and cast-in-place pile in zones with different ground temperatures in permafrost region, it takes a long time to freeze back to the natural ground

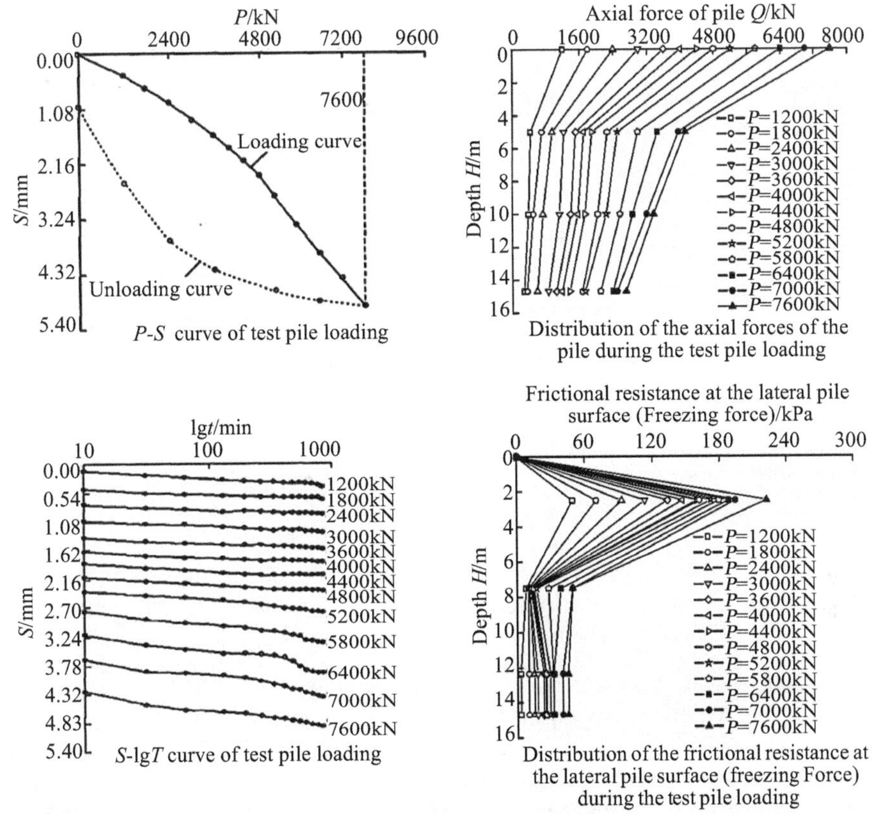

Fig. 4.24 Distributions of *P-S*, *S-log T*, axial force of the pile, and the frictional resistance at the lateral pile surface of the cast-in-place bored piles under loading at Kunlun Mountain Pass (Wang, 2013a, b)

temperature, usually more than 50 days. In cold permafrost area (with a mean annual ground temperature below −2.5 °C), the ground temperature adjacent to the pile surface is close to the natural ground temperature after 30 days, while in warm permafrost area (with a mean annual ground temperature higher than −1.0 °C), it takes 50–60 days to freeze back.

1. According to the two bearing capacity tests of the cast-in-place piles along the Qinghai-Xizang Highway in the cold permafrost area near Kunlun Mountain pass (Institute of Transportation Science of Heilongjiang, 2006), it demonstrates that after the ground temperature of permafrost adjacent to the pile drops to −0.5 °C, as the ground temperature continued to decrease, the bearing capacity of the cast-in-place pile increases. Table 4.2 shows the parameters of the test cast-in-place pile bored manually. The stratum situation is: the silty clay is 0–1.6 m, with 40% moisture content; the crushed stone and silty sand are 1.6–5.7 m, with 20–30% ice content; the crushed stone and silty clay are

Table 4.2 Parameters of cast-in-place piles test at Kunlun Mountain Pass of the Qinghai-Xizang Highway

Number of test pile	Diameter of pile /mm	Actual pile length /m	Effective pile length /m	Position of load box from the lower end of the pile/m	Pile construction method
1#	1200	16	15	1.5	Manually bored and cast-in-place pile
Buried depth of reinforcement meter	5.7 m, 7.6 m, 8.9 m, 12 m, 15 m				
2#	1200	16	15	1.5	Manually bored and cast-in-place pile
Buried depth of reinforcement meter	5.5 m, 7.3 m, 8.9 m, 12 m, 15 m				

5.7–8.9 m, with 15–25% ice content; the silty clay with peat is 8.9–12 m, with 10–20% ice content; the silty sand with peat is 12–17 m, with 60% ice content; the strongly weathered bedrock is 17 m below, with 4% moisture content.

In the first static load test, the concrete temperature at the center of the test pile and the ground temperature adjacent to the pile is ranging from −0.5 °C to −0.7 °C, and the ground temperature at the lower end of the pile is −1.1 °C to −1.2 °C, which has not yet reached the natural ground temperature (Fig. 4.25). Under the maximum load, it presents a shear stress in the interface between the pile and the permafrost and a freezing frictional resistance at the lateral pile surface, whose variations are presented in Table 4.3 and Fig. 4.26 respectively. The maximum freezing frictional resistance in permafrost layer is 148.69 kPa, with a corresponding displacement of 6.54 mm, and an average frictional resistance at the lateral pile surface for the upper part of the pile is 108.59 kPa.

During the second static loading test (in 90 days after the pouring), the concrete temperature in the center of the test pile and the average ground temperature adjacent to the pile were −1.5 °C, the ground temperature at 2.4 m away from the pile center was −1.7 to −1.8 °C (Fig. 4.27), and the ground temperature at the lower part of the pile was −1.8 to −1.9 °C, which could be considered as completely re-frozen. In the second test, the loading was 5000 kN, greater than that of the first test. In the second test, the upward and downward displacements were reduced by approximately 2 mm, compared to the first loading, which indicated that the bearing capacity of the pile foundation increased. Under the maximum load, it presented a shear stress in the interface between the pile and the permafrost and a freezing frictional resistance at the lateral pile surface, whose variations are presented in Table 4.3 and Fig. 4.28, respectively. The maximum freezing frictional resistance of the permafrost layer was 220.50 kPa, and the average frictional resistance at the lateral pile surface for the upper part of the pile was 125.4 kPa.

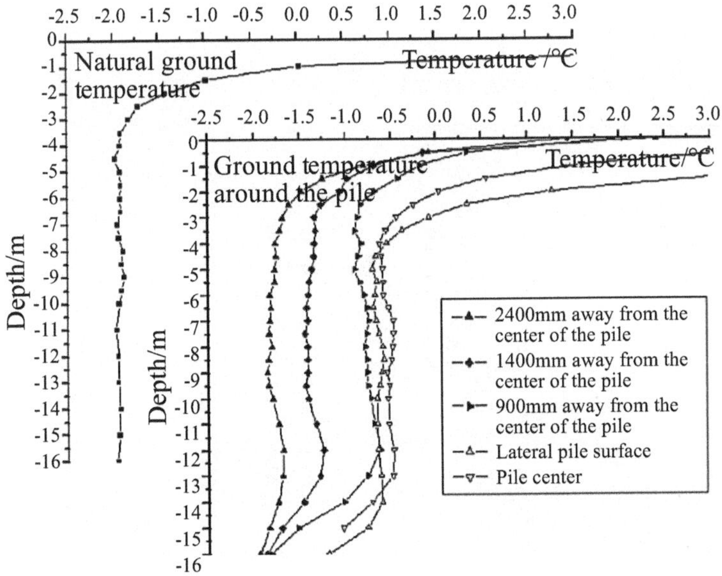

Fig. 4.25 Natural ground temperature and ground temperature around the pile in the first test (Liu, 2005)

At the Kunlun Mountain pass in cold permafrost region, the variations of ground temperature and freezing strength with time adjacent to the lateral pile surface at depths of 5 m and 10 m of the No.1 test pile (Fig. 4.29) showed that in 12 days after concrete pouring, the ground temperatures adjacent to the pile decreases to 0 °C, and the freezing strength reaches 3 kPa. On the 28th day, the ground temperature adjacent to the pile drops to −0.2 °C, and the freezing strength reaches 40 kPa. As the ground temperature adjacent to the pile continues to decrease, the freezing strength also increases accordingly. The temperature of the permafrost at the lower end of the pile also reaches 0 °C in 12 days after the concrete is poured, and the compressive strength reaches 500 kPa. With the decrease of ground temperature, the compressive strength continues to increase (Fig. 4.30).

Figures 4.31 and 4.32 show the shear strength on the lower part of the pile-soil interface. At the beginning of the test, as the shear stress (the load) increases, the shear displacement of the pile increases rapidly. When the shear stress increases to a certain value (a proportional limit) the shear displacement of thepile increases rapidly with the increase of the shear stress. After the shear stress reaches another limit (an ultimate load) the displacement of the pile enters the progressive flow stage. Generally, the shear stress at the proportional limit is called the calculated freezing strength of the pile-soil interface, and the shear stress corresponding to the ultimate load is called the ultimate freezing strength of the pile-soil interface. And the test results are listed in Table 4.4.

Figures 4.33 and 4.34 are the results of the loading test at the lower end of the pile in the permafrost region. The projected area at the lower end of the pile is

Table 4.3 Shear stress (kPa) and displacement (mm) on the interface between the upper part of the pile and the permafrost under the maximum load (Institute of Transportation Science of Heilongjiang, 2006)

Number of test pile	Test temperature/°C	Load/kN	12 m ~ 13.5 m		8.9 m ~ 12 m		7.6 m ~ 8.9 m		5.7 m ~ 7.6 m		5.7 m to the permafrost table		Average shear stress of frozen pile section
			Stress	Displacement	Stress	Displacement	Stress	Displacement	Stress	Displacement	Stress	Displacement	
1#	−0.6	4500	148.69	3.52	142.51	3.27	116.95	3.11	83.15	3.04	38.39	2.96	105.98
2#	−0.6	4500	167.20	3.60	131.19	3.36	106.35	3.19	85.26	3.12	38.64	3.04	105.72
1#	−1.5	5000	220.51	2.68	153.10	2.42	106.91	2.26	87.42	2.19	37.87	2.11	121.15
2#	−1.5	5000	243.11	2.61	133.80	2.25	108.47	2.09	86.96	2.01	39.41	1.93	122.35

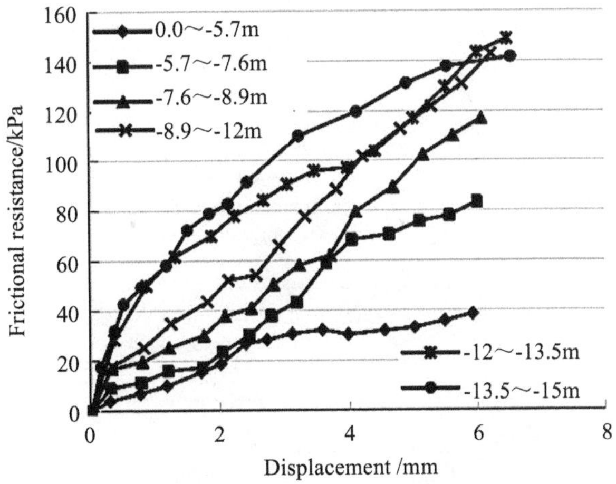

Fig. 4.26 Variations of frictional resistance at the lateral pile surface-displacement in the first test (Liu, 2005)

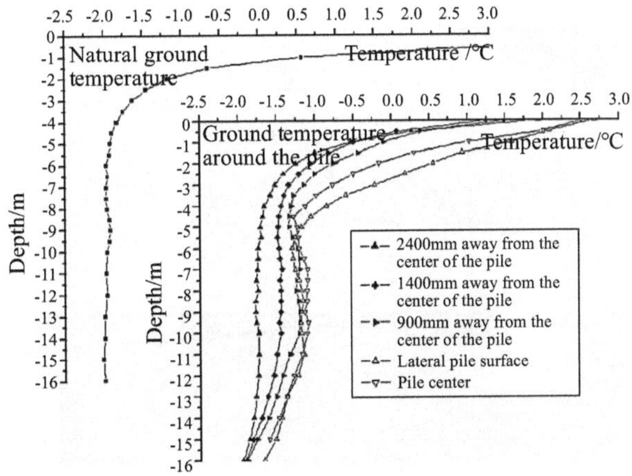

Fig. 4.27 Natural ground temperature and ground temperature around the pile in the second test (Liu, 2005)

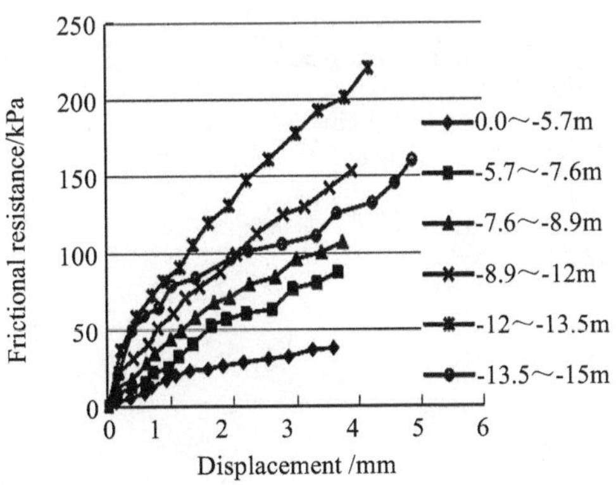

Fig. 4.28 Variations of frictional resistance at the lateral pile surface-displacement in the second test (Liu, 2005)

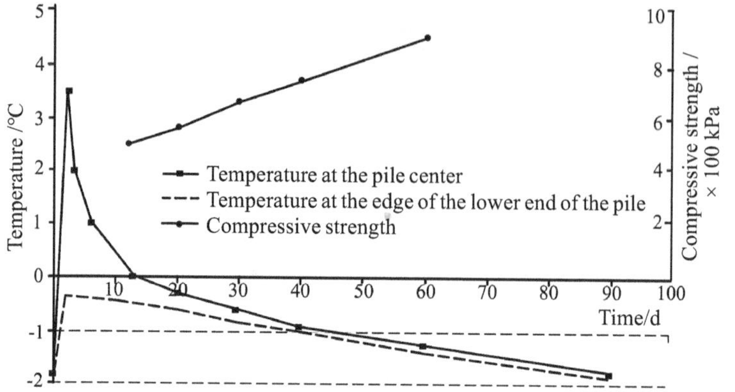

Fig. 4.29 Variations of the ground temperature and the freezing strength adjacent to the pile with time of No. 1 test pile at Kunlun Mountain Pass (Institute of Transportation Science of Heilongjiang, 2006)

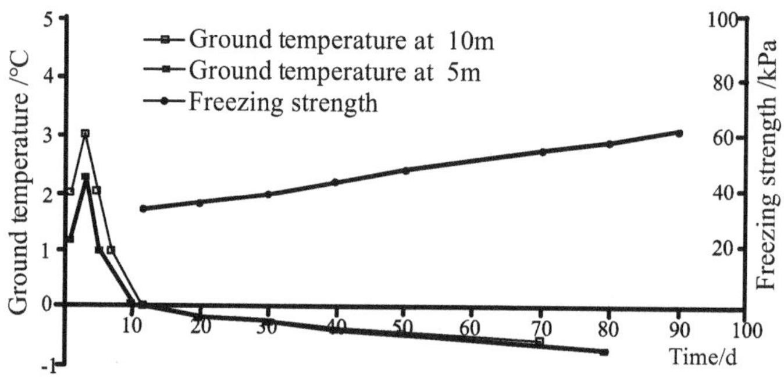

Fig. 4.30 Variations of temperature and the compressive strength at the lower end of the pile with time of No.1 test pile at Kunlun Mountain Pass (Institute of Transportation Science of Heilongjiang, 2006)

Fig. 4.31 Variations of τ-S at the Pile-soil Interface (Institute of Transportation Science of Heilongjiang, 2006)

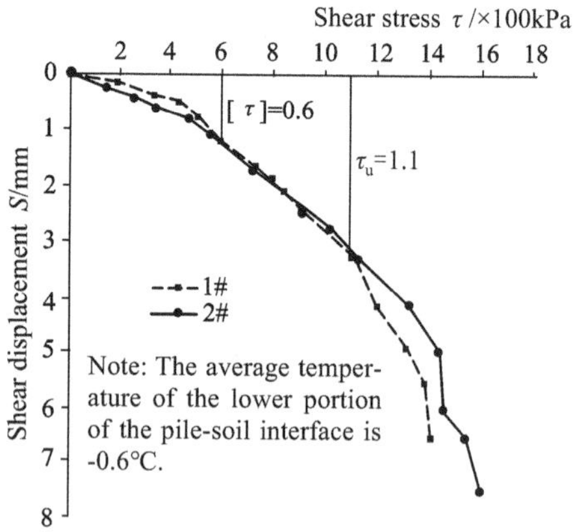

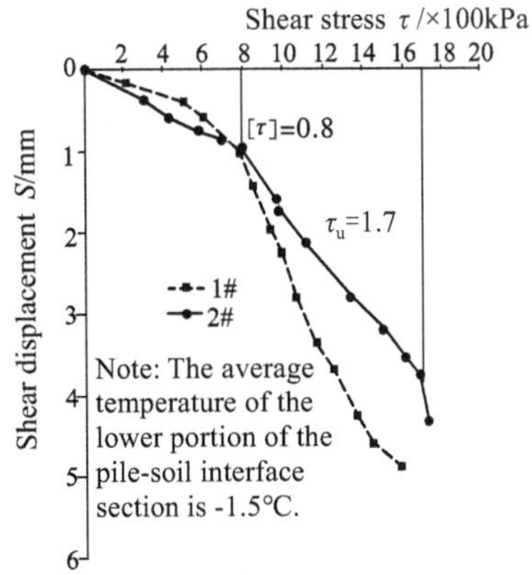

Fig. 4.32 Variations of τ-S at the Pile-soil Interface (Institute of Transportation Science of Heilongjiang, 2006)

Table 4.4 Calculated freezing strength and ultimate freezing strength and its corresponding displacements in the lower portion of the test pile (Institute of Transportation Science of Heilongjiang, 2006)

Test time	Average temperature of pile-soil interface /°C	Calculated freezing strength / kPa	Corresponding displacement at the calculated freezing strength /mm	Ultimate freezing strength / kPa	Corresponding displacement at ultimate freezing strength/mm
July,2004	−0.6	60	1.3	110	3.3
September,2004	−1.5	80	1.0	170	3.7

11,310 cm^2, and the ground temperature of permafrost during the test is −1.2 °C for the first time and − 1.8 °C for the second time. According to the variations of the load and settlement of the test, when the ground temperature of permafrost at the lower end of the pile is −1.2 °C, the allowable bearing capacity of the foundation is about 1240 kN, that is, the calculated strength is 110 kPa. When the ground temperature of permafrost at the lower end of the pile is −1.8 °C, the allowable bearing capacity of the foundation is about 1470 kN, that is, the calculated strength is 130 kPa.

When the pile length is 15 m, the calculated results of the allowable bearing capacity and ultimate bearing capacity of the pile are shown in Table 4.5.

The bearing capacity of pile foundation in permafrost is composed of the freezing strength adjacent to the lateral pile surface and the reacting force of the permafrost at the lower end of the pile. In general, when the pile length is relatively long and the pile diameter is relatively small, and also when the permafrost at the lower end of the pile has not yet borne the load or when the bearing load is small, the freezing force adjacent to the lateral pile surface will be destroyed. Therefore, when determining the bearing capacity of the pile foundation in the permafrost, the reacting force of the permafrost at the lower end of the pile should be determined according to the displacement of the freezing strength damage on the lateral pile surface, instead of according to the compressive strength of the permafrost at the lower end

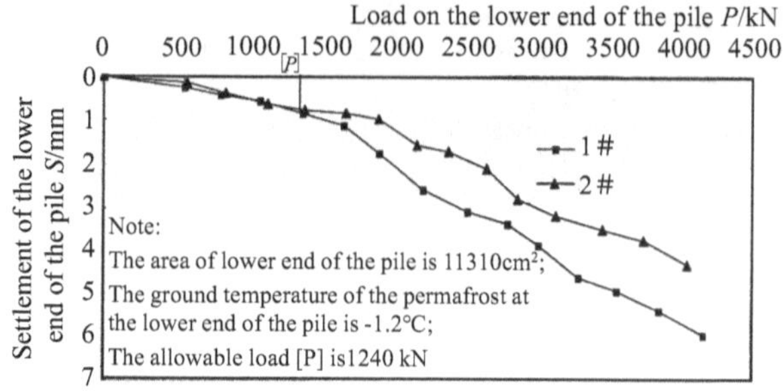

Fig. 4.33 Variations of the load-settlement for the first test (Institute of Transportation Science of Heilongjiang, 2006)

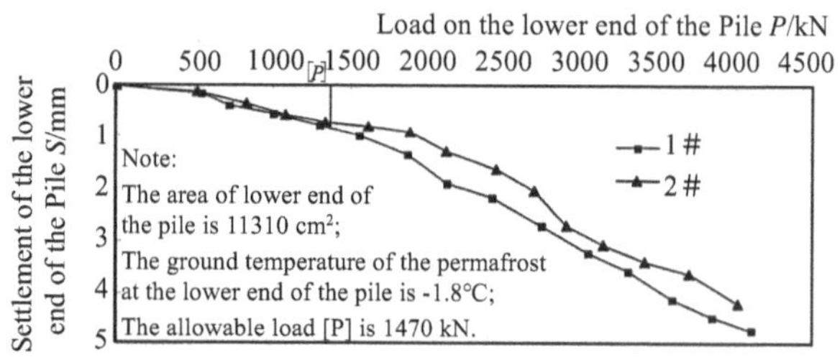

Fig. 4.34 Variations of the load-settlement for the second test (Institute of Transportation Science of Heilongjiang, 2006)

of the pile. The test at Kunlun Mountain pass shows that the freezing interface at the lateral pile surface with a large-diameter and the permafrost foundation of the lower end of the pile can cooperate, and the allowable and ultimate bearing capacity of the pile foundation can be determined by considering the freezing strength at the lateral pile surface and the bearing capacity at the lower end of the pile at the same time.

2. A Pile foundation test is also carried out in the permafrost of Jintao Town, eastern part of Mohe County, Daxing'anling Area, northeast of China. The site is located on the terrace of the Emur River, with a mean annual air temperature of −4.9 °C, which is swampy and has a large area of permafrost. The total design length of the pile foundation is 11.5 m. The exposed ground is 0.5 m, and the diameter of the pile is 1.0 m. The concrete used to pour is the concrete that is able to consolidate rapidly in a negative-temperature condition, and the pile is manually excavated and cast-in-place. The construction began in May, 2010 and was completed in June, 2010.

The first static load test is carried out in late November, about half a year after the pile foundation is poured. The temperature of the permafrost remains basically constant at a depth of 11 m below the ground surface (Fig. 4.35), and it is −1.8 °C. The ground temperatures of the permafrost within 0 to 3 m change greatly, with a highest ground temperature of −0.05 °C. And at a depth of 8 m, it has the lowest ground temperature, and it is −2.28 °C. Before construction, the highest natural ground temperature is −1.93 °C, and the lowest is at a depth of 3 m, which is −3.45 °C. It has not frozen completely back to its original natural state when testing.

Estimated according to the specification (JGJ118-2011), the ultimate bearing capacity of the pile is 2000 kN. The standard value of the load for a single pile in the vertical compressive static load test is 1000 kN.The first level is loaded according to

Table 4.5 Allowable bearing capacity and ultimate bearing capacity of the pile with a length of 15 m at Kunlun Mountain Pass (Institute of Transportation Science of Heilongjiang, 2006)

Number of test pile	Ground temperature of the permafrost at the lower end of the pile/°C	Calculated freezing strength /kPa	Ultimate freezing strength /kPa	Calculated freezing Force /kN	Ultimate freezing Force /kN	Calculated reacting force at the lower end of the pile /kN	Ultimate reacting force at the lower end of the pile /kN	Allowable bearing capacity/kN	Ultimate bearing capacity /kN
1#	−1.2	60	110	3030	5560	1500°	2570*	4530	8130
2#				3030	5560	1800	3170	4830	8730
1#	−1.8	80	170	3950	7900	1800	3150	5750	11,550
2#				3950	7900	2400	3600	6350	12,000

Note: The settlement of the foundation is determined according to the P-S curve. *Foundation reacting force is calculated by sinking 1.3 mm.; **The foundation reacting force is calculated by sinking 3.3 mm

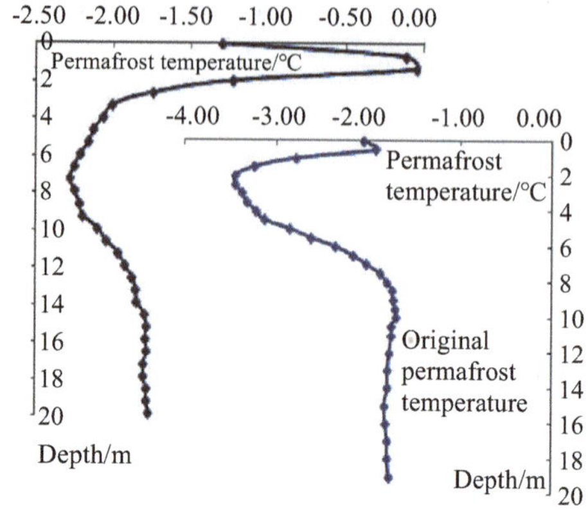

Fig. 4.35 Variation of ground temperature of the permafrost in the first test (Sun, 2012)

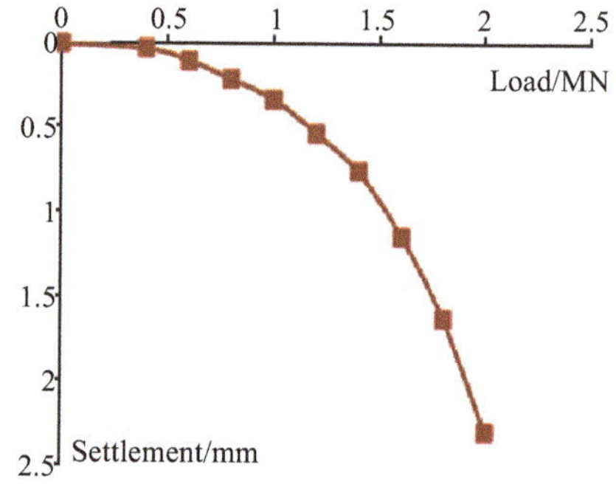

Fig. 4.36 Variation of *P-S* of the single pile in the first vertical static load test (Sun, 2012)

40% of the standard value.The static load is divided into 9 levels, and each level is 20% of the standard value; and the applying duration lasts for 60 min until it reaches 200% of the standard value.

In the variation of the *P-S* of the single pile in the vertical static load test (Fig. 4.36), the curve deforms gently, and there is no obvious steep dropping section. In the variation of *s-logt* of the single pile in the vertical static load test (Fig. 4.37), there is also no obvious downward bending found. When the pile top load reaches the design bearing capacity of 2000 kN, the total settlement of the pile top is only 2.30 mm. According to the settlement of 40 mm as the ultimate bearing capacity standard, the test pile is far from reaching the ultimate bearing capacity.

The second static load test is carried out in the late February of the following year, 9 months after the foundation piles are poured. The temperature of permafrost remains basically constant at depths below 11 m from the ground surface (Fig. 4.38), which is −1.9 °C. The maximum ground temperature of the permafrost adjacent to the lateral pile surface is −2.0 °C, which has completely refrozen. Affected by the air temperature the ground surface temperature is relatively low.

Estimated according to the specification (JGJ118-2011), the ultimate bearing capacity of the pile is 6000 kN. The standard value of the single pile load in the vertical compressive static load test is 3000 kN, the first level is loaded according to

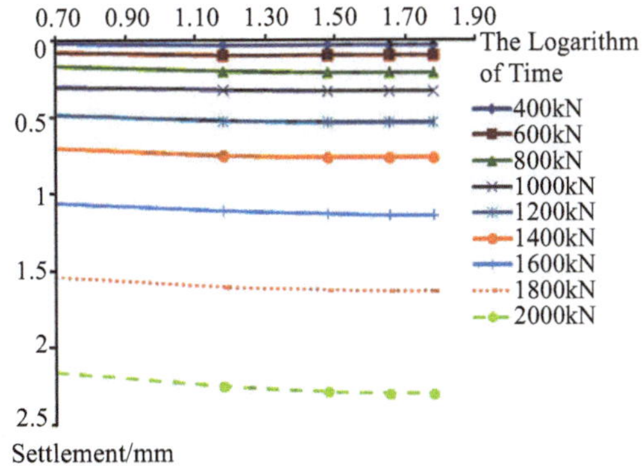

Fig. 4.37 Variation of *S-logT* of the single pile in the first vertical static load test (according to Sun, 2012)

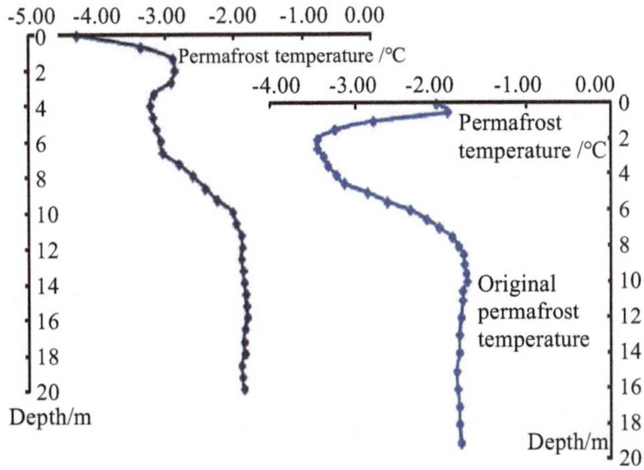

Fig. 4.38 Variation of ground temperature of the permafrost soil in second test (Sun, 2012)

40% of the standard value, and the static load is divided into 9 levels, each level is 20% of the standard value; the applying duration lasts for 60 min until it reaches 200% of the standard value. However, due to the low air temperature, the steel bars of anchor piles are pulled to break, and the test is terminated early at level 6.

In the variation of *P-S* of the single pile in the vertical static load test (Fig. 4.39), the curve deforms gently, and there is no obvious steep dropping section. In the variation of *s-logt* of the single pile in the vertical static load test (Fig. 4.40), there is also no obvious downward bending found. When the pile top load reaches the design test bearing capacity of 4200 kN, the total settlement at the lower end of the pile is only 2.21 mm. The load of this test increases, but the total settlement decreases, indicating that the temperature of the permafrost around the pile declines, and the bearing capacity of the pile is further improved.

According to the finite element calculation, when the test pile is continuously loaded till failure, the variation of the load displacement of the single pile is with a character of slow changing (Figs. 4.41 and 4.42). If the temperature of the permafrost is −2.0 °C, according to the formula of the specification (JGJ118-2011), the

Fig. 4.39 Variation of *P-S* of single pile in the second vertical static load test(according to Sun, 2012)

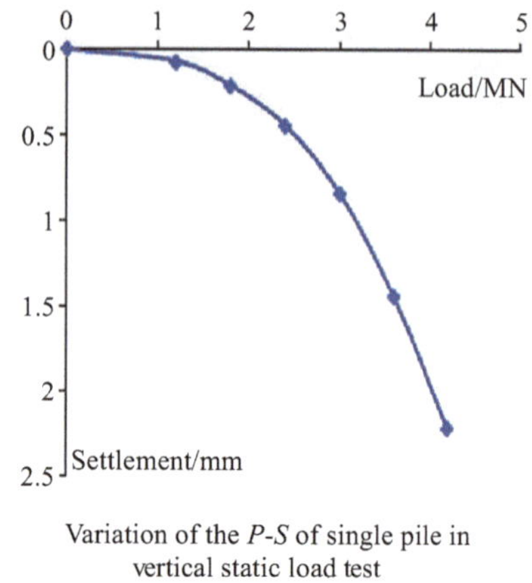

Fig. 4.40 Variation of *S-log T* of single pile in the second vertical static load test (according to Sun, 2012)

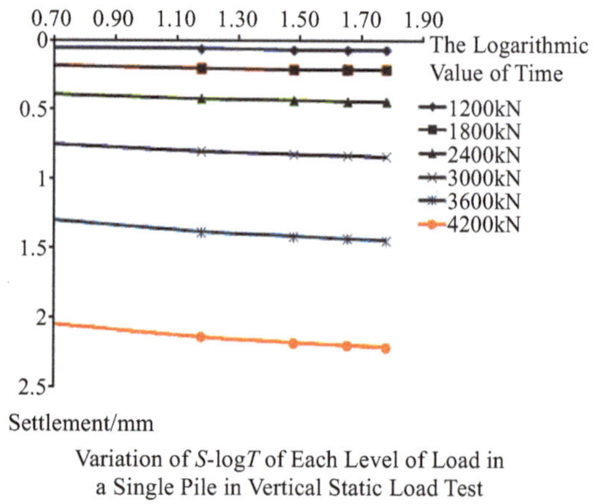

load at the pile top will be 5351 kN when the permafrost is damaged, and the corresponding settlement at the pile top is 15.14 mm from Fig. 4.41. Taking the pile top settlement of 40 mm as the standard, the ultimate bearing capacity of the pile is 7357 kN from Fig. 4.41. When the temperature of the permafrost at the lower end of the pile is −6 °C, according to the code formula for calculation, the load on the pile top will be 9315 kN when the permafrost is damaged, and the settlement at the pile top at this time is 13.69 mm from Fig. 4.42. With the standard settlement at the pile top of 40 mm, the ultimate bearing capacity of the pile is 13,125 kN from Fig. 4.42. It shows that the settlement at the pile top will be less than the standard specified for the static load tests when the permafrost at the lower end of the pile is crushed. The ultimate bearing capacity of the pile foundation determined by the finite element analysis is about 70% of the ultimate bearing capacity of the pile foundation determined by the current standard. The laboratory test shows that when the temperature of permafrost is −2 °C, under the vertical load of 5351 kN, the

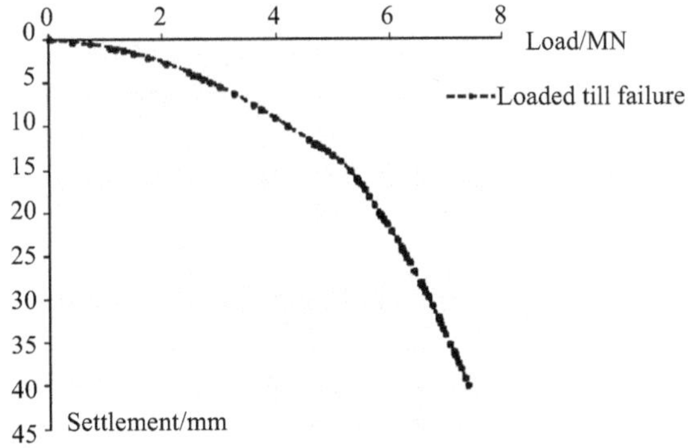

Fig. 4.41 Variation of *P-S* of an in-situ test pile loaded till failure (Sun, 2012)

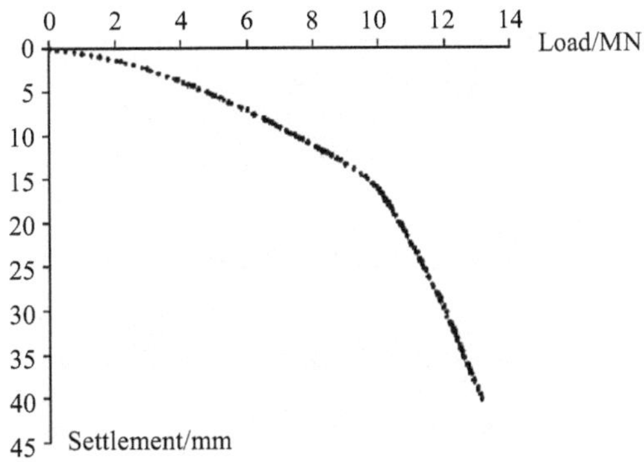

Fig. 4.42 Variation of *P-S* at the pile top when the ground temperature of the permafrost at the lower end of the pile is −6 °C (according to Sun, 2012)

permafrost failure will be a triaxial compression-shear failure. Therefore, the determination of the ultimate bearing capacity of piles in permafrost by the failure criterion of non-frozen soil area still needs to be discussed.

5. **Duration of Applying Load to the bored and Cast-in-place Pile**

The above test observations show that under the condition that the mean annual ground temperature of the permafrost is lower than −2.0 °C, in 12 days after pouring, the ground temperature around the pile will freeze back to 0 °C (or the initial ground temperature when starting freezing), and the temperature of the permafrost at the lower end of the pile will be colder than 0 °C. In warm permafrost area, it usually takes 30–60 days for the pile to freeze back to 0 °C. After the permafrost around the bored and cast-in-place pile is refrozen, the freezing strength around the pile is gradually formed, and the permafrost at the lower end of the pile also has a certain bearing capacity. Whether it can bear the applied load is related to the increase of the strength of concrete. Generally, under the normal maintaining conditions, the pile foundation cannot bear the load pressure within the 28-day maintaining period of concrete. Due to the low molding temperature of the bored and cast-in-place pile in the permafrost region and the relatively low temperature around the pile, the maintaining environment of the concrete in the permafrost cannot reach the standard of the regular maintaining environment. The concrete strength of the pile foundation should be based on the maintaining environment of the concrete in the permafrost area or the law of strength increase for the concrete obtained by calculation or experiment. As long as the concrete strength of the pile foundation meets the design strength requirements, the load can be applied.

4.1.2.3 Base Design of Bridge Foundation in Permafrost Regions

The location, structure type and base design principle of bridges in permafrost regions should be comprehensively analyzed, compared and determined according to the permafrost engineering geological conditions in site area of the bridge, such as the stratum lithology, thickness of permafrost, ice content, mean annual ground temperature, physical and mechanical properties, permafrost table variation, groundwater dynamics, unfavorable cryogenic phenomena (aufeis, pingos, thick ground ice, etc.), and the variations of permafrost foundation (rises and falls of permafrost table, ground temperature changes, changes of physical and mechanical properties, etc.) caused by the completion of bridge engineering.

1. **Design Principles for Bridge Foundation**

The design principles for permafrost bridge foundations are determined by the thermal stability of permafrost:

1. Design principles for maintaining the frozen state of permafrost foundations

   Keeping the foundation in the frozen state means that the permafrost foundation under the base is frozen during the construction and operation of the bridge foundation. This principle is applicable to cold permafrost area. According to the classification of frozen soil temperature in China, the mean annual ground temperature in cold permafrost area should be lower than −1.0 to −1.5 °C, and it should be a riverbed without constant water flow. It is also suitable for the sections, where bridges are built to replace the roads, in continuous permafrost regions and discontinuous permafrost regions.

2. Design Principles for Allowable Thawing

   The design principle for allowable thawing refers to the changes in thermal state of permafrost foundation during the construction and operation of bridge foundation. It is suitable for continuous or discontinuous permafrost regions, where the mean annual ground temperature higher than −1.0 °C in warm permafrost areas. In these areas, the thermal state of permafrost is unstable or extremely unstable, it will be difficult to recover and maintain the thermal state of permafrost under human or natural (climate and water flow) thermal disturbances. According to permafrost conditions, it can be designed in two situations:

① Natural thawing: it is applicable to the foundation base underlain the weak thaw settlement or non-thaw settlement permafrost foundation. As long as the total thaw settlement does not exceed the allowable value, regardless of the thickness of permafrost, the foundation of permafrost under the base is allowed to thaw naturally during the construction and operation. Or the bridge foundation goes through the permafrost layer and stands on the unfrozen soil layer as an end bearing pile.

② Pre-thawing: it is suitable for the permafrost whose thickness is thinner and permafrost table is shallow, and the permafrost foundation is thaw settlement or strong thaw settlement. During the period of foundation construction, the permafrost foundation will be excavated and replaced. When the permafrost table is deep, it will be designed on the basis of the principle for natural thawing.

When the above design principles are applied, the anti-frost heaving measures must be taken in the active layer to avoid the deformation of bridge foundation due to frost heave.

2. **Artificial Permafrost Table and Minimum Buried Depth of Bridge Foundation in Permafrost Regions**

The riverbeds, floodplains and bankside zones in permafrost regions are mostly taliks. Most of the riverbeds of the creeks are non-penetrating taliks and there is permafrost with ground ice. Whether there is permafrost or not the floodplains and bank side zones are frozen every year in the cold season, forming a seasonally frozen layer with a certain thickness and causing the occurrence of aufeis and pingos.

In recent years, the foundation of bridge engineering such as in the construction of the Qinghai-Xizang Railway has been mostly constructed by the high-power rotary drilling rigs, but in the constructions of some small bridge, the open-cut foundation is still used, and the buried depth of abutment foundation is mostly placed in permafrost foundation. During the construction of the open-cut foundation in warm seasons, in addition to the exposure of the permafrost, it often makes the foundation pit prone to water accumulation, and the consequential thermal disturbance will lower the permafrost table. The post-construction refreezing period will last several months, and even several years.

According to the experiences, the open-cut foundation would lead to great thermal disturbance to the permafrost base, resulting in the temperature rise and the thawing of the permafrost under the foundation base. For example, the Hongliang River bridge foundation on the Qinghai-Xizang Highway (the ground temperature at a depth of 4.0 m in the thawing season with the maximum thawing depth of the riverbed is −2.6 °C) is constructed in June and July, the permafrost of the bridge pier and the foundation pit of the bridge abutment thawed by 0.5–0.7 m, and the ground temperature rises to 1.0–1.6 °C. After backfilling, due to the low ground temperature, the foundation base started to rc-freeze after 1 month, but 1 year later (in June), the riverbed at the bridge pier thawed by 0.1 m, and when it reaches the maximum thawing depth in September, the artificial permafrost table is 0.9 m lower than the natural permafrost table. About half a year after the completion of the small bridge, the settlement of the abutment on the south side facing the water flow reaches 45 mm, and the settlement of the pier reaches 34 mm. After that, it gradually stabilizes. At the site of the Qingshui River highway bridge in the warm permafrost area (Fig. 4.43), the mean annual ground temperature at the depths from 2.5 m to 7.0 m below the riverbed is −0.4 °C. During the open-cut construction, under the I-shaped abutment, the foundation base is 0.2 m below the original natural permafrost table. After it is open to traffic, the abutment is broken at the middle. According to the drilling in the second year and the following temperature measurement in the third year, it is concluded that the permafrost table beneath the main river channel under the bridge has thawed to 4.0 m deep, which is 1.5 times the natural permafrost table, and the permafrost table at the center of the abutment base drops by 0.2–0.3 m.

Some types of pile construction including the inserted pile, the driven pile and the cast-in-place piles have less thermal disturbance on the permafrost under the foundation base than the open-cut foundation construction (Table 4.6). As long as the molding temperature of the concrete is controlled, even if the cast-in-place pile is adopted, the direct influence on the original ground thermal regime of the permafrost is about 0.5–1.0 m around the pile. The construction period of the open-cut foundation should be carried out in cold season, which can reduce the temperature rise of the permafrost caused by the thermal disturbance of the construction, and also make the permafrost foundation refreeze as soon as possible and reduce the thaw settlement of the foundation and accelerate the refreezing of the permafrost around the base and the foundation base. In this case, the artificial permafrost table rises, which provides sufficient freezing strength around the pile and ensure the thermal stability of the permafrost under the foundation base. The observation data show (Table 4.7) that for the buried bridge abutment, the permafrost table of the foundation is rising, and the permafrost table of the pier foundation is declining, and the variation of the depth is about 1.5 times of the natural permafrost table.

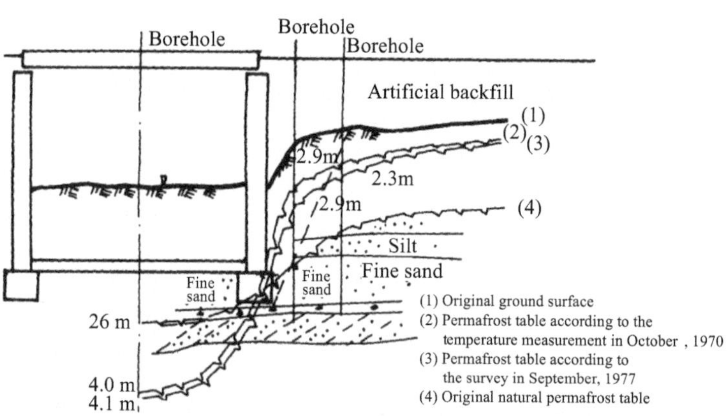

Fig. 4.43 Variations in artificial permafrost table in the Qingshui River Highway Bridge on the Qinghai-Xizang Plateau

Table 4.6 Comparisons of influence of construction methods and permafrost conditions on artificial permafrost table of foundation

Items	Hongliang River bridge pier	Dongdagou bridge pier of Fenghuo Mountain	Qingshui River bridge pier
Bridge site condition	The base are silt and silty clay	The base are pebble soil, silty clay and block stone soil	The base is fine sand, silty clay and marlstone
Natural permafrost table of riverbed/m	1.8	2.5	2.5
Construction date	Excavated in January, 1969, and backfilled in early August,1969	Poured on august 4, 1969	Poured on august 10, 1976
The average air temperature of the construction year (°C)	At −4.8 °C(in Wudaoliang)	At −4.9 °C(in Fenghuo Mountain)	At −6.2 °C(in Qingshui River)
Positive accumulated air temperature in the year of construction/°C·d	459	445	384
Bridge type	Concrete beam bridge with the height of 2 m to 6 m	Concrete beam bridge with the height of 2 m to 8 m	Concrete beam bridge with the height of 2 m to 6 m
Foundation construction method	Open-cut masonry foundation construction	No. 70 cast-in-place pile	No. 120 cast-in-place pile
Buried depth of the foundation calculated from the riverbed surface/m	2.9	8.5	7.5
Original ground temperature of the riverbed (at 40 m)/°C	−2.7	−2.2	−0.7
Internal initial temperature of the foundation upon completion (°C) (at 0.5 m to 5.0 m on average)	Higher than 10 °C (tested on July 26)	At 12.6 °C (tested on august 5)	At 6.2 °C (tested on august 11)
The temperature of the foundation base upon the completion of foundation (°C)	4.3(2.9 m)	1.5(8.5 m)	05(7.5 m)
Refreezing height and time required for the foundation base in the same year	It took 81 days to freeze back for 0.6 m (0.9 m away from the natural permafrost table)	It took 29 days to freeze back for 5 m (1.0 m away from the natural permafrost table)	It took 70 days to freeze back for 4.0 m (1.0 m away from the natural permafrost table)

(continued)

Table 4.6 (continued)

Items	Hongliang River bridge pier	Dongdagou bridge pier of Fenghuo Mountain	Qingshui River bridge pier
Refreezing rate of the foundation base (mm·d^{-1})	9 (in 81 days)	17.2 (in 29 days)	5.7 (in 70 days)
The temperature in the foundation in the second year of the bridge construction/°C; average temperature within the depths from 1.0 to 4.0 of the foundation base	0.9 −0.5	0 −2.4	0.5 −0.2
The artificial permafrost table of foundation/ m	2.7 (inside the foundation)	2.5 (inside the pile)	3.5 to 4.0 (inside the pile)
The deepening of the artificial permafrost table over the natural one/m	0.9	0	0.9 ~ 1.4
Artificial permafrost table/ natural permafrost table (folds)	1.5	1.0	1.4 ~ 1.5

Table 4.7 Comparison of the variations in artificial permafrost table in bridge pile foundation piers and abutments

Items	Qingshui River bridge		Dongdagou bridge of Fenghuo Mountain	
	Bridge abutment	Bridge pier	Bridge abutment	Bridge pier
Construction method	Driven pile in borehole	Cast-in-place pile	Cast-in-place pile	Cast-in-place pile
Natural permafrost table/m	1.6	2.6	2.85	2.5
Backfill height of abutment /m	1.46		1.50	
Thawing depth upon the completion of foundation /m	At the depth of 2.8 m (tested on September 7)	At the depth of 7.5 m (tested on August 11)	At the depth of 8.5 m (tested on July 24)	At the depth of 8.5 m (tested on August 5)
Thawing depth in late September in the same year/m	2.9	3.5	3.0	3.5
Artificial permafrost table (next year)/m	2.2	3.5 ~ 4.0	2.0	2.5
Variation of the natural permafrost table/m	+0.8	−0.9 ~ 1.4	+1.35	0

The artificial permafrost table (H_r) of the foundation should be determined by thermal calculation. In general, the empirical formula can be used for calculation for small bridges:

$$H_r = kH_t \tag{4.5}$$

where H_t is the natural permafrost table in the riverbed (m); k is the engineering thermal influence coefficient, dimensionless, as shown in Table 4.8.

In addition to considering the thermal and mechanical properties of the foundation, it should also consider the effects of air temperature on the ground temperature of permafrost when deciding the buried depth of the bridge foundation. If only the thermal influence is considered, the minimum buried depth of the foundation can be determined as follows:

$$h_{\min} = H_r + \Delta H \tag{4.6}$$

where $h_{\min}$ is the minimum buried depth of the foundation, m; H_r is the artificial permafrost table of the foundation, m; $\triangle H$ is the air temperature increment during the variation of the natural permafrost table, m.

The air temperature increment of the natural permafrost table is the upper limit increment caused by climate warming during the service life of the engineering, which should be determined by using the the centennial forecast data. According to statistics of the data from Wudaoliang and Tuotuo River, in a hundred years (design frequency is 1%), if the average air temperature increases by 1.0 °C, the deepening of natural permafrost table is generally about 0.2 time of the natural permafrost table. In addition, for the buried depth of the foundation, the deepening of the permafrost table caused by engineering influence should also be added. In general, the buried depth of the foundation can be considered as 1.0–1.5 m below the artificial permafrost table.

According to the design principle for protecting the frozen state of permafrost foundations, the *Interim Provisions for Engineering Design in Permafrost Regions of the Qinghai-Xizang Railway* stipulates that the minimum buried depth of bridge foundation should be calculated according to the following formula:

$$h_{\min} = H_r + \Delta H + \Delta h \tag{4.7a}$$

Table 4.8 Engineering thermal influence coefficient *k*

		Bridge pier			
		Warm permafrost area		General permafrost area	
Foundation type	Buried abutment	Open-cut masonry	Pile foundation	Open-cut masonry	Pile foundation
k values	1.0	2.0	1.6	1.6	1.2

Note: Refers to the permafrost layer with a mean annual ground temperature higher than −0.5 °C, and if the mean annual ground temperature is less than 0.5 °C, it is a regular permafrost

$$\Delta H = H_t \cdot A\left(K_p - K_H\right) \quad (4.7b)$$

where K_p is the annual average air temperature (°C) of the design frequency, which is taken as the guarantee rate of 1%; K_H is the annual average air temperature (°C) of the survey year (and the year that the design is based on); A is the maximum variation rate in air temperature fluctuation (%/°C) determined by the conditions including air temperature, stratum, vegetation, etc., %/°C, which is determined by the variation rules of annual average air temperature and thawing depth. If there is no data, the following statistics can be used:

In regular cohesive soil layer, without vegetation coverage: $A = 20\%/°C \sim 25\%/°C$.

In regular cohesive soil layer, with good vegetation coverage: $A = 7\%/°C \sim 10\%/°C$.

k is an empirical coefficient. When calculating the artificial permafrost table according to formula 4.5, it is 1.0 for the buried abutment. For the piers and abutments without scouring, it is 1.0–1.4. According to the degree of ground surface damage, it takes the smaller value for the minor damage, and the larger value for the major damage. For the piers and abutments with scouring, the value is 1.5–2.1. It takes the smaller value for the slow flow and shallow water, and the larger value for the fast slow and deep water. Δh is the safety value (m). For the open-cut foundation, the value of 1.5 m is taken, and 4.0 m is taken for the pile foundation.

For the discontinuous permafrost, when the active layer is relatively thick, the open-cut foundation can be placed in it. In the soil layer with frost heave, the open-cut foundation base should be buried not less than 0.25 m below the maximum freezing depth, and the base of the pile foundation platform should be not less than 0.3 m above the ground. In the soil layer with weak frost heave, the open-cut foundation base should be buried not less than 80% of the freezing depth, and anti-frost heaving measures should be taken on the foundation side.

The buried depth of the paved foundation on the conical slope of the bridge abutment should not be less than 1.25 m when there is no scouring, and the foundation soil with frost heave should be replaced with gravel stone around the foundation.

3. **Selection of Bridge Foundation Type**

The selection of bridge foundation types in permafrost regions should consider the structural characteristics of riverbed and the conditions of frozen soil engineering, and then design principles are determined. If the design principle for maintaining the frozen state of the permafrost foundation is adopted, the bridge foundation design should try to adopt the foundation types that can help to reduce the thermal disturbance of foundation construction to the permafrost foundation as much as possible.

The bridge foundation in permafrost regions of China mainly adopts the bored and cast-in-place pile foundation and the open-cut foundation. Engineering practices show that the open-cut foundation construction will cause great thermal disturbance to the permafrost foundation, and the refreezing time around the foundation will be longer. The open-cut foundation is suitable for the ice-poor or icy permafrost foundation with the buried depth of less than 5 m. When the open-cut foundation is used, it should be constructed in the cold season, and the molding temperature of concrete should be controlled during pouring, which should not be higher than 10 °C.

The pile foundation is a better foundation type for the bridge engineering in permafrost regions. The biggest advantage is that it has less thermal interference to the permafrost foundation, with some other advantages such as the faster borehole forming, the shorter refreezing time around the pile and the higher bearing capacity. The cast-in-place piles are suitable for all types of permafrost whose mean annual ground temperature of permafrost is lower than −1.0 °C. Compared to the bored and cast-in-place piles, the driven pile in borehole or the inserted pile has less thermal disturbance to permafrost foundations, which is suitable for warm permafrost foundations. The driven pile in borehole is more suitable for plastic permafrost foundation dominated by cohesive soil and sand.

The calculation shows that although increasing the length of the bored and cast-in-place piles in permafrost can improve the bearing capacity of pile foundation it will have fewer effects on the displacement of the load when the pile length exceeds to a certain value. When the pile length is 15 m, if the pile diameter increases by 100 mm, the ultimate bearing capacity will increase by 44%. If the pile diameter increases by 200 mm, the ultimate bearing capacity will increase by 111%. If the pile diameter increases by 350 mm, the ultimate bearing capacity will increase by 244%. The increase of pile length can only slightly improve the bearing capacity of the pile foundation, but it can't control the displacement of the lower end of the pile when the pile reaches the ultimate bearing capacity. Therefore, the increase of pile diameter has obvious influence on the bearing capacity, and the bearing capacity of the pile with a large diameter is greater than that with a small diameter.

4. **Calculation of Bearing Capacity of Pile Foundation**

1. Calculation of Vertical Bearing Capacity of Pile Foundation

After the thermal regime of the permafrost foundation around the pile is restored, the stress system between the pile and the soil is formed. In the calculation, the freezing strength within the active layer and at a certain range of depth under the permafrost table should not be considered. Because, during the warm season, the active layer frozen in the cold season will thaw completely, and coupled with the depth subjected to the thermal influence introduced by the pile foundation, which is generally not more than 0.5 m. For the sake of safety, the freezing force at depths that equals one-fold of the pile diameter blow the permafrost table should not be considered. Therefore, the effective pile length, by which the freezing force of the pile foundation can be calculated within a year, should be deducted the thickness of the active layer and the influence depth of the pile foundation (equals as the pile diameter) from the whole length of the pile. Within the effective pile length, the bearing capacity of the large-diameter pile foundation is calculated by the sum of the freezing force around the pile and the resistance at the lower end of the pile. It can be calculated as follows:

Railway Bridge specification (TB 10002.5-2005):

$$[P]=\frac{1}{2}U\sum f_i l_i m''+m_0 A[\sigma] \tag{4.8}$$

The highway bridge specification (TG/T D31-04-2012) does not make specific provisions for the permafrost. It can be calculated according to the allowable value formula for the bearing capacity of the bored or excavated and cast-in-place pile with the allowable value of the axial compressive bearing capacity of frictional piles for the single pile:

$$[P]=\frac{1}{2}U\sum_{i=1}^{n}f_il_i+A[\sigma] \tag{4.9}$$

The frozen soil building foundation specification (JGJ 118-2011): the last item in the formula is the active layer of permafrost. During the warm season, the active layer completely thaws, so its frictional resistance can usually be ignored in the calculation. If friction force generated during thawing is negative, which should be taken into account, and the value -10 kPa can be taken.

$$[P]=[\sigma]A+U\left[\sum_{i=1}^{n}f_il_i+\sum_{i=1}^{n}q_jl_j\right] \tag{4.10}$$

where U is the circumference of the completed pile diameter (m); l_i is the contacting frozen area between each soil layer and the lateral pile surface (m^2); f_i is the freezing force at each layer between each soil layer and the lateral pile surface (kPa), whose values are assigned according to Table 4.9; A is the sectional area of the lower end of the pile (m^2); $[\sigma]$ is the allowable value of the soil bearing capacity at the lower end of the pile (kPa), assigning the value according to the railway bridge specification; q_j is the characteristic values of resistance around the pile in each layer of thawed soil (kPa); l_j is the vertical pile length of each soil layer (m); m" is the correction coefficient of freezing force when different pile-laying methods are adopted,

Table 4.9 Freezing strength between permafrost and surface of concrete foundation *f*/kPa(JGJ 118-2011)

Soil type	Maximum monthly average ground temperature of soil layer /°C						
	−0.5	−1.0	−1.5	−2.0	−2.5	−3.0	−4.0
Cohesive soil	60	90	120	150	180	220	280
Sand	80	130	170	210	250	290	280
Crushed stone soil	70	110	150	190	230	270	250

Note: 1. The value for the non-thaw settlement permafrost is reduced by 10% to 20% in accordance with the values listed in the table. For the dry soil without obvious cementing force with the foundation, its freezing force can be ignored (calculated by the friction force of the thawed soil).The value for the permafrost with strong thaw settlement (ice-saturated permafrost) decreases by 20%. For the soil-bearing ice layer, the value is reduced by 50%. When 0.05 m to 0.1 m sand layer is backfilled around the foundation, the value can be taken as the value of the soil with strong thaw settlement (ice-saturated permafrost)

2. The value for the untreated and non-steel structures is reduced by 30% according to the values listed in the table

3. The values listed in the table are not applicable to the permafrost with salt content greater than 0.3%

which takes 0.7 to 0.8 for the bored and insert pile; take 1.1 to 1.3 for the bored and driven pile; take 1.3 to 1.5 for the bored and cast-in-place pile; m_0 is the reduction coefficient of the supporting force at the lower end of the pile toe, which is 0.5 to 0.9 according to the conditions of the bottom of the borehole.

2. Checking Calculation of Bridge Pile Foundation Stability under Frost-heave Force

Every year when the active layer above permafrost freezes in the cold season, the soil layer expands, and the pile foundation deforms under the impact of tangential frost-heave force. Therefore, the pile foundation in permafrost area needs the checking calculation for the stability under the impact of frost heave.

Under the impact of frost-heave force, the stability of pile and pier foundation should meet the requirements of the following formula:

$$\begin{gathered} \Sigma \sigma_\tau H_j U \le 0.9 G_k + R_m \\ R_m = \frac{1}{k} U \sum_{i=1}^{n} f_i \left(L - H_j \right) \end{gathered} \tag{4.11}$$

where σ_τ is the tangential frost-heave force during the freezing process in the active layer (kPa), values assigned according to Table 4.12. H_j is the thickness of the active layer (m). U is the circumference of the pile (m). G_k is the permanent load applied on the pile (kN). R_m is the anchoring force of pile and pier foundation in the permafrost (kN). f_i is the freezing strength between the permafrost and the foundation (kPa). L is the buried depth of foundation (m). k *is* the safety factor.

Through the above calculation, if the design requirements cannot be met, anti-frost heaving measures should be taken to replace the soil of the active layer on the pile side, or to apply anti-frost heaving chemical materials on the pile side (Fig. 4.44) to reduce and weaken the tangential frost-heave force. Besides, the diameter of the lower end of the pile and the buried depth of the pile foundation can be increased, but it should be re-checked until the design requirements are met.

4.1.3 Protection of Abutment Foundations in Permafrost Regions

The bridge head section includes the abutment, the conical revetment and the road-bridge transition section. The main cause of bridge diseases in the permafrost regions of Daxing'anling Mountains and the Qinghai-Xizang Plateau is the deformation of bridge abutments and piers, of which the disease at the head of the bridge accounts for the main part. The diseases are manifested as abutment settlement and frost heave, tilt (forward and backward), settlement and cracking of slope protection (see Fig. 4.8), bridgehead embankment sinking, etc., and even the phenomenon of

Table 4.12 Variations of annual maximum thawing depth of culvert foundations/m (according to Wu et al., 2005a, b, c)

Sections	Years	Natural permafrost table	The left opening	The left shoulder	The road center	The right shoulder	The right opening	Notes
Kunlun Mountains	1984	1.72 to 2.10	2.10	2.50	3.91	2.30	1.75	Completed in 1983. Silty clay, F, B. The permafrost table in the road center varied from 0.88 m to 0.92 m in 7 years
	1985		1.63		0.92	1.58		
	1987		1.68	0.81	0.88	<0.5	0.92	
	1990		1.63	0.60	<0.50	<0.50		
Hoh Xil	1984	1.60 to 1.7	1.70	3.10	3.25	3.10	1.60	Constructed in 1981. Silty clay, H. The thermosyphons were applied at the opening in 1989. In the middle of the culvert, it appeared the leakage from the cracks and the wing walls subsided
	1985		1.20		<0.5	2.30	1.75	
	1988		2.72	1.07	1.00	0.67	1.88	
	1993		1.80	1.50	1.18	1.37	1.64	
Wudaoliang	1984	2.05 to 2.1	2.10	3.34	3.67	3.32	2.05	Completed in 1981. Silty clay, B. In the middle of the culvert it appeared settlement and cracks and the wing walls appeared cracks and settlement
	1985		2.50	1.43	1.30	1.50	1.73	
	1988		2.17	1.5	1.40	1.25	2.00	
	1992		2.30	1.50	1.20	1.33		
Fenghuo Mountain	1984	1.80 to 2.4		3.45	3.50	3.55	2.45	Completed in 1984. Gravelly silty clay, S. Culvert bottom fractures and leakages. Cut-off walls at the outlets were hollowed out

Notes: S is Ice-poor permafrost; F is Ice-rich permafrost; B is Ice-saturated permafrost; H is Soil –bearing Ice Layer

Fig. 4.44 Bridge Abutment Coated with 10 mm Anti-Frost Heaving Oil Residue

backward tilt of the front and rear abutment.The deformation of the bridge abutment leads to the destruction of bridge components and the main forms are as follows:

① The cracks in the cushion stone;
② The upper flange or the lower flange of the beam end are pressed tightly, forming rigid contact;
③ The cracks or fractures of limit block;
④ The moves of bearing chair;
⑤ The failure of anti-seismic pile;
⑥ The pier sinking, etc. The essence of these diseases is the result of thaw settlement and frost heave of abutment foundation.

4.1.3.1 Deformation Characteristics of Bridge Piers and Abutments

Pile foundations are predominantly used for piers and abutments on the Qinghai–Xizang Railway. Pile foundations are buried below the permafrost table, but the base of the pile deck is in the thawed layer (Fig. 4.45). The main deformation types are:

1. Settlement deformation

It has the characteristics of slow heaving in the cold season and restoring in the warm season. The settlement deformation of the bridge abutment is very small, and the maximum frost heaving amount is 5–7 mm. In general, the vertical deformation is in a basically stable state.

2. Longitudinal deformation

The abutment has forward or backward tilts. Figure 4.46 shows the change curve of the inclination angle of an abutment (the curve angle is the angle between the side line of the front wall of the abutment and the vertical line). During the observation period, the abutments tilts either forward or backward (the settlement observation

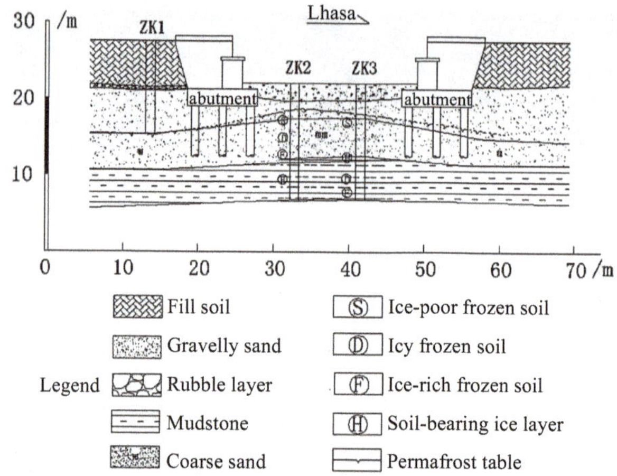

Fig. 4.45 Engineering geological section map of the middle bridge in Chiqu Valley of Qinghai-Xizang Railway (Xiong, 2012)

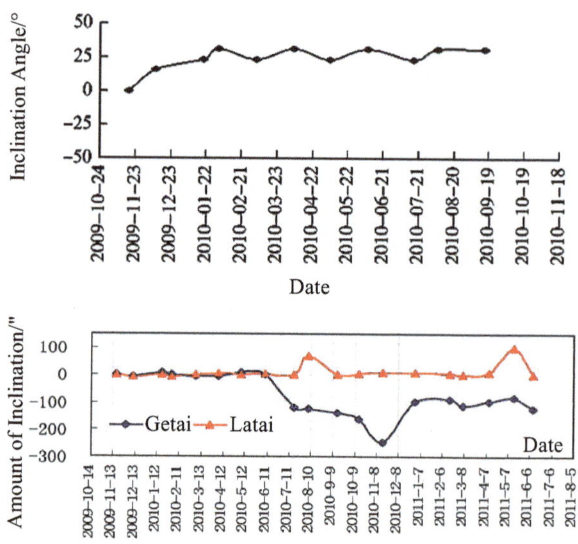

Fig. 4.46 Variations of the inclination angle of the bridge abutment (Xiong, 2012). Note: The left figure is the cold permafrost area, and the right figure is the warm permafrost area. The Getai is in the direction toward Golmud; Latai is in the direction toward Lhasa. The positive value means tilting forward, that is, inclined to Lhasa, and the negative value means tilting backward, that is, inclined to Golmud

point is set at the front wall). The initial time that Getai (toward Golmud direction) begins to tilt towards Lhasa (direction) mostly happens in November, and the oppositing tilt occurs in June to July. This is related to the thaw settlement in front of the bridge abutment and frost heave behind the abutment.

4.1.3.2 Reasons for the Deformation of the Bridge Abutment

The deformation of the bridge abutment in permafrost regions is mainly due to the frost heave and thaw settlement of the foundation soil in the active layer. In addition to the maximum frost heave and thaw settlement of the active layer, for the bridge abutment with pile foundation the changes in the bearing capacity of pile foundations due to the warming and thawing of the permafrost foundation are also the causes for deformation.

During the freezing process of foundation soil in permafrost regions, the bridge abutment will be affected by the normal frost-heave force on the base surface of the pile deck and the horizontal frost-heave force on the back of the platform (Fig. 4.47). Frost-heave forces depend on soil type, moisture, and temperature. The thaw settlement of permafrost foundation depends on its soil quality, ice content, ground temperature and load.

According to the bridge deformation survey (Xiong, 2012), although the causes for deformation are different, they are all related to the permafrost engineering geological and hydrogeological conditions at the bridge site. The causes for bridge abutment diseases are summarized as follows: ① When the content of silty clay in the foundation soil of the bridge abutment and subgrade filler is high, such as the content of silty clay in the foundation soil accounts for 17–25%, these foundation soils as well as the similar silt, have a 11–25% possibility of frost heaving. When the content of silt in the subgrade filler ranges from 12 to 26%, the possibility of frost heaving is 6–21.5%; ② The permafrost of the bridge abutment foundation soil is mostly the soil layer of ice-rich permafrost to soil-bearing ice layer; ③ The

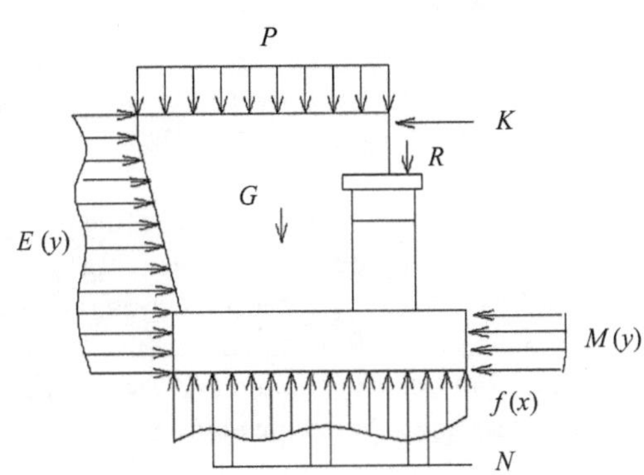

Fig. 4.47 Schematic diagram of the stress on bridge abutment on permafrost foundation. *P*—The manual transformation loads including ballast, sleepers, steel rails, and dynamic loads of train on the upper abutment; *R*—Loads from beam weight and the bridge deck system; *E*(*y*)—Earth pressure behind the abutment and horizontal frost-heave force; *G*—The self-weight of the bridge abutment; *N*—The reaction of the pile foundation; *f*(*x*)—The normal frost-heave force on the base surface of the pile deck; *M*(*y*)—The resistance force in front of the abutment; *K*—the constraint force from the beam body exists when the beam end contacts with the breast wall or when the support fails

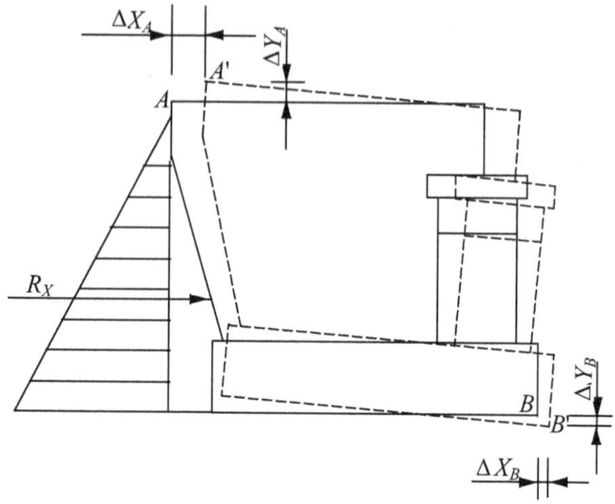

Fig. 4.48 Schematic diagram of abutment displacement under the impact of horizontal frost-heave force

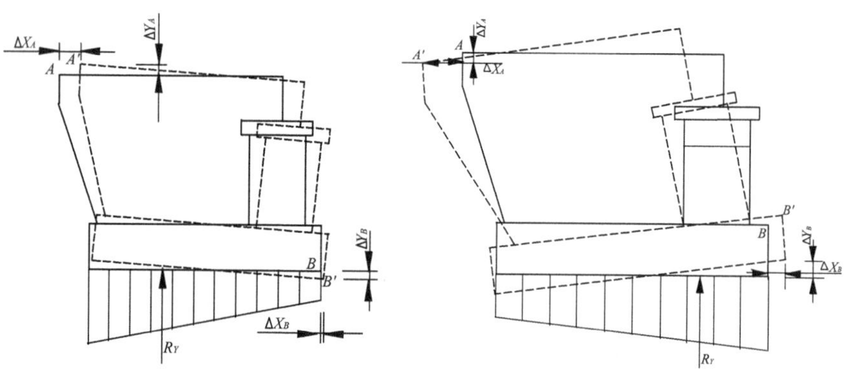

Fig. 4.49 Schematic diagram of abutment displacement under the impact of normal frost-heave force (the left: tilting forward; the right: tilting backward)

suprapermafrost water occurs in the active layer. Before freezing, there is water supply, and the aufeis appears in front of the abutment, leading to the tilting backward of the abutment, the frost heaving of the cone protection, and the cracking of the abutment. ④ The bridge site is heated from three dimensions, causing the thawing of the permafrost under the abutment foundation and the degradation of warm permafrost under the abutment. It shows that in the cold season, the foundation soil and the subgrade fill behind the abutment have great frost heave, sometimes reaching strong frost heave or even extremely strong frost heave. In the warm season, the permafrost foundation with high ice content under the bridge abutment and subgrade of the bridge back will experience degradation of permafrost and large thaw settlement under the influence of three-dimensional heat.

Therefore, the main reasons for the deformation of bridge piers and abutments in permafrost regions are: ① The horizontal frost-heave force caused by the strong frost heave of the soil on the bridge abutment and bridge back (Fig. 4.48); ② The normal frost-heave force generated by the foundation soil with strong frost heave under the pile deck of bridge abutment (Fig. 4.49); ③ The variations of permafrost conditions cause the rise and degradation of ground temperature of permafrost and the weakening of the bearing capacity of pile foundation.

4.1.3.3 Deformation Characteristics of Bridgehead Subgrade

The settlement deformation of bridgehead subgrade in permafrost regions is a typical and common disease in road engineering. The settlement deformation in the transition section of the bridgehead subgrade is often manifested in the settlement of the subgrade surface and the abutment cone protection (Fig. 4.50), cracking (Fig. 4.9), uplift, etc.

According to the disease investigation from164 road-bridge transition sections along the Qinghai-Xizang Railway (Niu et al., 2011), the settlement ratio is as high as 82.6%, and the average settlement is 70 mm. The settlement in plains, basins and other regions is larger than that in mountain and valley areas.

1) Relationship between the Settlement in the Transition Section and the North-south Ends of the Bridge and the Slope Aspect

According to statistics (Fig. 4.51), the total settlement at the north end of the bridge (toward Golmud direction) is 17.5% higher than that at the south end, and the average settlement is 11 mm higher than that at the south end. The total settlement on the sunny slope is about 65.9% higher than that on the shady slope, and the average settlement is 35 mm higher than that on the shady slope. This situation indicates that it receives different amounts of solar radiation in different directions: the north end receives more radiation heat, and the horizontal heat flow warms the embankment and increases the heat storage in the embankment, which aggravates the warming and thawing of the permafrost.

Fig. 4.50 Surface settlement of the abutment cone protection (Photo by Niu Fujun)

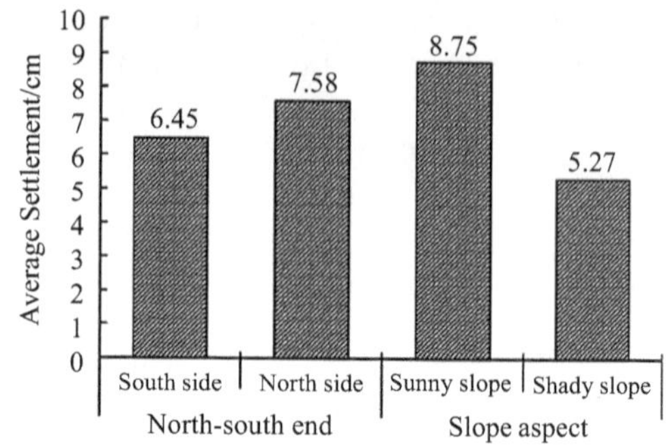

Fig. 4.51 Relationship between the settlement in transition section and north-south ends of bridge and slope aspect

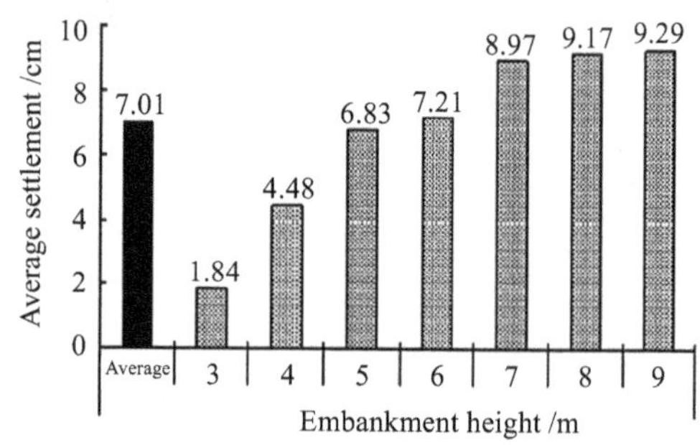

Fig. 4.52 Statistical relationship between settlement in transition section and embankment height

2) Influence of Subgrade (Abutment) Height on the Settlement in the Transition Section

The general trend is that the higher the subgrade, the greater the settlement in the transition section (Fig. 4.52). The total average settlement of the subgrade height less than 3 m is only 18 mm, and the average settlement of the subgrade height greater than 8 m is 93 mm. When the subgrade height is 4–6 m, the average settlement is close to the total average settlement (70 mm).

3) Relationship between Settlement in the Transition Section and Subgrade Structure and Soil Type of the Foundation

According to the statistics of survey data, the block stone and crushed stone slope protection are set as the thermal insulation measures; the subgrade is a plain soil entity; the bridgehead subgrade is thermally affected by the three-dimensional heat, which is under stronger impact than the subgrades under the two-dimensional heat condition in other sections. The characteristics of the settlement deformation in the transition section are still reflected in the subgrade structure of the plain soil. The average settlement of fine-grained foundation soil is larger than that of coarse-grained foundation soil, reaching 92 mm, and the settlement of weathered sandstone and rubble foundation soil is very small (Fig. 4.53). Obviously, the greater the strength of the foundation soil, the smaller the amount of settlement.

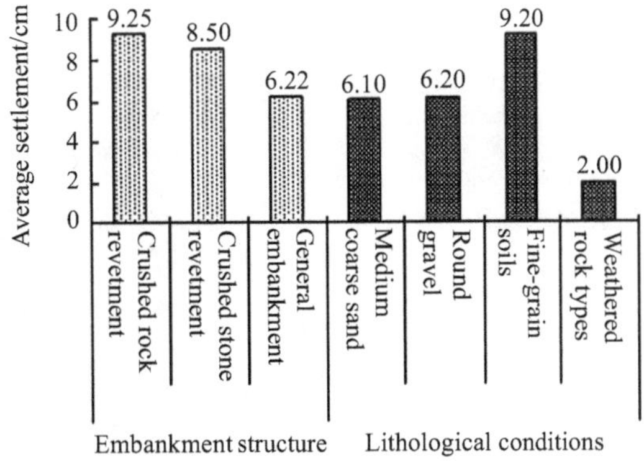

Fig. 4.53 Statistical relationship between settlement in transition section, embankment structure and engineering types of frozen soil

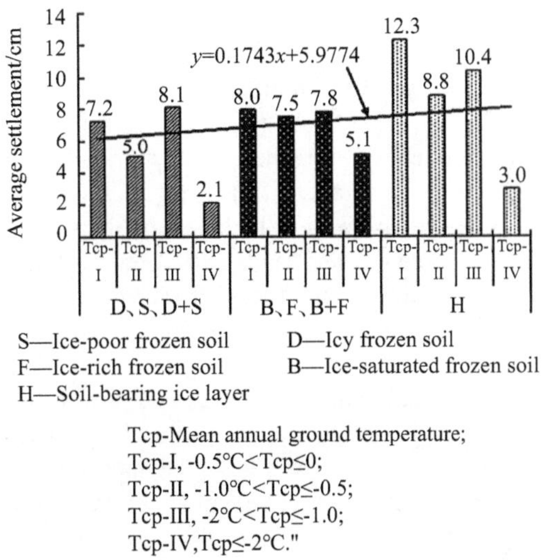

Fig. 4.54 Relationship between settlement in transition section, engineering types of frozen soil and mean annual ground temperature

4) Relationship between the Settlement in the Transition Section and the Types of Frozen Soil Engineering and the Mean Annual Ground Temperature

The types of frozen soil engineering are divided by the ice content. The higher is the ice content, the greater the settlement becomes, which is similar to a linear relationship. The average settlement in soil-bearing ice layer is 93 mm. In ice-rich and ice-saturated permafrost sections it is 90 mm, and in icy and ice-poor permafrost sections it is 72 mm. The average settlement in the transition section of warm permafrost areas is larger than that in the cold permafrost areas, and the average settlement in the warm permafrost sections with high ice content is the largest (Fig. 4.54). Regardless of the direction of the abutment and the shady and sunny slopes of the transition section, the average settlement in the transition section of the warm permafrost areas is larger, which is approximately 14 mm, than that in the cold permafrost areas (Fig. 4.55).

It indicates that the types of frozen soil engineering and mean annual ground temperature of the permafrost foundation in the bridgehead section are the decisive factors for the settlement in the road-bridge transition section, and the absorption of

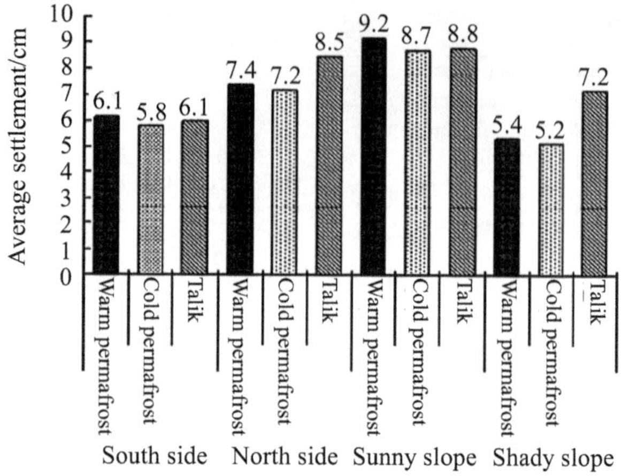

Fig. 4.55 Relationship between settlement in transition section with different embankment orientation and slope aspects and geotemperature zones

solar radiation heat in different directions and slope aspects is the induction factor for the settlement in the transition section. Due to an additional surface of the conical revetment of bridgehead that receives heat, which is an extra heat source, in the freeze-thaw dynamics of the bridgehead foundation, the heat is turned into a three dimensional process, which inevitably makes the thawing depth of bridgehead greatly increase, and leads to a deeper artificial permafrost table than that of the adjacent roadbed, and takes longer time to restore its stable state. By means of the subgrade structure and engineering measures, the thermal influence of solar radiation can be changed and reduced, and the subgrade settlement of the bridge abutment in the transition section can be prevented and controlled.

During the investigation on the Qinghai-Xizang Railway bridges, most of the bridgehead roadbeds were found to have large settlement deformation, and the maximum accumulative settlement (not a very strict conclusion based on the observation, for some were compared with the design, and some were compared with the situation after the opening to traffic) at the site with the most serious situation was 1.3 m. Since April, 2010, according to the investigation and statistics of five bridges with relatively serious bridgehead subgrade settlement within 1 year and the survey on the altitude differences between the rail surface and road shoulder, the annual settlement of most bridgeheads was greater than 20 mm. At a place 2 m away from the bridgehead, the maximum annual settlement was nearly 50 mm, and the height difference between the subgrade surface and the rail surface was 1.9 m. At a place12 m away from the bridge, the maximum annual settlement was 44 mm, and the height difference between the subgrade surface and the rail surface was 1.83 m. According to the maintenance data, the influence range of bridgehead subgrade settlement is generally 1.5–2.0 times of the height of the abutment.

4.1.3.4 Treatment of Deformation for Bridge Abutment and Road-Bridge Transition Section

According to the deformation characteristics of abutment and bridgehead subgrade, the horizontal frost-heave force on the contacting area between the bridge abutment and the embankment and the normal frost-heave force at the abutment base are the main causes for abutment deformation during the freezing process. The thawing of permafrost foundation is an important factor causing the settlement of bridge pier, abutment and subgrade in the transition section. Accordingly, corresponding engineering measures can be adopted to control or eliminate the deformations of bridge piers and abutments. The following basic principles should be followed in design and construction:

1. Strengthen the waterproofing and drainage measures to prevent the erosion and influence of ground surface water and suprapermafrost water on piers, abutment foundations and fill soil.
2. Adopt cooling measures to reduce the thawing depth of the permafrost foundation, raise the artificial permafrost table under the base of the pile deck, and improve the bearing capacity of the permafrost foundation.
3. Strictly control the silty clay content and moisture content of coarse-grained fillers to reduce the frost heave.
4. Adopt thermal protection measures to shield or isolate the intrusion of external heat.
5. Change the subgrade structure to reduce the thermal disturbance to the permafrost foundation.

A large number of tests and engineering practices show that the thermosyphon, block stone roadbed, crushed rockslope protection and berm protection, ventilation tube, thermal insulation material, shading board, water-stop ridge, water-stop board, the controlling of silty clay content, and so on, are effective engineering measures for thermal protection and waterproofing and drainage to protect permafrost foundations from thaw settlement and frost heave.

At the abutment back, pile deck and conical revetment, the thermal insulation layer should be applied. In the bridgehead transition section, it adopts the block stone roadbed structure (Fig. 4.56). For the ventilation tube embankment, the crushed stone slope protection and berm protection can be set on the embankment slope (the bottom is firstly backfilled with clay, preventing surface water from intrusion), or the ventilation tube can be laid on the embankment slope. The thermosyphon should be buried on both sides of the bridgehead and embankment. The silty clay particles in coarse-grained filler in the contacting area between the bridge abutment and the embankment or under the pile deck should be controlled within

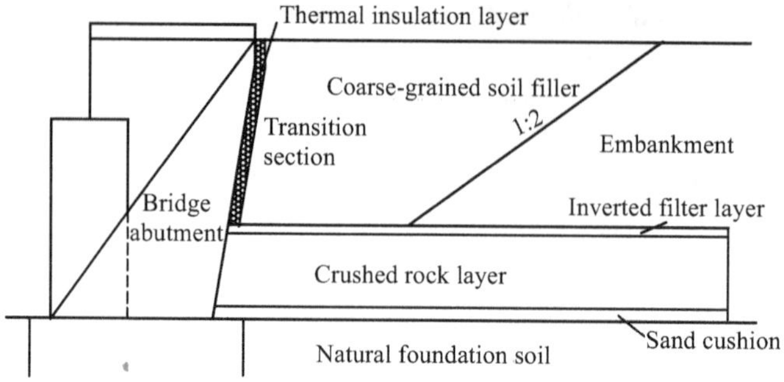

Fig. 4.56 Schematic diagram of the setting of road-bridge transition section

5–10%, and waterproofing and drainage measures should be taken to reduce the frost-heave force. The thermosyphon can be embedded in the pile foundation (or inside) under the pile deck, which can decrease the ground temperature and increase the bearing capacity.

4.1.4 *Bridge Foundation Construction in Permafrost Regions*

Several aspects should be considered for bridge foundations and their construction methods in permafrost regions: ① Thermal and mechanical characteristics of buildings; ② Large permanent bridge loads and small allowable displacements require high strength and stability of the foundation; ③ Foundation conditions, characteristics of permafrost, frost heave and thaw settlement, groundwater, etc.; ④ Construction methods, construction conditions, equipment, requirements for construction time, building materials, etc.

The domestic and abroad experiences indicate that the bored pile foundation is the optimal foundation type for bridges in permafrost regions, which has little thermal disturbance to permafrost and suits the requirements for foundations with large buried depth. In addition to the requirements of topographic conditions, it can adapt to various geological and permafrost conditions. Besides, the operation of the bored pile foundation is simple and it is beneficial to the environment protection of permafrost. However, under some topographic, geological, hydrological and engineering requirements, the open-cut foundations can also be used.

4.1.4.1 Foundation Construction Method

1. **Bored Pile Foundation**

The use of bored pile foundation has obvious advantages. It is not limited by warm or cold permafrost areas, high-ice content or low-ice content permafrost areas, the thickness of seasonal active layer, lithology of foundation, ground surface water and groundwater conditions. For the use of bored pile foundation, its high degree of construction mechanization has less thermal interference to the permafrost and less damage to the environment of permafrost. The refreezing time for permafrost around the pile is shorter, which has high bearing capacity and shortens the construction time.

During the construction of the Qinghai-Xizang Railway, the domestic-made of abrasion drill rigs/whirling driller, the percussion drill rigs, the reverse rotary rigs and the imported rotary drilling rigs were used together for pore-forming drilling of the pile foundation. The rotary drilling rigs had more advantages.

The construction of borehole with a constant diameter adopts the medium or high speed, low torque and low progress rate technology, which can improve the speed and quality of pore-forming drilling. Selecting drill bits according to the local conditions can greatly improve the drilling speed and pore-forming quality. Rapid pore-forming and rapid concrete pouring can minimize the thermal disturbance to permafrost and avoid borehole collapse accidents.

Select the drill bits reasonably according to the lithology of the formation.

For the bored and expanded-base pile foundation, it is a fact that in permafrost region, the high-power rotary drilling rigs are used to construct the expanded-base piles, but there isn't any machine particularly used in the construction yet. The bored and expanded-base pile can enlarge the section of the borehole bottom, improve the bearing capacity of the pile foundation and the ability to resist frost heave, and shorten the pile length.

The reaming bit (Fig. 4.57) can enlarge the bottom of the original straight borehole with a constant diameter of 0.4 m. For the construction of 446 bored and expanded piles through Budongquan Bridge, Chumar River Super Bridge and Hoh Xil No.1 Grand Bridge, the reaming time should be controlled within 1.5 h. The use of rotary drilling rigs for pore-forming boreholes is several to dozens of times faster than those of the abrasion drill rigs and percussion drill rigs, shortening the construction period by 50–65% and saving costs by 30–42%.

The bored and inserted pile is suitable for the construction of small and medium size bridge foundations. The borehole diameter of Qingshui River Bridge was 0.7 m (the diameter of the pile foundation was 0.6 m), and the surrounding of the pile was backfilled with clay mortar (the volume ratio was 1:8). After the pipe was inserted, the borehole was filled with C30 concrete, and the top of the pile was pre-buried with steel bars connecting the pile and the pile deck. The refreezing time was 7 days, the load was 73 tons (the design load was 71 tons), the settlement of the two piles was 1.97 mm and 2.10 mm respectively, and the residual settlement of 1.00 mm and 1.11 mm was maintained after unloading, which met the design requirements. The bored and inserted pile was suitable for the water-free and low-water/water-limited conditions. It should be constructed when the temperature is low in the warm season. The molding temperature of the concrete is needed to be controlled below 5 °C, and its bearing capacity is small, which is suitable for engineering with small load requirement.

2. **Open-cut Foundation**

It is generally used for foundation whose buried depth does not exceed 5 m, but in cold permafrost areas, the deep excavated pile foundations can also be used, such as in the Kunlun Mountains, the highway bridge foundation was buried down to 15 m. The construction should be conducted in cold season in order to avoid borehole

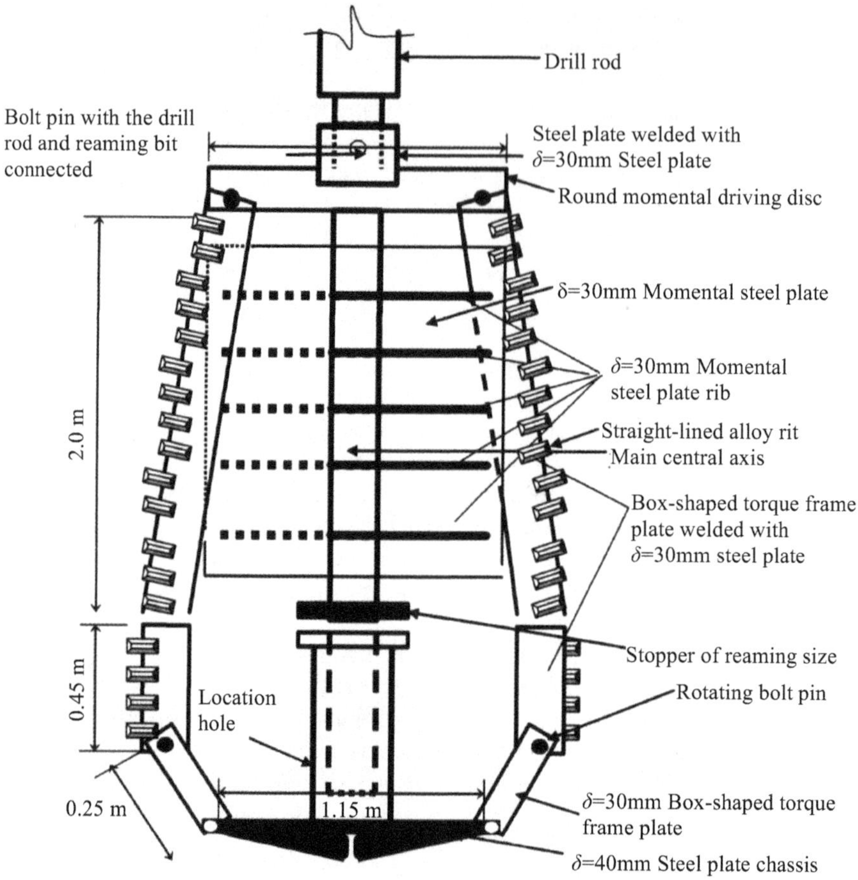

Fig. 4.57 Schematic diagram of reaming bit (Editorial committee of the Qinghai-Xizang Railway construction technology of China Railway Engineering Corporation, 2016)

collapse. If the construction is carried out in warm season, several preparations should be made in advance: ① Set up a sunshade to avoid direct sun exposure; ② Protection measures for the borehole wall within the depth range of the seasonally thawed layer should be taken, such as protective tubes, and at its bottom should go through the permafrost table by 0.5 m; ③ Foundation pit drainage should be equipped to prevent the infiltration of super permafrost water or the thawed water from permafrost in the borehole wall, and the accumulated water should be drained out in time; ④ Blasting parameters should be optimized and the smooth blasting technology should be used, which is better to be conducted in loose permafrost layer. The general principle is to adopt a quick construction with a rapid pouring to reduce the exposure time of the permafrost foundation pits, which should generally not exceed 15 days.

Table 4.10 Strength loss of concrete frozen immediately after pouring

Precured time/hours	Freezing time/hours	Standard maintaining stage after freezing / days	Ambient temperature/°C	Strength loss of labeled concrete/%		
				C20	C30	C40
0	24	28	−5	38.0	48.9	48.1
			−10	32.5	41.8	43.3
			−20	27.8	16.8	20.1
0	24	28	−5	35.9	44.1	47.8
			−10	30.5	37.0	41.8
			−20	19.4	22.5	20.8

The influences of repeated freeze and thaw cycle on the strength of concrete are mainly manifested in the following aspects.

1) Impact of Freeze-thaw on fresh concrete

When the fresh concrete freezes before the initial setting, except for some of the mixed water participating in the hydration reaction, the rest of the water will freeze into ice. On the one hand, the ice volume expands, on the other hand, the solidification process of concrete stops for lack of hydration reaction. When the concrete is heated and the ice thaws, the hydration reaction will recover. Generally, the shorter the process, the better the situation, and it will not affect the solidification process and strength increase of concrete.

2) Impact of Freeze-thaw during the maintaining period of concrete

The freezing of concrete after initial setting during the maintaining period has a great influence on the strength of concrete. The freezing and expanding of water in the concrete pores cause micro-cracks inside the concrete, which reduces the cohesion between the cement and the sandstone. After heating up, although the concrete continues to maintain the following stage, it often fails to fully reach the design strength and has a great loss of durability. The later the freezing occurs during the maintaining period, the smaller the effect on the concrete strength. After the concrete is poured, it is immediately frozen in freezing chambers with different negative temperatures. After 1 day of freezing, it then maintains for 28 days in accordance with the maintenance standard. Compared with the standard maintaining specimens, the loss of concrete strength is measured (Table 4.10).

It shows that if the concrete after the initial setting is not insulated and maintained in time, the strength loss of concrete after freezing is very large, which could be up to 50%. It can be seen from the table that the higher the ambient temperature, the slower the freezing rate, the stronger the water migration in the cement slurry, the more micro-cracks in the concrete caused by the freezing and expanding, and the greater the ascertain loss. On the contrary, the lower the ambient temperature, the faster the freezing rate, so the water in the cement slurry has no time to migrate, causing less micro-cracks in the concrete and less strength loss.

4.2 Culvert Engineering

In the railway and highway engineering in permafrost regions, culverts are indispensable small-scale projects, with a large number and distributing in various geomorphic areas with varying lithology and frozen soil engineering types. The types of culverts are mostly the segmented reinforced concrete circular pipe culvert or the rectangular culvert, and the foundation is the cast-in-place one-piece foundation or the assembled foundation. Culverts in abroad permafrost regions are usually made as the corrugated metal pipe culverts. In the early days, in China, it also adopted the corrugated metal pipe culverts (in the seasonally frozen soil areas), and recently this type of culvert has been widely used in permafrost regions.

4.2.1 Culvert Diseases and Variations of Permafrost Table

According to a survey on the working state of 772 culverts along 550 km of the Qinghai-Xizang Highway in permafrost region in October, 1990 (Zhang, 2000), 15.4% of the total culverts were severely damaged, 21.1% were damaged in a medium level and 63.5% were slightly damaged. The damages at the culvert inlet and the culvert bottom paving were the most serious, accounting for more than 86% of the damage types of the total damaged culverts. When the renovation project of the Qinghai-Xizang Highway started in August, 1998, the survey on the 368 newly built and reinforced culverts along the 240 km highway section in permafrost regions from Kunlun Mountains to Amdo was carried out. According to the survey, 8.7% of these culverts were severely damaged, 30% of them were damaged in a medium level, and the most seriously damaged parts were the culvert inlets (conical slope) and the culvert bottom paving. The culverts with the tilting, sinking and cracking straight end wall and wing wall account for 19.6%. The survey data of the Qinghai-Xizang Highway from Kunlun Mountain to Wudaoliang in 2011 (Fu et al., 2014) showed that the proportion of the reinforced concrete slab culvert (old culvert) with basic service ability was 41%, the proportion of culvert with medium level of damages was 21.7%, and the proportion of culvert with severe level of damages was 37% %. The service ability rate of the newly built corrugated metal pipe culvert was 100%.

In 2004, the investigation on the use of 8 culverts in the permafrost section of the Tianjun-Muli Highway was conducted. The main culvert diseases were the collapse and cracks of the inlet and outlet wing walls, and the settlement and inclination of the culvert body.

In the permafrost section from Jiangluling to Maduo along the National Highway 214, the service ability rate of all parts of the original culvert before the renovation was only 36%, the culverts with a basic service ability was 55%, and the damage rate of the important parts of the culverts were 9%. In the medium-temperature permafrost transition zone, the service ability rate was only 12%, and the damage rate of culverts at the important parts was 22%. The damage rates of important parts in warm permafrost zone and seasonally frozen soil area were 13% and 5% respectively. The more developed the permafrost, the lower the service ability rate of culverts.The damage types of culvert body include: abutment body tilts or collapses, abutment body cracks, culvert roof subsidence and culvert body paving damage, as well as siltation, water accumulation, ice accretion, etc. The damage types at the culvert inlet include: wing wall tilting or separated from the culvert body, wing wall cracks, chute paving and the opening paving hollowed out, etc. The damage rate at the inlets such as damages on wing walls was 40%. Three years after the reconstruction of the highway (in 2005), the service ability rate of the culvert was significantly improved, and the diseases such as cracking, paving damage and other diseases were eliminated. In 2011, the survey on the use of reinforced concrete slab culverts and corrugated metal pipe culverts from the Yellow River Source to Chalaping showed that the basic service ability rates were 54.79% and 75.68% respectively; and the medium level rates of damages were 28.78% and 10.81% respectively, and the serious damage rates were 16.43% and 13.51% respectively.

4.2.1.1 Culvert Damage Types and Causes

The types of culverts in the railway and highway engineering in permafrost regions are mainly the reinforced concrete slab culverts, the reinforced concrete circular pipe culverts and the concrete arch culverts. In summary, the main types of culvert damages are as follows:

1. **Frost-Heaving Damages**

At the inlet and outlet of culvert, it appears frost-heaving uplifts and frost cracks (Fig. 4.58), and the cracks on the end wing walls of culvert (Fig. 4.59), as well as the tilting outwardly of the wing walls of culvert (Fig. 4.60). At the inlet and outlet paving of culverts, it appears frost cracking, crushing, and scouring collapse (Fig. 4.61). The deformation and failure of culvert inlet and outlet is the most common and serious disease of culvert engineering in permafrost regions. In the design, the buried depths of the foundation at the inlet and outlet of the culvert are in accordance with the requirements of the specification, and the foundation base is buried at a depth of 0.25 m deeper than the maximum depth of the active layer (seasonal freezing depth or thawing depth). However, operations such as foundation excavation and masonry foundation and other works, coupled with the intrusion of ground surface water, increase the heat storage of the foundation, resulting in a deepening in the permafrost table. In the cold season, there is a three-dimensional freezing (from the slope, tunnel wall, and the ground) at the inlet and outlet of the culvert, and the freezing depth is much deeper than the depth of the active layer in natural state. The result is that not only the culvert foundation frost heaves, but also the slope soil body on the back of the wing wall frost heaves as well. Under the impact of the normal frost-heave force of the foundation base and the horizontal frost-heave force on the back of the wing wall, the frost heave at the inlet and outlet of the

Fig. 4.58 Frost heaving and frost cracks at the inlet and outlet of the culvert (Photo by Li Qingwu)

Fig. 4.59 Cracks in wing wall of culvert (Photo by Tong Changjiang)

culvert is frost heaved and damaged, and the wing wall is inclined or broken. During the warming-up period of the following year, the fine-grained soil invades the foundation soil, continuously increasing the silt-clay content of the original foundation soil, and increasing the frost heave of the foundation soil year after year. At the same time, the intrusion of the three-dimensional heat source also makes the thawing depth at the culvert inlet and outlet increase, permafrost table decline, and thaw settlement appears, and then, the water erosion is intensified, causing collapses.

2. **Damages of Thaw Settlement**

The so-called waist downward bowing of culvert means that after thaw settlement of the culvert foundation, the relative positions of the inlet and outlet of the culvert are high while the middle part of culvert is low, which makes the middle part

Fig. 4.60 Wing wall of culvert tilting outwardly (Photo by Tong Changjiang)

Fig. 4.61 Scouring collapse of the culvert outlet (Photo by Tong Changjiang)

depressed (Fig. 4.62). The relatively large amount of depressing of the culvert in the permafrost region of Daxing'anling Mountains could reach 500 mm (The Third Survey and Design Institute of the Ministry of Railways, 1994), resulting in the failure of flood discharge capacity of the culvert. Culvert foundation excavation and foundation concrete pouring have thermal disturbance to the permafrost foundation, increasing the thawing depth of permafrost under the foundation base. During operation, the middle part of the culvert bears a relatively large load, leading to a larger settlement. The two ends are affected by the frost heave of the foundation soil, which makes the downward bowing of culvert more prominent.

Fig. 4.62 Culvert body sinking with the bowing and leaking at the middle part (Photo by Li Qingwu)

According to the data of three test sections of the Qinghai-Xizang Railway, the influence depth of culvert foundation construction on the permafrost under the foundation base is 1.0–2.0 m (Editorial Committee of *Qinghai-Tibet(Xizang) Railway*, 2016). The foundation of the Qingshui River culvert is cast-in-place concrete. Although the foundation was constructed in winter, the influence depth was 1.2–1.5 m due to the thermal interference of the hydration heat generated after the pouring of the foundation concrete.The culvert foundations of the Beilu River and the Tuotuo River adopted the precast assembled foundation, and the heat source affecting the permafrost under the foundation base came from the intrusion of atmospheric thermal energy. However, because the foundation assembly needs to set up the longitudinal slope and slope adjustment, the joint mortar needs to be poured, and the construction technology is more complicated than the cast-in-place concrete, for which the exposure time of the foundation pit is the same as that of the cast-in-place foundation. According to the ground temperature observation, the influence depth of the culvert foundation of the Beilu River on the permafrost under foundation base was 1.0–1.3 m. Tuotuo River culvert was constructed during the transition period from the warm season to the cold season (in September), and the heat source intrusion was relatively large, and the influence depth on the permafrost under the foundation base was about 1.7 m. The foundation of the Qingshui River culvert stayed in cold season until the culvert section was assembled in April of the following year. The subgrade was not filled in the whole cold season, which was conducive to the refreezing of the culvert. After 45–50 days, the permafrost below the culvert foundation could be refrozen.

According to the observation results of foundation settlement after the culvert was completed (Table 4.11), there was basically no frost heave in the culvert during the cold season. The settlement in the center of the culvert was larger than that at the

Table 4.11 Settlement and average settlement rate of culverts in test sections

Items	Qingshui River	Beiluhe River	Tuotuo River	Project
Numbering	Circular culvert	Rectangular culvert	Rectangular culvert	Rectangular culvert
Minimum settlement/mm	−72.9	−51.8	−48.5	−80.6
Maximum settlement/mm	−120.2	−113.0	−84.0	−210.2
Average settling velocity/mm·d^{-1}	0.0227	0.0193	0.0224	0.0290

two ends, and the settlement at the inlet (on the shady side) was smaller than that at the outlet (on the sunny side). The large settlement in the middle of the culvert was affected by the impact of the filling and the loads from vehicles, and the additional stress of the foundation was greater than that at both ends. The thawing depth on the shady side was smaller than that on the sunny side, and the settlement was also smaller than that on the sunny side. The causes for the settlement of the culvert are as follows: ① The thawed soil layer under the foundation base is gradually compacted under the impact of the external loads; ② The rainfalls and flowing water form a ponding with a large amount in front of the culvert, which accelerates the thawing depth of the permafrost under the foundation base; ③ The replacement layer of the foundation base (with the thickness of 300 mm) fails to reach the standard degree of compaction during the construction period in the cold season, and then, the compaction settlement occurs; ④ The secondary construction of the wing wall in the Tuotuo River had an impact on the permafrost, so the settlement was larger than that of other sections; ⑤ The temperature observation in the culvert shows that the ground temperature under the foundation was in an unstable state, swinging around 0 °C (Fig. 4.64).

3. **Culvert Ice Jam**

The culvert ice jam is the phenomenon that the culvert is filled with ice (Fig. 4.63). Some are partially icing and some are mostly icing, resulting in ice jam. The degree of ice jam is related to the topography, geomorphology and hydrogeological conditions in which the culvert is located. The section with much developed local suprapermafrost water or exposed springs may cause the culvert ice jam, or icing formed by flowing water gradually blocks the culvert, or the icing may glaciate the road as the ice accretion is over culvert. In the warm season, the ice in the culvert thaws slowly, and water accumulates outside the culvert into a "pond".

4.2.1.2 Variations in Artificial Permafrost Table of Culvert Foundation

After the construction of the culvert, the thermal state of the permafrost will be changed, and the morphological variation of the permafrost table (the artificial permafrost table) under the culvert foundation is the most obvious. In the second

Fig. 4.63 Ice jam in the culvert (Photo by Northwest Research Institute of China Railway)

reconstruction of the Qinghai-Xizang Highway, the culverts with high ice content in permafrost zones with different ground temperature were selected for on-site observation and the data (Table 4.12) showed that the buried depth of the artificial permafrost table in the culvert foundation changed greatly, and the artificial permafrost table at the inlet of the culvert also decreased, and the artificial permafrost table at the shoulder of the culvert increased (Fig. 4.65), forming arched features.The depth of the zero isotherm at the inlet of the culvert was 0.5–1.8 m deeper than that in the middle of the culvert, and the thawing time of permafrost foundation was 20–40 days earlier than that in the middle of the culvert. When the culvert foundation broke and leaked, the artificial permafrost table would decrease, and the culvert would settle. In the warm season, when the embankment height was high, there were taliks occurring in the transverse section of the culvert. When the embankment height is small, the frozen soil in the embankment above the culvert would completely thaw, and the culvert foundation would form a basin-shaped talik.

4.2.2 *Culvert Design in Permafrost Regions*

For the applying of culverts in the permafrost, in addition to the key building of the subgrade drainage system, in the thermal system of the subgrade, the culvert is also a ventilation tunnel, which is closely connected with the outside. The temperature state is not only affected by the air temperature, but also by the heat and infiltration of the water flow. Culverts can provide shade, increase wind speed, reduce solar radiation and decrease temperature, which is beneficial to cool the ground

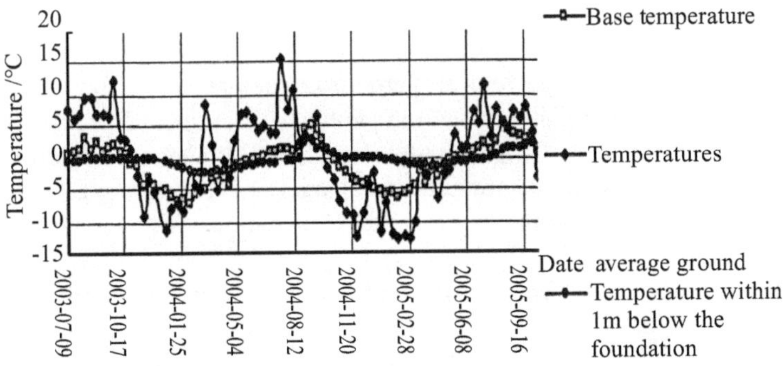

Fig. 4.64 Variation of temperature, base temperature and foundation temperature in the center of Beilu River culvert (Editorial Committee of *Qinghai-Tibet(Xizang) Railway*, 2016)

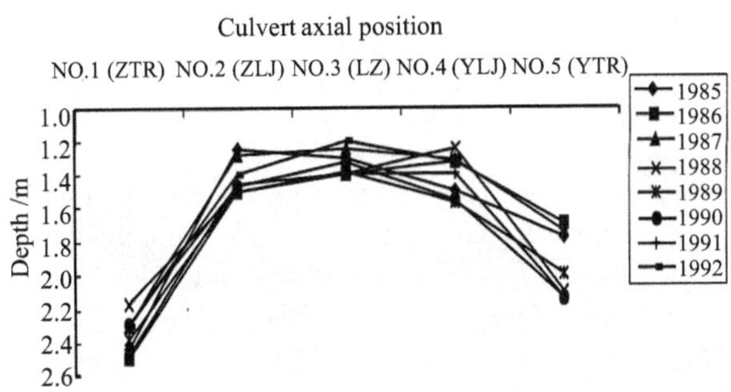

Fig. 4.65 Depth variation of zero-temperature line at different locations of the Wudaoliang culvert foundation (According to Zhang, 2000)

temperature of subgrade. However, due to the inflow of hot air and overflow in the warm season, the temperature in the culvert rises, which affects the thermal stability of the permafrost foundation of the culvert. Therefore, culvert engineering is actually an interference and interactive system of culvert structure, foundation, subgrade and environment. For permafrost regions, its essence is a thermodynamic problem of heat change. In the design of culverts, the structure of the construction, the engineering properties of the permafrost foundation, the freeze-thaw characteristics of the culvert foundation and the subgrade soil layer, the frost resistance of the building materials, the environmental conditions of the permafrost (atmosphere, hydrology and hydrogeology, etc.), the construction methods and the operation management, etc., should be considered comprehensively and constitute a culvert-permafrost thermodynamic system to determine the culvert design principles, structural types, foundation types, thermal protection measures, construction methods, and environmental protection measures.

4.2.2.1 Design Principles for Permafrost Foundation

The design principles for culverts in permafrost regions should be based on the characteristics of frozen soil engineering geology (mean annual ground temperature, ice content, lithology), permafrost environment (temperature, hydrology and hydrogeology), and proposed structures (mechanics and thermology). According to the characteristics, the principle for permafrost foundation design should be selected reasonably.

The *Specification for Design of Building Foundations in Frozen Soil Areas* proposes three principles for using permafrost as a foundation. Considering that the culvert projects are small, with shallow foundations and are also cold buildings, in the case of permafrost with a low mean annual ground temperature, a thick permafrost layer, and a high-ice content fine-grained soil, it is basically needed to keep the permafrost foundation in a frozen state and maintain the natural state of it, that is, the design principle for maintaining the permafrost foundation in a frozen state should be adopted. In the taliks, the design principle for non-frozen soil area can be adopted. The design principle of pre-thawing can be used in the extremely unstable permafrost zone, where the permafrost layer is thin, the frozen coarse-grained soil is ice-poor or the buried depth of the bedrock is shallow. The engineering geological conditions of permafrost in actual engineering areas are often not ideal combinations. When the ground temperature of permafrost is high, the thickness of permafrost is more than 5 m, and is the coarse-grained soil with high ice content, the design should be adapted to the deformation of the permafrost foundation according to the situations of culvert structure, foundation types and auxiliary engineering measures. The design principles are determined after technical and economic comparison. Therefore, the selection of culvert design principles in permafrost regions should be based on the ground temperature of permafrost zone, the thickness of permafrost and buried depth of permafrost table, the ice content (thaw settlement), and the lithology to determine the adaptability of culvert structure to deformation of the permafrost foundation.

No matter which design principle is adopted, the destructive effects of thaw settlement and frost heave of permafrost foundation on culvert must be taken into consideration and corresponding engineering measures must be taken, such as the prevention and control of thaw settlement and normal frost-heave forces under the culvert baseplate, tangential and horizontal frost-heave forces on the base side and culvert side. The commonly used methods include: appropriately increasing the buried depth of the foundation, adapting to types of thaw settlement of permafrost foundation, replacement with coarse-grained soil, setting up thermal insulation layer, coating the wall of the culvert with residual oil or chemical anti-freezing coat, using waterproofing and drainage measures and anti-seepage measures, and the applying of the composite measures of thermosyphons and thermal insulation layer, etc.

4.2.2.2 Selection of Culvert Foundation Types

The main types of culvert foundation in permafrost regions are: the open-cut foundation and the pile foundation.

The commonly-used open-cut foundation types include: the cast-in-place reinforced concrete integral plate foundation (Fig. 4.66), the strip or segmental concrete foundation, and the precast assembled foundation. In the area of Da xing'anling Mountains, the composite foundation of wood raft (Fig. 4.67) and ligno-concrete had been used.

The advantages of the open-cut foundation are that the construction method is simple and flexible, and it is suitable for scattered culvert projects, especially suitable for the use in the permafrost area with low ice content and coarse grain. The disadvantages are that large-scale excavation is required, and the foundation pit is exposed to the atmosphere, which is easy to heat up and thaw the permafrost foundation, accumulate water in the foundation pit, and accelerate the warming and thawing of the permafrost foundation by the hydration heat of the cast-in-place concrete. The heat makes the ground temperature of permafrost foundation takes a longer time to recover, and it may be difficult for some areas to recover. Due to the large disturbance area of the open-cut foundation and the high moisture content of the backfill, the greater frost heave may occur during the refreezing process.

The short pile foundations (Fig. 4.68), the explosive-expanded pile foundations (Fig. 4.69), and the expanded-base pile foundations are often used in permafrost foundations with high ice content. In areas with large thickness and high ground temperature, the lower end of the pile should be placed within the soil layers with weak thaw settlement such as the icy or ice-poor permafrost, etc. The advantage is

Fig. 4.66 Cast-in-place reinforced concrete foundation of circular culvert (Photo by Li Yongqiang)

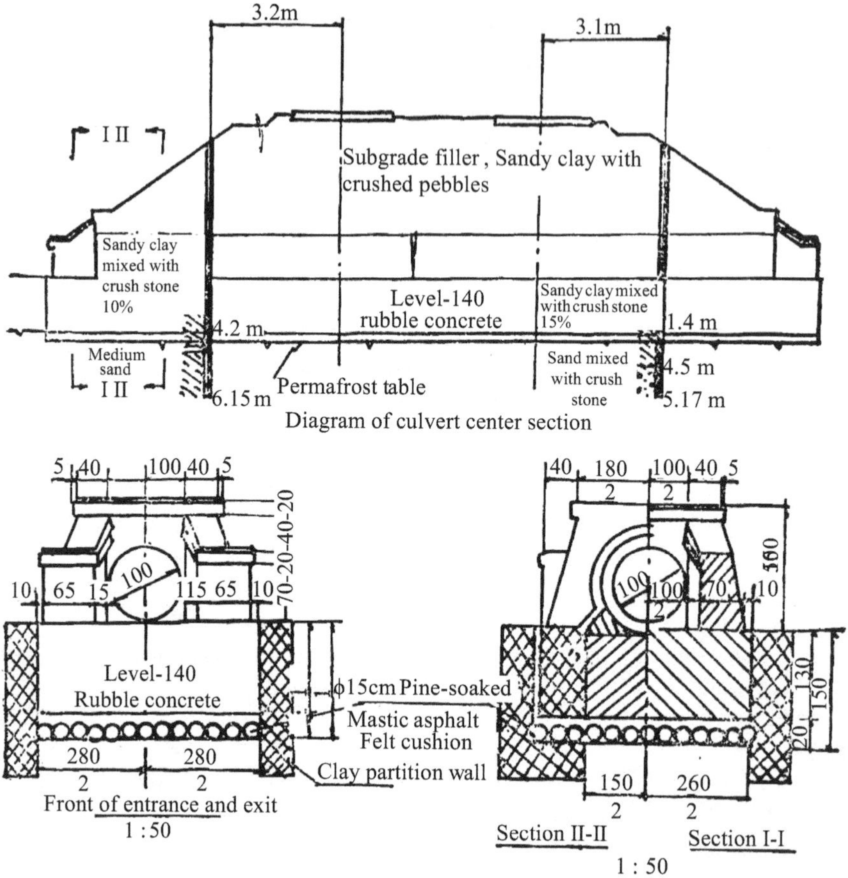

Fig. 4.67 Design of wood raft foundation in the permafrost area of Da Xing'anling Mountains (Unit: cm) (The Third Survey and Design Institute of the Ministry of Railways, 1994)

Fig. 4.68 Schematic diagram of short pile foundation (Unit: cm) (Wu, 2003)

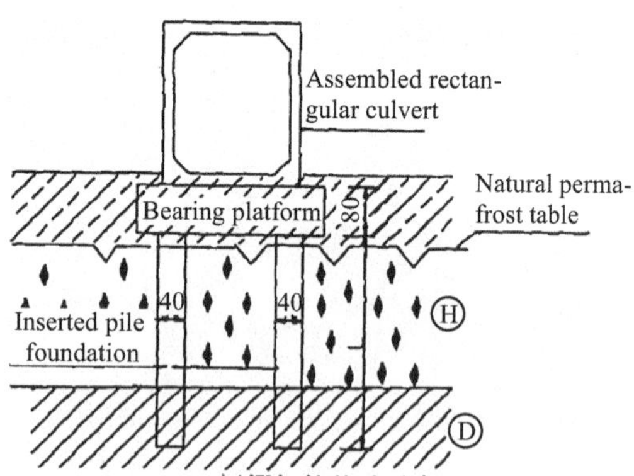

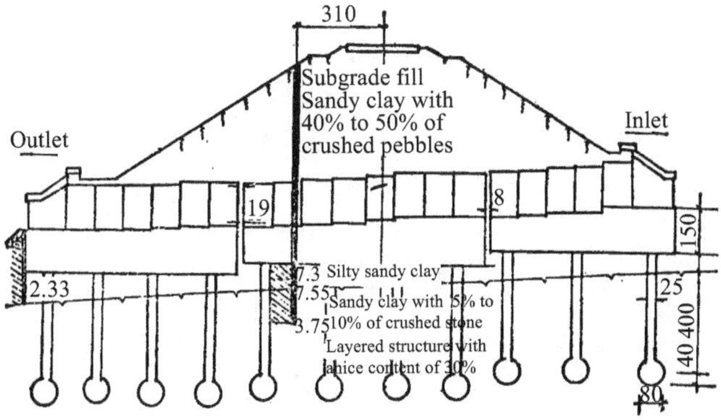

Fig. 4.69 Design of explosion-expanded pile foundation of culverts in permafrost area in Da Xing'anling Mountains (Unit: cm) (The Third Survey and Design Institute of the Ministry of Railways, 1994)

that the thermal disturbance to the permafrost foundation is small, the ground temperature is easy to recover in a short time, the lower end of the pile can be placed within a stratum with better foundation conditions, and the anti-frost heaving measures in the active layer are easier to take, which is beneficial to the stability of the culvert. In designing, the collaborative works of the pile foundation, pile cap, culvert body and the opening should be considered when the permafrost foundation varies, and the anti-seepage measures for the foundation and culvert joints should be taken into account to avoid the culvert deformation and destruction caused by the thaw settlement of the permafrost foundation.

Generally, in culvert projects the pile foundations are rarely used but the open-cut foundations are commonly adopted.

4.2.2.3 Artificial Permafrost Table under Culvert

The analysis of the causes for culvert engineering diseases shows that the frost heave of foundation soil and the thaw settlement of permafrost foundation are the fundamental causes for culvert engineering diseases. The factors that lead to frost heave and thaw settlement of culvert foundations are as follows: ① The buried depth of the foundation is insufficient; ② The anti-frost heaving measures are not in place, and some anti-frost heaving measures are not taken, especially under the foundation plate; ③ The thermal protection measures during foundation excavating and pouring are insufficient so that the temperature of the permafrost foundation rises and then thaws, and the differential settlement deformation increases; ④ During the construction and operation periods, the culverts take insufficient waterproofing and drainage measures, and the water accumulation and water leakage increase the moisture content of the foundation soil and the thawing of the foundation soil; ⑤Some culvert projects are not constructed according to the design, or the construction quality does not meet the design requirements; ⑥ The bearing capacity of original culvert could not meet the current vehicle-load or overload.

According to the investigation data of culvert usage, preventing and eliminating the normal frost-heave force within the foundation soil under the baseplate is an important link to avoid the damages to culvert engineering. And two aspects can be considered: one is reasonably to determine the buried depth of the foundation; the other is to strengthen the thermal protection and waterproofing and drainage measures.

After 1–3 freeze-thaw cycles after the completion of the culvert, the variations of the 0 °C isotherms of ground temperature in different parts of the culvert foundation are shown in Fig. 4.67. The artificial permafrost table under the culvert foundation has increased to different degrees, forming an arched curve with deep openings and shallow culvert in the middle part. According to the observation data of the Wudaoliang Culvert, the artificial permafrost table at the inlet of the culvert was about 0.5 m higher than the natural permafrost table, and the artificial permafrost table under the culvert was about 0.5 m higher than the natural counterpart. The permafrost table under the inlet was 1.0 m deeper than that in the middle of the culvert.

The artificial permafrost table under culvert foundation is related to the diameter of culvert, the permafrost engineering types, the thermophysical properties and the thickness of foundation material, and also the air temperature thawing index at the site.

(1) Theoretical calculations are basically modified based on the Stephen's equation. The former Soviet Union Rukyanov put forward the calculation formula for the permafrost thawing depth (h) of the culvert foundation:

$$h = \sqrt{\frac{2\lambda_1\theta_0}{Q_0 + \frac{1}{2}C_1\theta_0}(t - t_1) + S^2} - S \tag{4.12}$$

where S is the equivalent foundation thickness, its thermal conductivity is λ_2. According to the equivalent thermal resistance, the foundation is converted into the equivalent thickness of the foundation soil, namely $S = \delta\frac{\lambda_1}{\lambda_2}$. δ is the thickness of culvert foundation (including the wall thickness of the culvert pipe), m; t is the cumulative thawing time (h); t_1 is the time required for the temperature of the culvert foundation base to reach 0 °C, that is, $t_1 = 26.4\sqrt{\frac{1}{\alpha_2}}\delta$ (h), where α is the thermal diffusivity of the foundation materials (m^2/h).

Substitute the equivalent foundation thickness and the time when the temperature of the culvert base surface reaches 0 °C into Eq. (4.12), there is:

$$h = \sqrt{\frac{2\lambda_1\theta_0}{Q_0 + \frac{1}{2}C_1\theta_0}\left(t - 26.4\sqrt{\frac{1}{\alpha_1}}\delta\right) + \left(\delta\frac{\lambda_1}{\lambda_2}\right)^2} - \delta\frac{\lambda_1}{\lambda_2} \tag{4.13}$$

Q_0 is the latent heat of the thawing of frozen soil (kcal/m^3); $\lambda_1$is the thermal conductivity of the base permafrost (kcal/(m·h·°C)); C_1 is the volume thermal capacity of the base permafrost (kcal/m^3); θ_0 is the ground surface temperature where the waterflow goes through on the culvert bottom (°C).

For intermittent water flow, θ_0 is recommended to use the average positive temperature from the local weather station; for a culvert with perennial flow, it is recommended to use the average ground surface temperature from the weather station during the thawing period.

When h is 0, δ is the artificial upper limit depth of the frozen soil in the culvert, which is the minimum buried depth of the culvert foundation.

(2) Empirical formula. According to the observation data from the Fenghuo Mountain tests (China Railway Northwest Research Institute Co., Ltd., 2006), the artificial permafrost table in the culvert was related to the thawing index of local air temperature, the thermal conductivity of the permafrost foundation, the thermal conductivity of the culvert foundation, the pipe diameter, and the flow rate of water. The empirical formula for the artificial permafrost table under the base of the culvert was obtained:

$$h = K\lambda\sqrt{\Sigma I_t} \tag{4.14}$$

where h is the thawing depth of the culvert foundation, including the thickness of the culvert pipe wall, the thickness of foundation and the thawing depth of the permafrost foundation, m; λ is thermal conductivity, and the weighted average of the culvert pipe, the base and the permafrost foundation should be taken in the calculation (kcal/(m·h·°C)); I_t is the temperature thawing index, taking the 10-year average of the air temperature thawing index from the weather station near the culvert location (°C·d); K is the empirical coefficient related to the diameter of the culvert,which is calculated by inverse computation according to the measured values of culverts with different diameters (Table 4.13).

Table 4.13 *K* values of culverts with different diameters in northeast China and the Qinghai-Xizang Plateau

	Diameter/m		
K values of culverts with different diameters in the Daxing'anling Mountains (The Third Survey and Design Institute of the Ministry of Railways, 1994)			
Runoff conditions	1.0	1.5	2.0
Small waterflow (when it rains heavily, there is a waterflow, but generally there is not)	0.025	0.050	0..075
Seasonal intermittent waterflow with low water amount	0.030	0.060	0.090
Water flows all year round (except the freezing period), and the water accumulates in front of the culvert to form a pond	0.025	0.070	0.105
The *K* values of culverts on the Qinghai-Xizang plateau (China railway northwest research institute co., ltd., 2006)			
Diameter of culvert/m	0.75	1.0	2.0
K values	0.040	0.050	0.075

4.2.2.4 Buried Depth of Culvert Foundation

The buried depth of the culvert foundation should be determined according to the types of frozen soil engineering (the degree of ice content), the permafrost table, the adaptability of the culvert structure to the deformation of the foundation, and the conditions of waterflow. In principle, the culvert foundation should be buried 0.25–0.5 m below the artificial permafrost table in the stable permafrost area, so as to eliminate the normal frost-heave force under the foundation base. Only when the permafrost layer under the foundation base is coarse-grained soil with low ice content (non-thaw settlement or weakthaw settlement soil layer), the culvert foundation is allowed to be buried at the permafrost table.

According to the design principle for maintaining the frozen state of the permafrost under the foundation base, for the culvert with seasonal small waterflow, the permafrost table under the foundation is on the rise, but those under the inlet and outlet sections are still going downwardly. The buried depth of the foundation in the middle section of the culvert can be considered according to the distribution characteristics of frost heave along the freezing depth, that is, the buried depth is considered to be 2/3–3/4 of the natural permafrost table. The buried depth of the foundation in the inlet and outlet sections of the culvert should be 1.0 m below the natural permafrost table, and the buried depth in the transition section between the culvert body and the openings should be 0.4–0.8 m deeper than that in the middle section. However, in culverts with perennial waterflow, the buried depth of the foundation should be deepened to 0.25 m below the natural permafrost table.

The circular culvert with a cast-in-place concrete foundation was used in the Qingshui River along the Qinghai-Xizang Railway, and the construction began in winter (in late October), and its influence depth reached 1.2–1.5 m below the foundation base. The Beilu River and Tuotuo River adopted the precast assembled rectangular culverts. Because the thermal effect of the exposure time of the foundation pit was not significantly different from that of the cast-in-place foundation, the influence depth of foundation construction on the permafrost under the foundation base was1.0–1.3 m. The Tuotuo River culvert foundation was constructed in September during the transition period from the cold to the warm season, and the influence depth on the permafrost under the foundation base was about 1.7 m (Table 4.14). According to the natural permafrost table at the culvert site, the buried depth of culvert foundation could be determined, which was divided into three sections: the section of inlet and outlet, the transition section and the culvert body section. The test proved that it was reasonable (Table 4.15).

When employing the design principle for allowable thawing state of permafrost foundation, the permafrost foundation belongs to detachment of frozen ground, and the buried depth of the foundation should be 0.25–0.5 m below the maximum seasonal freezing depth. For the permafrost with coarse particles and low ice content, according to the stability requirements for culvert engineering, the buried depth of culvert foundation can be determined through thermal calculation, or by the value of 2/3–3/4 of the calculated artificial permafrost table.

Table 4.14 Summary of ground temperature tests of permafrost foundations of culverts (Editorial Committee of *Qinghai-Tibet(Xizang) Railway*, 2016)

Items	Qingshuihe River		Beiluhe River	Tuotuo River
Location of culverts	Circular culverts	Rectangular culverts	Rectangular culverts	Rectangular culverts
Ground temperatures of benchmark borehole before construction/°C	−0.87	−1.64	−0.28	−0.37
Construction season	In October the cold season	In October the cold season	In October the cold season	In September the transition period from the warm to the cold season
Ground temperature zones (*Tcp*)	Tcp-II	Tcp-II	Tcp-I	*Tcp*-I (isolated permafrost)
Types of foundation	Cast-in-place	Cast-in-place	Precast	Precast
Influence depth of foundation construction/m	3.5	3.6	3.5	3.3
Natural permafrost table/m	2.0	2.6	2.2	Permafrost table of 2.1 m, and the permafrost base of 10.5 m
Influence depth of construction below the natural permafrost table/m	1.5	1.2	1.3	1.7
Refreezing time/d	45	50	50	70
Artificial permafrost table in 2002/m	Outlet of 1.9 m, the middle of 1.2 m and inlet of 1.3 m	Outlet of 1.9 m, the middle of 1.2 m and inlet of 1.3 m	Outlet of 3.5 m, the middle of 3.1 m and inlet of 3.4 m	
Artificial permafrost table in 2003/m	Outlet of 1.8 m, the middle of 0.8 m and inlet of 1.1 m	Outlet of 2.1 m, the middle of 1.2 m and inlet of 1.5 m	Outlet of 2.5 m, the middle of 2.3 m and inlet of 2.4 m	

4.2.2.5 Selection of Culvert Structure Types

The selection of culvert structure types in permafrost regions should be based on the comprehensive consideration of the permafrost characteristics (ground temperature, ice content, lithology, etc.), the permafrost table, the environment of permafrost (topography, hydrology and hydrogeology, etc.), the design principles, the construction conditions (construction methods, construction seasons, etc.) of the culvert site. And the structures, which have strong ability to adapt to differential deformation of permafrost foundation and have small thermal disturbance to permafrost

Table 4.15 The Permafrost table and the Buried Depth of the Foundation of the Test Culverts along the Qinghai-Xizang Railway

Test culverts		Circular culvert at Qingshui River	Rectangular culvert at the Qingshui River	Rectangular culvert at Beilu River
Artificial permafrost table/m	Inlet	1.1	1.5	2.4
	Culvert body (in the middle)	0.8	1.2	2.3
	Outlet	1.8	2.1	2.5
Buried depth of the foundation/m	Inlet and outlet	2.2	3.1	2.7
	Culvert body	1.5	2.0	1.5
Natural permafrost table		2.0	2.6	2.2

environment and can meet the requirements for rapid construction, can be selected as the main option. The culvert structure is required to have strong rigidity, integrity and stability, suitable for rapid construction. The investigation and test data show that the culvert structure types commonly used in permafrost regions are the reinforced concrete slab culvert, the reinforced concrete framed slab culvert, the reinforced concrete rectangular culvert, the reinforced concrete circular culvert, the corrugated metal pipe culvert, and etc.

1. Reinforced concrete slab culvert: it is composed of two parts: the reinforced concrete slab and the culvert abutment (pier). It is a simple structure with even stress distribution, good stability and adaptability to the deformation. The integral foundation is adopted, and the culvert is longitudinally and reasonably segmented, which is beneficial to improve the ability to adapt to the heterogeneous permafrost.
2. Reinforced concrete framed slab culvert: it is a new type of structure, with the reinforced concrete slab culvert as its basic structure, and it is formed by adding a reinforced concrete rigid steel frame to the culvert body. It has a good integrity, a strong resistance to deformation, and can appropriately reduce the buried depth of the foundation.
3. Reinforced concrete rectangular culvert: it is a closed framework structure of reinforced concrete. The cross section of the culvert has high rigidity, good integrity and strong resistance to deformation. It can be prefabricated, assembled with culvert sections, and constructed with machines (Fig. 4.70).
4. Reinforced concrete circular culvert: with a high cross-sectional rigidity and a good integrity, it can be prefabricated and assembled (see Fig. 4.66).
5. Corrugated metal pipe culvert: with a high strength, a large cross-sectional rigidity and a good stability, it is made in a whole piece and jointless. When the foundation is not properly treated, the longitudinal deformation of the corrugated metal pipe culvert will also produce differential deformation, which will affect the normal use of the culvert.

Fig. 4.70 Assembling of prefabricated reinforced concrete rectangular culverts (Photo by Northwest Research Institute of China Railway)

The precast reinforced concrete rectangular and circular culverts can be constructed by machines, and their foundations are the types of the open-cut and the cast-in-place integral foundations. The foundation treatment needs to be handled properly to avoid excessive differential deformation which affects the normal use of culverts.

The test conclusion of the Qinghai-Xizang Railway culvert shows that the problems of circular culverts are obviously more serious than that of rectangular culverts because the resistance to differential deformation of circular culvert structures is not as good as that of rectangular culverts. When the permafrost is high in ground temperature, ice content and subgrade fill, the culvert will leak when it has small and differential settlements and the waterproofing of culvert sections is staggered. According to the record two culverts with subgrade fills exceeding 4 m had subsided significantly and it is greater than that of the rectangular culvert after 1 year, indicating that the circular culverts should not be used in the case of poor engineering geological conditions of permafrost. And at the same time, the rectangular culvert should be used with caution.

Although the structure of the tunnel openings is an ancillary part, it has an important impact on the regular use of the culvert. At present, there are several forms for the construction of permafrost culvert openings: the straight-lined wall type (or end wall type), the wing wall type, and the stone gabion openings.

1. The straight-lined wall type (or end wall type): due to the reduction of the horizontal and normal frost-heave forces on the subgrade fill, the measures to eliminate the tangential frost-heave force on the lateral surface of the wall are relatively simple. The straight-lined wall is a low wall perpendicular to the axis of the culvert, supporting the fill of side slope of the subgrade, and at two sides of the opening in front of the wall. They are mortar pitching (dry-laid) conical

revetment, with simple structures. The end wall and the culvert are basically built as one piece, whose overall strength is strong. But the conical revetment on both sides of the opening in front of the wall has poor resistance capacity to deformation. If the precast concrete sunken fascine mattress, non-woven bag cage or grass protection is adopted, the resistance capacity to deformation may be better.

2. The wingwall: the sprayed wing walls on both sides of the end wall of the opening are rigid structures with strong resistance to deformation. It is easy to be overturned, fractured and heaved by the horizontal frost-heave force of the embankment side slope behind the wall and the normal frost-heave force at the bottom of the wall, so it requires a deeper buried depth of foundation and high crack resistance of the wall.
3. Stone Gabion style: the net cage is made of round steel bars with the diameter of 8 mm and filled with stones about 0.3 m. The bottom of the gabion should be filled with mortar to fill the pores between the block stones, and the composite geotextiles should be laid on the gabion.

4.2.3 Application of Corrugated Metal Pipes Culverts in Permafrost Regions

The application of corrugated metal pipe culverts in road engineering has a history of nearly 100 years and the method had also been applied in China. In the 1950s, it was used for rush-repair work when the Qinghai-Xizang Highway was built. In the 1960s, the corrugated metal pipe culvert was used in the construction of the Baoji-Lanzhou Railway, and it was also used in other parts of China. In the late 1990s, corrugated metal pipe culvert was used in permafrost regions and there were three corrugated metal pipe culverts along the Qinghai-Xizang Highway, there was no unusual phenomenon, and the requirements for transverse drainage of the subgrade were met. For the two tests corrugated pipe culverts along the Qinghai-Xizang Railway, under the train load (with an even load of 57.5 kPa), the maximum disturbance of the wave crest of the pipe top was 5.5 mm–5.8 mm, and there was no differential settlement except for peeling of the zinc-rich coating at the inlet and the outlet. At the beginning of the twenty-first century, there had been accumulated 418 corrugated pipe culverts in the four projects of National Highway 214, reaching 8231.2 m, with a pipe diameter of 1.5–2.5 m, a wall thickness of 3–5 mm, and a fill height of 1.0 m–17 m at the top of the tunnel. There were more than 100 corrugated pipe culverts built along the Jiagedaqi-Mohe Highway in the Daxing'anling Mountains of Northeast China, with overall or partial settlement, and inclination and collapse of the wing wall at the openings. The diameter of the corrugated pipe culvert used in the Qinghai-Xizang Plateau was 1.0–2.5 m, the wall thickness was 3–5 mm, the wave height was 70 mm, and the pitch of wave was about 140 mm. After investigation, except for a very small number of differential settlement

Fig. 4.71 Corrosion of corrugated pipe culvert (Left), revetment damages (Middle), and waist downward bowing (Right) on National Highway 214 (Cui, 2013)

deformation due to improper construction, foundation treatment and waterproof treatment, there was no such unusual phenomenon.

According to the post-earthquake survey in Yushu, Qinghai Province, the main diseases occurred to the corrugated pipe culverts in National Highway 214 were: the rusting corrosion, the destruction of slope protection and the waist downward bowing (Fig. 4.71), and the number of the corrugated pipe culvert damage accounted for about 65.4%, 44.7% and 36.2% of the total number of corrugated pipe culverts surveyed (Cui, 2013). The Corrosion damage was due to the abrasion of culverts caused by a large amount of sediment in the flowing water, and the action of rusting of saline water in salinized soil. The destruction of slope protection was caused by the impact of repeated freeze-thaw, making the slope uneven or even fall off. The high subgrade fill and excessive load caused waist downward bowing.

The biggest advantage of corrugated metal pipe culverts is that it has greater flexibility and relatively high rigidity, which can adapt to the differential deformation caused by the repeated freeze-thaw process of the active layer in permafrost regions. It does not need foundation excavating and rigid foundation pouring, but it meets the requirements for rapid construction, and has little thermal disturbance to permafrost. The corrugated pipe culverts can reduce culvert engineering diseases, improve the service life of culvert engineering in permafrost regions and improve subgrade drainage conditions.

4.2.3.1 Thermal Characteristics of Corrugated Pipe Culverts

The heat exchange in the culvert is mainly by means of air convection. Because the air circulation speed is fast and the heat exchange with the foundation soil is also fast, it makes the permafrost table under the culvert rise, forming an "arch" shape. But compared with the reinforced concrete culverts, the top of "arch" is relatively flat. The permafrost table under the inlet and outlet decreases little. The corrugated pipe culvert has small thermal resistance (Table 4.16) and better heat dissipation performance, which is beneficial to the rising of the permafrost table at the bottom of the culvert. The wall of corrugated pipe culvert is in the corrugated form, which can increase the contact area between the culvert and the air, and also increase the

Table 4.16 Thermal properties of different culvert materials

Materials	Density/ kg·cm^3	Thermal conductivity/J·(m·S·°C)$^{-1}$	Specific heat/ kJ·(kg·°C)$^{-1}$
Low carbon steel	7833	54	0.465
Reinforced concrete	2400	1.54	0.84

frictional resistance of the surface of the pipe wall so that the air in the culvert functions as the turbulent flow and enhances the heat exchange capacity.

4.2.3.2 Mechanical Properties of Corrugated Pipe

Vertical deformation capacity: under the force in a same direction (1000 N), the corrugated pipes have more obvious effects of axial displacement compensation, better axial expansion deformation capacity and stronger adaptability (Table 4.17).

Deformation resistance: under the same vertical force (1000 N), the "flattening" of the corrugated pipes (the quotient of the absolute value of the maximum deformation along the radial direction and the radius) is only 1/845 of that of the circular pipe. Because the corrugation of the corrugated pipe effectively enhances its vertical deformation resistance, it is far greater than the deformation resistance of the circular pipe.

Comparison of Von Mises stress: when the confining pressure is the same, the displacement in the cross-section (radial) direction of the corrugated pipe is 0.01922 mm, and that of the ordinary circular pipe is 0.0178 mm, so basically there is no significant difference. However, if the Von Mises stress is compared as a strength condition, the maximum Von Mises stress of the corrugated pipe will be concentrated at the axial end of the pipe (the maximum stress is 2768.22 kN/m^3, and the minimum is 1420.22 kN/m^3) (Institute of Transportation Science of Heilongjiang, 2006), equivalent to the circular pipe. It also reflects that the ordinary circular pipe is uniformly stressed, and the combination of principal stress at each point is equivalent.

The permafrost table under the culvert in permafrost regions presents an "arch" shape. Under the impact of repeated freeze-thaw, the deformation at both ends of the culvert is larger than that of the culvert body, and a certain tensile and compressive stress is generated in the culvert body. The use of corrugated steel pipe culverts in permafrost regions can exert its excellent deformation resistance capacity and the characteristics of steel tensile and fatigue resistance, reducing or eliminating the freeze-thaw diseases of culverts.

4.2.3.3 Application Conditions of Corrugated Pipe Culvert

From the domestic use experience of the corrugated steel pipe culverts, the stress characteristics, and the anti-corrosion conditions, the use of corrugated pipe culverts should meet the following requirements:

Table 4.17 Mechanical properties of corrugated pipe and circular pipe under 1000 N radial force (Institute of Transportation Science of Heilongjiang, 2006)

Categories	Maximum radial displacement/mm	Maximum axial displacement/mm	Maximum equivalent stress/MPa	Maximum equivalent strain/mm	Maximum stress Intensity/MPa
Corrugated pipes	56.3	57.3	1.790	3.03	1.926
Circular pipes	2.39	0.12	0.5416	3.975	5.548
Corrugated pipes/circular pipes	23.56	477.5	3.30	0.76	0.35

1. The wave form and wall thickness should be suitable for the fill height and live load, to ensure that the stress of the corrugated steel pipe under the maximum load is lower than the allowable stress (yield strength) of the steel structure.
2. The anti-corrosion treatment for the corrugated steel pipe should ensure that it has sufficient corrosion resistance within a certain period of time, and there is no obvious corrosion phenomenon in hot-dip galvanizing. When determining the wall thickness of the steel plate, the surplus amount of thickness for corrosion prevention should be properly considered;
3. It should be able to withstand the deformation caused by the standard load, without causing nonlinear failure, and without affecting the function of the upper structural layer;
4. It is suitable for special areas such as permafrost regions and the areas where sand and gravel materials are lacking, and also the project that requires accelerated construction progress.

4.2.3.4 Designs of Corrugated Steel Pipe Culvert

The Structure design of corrugated steel pipe culverts refers to determine the parameters of the pipe diameter and parameters of corrugation (wavelength, wave height and wall thickness).

The basic design contents include: the culvert hydraulic calculation and the diameter design, the design of culvert foundation, the buried depth and oblique angle design, the culvert structure calculation, the culvert corrosion prevention design, the culvert detailed structure design and the pipe strength requirements, the design of culvert inlet and outlet, and the combined design of culvert body and openings.

Generally, compared with the round pipe culvert, the influence of wave form on characteristics of the water flow in the culvert should be considered in the hydraulic calculation of the corrugated pipe culvert. The corrugated steel pipe culvert has relatively low requirements for the foundation, and the use of replacement foundation can meet the design requirements.

1. **Pipe Diameter Design**

The size of the pipe diameter is determined according to the fill height of the roadbed where the corrugated steel pipe culvert is installed, and it is also necessary to comprehensively consider the local hydrological conditions, the precipitation, the topographic conditions and the construction techniques. The design of the diameter of the corrugated pipe culverts should meet the following requirements (Ming, 2011):

1. It is necessary to ensure the safe passing of all levels of floods and water flow within the design of flood level, and consider the impact of backwater and scouring on the upstream to ensure the stability of the embankment near the bridge and culvert.
2. The conditions of the upstream and downstream bridges and culverts and their influences on the evolution of the riverbed should be considered.
3. Attention should be paid to the topography of the riverbed, and it is not advisable to over-compress the river channel and change the natural state of water flow.
4. It should be determined according to the conditions of the design flood discharge, the geology of riverbed and the reinforcement form of riverbed and conical slope, etc. When the upstream conditions permit water accumulation, the flow can be considered to be reduced according to the rainstorm runoff, but the reduced flow should not be greater than 1/4 of the total amount.
5. It should be designed as a pressureless type, and the clear height from the roof of the culvert to the standard water level of the design flood frequency in the culvert meets the regulations.

According to the using experience of the corrugated steel pipe culverts used in the highways and the railways in permafrost regions such as the Qinghai-Xizang Plateau, the National Highway 214, and Daxing'anling Mountains in Northeast China, the diameter of the culvert is 1.0–2.0 m, and the tubes are connected by flanges.

2. **Determination of Wall Thickness of Corrugated Pipe Culverts**

According to the research report on the corrugated pipe culverts, the standard yield strength of the steels used is 245 MPa, but the actual thicknesses of the steel are different. According to the comprehensive consideration of the foreign experience, the safety factor is taken as 1.5, and the allowable stress is above or equal to 164 MPa, that is, $[\sigma] \geq 164$ MPa.

According to calculations, when the corrugated pipe culvert with the pipe diameter of 1.5 m, the wall thickness is 3 mm, and the fill height on the pipe top is less than 5 m (including 5 m), it is the maximum equivalent stress of the fill load plus the car of Super-20 or the Trailer-120, which does not exceed 164 MPa, and the maximum value is 118.71 MPa.When the fill height is 10 m, the maximum equivalent stress of wall thickness 3 mm and 4 mm thick will be more than 164 MPa. And when the wall thickness is 5 mm, the maximum equivalent stress under the application of Trailer-120 will be 176.58 MPa. And it will be 151.33 MPa when the wall thickness is 6 mm. The determination of the wall thickness of the corrugated pipe depends on the overburden (dead load plus live load).

Appropriate consideration of the surplus of the thickness of the wall should be given to the pipe for the case of the wall corrosion.

3. **Minimum Fill Thickness on the Top of Corrugated Pipe**

The thickness at the center of the fill on the top of the corrugated pipe culvert is related to the pipe diameter and the level of the live load. The larger the pipe diameter, the heavier the live load and the thicker the central fill. Based on the relationship between the structural characteristics of the corrugated pipe culvert and its function and its stress characteristics, the top of the pipe should be filled with soil, especially during the construction process.The thickness of the first layer of the fill soil above the pipe top should be controlled properly. If the thickness of the fill is too small, it will affect the stress and deformation of the corrugated pipe culvert, and even cause damages. According to the survey data, the minimum fill thickness above the top of the pipe should not be less than 1/10–1/6 of the pipe diameter (Xiong, 2012), and the upper limit can be applicable to the entirely flexible culverts, generally not less than 0.6 m. The following values can be referenced:

According to *Canadian Highway Bridge and Culvert Design Specification*, the maximum value should be taken.

① 0.6 m;

② $$\frac{D_h}{6}\left(\frac{D_h}{D_V}\right)m;$$

③ $$0.4\left(\frac{D_h}{D_V}\right)m$$

where D_h and D_V are the horizontal span and vertical height of the culvert.

4. **Dead Load**

The dead load of the corrugated steel pipe culvert includes the weight of culvert structure and the vertical earth pressure on the top of the culvert. The self-weight of culvert structure is the product of weight density of the culvert structure material and the wall thickness, namely $\rho' t$(ρ' is the weight density of the corrugated pipe material; $t$is the wall thickness of the corrugated pipe structure).

The vertical earth pressure on the top of culvert: ρH (ρ is the density of soil; H is the fill height on the top of the corrugated pipe).

5. **Live Load**

It is suggested that the design and calculation should be carried out according to *General Code for Highway Bridge and Culvert Design* and *Basic Code for Railway Bridge and Culvert Design* (TB10002.1/J460-2005). The total design load is that the dead load plus the live load.

6. **Layout and Treatment for Corrugated Pipe Culvert Openings**

(1) Orthogonal Opening Types and Adaptability

The commonly-used types of culvert openings are the wing wall type, the end wall type, the conical slope type, the corridor type, the flat head type, the straight pipe extension type and the dustpan type, etc. (Fig. 4.72). The adaptability, advantages and disadvantages of various opening types are shown in Table 4.18.

Fig. 4.72 Types of corrugated pipe culvert openings: wing wall type (the Upper Picture) and dustpan type (the Lower Picture)

(2) Layout and Treatment for Oblique Openings

When the culvert is obliquely intersected with the line, the building at the opening is basically the same as the orthogonal structure. According to the different structure of the culvert barrel, two treatment processes can be adopted (Fu Jin etc. 2008):

① Oblique intersection and oblique construction: it strives to be beautiful in appearance and adapts to the water flow conditions so that the end of the culvert barrel is parallel to the line, that is, the oblique construction (Fig. 4.73 a and b).

② Oblique intersection and orthogonal construction: the culvert body is the same as the orthogonal one, and the height of the end wall at the opening is adjusted. Generally, the end wall is designed as a sloping form or a sloping staircase type (Fig. 4.73 c and d).

Table 4.18 Comparison of adaptability and advantages and disadvantages of various culvert construction types (Li, 2006)

Forms of tunnel opening	Applicability	Advantages and disadvantages	Notes
The wing type	The stream is flat and straight, with small height difference in longitudinal section. With settings on the embankment slope, it is widely used in the section that is capable to store water flow or diffuse the holding water	Good hydraulic performance, simple construction and less engineering works	
End wall type	Rivers and canals with low flow velocity and small runoff in plain areas	Simple structure, low cost, but poor hydraulic performance	
Conical slope type	Wide and shallow streams, culverts that can relatively compress the water flow. Often used with taller and larger culverts	The hydraulic performance is good, and the stability of the openings and the tunnel trunk of the high embankment can be enhanced. But the engineering works are much	
Flat head type	The lateral extrusion of water flowing through the culvert is not large, and the flow velocity is small	This type saves material, but the process is more complicated and the hydraulic performance is slightly bad. In addition, the tube couplings at the opening can be prefabricated	
Straight pipe or straight pipe extension type	This type of culvert is suitable for streams and water channels with high flow velocity. The culvert with a large diameter is suitable for the wide stream, with a straight pipe that does not need to be extended. The culvert with a small diameter is suitable for the narrow stream, which needs to be appropriately extended	The structure is simple, the prefabrication is convenient, and it is convenient to extend the culvert, but the drainage capacity of the opening is relatively poor when the diameter of culvert is small, and the hydraulic performance is not good	For arid and semi-arid areas, this type of culvert is mostly used with short-term flowing water and the culvert diameter is smaller than that of streams and river channels
Dustpan type	This type of culvert is suitable for the wide and shallow streams with large flow velocity	Relatively complex structure, and poor hydraulic performance	Corrugated or flat

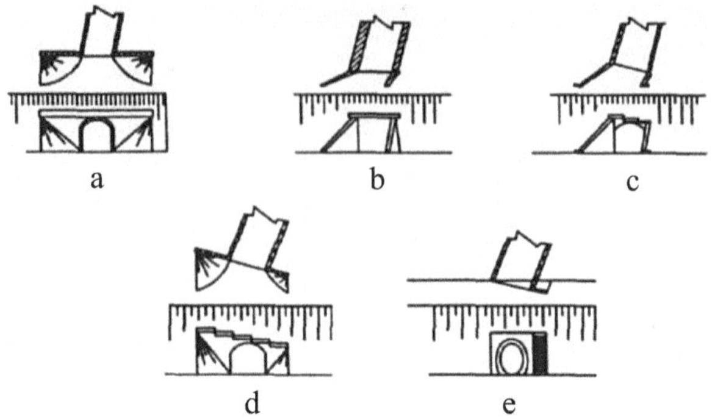

Fig. 4.73 Layout for oblique culverts

When the oblique culvert is at a flat-head opening, it protrudes from the triangular platform outside the subgrade, and is reinforced by paving the expanded slope (Fig. 4.73e).

When the method oblique intersection and oblique construction is used, and when the oblique angle is less than or equal to 20°, the end-section culvert pipe can be gas-cut into an inclined plane parallel to the center line of the highway, and the oblique cutting slope should not exceed 2:1. However, when the oblique angle is more than or equal to 20°, it is advisable to adopt orthogonal construction for the end-section culvert and extend out of the embankment slope. The minimum width of the oblique embankment should be 1.5 times the sum of the rise of the culvert and the thickness of the overlaying layer (Fig. 4.74).

The Culvert openings are usually the wing type or straight-lined type. Because the corrugated pipe culvert has a strong ability to adapt to deformation, the corrugated tube couplings can be directly extended outside the embankment slope, and the advantages of the overall tensile strength, shear resistance and adaptivity to deformation of the corrugated pipe can be exerted. But the anti-seepage and scour protection measures should be taken at the entrances of the culvert (Fig. 4.75).

(3) Transverse Spacing of Multi-Barrel Corrugated Steel Pipe Culverts

When the culvert diameter is less than or equal to 2.0 m, the clear spacing between the two pipes can be determined by referring to Table 4.19.

7. **Base and Foundation of Corrugated Steel Pipe Culvert**

Although the corrugated steel pipe culvert has strong deformation adaptability, it still needs foundation to keep the stability of the culvert. Generally, the sand and gravel materials are used, the thickness is 0.5–0.55 times the pipe diameter of the culvert, the minimum thickness is not less than 0.6 m, and the maximum thickness is not more than 1.5 m. In order to prevent frost heave, the soil content in the gravel should not be more than 5%. For avoiding the damage to the culvert, the maximum particle size of the gravel should not exceed 50 mm, and the degree of compaction should be 85–90%. Both sides of the pipe should be backfilled with coarse-grained soil with a soil content less than 5%, and the degree of compaction should be the same as that of the subgrade fill at the same depth.

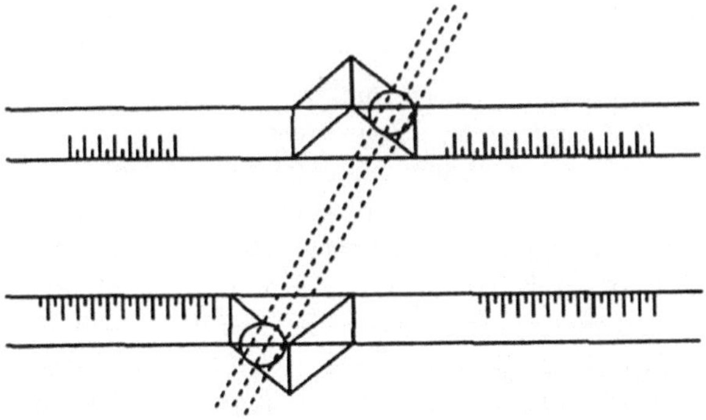

Fig. 4.74 Schematic diagram of oblique intersection and orthogonal construction of corrugated pipe culvert openings

Fig. 4.75 Treatment for corrugated pipe culvert openings

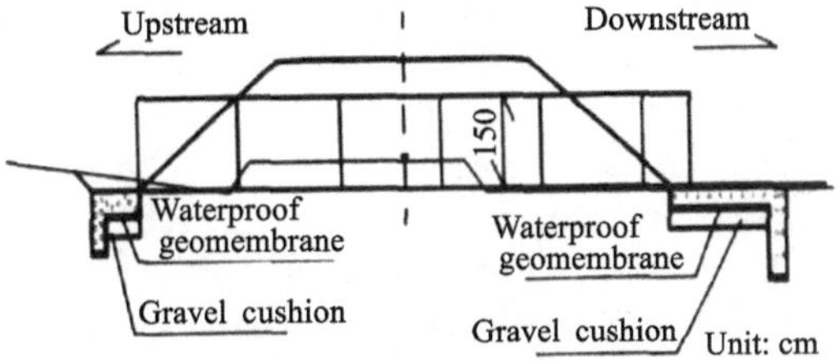

Table 4.19 Spacing between two corrugated steel pipe culverts

Diameter of corrugated steel pipe culvert D/m	Clear spacing between two pipes e/m
$D \leq 0.6$	0.3
$0.6 < D \leq 2.0$	0.5*D*, No larger than 1.0
$D > 2.0$	1.0

The pipe base should be constructed with the same radian as the pipe culvert to make the pipe body and the pipe base closely fit. The materials used for pipe base must be homogeneous, without rock blocks or other sundries, and durable.

4.2.3.5 Requirements for Anticorrosion of Corrugated Steel Pipe Components

The Common corrosion circumstances of the metal pipes include the atmospheric corrosion, the soil corrosion, the water corrosion and the chemical medium corrosion. The outer side of the corrugated steel pipe culvert is mainly the soil corrosion, and the inner side is mainly the water corrosion and the atmospheric corrosion.

The main factors for soil corrosion are the salt content of soil, the water content, and the soil resistivity. Generally, the greater the salt content and water content of the soil, the smaller the soil resistivity and the stronger the soil corrosion, and the better the correlation between the minimum resistivity and the maximum corrosion rate. The factors, such as the organic matter content, the chloride ion content, the average water content, the sulfate ion content, the total salt content and the pH value, etc., have weak correlation with the maximum corrosion rate. Under a regular water conditions, most of the soils in China belong to the categories of the strong corrosion or the moderate corrosion. The water content of the fluvo-aquic soil with the maximum corrosion rate is around 30%.

The main factors for atmospheric corrosion are the critical humidity of steel and the harmful substances in the atmosphere. The average corrosion rates of carbon steel, zinc, and aluminum in atmosphere are 4 μm/a to 65 μm/a, 2 μm/a to 3 μm/a, and 0 μm/a to 0.1 μm/a respectively.

The factors for aqueous solution corrosion include the dissolved oxygen in fresh water, the water hardness, the salt content of water, the sulfur content, the temperature, the flow velocity and the pH value, etc.The fresh water corrosion of the low carbon steel and the galvanized steel in the electrochemical process is the main reason. And the increase of salt content in water will increase the capacity of conductivity and accelerate the corrosion rate.

The abrasion of corrugated steel pipe culverts: in most cases, the water flow with low velocity basically does not bring the effect of abrasion from the scouring sediment. When the flow velocity is greater than 5 m/s, the abrasion of sediment will become obvious, and the sand stones will scour on the inner wall of the culvert, and the amount of abrasion will increase.

The types of surface protective layer of steel components include the galvanizing, the aluminizing, and the galvanizing (or aluminizing) before plastic-coated. Both the galvanizing and aluminizing are hot-dip galvanizing, and the quality of galvanized and aluminized coatings for steel plates and steel pipes (with the wall thickness more than 3 mm) is 600 g/m^2 and 120 g/m^2 respectively. For fasteners and connectors, the quality of galvanized and aluminized coatings is 350 g/m^2 and 110 g/m^2, respectively. For steel plates and steel pipes, the second layer is coated with plastic after galvanizing (270 g/m^2) or aluminizing (61 g/m^2), the thickness of polyvinyl chloride and polyethylene is greater than 0.25 mm, and the thickness of polyester is greater than 0.076 mm. For fasteners and connectors, the second layer is coated with plastic after galvanizing (120 g/m^2) or aluminizing (61 g/m^2), the thickness of polyvinyl chloride and polyethylene is greater than 0.25 mm, and the thickness of polyester is greater than 0.076 mm. The thickness of coating for the single plastic-coated steel plates, steel pipes, fasteners, and connectors, etc. is required to be 0.38 mm.

4.3 Tunnel Engineering

In addition to the engineering problems in regular areas, the characteristics and technical problems of tunnel construction in alpine and permafrost regions are more complicated than those in non-frozen soil areas. Especially, the complexity arises from the impacts of the repeated freeze-thaw and freezing on the surrounding rocks caused by the variations of engineering geological and hydrogeological conditions of the frozen soil environment, the frost heave and frost-heave force on the surrounding rocks of the tunnel during the freeze-thaw process and the freezing of the tunnel, etc. These issues impact the anti-freezing capabilities, operational safety, and long-term durability of the main structures. Reducing the freeze-thaw process and thickness of tunnel surrounding rocks and maintaining the well drainage of groundwater in frozen soil can ensure the long-term stability of tunnel engineering in permafrost and cold regions.

4.3.1 Particularity of Tunnel Engineering in Permafrost Regions

In permafrost and cold regions more than 43 tunnel projects have been built or been under construction. With the economic development in cold regions, the tunnel engineering will continue to be carried out. Among the 43 tunnel projects, 33.3% of the tunnels are in permafrost regions, 45.2% in seasonally frozen soil area with high altitude and cold climate, and 21.4% in the deep seasonally frozen soil area. Due to the influence of cold climate and permafrost, there are many frost damages, such as: ice hanging on the top and side wall of the tunnel (Fig. 4.76a), ice accretion on the ballast bed (Fig. 4.76b), ice jam in the tunnel (Fig. 4.76c), cracks in the tunnel portal (Fig. 4.76d), lining cracking, crumbling and spalling of vault side wall (Fig. 4.76 e and f) and so on. The damage of lining is caused by the pressure from frost heave, the impact of repeated freeze-thaw and the water accumulation in the surrounding rock of the tunnel. Hanging ice, ice mantles, and ice jams primarily result from inadequate waterproofing and drainage, which permit water seepage that subsequently freezes. The tunnel portal cracking cause by frost heave mainly occurs in permafrost regions, and 56% of tunnels in Da xing'anling Mountain had such problem of tunnel portal cracking. The thaw settlement is mostly caused by the subsidence of permafrost layer in the surrounding rock.

The basic conditions for the occurrence of the tunnel frost damages in cold areas include the cold ambient temperature and the groundwater in surrounding rock. When the air temperature in the tunnel is lower than 0 °C, it will not only cause the lining wall to seep and freeze, forming hanging ice and ice mantle, resulting in the occurrence of ice jam, but also play a more important role in the freezing of the water-containing tunnel surrounding rocks, producing a large frost-heave force on

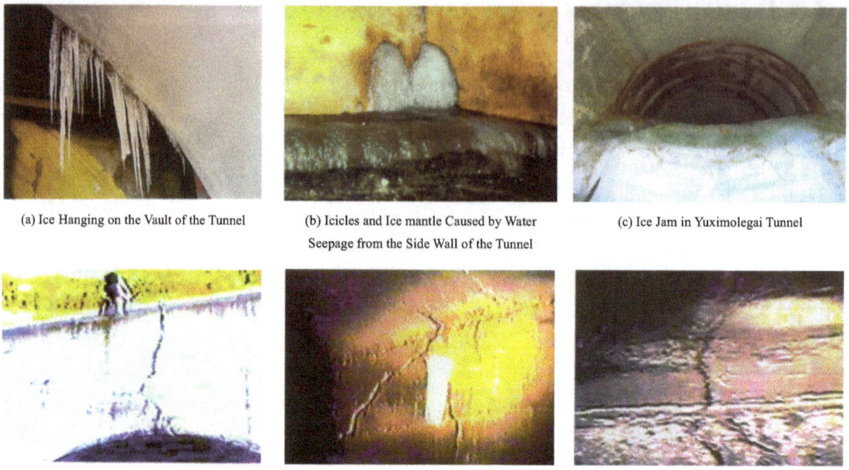

(a) Ice Hanging on the Vault of the Tunnel

(b) Icicles and Ice mantle Caused by Water Seepage from the Side Wall of the Tunnel

(c) Ice Jam in Yuximolegai Tunnel

(d) Cracking in the Tunnel Portal

(e) Lining Cracking in the Side Wall of the Tunnel

(f) Lining Cracking in the Vault of the Tunnel

Fig. 4.76 Frost damages of tunnels in permafrost regions (Wu et al., 2005a, b, c)

the main structures of the tunnel, resulting in cracks in the lining body. Under the impact of the repeated freeze-thaw, the lining body cracks and breaks.

The tunnel construction in permafrost and cold regions has changed the original mountain structure, increasing the heat exchange between the air and the ground in the mountain and changing its heat balance, which has caused many special engineering problems.

4.3.1.1 Formation of Freeze-Thaw Belt in Surrounding Rocks of the Tunnel

1. **Air Temperature Variation and Distribution Characteristics in the Tunnel**

Regardless of whether the tunnel passes through the mountain on permafrost layer, or the entrance and exit sections are in permafrost layers, after the construction of the tunnel in the cold region, the air temperature in different periods and different sections will change, which will change the original temperature state, link to the air temperature outside the tunnel, and vary periodically with the air temperature outside the tunnel.

Under the natural conditions, during tunnel excavation, the air temperature inside the tunnel and the surface temperature of the surrounding rock are higher than the air temperature outside the tunnel in the cold season, and the temperature inside the tunnel is lower than that outside the tunnel in the warm season (Fig. 4.77). After the tunnel is completed, the average annual air temperature inside the tunnel and the surface temperature of the surrounding rock are generally equal to or lower than the ground surface temperature and air temperature of the corresponding section.

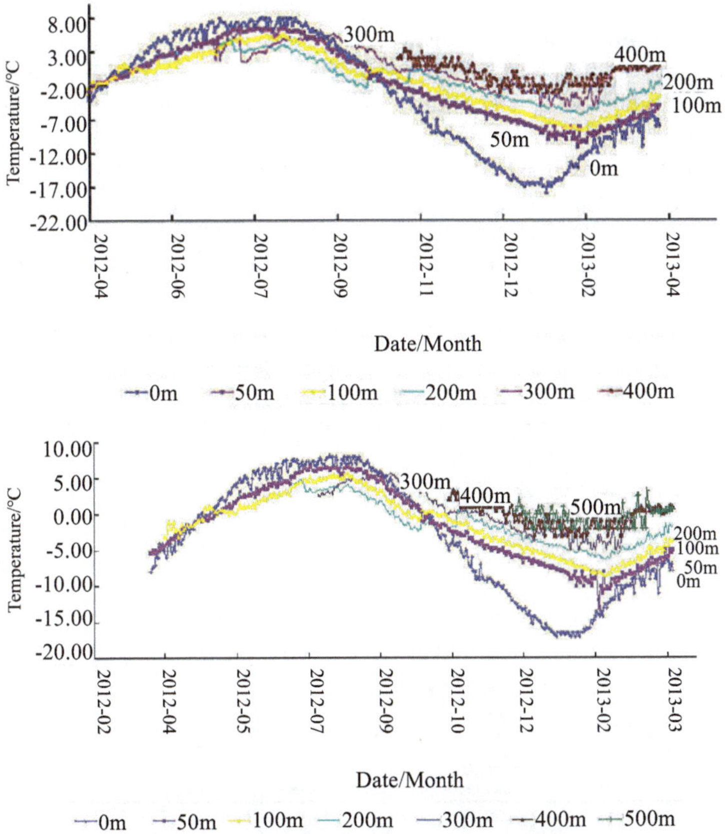

Fig. 4.77 Air temperature variation at the entrance (on the left) and exit (on the right) of Jiangluling Tunnel of National Highway 214 (Li, 2014)

Although the temperatures are different, they all change synchronously with the change of the four seasons of the atmosphere, indicating that the temperature inside the tunnel is lower than that outside the tunnel in the warm season and higher in the cold season, and it is so-called "warm in winter and cool in summer". The air temperature variation law at the tunnel entrance and exit are as follows: ① The closer the distance to the entrance of the tunnel, the closer the temperature inside the tunnel to the temperature outside the tunnel, and the greater the range of variation; ②The air temperature inside the tunnel changes periodically along with the air temperature outside the tunnel. Generally, the temperature inside the tunnel reaches the highest in August and the lowest in January; ③ The closer to the center of the tunnel, the smaller the amplitude of variation. At a distance of 500–600 m from the entrance of the tunnel, the temperature inside the tunnel varies very little along with the temperature outside the tunnel, almost close to the "stable state".

The monthly variation characteristics of the temperature inside KunlunshanTunnel and Fenghuo Mountain Tunnel in the permafrost region of the Qinghai-Xizang Plateau (Fig. 4.78) shows that during the construction period, the impact of the construction on the air temperature inside the tunnel was very obvious, which was significantly higher than the air temperature outside the tunnel. After completion, the temperature inside the tunnel showed obvious seasonal variations: the positive temperature period was from June to September (the hottest month was in August), and the negative temperature period was from October to May (the coldest month was in January). After opening to traffic, the air freeze-thaw index in the tunnel decreased by 30% as a whole.

The variation and distribution of the annual average air temperature in Kunlun Mountain Tunnel and Fenghuo Mountain Tunnel have several characteristics: after the completion of the construction, the air temperature in the entire tunnel had dropped significantly compared with that during the construction period, and this dropping of air temperature gradually slowed down. In the predominant wind direction, the air temperature decline inside Kunlun Mountain Tunnel was lower than that inside Fenghuo Mountain Tunnel, and it was also close to the temperature outside the tunnel (the average annual temperature outside the Fenghuo Mountain Tunnel was below −6 °C). Before the opening to traffic, the annual average air temperature in the tunnel presented a linear decrease from north to south, and after the opening to traffic, the tunnel temperature overall increased by about 1 °C.

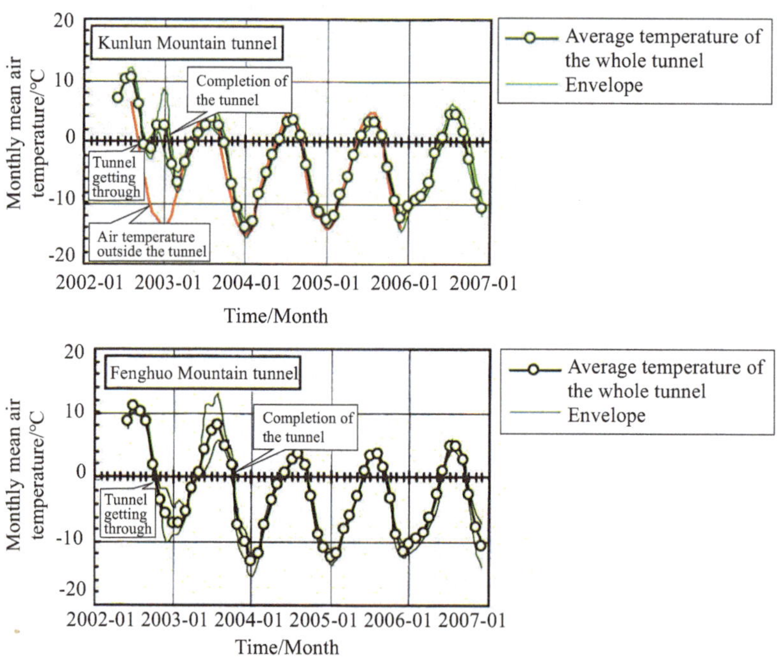

Fig. 4.78 Monthly variation of air temperature in Kunlun Mountain tunnel and Fenghuo Mountain tunnel (China Railway First Survey and Design Institute, 2007)

Based on the longitudinal distribution of the monthly air temperatures in the annual hottest and coldest months of Kunlun Mountain Tunnel and Fenghuo Mountain Tunnel, before the opening of the tunnel the annual highest and lowest air temperatures in the tunnels also had the characteristics of high air temperature in north and low temperature in south. The hottest monthly air temperature at the entrance of the north slope of the Kunlun Mountain Tunnel and Fenghuo Mountain Tunnel were both 3.9 °C, while at the exits of the south slope of the two tunnels were 3.0 °C and 3.5 °C respectively. According to the annual variation of air temperature in the tunnel, after the completion of construction, the maximum variation of annual air temperature was located at the entrance of the tunnel, especially at the south opening, which was more intense. After opening to traffic, the temperature variation at the south opening was over 5 °C higher than that at the north entrance (during the observation period). And the minimum temperature variation was located at 1/4–1/3 of the tunnel length near the north entrance.

When the tunnel passed through the sections of partial permafrost, the temperature inside the tunnel changed with the depth and basically showed an increasing trend (Fig. 4.79).The longer the tunnel, the longer the distance between the entrance

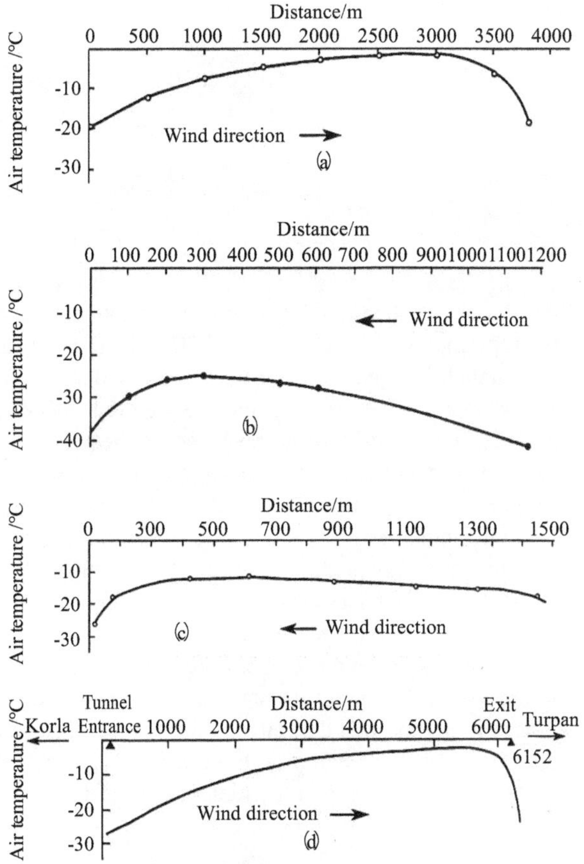

Fig. 4.79 Air temperature distribution along the distance in the tunnel (Wu et al., 2003a, b, c). (**a**) Ducao Tunnel in Xiao Xing'anling Mountains (in January, 1979); (**b**) No.2 Tunnel of Xiluoqi in Da Xing'anling Mountains (in January, 1979); (**c**) Daban Mountain Tunnel in Qinghai Province (in January, 2000); (**d**) Kuixiandaban Tunnel in Xinjiang, China (in January, 1979)

Table 4.20 Relationship between air temperature difference inside and outside the tunnel and the tunnel length in January (Wu et al., 2003a, b, c)

Tunnel name	Tunnel length/m	The Maximum daily temperature difference between at the tunnel entrance and inside the tunnel in January/°C
Kuixiandaban tunnel in Xinjiang, China	6150	24
Ducao tunnel in Xiaoxing'anling mountains	3849	18
No.2 tunnel of Xiluoqi in Daxing'anling mountains	1160	15
Dabanshan tunnel in Qinghai Province	1530	15
No.2 tunnel of Cuiling in Daxing'anling mountains	420	5

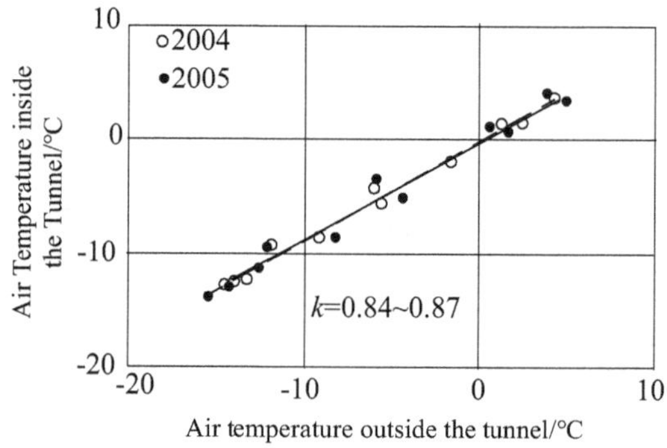

Fig. 4.80 Correlation of air temperature inside and outside the Kunlun Mountain Tunnel

and the center of the tunnel, and the greater the maximum temperature difference in the coldest month (January) (Table 4.20). The temperature difference at the opening of the tunnel was the largest, and gradually decreased with the increasing of the distance between the opening and the location inside the tunnel. The temperature difference in the middle section was the smallest. Due to the influence of ventilation the temperature at the air inlet was generally lower.

Figure 4.80 is the fitting line of the air temperature inside and outside the Kunlun Mountain Tunnel before the opening of the tunnel from 2004 to 2005, indicating that there was a very good correlation between them. When the temperature outside the tunnel changed by 1 °C, the average temperature inside the tunnel would change by 0.8 °C, and the average temperature would decrease by nearly 0.2 °C during the transfer process of the air temperature in the tunnel. After completion, the air temperature inside the tunnel was still lower than that outside the tunnel in the warm season, and higher than that outside the tunnel in the cold season. According to the average and envelope statistics of the whole Kunlun Mountain Tunnel, the absolute value of the maximum monthly temperature difference was about 2 °C in the warm

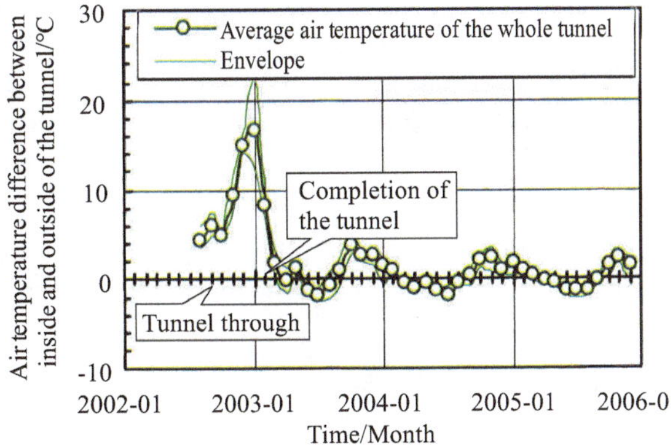

Fig. 4.81 Air temperature difference between inside and outside Kunlun Mountain Tunnel (China Railway First Survey and Design Institute, 2007)

season (located at the south opening of the tunnel, that was, the exit), and about 3 °C in the cold season (located in the middle of the tunnel) (Fig. 4.81).

2. **Variation and Distribution Characteristics of the Temperature of the Surrounding Rock of the Tunnel**

After the excavation of the tunnel, an air-ground heat exchange interface was added in the permafrost layer. At the beginning of construction, the heat exchange between the air temperature in the tunnel and the surrounding rock also began. With the opening and operation of the tunnel, in 2–3 years, a new heat exchange system between the surrounding rock and the air temperature, and a stable thermal regime were gradually formed in the tunnel, which was characterized by no solar radiation and with open ventilation and convection. Regardless of the tunnel in the fully permafrost, partial permafrost or seasonally frozen area the new heat exchange system results in the repeated freeze-thaw. The freeze-thaw action at the opening of the tunnel is the most severe, but it gradually weakens with the length of the tunnel and gains a relative stability.

(1) Ground Temperature of the Surrounding Rocks of the Tunnel during the Construction Period

Take the Dabanshan Mountain Tunnel in Qinghai Province (non-fully permafrost) as an example. In the early stage of construction, due to construction requirements, the ambient temperature in the tunnel should not be too low, and the engineering measures needed to be adopted to maintain the requirements for the working temperature in some sections of the tunnel. In the coldest month, the freezing length of the surrounding rocks at the entrance of the tunnel was 300 m, and the freezing depth was generally 0.5 m. The maximum freezing depth near the opening of the tunnel was about 2 m, and there was a 200 m freezing section at the exit end, with a freezing depth less than 0.5 m. In the warmest month, the isotherm of 5 °C in the tunnel could reach a length of 500 m (Fig. 4.82 a and b).

After completion of the tunnel, in the first cold period, the freezing length at the exit end reached 500 m and the freezing length at the entrance end reached 400 m

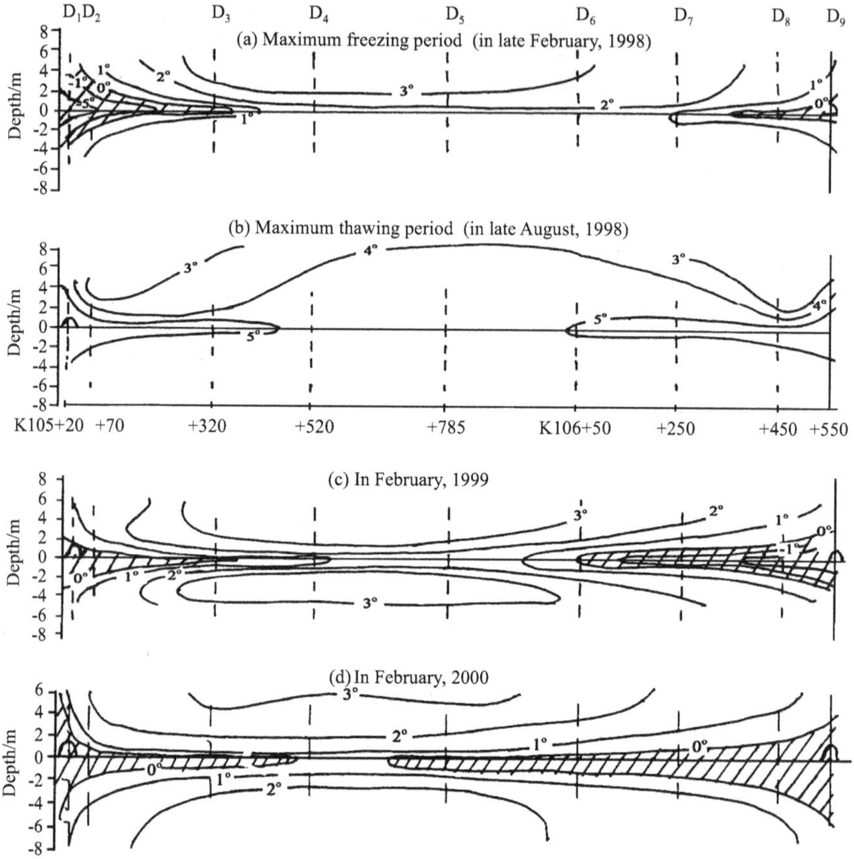

Fig. 4.82 Ground temperature variation of surrounding rocks in early stage of construction (**a** and **b**), after the getting-through (**c**) and after the operation (**d**) in the groundwater tunnel (Wu et al., 2003a, b, c)

(Fig. 4.82c). The freezing depth at the opening of the tunnel was slightly deeper, the freezing depth of the terrace was less than 2 m, and the freezing depth of the side wall and the vault was generally less than 0.5 m. The negative accumulated temperature in the first cold month after the operation was much higher than that during the survey period, with a temperature difference of 25%. The maximum freezing depth of the terrace was 2.1 m (K106 + 450), and the freezing length was 750 m (Fig. 4.82d), and when it warmed up and thawed during the summer time, the ground temperature could be 4–5 °C (in September).

(2) Temperature Variation and Distribution inside and outside the Thermal Insulation Layer.

The tunnels in cold regions adopt the construction method of conducting the lining while excavating, and before the secondary lining, the water-stop board and thermal

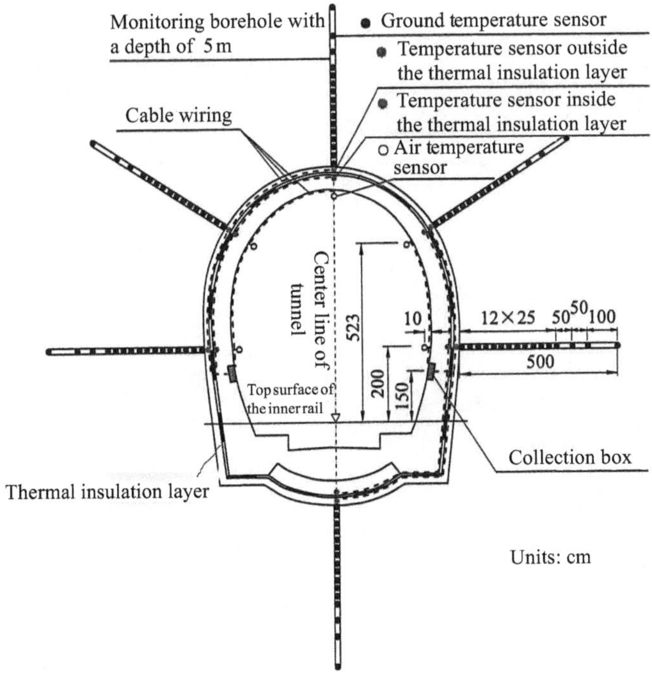

Fig. 4.83 Layout of observation elements for air temperature and ground temperature in the tunnel (China Railway First Survey and Design Institute, 2007)

insulation layers (with the thickness of 50 mm) are laid to reduce the destruction of the repeated freeze-thaw on the main structure of the tunnel.

The Kunlun Mountain Tunnel and Fenghuo Mountain Tunnel of the Qinghai-Xizang Railway are full permafrost tunnels. The Kunlun Mountain Tunnel is 1686 m long, with a maximum buried depth of 106 m; the average annual air temperature of the tunnel is −5.1 °C, and the extreme maximum temperature and minimum temperature are 18.3 °C and −26.5 °C respectively. The Fenghuo Mountain Tunnel is 1338 m long, with a maximum buried depth of 100 m; the average annual temperature of the tunnel is −6.1 °C, and the extreme maximum temperature and minimum temperature are 23.2 °C and −37.7 °C respectively. There are 11 air temperature and ground temperature observation sections in each of the two Tunnels (Fig. 4.83).

Annual average temperature: the annual average temperature inside and outside the thermal insulation layer in the main structure of the tunnel was distributed along the longitudinal direction of the tunnel (Fig. 4.84). It shows that after the completion of the tunnel and before the opening of the tunnel to traffic, the average annual temperature inside and outside the thermal insulation layer was declining year by year, and the declining rate gradually slowed down.

The temperature inside the thermal insulation layer was closer to the air temperature distribution in the tunnel than the temperature outside the thermal insulation

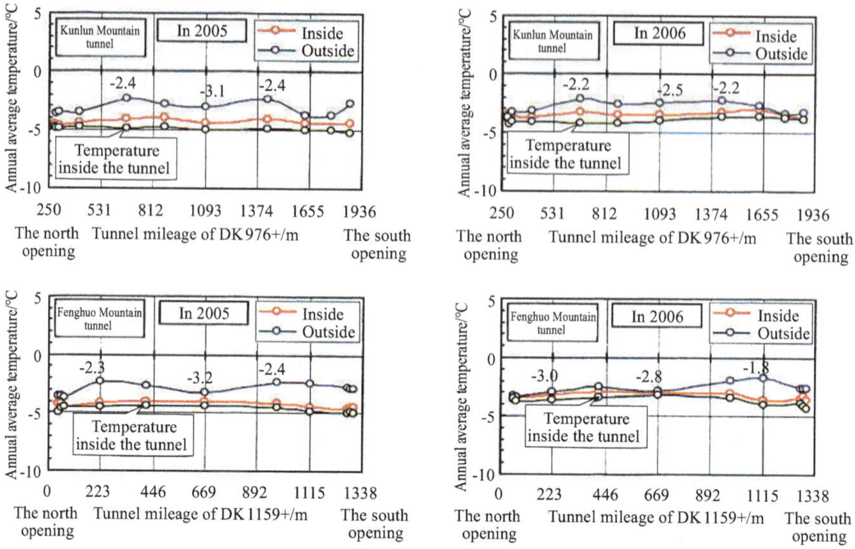

Fig. 4.84 Longitudinal distribution of annual average temperature inside and outside thermal insulation layers in Kunlun Mountain Tunnel and Fenghuo Mountain Tunnel (China Railway First Survey and Design Institute, 2007)

layer, and there was a large temperature difference between the inside and the outside. After opening to traffic, due to the influence of operation, the average annual temperature of the inside increased significantly (by 1 °C overall), and the average annual temperature of the outside changed little (by 0.2–0.3 °C overall), indicating that the thermal insulation layer could block heat transfer.

The annual highest and lowest temperatures: the tests showed that the temperatures outside the thermal insulation layers in Kunlun Mountain Tunnel and Fenghuo Mountain Tunnel in the hottest month was lower than that inside the thermal insulation layer by more than 5 °C, and the temperature of the inside (inner side) in the hottest month was closer to the distribution of air temperature. The thermal insulation layer could inhibit the heat transfer of the air temperature in the hottest month to a certain extent, prevent the heat exchange between the tunnel and the surrounding rocks, and reduce the thawing of the permafrost layer in the surrounding rocks behind the lining layer.

After the opening to traffic, as the air temperature in the tunnel increased, the temperature inside the thermal insulation layer in the hottest month also rose accordingly. The thermal insulation layer laid behind the lining layer played a role in heat resistance, and the influence on the temperature outside the thermal insulation layer was much smaller (Fig. 4.85).

Monthly variation of the temperature: the temperature inside and outside the thermal insulation layer in the Kunlun Mountain Tunnel and Fenghuo Mountain Tunnel changed with the air temperature within the tunnel (Fig. 4.86). The temperature inside the thermal insulation layer was close to the air temperature within the

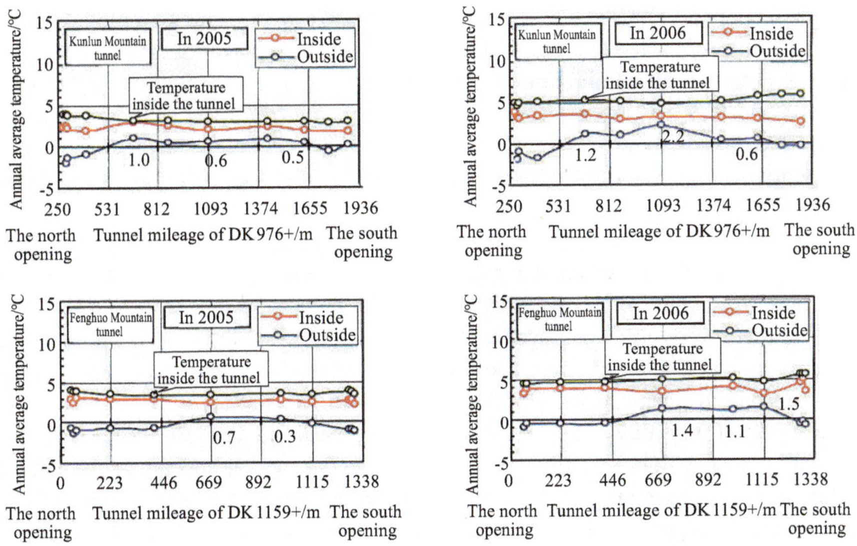

Fig. 4.85 Longitudinal distribution of temperatures inside and outside the thermal insulation layers in Kunlun Mountain Tunnel and Fenghuo Mountain Tunnel in the hottest month (China Railway First Survey and Design Institute, 2007)

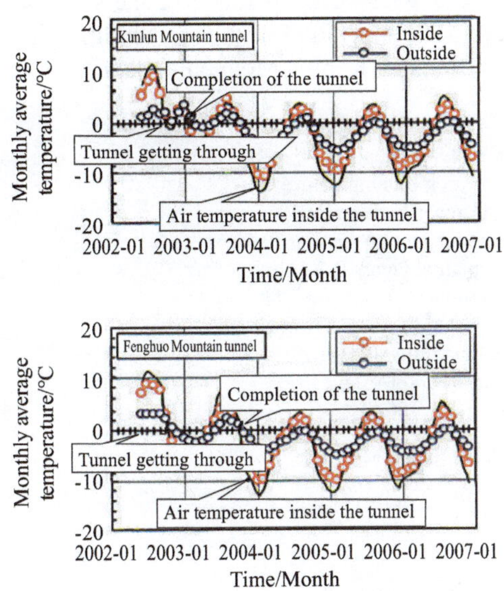

Fig. 4.86 Monthly variation of temperatures inside and outside the thermal insulation layers in Kunlun Mountain Tunnel and Fenghuo Mountain Tunnel (China Railway First Survey and Design Institute, 2007)

tunnel, and the temperature outside the thermal insulation layer had a significant attenuation and also had a trend of decreasing year by year. The performance of Fenghuo Mountain Tunnel was more obvious, dropping to 0 °C 1 year earlier than that of the Kunlun Mountain Tunnel. After the completion, the monthly temperature variation characteristics inside the thermal insulation layer of the two tunnels were basically the same, but the amplitude of variation outside the thermal insulation layer of Fenghuo Mountain Tunnel was relatively small, indicating that the thermal insulation effect was more obvious. After the opening to traffic, the temperature inside and outside the thermal insulation layer of the two tunnels had increased, but the annual thawing index of the temperature outside was still zero throughout the year.

The characteristics of temperature difference: the temperature difference between inside and outside the thermal insulation layer in Kunlun Mountain Tunnel and Fenghuo Mountain Tunnel changed with the air temperature variations in the tunnel, and there were obvious seasonal changes after the completion of the tunnels. In the transition period of the cold and the warm seasons, the temperature difference was zero, and the absolute value of the temperature in the hottest and coldest months reached the maximums. The temperature difference in the coldest month was significantly greater than that in the hottest month (Fig. 4.87), and the accumulated temperature of the temperature difference in the cold season was 2–3 times of the accumulated temperature in the warm season. It showed that the thermal insulation layer of the two tunnels in permafrost regions could reduce the range of the thawing rings around the tunnels.

(3) Temperature Variation and Distribution of the Surrounding Rocks

Annual average temperature of the surrounding rocks at different depths: annual average temperature of the surrounding rocks within a depth range of 5 m behind the lining in Kunlun Mountain Tunnel and Fenghuo Mountain Tunnel rose with the

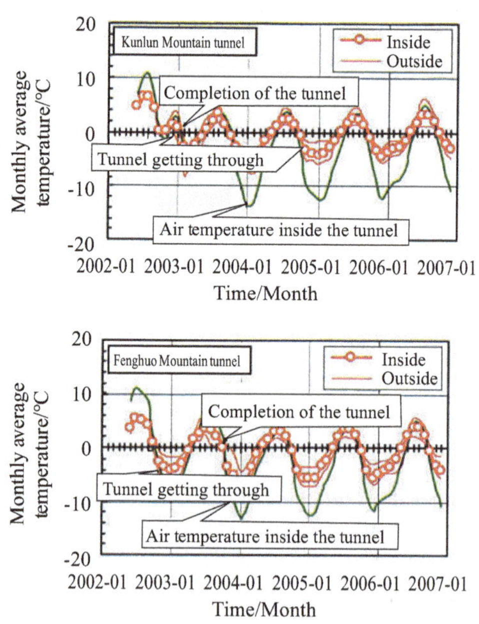

Fig. 4.87 Monthly variation of temperature difference between inside and outside the thermal insulation layers in Kunlun Mountain Tunnel and Fenghuo Mountain Tunnel (China Railway First Survey and Design Institute, 2007)

increase of depth (Fig. 4.88). The figure showed that affected by the construction the surrounding rocks in the tunnel wall warmed up. Since the second year, the surrounding rocks in the tunnel wall had begun to cool down and refreeze, and the cooling rate slowed down in the third year. With the increase of air temperature in the tunnel, the temperature of the surrounding rocks at a depth of 2 m increased slightly, and the annual average temperature of the surrounding rocks at depths of 2–5 m still decreased, but the rate slowed down.

Along the longitudinal direction of the tunnel, the annual average temperature of the surrounding rocks at the opening of the tunnel was lower than that in the central part, but the annual average temperature of the surrounding rocks at the entrance of the north slope was significantly lower than that at the exit of the south slope, namely, the characteristic of temperature distribution was that the temperatures in the north were lower but higher in the south, and this characteristic was more obvious in Kunlun Mountain Tunnel with relatively steep terrains (Fig. 4.89). The deeper the buried depth of the tunnel, the higher the temperature of the surrounding rocks. That was, the annual average temperature of the surrounding rocks within a depth of 5 m was also higher than that of the surrounding rocks with shallower buried depth, which was related to the influence of different terrains and slope aspects on the temperature in the permafrost.

The temperature of the surrounding rocks in the hottest month: affected by the construction during the hottest month, the maximum depth of seasonal thawing in Kunlun Mountain Tunnel and Fenghuo Mountain Tunnel reached 5 m. After completion, the thawing depth of the surrounding rocks behind the tunnel lining decreased, and the thawing depth of Fenghuo Mountain Tunnel decreased faster. In

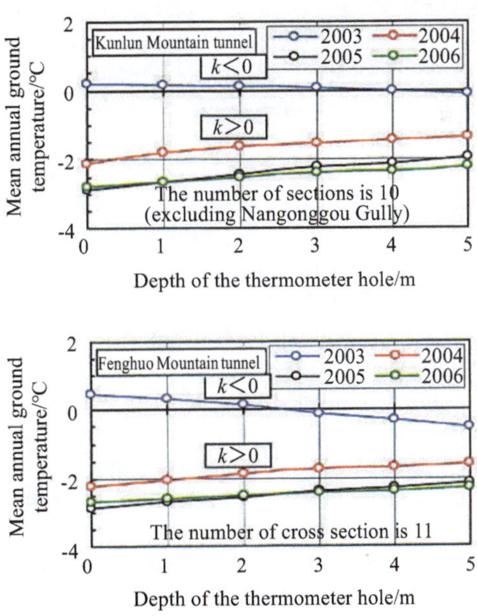

Fig. 4.88 Distribution of annual average temperature of surrounding rocks along the depth within a range of 5 m behind tunnel lining (China Railway First Survey and Design Institute, 2007)

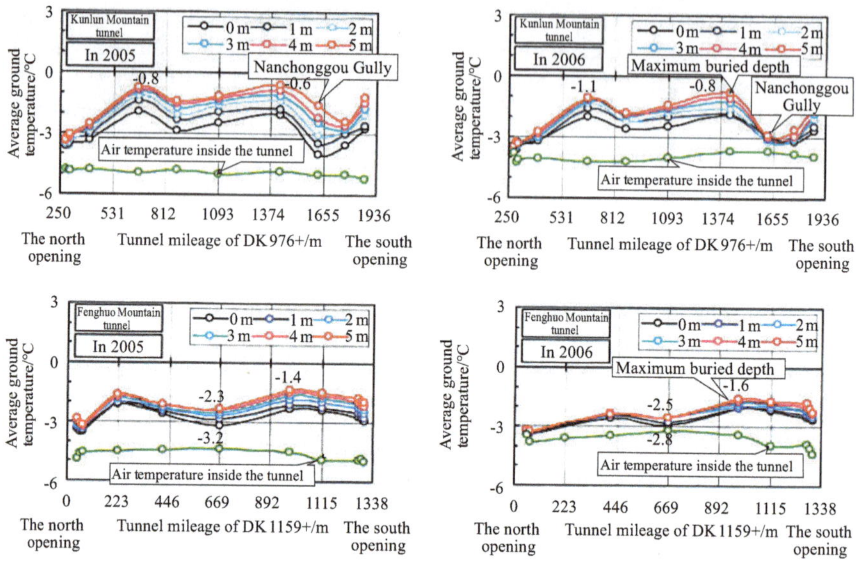

Fig. 4.89 Longitudinal distribution of average annual temperature of surrounding rocks within 5 m depth behind tunnel lining (China Railway First Survey and Design Institute, 2007)

the fourth year, most of the surrounding rocks of the two tunnels refroze, and only a few sections thawed in some parts during the warm season. The thawing depth of Kunlun Mountain Tunnel in the two gullies was about 1.0 m, and the maximum thawing depth was less than 1.5 m. The maximum thawing depth of Fenghuo Mountain Tunnel was less than 1 m.

The monthly variation of temperature of the surrounding rocks: according to the variations of the monthly rock temperature at different depths of the surrounding rocks behind the lining in Kunlun Mountain Tunnel and Fenghuo Mountain Tunnel (based on the average temperature of the whole tunnel) (Fig. 4.90), the temperature of the surrounding rocks at a depth of 5 m behind the tunnel lining varied with the variations of air temperature, and the varying rate decreased year by year, and the variation cycle exhibited different degrees of lag with depth. From the comparison, the monthly variation cycle of the surrounding rock temperature in Fenghuo Mountain Tunnel obviously lags behind that in Kunlun Mountain Tunnel, especially in deep surrounding rocks. The reason was that the stratum of Fenghuo Mountain Tunnel was the mudstone sandwiched with sandstone, and the thermal conductivity was lower than that of the slate sandwiched with schist stratum in Kunlun Mountain Tunnel. In addition, the monthly variation amplitude of the surrounding rock temperature in Fenghuo Mountain Tunnel was smaller than that in Kunlun Mountain Tunnel. Therefore, the temperature variation of the surrounding rocks in Fenghuo Mountain Tunnel was relatively stable.

The observation data of the temperatures of surrounding rock of the two tunnels show that within the depth of 5 m around the clearance, the differences of the annual

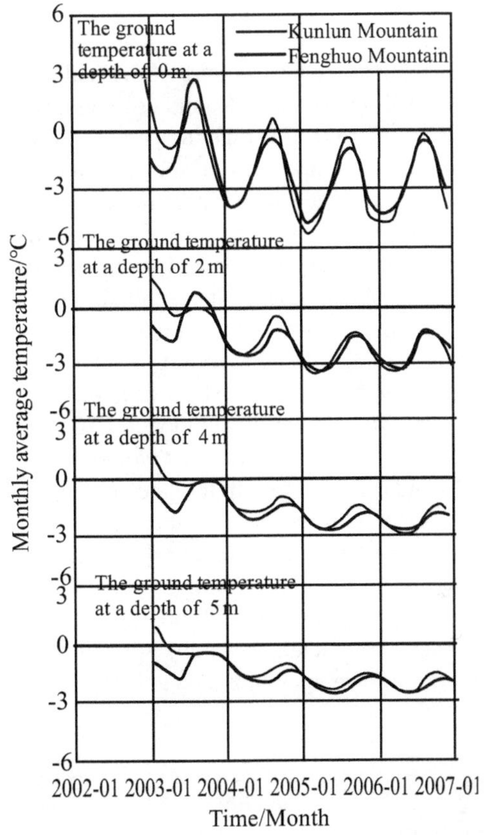

Fig. 4.90 Monthly variation of the surrounding rock temperature in tunnel (China Railway First Survey and Design Institute, 2007)

average temperatures between the side wall and the surrounding rock in the vault is not obvious and the maximum difference is less than 0.5 °C.

3. **Influence of Temperature inside and outside the Tunnel on the Freeze-thaw Ring of Surrounding Rocks**

The annual average temperature: after the tunnel in permafrost regions has been constructed, the atmospheric temperature outside the tunnel is the main factor affecting the temperature of the freeze-thaw ring of the surrounding rocks in the tunnel, and besides, the traveling cars also have an impact on the temperature. According to the observation of Kunlun Mountain Tunnel and Fenghuo Mountain Tunnel along the Qinghai-Xizang Railway, the amplitudes of the temperature variation from the air temperature outside the tunnel, to the air temperature inside the tunnel, then to the inside of the thermal insulation layer, and finally to the outside of the thermal insulation layer were in the following influential strength: 1 °C → 0.8 °C → 0.6 °C → 0.3 °C (Kunlun Mountain) or 0.2 °C (Fenghuo Mountain), and the temperature of surrounding rock changed with the depth in a certain gradient.

According to overall variation and distribution characteristics of the air temperatures inside and outside the Kunlun Mountain Tunnel and Fenghuo Mountain Tunnel, and the temperature of thermal insulation layer and the temperature of the surrounding rocks in Kunlun Mountain Tunnel and Fenghuo Mountain Tunnel (according to the average temperature of the whole tunnel) (Fig. 4.91), after completion, the temperature of the surrounding rocks decreased significantly with the cooling of the air temperature inside the tunnel, and the refreezing occurred in the second year. The temperature of the surrounding rocks within a depth of 5 m behind the lining changed from an overall heat absorption state during the construction period to a heat release state, and all the temperatures of Kunlun Mountain Tunnel turned into negative temperatures in the third year. After the opening to traffic, the temperature of the surrounding rocks in the two tunnels had increased, but there was still a downward trend. The mean annual ground temperature of the permafrost in Kunlun Mountain Tunnel (about −2 °C) was higher than that in Fenghuo Mountain Tunnel (−3 °C to −2 °C), so it would be a little later for the temperatures in Kunlun Mountain Tunnel to turn into the negative temperatures.

The monthly average temperature variations in the hottest and coldest months: the hottest temperature causes the permafrost in the surrounding rocks in the tunnel to warm up and thaw, forming a thawing ring, and the coldest temperature can make the temperature of the surrounding rocks in the tunnel drop and refreeze. After the tunnel is completed, the temperature of the surrounding rocks in the tunnel increases with the air temperature, and a thawing ring appears. With the passing of time, the

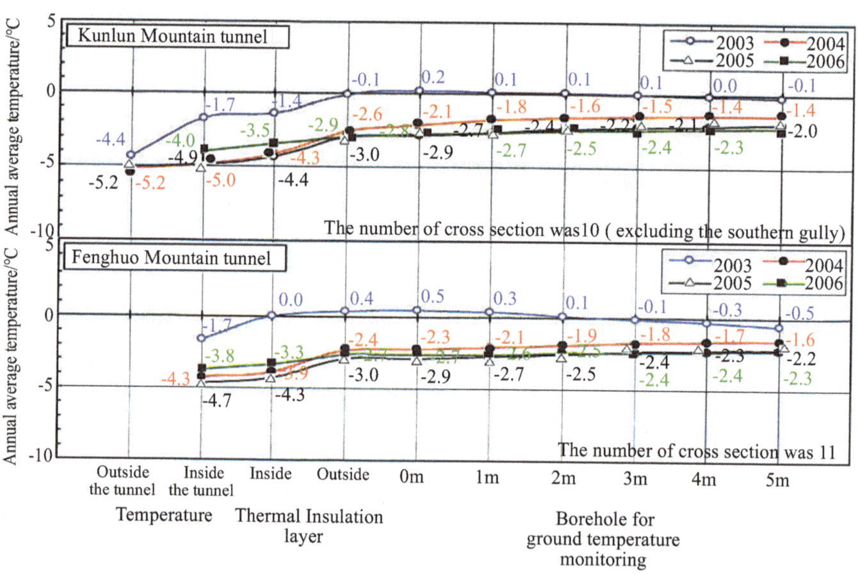

Fig. 4.91 Annual average temperature variation and distribution of temperature inside and outside the tunnel, in thermal insulation layer and surrounding rocks in Kunlun Mountain Tunnel and Fenghuo Mountain Tunnel (China Railway First Survey and Design Institute, 2007)

long-term severe cold temperature on the Qinghai-Xizang Plateau makes the surrounding rocks in the tunnel gradually refreeze (Fig. 4.92). It can be seen from the figure that the thermal insulation layers of the two tunnels worked well, and the influence of the temperature outside the tunnel and the operation on the ground temperature of the surrounding rocks in the tunnel had been minimized.

4. **Numerical Simulation Analysis**

According to *Prediction of Environmental Change in Western China* (Ding, 2002), under the influences of climate and human activities, the mean annual air temperature of the Qinghai-Xizang Plateau will increase by 2.6 °C in the next 50 years. Considering the water migration, ice-water phase transition and the effect of air temperature warming, the control differential equation and finite element formulation of the two-dimensional thermal regime are used to predict and analyze the thawing ring of the surrounding rocks in two tunnels (Fig. 4.93).

DK976 + 300 (construction mileage of 976 km and 300 m) with a buried depth of 13 m from Kunlun Mountain Tunnel, and DK1159 + 671 (construction mileage of 1159 km and 671 m) with a buried depth of 75 m from Fenghuo Mountain Tunnel were selected for the simulation, and the thermal insulation layers with a thickness of 50 mm were laid (the thermal conductivity was0.03 W/(m·K)). Considering the temperature increases by 2.6 °C in 50 years, the thawing ring of the surrounding rocks behind the lining of the two tunnels had been predicted. The surrounding rocks in Kunlun Mountain Tunnel had a seasonally thawed layer with a thickness of

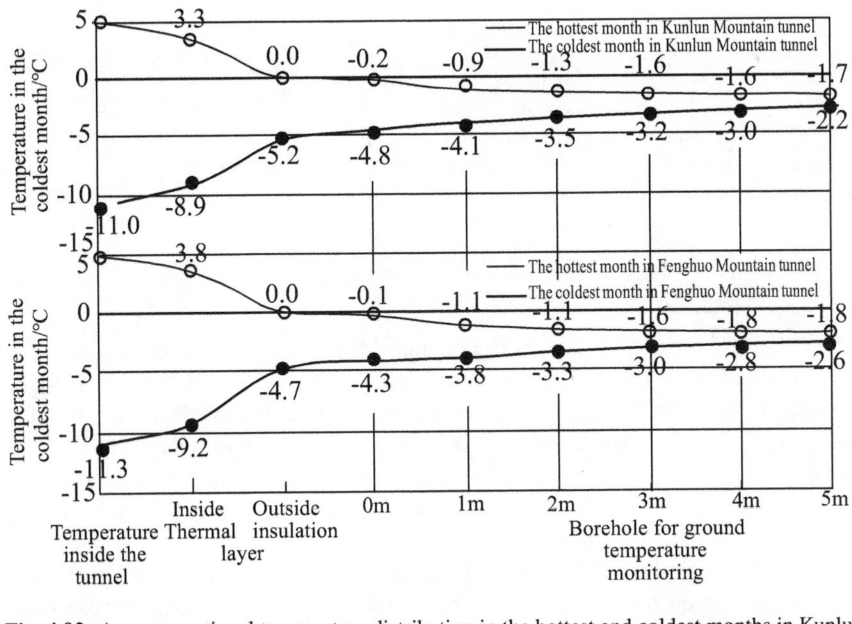

Fig. 4.92 Average sectional temperature distribution in the hottest and coldest months in Kunlun Mountain Tunnel and Fenghuo Mountain Tunnel (China Railway First Survey and Design Institute, 2007)

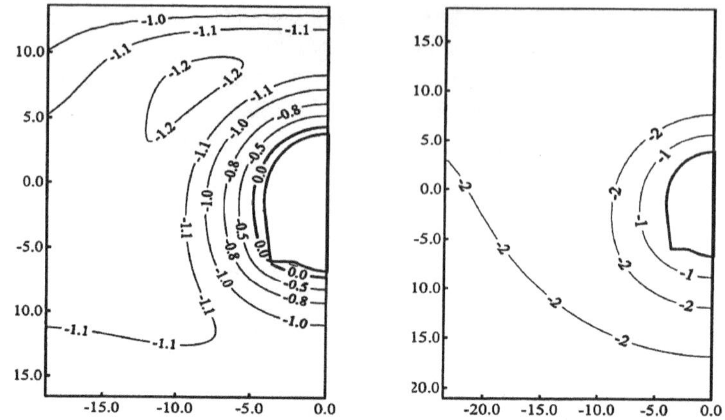

(a) On September 30 of the 50th year (b)On September 30 of the 50th year

Fig. 4.93 Distribution of temperatures of surrounding rocks in Kunlun Mountain Tunnel (**a**) and Fenghuo Mountain Tunnel (**b**) in the 50th Year (China Railway First Survey and Design Institute, 2007)

Table 4.21 Prediction of maximum thawing depth in Kunlun Mountain Tunnel (Construction Mileage of DK976 + 300) and Fenghuo Mountain Tunnel (Construction Mileage of DK1159 + 671)

Location	Time	In the first year	In the 5th year	In the 10th year	In the 30th year	In the 50th year
Fenghuo Mountain	No thermal insulation layer is laid	/	0	Very little	0.25	0.47
	Laying 50 mm thick thermal insulation layer(thermal conductivity was 0.03 W/(m·K))	0	0	0	0	0
Kunlun Mountain	No thermal insulation layer is laid	0.61		0.87	1.05	1.28
	Laying 50 mm thick thermal insulation layer(thermal conductivity was 0.03 W/(m·K))	0.15		0.21	0.35	0.50

0.5 m, and the surrounding rocks in Fenghuo Mountain Tunnel had no seasonally thawed layer (Table 4.21) (China Railway First Survey and Design Institute, 2007). In general, the influence depth of air temperature variations inside and outside the tunnel on the ground temperature of the surrounding rocks was more than 5.7 m. The depth of the non-shallow buried section with an annual amplitude of 0 °C was generally about 9–10 m, which was shallower than the depth with an annual amplitude of 0 °C outside the tunnel.

In the plateau and the cold regions, in the coldest month, for the severe cold areas with an average air temperature lower than −10 °C and the seasonally frozen soil areas in the cold regions with an average air temperature of −3.0 °C to −1.0 °C, the external temperature has a great influence on the surrounding rocks in the cold

Table 4.22 Length of tunnel affected by freezing (Tan Xianjun etc., 2013)

	Entrance end				Exit end			
	Influence length of freeze-thaw action under different wind speeds/m							
Length of insulation material/m	6 m/s	8 m/s	10 m/s	12 m/s	6 m/s	8 m/s	10 m/s	12 m/s
0	600	700	780	850	430	450	500	600
300	450	550	630	700	0	0	450	500
400	0	530	580	650	0	0	0	0
500	0	0	0	600				
600	0	0	0	0				

season and the freeze-thaw cycles often occur at the entrance and exit of the tunnel, causing destructive effects on the lining. In the middle of the tunnel, due to the geothermal effect, the temperature of the surrounding rocks is less affected by the outside air temperature. After laying a certain length of thermal insulation layer as an anti-frost measure, the air temperature in the middle of the tunnel is often higher than 0 °C, except that the air temperatures within a certain length at the tunnel entrance and exit is below 0 °C. Through the finite element calculation of the control equation of the thermal regimes in the surrounding rock area and of the tunnel clearance, the influence length in the longitudinal distribution of air temperature (0 °C) in the coldest month on the surface of the secondary lining in the tunnel under different wind speed can be obtained, so as to determine the length of the thermal insulation layer. For example, under a certain wind speed, when the tunnel is not laid with thermal insulation layer, the length of the tunnel affected by the freeze-thaw action will be about 600 m. But after laying 300 m long thermal insulation layer, the length of the freeze-thaw effect on the surrounding rocks will be 450 m. When the 400 m long thermal insulation layer is laid, except for a small section of surrounding rocks at the openings of the tunnel, the tunnel will basically not be affected by freeze-thaw action. When the 500 m and 600 m long thermal insulation layers are laid the temperature in the tunnel will increase. Table 4.22 shows the length affected by freeze-thaw action at the entrance and exit of Galongla Tunnel in Xizang Autonomous Region. Within the range of 600 m at the entrance end and 400 m at the exit end of the tunnel, the surface of the secondary lining has been paved with a 6 mm-thick polyphenolic insulation layer, which could effectively protect the tunnel lining and the surrounding rocks from freeze-thaw damages (the maximum freezing depth of the vault was 0.5 m, which was reduced to 0.21 m after the insulation).

4.3.1.2 Effects of Frost-Heave Force

1. **Frost Heave of Tunnel Surrounding Rocks**

In permafrost regions and alpine seasonally frozen soil areas, when the temperature of the soil (tunnel surrounding rocks) drops below the freezing temperature, frost heave will occur. The tunnel has earthy (or weathered layer) and rocky surrounding rocks. Usually, there are some weathered layers in the inlet and outlet sections, especially in fault zones, which are earthy or semi-rocky in nature, with well-developed fissures and joints to form channels for groundwater.

The normal frost-heave force of the tunnel is related to the properties of the surrounding rocks, the water content and the water recharge (hydrological conditions), the freezing depth and the engineering measures (the stiffness of the lining layer and the thermal insulation layer). The greater the frost heaving amount of soil body, the greater the frost-heave force. Under the condition of water migration, the frost heaving amount of soil (h_{fh}) (Tsytovich, 1985) is:

$$h_{fh} = \alpha\omega\rho_d h_f i_0 + (1+\alpha)\int v_s dt \tag{4.15}$$

where α is the volume expansion coefficient when water freezes; ω is the total mass of water content in the freezing layer of soil; ρ_d is the dry density of soil; h_f is the freezing depth of the soil layer; i_0 is the relative ice content of the frozen soil layer obtained according to the unfrozen water content; v_s is the water migration rate; t is the water migration time.

The magnitude of the normal frost-heave force can be estimated according to the pressure value generated by the growth of ice crystals when the water freezes under constraint conditions, and the maximum value occurs only when the water freeze and cannot expand at all (rigid airtight). When the temperature reaches −22 °C, the maximum value is about 211 MPa, and when the temperature is above −22 °C, the pressure value will decrease significantly. However, when the restraining pressure increases, the freezing point of the ice will decrease, and the unfrozen water in the frozen soil will increase accordingly. The empirical relationship can be expressed by the Clapeyron formula:

$$\Delta p = 1 + 127\theta - 1.519\theta^2 \tag{4.16}$$

where Δp is the pressure (kg/cm^2); θ is the absolute value of the negative temperature (°C).

Under the standard atmospheric pressure, water freezes at 0 °C, and as the temperature decreases, unfrozen water and extraneous water in the soil gradually freeze, resulting in frost-heave force. The dynamic equilibrium of unfrozen water and ice occurs constantly during the occurrence of water freezing and the generation of frost-heave force. The variations of frost-heave force when increasing doesn't follow a shape in a circle smooth curve as the temperature decreases, but increases in a serrated manner. When the temperature doesn't decrease (usually around March) the frost-heave force decreases immediately.

The frost-heave force between the tunnel support and the surrounding rocks is essentially the normal frost-heave force perpendicular to the surface of the support body. At present, the frost-heave force measured in the tunnel refers to the surrounding rock pressure that achieves a balance between the support and the surrounding rock, including the swelling force and the rock pressure when the water in the surrounding rock freezes, or defined as "generalized frost-heave force" (Wang et al., 2004a, b, c).

There are two situations for the freezing of the surrounding rocks behind the tunnel lining in cold regions. One is that the tunnel passes through the unfrozen surrounding rock, and after the completion of the tunnel (lining), under the influence of cold air, a new freezing ring is formed in the surrounding rocks in the tunnel, that is, the newborn frozen soil. The other is that the tunnel passes through the frozen surrounding rocks (frozen soil). During the construction, the frozen surrounding rocks within a certain range around the tunnel thaws. After the completion of the tunnel (lining), the thawed surrounding rocks refreezes, leading to the formation of regenerated frozen soil. In the former, the water in the surrounding rocks of the tunnel infiltrates (flows) into the freezing ring from the unfrozen surrounding rocks, and gradually freezes and expands in volume, which produces frost-heave force on the support lining.And the magnitude of the frost-heave force depends on the amount of water storage between the lining and the surrounding rocks during the unfrozen period and the amount of infiltration from rock fissures. During the freezing process, under the impact of pressure, the free water will continue to infiltrate and freeze. In the latter, in addition to the groundwater recharge in the fault zone, when the frozen surrounding rocks thaw the water will gradually lose. During the refreezing process, the existing water in the thawing ring and the newly thawed water of frozen soil caused by the lining body freeze and produce volume expansion, which poses frost-heave force on the support lining.

2. **The Effect of Frost Heave of Surrounding Rocks on the Lining**

The frost-heave force caused by the freezing of the water-containing surrounding rocks behind the tunnel lining layer is the essential cause for the tunnel frost damages. The fissure water of rock tunnel intrudes (permeates) and freezes into fissure ice, which produces frost-heave force on the surrounding rocks, and the deformation of the surrounding rocks causes the deformation of lining layer. When the fissure water accumulates between the surrounding rocks and the lining layer, the free water freezes into ice and produces frost-heave force, which causes the deformation of the lining layer. Sometimes, the deformation caused by the free water is comparable to the strength of the surrounding rock on the lining layer. In the earth tunnel (in completely or highly weathered layer), the pores of the surrounding rock are rich in water or saturated, and the water in the thawed surrounding rocks continues to infiltrate, and some water migrates to the freezing front. The ice exists not only in the pores of the surrounding rocks, but also is sandwiched in the weathered layer in the form of a free and thin layer, and an ice layer is also formed on the surface of the lining layer. When there is an external water supply, the ice may continue to freeze on the surface of the lining layer. Regardless of whether it is a rocky tunnel or an earthy tunnel, in the absence of drainage, the frost-heave force increases with the increase of the rigidity of the surrounding rocks and lining layers, and also increases with the increase of freezing depth and the negative temperature of surrounding rocks in the tunnel, and almost increases proportionally with the increase of water content in the surrounding rocks. During the freezing process, when the water in the surrounding rocks has a drainage channel the frost-heave force value is greatly reduced or even disappears, and it also decreases with the weakening of the constraint degree (rigidity) of the lining layers and the surrounding rocks (Figs. 4.94 and 4.95).

At present, the measurements of frost-heave force in the tunnel are typically conducted by using the pressure cells (strain-gauge pressure cell and vibrating-wire pressure cell). When using the pressure cell for earth to measure the pressure of hard rock stratum, the force system should be converted to meet the requirements for the earth pressure cell (pressure membrane) to measure the uniform load. Otherwise, the measured value may not able to fully reflect the authenticity of the pressure value, and the frozen soil also belongs to the quasi-hard rock. The deformation of the constraint body of the lining layer (and other factors) may cause the release and decrease of frost-heave force. In this circumstance, the frost-heave force measured in the tunnel could be smaller than those measured in the laboratory. The frost-heave force observed in No.2 Tunnel of Chaoyang in Nenlin Line of Daxing'anling Mountains in Northeast China was 25 kPa to 200 kPa (Fig. 4.96).The observed frost-heave force in the Daban Mountain Tunnel in Qinghai Province was about

Fig. 4.94 Relation between frost-heave stress and strain

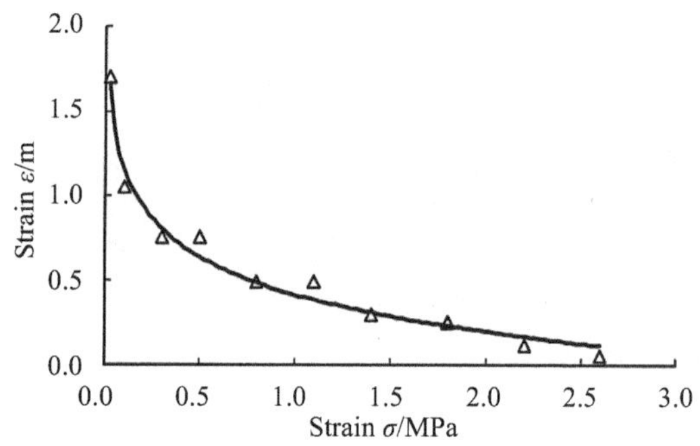

Fig. 4.95 Normal frost-heave force under different constraint degrees (Tong & Guan, 1985). Silty Clay, with *Ip* of 7.7 and *W* of 22–25.8%, and the area of test loading plate is 491 cm^2

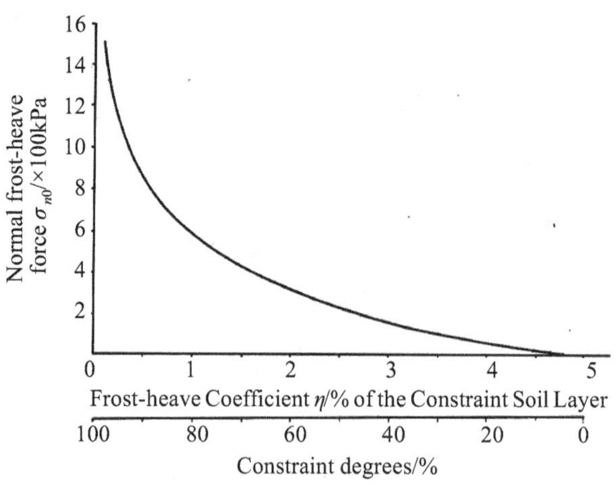

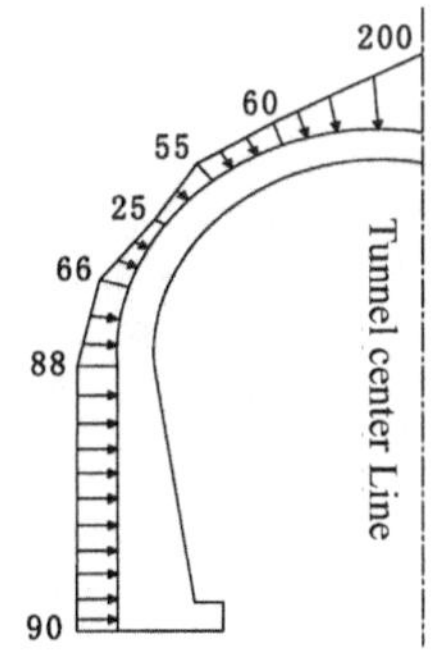

Fig. 4.96 Distribution of frost-heave force in Chaoyang Tunnel of Nenlin Line (China Railway First Survey and Design Institute, 2007)

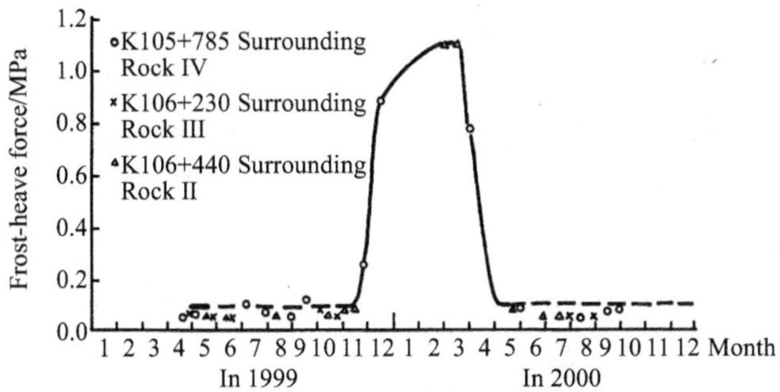

Fig. 4.97 Observed values of frost-heave force behind the lining layer of Daban Mountain Tunnel in Qinghai Province (Wu et al., 2003a, b, c)

1.1 MPa (Fig. 4.97). The peak value of frost-heave force in the surrounding rocks at the entrance of Qingsha Mountain Tunnel along Ping'an-A'dai Expressway in Qinghai Province was 1.3 MPa (at the intersection of the bottom of the side wall and the inverted arch), and the value was 0.35 MPa at the exit side (Lai, 2008). According to relevant data from the former Soviet Union, when there is water intrusion in rock fissures, the possible frost-heave force generated during freezing could be 15 MPa (the observation method is unknown) (Edited by China Railway First Survey and Design Institute, 2007). The observed values of generalized frost-heave force in deep and shallow buried sections of Kunlun Mountain Tunnel were 0.12–0.3 MPa (Fig. 4.98). Since the frost-heave force varies with temperature, the envelope diagram of frost-heave force with the most unfavorable load acting on the supporting structure could be plotted by using the theory of structural mechanics and the statistics of frost-heave force test values under different temperature conditions (Fig. 4.99).

The calculation of frost-heave force of the tunnel is based on three assumptions. At present, these three assumptions are: ① Assuming that the entire surrounding rock within the freeze-thaw ring around the tunnel frosts (Lai et al., 2009), that is, assuming that the uniformly distributed pores in the surrounding rock in the frozen area are filled with groundwater, and the water is frozen into ice, then its volume

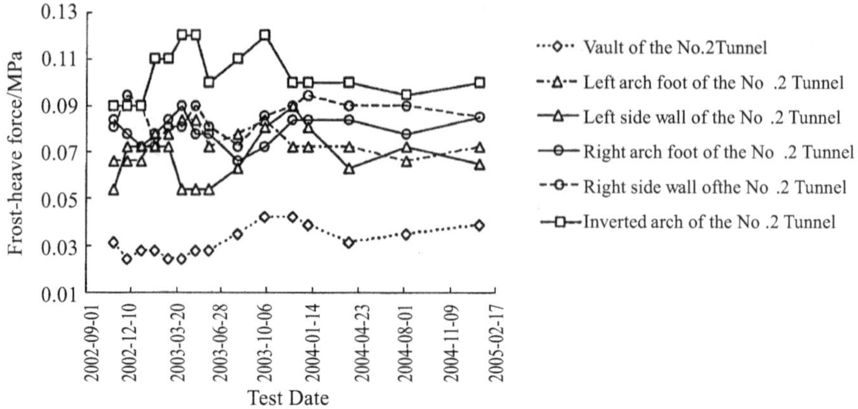

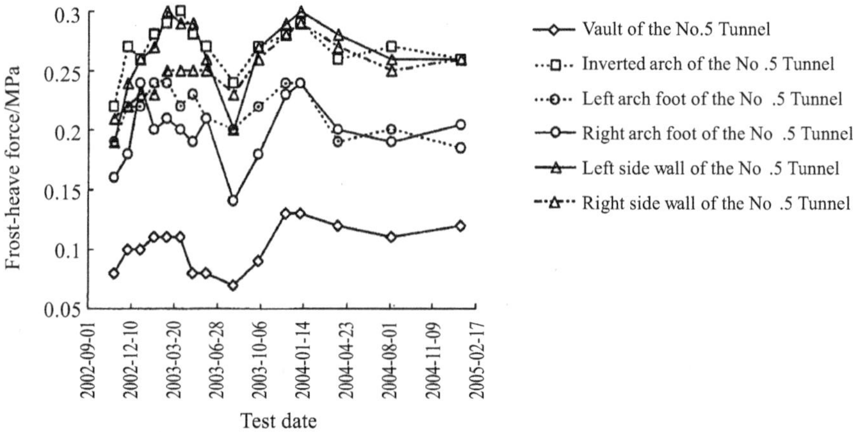

Fig. 4.98 Variation of Frost-heave force with time between Primary Support and Secondary lining (the Upper) and the surrounding rocks and primary support (the Lower) in Kunlun Mountain Tunnel (Editorial Committee of *Qinghai-Tibet(Xizang) Railway*, 2016)

Fig. 4.99 Envelope of frost-heave force behind the tunnel lining

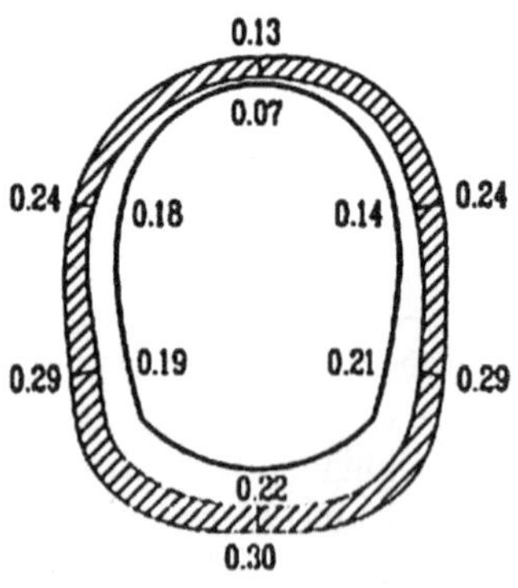

expands as a whole, producing frost-heave force on the tunnel lining. In this case, the frost-heave force should be calculated as an elastic body. When the freezing depth is less than 2 m and the porosity is less than 5%, the frost-heave force of the lining does not exceed 2 MPa. Generally, the compressive strength of the rock is relatively large, and the freezing of the fissure water is not enough to cause the overall expansion of the surrounding rocks. ② Assuming that the accumulated water causes frost heave (Wang et al., 2004a, b, c), that is, the accumulated water behind the tunnel lining freezes and then expands, generating the frost-heave force. Depending on the different locations of waterlogged space, the local frost-heave force can reach 1.0–2.79 MPa. Usually, the strength of the shotcrete of the surrounding rocks in the initial stage should reach the strength of C20, and a waterproof board should be installed between the secondary lining to introduce water seepage into the drainage system. Even if the initial spray layer can't contact tightly with the secondary lining, which could cause a little "cavity" left in between. However, unless there is a water supply, or only by the freezing and expanding of the water body from the "cavity", it will be difficult to pose a threat to the secondary lining support. ③ Assuming that the water-containing weathered layer frosts (Zhang & Wang, 2004). A large number of frost damage surveys in Hokkaido, Japan and Northeast China showed that for any tunnel with frost damages, the surrounding rocks around the lining had a 0.1–0.2 m thick weathered layer, and most of them were intercalated with free and thin layer of ice when frozen. According to the test, the frost heaving ratio was generally 20–50%, and when there was abundant groundwater, the frost heaving ratio could reach 100–125%. In fact, the weathered layers of the surrounding rocks behind the tunnel lining are mostly loose and broken rock blocks, and there is no water migration during the freezing process. The frost-heave force under drainage conditions is only 9% of the expansion ratio of water freezing. However, when the drainage channel is firstly frozen and then blocked, the frost heaving amount should be calculated according to Eq. 4.15. The overall displacement may appear in the weathered layer. According to Zhang Zhidao, when the thickness of frost-heaving layer is 100 mm, the frost-heave force will be 1.2 MPa to 3.1 MPa. Under the condition of cold-proof measures, the frost-heave force can be weakened. It is suggested that the frost-heave force at the opening of the tunnel should be 0.9 MPa and that at the inner section of the tunnel should be 0.6 MPa.

Practice has proved that the influence range of the cavern excavation on the surrounding rocks is limited, and the influence range is about 3–5 times of the cavern diameter from the center of the cavern. According to the results of the indoor model test, the greater the freezing depth of the surrounding rocks, the greater the frost-heave force, and the stronger the top constraint, the greater the frost-heave force. For the curved batter and vertical wall lining, the top constraint has a great influence on the frost-heave force on the arch and inverted arch, and has little influence on the frost-heave force of the side wall. The influences of the top constraint of the circular lining on the frost-heave force of the whole lining are not much different in each part. The maximum frost-heave force always occurs in the section of the lower arch foot. For the frost-heave force on the vertical wall lining, it is the largest, bit for that on the circular lining, it is the smallest.

3. **Frost-Heave Force Levels of Tunnels in Frozen Soil Areas**

For the tunnels, most of the surrounding rocks belong to the rock (hard rock and soft rock) and fracture zone (weathered layer and fault zone, etc.). Water and negative temperature become the dominant factors for the frost-heave force in the surrounding rocks of the tunnel. During the freezing process, the water increment (supply) and the constraint conditions can make the frost-heave force vary greatly, from several kPa to dozens of MPa. Under the condition that the frost-heave force is small, the secondary lining of the tunnel can be designed according to the standard condition, and the design requirements can be met without using reinforced plain concrete. According to the tunnel design specification, the ultimate compressive strength and ultimate tensile strength are used to control the bearing capacity and calculate the minimum safety factor. The grade IV, V and VI surrounding rocks are selected to calculate the internal forces of structure corresponding to different frost-heave forces under the conditions of deep burial and shallow burial, so as to determine the threshold values of frost-heave forces corresponding to different surrounding rocks at different buried depths (Table 4.23). The threshold values of frost-heave force in the table are the active forces. The force values measured on site are the surrounding rock pressures when the support and the surrounding rocks reach a balance after the surrounding rocks freeze, that is, the generalized frost-heave force. After the corresponding transformation of the calculation model, the critical value of frost-heave force grade under the impact of generalized frost-heave force is obtained (Table 4.24).

For the lining section in Kunlun Mountain Tunnel along the Qinghai-Xizang Railway, through calculation and analysis, and combined with the model test and the field test results, the classification standard of frost-heave force for the tunnels in frozen soil areas is determined.

Table 4.23 Threshold values of frost-heaving of varying levels under the impact of generalized frost-heave force (Bai et al., 2007)

Working conditions	Buried depth	Surrounding rocks	Average value of surrounding rock pressure at the threshold value/MPa	
			Weak to moderate frost heave	Moderate to strong frost heave
1	Deep burial	Level IV	0.070	0.382
2	Deep burial	Level V	0.073	0.353
3	Deep burial	Level VI	0.076	0.334
4	Shallow burial	Level IV	0.341	0.707
5	Shallow burial	Level V	1.074	1.441
6	Shallow burial	Level VI	1.141	1.800

Table 4.24 Threshold values of frost-heave force levels under different working conditions (Bai et al., 2007)

Working conditions	Buried depth	Surrounding rocks	Concrete	Stress control point	Threshold values of weak and medium frost heave/MPa		Threshold values of frost heave and strong frost heave/MPa	
1	Deep burial	Level IV	C30	Side wall	0.04 to 0.05	When the frost-heave force is greater than this threshold value, the supporting structure must consider reinforced concrete	0.2 to 0.3	When the frost-heave force is greater than the threshold value, the structure of frost resistance should be considered to strengthen
2	Deep burial	Level V	C30	Side wall	0.04 to 0.05		0.2 to 0.3	
3	Deep burial	Level VI	C30	Side wall	0.04 to 0.05		0.2 to 0.3	
4	Shallow burial	Level IV	C30	Side wall	0.4 to 0.5		0.5 to 1.0	
5	Shallow burial	Level V	C30	Vault, inverted arch	1.4 to 1.5		1.5 to 2.0	
6	Shallow burial	Level VI	C30	Vault, inverted arch	1.5 to 1.6		2.0 to 2.5	

Note: Under the condition of shallow burial, the threshold value of frost-heave force is the peak value of frost-heave force distribution

1. Under the condition of deep burial, when the generalized frost-heave force is less than 0.07 MPa or within the range of 0.07 MPa to 0.3 MPa, it is moderate frost heave. Generally, the safety requirements for the lining can be met by increasing the strength of the concrete or increasing the amount of reinforcement.When the generalized frost-heave force is greater than 0.3 MPa, it is strong frost heave, and the heat insulation and heat preservation measures need to be taken to prevent frost damages.
2. Under the condition of shallow burial, when the peak value of generalized frost-heave force distribution is less than 0.3 MPa, it is weak frost heave, and the plain concrete can meet the safety requirements for the lining. When the peak value of generalized frost-heave force distribution is within the range of 0.3–0.7 MPa, it is moderate frost heave, and the safety requirements for the lining can be met by improving the strength of concrete or increasing the amount of reinforcement. When the peak value of generalized frost-heave force distribution is greater than 0.7 MPa, it is strong frost heave, and the heat insulation and heat preservation measures should be taken.

4.3.1.3 Freezing of Waterproofing and Drainage System

The essential cause for tunnel frost damages in frozen soil area is the impact of the repeated freeze-thaw of water seepage in surrounding rocks behind the lining, which causes the lining frost cracking, crumbling, spalling, ice hanging, and water seeping and icing on the ballast bed (Fig. 4.76). The water temperature of the surrounding rock aquifer of the tunnel in the frozen soil area is relatively low. Generally, the water temperature of the aquifer in the regular seasonally frozen soil area is 6–8 °C, in the deep seasonally frozen soil area is 4–6 °C, and in permafrost regions is 0–4 °C. In freezing period, the water temperature in surrounding rocks is stable. For example, the annual variation amplitudes of water temperature in Daban Mountain Tunnel of Qinghai Province is less than 1.6 °C, and in Baikaer Tunnel of Daxing'anling Mountain is less than 1.7 °C. Generally, the groundwater temperature of the tunnel in frozen soil area is affected by the ground temperature of the surrounding rocks. The water temperature with the depths of the active layer at the opening of the tunnel is easily affected by the air temperature and fluctuates with the air temperature. As going into the tunnel, the water temperature will be affected by the ground temperature of the frozen soil layer and decrease. However, the water temperature will increase in the middle of the tunnel. When the tunnel is getting through, the air temperature in the tunnel also has an impact on the water temperature in the surrounding rocks, especially in the opening section, which fluctuates with the air temperature. During the cold season, the low-temperature tunnel groundwater (including the water flow of the drainage system) is easily affected by the negative temperature in the tunnel and freezes, causing a series of engineering problems related to the freezing of the water.

The frost damages during the operation stage of tunnels in permafrost regions and in the deep seasonally frozen soil area are almost caused by the infiltration of groundwater in the surrounding rocks. Some are caused by the seepage at the lining joints (Fig. 4.100), some are caused by water along the lateral drainage ditch of the fault zone emerging and submerging the ballast bed, and some are arch foot joints or cracks with seepage-curtain hanging on the side wall.

Permafrost exists as an impermeable layer, making the types and distributions of groundwater more complicated. Generally, according to the relationship between the groundwater and the permafrost layer, the groundwater is classified as the suprapermafrost water, the interpermafrost water and the subpermafrost water.

1. Suprapermafrost water: it occurs in the active layer above the permafrost table, and it is recharged by atmospheric precipitation, ice meltwater and snow meltwater, and infiltrated soil water. The water volume and water temperature have seasonal variations and they are also abundant, which have a significant impact on the tunnels with shallow burial and the inlet and outlet sections;
2. Interpermafrost water: it occurs in the permafrost layer, and it could be with high salinity, or it is in a special geological structure such as a fault zone, usually closed or semi-closed in the surrounding rocks.The water temperature is close to 0 °C, and it has pressure-bearing properties, and its degree of influence depends on the passing conditions of the tunnel and the groundwater recharge.

Fig. 4.100 Water seepage freezing and ice covered the road surface in the tunnel

3. Subpermafrost water: it occurs in the aquifer at the permafrost base, and it may be sealed under the permafrost layer, or exposed to the surface through the fault zone. The water is abundant, high in temperature or is hot spring, and it has pressure-bearing properties and exerts a great and even serious impact on the tunnel crossing the permafrost layer.

4.3.1.4 Frost Damage Level of Tunnels in Cold Regions

1. **Frost Damage Level of Tunnels in Cold Regions**

Generally, when the annual average temperature of the ground surface is lower than 0 °C, there are conditions for the formation of permafrost, and when the daily average surface temperature is lower than 0 °C, there will be an environment for the formation of frozen soil. In the Daxing'anling Mountains, the freezing index is greater than 2900 °C·d, and when the annual air temperature is 0 °C to 1 °C or −1 °C to 0 °C (in the wilderness area), the permafrost layer exists. The Qinghai-Xizang Plateau is affected by the altitude, with the freezing index greater than 1700 °C·d, and the permafrost layer exist when the annual air temperature is −2.0 °C. When the freezing index in the seasonally frozen soil area is greater than 500 °C·d, the freezing depth can be greater than 1 m, which will cause mild to moderate frost damages.

The monthly air temperature in January is selected as one of the first-level zoning indexes, which can better reflect the degrees of cold and heat in a region and reflect the characteristics of the seasonal freezing depth. From the statistical relationship between the monthly air temperature in January, the coldest period, and the maximum seasonal freezing depth of 108 weather stations in China (Fig. 4.101), it indicates a good correlation. In fact, the monthly air temperature in January can also better reflect the annual air temperature and the climate during the cold period in the

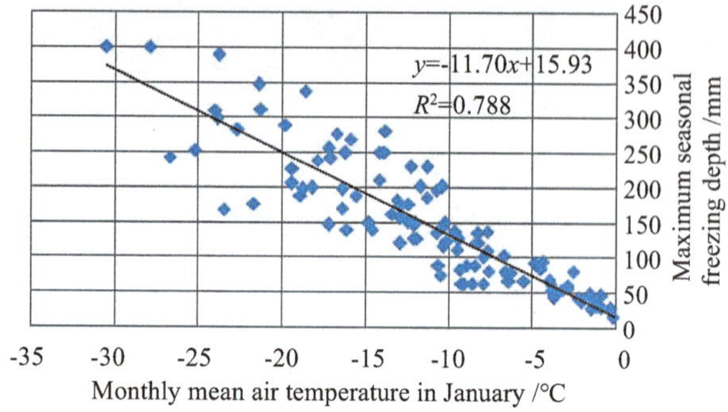

Fig. 4.101 Relationship between the monthly mean air temperature in January and seasonal freezing depth

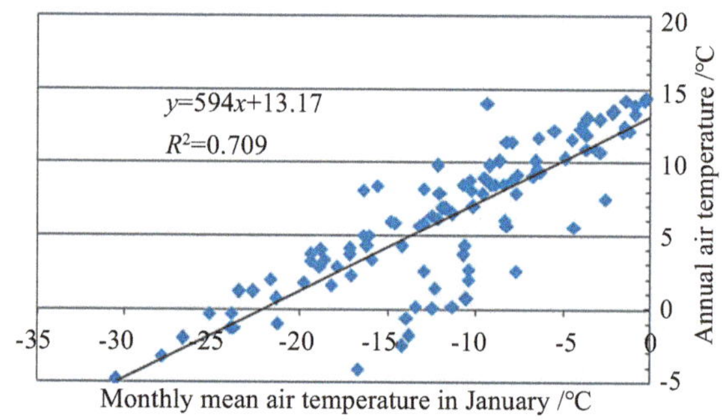

Fig. 4.102 Relationship between the monthly air temperature in January and the annual average temperature

region (Fig. 4.102). January mean air temperature correlates well with maximum seasonal freezing depth and is suitable as a first-level zoning index. Although it is not as good as the correlation between the freezing index and freezing depth, it can meet the engineering design requirements. Moreover, it is easier to directly obtain the data from the weather station, and there is no need to count the daily negative temperature values during the negative temperature period throughout the year.

The main factors for tunnel frost damages are the negative temperature and the water. The former involves water freezing and freezing depth of the surrounding rocks, while the latter involves water occurrence and water supply in surrounding rocks. The lower the negative temperature is, the deeper the freezing depth and the higher the possibility of frost damage. The water occurred in the surrounding rocks is mostly fissure water, and the amount of water supply during the freezing process will affect the severity of the tunnel frost damages. Generally, except that there are loose soil layers in the opening section or the fault zones of some tunnels, the main body of the surrounding rocks in the tunnel is mostly rock (hard rock or soft rock). According to the influencing factors of frost heave of foundation soil, the factor "soil quality" can be ignored, and only water and negative temperature need to be considered as the main factors for tunnel frost damages. The tunnel frost damages are mainly manifested in lining concrete fractures, cracking, spalling, water leakage, ice hanging, seeping and freezing on the bottom of the tunnel, ice jam in the tunnel, cracking in the tunnel portal and thaw slumping, etc. When classifying the tunnel disease grades, two aspects should be considered: lining deformation (failure) caused by frost-heave force and water seeping and freezing (water seepage from lining cracks and unfavorable drainage of seepage). With reference to the data of Sun Bing (in 2012) and Luo Yanbin (in 2010), the revised classification of frost damage grades for tunnels in cold regions (Table 4.25) is the following.

The frost damages of tunnels in cold regions are characterized by the frost cracks in the lining layer and the water seeping and freezing in the tunnel. Based on this, different levels of frost damages are determined:

Level I: there are no cracks in the lining layer, without freezing or with weak freezing, which doesn't affect the smooth drainage.

Level II: there are micro-cracks in part of the lining layer, and water seeping and freezing, which doesn't affect the overall strength of the lining layer and the traffic.

Level III: there are frost cracks in the lining layer, and water seeping and ice hanging in the tunnel, and the drainage ditch overflows and the water freezes on the road surface, which mostly occurs from December to February of the following

Table 4.25 Frost damage levels of tunnels in cold regions

Frost damage levels					Groundwater				
					Dry	Dripping	Seepage	Small gushing water	Large gushing water
					Frost-heave force at lining back				
					Without frost heave	Weak frost heave	Frost heave	Strong frost heave	Extremely strong frost heave
Air temperature	Air temperature zoning	Monthly air temperature in January/°C	Maximum freezing depth/m	Degree of damage	None	Light	Moderate	Heavy	Severe
	Temperate	0 to -5	0.2 to 0.8	None	I	I	I	I	I
	Cold	-5 to -10	0.8 to 1.4	Light	I	II	II	III	IV
	Frigid	-10 to -15	1.4 to 2.0	Moderate	I	II	III	IV	IV
	Very frigid	-15 to -25	2.0 to 3.0	Heavy	I	III	IV	IV	V
	Severe frigid	<-25	>3.0	Severe	I	IV	IV	V	V

Note: 1. Frost damage grade: I means non-frost damage or slight frost damage; II means mild frost damage; III means moderate frost damage; IV means heavy frost damage; V means severe frost damage

2. Monthly air temperature in January: < −25°C indicates the permafrost regions; −15 ~ −25°C indicates the isolated permafrost regions or deep seasonally frozen soil areas; −10 ~ −15°C indicates the deep or medium seasonally frozen soil areas;−5 ~ −10°C indicates the moderate or shallow seasonally frozen soil areas; 0 ~ −5°C indicates the shallow seasonally frozen soil areas

3. Groundwater conditions (Ma, 1990): the dripping water refers to the water inflow less than 0.01; the seepage refers to the water inflow of 0.01–0.1; the small gushing water refers to the water inflow of 0.1–1; the large gushing water refers to the water inflow of 1–10; unit: $\times 10^4$ m^3/d

4. Classification of lining frost-heave force: non-frost damage refers to the frost-heave force less than 0.01; weak frost heave refers to the frost-heave force of 0.01–0.5; frost heave refers to the frost-heave force of 0.5–1.0; strong frost heave refers to the frost-heave force of 1.0–2.5; extremely strong frost heave refers to the frost-heave force greater than 2.5; unit: MPa

year, affecting the safety of driving. It is necessary to adopt reinforced concrete, thermal insulation layer structure, and deep buried water ditch measures.

Level IV: the freezing period is more than 4 months, there are severe cracks in the lining layer, large fractures and leakages in the tunnel portal, spandrels and joints, and serious ice hanging (on the vault and the side wall), and large area of the road surface is covered with ice, seriously affecting the traffic safety. It is necessary to use reinforced concrete and thermal insulation layer structure, as well as the deep buried ditch with double-layer insulation or central drain cavern.

Level V: the freezing period is more than 5 months, the lining layers are severely damaged, the spalling occurs, the water leakage happens and freezes in large scale, the hanging ice springs up like a forest, the drainage system is blocked by ice, serious freezing occurs on the road surface and the ice jam happens in the tunnel, and the traffic in the tunnel has to be stopped. It is advisable to adopt reinforced concrete and thermal insulation layer structure, as well as the grouting for water blocking and the central drain cavern measures.

For the tunnel, when it is in an area where the monthly air temperature in the coldest month is higher than −5 °C and the freezing depth is higher than 0.8 m, whether the tunnel contains water or not, the frost damages will basically not occur. The dry tunnels do not suffer from frost damages due to severe cold air temperatures. When crossing the permafrost layer, especially under the condition of the surrounding rocks with thick layer of ice, it should be treated differently according to the mean annual ground temperature. In cold permafrost regions, a small amount of water dripping and seeping occurs during the excavation process due to human activities. In the warm permafrost regions, there may be a large thawing ring in the surrounding rocks and a large impact of freeze-thaw cycle, especially when there are fault zones and subpermafrost water, it may cause serious frost damages.

2. **Tunnel Zoning in Cold Regions**

The cold regions include the permafrost regions and the seasonally frozen soil areas. When the surrounding rocks through which the tunnels pass are the permafrost layers, the medium or deep seasonally frozen soil layers, the frost damages could occur. According to the condition of the surrounding rocks in the frozen soil layer that the tunnels pass through, the tunnels can be divided into the following types.

(1) Tunnels in full Permafrost Regions

The surrounding rocks that the tunnels pass through are permafrost layers, and there are seasonally thawed layers in the inlet and outlet sections in warm season. In large continuous permafrost regions, the thickness of permafrost in cold permafrost region is more than 80–100 m, and the thickness of permafrost in isolated permafrost region is generally 40–50 m. The characteristics of the Qinghai-Xizang Plateau and mountainous permafrost regions in western China, with the increase of altitude and the influence of snow cover and glaciers, the lower is the air temperature, the greater is the thickness of permafrost. The permafrost region in the Daxing'anling Mountains belongs to the high-latitude permafrost region. The permafrost with the largest thickness is distributed in low-lying marsh wetlands. Affected by the temperature inversion layer, the permafrost in the high mountains is thin. The Kunlun Mountain Tunnel and the Fenghuo Mountain Tunnel on the Qinghai-Xizang Plateau are the full permafrost tunnels.The mountain that Kunlun Mountain Tunnel passes through belongs to Triassic slate sandwiched with schist. The maximum buried depth of Kunlun Mountain Tunnel is 106 m, the maximum permafrost base within the tunnel section is 120 m, and the maximum depth of the permafrost table is about 3 m. At a depth of 1.5 m at the entrance section and 1.8 m at the exit section, they are ice-saturated permafrost, and the main tunnel is located in the icy or ice-poor permafrost.The mountain that Fenghuo Mountain Tunnel passes through is Tertiary sandstone and mudstone. The maximum buried depth of the tunnel is 96 m. At a depth of 1.5 m at the entrance section and 2.4 m at the exit section are the ice-rich and ice-saturated permafrost, and the tunnel is located in the icy and ice-poor permafrost and 1.0–2.5 m thick ice-rich and ice-saturated permafrost occurs in the weathered layer of local bedrock.

Another case is the sidehill tunnel in permafrost regions. Due to the shallow buried depth, the tunnel passes through between the permafrost table and the permafrost base, the opening sections of the tunnel are in seasonally thawed layers, and the main tunnel is in permafrost.

Generally, the surrounding rocks of the full permafrost tunnel are in a frozen state all year round, which blocks the supply of external water sources. Except for the water seepage in the thawing ring during the excavation process, the possibility of water leakage is small, so the frost-heave force of the surrounding rock behind the lining is also small. When there are broken rocks in the fault zone, there may be interpermafrost water or hydraulic connection with the surface water in the valley, and there could be a large water seepage or gushing water, such as the normal water inflow in Kunlun Mountain Tunnel reaches 222.55 m^3/d. In the warm permafrost sidehill tunnel with a shallow burial, when the surrounding rocks are broken or the fissures and joints are relatively developed, the ground surface water may infiltrate, causing the tunnel to seep and freeze. For example, in a valley of Kunlun Mountain, under natural conditions, the water flowed regularly, the permafrost table surveyed was 3–4 m deep, and the permafrost table decreased to 10 m ~ 13 m after the dam was built to accumulate water. Therefore, it is necessary to determine the tunnel design principles and schemes according to the geological and hydrogeological conditions of the tunnel in full permafrost regions.

(2) Tunnels in Partial Permafrost Regions

There is seasonally thawed layer, permafrost layer and non-permafrost layer in the surrounding rocks that the tunnels pass through, and the annual air temperature is lower than 0 °C in cold regions. A large number of tunnels in permafrost regions belong to this type. The tunnels in the permafrost regions of Daxing'anling Mountains in Northeast China are almost partial permafrost tunnels. The deep buried tunnels in the cold permafrost regions, and the middle part of the tunnel in isolated permafrost regions are often in non-permafrost layers.The strong freeze-thaw cycles often occur in tunnels and at the exit sections. The surface water erosion, the water infiltration and the intrusion of confined underground water could cause permeation of surface water into the tunnel through the fault zone, so the water seepage should be drained by thermal insulating deep-buried ditches or drain cavern, and the thermal insulating outlets should be used to avoid freezing. For example, for the Lingding Tunnel in the permafrost of Daxing'anling Mountain and No. 2 Tunnel of Cuiling in Huzhong branch line, under the influence of cold air temperature, the minimum temperature in the tunnel and the drain cavern can still reach −2.5 °C, so that the ice jam phenomenon occurs in the drainage ditches and the drain caverns. In such tunnels, it may form the epigenetic permafrost after the tunnels are getting through.

(3) Non-permafrost Tunnels

The surrounding rocks that the tunnels pass through are totally non-frozen soil layers, which are positive temperature rock stratum. Such tunnels are distributed in the isolated permafrost regions, the alpine mountainous areas and the medium to deep seasonally frozen soil areas. There are seasonally frozen soil layers in the opening sections of the tunnels, and non-frozen soil layers in the tunnel. The annual air temperature is close to 0 °C, and the monthly air temperature in the coldest month is below 0 °C. After the tunnels are getting through, if the seasonal freezing depth in the opening section of the tunnel is relatively large, the epigenetic permafrost may form. Under the regular circumstances, the surface water or groundwater seeps into the tunnel and can be drained through the drainage gutter, or, if necessary, through the deep buried ditch whose outlet has been set up with thermal insulation measures.

3. **Precautionary Measures for Tunnel Frost Damages in Cold Regions**

According to the freezing degree, the frost-heave force of the lining and durability or the fatigue strength of the concrete under freeze-thaw cycles in the tunnel in cold regions, the corresponding precautionary measures should be carried out according to the level of tunnel frost damages. The degree of frost damage in lining structure can be determined by the safety factor of frost-heave force.When the frost-heave force is very small, the design requirements can be achieved by using the plain concrete, which belongs to non-frost damage or weak frost damage. But when the frost-heave force reaches a certain extent, the plain concrete cannot meet the safety requirements, and the lining needs to be reinforced to meet the safety requirements, which belongs to moderate frost damage. When the frost-heave force is further increased, the reinforced concrete still can't meet the safety requirements, and certain measures need to be taken to make the lining in a safe state, such as adding an additional insulation layer etc., which belongs to strong or extremely strong frost damage. The precautionary measures for different levels of frost damages in tunnels can be determined by referring to Table 4.26.

(1) For the durability of the frost resistance of concrete lining, as the concrete has a damage extent and residual strength within a specified time, when the damage extent reaches the failure criterion within the specified period of time, some

Table 4.26 Levels of frost damages and corresponding precautionary measures for tunnels in cold regions (Sun, 2012)

Level of frost damage in the tunnel	Degree of lining damage		
	Non-frost damage or weak frost damage	Moderate frost damage	Strong or extremely strong frost damage
Non-frost damage I	It can be designed according to the regular tunnel. When the frost-heave force is close to the weak and medium frost-heave force, the reinforced concrete structure should be adopted, and other precautionary measures are not needed	The reinforced concrete structure must be used. When the frost-heave force is close to medium and strong frost-heave force, the thermal insulation measures should be adopted	The reinforced concrete structure must be used. When the frost-heave force is close to medium and strong frost-heave force, the thermal insulation measures should be adopted
Weak frost damage II	The cold-proof ditches need to be used. When the frost-heave force is close to weak and medium frost-heave force, the reinforced concrete structure should be adopted	It has to use cold-proof ditches, the reinforced concrete structure and the thermal insulation measures	The cold-proof ditches need to be used. The reinforced concrete structure must be used. When the frost-heave force is close to medium and strong frost-heave force, the thermal insulation measures should be adopted
Moderate frost damage III	The deep-buried seepage ditch should be used. When the frost-heave force is close to weak and medium frost-heave force, the reinforced concrete structure should be used	When the freezing depth is shallow, the deep-buried seepage ditch should be used. When the freezing depth is relatively deep, the drainage ditch should be used, and the reinforced concrete structure must be used. When the frost-heave force is close to medium and strong frost-heave force, the thermal insulation measures should be adopted	When the freezing depth is shallow, the deep-buried seepage ditch should be used. When the freezing depth is relatively deep, the drainage ditch must be used, and the reinforced concrete structure and thermal insulation measures must be adopted

(continued)

Table 4.26 (continued)

Level of frost damage in the tunnel	Degree of lining damage		
	Non-frost damage or weak frost damage	Moderate frost damage	Strong or extremely strong frost damage
Strong frost damage IV	The deep-buried seepage ditch should be adopted. When the construction joints are leaking, the grouting for water-blocking should be carried out in the corresponding location. When the frost-heave force is close to the weak and medium frost-heave force, the reinforced concrete structure should be adopted	When the freezing depth is shallow, the deep-buried seepage ditch should be used. When the freezing depth is deep, the drainage ditch should be adopted. When there is water seepage from cracks in construction joints or the local structure, the grouting for water-blocking should be carried out in the corresponding parts, and the reinforced concrete structure must be used. When the frost-heave force is close to medium and strong frost-heave force, the thermal insulation measures should be adopted	When the freezing depth is shallow, the deep-buried seepage ditches and the cold-proof ditches should be used. When the freezing depth is deep, the drainage ditch (drain cavern) and cold-proof ditch should be used. When there is water seepage from cracks in construction joints or the local structure, the grouting for water-blocking should be carried out in the corresponding parts, and the reinforced concrete structures and the thermal insulation measures must be adopted
Extremely strong frost damage V	The cold-proof and thermal insulating doors or the snow-proof and thermal insulating sheds should be used at the openings of the tunnels, and the deep-buried seepage ditches should be used. When water seeps in the construction joints, the grouting for water-blocking shall be carried out in the corresponding parts. When the frost-heave force is close to weak and medium frost-heave force, the reinforced concrete structure should be used	The cold-proof and thermal insulating doors or the snow-proof and thermal insulating sheds should be used at the openings of the tunnels, and when the freezing depth is shallow, the deep-buried seepage ditch should be used. When the freezing depth is deep, the drainage ditch (drain cavern) should be used. When there is water seepage from cracks in construction joints or the local structure, the grouting for water-blocking should be carried out in the corresponding parts, and the reinforced concrete structure must be used. When the frost-heave force is close to medium and strong frost-heave force, the thermal insulation measures should be adopted	The cold-proof and thermal insulating doors or snow-proof and thermal insulating sheds should be used at the openings of the tunnels. When the freezing depth is shallow, the deep-buried seepage ditch and the cold-proof ditch should be adopted. When the freezing depth is deep, the drainage ditch (drain cavern) and cold-proof ditch should be adopted. When water seepage occurs in construction joints or the local structure, the grouting for water-blocking should be carried out in the corresponding parts, and if necessary, the local heating measures should be adopted, and the reinforced concrete structures and thermal insulation measures must be adopted

corresponding measures should be taken to strengthen the frost resistance durability of concrete, such as reducing the water-cement ratio, adding air-entraining agent, and applying thermal insulation layer to reduce the time of freezing, etc.

(2) For the fatigue damages of frost-heave force in concrete lining, when the damage extent reaches the failure criterion within the specified time, corresponding measures, such as applying thermal insulation layers to reduce the magnitude of frost-heave force and the time of fatigue cycles, improving concrete strength and so on, should be taken.

4.3.2 *Thermal Insulation Design for Tunnels in Cold Regions*

The particularity of tunnel engineering in cold regions has given rise to the special frost damage characteristics of tunnels in frozen soil areas (in permafrost regions and in seasonally frozen soil areas), which can be summarized into two categories: one is the damage of lining concrete structure; the other is the freezing inside the tunnel. Therefore, the basic countermeasures for tunnel engineering design in permafrost regions are as follows:

① the time of freeze-thaw cycles of the surrounding rocks of the tunnel and the thickness of the active layer in the freeze-thaw ring should be controlled (or reduced), that is, to control the variation of thermal regime of the surrounding rocks of the tunnel;

② the waterproof and drainage system of the tunnel should be designed reasonably to ensure the smooth and safe operation of the system;

③ the strength, impermeability and durability of lining concrete structure should be enhanced.

4.3.2.1 Design for Thermal Insulation Layer in Tunnel

Whether it is a full permafrost tunnel, or a partial permafrost tunnel, or a non-permafrost tunnel, there is a certain depth of freeze-thaw ring in the surrounding rock in the whole section of the tunnel or in the local section from the exit. Under the influence of repeated freezing and thawing cycles caused by the fluctuating external air temperatures, it leads to damage to the tunnel lining and the occurrence of frost damage.

One of the designs for tunnel engineering in cold regions is to minimize the range of freeze-thaw ring in the surrounding rocks of tunnel. The way is to use thermal insulation techniques, that is, to set up the thermal insulation layer inside the tunnel lining layer (or outside the tunnel lining layer) to reduce the heat exchange between the air temperature and the surrounding rocks in the tunnel.

Several factors should be considered in the design for thermal insulation layer in tunnels:

1. The maximum freezing (thawing) depth of the surrounding rocks in the area where the tunnel is located, the monthly air temperatures in the coldest month and in the hottest month, and the annual range of the air temperature is the main basis.
2. The thermal insulation layer should be set between the two layers of lining, and its thickness should not only meet the thermal insulation effects, but also have sufficient strength to ensure the stability of the thermal insulation layer itself.
3. The thermal insulation layer should have low toxicity, high flame retardancy, and be easy to operate during construction.
4. The tunnel structure design should provide conditions for laying thermal insulation layers.
5. The aging of the thermal insulation layer which increases its thermal conductivity should be fully considered so that there are still remedial conditions when it fails.

The thermal insulation layer can be set outside the lining layer or inside the lining layer, which is determined according to the conditions and requirements. It is generally set between the primary support and the secondary lining, such as in the Kunlun Mountain Tunnel and Fenghuo Mountain Tunnel, and it also can be set outside the secondary lining, such as in Daban Mountain Tunnel.

4.3.2.2 Selection of Thermal Insulation Materials

Due to the special climate and geological conditions in cold regions, especially for the full permafrost tunnel or the partial permafrost tunnel, the construction should be carried out at a low temperature or near 0 °C. The selected thermal insulation materials should have good low-temperature performance, and the bonding strength and setting time at low temperature have to meet the construction requirements. The industrial thermal insulation materials currently used include EPS (expanded polystyrene board), XPS (extruded polystyrene foam board) and PU (foaming polyurethane board). From the comparison of basic physical properties of these materials, PU has advantages in thermal conductivity and temperature resistance, XPS has advantages in compressive strength and humidity resistance, EPS has no advantage in other aspects but a slight advantage in price.

4.3.2.3 Determination of the Thickness and Thermal Conductivity of the Thermal Insulation Layer in Tunnel

1. **Calculation Method for the Thickness of the Thermal Insulation Layer in Tunnel**

Based on the calculation methods for the thickness of thermal insulation layer for the frost prevention of tunnel, there are the equivalent thickness conversion method, the meteorological analytical method and the finite element modelling calculation method. The equivalent thickness conversion method is more suitable for the engineering practice. In tunnel design, measures to increase the lining or thermal insulation can be taken to avoid or reduce the seasonally thawed layer of lining in frozen surrounding rock, which can prevent the occurrence of frost damages in tunnel. The thickness of seasonally thawed layer of the lining in surrounding rocks can be measured, or the thickness of the seasonally thawed permafrost layer can be calculated by Stephen's formula.

The concrete (or reinforced concrete) lining is used to avoid the damage of freeze-thaw cycle, and its thickness must be equal to or greater than the depth of the seasonally thawed layer, namely:

$$H_t = \sqrt{\frac{2\lambda_u\left(\frac{365T_0}{4} + \frac{365}{\pi}\sqrt{A^2 - T_0^2}\right)}{L\rho_d\left(W - W_u\right)}} \leq H_c \tag{4.17}$$

where H_t is the maximum seasonal thawing depth of permafrost (m); H_c is the thickness of the concrete lining (m); λ_u is the thermal conductivity of the soil body under the thawing state (W/(m·K)); T_0 is the average temperature of air (°C); A is the half of the annual range of air temperature (°C); L is the latent heat of water freezing or thawing, generally 33.456 J/kg in thermal calculation; ρ_d is dry density of soil (kg/m^3); W is the moisture content in soil, in decimals; W_u is the content of unfrozen water in frozen soil, in decimals.

Assuming that λ_c is the thermal conductivity of the concrete lining thickness (W/(m·K)), H_b is the thickness of the lining laid in the tunnel in cold region, and the thermal insulation layer with the thermal conductivity of λ_b, the concrete thickness H_{c1} with the equivalent thermal resistance of the thermal insulation layer can be obtained:

$$H_{c1} = \frac{\lambda_c}{\lambda_b} H_b \tag{4.18}$$

Assuming that H_0 is the thickness of the concrete in the thermal insulating lining of the tunnel in cold region, and H_b is the thickness of the thermal insulation layer, then the thickness of the entire thermal insulating lining is $H_0 + H_b$, and from the formula 4.18, the concrete thickness H_c equivalent to thermal resistance of the thermal insulating lining with the thickness of $H_0 + H_b$ can be obtained:

$$H_c = H_0 + \frac{\lambda_c}{\lambda_b} H_b \tag{4.19}$$

Substitute Eq. 4.19 into Eq. 4.17, and then:

$$\sqrt{\frac{2\lambda_u\left(\frac{365T_0}{4} + \frac{365}{\pi}\sqrt{A^2 - T_0^2}\right)}{L\rho_d\left(W - W_u\right)}} \le H_0 + \frac{\lambda_c}{\lambda_b} H_b \tag{4.20}$$

When the thickness of the thermal insulation layer laid in the tunnel in cold region is constant, the optimal thermal conductivity will be:

$$\lambda_b = \frac{\lambda_c H_b}{\sqrt{\frac{2\lambda_u\left(\frac{365T_0}{4} + \frac{365}{\pi}\sqrt{A^2 - T_0^2}\right)}{L\rho_d\left(W - W_u\right)}} - H_0} \tag{4.21}$$

When the thermal conductivity of the thermal insulation layer laid in the tunnel in cold region is constant, the optimal thickness will be:

$$H_b = \frac{\lambda_b}{\lambda_c}\left(\sqrt{\frac{2\lambda_u\left(\frac{365T_0}{4} + \frac{365}{\pi}\sqrt{A^2 - T_0^2}\right)}{L\rho_d\left(W - W_u\right)}} - H_0\right) \tag{4.22}$$

The maximum freezing depth from local meteorological data can also be used for equivalent thickness conversion (Chen & Luo, 2007). However, the air temperatures inside and outside the tunnel are different. The air temperature inside the tunnel is higher than that outside the tunnel, but the difference at the openings of the tunnel is small. The freezing depth of the surrounding rocks in the tunnel is slightly larger and the degree of safety is greater.

The basic principles for the equivalent thickness conversion are as follows:

$$\frac{H_0}{\lambda_0 A} = \frac{H_w}{\lambda_w A} \tag{4.23}$$

where H_0 is the maximum freezing depth form the meteorological data (m); λ_0 is the thermal conductivity of the loose rock or soil on the ground surface (W/(m·K)); H_w is the converted (or measured) maximum freezing depth of the surrounding rocks (m); λ_w is the thermal conductivity of the surrounding rocks (W/(m·K)); A is the heat transfer area (m^2). For a certain section of the tunnel, it can be approximately considered that the temperature only transfers along the radial one-dimensional cylinder wall.

The thickness of the thermal insulation layer set on the lining surface is:

$$\frac{1}{2\pi\lambda_w}\ln\left[\frac{r+H_w}{r}\right]=\frac{1}{2\pi\lambda_b}\ln\left[\frac{r+H_b}{r}\right] \tag{4.24}$$

where λ_b is the thermal conductivity of the thermal insulation layer (W/(m·K)); H_b is the thickness of thermal insulation layer (m); r is the equivalent radius of the tunnel.

The thickness of the thermal insulation layer installed between the primary support and the secondary lining:

$$\frac{1}{2\pi\lambda_w}\ln\left[\frac{r+H_w}{r}\right]=\frac{1}{2\pi\lambda_b}\ln\left[\frac{r+H_c+H_b}{r+H_c}\right]+\frac{1}{2\pi\lambda_c}\ln\left[\frac{r+H_c}{r}\right] \tag{4.25}$$

In this formula, the physical meanings of the symbols in the formula are the same as the above formulas.

The length of the thermal insulation layer laid in the tunnel. The known conditions, such as the annual average temperature at the entrance and exit of the tunnel, the annual temperature amplitude, the function of the tunnel changing with the buried depth, and the thermophysical parameters of the materials etc., need to be determined at first. In permafrost regions, at a certain distance of l from the tunnel entrance or exit, the interface temperature between the primary lining and the surrounding rocks is less than or equal to 0 °C, and this length can be taken as the laying length of the thermal insulation layer. For non-frozen soil areas, when the temperature at l is higher than 0 °C, l can also be taken as the laying length. After the thermal insulation layer is laid, the air temperature in the tunnel changes along the length of the path, and a longer path is required to achieve the state where there is no heat exchange between the thermal insulation layer and the surrounding rocks. At this time, according to the basic equation of heat transfer, the relevant correction coefficient of the laying length of the thermal insulation layer can be derived.

2. **Usage of Tunnel**

(1) Analysis of Mechanical Properties of Thermal Insulating lining

The section of Kunlun Mountain Tunnel with the maximum buried depth of 106 m was constructed in two stages. The first stage was the full section excavation, and the C25 moulded reinforced concrete support with the lining thickness of 0.25 m was used. The second stage was to lay a waterproof layer at first, to lay a 50 mm thick thermal insulation layer after that, and then to lay a waterproof layer again, and finally to lay the C30 moulded reinforced concrete lining with different thickness. A finite element analysis was carried out based on the mechanical parameters of V-class surrounding rocks, the concrete and the thermal insulation materials. In the case of laying thermal insulation layer on the section, the maximum compressive stress of the thermal insulation layer was 0.234 MPa, which was located on the bottom of the inverted arch of the tunnel. And there would be no tensile stress on the thermal insulation layer, the stress of the thermal insulation layer on the vault and the side wall was small, the stress on the bottom of the tunnel was large, and the stress on the inverted arch of the tunnel was the largest. At this time, the strain was 2.85%, the density of the thermal insulation layer increased by 2.28 kg/m^3, and the thermal conductivity increased by 0.00083 W/m·K. From the strength analysis (Fig. 4.103), it was feasible to use polyurethane plates with a density of 80 kg/m^3 in Kunlun Mountain Tunnel.

The section of Fenghuo Mountain Tunnel with the maximum buried depth of 96 m was constructed in two stages. The first stage was the full section excavation, and the C25 moulded reinforced concrete support with the lining thickness of 0.25 m was used. The second stage was to lay a waterproof layer at first, and to lay a 50 mm thick thermal insulation layer after that, and then to lay a waterproof layer again, and finally to lay the C30 moulded reinforced concrete lining with different thickness. The tunnel structure diagram is shown in Fig. 4.104. The finite element analysis was carried out according to the mechanical parameters of V-class surrounding rocks, the concrete and the thermal insulation materials. In the case of laying thermal insulation layer on this section, the maximum compressive stress of the thermal insulation layer was 0.14 MPa, which was located on the bottom of the inverted arch of the tunnel. And there was no tensile stress on the thermal insulation layer, the stress of the thermal insulation layer on the vault and the side wall was small, the stress on the bottom of the tunnel was large, and the stress on the inverted arch of the tunnel was the largest. At this time, the strain was 2.2%, the density of the thermal insulation layer increased by 1.76 kg/m^3, and the thermal conductivity increased by 0.00064 W/m·K. From the strength analysis (Fig. 4.105), it was feasible to use polyurethane plates with the density of 80 kg/m^3 in Fenghuo Mountain Tunnel.

(2) Determination of Thermal Parameters of Thermal Insulation Layer in Tunnel

For the section of Fenghuo Mountain Tunnel with a buried depth of 15 m, the thermal insulation layer with a thickness of 50 mm and a thermal conductivity of 0.03 W/m·K has been adopted. Comparing the calculation results in the first year (in 2004) with the field observations at the corresponding time, the thick dash line at −1.5 °C in the warm season basically coincides with the solid line (Fig. 4.106a). And during the cold season, the difference between the thick dash line and the solid line of −3.5 °C and −4.0 °C is very small (Fig. 4.106b). Considering the climate warming in the next 50 years, the impacts of the increase of air temperature on the Qinghai-Xizang Plateau by 2.6 °C and the water migration, the predictions of temperature distribution of the tunnel in the 10th year and the 50th year are shown in Fig. 4.106 c and d. There is no seasonal thawing of the surrounding rocks in the tunnel, indicating that the thickness and thermal conductivity of the thermal

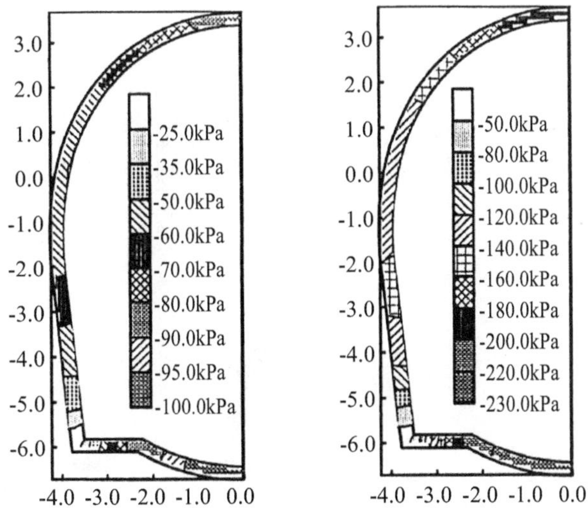

Fig. 4.103 Principal stress distribution of the thermal insulation layer in Kunlun Mountain Tunnel

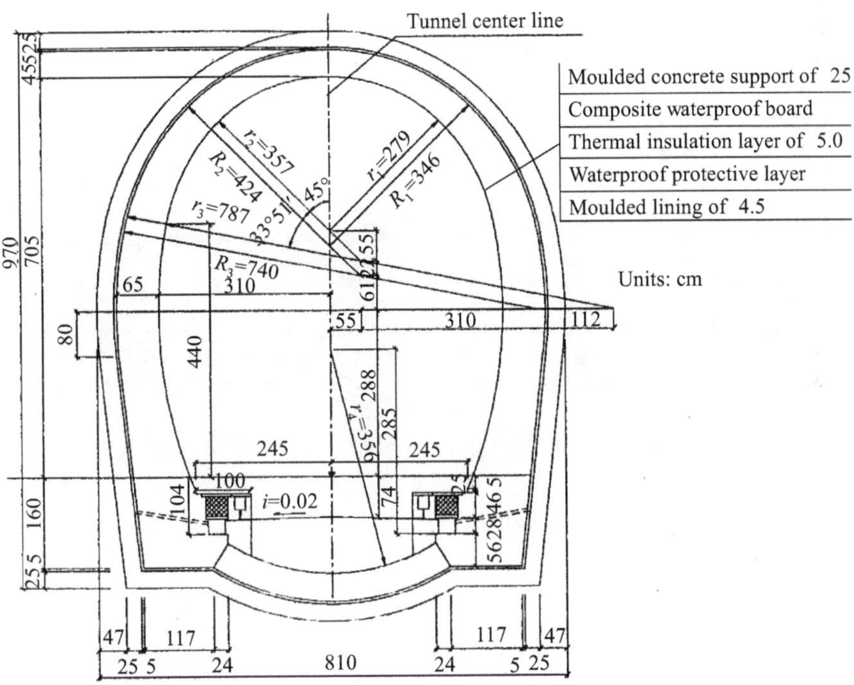

Fig. 4.104 Structural section of the level-IV surrounding rock lining in Fenghuo Mountain Tunnel (Editorial Committee of the *Qinghai-Xizang Railway*, 2016)

insulation layer used in the tunnel are feasible. Similarly, in the section of the tunnel with a buried depth of 75 m, there is no seasonal thawing of the surrounding rocks in the tunnel when using the thermal insulation layer with the same thickness and thermal conductivity (Fig. 4.107).

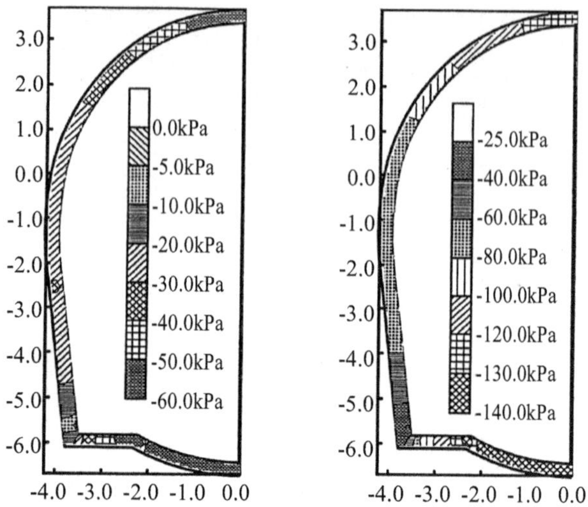

Fig. 4.105 Principal stress distribution of thermal insulation layer in Fenghuo Mountain tunnel

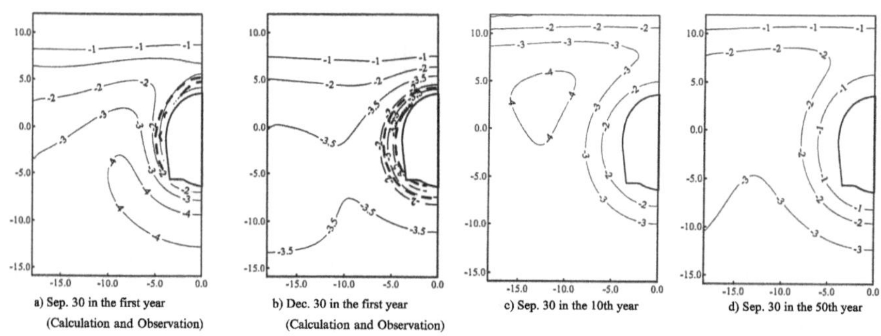

Fig. 4.106 Temperature distribution of the Fenghuo Mountain Tunnel section with the buried depth of 15 m in the 1st, 10th and 50th year (Editorial Committee of *the Qinghai-Tibet(Xizang) Railway*, 2016). (**a**) On September 30 in the First Year (Calculated Value and Field Observation Value). (**b**) On December 30 in the First Year (Calculated Value and Field Observation Value). (**c**) On September 30 in the 10 th Year. (**d**) On September 30 in the 50 th Year

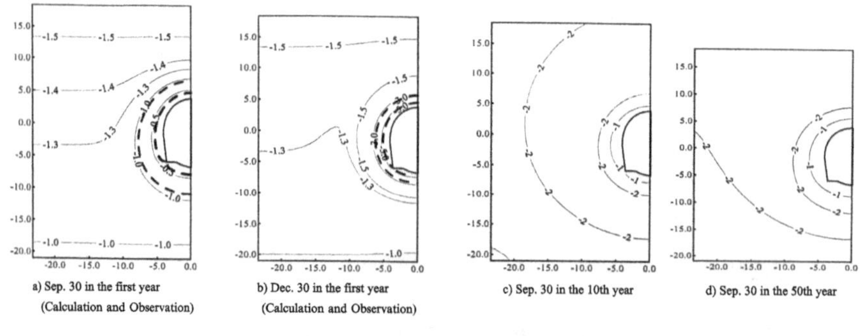

Fig. 4.107 Temperature distribution of the Fenghuo Mountain Tunnel section with a buried depth of 75 m in the 1st, 10th and 50th year (Editorial Committee of*the Qinghai-Tibet(Xizang) Railway*, 2016). (**a**) On September 30 in the First Year (Calculated Value and Field Observation Value). (**b**) On December 30 in the First Year (Calculated Value and Field Observation Value). (**c**) On September 30 in the 10th Year. (**d**) On September 30 in the 50th Year

For the section of Kunlun Mountain Tunnel with a buried depth of 13 m, the thermal insulation layer with the thickness of 50 mm and the thermal conductivity of 0.03 W/m·K has been adopted. Considering the influence of water migration and the increase of air temperature by 2.6 °C in the next 50 years, the distribution of numerical prediction temperature of the tunnel surrounding rocks in the 5th, 10th, 30th and 50th year are shown in Fig. 4.108. The results show that the maximum thawing depth of the freeze-thaw cycle in the surrounding rocks in the 5th year will be 0.15 m; the maximum thawing depth of the freeze-thaw cycle in the surrounding rocks in the 10th year is 0.21 mm; the maximum thawing depth of the freeze-thaw cycle in the 30th year is 0.35 m; the maximum thawing depth of the freeze-thaw cycle in the 50th year is 0.50 m. When the thickness of the insulation layer is changed to 120 mm, there is no seasonal thaw circle in the frozen surrounding rocks in the 50th year, but the temperature of the surrounding rocks near the tunnel lining is close to 0 °C. If the thickness of the thermal insulation layer is less than 120 mm, or the thermal conductivity is greater than 0.03 W/(m·K), a seasonal thawing ring will occur in the tunnel.

Similarly, if the Kunlun Mountain Tunnel section with a buried depth of 106 m also uses a 50 mm thick thermal insulation layer with a thermal conductivity of 0.03 W/(m·K), the numerical analysis results of freeze-thaw conditions of the tunnel within 50 years are shown as follows (Fig. 4.109): the maximum thawing depth of the freeze-thaw cycle in the 5th year is 0.16 m; the maximum thawing depth of the freeze-thaw cycle in the 10th year is 0.20 m; the maximum thawing depth of the freeze-thaw cycle in the 30th year is 0.31 m; the maximum thawing depth of the freeze-thaw cycle in the 50th year is 0.42 m. It can be seen that after climate warming, the thawing depth of the surrounding rocks in the tunnel increases year by year. When the thickness of the insulation layer is changed to 100 mm, there is no thawing ring in the frozen surrounding rocks of the tunnel, but the temperature of the surrounding rock near the tunnel lining is close to 0 °C. If the thickness of the thermal insulation layer becomes thinner, or the thermal conductivity is larger, the surrounding rocks near the tunnel lining may have a thawing ring.

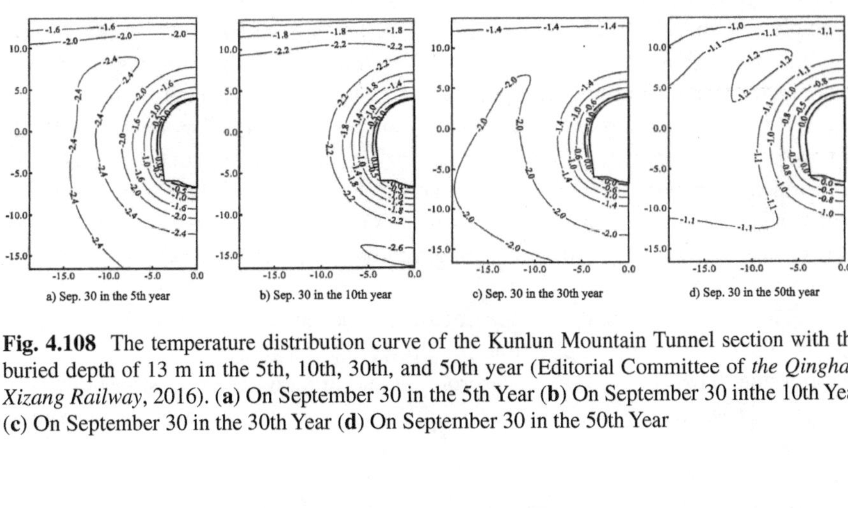

Fig. 4.108 The temperature distribution curve of the Kunlun Mountain Tunnel section with the buried depth of 13 m in the 5th, 10th, 30th, and 50th year (Editorial Committee of *the Qinghai-Xizang Railway*, 2016). (**a**) On September 30 in the 5th Year (**b**) On September 30 inthe 10th Year (**c**) On September 30 in the 30th Year (**d**) On September 30 in the 50th Year

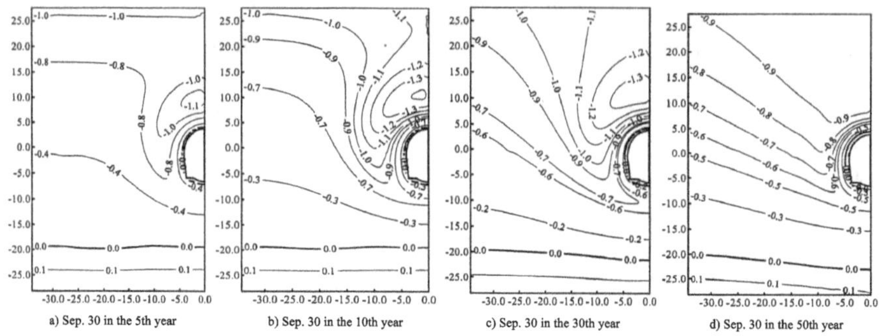

Fig. 4.109 Temperature distribution of the Kunlun Mountain Tunnel section with a buried depth of 106 m in the 5th, 10th, 30th and 50th year (Editorial Committee of *the Qinghai-Tibet(Xizang) Railway*, 2016). (**a**) On September 30 in the 5th Year (**b**) On September 30 in the 10th Year (**c**) On September 30 in the 30th Year (**d**) On September 30 in the 50th Year

4.3.3 Waterproofing and Drainage Design for the Tunnel in Cold Regions

The practice of tunnel engineering in permafrost regions shows that water plays a dominant role in the stability of tunnels. Generally, under conditions without water, the freeze-thaw cycle of surrounding rocks in the tunnel does not cause serious diseases or affect for the regular operation of the tunnel. When groundwater appears in permafrost tunnels, it leads to various diseases, of which 60% are caused by leakage, such as icicles and ice mantle, and 24% are caused by the break of concrete lining, etc.

4.3.3.1 Principles of Waterproofing and Drainage Design for Tunnel in Cold Regions

The waterproofing and drainage design for permafrost tunnels should follow the principle of "emphasis on drainage (or plugging), supplemented by plugging, draining, preventing, intercepting, and insulation", with a combination of insulation and drainage measures forming a comprehensive system. The measures for ground surface are mainly draining and intercepting, and the measures for tunnel is mainly draining. The groundwater and drainage channels should be kept unblocked according to local conditions, and the storage and leakage of groundwater in the tunnel should be eliminated. Before tunnel construction, the principle of "drain and intercept surface water first, guide and drain groundwater, and then proceed with tunnel construction" should be implemented.

The waterproofing and drainage measures for tunnels in permafrost regions should consider the design principle of "protecting permafrost" and minimize the scope of the thawing ring. The waterproofing and drainage measures for tunnels in non-permafrost regions should aim to reduce the freezing circle of the surrounding rocks.

The waterproofing and drainage for tunnels in cold regions should adopt the self-waterproof concrete structure, with permeability not lower than S8, and a frost resistance grade F = 300. Reliable waterproofing measures should be taken for the construction of the joints, the expansion joints, the settlement joints and the holes for equipment installation to ensure that there is no water seepage and no water accumulation behind the lining layer and on the road surface. Additionally, the drainage ditch needs to be kept unfrozen.

When the thickness from the outer edge of the thawing ring on the top of the shallow buried tunnel to the ground surface is less than or equal to the maximum seasonal thawing depth (such as valleys and low-lying areas), measures such as surface water guiding, blanketing, leveling, and adding thermal insulation layers should be taken on the ground surface to prevent suprapermafrost water entering the drainage system of the tunnel. The guiding measures should not reduce the thickness of the soil layer at the original thawing depth. If necessary, the anti-seepage grouting treatment can be carried out on the base of suprapermafrost water.

Thermal insulation and waterproofing measures should be set up between the primary support and the secondary lining. Attention should be paid to the anti-seepage measures for the connection between the transverse drainage and the longitudinal drainage pipe.

The drainage ditch in the tunnel should be set as a thermal insulating blind ditch. If necessary, heating cables should be added. The blind ditch connected to the outside of the tunnel also deserves attention regarding its thermal insulation measures, whose water outlet should be set up with a thermal insulating layer to avoid freezing. Water intercepting ditch and water retaining facilities should be set up at the opening of the tunnel and the front slope, and the flowing water in the road cutting should not be allowed to enter the tunnel.

It's necessary to control overbreak and collapse during construction. The overbreak and hollow parts behind the lining should be backfilled compactly with concrete at the same level to eliminate the longitudinal migration of accumulated water or groundwater along the tunnel.

4.3.3.2 Waterproofing and Drainage System of the Tunnel

The tunnel drainage system usually has two components: the waterproofing measures and drainage measures. The waterproofing measure is to plug the water outside the structural layer, and the drainage measure is to drain the water out of the tunnel (Fig. 4.110).

1. **Waterproofing Measures**

Waterproofing measures in tunnel: the measures of full-section waterproofing and thermal insulation should be taken, that is, the laying of a composite and all-in measure consisting of "a waterproof board and a thermal insulation layer and a waterproof board".

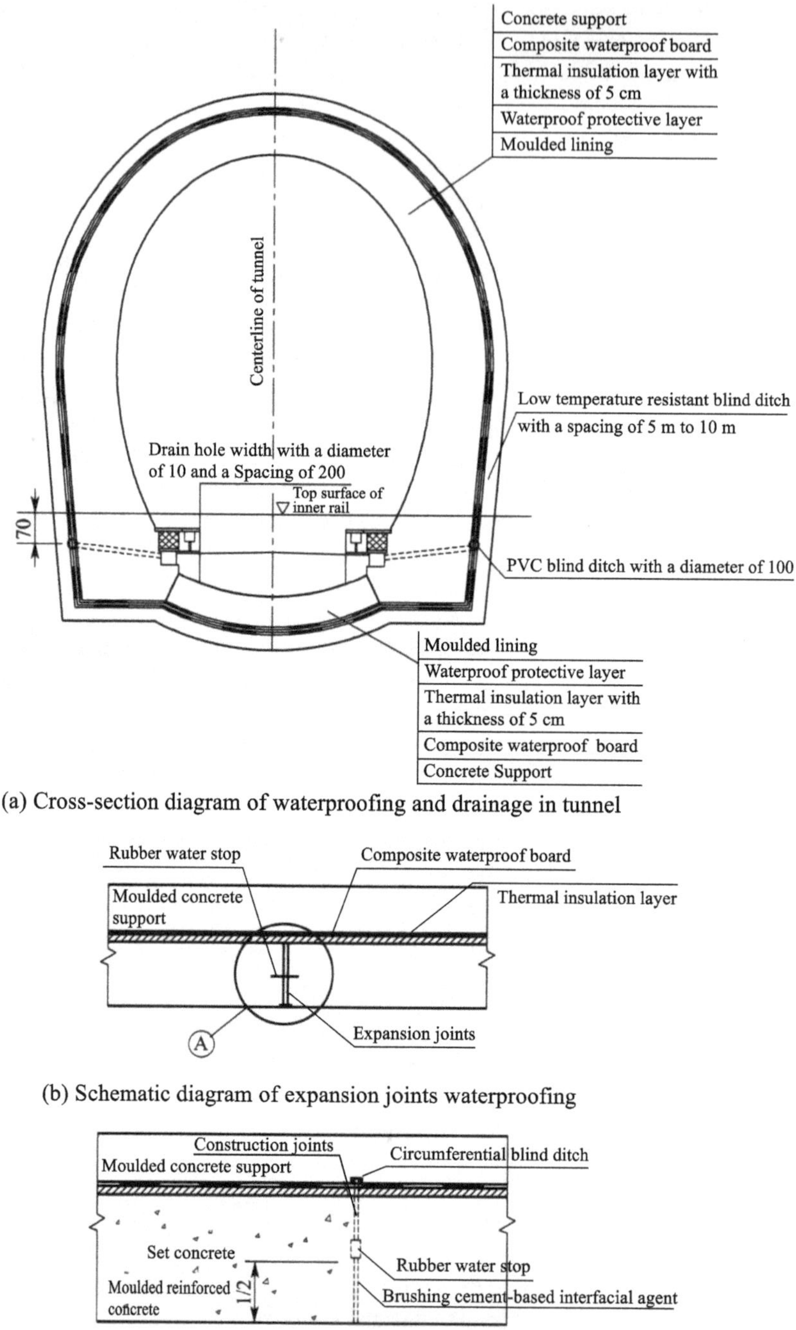

(a) Cross-section diagram of waterproofing and drainage in tunnel

(b) Schematic diagram of expansion joints waterproofing

(c) Schematic diagram of construction joints waterproofing

Fig. 4.110 Waterproofing and drainage design of permafrost tunnel along the Qinghai-Xizang Railway (Editorial Committee of the *Qinghai-Tibet(Xizang) Railway*, 2016)

① Waterproofing of primary support: for most of the tunnels in China, after steel arch centering has been installed and the anchor bolting steel mesh reinforcement has been laid, the concrete will be sprayed directly at one time. In alpine frozen soil areas, improving the quick-setting time and the early strength of shotcrete to ensure the bonding strength with the frozen soil to avoid the falling off has become the key of the construction technology. The liquid accelerator selected should meet the requirements that after the heating (at 20–40 °C), the initial setting time is within 5 min and the final setting time is 15–20 min at the ambient temperature of 0–5 °C. For the shotcrete, the thickness of one shot needs to be 150 mm for downward shot, 100 mm for horizontal shot and 70 mm for upward shot.

② Adding waterproof board. The composite waterproof board coil is added between the primary support and the secondary lining. During the construction, the composite waterproof board is laid together with the thermal insulation board.

③ For the secondary lining, it adopts the self-waterproofing concrete lining. The impermeability grade of the secondary moulded lining concrete of the tunnel in cold regions should be not lower than S8 (the level of 8 kg/cm^2). Advanced expansive agent such as HEA and AEA etc., could be added to improve the density and construction quality of the concrete, reduce the occurrence of cracks, and enhance the self-waterproofing capability of the secondary lining.

④ Waterproofing treatment for the two types of joints (construction joint and deformation joint). The construction joint and deformation joint are the most difficult problems to solve in tunnel waterproofing system. In the selection of waterproof materials for tunnels in permafrost regions, the impacts of low temperature and freeze-thaw cycles on the performance of waterproof materials must be considered. It is advisable to adopt the partitioned waterproofing method to divide the waterproofing section of the tunnel into smaller areas which are isolated from each other.

2. **Drainage Measures**

Anti-freezing of drainage system in the tunnel is the focus of tunnel drainage design in cold regions and also the key to avoiding frost damage. At present, the anti-freezing measures for tunnel drainage in cold regions and in permafrost regions of China mainly include: the central deep-buried ditch, the drain cavern, the thermal insulated and cold-proof ditch, the heated ditch, the thermal insulated outlet and ancillary facilities, such as the inspection well/manhole, etc.

(1) Drainage System in Permafrost Section

The sources of groundwater in tunnel in the permafrost section are mainly the thawed water from the frozen surrounding rocks in tunnel, the groundwater (or ground surface water) infiltrated through the fault zone or the thawing zone, and the groundwater longitudinally flowing from the non-permafrost section.

The thawed water from the frozen surrounding rocks is mainly the pore water and the fissure water. Generally, the amount of this type of water is limited, and there may still be a small amount of groundwater within a certain period of time after the tunnel is constructed, which can be drained by setting up the thermal insulated ditch.

For the groundwater (or ground surface water) infiltrated from the fault zone or the thawing zone, the water volume depends on the source of the water, such as the subpermafrost water or the ground surface water in the gully with a large amount of water. In this case, it should be drained by the drain cavern or other forms, and also should be dealt with comprehensively by the measures of intercepting and drainage on the ground surface.

For the groundwater longitudinally flowed from the non-permafrost section, the measure of plugging should be adopted at the junction of the two zones, such as a curtain grouting to prevent the inflow of groundwater. At the same time, the groundwater in the non-frozen soil zone should be discharged by the drainage measures.

The main drainage measures for tunnels in permafrost regions are mainly the drain cavern, the deep-buried central drainage ditch, the thermal insulated (heated) ditch and the thermal insulated outlet.

① Deep-buried central drainage ditch: it is to bury the drainage ditch within or below the corresponding freezing depth under the road surface in the tunnel, and to utilize the initial temperature of groundwater to achieve a freeze-thaw balance to prevent freezing (Fig. 4.111) (China Communications Construction Company First Highway Consultants CO. LTD etc., 2015). However, when the buried depth of the ditch is within the freezing depth, in the coldest month, the side wall of the ditch may freeze. However, it is not easy to completely freeze due to the initial temperature of the running water at the beginning of the month and over a period of time it may cause frost damages in the ditch. During the warm season, after removing the cover of inspection well, the ice in the seepage ditch will thaw quickly. Alternatively, a thermal insulation layer can be laid under the tunnel pavement (in case of meeting the strength requirements) to reduce the freezing depth. The thermal insulation layer can be laid around the ditch to reduce the negative temperature in the ditch. At the same time, it is necessary to build inspection wells and thermal insulated outlets.

The No.1 Tunnel of Chaoyang along the Nen-lin Line in permafrost region of the Daxing'anling Mountains has an annual air temperature of −3.2 °C and a monthly air temperature of −27.9 °C in the coldest month. The measured freezing depth in the tunnel is 5 m, and the buried depth of the central ditch is 3.5–4 m, which is running well. The practice has shown that the deep-buried central drainage ditch in the tunnel is more suitable for non-permafrost regions, or permafrost regions where the freezing depth of rocks is less than 5 m. In permafrost region, for a tunnel where the air temperature in the coldest month is much lower, in addition to strengthening the thermal insulation measures in the tunnel, the thermal insulation measures should also be set up around the central drainage ditch, and the heating cables with constant temperature controlling should be arranged as well. Or the deep-buried central ditch and the drain cavern should be used in combination, that is, in the section with a high temperature and shallow freezing depth with a certain distance from the

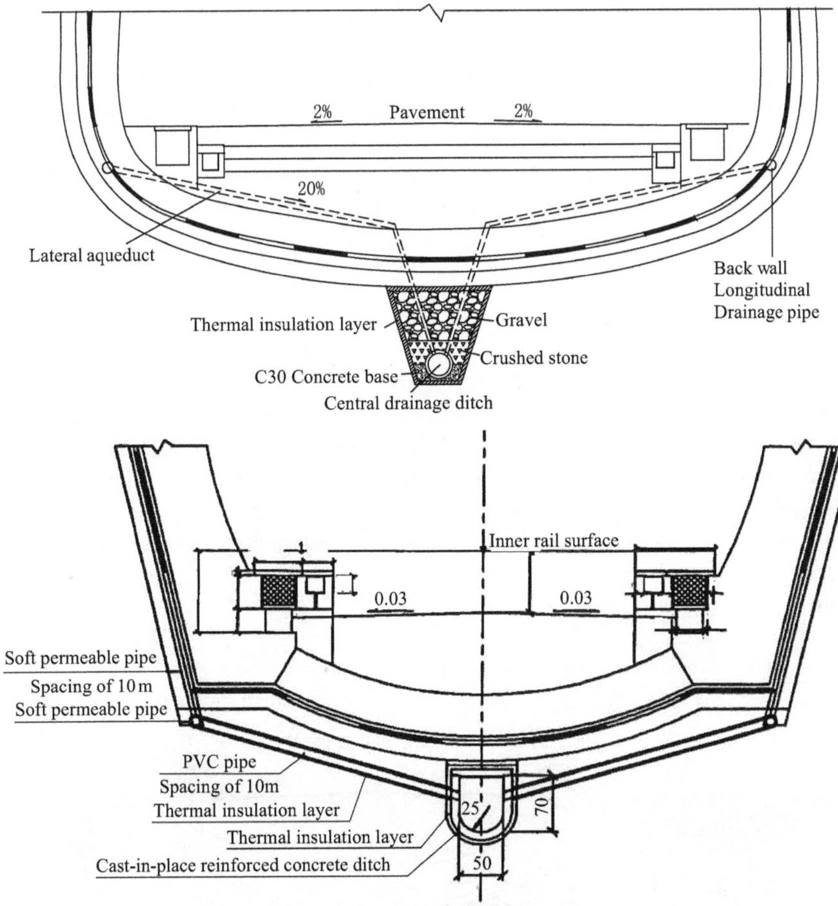

Fig. 4.111 Cross-section schematic diagram of the deep-buried central drainage ditch in the tunnel (Units:cm)

opening of the tunnel, the deep-buried central drainage ditch should be adopted, and at the opening of the tunnel, the drain cavern should be used.

The buried depth of the central deep-buried drainage ditch can be estimated by the empirical formula:

$$h_x = Kh_0 t_x / t \tag{4.26}$$

where h_x is the minimum buried depth of the drainage ditch with a distance of X meters from the opening of the tunnel (m); K is the coefficient of freezing depth related to lithology, 1.0 for cohesive soil, 1.1–1.3 for sandy soil and 1.3–2.0 for rock; h_0 is the maximum freezing depth in the ditch is located (m); t_x is the monthly air temperature in the coldest month at a distance of X meters from the opening (°C); t is the monthly air temperature in the coldest month in the area where the ditch is located (°C).

Under a regular circumstance, the deep-buried central drainage ditch should be excavated at the same time when the tunnel is constructed, and the excavation should be carried out before the construction of the primary support of the inverted arch in the tunnel. If the deep-buried central drainage ditch is excavated after the completion of the tunnel, there are some problems in the construction. Firstly, when the bottom of the tunnel is grooved, which is close to the lining of the side wall, the construction vibration is likely to have an adverse effect on the stability of the lining foundation and the lining structure. Secondly, it is difficult to ensure the quality of the ditch construction.

② Drain cavern: it is one of the important measures for tunnel drainage in permafrost regions. Its shape is similar to a small-sized tunnel, located directly under the main tunnel (Fig. 4.112) (Editorial Committee of the *Qinghai-Tibet(Xizang) Railway*, 2016). The drain cavern is connected with the vertical blind ditch, the drainage opening, the branch guide adit, the inspection well, and the thermal insulated outlet, etc., constituting a complete drainage system. The drain cavern is commonly used in the sections where the freezing depth is large and the groundwater is abundant in the tunnel. The drain cavern can gather the groundwater from

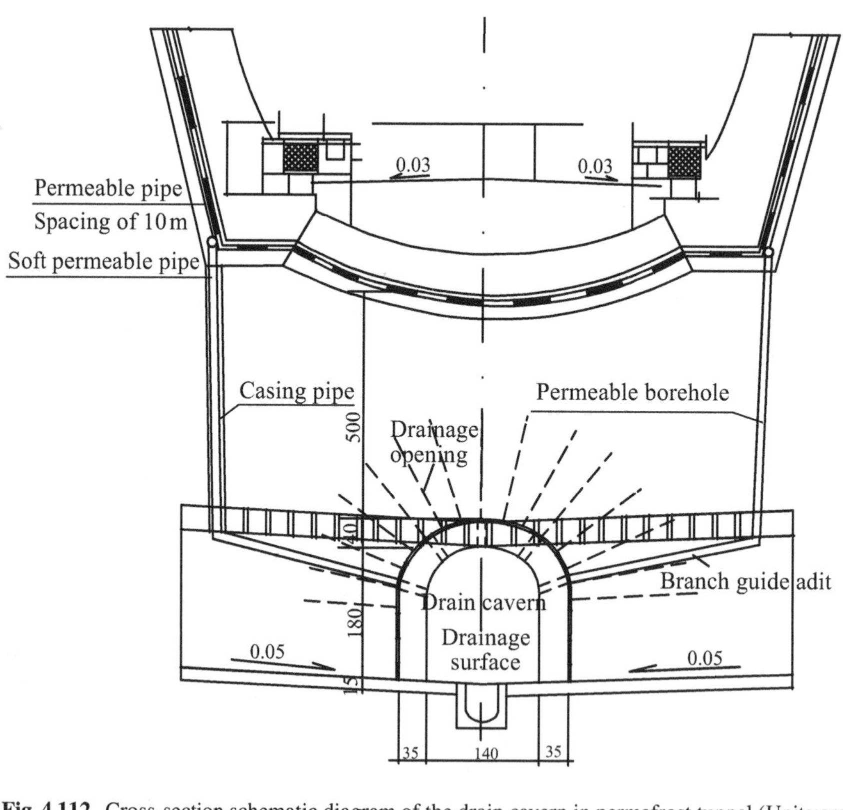

Fig. 4.112 Cross-section schematic diagram of the drain cavern in permafrost tunnel (Units:cm)

the surrounding rocks in the tunnel to lower the groundwater level and prevent the frost-heaving effects of the surrounding rock on the lining. The use of drain cavern in permafrost regions is a very successful drainage measure and it is suitable for severe cold regions where the monthly air temperature in the coldest month is lower than −20 °C and the freezing depth of surrounding rock reaches 5–6 m.

The buried depth of the drain cavern mainly depends on the maximum freezing depth. If the buried depth is shallow, the drain cavern is prone to freezing. If the buried depth is deep, the outlet needs to extend too long before it can go through to the ground surface (also related to the slope of the terrain). When the thickness of the overlaying soil layer above the drain cavern outside the tunnel is less than the maximum freezing depth, the thickness of the overlaying soil layer should be increased by adding an embankment.

The cross-section form of the drain cavern is determined according to the construction conditions. When it is inside the tunnel, it can be constructed by a form of underground excavation and choose the integral type. When it is outside the tunnel, it can be constructed with a form of open-cut excavation and choose the assembly type.

The branch guide adit is connected with the vertical drainage hole and can quickly discharge the groundwater into the drain cavern. At the same time, many drainage holes are drilled in the upper part of the drain cavern to gather more groundwater and lower the groundwater level. The practices have shown that most of the groundwater flows into the drain cavern through rock fissures and drainage holes. The inspection well can be used for ventilation, inlet and outlet, material hauling during the construction of the drain cavern, and also used for ventilation and inspection during the operation period, etc. A sedimentation tank with a depth of 0.3 m should be preserved at the bottom of the inspection well. In order to prevent cold air from entering the drain cavern, a double-layer thermal insulation cover plate should be set over the top of the inspection well. During the warm season, the well should be opened for ventilation to thaw part of the ice in the drain cavern.

The design for the drain cavern in permafrost regions should also have anti-freezing measures, and the outer wall of the drain cavern should be equipped with thermal insulation layer. And the thermal insulation materials for the outer wall should be the same as the thermal insulation layer in the main tunnel, with a thickness over 50 mm. If necessary, the heating facilities should be reserved in the drain cavern. The heating cables with constant temperature controlling can be arranged in areas with electricity supply, or the heating pipes can be reserved in areas without electricity supply to harness the locomotive steam to thaw the ice.

The issues of mutual influence and construction sequence between the drain cavern and the main tunnel: if the cold-proof drain cavern is constructed at first, due to its small area of the cross-section and a slow construction progress, the construction progress of the main tunnel will be affected. If the main tunnel is constructed at first and then followed by the construction of the drain cavern, it will have some influence on the structural safety of the main tunnel. According to relevant data (Liu,

2014), in the case that the cold-proof drain cavern is constructed at first, as the rock is not disturbed in advance, the bias effect could hardly exist and the coupling equivalent stress on the vault would change little, which will cause almost no impact on the side walls of both sides. In the case that the main tunnel is constructed at first, the rock would experience a disturbance before the excavation of the cold-proof drain cavern. After the two disturbances, the surrounding rocks would produce a relatively obvious bias effect. At the point where the included angle of the horizontal line and the connection line between the top of the cold-proof drain cavern and the inverted arch in the main tunnel is approximately 51°, the bias effect is the most pronounced, and the equivalent stress on the vault in the drain cavern also reaches its maximum value. Accordingly, the cold-proof drain cavern should be constructed at first, which has little impact on the safe operation of the tunnel. Generally, the construction progress of the cold-proof drain cavern should be ahead of the main tunnel by at least 100 m.

③ Cold-proof and thermal insulated (heated) ditch: in the tunnel drainage system, the longitudinal and transverse drainage pipes are usually arranged behind the lining to collect groundwater into the drainage ditches on both sides of the tunnel and discharge the water out of the tunnel (Fig. 4.113) (China Communications Construction Company First Highway Consultants CO. LTD., etc., 2015). In cold and also in permafrost regions, the buried depth of drainage ditch should usually be below the maximum freezing depth, so as to keep the flowing water in the drainage ditch from freezing. For this reason, the cold-proof insulating or heating measures should be adopted to reduce the buried depth of the drainage ditch and meet the drainage requirements.

Most of the tunnels in warm permafrost regions such as Da Xing'anling Mountains in Northeast China, Xinjiang and the Qinghai-Xizang Plateau have adopted the thermal insulated drainage ditches. Practices have proven that the thermal insulating drainage ditches are effective, but there are also some failures. The reasons are the following. Firstly, the air temperature in the tunnel is too low. The thermal insulated ditches should be used in the tunnels where the monthly temperature in the coldest month is −5 °C to −15 °C, with a freezing depth ranging from 1 m to 1.5 m, and there are water or may have water in winter. When the average air temperature in winter is colder than −15 °C, the thermal insulated drainage ditch freezes in these areas. Secondly, the length of the thermal insulating drainage ditch is too long. Because the initial water temperature of the groundwater in the tunnel is low, and if the flowing path is too long, the temperature in the thermal insulated drainage ditch will also decrease because of the cold air temperature in the tunnel, and the groundwater could freeze because of the temperature decreasing in the ditch. The buried length of the ditch depends on factors such as the air temperature and the length of the tunnel, the water content and the initial water temperature of the surrounding rocks, and the slope of the ditch, etc.Thirdly, the thermal insulation measures for the outlet of the thermal insulated drainage ditch are not effective. If the outlet of the thermal insulated drainage ditch freezes, it will lead to the drainage failure of the thermal insulated drainage ditch. The outlet of the thermal insulating drainage ditch

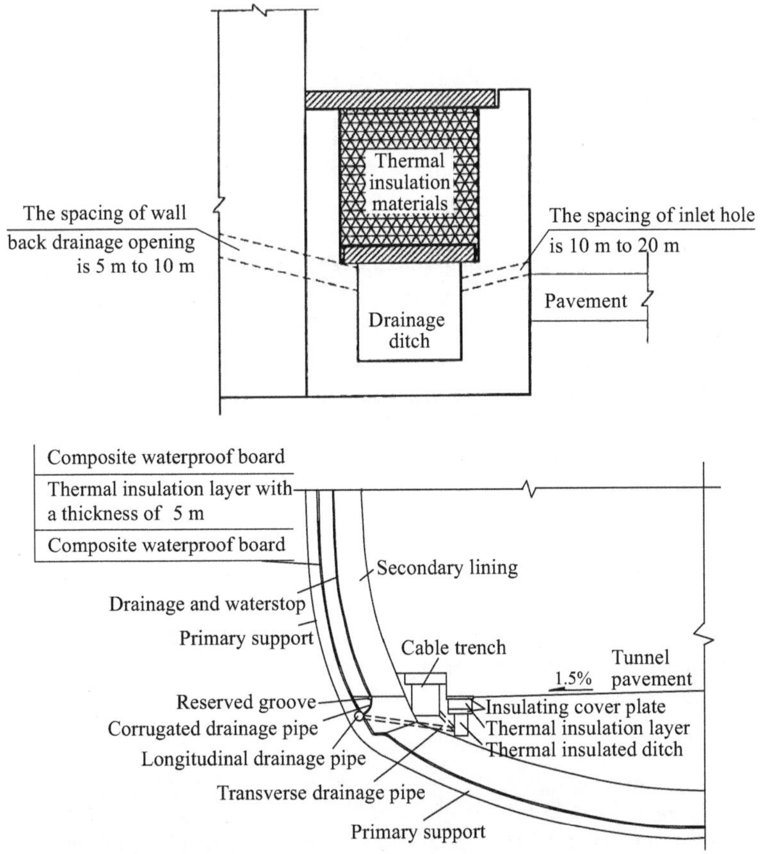

Fig. 4.113 Schematic diagram of the structures of drainage and thermal insulated (heated) ditch

in Kunlun Mountain Tunnel freezes soon after the winter is coming, resulting in the drainage failure and the overflow of the thermal insulating drainage ditch.

When setting up the thermal insulated drainage ditches in tunnels in severe cold regions, the both-sided and full-section thermal insulation should be adopted, aiming to maintain the structure as sealed as possible and reduce air convection inside the ditch. Meanwhile, the damp-proof treatment should be taken for thermal insulation layers (such as coating asphalt layers inside and outside the thermal insulation layers), or working together with the drain caverns.

In areas where it is available (with power supply), a heating cable of constant-power or self-regulating heating cable can be pre-buried (China Communications Construction Company First Highway Consultants CO. LTD., etc., 2015).

The constant-power heating cable includes a double-conductor thermal resistance wire, a polyester insulating layer, a high-density polyethylene inner sheath, an all-metal meshgrid shielding layer and an outer sheath. After the power is turned on, the power output is constant and does not change alongside the variations of the external environment, thermal insulation materials or heating materials. The output or stop of the power is controlled by the temperature sensor.

The self-regulating heating cable is composed of a conductive wire, a positive temperature coefficient heating body, an insulating layer, a shielding layer and an outer sheath. The heating body of the self-regulating heating cable is a functional material with a positive temperature coefficient. The conductivity of the material decreases with the increase of temperature, exhibiting a nonlinear relation. When it reaches a certain temperature range, the rate of decline of the conductivity accelerates rapidly. When working, the environmental temperature changes can cause thermal expansion and contraction in materials with positive thermal coefficients. When the environmental temperature is high, the expansion can cause the conductive heating fiber become less interconnected, leading to a reduction in output power. Conversely, when the environmental temperature decreases the contraction can cause the conductive heating fiber more interconnected, resulting in an increase in output power.

④ Thermal insulated outlet. The water outlet is a key construction step for the entire tunnel. If the water outlet is plugged by freezing, the entire drainage system becomes ineffective, leading to potential frost damages in the tunnel. The site of the water outlet of the tunnel should be selected on the leeward, sunny position with steep slope and smooth drainage. Figure 4.114 is the structural diagram of the buried thermal insulated outlet.

The slope of the flowing path at the water outlet should be increased generally greater than or equal to 5–8% to accelerate the water flow velocity and prevent the water temperature from falling to freeze. The water flows should be centralized to drain at the outlet to increase the flow rate. The outlet should be built with dry-laid rubble, which should be unweathered and with clean surface. The stacking width of the outlet should exceed the width of water flow by 0.5–1.0 m to prevent sundries intrusion.

The height of the composite thermal insulation layer stacked at the outlet should not be less than the local freezing depth to ensure that the water flow is not frozen. Geotextiles should be laid between the upper aquifer of the dry block rubble and the crushed stone layer to prevent the fine-grained soil from falling into the crushed stone and rubble layer and reduce its permeability.

A scarp should be excavated at the outlet, with a height not less than 1.5 m, to prevent the water from freezing due to the low environmental temperature after flowing out, leading to a plugging at the water outlet.

When the terrain is large, the end wall-type thermal insulated outlet can be built (Fig. 4.115), with a deep scarp dug at the outlet, and an aquifer with dry crushed stones and rubbles backfilled.The surface layer of the end wall-type thermal insulated outlet should be covered with sufficient thermal insulation layer to avoid freezing.

Based on the experiences, although the thermal insulated ditches and the central deep-buried ditches have been constructed in the tunnel in cold regions, ice jam still

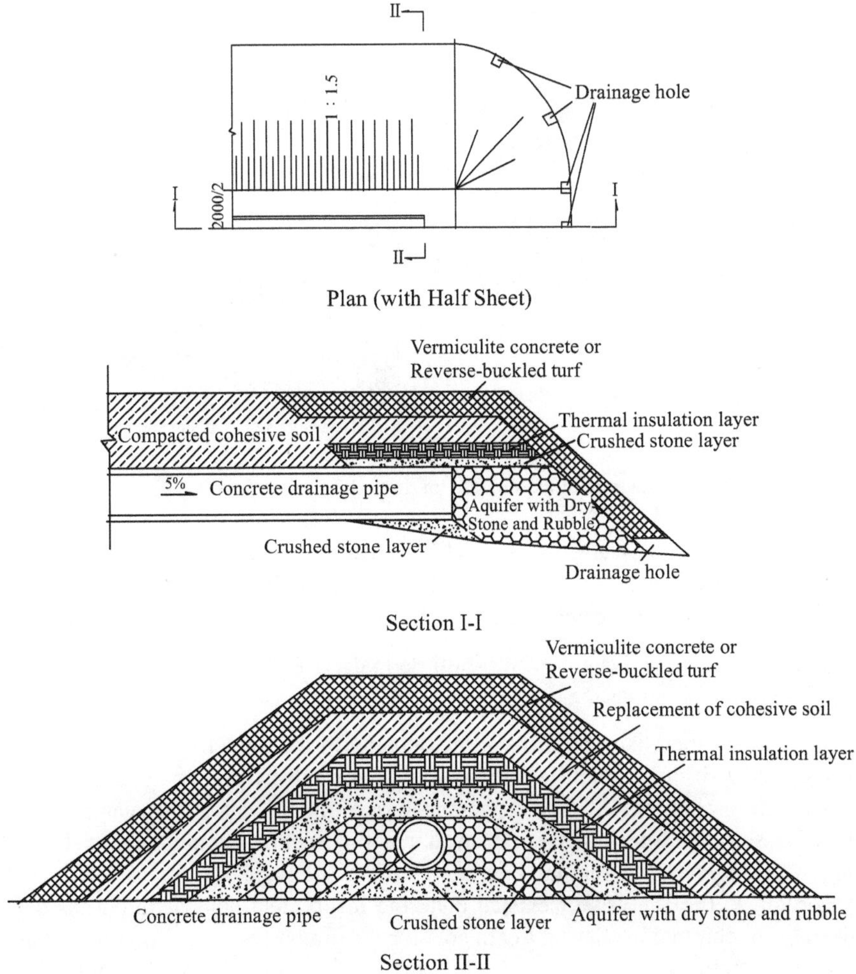

Fig. 4.114 Schematic diagram of the structure of the buried thermal insulating outlet

occurs in the drain cavern. One of the important reasons for this situation is that the water outlet freezes, resulting in the occurrence of ice jam and the further failure of the drainage system in the tunnel. A careful treatment for the outlet in the tunnel is a key to the effectiveness of the tunnel drainage system.

In the waterproofing and drainage measures for tunnel in permafrost regions, the heating in the drainage ditch is commonly adopted. For example, in Russia, the tubular electric heaters or steam heaters are used in drainage ditches, with additional thermal insulation measures, etc. In Norway, the heating cables are installed in the tunnel drainage system, and sometimes with double cold-proof doors installed. In the United States, every drainage ditch needs to be carefully inspected before winter, and the thermal insulation measures are added to the outlet of the drain cavern,

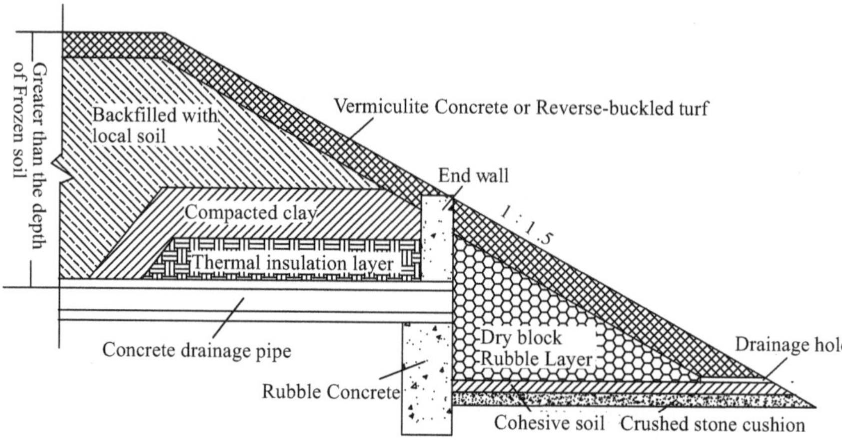

Fig. 4.115 Schematic diagram of the structure of the end wall-type thermal insulated outlet

and the heating cables are placed in the drainage pipe. In Hokkaido, Japan, the thermal insulation measures are adopted by putting the thermal insulation materials on the lining surface to prevent geotherm from releasing as much as possible.

(2) Tunnel Drainage System in Non-Full Permafrost Regions and Deep Seasonally Frozen Soil Regions

In some areas, there is permafrost at the opening of the tunnel. However, for most of the section in the tunnel, it belongs to non-permafrost part or deep seasonally frozen soil in cold regions. The biggest feature of the tunnels in these areas is that there is a seasonal freezing layer occurred in a long section around the tunnel opening, which affects the regular operation of the drainage system of the tunnel. According to the analysis of existed data, the insufficient consideration has been given to the thermal insulation treatments for the circumferential, longitudinal and transverse drainage pipes in the design for tunnels in non-permafrost regions, which often causes problems.

① The thickness of thermal insulation layer in the tunnel is insufficient, causing the freezing of the circumferential, longitudinal and transverse drainage pipes. As a result, the groundwater behind the lining freezes and the frost-heave force on the lining structure increases. During the thawing period in spring, the groundwater behind the lining gathers and infiltrates into the structural layer of the lining, resulting in the leakage of the construction joints and the deformation joints.

② The anti-freezing measures for arch foots are insufficient. Limited by the construction conditions, the thermal insulation layer for the side wall and the inverted arch cannot be constructed uniformly, which will cause the convective heat transfer between the construction site and the outside cold air, resulting in the freezing of the longitudinal drainage pipe, the plugging of the regular operation of the drainage system, the thawing of the circumferential drainage in the vault of the tunnel when the longitudinal drainage pipe in the arch foot is still in a frozen state.

③ The drainage system under the subgrade is frozen. The improper selection or laying of the thermal insulation layer, or the lacking of thermal insulation measures could cause the circumferential drainage pipes under the pavement to freeze within the range of seasonal freezing depth, leading to the situation that the groundwater concentrated in the circumferential and longitudinal drainage pipes cannot be drained through the central ditch.

④ The underestimation of the groundwater volume in the drainage system under the non-frozen soil section of the tunnel could cause water gushing and freezing at the tunnel opening. Therefore, the following measures need to be taken to solve these problems.

Firstly, tunnels in cold regions need to be laid with thermal insulation layers, either between the primary support and the secondary lining, or on the surface of the secondary lining. The thickness of thermal insulation layer should be calculated to ensure that the circumferential and longitudinal drainage pipes behind the lining structure don't freeze.

Secondly, insulation treatment for the longitudinal drainage pipe in the arch foot should be treated separately or enhanced, and it is critical that the thermal insulation material should be treated with waterproofing to reduce the impact on its effectiveness of heat preservation.

Thirdly, combined with the content of water inflow in the tunnel and the monthly air temperature in the coldest month, one of the above-mentioned drainage systems in permafrost regions should be selected to design the drainage system.

In cold regions, in order to prevent the damages caused by the freezing of water flow inside the tunnel, in addition to strengthening the thermal insulation measures for the lining (inside or outside) in the tunnel, the drainage measures inside and outside the tunnel should be enhanced to form a complete and effective drainage system, which requires the drainage facilities to be well-established and the drainage to be unobstructed. The drainage system can be roughly considered as Table 4.27.

(3) Connection between Drainage Systems

For the designs of tunnel drainage in cold regions, in addition to determining the segmented drainage forms according to the types of frozen soil, the interconnection of drainage systems in different types of frozen soil should also be considered. There are usually three situations that can be encountered. Firstly, for fully permafrost tunnel, there should be connection between the tunnel drainage systems in the warm central section and in the cold permafrost section at the tunnel opening. Secondly, for non-full permafrost tunnel, there should be connection between tunnel drainage systems in permafrost section at the tunnel openings and in the middle of the tunnel of non-permafrost section or of seasonally frozen soil section. Thirdly, for tunnels in deep seasonally frozen soil areas, there should be connection between the tunnel drainage systems in deep seasonally frozen soil section at the tunnel openings and in non-frozen soil section in the middle of the tunnel.

① Connection between drainage systems at Openings and of Central Drainage Ditch in Full Permafrost Tunnel

Under the influence of geothermal energy, the air temperature in the full permafrost tunnel will gradually increase along with the increase of the depth of the tunnel, but the difference is minor. The monthly air temperatures in the coldest month in Kunlun Mountain Tunnel (1686 m long) and Fenghuo Mountain Tunnel (1338 m long) on the Qinghai-Xizang Plateau are about −12.5 °C, the difference between the air temperatures at the opening and in the middle of the tunnel is about 0.9 °C. In the case of super-long tunnels, there may be a large temperature difference. Both Kunlun Mountain Tunnel and Fenghuo Mountain Tunnel are equipped with the thermal insulating drainage ditches for both sides, and the connection between them is from a thermal insulating ditch to another thermal insulating ditch. In permafrost regions of Daxing'anling Mountains in Northeast China, due to low air temperatures, the monthly air temperature in the coldest month is lower than −25 °C, or even below −30 °C, so the drain cavern has been adopted.

② Connection between the Drainage Ditches in Permafrost Section and Non-permafrost Section in Non-Full Permafrost Tunnel

The tunnels in the non-full permafrost regions, that is, the permafrost layer is buckled on the mountain top like a "pot", and the inlet and outlet are in permafrost section, but the middle of the tunnel is in non-permafrost section or seasonally frozen soil section. The layout for the tunnel drainage system can be determined according to the monthly air temperature in the coldest month and the development degree of groundwater in the tunnel (Table 4.28). There could be three forms of connection.

A. From insulated ditch to insulated ditch connection. The air temperature in the tunnel indicates that the tunnel is located in a cold region, where the monthly air temperature in the coldest month is lower than −15 °C. The entire drainage system inside the tunnel can adopt the thermal insulated ditches in both sides, which are connected to each other.
B. Connection between the thermal insulated ditch and the deep-buried central ditch. For a tunnel, there is a permafrost section with large thickness at the open-

Table 4.27 Forms of tunnel drainage ditch under different environmental conditions (according to Ye, 2010)

Regions	Monthly air temperature in the coldest month /°C	Maximum freezing depth of cohesive soil /m	Forms of main drainage ditch
Cold regions	-5 to −10	≤1.0	General drainage ditch
	−10 to −15	1.0 to 1.5	Insulated drainage ditch
Severe cold regions	−15 to −25	1.5 to 2.5	Central deep-buried drainage ditch
	< −25	>2.5	Cold-proof drain cavern

ing of the tunnel and a non-frozen soil section in the middle part. After the tunnel has been completed, the cold air enters and freezes the surrounding rocks, forming a seasonal freezing layer in the middle of the tunnel. Generally, the permafrost sections at the openings are mostly the warm permafrost with a thin thickness and the monthly air temperature is not lower than −25 °C in the coldest month in the area of the tunnel. In the non-frozen soil section in the middle of the tunnel, the thermal insulated ditches in both sides can be used for drainage. For the permafrost sections at the opening of the tunnel where the freezing depth is relatively large, it is necessary to use the deep-buried central ditch with thermal insulation to meet the requirements that the groundwater does not freeze. In the non-frozen soil section at the transition part of the permafrost and non-frozen soil in the tunnel, a water diversion tunnel should be excavated (in case of dual tunnels, it is necessary to excavate inclined water diversion tunnels under the thermal insulating ditches in both tunnels to concentrate the water and divert them together) to divert the water from the insulated ditches in both sides into the deep-buried central ditch with thermal insulation, and then the water will be discharged from the permafrost section through the buried thermal insulated outlet outside the tunnel.

C. Connection between the thermal insulated ditch and the drain cavern. The tunnel is low-lying and relatively long, located in a permafrost region with medium temperature, where the permafrost thickness is relatively large and the monthly air temperature in the coldest month is lower than −25 °C. However, there are still some non-permafrost sections in the middle of the tunnel. In this circumstance, the thermal insulated ditches in both sides can be used. In the permafrost section, two water diversion tunnels are excavated to divert groundwater into the drain cavern under the permafrost section, and then, the water will be discharges through the buried thermal insulated outlet outside the tunnel

Table 4.28 Designs for tunnel drainage systems in cold regions

Degree of coldness	Developmental conditions of groundwater	Monthly air temperature in the coldest month (°C)	Forms of drainage system
Cold zone	Groundwater is developed and with long-term recharge	−10 to −15	Insulated ditches in both sides
		−15 to −25	Central deep-buried insulated ditch
Severe cold zone		< −25	Cold-proof drain cavern
Cold zone	Groundwater is underdeveloped	−10 to −15	Insulated ditches in both sides
Severe cold zone		< −25	Insulated ditches in both sides (appropriately increase the thickness of the insulation materials or heating)

4.3.3.3 Examples of Tunnel in Permafrost Region in China

Among 45 railway and highway tunnels in cold regions in China, the tunnels in permafrost regions (33.3%) and in alpine cold regions (45.2%) account for 78.5%. The tunnels in permafrost regions have the most frost damage problems, which are followed by the tunnels in cold regions. In summary, the fundamental cause for frost damages is the freezing of "water". Ensuring proper insulation and drainage in tunnel is the core measures to guarantee the safety and stability of tunnels in cold regions.

1. **Cold-proof Drain cavern**

(1) Lingding Tunnel is the first tunnel in permafrost regions in China. The local annual air temperature is −6.7 °C, with a lowest temperature of −50 °C. The tunnel is 936.8 m long, with a one-direction slope. The surrounding rock is tuff breccia and andesite with well-developed joint and severely weathering. The groundwater inflow of the tunnel is 80 T/d, which belongs to subpermafrost water. The exit section is located in the continuous permafrost, with a mean annual ground temperature of-1.0 °C, where the maximum seasonal thawing depth is 9 m. The drainage system in this tunnel is designed with a common ditch with grouting behind the lining.

The tunnel was completed in September 1961, and suffered from frost damages in November of the same year. The arch was generally leaking, forming hanging ice. The ice stalagmites appeared at the bottom, with an icicle diameter of 1 m on the side wall. Circumferential cracks appeared in the lining, with the largest one of 3 mm. The thickness of the ice accretion at the bottom of the tunnel ranged from 0.3 to 1.3 m.

Treatment: a cold-proof drain cavern had been constructed under the tunnel, with a length of 720 m and a buried depth of 5.5 m (from the bottom of the tunnel to the bottom of the drain cavern). The blind ditches, branch-guide adits, blind drains outside the tunnel, thermal insulated outlets, and inspection wells, etc., had also been constructed. After the construction, the icicles, hanging ice in the tunnel, and the accumulated water on the bottom of the tunnel, etc., had completely disappeared, and meanwhile, the water had been flowing in the drainage tunnel all years round.

Flaws: it had no thermal insulation layer in the drain cavern. In the permafrost section (with a length of 190 m), according to the data in July 1974, it was found that the drain cavern was frozen, with an ice thickness of 0.8 m, and the water was flowing under the ice (the water temperature in the non-frozen soil section in the tunnel ranged from1 °C to 3 °C). After manual treatment and natural thawing, no new freezing phenomena have been observed.

Inspiration: the drain cavern is one of the effective drainage measures for permafrost tunnels, but effective insulation measures have to be arranged properly to prevent the discharge water from freezing.

(2) No.2 tunnel of Cuiling, built in 1966, is located in Linhai-Bizhou branch line of Nenlin line in Da Xing'anling Mountains. The mean annual ground tempera-

ture of the permafrost around the site of the tunnel was −3.0 °C, the lowest temperature in the drain cavern was −2.3 °C, and the amount of groundwater gushing out was 73 m^3/d, with water temperature of 0.4 °C. The section area of the drain cavern was large, and the slope of the water flow was steep, but both were in low temperature all years round and with a layer of ice hanging year by year. After 7 years of service, freezing and ice accretion had appeared in the drain cavern in the permafrost section. In summer, the hot air outside had been pumped into the tunnel with a 75-kilowatt ventilator, and more than 400 m^3 of ice accretion were thawed in 7 days. Since then, the inspection and ventilation had been strengthened, and the drainage effectiveness of the drain cavern has not been affected.

Inspiration: in addition to strengthening inspection and ventilation, the thermal insulation measures for drain cavern should be added to the tunnels in permafrost regions.

(3) Yong'an Tunnel, built in 1974, with a length of 1244 m, was located in Tahe-Hanjiayuanzi branch line of Nenlin line in Daxing'anling Mountains. The surrounding rock was granite coarse-grained structure, with well-developed joints and groundwater.The annual air temperature in the area of the tunnel was −2.7 °C, with a monthly air temperature of −31 °C in the coldest month and a freezing depth of more than 4 m in the surrounding rock. The construction of the tunnel was started in August 1979, but without the construction of the blind ditch and the drain cavern, there was water seepage in the lining arch and side wall, with ice hanging on the arch and also accumulated on the bottom in winter. But after the construction of the blind ditch and the drain cavern had been completed, the above phenomena disappeared, although with an amount of drainage water of more than 150 tons per hour.

Inspiration: the drain cavern is an ideal drainage measure for tunnels in permafrost regions.

The Ela Mountain Tunnel was constructed in 2011, located in 214 national highway in Qinghai Province. It was a separated tunnel, whose left tunnel was 4695 m long and right tunnel was 4635 m long, with an altitude of 4200 m at the opening and a maximum of 137 m buried depth. The annual air temperature in the area of the tunnel was −4.2 °C, with a monthly air temperature of −10.3 °C in the coldest month. The freezing period was as long as 7 months, the maximum freezing depth was 277 mm and the average annual precipitation was 369.2 mm. The surrounding rocks were Triassic clastic rock, volcanic rock, volcano clastic rock and Tertiary glutenite sandwiched with mudstone.

The permafrost was distributed at the entrance and exit on the north and south slopes, as well as at the top of the mountain. The permafrost table was 1.4–1.8 m, and the permafrost base was 35–56 m (Wang, 2013a, b), with a mean annual ground temperature of −0.5 to −1.0 °C. The upper ice-water deposits were silty clay mixed with gravel, mainly composed of icy, ice-rich and ice-saturated permafrost, and there was also the layered ice and granular ice in joint fissures of frozen rocks. The

Ela Mountain Tunnel was a non-full permafrost tunnel, and the main part of the tunnel was in non-frozen soil.

There was suprapermafrost water in loose rock stratum in the entrance and exit sections of Ela Mountain Tunnel with a large amount. There had thawed water from ground ice in permafrost. During excavation, there were dripping and linear flowing water from the fissures in the bedrock, while significant flowing and gushing water occurred from the fault fissures.

The waterproofing measures adopted in Ela Mountain Tunnel included the primary support, the waterproof board, the thermal insulation layer and the secondary lining. High-efficiency expansive agents such as HEA, AEA, etc., had been added to the secondary lining concrete to enhance the self-waterproofing capacity of the secondary lining. For the waterproof structure of the construction joints and horizontal construction joints, it had been laid with a cement-based interface agent and waterstop. For the circumferential construction joints, it had been painted with a 2 mm thick of WJ interface adhesive agent before pouring new concrete, and hydrophilic waterstops had also been installed in the middle of the lining interface. Additionally, the segmented waterproofing measures had also been applied in Ela Mountain Tunnel.

(4) According to the requirements of relevant specifications, two drainage measures, the central drainage ditch and cold-proof drain cavern, had been set up at the entrances of Ela Mountain Tunnel. In the main tunnel, the water behind the lining was first discharged to the central drainage ditch through the blind drainage pipes, and then discharged from the central drainage ditch to the cold-proof drain cavern (Fig. 4.116). Due to the gentle terrain at the entrance of the tunnel, the drain cavern outside the tunnel was long. To avoid freezing, anti-freezing measures for the drain cavern had been added to reduce the length of the drain cavern.

Inspiration: the tunnel has been running well so far. Thermal insulation measures should be taken for the central drainage ditch.

(5) The Daban Mountain Railway Tunnel is the second dual-tracks high-speed railway tunnel in Lanxin Railway, with a total length of 15,897 m. It is located in

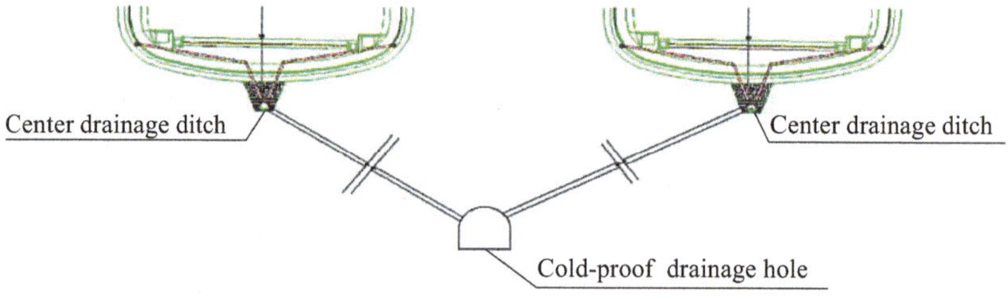

Fig. 4.116 Cold-proof drainage system of Ela Mountain Tunnel (Xu, 2013)

the main ridge of Daban Mountain at an altitude of 4060 m, in Datong County in Qinghai Province. The entrance altitude of Daban Mountain Railway Tunnel is 2600 m and the exit altitude is 4300 m. Permafrost has occurred in the areas higher than 3800 m. The annual air temperature at the entrance is 0.48 °C, with a monthly air temperature of −13.1 °C in the coldest month. The surrounding rock in the tunnel is mainly gneiss sandwiched with schist, diorite, andesite, slate, sandstone sandwiched with shale. Near the fault zone in Daban Mountain, the rock stratum is broken, 60% of which belongs to the water-rich section with high-ground stress, where the maximum water inflow during construction is more than 53,229 m^3/d.

The waterproofing and Drainage System of the Tunnel (Dong, 2014):

Firstly, the drainage ditches and central drainage ditches have been set up on both sides of the bottom of the tunnel lining. The thermal insulated ditches with double-layer cover plate have been constructed within the range of 1000 m at the entrance and 1500 m at the exit. The cold-proof drain cavern, with an inner dimension of 220 × 250 cm, has been constructed below the main tunnel at a depth of 5 m, with a length of 1000 m from 50 m inside the entrance and 30 m to the exit. At the side of entrance, the water is diverted through the drain cavern to the Santanggou gully on the left side of the line. At the side of exit, the water is diverted through the drain cavern to the slope on the right side of the line. And the end wall-type thermal insulating outlets have been set up at the outlets of the drain caverns.

Secondly, in the regional fracture zone with developed groundwater, the arch wall has been applied with the radial grouting to reinforce the stratum for water plugging. Within the lengths of 1000 m from 20 m inside the entrance and 965 m from 35 m inside the exit, the treatment of grouting has been carried out behind the primary support.

Thirdly, the EVA waterproof board (ethylene-vinyl acetate copolymer) (with a thickness of 1.5 mm) and backing non-woven fabrics (density greater than 400 g/m^2) have been set up behind the lining arch wall, the perforated corrugated pipes (with a diameter of 50 mm and a spacing ranging from 6 m to 12 m) have been used for the circumferential blind drainage pipe, and the perforated corrugated pipes (with a diameter of 80 mm) have also been adopted for the vertical blind drainage pipe at the foot of the wall, segmented in sections of 12–20 m, concentrating the drainage pipe from the inlet in the side wall to the side ditch.The transverse blind drainage pipes have been set up every 30–50 m to discharge the water from the drainage ditches on both sides into the central drainage ditch, and an inspection well have been arranged every 30–50 m in the central drainage ditch.

Fourthly, the waterproof design for construction joints and deformation joints. For the circumferential construction joints, the composite waterproof structure of "medium-buried rubber waterstop and externally bonded waterstop" has been adopted. For the longitudinal construction joint, the composite waterproof structure of "medium-buried rubber waterstop with steel edge and externally bonded waterstop" has been adopted. For the deformation joint, the composite waterproof structure of "medium-buried rubber waterstop and externally bonded waterstop and joint material" has been adopted.

Fifthly, the self-waterproofing capacity of concrete structure has been fully utilized. The impermeability grade of lining concrete is not lower than P10–P12. In some lining sections, it's necessary to mix it with WQ preservative with a dosage of 5% of the cement consumption.

The Daban Mountain Highway Tunnel is located 5–10 km west of the second dual-tracks high-speed railway tunnel in Lanxin Railway and south of the Datong River in the Xining-Zhangye Highway, with an altitude of 3800 m. The annual air temperature is −3.1 °C and the coldest air temperature is −34 °C in the area of Daban Mountain Highway Tunnel. The design of "rigid polyurethane foam and dry-process silicon-aluminum fiberboard and glass fiber reinforced plastic" has been adopted in the tunnel, with a composited rain cavern used for drainage.

2. **Thermal Insulating Ditches**

The Kuixian Railway Tunnel is located in the southern Xinjiang Railway from Turpan to Korla. It is a cross-mountain tunnel that passes through the watershed in the middle of the Tianshan Mountains. The length of the tunnel is 6154.16 m, with an elevation of 2982 m for the entrance. The surrounding rocks in the tunnel are granite, gneiss and schist, and fissure, with well-developed fissure water. There is a range of 700 m of the tunnel that is going through a permafrost layer in the section close to the exit. In Kuixian area, the annual air temperature is −3.8 °C, with a monthly air temperature of −15.8 °C in the coldest month. For the mean annual ground temperature, it could be as high as −0.1 °C to −0.2 °C and as low as −2.0 °C to −2.4 °C. The maximum thawing depth is ranging from 1.4 m to 4.0 m, with an annual precipitation of 300–350 mm. During construction, as the excavation is in progress, the amount of groundwater is larger and larger and the fault zone is filled with a large amount of detritus, and at the meantime, there is ground surface water infiltrating to recharge groundwater.

The cold-proof ditches have been set up at positions 500 m away from the entrance and exit of the tunnel, extending to a certain distance outside the tunnel. The third channel at the west end 570 m away from the entrance diverts the water in the main tunnel into the parallel adit, and a cold-proof ditch has been constructed in the center of parallel adit, extending by 25 m outside the tunnel.

In the winter of 1978, when the construction of the outlet section of the cold-proof ditch has not completed, the outlet is frozen in late November, and the frozen part has been gradually developing towards the inside of the tunnel, which has resulted in the situation that the parallel adit is frozen by 1250 m and the main tunnel is frozen by 2870 m. The ditch water overflowing on the ice and the ballast bed is frozen, with an ice accretion of about 0.33 m thick. Due to the ventilation effect during operation, the air temperature inside the tunnel is decreased, making it difficult for the 500 m-long cold-proof ditch originally designed at both ends of the tunnel to cope with the temperature drop.

Treatment: the treatment begun in 1980. Firstly, a 3859 m-long cold-proof ditch was constructed, including a main tunnel of 1820 m and a parallel adit of 1399 m. In the permafrost section of 740 m, an asphalt-glass wool was used for insulation, reducing the heat dissipation of the groundwater (the temperature of groundwater ranges from 5 °C to 5.6 °C) when flowing, so that the water with positive temperature in the side ditch in the main tunnel could be diverted to the parallel adit. Secondly, the longitudinal slope at the bottom of the ditch had been increased and the cross section area had been reduced in the susceptible freezing section, which could increase the flow velocity of the water. Thirdly, the cold-proof ditch in the permafrost section had adopted the circumferential thermal insulation. Fourthly, the thermal insulating outlet had been constructed.

Operation: there had been no frost damages for three winters and the water temperature at the outlet was above 1.2 °C. However, the ditch was frozen and jamed in January, 1984. This was because that the oil-impregnated wooden board shad missed to be installed in the inner layer of the ditch, and the diameter of the horizontally connected drainage pipes was too small (only 200 mm) and without filter screens, which resulted in the clogging. In June, the drainage ditch had been cleaned and replaced with an insulation layer with high thermal insulation performance. And the horizontally connected drainage ditch had been renovated and changed to a U-shaped drainage ditch. 245 kW electric heaters had been divided into three groups and installed in the cold-proof ditch in February, 1985.

Chapter 5
Hydraulic Engineering in Frozen Soil Regions

Hydraulic structures in permafrost area are the most complex and important engineering structures in cold regions, including those in seasonally frozen soil regions and permafrost regions. The frost damage of hydraulic structures in seasonally frozen soil region is mainly caused by the frost heave from the freezing of foundation soil in cold season, and it sinks due to the thaw settlement of frozen foundation in warm season. The frost damage of hydraulic structures in permafrost area is mainly caused by the subsidence of frozen soil foundation from hydrothermal erosion, and also from the frost heave caused by the freezing of foundation soil. The former is disturbed by "epigenetic" freezing, while the latter is damaged by thawing of "protogenetic" portion. There are similarities and differences in the causes of frost damages in the two frozen soil regions, which should be treated with caution.

At present, the constructions of hydraulic engineering projects in frozen soil regions in China are mainly concentrated in seasonally frozen soil regions, and most of them are small and medium-sized abstraction works and existing dams. The permafrost regions have not been involved so far. However, in the future, in the west line project of the South to North Water Diversion Project and the water conservancy development in Heilongjiang, Argun River and Xizang will encounter challenges related to permafrost foundations.

Based on the causes and characteristics of freeze damage of hydraulic structures in frozen soil area, they mainly occur in the foundation part. Therefore, the freeze damage prevention measures for hydraulic structures will be discussed from the perspective of the damages associated with the foundation.

Q. Wu, C. Tong, *Frozen Soil Engineering in Cold Regions*,
https://doi.org/10.1007/978-981-95-2527-0_5

5.1 Particularity of Engineering Design

During the construction and operation of earth dams and hydraulic structures in frozen soil areas, complex cryogenic processes occur on the foundation of buildings due to the impact of freezing and thawing. On the basis of complying with the design criteria of non-frozen-soil areas, the design of hydraulic buildings should incorporate some special requirements and research works specific to the environment and frozen soil foundation characteristics in frozen soil areas.

1. Study the state of regional permafrost and frozen soil engineering types and the complexity and diversity of their burial conditions, and add research contents and measuring scales of permafrost according to different design sections. Study the states and changes of permafrost within the scope of hydro junctions, related projects, transportation and residential areas.
2. Study the property changes and corresponding indexes of frozen soil foundation in the processes of freezing and thawing of soil and rock foundation, such as frost heaving potential, thawing settlement, permeability and strength characteristics.
3. Study the temperature field changes and calculations of hydraulic structures and their foundations, and predict the development trend under global climate change and hydraulic actions.
4. Study the mechanical and hydraulic stability of dam structures, facilities and water diversion tunnel, and the interaction between foundations. Study the prevention and control measures and methods of foundation frost heave and thaw settlement.
5. Study the interaction and function of the joint parts of various buildings (especially considering two different design principles).
6. Study the cryogenic weathering of large gravel, bedrock gravel and earth rock filling materials in the dam and its foundation under the condition of repeated freezing-thawing and wetting-drying cycles.
7. Study the stability of declivity (slope) and displacement due to freeze-thaw effect and ground temperature variations in soil and rock.
8. Study the seepage and exposure of surface water and groundwater (usually pressurized), including springs and water flows etc., and the possibility and harmfulness of seepage through dam bodies, foundations, slopes and other parts.
9. Study the reformation of reservoir banks, riverbanks, and other slopes caused by hydrothermal erosion and their stabilities.
10. Study the occurrence and treatment of cryogenic phenomena caused by the changes of frozen soil environment due to construction.
11. Study the changes of properties and strength of building materials under repeated freeze-thaw cycles, and also study building material selection and treatment.

The above problems are all related to the foundations of permafrost and non-frozen ground under the influences of cold climate and hydrothermal erosion and they need to be studied before and during the design. Comprehensive and detailed engineering geological surveys and researches of frozen soil are essential in the design of hydraulic structures in frozen soil areas to avoid construction and operation accidents.

At present, it is difficult to solve or predict the cryogenic phenomena of frozen soil in the dam bodies, dam foundations and other hydraulic structures in permafrost areas by numerical calculations. Despite the availability of various sophisticated computational software, it is difficult to obtain the required parameters or the reliability may be insufficient, so the calculation results are not easy to be accurately consistent with the real-world engineering conditions, and can only approximately evaluate some of the most dangerous or important cryogenic phenomena of frozen soil. Nowadays, the following prevailing approaches are adopted to minimize the adverse effects of cryogenic phenomena on engineering buildings. Using non-frost heave materials to modify foundation soil; taking the method of closed and sun-dried foundation soil (in areas lacking coarse-grained soils) to modify the foundation of small-scale projects; using industrial insulation materials to reduce the thermal erosion of frozen soil foundation or reduce the freezing depth of foundation; integrating the use of thermosyphons (heat pipes) and insulation layers with cooling measures to increase the cold storage capacity and intensify the frozen state of frozen soil foundations; modifying the foundation structure design to adapt to the cryogenics of frozen soil.

Different areas have different considerations in engineering designs. The key point of engineering design in seasonally frozen soil areas is the frost heave prevention of building foundation. The emphasis of engineering design in permafrost areas should be preventing the foundation of frozen soil not only from thaw settlement but also from frost heave. The key points of engineering design are "prevention" and "isolation" or "resistance". The key points of frost heave prevention design are preventing freezing, isolating water and resisting external forces, and the key points of thaw settlement prevention design are insulating heat, isolating water and cooling foundation.

5.2 Hydraulic Structures with Flat Foundation

The common flat foundation hydraulic structures include water intake, channel, water conveyance tunnel, culvert apron, pressure forebay and etc. Most foundations are shallow flat foundations. Many frost damages related to hydraulic structures with flat foundation commonly occur in seasonally frozen soil regions.

5.2.1 Normal Frost Heave Force under Flat Foundation

The foundations of water diversion projects in cold regions include both those constructed on seasonally frozen soil and those constructed on permafrost. In winter, the water in soil is frozen and transforms from liquid state to solid state, and the groundwater (or surface water infiltration) migrates towards the freezing front, causing frost heave in the foundation soil, resulting in differential deformations and damages to buildings.

A large number of frozen soil tests and profile data show that there is abundant ground ice within the range from permafrost table to a certain depth. Under the influence of heat (thermal erosion of buildings and water bodies), the frozen soil foundation will undergo different degrees of thawing settlement deformations, affecting the stability of buildings.

The foundations of engineering buildings such as water intake, sluice chamber and pumping house are mostly in the form of piles. The interaction forces between the foundation and frozen soil involve tangential frost heaving forces, freezing forces, and frictional resistance at the thawed soil-foundation interface. The working environment of hydraulic structures results in higher values of tangential frost heaving forces compared to conventional building projects, which is due to enhanced frost heave caused by moisture supply during the freezing process.

In hydraulic structures, the bottom slabs of culverts and sluices, the fore-bay cover, the after-bay apron, various pool bottom plates and channel linings etc. are all subjected to frost heave of the foundation soil. Due to the different surrounding conditions of the foundation and the variations to the extent where the upper structures constrain the foundation, the flat foundation will be subjected to bending, torsion, shearing and uplift under differential freeze and thaw of the foundation soil. Among these, the normal frost heave force of the foundation soil plays a decisive role.

The internal causes of frost heave forces are the natural factors such as soil, water and temperature, and the external causes are the constraint conditions on soil frost heave, for example, the engineering factors such as the embedded depth, shape, rigidity, foundation and upper load. The total force of the normal frost heave force constrained by the foundation plate increases with the increase of the area of the foundation plate, while the cooperative frost heave force (or additional normal frost heave force) of the soil around the foundation decreases gradually due to the cumulative effect of the tangential forces between the side soil masses. The smaller the area of the flat foundation is, the larger the relative cumulative effect of the surrounding soil masses. Figure 5.1 is a sketch of the test pit profile by Orlov, Russian scholar. The ground ice distribution within the foundation soil under the flat foundation can be observed when frost heave happens. The soil column is formed under the foundation, the ground ice layer is basically horizontally distributed, and the ground ice layer on both sides is curved along with the frost heave of the soil. A comparative test at Heilongjiang Wanjia frozen soil test site with a flat foundation having an area of 1 m^2 (Jia, 2000) shows that the normal frost heave force of the foundation

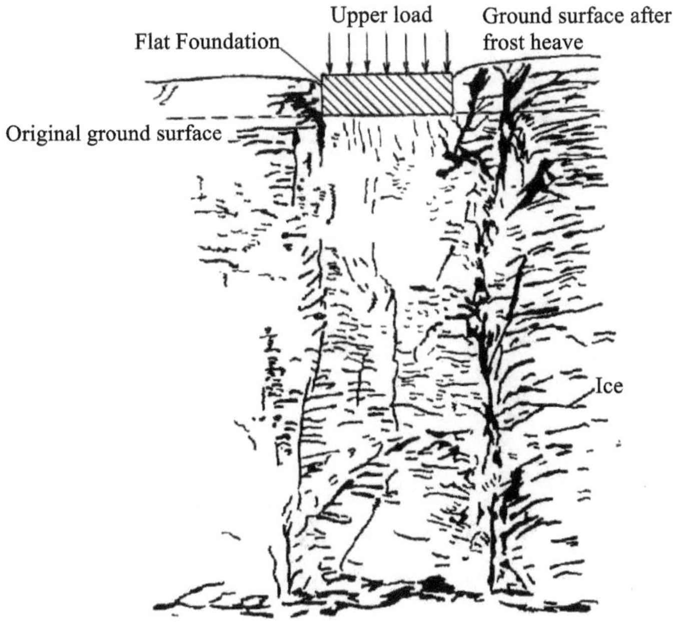

Fig. 5.1 Distribution of ice in soil under flat foundation

plate without lateral restraint is 387 kPa (with a deformation of 70 mm), and the normal frost heave force of the soil column under the restraint of the foundation plate is 79 kpa (with a frost heave deformation of 20 mm), which only accounts for 22% of the former, and the additional normal frost heave force accounts for 78%. However, Tsytovich, a Geocryologist in the former Soviet Union, conducted experiments using fully water-filled sealed all-steel plate cubic molds, and it showed that the expansion pressure of frozen ice on steel plates reached 211.5 MPa (H.A. Tsytovich, 1985), which was the normal frost heave force value under full constraint conditions. The actual engineering conditions are not fully constrained. There are varying degrees of displacement during freezing processes, and the normal frost heave force acting on the foundation plate could be released to a certain extent. Moreover, under the load on the foundation plate, the saturated water body does not completely freeze under the foundation plate during the freezing process, and will migrate to the direction of lower stress, therefore, the normal frost heave force under the foundation plate measured in the field is not as great as that measured in the laboratory environment (the measured normal frost heave force at Wanjia frozen soil test site is 145 kPa).

According to the experimental research data, the normal frost heave force per unit acting on the normal direction of the foundation bottom when the foundation soil frost heaves is listed in Table 5.1 under the condition that the building

Table 5.1 Normal frost heave force per unit σ_0 (kN/m^2) (NB/T 35024–2014)

Frost heave level of foundation soil	Frost heave amount of foundation soil h(mm)	Area of a single foundation plate(m^2)			
		5	10	50	100 and above
I	$0 < h \leq 20$	50 ~ 100	30 ~ 60	20 ~ 50	10 ~ 30
II	$20 < h \leq 50$	100 ~ 150	60 ~ 100	50 ~ 80	30 ~ 60
III	$50 < h \leq 120$	150 ~ 210	100 ~ 150	80 ~ 130	60 ~ 100
IV	$120 < h \leq 220$	210 ~ 290	150 ~ 220	130 ~ 190	100 ~ 150
V	$h > 220$	290 ~ 390	220 ~ 300	190 ~ 260	150 ~ 210

Note: 1. In a same frost heave level, the values in the table should be interpolated according to the frost heave amount, and then according to the area of the foundation
2. This table is applicable to the flat foundation with the size of a shorter side not less than 2.0 m. If the area of the foundation plate is less than 5 m, the value should be taken as 5 m^2

foundation cannot be displaced in the vertical direction and there is no frost heave effect from the surrounding soil.

For the design of flat foundation in frozen soil area, the general method is to follow the conventional design, and then check and calculate the overall stability, foundation deformation and displacement and structural strength of the building under the condition of frost heave soil, so as to determine whether it is necessary to adopt the measures to prevent frost damage.

The anti-frost heave design of hydraulic structures needs to obtain the geological data, the maximum freezing depth and the frost heave amount of the construction site, and obtain the freezing depth and the frost heave amount under the foundation plate through calculation. Determine the frost heave category and normal frost heave force according to the frost heave amount. When the frost heave amount under the foundation plate is greater than the allowable deformation of the upper load of the foundation and the structure, it is necessary to take engineering measures to ensure the stability of the building. In Northeast China, the frost heave h' under the foundation of buildings on cohesive soil foundation can also be obtained according to the empirical formula (Xie and Wang, 1992):

$$h' = 45\alpha_\rho \alpha_p \exp\left(0.017H_d - 2.3\right)^2 - 0.007Z \tag{5.1}$$

where H_d is the engineering design freeze depth, cm; Z is the buried depth of groundwater before freeze, cm; $\alpha_\rho = e^{-Z}(\rho_d - 1.35)$, α_ρ is the density correction coefficient of foundation soil; when $\rho_d < 1.35$ g/cm^3, take 1.35 g/cm^3; α_p is the load correction coefficient, $\alpha_p = e^{-\beta p_i}$, p_i is the load strength of the foundation on the calculation section (kPa), β is coefficient related to foundation soil density, $\beta = 0.034 \exp[-6.0(\rho_d - 1.35)]$; ρ_d is the dry density of foundation soil, g/cm^3.

In actual engineering, the value of normal frost heave force of flat foundation should be based on the area of foundation plate and the thickness of frost heave layer (i.e., the embedded depth of foundation plate, H_d–δ, (δ is the thickness of the foundation plate), displacement and other corrections, and the obtained value is the

Table 5.2 Allowable vertical displacement of flat foundation

Building type and structural position	[S]/mm
Slab foundation at culvert gate inlet and outlet	20
Reinforced concrete foundation plate of sluice chamber section	25
Bottom slab apron of steep slope section (reinforced concrete)	25
Reinforced concrete integral U-shaped trough	30

actual normal frost heave force value of the foundation plate. Table 5.1 is mainly used to determine the stability of plate-shaped buildings. The design value of normal frost heave force should be calculated as follows (Jia, 2000; Qu and Zhang, 2008).

$$\sigma_n = m_\sigma \alpha \sigma_0 \tag{5.2}$$

where, m_σ is the frost heave force attenuation coefficient, $m_\sigma = 1 - ([S]/h)^{0.5}$, when the foundation plate is not allowed to move upward, $m_\sigma = 1$; α is the influence coefficient of frost heaving layer thickness,$\alpha = [1 - (0.7\delta/H_d)]^{3/2}$; σ_0 is the normal frost heave force value per unit, which should be taken according to Table 5.1; $[S]$ is the allowable vertical displacement value of foundation plate, which should be taken according to Table 5.2; h is the frost heave amount of foundation soil at its location, cm; H_d is the engineering design freeze depth, cm; δ is the thickness of foundation plate, cm.

5.2.2 *Prevention and Remedy Measures for Frost Damage of Flat Foundation*

During the foundation design of hydraulic structures in cold regions, preventions for frost heave and thaw settlement should be considered. In seasonally frozen soil area, emphasis is placed on prevention for frost heave. Both frost heaving and thawing of foundation in permafrost area will endanger the safety of buildings. Both of them need to be considered in design. For shoulder flat foundation, both preventions for frost heave and thaw settlement are needed to be considered in design.

Because the size of flat foundation is large and the upper load is relatively small, it may cause displacement. Therefore, it may not be suitable for the domestic conditions to entirely adopt the "resistance" method of reinforced concrete structure used in foreign countries. Deep foundation is usually used to eliminate the frost damage for flat foundation that is not allowed for frost heave deformation. It belongs to the form of pile foundation. The measures for preventing frost heave and thaw settlement will be discussed in "hydraulic structure of pile foundation". For flat foundations that allow for frost heave deformation, the frost heave or thaw settlement of foundation soil can be reduced from the aspects of structure and foundation treatment to meet the adaptation requirements of flat foundation.

5.2.2.1 Measures for Improving Foundation Soil

The freezing and thawing sensitivity of foundation soil depends on the interaction between matrix soil and water. Different particle size, gradation, mineral composition and salt content of solution in soil determine the water migration, ice segregation and frost heave during the freezing process. The thawing settlement of permafrost region depends on the ice content of frozen soil foundation, the distribution and state of ice, and the interaction between matrix soil and water during the freezing process. When the clay content of soil with particle sizes of 0.05 ~ 0.002 mm is greater than 60%, as well as with presence of Kaolin, it is most susceptible to frost heave. As the salt content of the solution in the foundation soil increases, its frost heave amount decreases. The existence of salt affects the freezing temperature and water migration in soil, thus inhibiting the frost heave of foundation soil (Table 5.3)

Table 5.3 Classification of frost heave of natural foundation soil (according to the data from Heilongjiang Province Hydraulic Research Institute)

Frost heave classification	Soil texture	Moisture conditions	Frost heave coefficient /%	Total frost heave amount /mm	Remarks
Slightly frost-susceptible soil	Coarse grained soil coarser than fine sand (including medium coarse sand, gravel cobble, etc., with a particle content of 0.05 ~ 0.005 mm less than 10%), heavy clay (undisturbed)	Unconfined groundwater; although the groundwater level is high, there is drainage during the freezing period	1 ~ 2	20 ~ 40	Generally, small to medium-sized hydraulic structures may not need to consider frost heave effects.
Intermediate frost-susceptible soil	Silt, silty clay, clay	During the whole freezing period, the groundwater level is at least 1 ~ 1.5 m lower than the maximum freezing depth	5	100	There may be zones with strong frost heave and zones where freezing occurs without heaving.
Strongly frost-susceptible soil	Various silty soils, in which the particle content of 0.05 ~ 0.005 mm is more than 30%	Soil is saturated; the groundwater level remains close to the freezing front	>5	>100	Zones where freezing occurs without heaving do not exist.

1. **Foundation soil replacement**

The foundation soil replacement is to replace the frost-susceptible soil with non-frost-susceptible soil within the freezing depth under the foundation plate to reduce the frost heave of the foundation plate. Non-frost-susceptible materials are usually pure gravel, sand gravel, coarse sand etc. In general, the replacement rate is 60% ~ 80%, about 1/2 ~ 2/3 of the freezing depth. When the foundation soil is strongly frost-susceptible soil, as well as high groundwater level, the replacement rate can reach 100%. In actual construction, coarse sand is used as the reversed filter layer to avoid the fine particle soil infiltrating into the replacement material, but it is difficult to conduct; therefore, coarse sand in fiber bags can be used as the reversed filter layer. The fiber sand bag has function of isolation, drainage and reverse filtration, which can prevent the water from migrating to the replacement layer. According to 3 years actual application of the fiber sand bag in Hetao Irrigation Area of Inner Mongolia, the frost heave amount is reduced by 60% ~ 100%.

Two conditions need to meet when gravel is used as replacement filler to reduce frost heave of foundation soil. First, the content of silt and clay in gravel must be less than 5% ~ 10%, and the less is the content, the better is the prevention of frost heave. Second, the water in the sand and gravel should have a drainage outlet during the freezing process. Otherwise, the frost heave caused by water freezing cannot be avoided. The frost heave test of saturated sand gravel in open (with unobstructed drainage outlet) and closed (without drainage outlet) systems shows (Fig. 5.2 and Table 5.4), under the same conditions the total frost heave amount of the open system is only 50% ~ 60% of that of the closed system, that is, it is almost reduced by half.

Adding lime, sand, cement and other materials into the foundation soil to form tabia or four compound soil can not only improve the foundation soil, increase its density, but also form an impermeable layer, which is water-resisting and anti-seepage and can reduce the frost heave of the foundation soil.

2. **Density of compacted foundation soil**

Under the condition that the water content of the tri-phase system remains unchanged, increasing the density of the soil only changes the porosity and saturation of the soil. The initial frost heave ratio of the soil increases with the increase of the density, and after reaching a certain "limit", the frost heave ratio rapidly decreases (Fig. 5.3). At this time, the "free water" in the pores is squeezed out, and only the interfacial water remains, which makes it a two-phase system. When both of the water content and water migration amount decrease, the frost heave susceptibility also decreases, with a dry density of 1.65 g/cm^3 for the "limit" soil. In the two-phase system, the density of soil increases, resulting in the decrease of water content. In the process of soil freezing, the water migration is blocked, following a reduced migration amount, and then leading to a reduced frost heave ratio (Fig. 5.4).

Therefore, the frost heave susceptibility of foundation soil can be reduced by increasing the density of foundation soil through dynamic compaction. In actual practice, by using the layer-by-layer compaction method in earth dam construction,

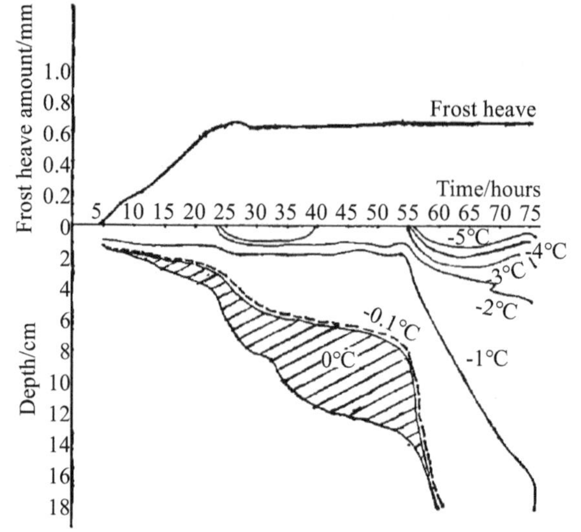

(a) Open system (with unobstructed drainage)

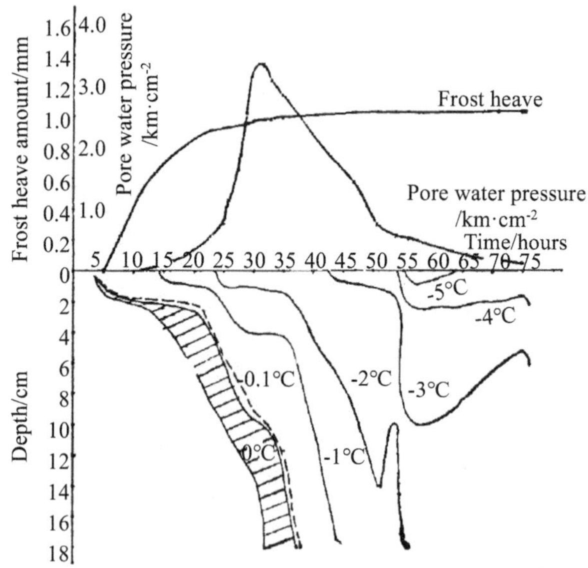

(b) Closed system (without drainage)

Fig. 5.2 Process of sand gravel freezing test (Chen et al., 1979)

Table 5.4 Frost heaving of sand gravel under open and closed system

Open system				Closed system			
Maximum frost heaving amount(mm)	0.69	0.65	0.59	Maximum frost heaving amount(mm)	1.38	1.25	1.06
Maximum amount of discharge water(ml)	90	95	87	Maximum pore water pressure(kg/cm)	3.73	3.63	1.67

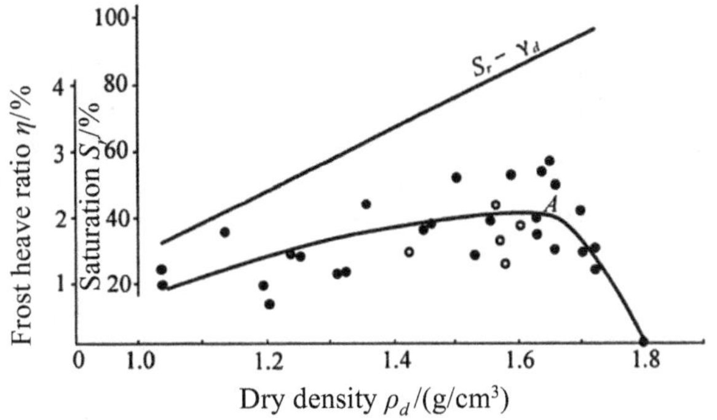

Fig. 5.3 Relationship between frost heave ratio and soil density under same water content (Tong & Guan, 1985)

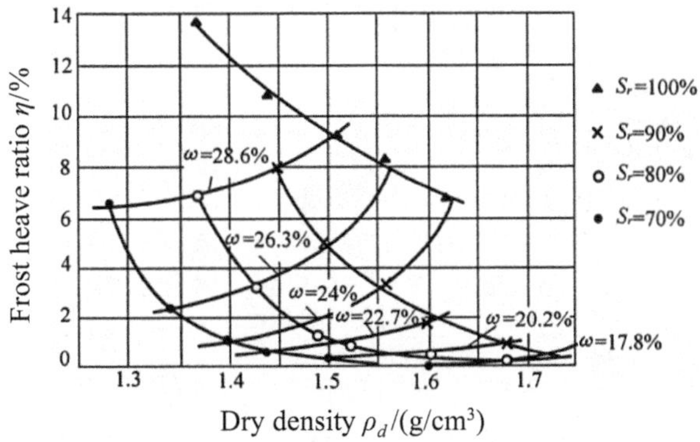

Fig. 5.4 Relationship among frost heave ratio and soil density, saturation and water content (Qu & Zhang, 2006)

Table 5.5 Compaction soil measures design parameters (Na & Yuan, 2007)

Frost heave level	Design thickness δ(cm)	Design dry density ρ_d(g/cm³)
≤III	(0.75 ~ 0.8)Z_f	≥1.55
>III	(0.95 ~ 1.0)Z_f	>1.60

Note: Z_f is the design freezing depth for engineering

achieving a dry density of 1600 kg/m³ or higher is necessary to transform frost-susceptible soil into non-frost-susceptible or slightly-frost-susceptible soil, thereby achieving the goal of preventing frost heaving. When the bearing load of the foundation is less than 10 kPa, the thickness and density within the design freezing depth can be determined by referring to Table 5.5.

3. **Chemical treatment**

The chemical treatment method is to inject or mix chemical materials into the foundation soil, so as to reduce the freezing point of the foundation soil or enhance the hydrophobicity of the foundation soil, and reduce the water migration and frost heave amount during the freezing process. Commonly used materials include table salt, magnesium chloride, calcium chloride and other soluble salts. NaCl is the most admixture, which can reduce the frost heave of foundation soil. The mixing amount can be calculated according to the following formula (Pei and Lu, 1982)

$$Q_{Nacl} = K\frac{10\rho W_{so}C_\theta V}{(1+0.01W)(\rho_s - C_\theta)} \tag{5.3}$$

where Q_{Nacl} is the added amount of NaCl crystal salt, kg; V is the volume of soil before treatment, m^3; ρ is the density of soil before treatment, T/m^3; W is the water content of the soil before treatment,%; K is the impure coefficient of salt (the ratio of the weight of salt mixed with water to its pure NaCl content); $W_{so} = W - W_{sb}$ (W_{sb} is the strong interfacial water of soil particles); ρ_s is the density of pore water solution in soil at the concentration of C_θ, $\rho_s = 1 + 0.65C_\theta$ g/cm^3; When C_θ is the concentration of pore water solution that remains unfrozen at $-\theta$ °C, g/cm^3.

Pay attention to "quantity" when increasing the salinity of the foundation soil. Otherwise, the foundation soil will be salinized, and the deformation and damage resistance capacity will be significantly reduced. Even in the case of slight salinization (the salinity is 0.5% ~ 1.0%), the cohesion will be reduced by 2/3 ~ 5/6, the bearing capacity will be correspondingly reduced, and the deformation will be relatively increased. However, the validity period of adding salt to prevent frost heave is short, and desalination usually occurs in 2 ~ 4 winters.

In order to prolong the effective time of salinization, the comprehensive measures of salinization and foundation layered dynamic compaction can be adopted to form a stable high-strength structure in the soil, resulting in the reduction of the freezing point of pore water solution and the strong change of the interaction capacity between mineral skeleton and groundwater. When the soil is frozen, the water does not migrate or slightly migrates, and the frost heave of the soil can be suppressed for a long time. The dry density of soil required to form high-strength structure is 0.9 ~ 1.03 times of the optimal dry density (Pei & Lu, 1982).

Hydrophobic substances are commonly used as petroleum products, i.e. heavy oil or diesel oil and cationic surfactants, or wood sulfate with sodium dichromate, polyester, polyamide, polyene slurry, polyurethane foam plastic, silicone organic colloid compound (such as ethyl sodium silanol solution, methyl sodium silanol solution, etc.), hydrophobic surfactants etc., which can prevent surface water from penetrating and groundwater from capillary elevation, and weaken water migration during freezing, and greatly reduce frost heave.

Soil solidification technology is to activate soil activity and cement soil particles through soil solidification agent, condense loose soil into materials with overall strength, change the instability of original soil after water absorption, and form consolidated soil with certain bearing capacity, impermeability, frost resistance and durability. Commonly used soil solidifying agents include inorganic soil solidifying agents, organic soil solidifying agents, organic-inorganic composite soil solidifying agents and biological enzyme soil solidifying agents.

5.2.2.2 Measures to Improve Water Conditions

Free water in the soil is the core factor causing frost heave in the freezing process. The continuous increase of ice crystals causes the displacement of soil particles, resulting in frost heaving. When the soil contains only bound water or slightly less

Table 5.6 Initial and safe frost heave water contents of several typical soils

Water content	Soils				
	Clay	Clay	Silty clay	Silty soil	Silty soil
Plastic limit water content, %	19.1	15.7	21.0	10.5	10.2
Initial frost heaving water content, %	13.0	12.0	18.0	10.0	8.0
Safe frost heaving water content:, %	20.0	17.0	22.0	13.0	12.0

than the plastic limit, it exhibits the phenomenon of freezing without heaving, which is known as the initial frost heave water content. When the moisture content in the soil slightly exceeds the initial frost heaving moisture content, weak frost heaving may occur. However, as long as it does not exceed the allowable deformation of the structure, this moisture content can still be considered as safe water content (Table 5.6). The frost heave caused by exceeding the "safe frost heaving water content" poses a threat to the building and causes frost damage. However, the initial frost heave water content and "safe water content" of soil are not fixed values, but depend on the physical properties of soil.

The water in the soil comes from the infiltration of precipitation and surface water, as well as the wetting of groundwater through capillaries, so that there are bound water and free water in soil pores. During the freezing process, free water in soil is the first to undergo freezing and ice segregation. With the decrease of the soil temperature, the weakly bound water also gradually freezes. This process extracts free water from the soil below the freezing front. The ice crystal gradually increases, forming an ice layer, expanding the volume, and the soil particles appear displacement (frost heave). When there is no more energy at the frozen front to extract free water at the lower part or no free water supply, it will move down and continue to repeat the ice segregation process. The cooling transmitted from the atmosphere through the ground decreases, the ice segregation (frost heave) ends, and the freezing process ends immediately. The frost heave amount of the soil in the closed system depends on its water content. With the increase of the water content, the frost heave amount of the soil also increases, and finally stabilizes at a relative value. In general, the frost heave amount of soil in a closed system is weak frost heave (II) or frost heave (III), and it may have greater frost heave when saturated. Under the condition of groundwater recharge, that is, an open system during the freezing process, the groundwater continuously supplies capillary water to the process of ice segregation at the freezing front, and ice crystals accumulate and increase continuously, forming a thick ground ice layer (or intrusion ice). The displacement of soil particles increases and the voids created by this displacement are subsequently filled by the ice layer. In this circumstance, when the soil is strongly frost susceptible, it forms pingos.

In addition to surface water infiltration, the rising height of capillary water plays an important role in the distribution of soil water content. In the range of capillary water height, from top to bottom, the closer to the groundwater level, the higher the water content of the soil. During the freezing process, the shoulder is the groundwater level, the greater the frost heaving of the soil becomes. According to the frost heave test observation of Liaoning Institute of Water Resources Research in the

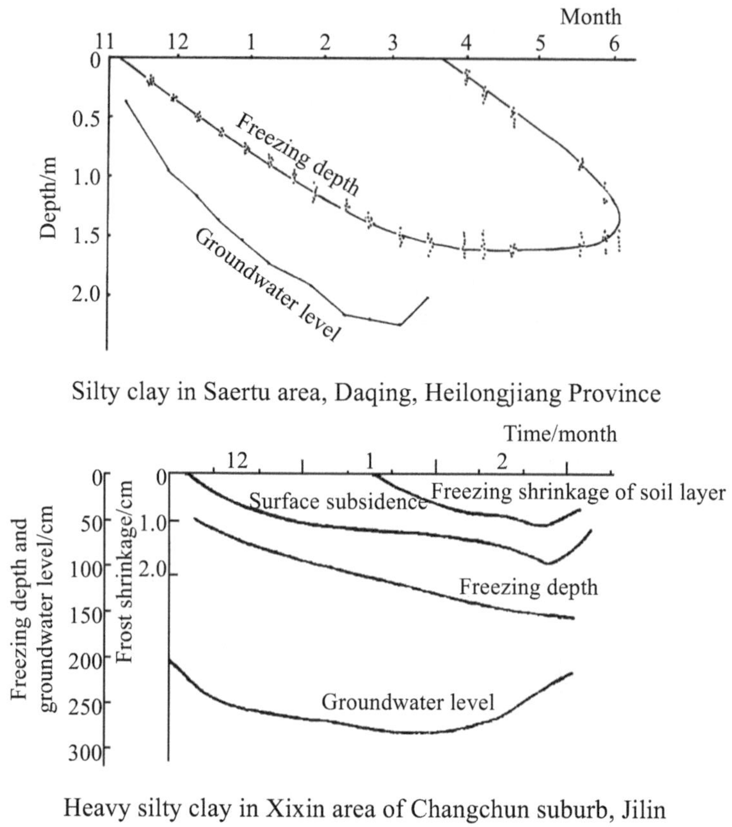

Fig. 5.5 Freezing depth and variation of groundwater level in silty clay (Xu & Lu, 1986)

seasonally frozen soil area of Inner Mongolia, the height of capillary water within the non-frost susceptible soil layer from the groundwater level is: in coarse sand, it is 0.4 m; in fine sand, it is 0.6 m; in silty soil (light loam soil), it is 1.20 m; and in silty clay (heavy loam soil), it is about 1.60 m. When the distance between the freezing front and the groundwater level is less than or equal to the rising height of capillary water, it will have a positive effect on the frost heave of soil. However, with the increase of freezing depth, the ground water level will drop (Fig. 5.5), with a range of 1.0 ~ 1.3 m (Wang, 1982), and the ground water level will start to rise in March next year. During the dry season in winter, the groundwater gradually decreases. In addition, the pore water of the soil freezes and expands in volume to push the excess pore water out of the pores. Under the effect of pore water pressure gradient, the groundwater level decreases.

Based on the research of Russian scholar, B. O. Orlov (1980), in the circumstance that the soil freezes without leading to frost heave, the minimum distance between the maximum freezing depth and the groundwater level before winter is listed in Table 5.7. Compared with the observation data of frost heaving test of

Table 5.7 Minimum distance between the maximum freezing depth and the groundwater level /m

Clay containing montmorillonite and hydromica	3.5
Kaolinite containing matrix clay, silty clay (loam)	2.5
Silty clay, silty soil (sandy soil)	1.5
Fine sand, silty sand	1.0
Fine sand	0.6
Coarse sand	0.4

Table 5.8 Harmful capillary rising height of highway subgrade in seasonally frozen soil area (Chen et al., 2008a, b)

Soil name	Plasticity index	Particle size composition (%)			Threshold height for frost damage	Compactness
		>0.075	0.075–0.005	<0.005		
Low liquid limit clay with lime	14.7	6.37	90.31	3.32	2.00	0.8
Low liquid limit clay	18.1	1.31	79.95	18.74	2.45	0.8
High liquid limit clay	26.4	1.19	74.1	24.71	1.70	0.8
Low liquid limit silt	9.24	23.45	64.19	1.33	2.25	0.9
Silty sand		68.04	31.48	0.48	1.25	0.9
Fine sand		86.45	13.41	0.48	0.60	0.9
Low liquid limit clay	12.8	25.3	70.13	4.87	2.40	0.8
Medium sand with mud (C = 25%)		75.0	20.25	4.75	1.95	0.9
Medium sand with mud (C = 10%)		90.0	8.1	1.9	0.80	0.9
Silty coarse sand (C = 10%)		90.0	8.1	1.9	0.4	0.9

Note: The sample was taken from the front edge of Songhua River first terrace and floodplain in Taolaizhao Town, Jilin Province and the second terrace of Yitong River in Changchun City

Liaoning Institute of Water Resources Research in Inner Mongolia seasonally frozen soil area, the values in Table 5.7 are larger. Table 5.8 is the test data of harmful capillary rising height of compacted soil of highway foundation in seasonally frozen soil area. Because cohesive soil includes a wide range and the capillary rising heights of them are varying, it is feasible to judge the possibility of frost heave according to whether the groundwater level before freezing is greater than the capillary rising height. Attention should be paid to the frost heave of medium coarse sand.

1. **Drainage measures**

The essence of the drainage measures is to reduce the ground water level and capillary rising height, drain the water in the foundation soil, remain "safe water content of frost heave", and reduce the frost heave of the foundation soil. Drainage measures should be designed in combination with topography, hydrogeology and specific conditions of the project. For the buildings with culverts, gates and molten iron at the outlet with height difference, vertical and horizontal drainage ditches can be set up to discharge the groundwater in the foundation soil under the plate foundation, taking advantage of the low drainage outlet conditions on the steep slope where the outlet lies (Fig. 5.6). When there is an unconfined permeable layer lying not deep under the plate foundation, several vertical sand wells can be arranged to drain the water of the foundation soil (Fig. 5.7). Under similar conditions at the bottom of a channel, one seepage well can be arranged every 20 m or so (Fig. 5.8). Under the maximum seasonal freezing depth, vertical and horizontal underground drainage pipes can also be arranged to drain the water in the soil layer under the foundation plate. When the drainage facilities are arranged at the bottom of culverts, gates and channels, they must meet the seepage prevention requirements and be designed in combination with the seepage prevention measures.

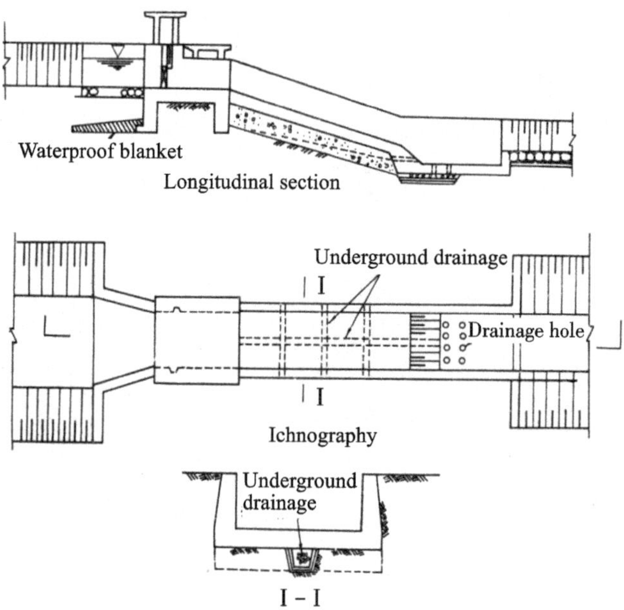

Fig. 5.6 Drainage measures for foundation soil at outlet of channel sluice chamber (Tong & Guan, 1985)

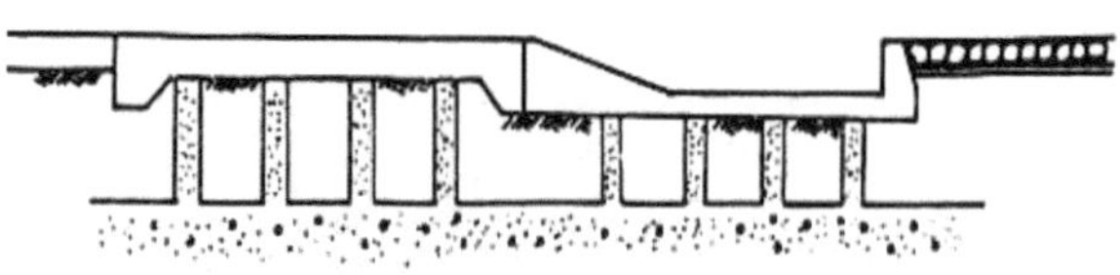

Fig. 5.7 Layout of sand well drainage under flat foundation (Tong & Guan, 1985)

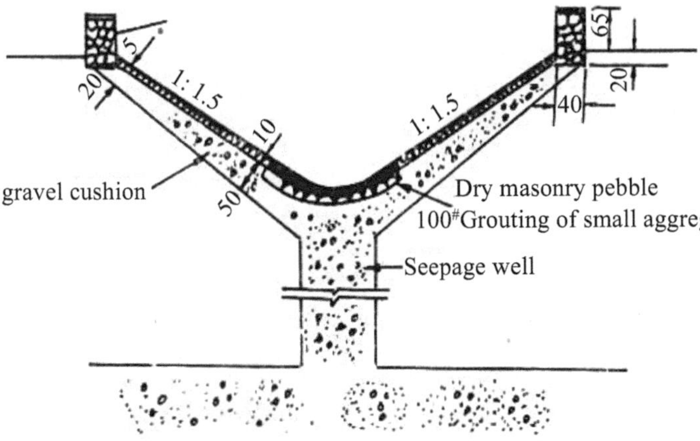

Fig. 5.8 Seepage wells at bottom of channels (Tong & Guan, 1985)

2. **Waterproofing measures**

The water in the soil comes from the infiltration of surface water and groundwater. From the perspective of frost heave, measures should be taken to ensure that the water content of the foundation soil within the freezing depth under the foundation plate is within the "safe frost heave water content" to avoid frost damage caused by the freezing of the foundation soil.

The measures to prevent water seepage from the upper part and reduce the water content of the foundation soil are usually to lay a layer of polyethylene film or geomembrane or geotextile (one geotextile and one film, two geotextiles and one film) under the foundation plate, and to fill the anti-seepage materials for the construction joints and deformation joints between the plates. The waterproof membrane is simply laid under the foundation plate, but the seepage prevention of the construction joints and deformation joints between the foundation plates is not properly handled, and there is still water under the foundation plate, which causes deformation. The filling materials in the construction joints and deformation joints should be anti-seepage mastic with good elasticity, strong adhesion and expansion in water, such as tar plastic mastic, polyvinyl chloride mastic, elastic polyester grease, polysulfide sealing paste, building grease, water-swelling rubber waterstop strip, etc.

Generally, in areas with shoulder groundwater level, the freezing depth is also small. In combination with the topographic and hydrogeological conditions, drainage measures such as interception, diversion and drainage are adopted, and the groundwater level reduction is the preferred scheme. If there is no drainage condition, fill should be used to raise the building base, increase the distance between the ground water level and the freezing line, and combine the measures of base water-resisting (or soil curing agent) and thermal insulation to reduce the freezing depth and reduce the frost heave amount of foundation soil. Under the foundation of the culvert gate, the foundation soil should be excavated, dried to the plastic limit water content, and then backfilled into the impermeable and anti-seepage material for sealing to form an artificial foundation and eliminate the frost heave of the foundation soil. This method is called the waterproof sealing method.

The water-resisting sealing depth is generally not less than the depth of the main frost heave zone, that is, more than 2/3 of the maximum freezing depth. If the building is not allowed to be deformed, it needs to reach 4 / 5 of the maximum freezing depth. The plane size should be greater than 0.5 m around the foundation plate, and the elevation of the foundation pit should be vertical. Any soil can be used as backfill, but it should not contain sharp objects such as block gravel and tree roots. The water content should be slightly less than the plastic limit moisture content of the soil. The density of the compacted soil layer should reach the dry density of the natural foundation soil. Thick geomembrane should be used as the sealing material, and 0.1 ~ 0.2 m coarse sand and gravel should be filled between the outer side of the sealing geomembrane coated with grease or in contact with the foundation pit wall. It can be sealed in layers, and the thickness should be about 0.5 m.

5.2.2.3 Measures for Changing Temperature Conditions

Lay thermal insulation layer under and around the foundation plate to increase the thermal resistance of the foundation plate, increase the temperature of the foundation soil, delay its freezing time, and reduce the freezing depth. At present, industrial thermal insulation materials such as polystyrene (EPS), polyurethane (PU) and extruded polystyrene (XPS) foam are widely used in actual practices. They have low thermal conductivity, no water absorption (or low) and strong compression resistance, and XPS has high strength. In addition, there are other thermal insulation materials, such as slag, vermiculite, foam concrete, glass fiber, expanded pearl cotton etc. Once these thermal insulation materials are soaked, the thermal insulation performance will be greatly reduced, resulting in opposite results. Water is also used as insulation material in water conservancy projects.

The general method is to lay the insulation board under the foundation plate or at the appropriate position in the foundation soil, or add insulation materials to the mixed concrete to produce a new type of insulation board. When adopting thermal insulation foundation, in addition to reaching thermal insulation index according to thermal calculation, it should also meet the requirements of buildings for foundation bearing capacity.

For thermal insulation foundation of lightweight buildings, such as flat foundation of culvert, canal and apron in front of gate, the thermal insulation plate can be directly laid at the appropriate position under the foundation plate (Fig. 5.9) or combined with the foundation plate to form a composite foundation plate (Fig. 5.10). Other thermal insulation materials can be used for the thermal insulation layer sandwiched in the composite foundation plate, such as slag cement thermal insulation plate, cement thermal insulation plate, expanded perlite concrete thermal insulation plate, polystyrene granule (organic waste) cement concrete thermal insulation plate, ceramsite concrete thermal insulation plate and other new composite thermal insulation plates. It has the properties of thermal insulation, low water absorption,

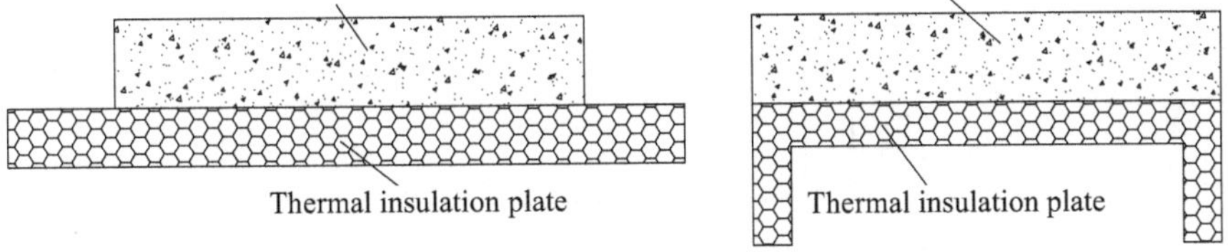

Fig. 5.9 Schematic diagram of placing insulation board under the foundation plate (Qu & Zhang, 2008)

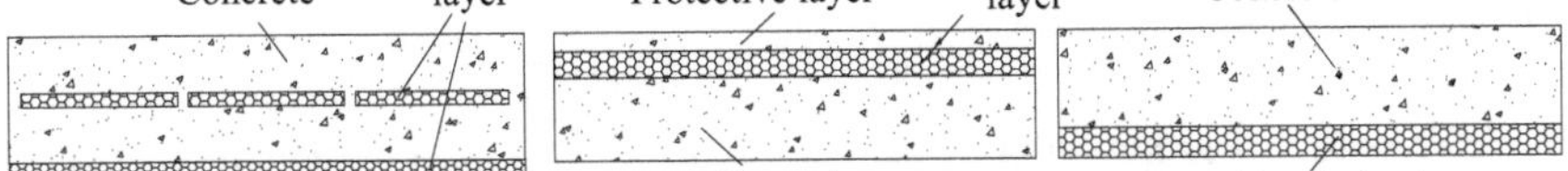

Fig. 5.10 Schematic diagram of composite foundation insulation plate (Qu & Zhang, 2008)

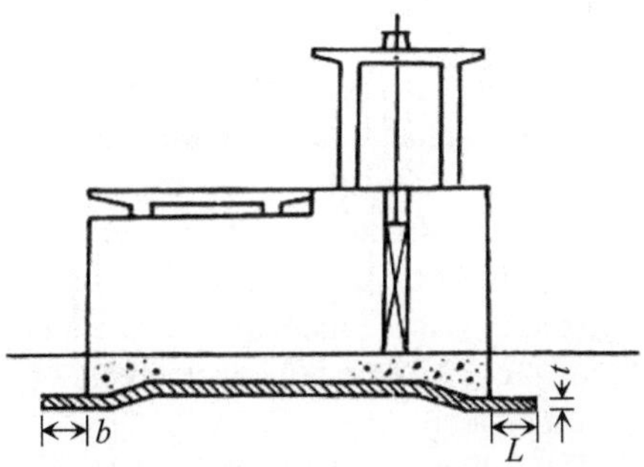

Fig. 5.11 Schematic diagram of thermal insulation foundation

meeting the freezing resistance of 25 freeze-thaw cycles, good integrity, strong toughness and light weight, and can also be produced by waste polystyrene.

In construction projects, polystyrene foam plates are often laid under the foundation plates to reduce the freezing depth (Fig. 5.11). For the sluice gates in many irrigation areas in the northeast, the open box type horizontal type is adopted at the gate foundation, and the bottom plate of the sluice body is 1.0 m. In winter, the water depth of 1.0 m is kept for thermal insulation. The stilling pool is used for water storage and thermal insulation to prevent foundation freezing (Fig. 5.12). The thickness of the water storage and thermal insulation layer should be greater than 0.5 times of the local freezing thickness or the maximum freezing depth (Xu and Gu, 1993), or the thickness of the single-layer thermal insulation material(δ)according to the following empirical formula:

$$\delta = \frac{H_f}{K} \tag{5.4}$$

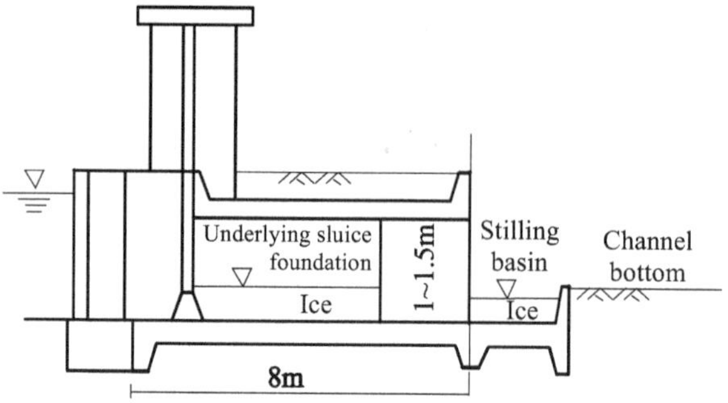

Fig. 5.12 Schematic diagram of water storage and anti freezing of 1–3 sluices in pear irrigation district, Jilin Province

where H_f is the local maximum freezing depth, m; K is the coefficient determined by the type of foundation soil and thermal insulation material, Table 5.9.

5.2.2.4 Measures to Improve Stress Conditions

The frozen expansion of foundation soil is constrained to generate frost heaving force. The greater the constraint is, the greater is the frost heaving force. If deformation is allowed, the frost heaving force will be released quickly (Fig. 5.13). Change the building structure, such as anchor base plate, inverted arch base plate, I-shaped sluice, arc bottom trapezoid, arc slope foot trapezoid, U-shaped groove, etc., to weaken or release the restraint direction and degree of the foundation plate on the foundation soil, and effectively avoid or weaken the frost heave damage of the foundation plate.

The culvert gate building has a large flat foundation. According to the application requirements, the foundation structure can be appropriately changed according to the integrity, rigidity and strength characteristics of the structure to resist or adapt to the frost heave of the foundation soil.

In terms of general arrangement, in addition to meeting the basic requirements of water flow, seepage prevention, application and structure, the general arrangement should be carried out from the perspective of antifreeze damage, and strive to be simple, so as to shorten the range of antifreeze expansion and uneven deformation.

Structurally, the wing wall at the inlet and outlet adopts I-shaped wing wall instead of splayed wing wall.

For anti-seepage arrangement, for example, the vertical anti-seepage with deep teeth is adopted for the main groove, and the I-shaped wall or keywall is adopted at both sides for anti-seepage.

On the premise of meeting the requirements of sluice chamber of opening, closing and other traffic functions, the length of gate chamber section should be shortened as far as possible, the stilling basin should be placed under the traffic bridge, and the gate chamber should adopt a narrow weir (Fig. 5.14). In Xinjiang and Hailun

Table 5.9 Coefficient K (Tong & Guan, 1985)

Foundation soil	Leaf	Shavings	Sawdust	Moss	Peat residue	Chopped straw	Dry stove	Damp slag	Loose snow	Accumulated snow	Loosing soil	Reed	Concrete	Reinforced concrete	Ice
Sandy soil	3.3	3.2	2.8	2.8	2.8	2.5	2.0	1.6	3.5	2.5	1.4	2.1	1.4	1.2	1.5
Silty sand	3.1	3.1	2.7	2.7	2.7	2.4	1.9	1.6	3.0	2.0	1.3	2.1	1.3	1.1	1.4
Silty clay	2.2	2.6	2.8	2.3	2.3	2.0	1.6	1.3	3.0	2.0	1.2	1.7	1.1	1.0	1.3
Clay	2.1	2.1	1.9	1.9	1.9	1.6	1.3	1.1	3.0	2.0	1.2	1.4	1.0	0.8	1.2

Fig. 5.13 Relation curve between frost heaving force and deformation (Tong & Guan, 1985; Qu & Zhang, 2006)

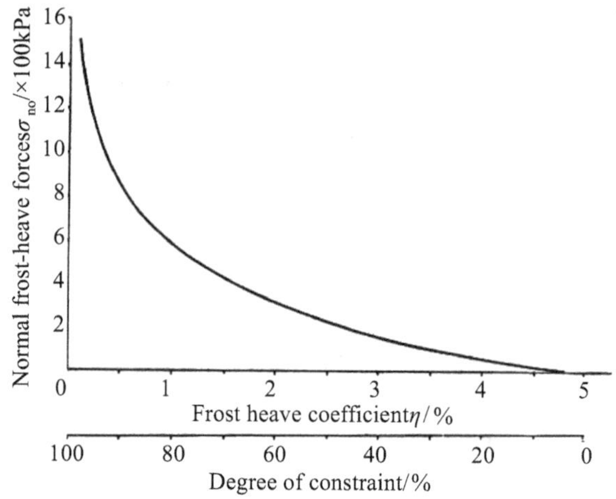

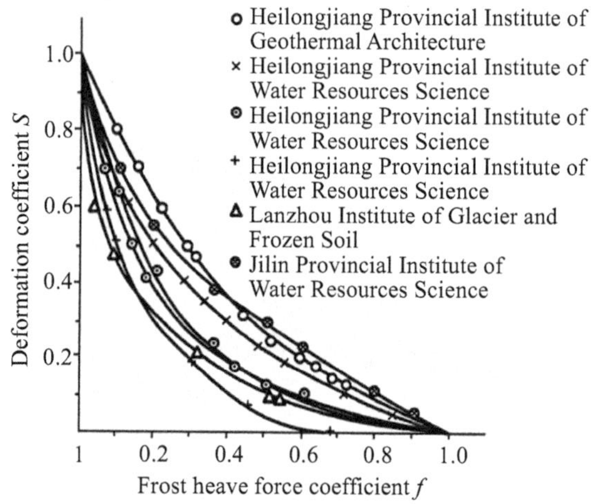

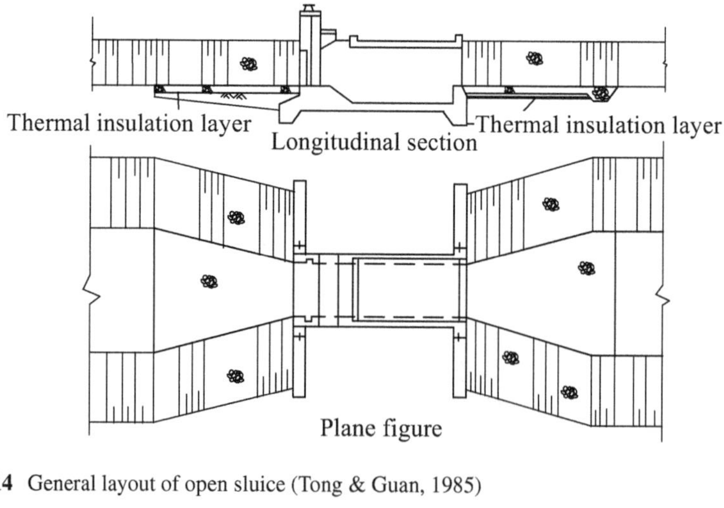

Fig. 5.14 General layout of open sluice (Tong & Guan, 1985)

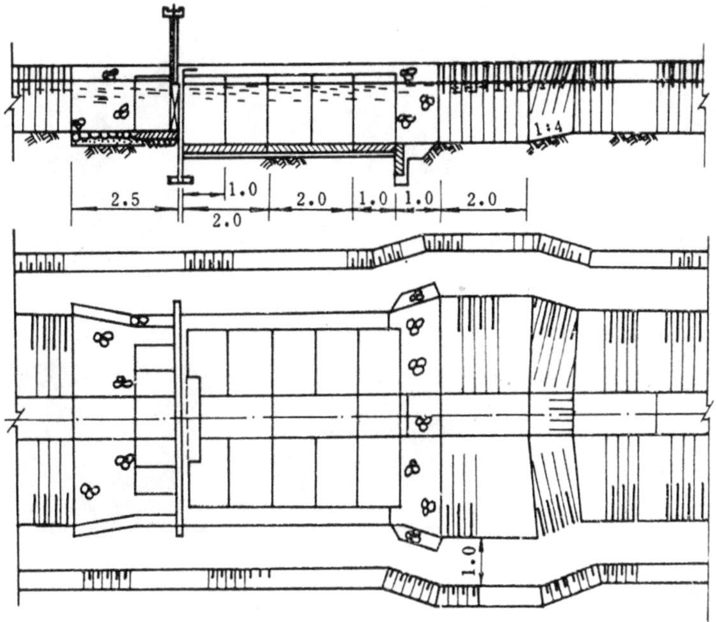

Fig. 5.15 Finalized design drawing of I-shaped sluice (Tong & Guan, 1985)

County of Heilongjiang, the I-shaped wall is used, which has been proved to be in good operation for many years. Heilongjiang Provincial Water Resources Survey and Design Institute has prepared the finalized design of the fabricated I-shaped sluice (Fig. 5.15). The splayed wing wall, which is vulnerable to frost damage at the inlet and outlet, can be replaced by the slope protection connection which is capable to adapt to frost heaving deformation. The sluice chamber is I-shaped wall, and both sides of the gate extend into the canal embankment. The inlet and outlet of the I-shaped sluice are trapezoidal sections with poor water flow conditions. The addition of a W-shaped energy dissipater placed at its outlet can significantly improve the energy dissipation effect at the outlet and shorten the length of the energy dissipation and erosion prevention section at the outlet.

The canal lining is a light flat structure, and the frost heaving force of the foundation soil can easily cause the lining of the canal to suffer from frost damage. The lining plates at the bottom of the canal and at the toe of the slope rise, dislocate, and crack along the axis of the canal, and 1/3 of the bottom, toe, and slope of the large slab lining are subject to frost heave and fracture (Fig. 5.16). Some large slab linings on canal slopes are divided into two or three parts, and the rigidity is relatively high, so the phenomenon of "climbing" occurs.

According to the characteristics of canal frost damage, the causes of frost heave are soil texture, moisture content and soil temperature. The soil water content of canal bed foundation plays a key role. When there is no groundwater recharge, channel leakage is the main source of foundation soil wetting, and the maximum frost heaving ratio is mainly distributed in the upper part, under the lining plate, and

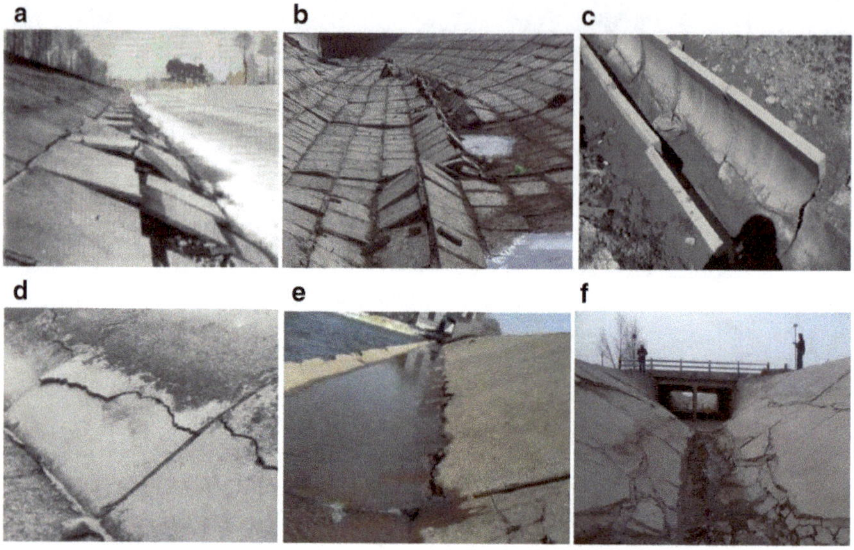

Fig. 5.16 Frost damage of canal bottom lining. (**a**) Frost damage of canal slope toe lining uplift; (**b**) The lining blocks at the bottom of the canal and at the toe of the slope bulge and crack (Photo by Gongli); (**c**) Fracture and dislocation of U-shaped drainage ditch (Photo by Tian Yahu); (**d**) Fracture of large slab lining of canal slope (Photo by Sun Xiuyan); (**e**) Cracking of large plate lining at slope toe (Photo by Xiulu); (**f**) Frost cracking and crushing of the lining at the bottom slope toe of the canal (Photo by Gongli)

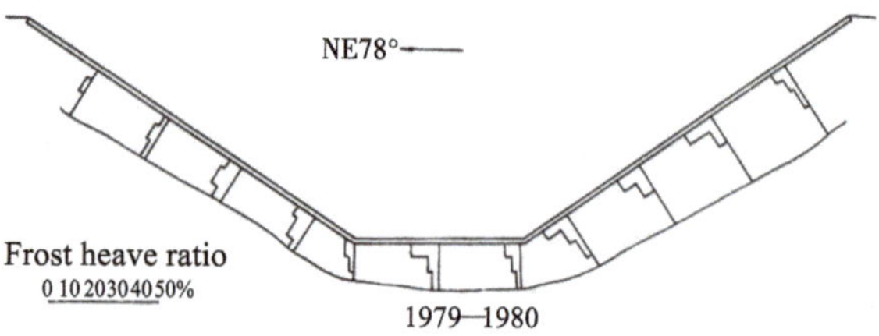

Fig. 5.17 Stratified frost heave ratio of Jinghui Main Canal in Gansu Province

the frost heaving ratio in the lower part is very small (Fig. 5.17). When there is groundwater recharge, the maximum frost heaving ratio at each point of the canal bed often appears in the middle and lower parts, and there is basically no area where there is frost but no heaving (Fig. 5.18). It can be seen from the distribution of water in the foundation soil of the canal bed that the water content at the bottom of the canal and the lower part of the canal slope is the largest (Fig. 5.19), so the maximum frost heaving ratio occurs at the bottom of the canal and the lower part of the canal slope. These parts are just the places where the interaction between lining plates has greater constraints, and the frost heaving force is the largest, so the frost damage is concentrated and prominent.

It can be seen that the double-layer plastic film (Fig. 5.20) is laid under the lining plate of the canal bed. First, it plays the role of anti-seepage and reduces the seepage

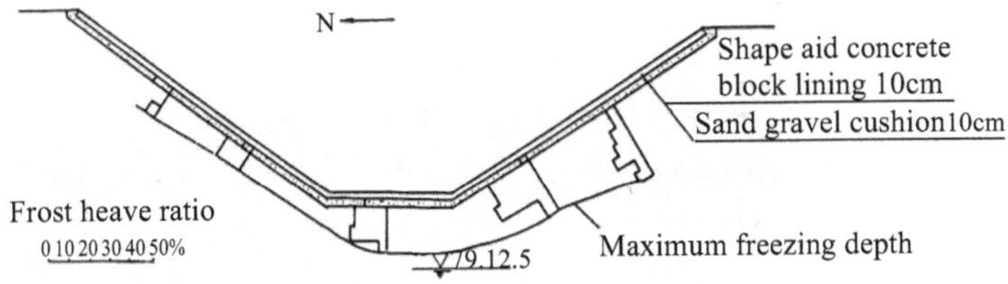

Fig. 5.18 Frost heave ratio of section II of Jingtai Test in Gansu Province

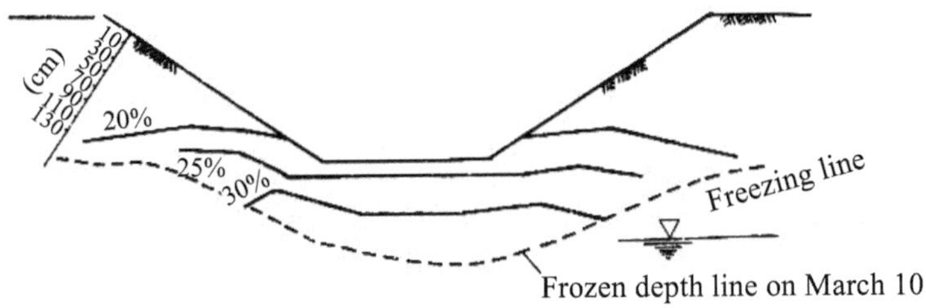

Fig. 5.19 Contour map of water content of canal bed

Fig. 5.20 Canal bed paved with double-layer film plastic for seepage prevention (Photo by Wang Jianjun)

of the canal. Secondly, it can isolate the lining plate from the foundation soil, reduce the freezing constraint on the foundation soil, and achieve the goals of seepage prevention and frost heave prevention. The asphalt concrete is made of flexible

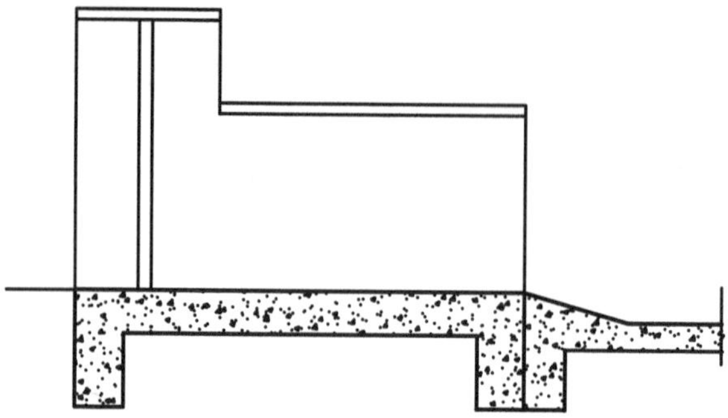

Fig. 5.21 Schematic diagram of section of trough foundation

materials mixed with asphalt, sand and crushed stone as the channel lining, which can not only prevent seepage, but also adapt to frost heave deformation. This method can achieve the goal of flexible anti frost heaving in the channel with or without groundwater recharge, especially suitable for the area with deep groundwater level in cold regions. In addition, waterproofing materials such as asphalt, glass fiber, asphalt felt can also be used, but a protective layer should be laid on the surface.

The channel lining plate adopts a special structural form, which can change the mechanical performance of the interaction between the concrete lining plate and the foundation soil, and can also prevent frost heave. The structural forms include rectangular, arc bottom trapezoid, arc toe trapezoid, U-shaped and gate hole shape, box shape, reverse arch shape and circle shape of unpressurized anti-seepage culvert. Modified PP fiber can be added in the materials to change the performance of the concrete slab.

1. **Trough plate foundation**

Trough plate foundation, also known as inverted-box-shaped foundation (Fig. 5.21), is widely used in small culvert and sluice foundations in northern China, with good frost damage prevention effect. The reasons are as follows: first, it has greater rigidity and can adapt to the action of downstream and vertical flow. Second, it can change the direction of heat flow and the direction of water migration during freezing. If the thermal insulation layer and geomembrane are laid under the trough plate, the freezing depth and normal frost heave force under the foundation can also be reduced, and the goal of anti-seepage can be achieved.

The channel lining plate can also be made into the shape of trough plate, shortening the vertical plates at both ends to form a "ribbed plate", which becomes the rib beam of the lining plate and enhances the rigidity of the lining plate. Rib beams can be added under the lining plate of any shapes or U-trough lining plate. The spacing depends on the size of the plate, but should not exceed 1 m. The rib height is 2 ~ 3 times the thickness of the plate, and it is better in wedge-shaped or with the lower end of semicircular.

Fig. 5.22 Section of U-shaped channel (GB T 50600-2010)

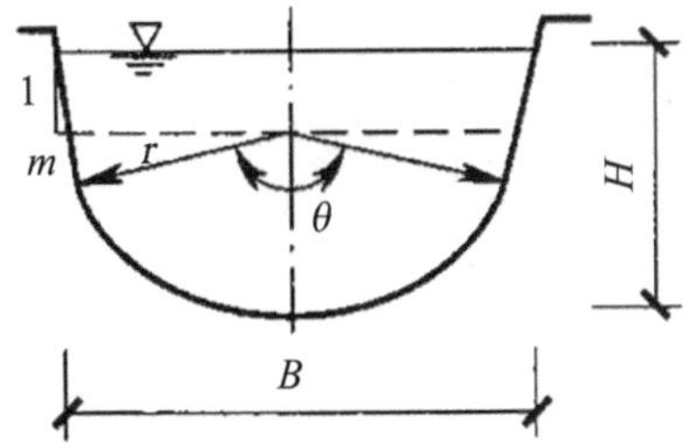

2. **Whole concrete U-shaped channel**

The whole concrete U-shaped channel (Fig. 5.22) has a relatively high rigidity. In order to improve the impact resistance, crack resistance and frost resistance of the concrete, modified PP fibers are mixed in the concrete. At the same time, anti-seepage geotextiles can be laid under the U-shaped channel. U-shaped channels are mostly used for channels with small section area. Its advantages are the following. ① The anti-seepage effect is remarkable, and the water utilization coefficient reaches 0.97 ~ 0.98. ② Good hydraulic conditions lead to high velocity of flow, which increases the sediment carrying capacity of the flow. ③ It is good in external force resistance, frost resistance and impact resistance. ④ It occupies small land, only 1/2 ~ 1/4 of trapezoidal channel. ⑤ It saves labor and materials, saving 25% ~ 30% of materials than trapezoidal channels, reducing hydraulic drop, steeping slope and other connecting buildings, and generally saving 40–50% of the cost of building materials. ⑥ It is easy to manage, saving 60 ~ 70% of management labor compared to trapezoidal channels.

According to the investigation of the branch canals in the irrigation area in Yellow River basin in Ningxia, there were 16 typical U-shaped sections, which were lined with precast concrete slabs, with an intact rate of 43.75% (Wang & Lu, 2007). The rest had fractures, collapses, cracks and displacements. The reasons are that the section was spliced with two half-U-shaped plates (small flow rate), or an arc plate was added at the bottom of the channel (namely, triple splicing type, for slightly larger flow rate), with a plate thickness of 50 ~ 70 mm, which was related to the slope of the inner slope, construction technology and groundwater. Relatively speaking, the three-splicing type has better frost heave resistance than the two-splicing type. The integrated U-shaped channel with well-designed width of the upper opening and depth of the channel, that is, the parabola form with an outward inclination of the opening and a large bottom depth, has good stress conditions.

The calculation of "U-shaped" channel section size can refer to the method introduced by Wang (2016).

3. **Arc bottom trapezoidal channel**

The concrete lining plates of trapezoidal channel section with curved bottom (Fig. 5.23) transit to the circular tangent of the arc at the junction of the straight slope and the arc channel bottom. The hydraulic performance and the force bearing performance of the slope are improved, the stability of the channel slope is enhanced,

Fig. 5.23 Section of trapezoidal channel with arc bottom

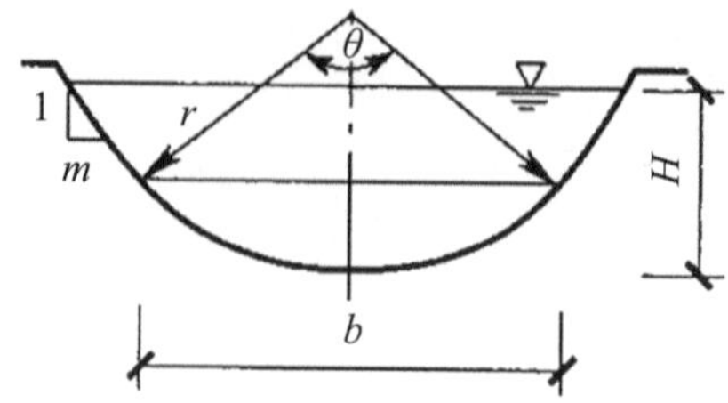

Fig. 5.24 Section of trapezoidal channel with arc slope toe

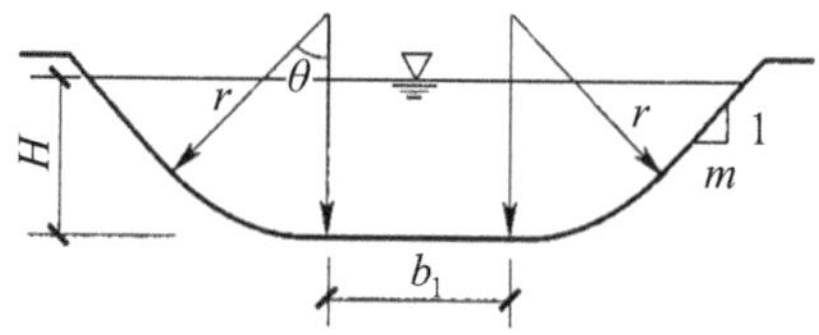

and the defects of easy destruction and fracture above the U-shaped section channel slope are avoided.

According to the investigation of the branch canals in the irrigation area in Yellow River basin in Ningxia, 15 arc bottom trapezoidal canals were lined with precast concrete slabs. Collapse and fracture were the most serious damages, accounting for 53.4% (Wang & Lu, 2007), which was less serious than that occurred for U-shaped section. Reason: The thickness of the lining slab in the straight section of canal slope was normally 45 ~ 70 mm, which was spliced with precast concrete flat plates with a size of 0.4 m × 0.4 m, and with an arc canal bottom, with a strength of more than C10. The thicker the plate is, the milder the damage becomes; the slower the slope is, the lighter the damage is. The thickness of 70 mm, strength of C15 and gradient of 1: (1 ~ 2) is preferable. The damage of this kind of channel is related to construction quality and excess water in the canal.

4. **Trapezoidal channel with arc slope toe**

On the basis of trapezoidal section, trapezoidal channel section with arc slope toe (Fig. 5.24) changes the linear angle of slope toe that is easy to be damaged into arc shape, increases the width of channel bottom plate, improves the anti-destruction ability and adapts to larger flow. Three typical channels were investigated, and the overall damage was the least.

5. **Trapezoidal compound channel**

Trapezoidal compound channel, that is, the straight channel slope of trapezoidal section is changed into broken line channel slope, which is called "quasi trapezoidal" section channel (Fig. 5.25). The frost heaving force on the bottom plate and the lower 1/3 of the slope plate of trapezoidal channel is the largest. The trapezoidal composite section reduces the stress on the center of the canal bottom plate, and the shape of the canal slope changes its stress state, thus enhancing the integrity of the trapezoidal composite canal lining.

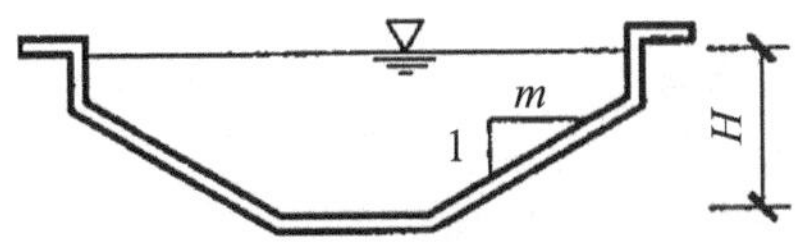

Fig. 5.25 Trapezoidal composite channel section

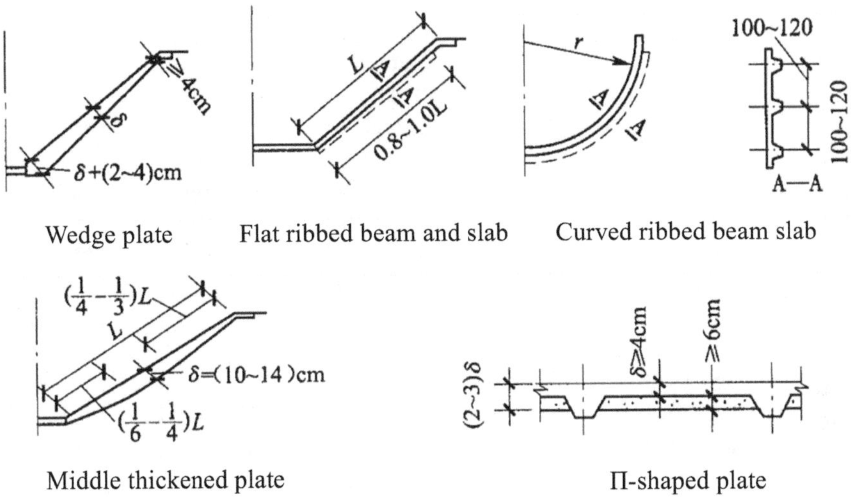

Fig. 5.26 Structural form of concrete slab for canal lining (GB T 50600-2010)

In the irrigation area in Yellow River basin in Ningxia, the trapezoidal composite section of lining plate and masonry toe protection was adopted. After investigation of five typical channels, common diseases such as cracks, bulges, fractures and collapses were found. The cracks and bulges occurred on the slope of the canal, and the fractures occurred at the bottom of the canal, causing little harm. However, the canal slopes were seriously damaged by collapse. On the whole, the damages to the channels were very light, and only leakage occurred without endangering safety. Comparatively, trapezoidal composite channel section lining structure had the longest preservation time, with the best integrity and frost heaving resistance, but also with the highest cost of construction.

6. **Lining concrete slab structure**

According to the frost damage characteristics of lined channels, the measures such as changing the structural section form (Fig. 5.26) at the position with the maximum frost heave force or increasing the thickness and rigidity of the concrete slab structure will enhance the frost-heave resistance of the concrete slab.

7. **New materials and structures**

The canal anti-seepage and anti-frost heaving are unified. The canal lining structure has a good anti-frost heave performance, so it has a good anti-seepage effect. The flexible structure is used to adapt to the frost heave of the foundation soil. The new flexible canal lining structure combining "internal prevention and external flexibility" (Wang et al., 2009; Li, 2006; Dou et al., 2009) can achieve the purpose of anti-seepage and anti-frost heave.

PP fiber reinforced concrete slab + composite geomembrane structure: It is used for filling, semi filling or excavation with low frost heave and deep groundwater level, and for channels without water in winter. The inducement of foundation soil frost heave comes from channel seepage and atmospheric precipitation. PP fiber concrete is a cement-based composite material made of concrete as the matrix, mixed with various discontinuous short fibers, to improve the internal structure of concrete and maintain the durability of concrete. PP fiber reinforced concrete slab has good plasticity and adaptability to frost heaving deformation of foundation soil, and also has flexural strength, impact resistance and tensile strength. According to data, the ultimate tensile rate of polypropylene fiber is 6% ~ 16% higher, the elastic modulus is 5% ~ 13% lower, and the toughness index is 21% ~ 35% higher than those of ordinary concrete (Wang et al., 2009). The geomembrane is laid under the PP fiber concrete slab, which can not only prevent seepage, but also adapt to frost heaving deformation, and has additional thermal insulation effect. It can reduce the frost depth and frost heaving amount. This structure, with a relatively low cost to construct, is raised by frost heave in winter, and it is generally restored to its original shape in summer with a flat and straight surface and without obvious cracks.

Cellular concrete slab + composite geomembrane structure. The structure is used for excavation or semi-excavation with high frost heaving, and cohesive soil channel soaked by groundwater. Cellular has a three-dimensional honeycomb structure with high strength, light weight, wear resistance and aging resistance. The concrete slab of cellular is made of cellular frame with cast-in-place concrete, so that the concrete slab has interlocking and flexible structure like plates, which can better adapt to differential frost heaving deformation and also has frost heaving resistance. Geomembrane is laid under the concrete slab of cellular to prevent seepage and weaken the frost heaving of foundation soil.

Armored fiber concrete composite coiled material structure. As a new type of channel protection, it adopts combined masonry materials of rigid, flexible and modified polypropylene fiber, modified polypropylene fiber concrete and high-strength composite anti-seepage membrane, which is anti-seepage and also has adaptability to differential deformation.

Geosynthetic clay liner (GCL). It is a thin layer of bentonite sandwiched between two layers of geotextile. GCL has the following advantages (Dou et al., 2009). ① Good flexibility, no increase in permeability when the tensile strain reaches 20%. ② Small amount of excavation when it is used as channel liner. ③ Strong self-healing ability when the geotextile is punctured. ④ Strong tensile strain resistance, which can resist 5% or more tensile strain without significantly increasing the hydraulic permeability coefficient. ⑤ Strong ability to resist dry and wet circulation. After dry shrinkage dehydration and re-injection, the cracks will be closed again, and the hydraulic permeability coefficient will return to the original low value. ⑥ Strong anti-freezing and thawing cycle capacity. The hydraulic permeability coefficient will not increase significantly after three freeze-thaw cycle tests. ⑦ Convenient splicing, simple installation and fast construction.

Geosynthetics are used in the above-mentioned new anti-seepage and anti-frost heave materials. After testing, there exists more or less leakage, which is related to the thickness of the membrane, the construction and operation protection. The probability of occurrence is 5% in its own hole, and 69% in the weld seam. The liner under the geomembrane should be made of permeable sand gravel instead of cohesive soil.

Basis for selecting lining structure for canal slope protection project in frozen soil area (Xiu, 2015): ① flexible structure that can improve the soil temperature field of canal foundation; ② canal body and lining structure with strong drainage performance that can reduce the water content of the lower part of the lining body; ③ The lining structure with high integrity that can reduce the ice thrust in the canal to the slope and the slope toe; ④ lining structure with ecological and environmental protection; ⑤ lining structure with frost and thaw resistances.

At present, the new structures of channel slope protection include: articulated concrete slope protection technology; Reno cushion structure slope protection technology; Penn's cage slope protection structure technology; structural technology of precast concrete slab combined with thermal insulation measures; honeycomb ecological slope protection structure technology; crushed stone polyurethane slope protection structure technology, etc.

5.2.2.5 Comprehensive Measures

The basic factors affecting frost heave potential of foundation soil are soil texture, moisture content, soil temperature and stress conditions. Generally, improving two or more of these factors can achieve the goal of frost heave resistance. A large number of engineering practices show that it is often difficult to achieve the goals of preventing and controlling frost damage by adopting a single treatment or structural measure to eliminate or weaken just one factor affecting frost heave potential. Therefore, the comprehensive anti-frost heave measures, mainly one major measure combined with other one or several measures can achieve satisfactory goals.

Comprehensive anti-frost heave measures should be determined according to local conditions in combination with the grade of buildings, application requirements, foundation conditions, local materials etc. in the actual practices. The key points of frost heaving prevention of flat foundation are waterproof and thermal insulation, combined with structural form to improve the stress conditions. For example, canal engineering, composite structure of lining plate + anti-seepage membrane composite structure; composite structure of lining plate + anti-seepage membrane + thermal insulation layer; composite structure of lining plate + double-layer anti-seepage membrane (upper anti-seepage membrane and lower anti-groundwater membrane); composite structure of lining plate + anti-seepage membrane + fiber sand bag; new anti-seepage and thermal insulation integrated composite structure etc. The flat foundation adopts the slab foundation + anti-seepage + thermal insulation composite structure; flat foundation segmentation + anti-seepage + gap + anti-seepage filler + thermal insulation composite structure; flat foundation + anti-seepage + drainage composite structure; flat foundation + anti-seepage + replacement + tamping composite structure, etc.

5.3 Hydraulic Structures with Pile, Column or Pier Foundation

Pile, column and pier foundations are often used in hydraulic structures such as sluice chamber, aqueduct, bridge, pumping house and so on for purpose of controlling. Relatively, their foundations are "deep" foundations. In seasonally frozen soil areas, the foundation is jacked due to the tangential frost heaving force of the foundation soil, causing damages to pile structures, such as Luoguo Bridge (a bridge having an arch shape due to frost heave of its foundation) (Fig. 4.5) or pile jacking (Fig. 4.6), or the pile foundation on the slope is broken due to the horizontal frost heaving force (left of Fig. 5.27, the pile foundation of Sidaogang aqueduct is broken), or the foundation of the aqueduct where water is unable to pass through due to the frost jacking of its foundation (middle of Fig. 5.27, the aqueduct in Zhoujiazhuang is frost jacked), or the retaining wall is inclined forward, bent and broken because of the horizontal frost heaving force (on the right of Fig. 5.27, the retaining wall behind the factory of a pumping station in Nehe County is inclined forward and then broken).

5.3.1 Anti-Frost Heave Checking Calculation for Hydraulic Structures with Pile, Column and Pier Foundations

The foundations of small and medium-sized aqueducts, sluice chambers, pumping houses and other hydraulic structures in the northeast, north and northwest of China are mainly pile, column or pier type, and experience different degrees of frost

Fig. 5.27 Deformation and damage of hydraulic structures due to frost heave

jacking, ranging from a few centimeters to dozens of centimeters, leading to bulge, crack, tilt, fracture, and even instability and damage of the structures.

From the perspective of supporting the building load and frost heaving prevention, there are two types of structures with pile foundation. One is the friction pile type with constant diameter of the whole pile foundation and the other is the expanded foundation type. In any case, the stability of frost heave resistance of pile, column or pier foundation under frost heaving force should be checked.

5.3.1.1 Checking Calculation of Frost Heave Resistance Stability of Friction Pile

The design of friction pile in frozen soil area should not only meet the design requirements of pile foundation in non-frozen soil area, but also meet the requirements of the stability of the overall building and the strength of the pile itself under the action of the tangential frost heaving force of the foundation soil, so as to determine the embedded depth of the pile and the number of reinforcements. In the stress state of pile foundation (Fig. 5.28), the stability of buildings should meet the following conditions:

$$P + G + \pi D\left(H - h_f\right) f_j \geq k\pi D h_f \sigma_\tau \tag{5.5}$$

where P is the upper load of pile foundation, kN; G is the weight of the pile foundation, kN; D is the pile diameter, m; H is the buried depth of the pile foundation, m; h_f is the thickness of frozen soil layer, m; f_j is the ultimate friction force, kPa; σ_τ is the tangential frost heaving force, kPa; k is the safety coefficient, which is 1.1 for statically determinate structures and 1.2 for statically indeterminate structures.

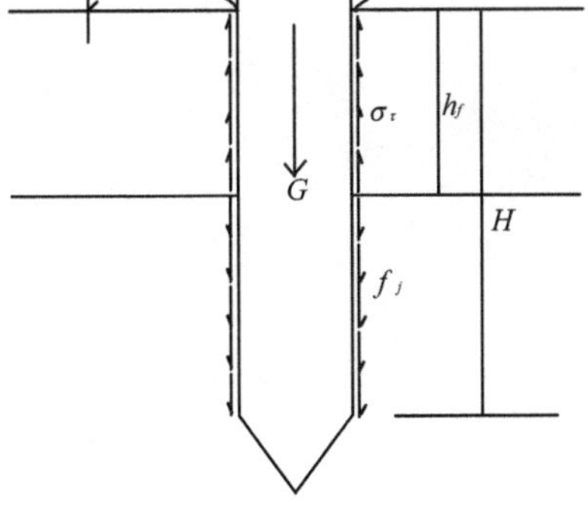

Fig. 5.28 Stress diagram of pile foundation in frozen soil

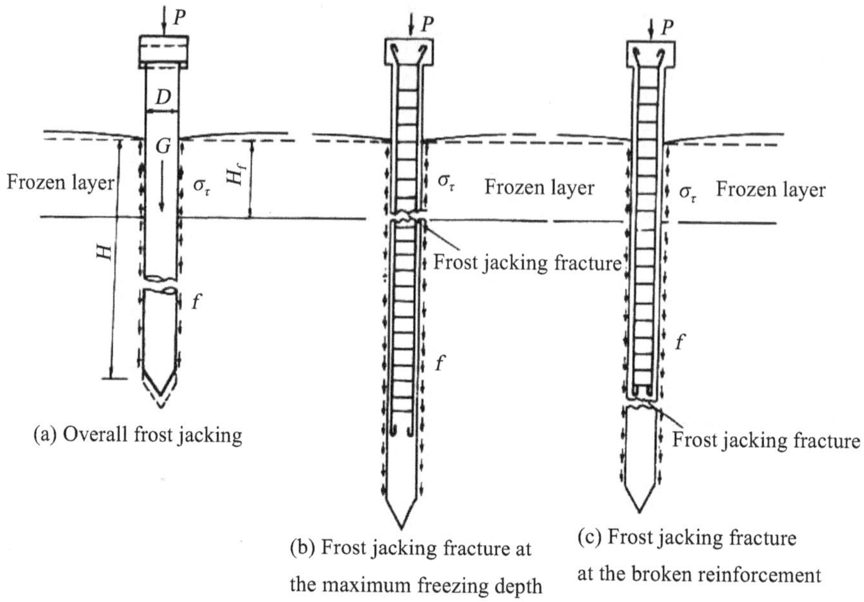

Fig. 5.29 Frost jacking of pile foundation in frozen soil

If the embedded depth of the pile foundation is insufficient, the pile foundation will be completely jacked out under the action of tangential frost heaving force (Fig. 5.29a), or if the pile section size or reinforcement does not meet the tensile strength requirements, it will result in pile break, which mostly occurs in the section with maximum frost jacking force (Fig. 5.29b) or the section at the lower end of the reinforcement (Fig. 5.29c). During the checking calculation, the tangential frost heaving force of frozen soil and the friction force of non-frozen soil should be selected according to actual situation on site (Tables 5.10, 5.11, 5.12 and 5.13) (Qu & Zhang, 2008).

Additionally, the tensile strength of components should be checked in the checking calculation of frost heave resistance of pile, column and pier foundations in frozen soil area:

Reinforced concrete pile

$$\frac{k\pi Dh_f\sigma_\tau-\left(P+G+\pi D\left(H-h_f\right)f_j\right)}{A_g}\le\left[R_g\right] \tag{5.6}$$

Concrete pile

$$\frac{k\pi Dh_f\sigma_\tau-\left(P+G+\pi D\left(H-h_f\right)f_j\right)}{A}\le\left[R\right] \tag{5.7}$$

Table 5.10 Ultimate friction of soil around pile f_j

Soil type	State	Ultimate friction resistance f_f /kpa	Soil type	State	Ultimate friction resistance f_f /kpa
Clay soil	$1.5 \geq I_L \geq 1$	15 ~ 30	Silty fine sand	Loose	20 ~ 35
	$1 > I_L \geq 0.75$	30 ~ 45		Moderately dense	35 ~ 65
	$0.75 > I_L \geq 0.5$	45 ~ 60		Dense	65 ~ 80
	$0.5 > I_L \geq 0.25$	60 ~ 75	Medium sand	Moderately dense	55 ~ 75
	$0.25 > I_L \geq 0$	75 ~ 85		Dense	75 ~ 90
	$0 > I_L$	85 ~ 95	Coarse sand	Moderately dense	70 ~ 90
				Dense	90 ~ 105

Note: I_L is the liquid index of soil, measured by the balance cone test with a weight of 76 g

Table 5.11 Ultimate friction f_j of soil around bored pile

Soil type	Ultimate frictional resistance f_j /kPa	Soil type	Ultimate frictional resistance f_j /kPa
Backfilled medium density slag and fly ash	40 ~ 60	Stiff plastic silty clay and silty sand	55 ~ 85
Flow plastic clay, silty clay and silty sand	20 ~ 30	Silt, fine sand	35 ~ 55
Soft plastic clay	30 ~ 50	Medium sand	40 ~ 50
Stiff plastic clay	50 ~ 80	Coarse sand, gritty sand	60 ~ 140
Stiff clay	80 ~ 120	Gravel (round gravel, angular gravel)	120 ~ 180
Soft plastic silty clay and silty sand	35 ~ 55	Crushed stone, cobble	160 ~ 400

Note: 1. Cobble and Boulder (constituting 40% ~ 50% of the mixture), with a grain size of 300 ~ 400 mm, can be used with a resistance of 600 kPa
2. The maximum or minimum value of resistance of sand can be selected according to its compactness
3. The maximum or minimum value of resistance of round gravel, angular gravel, crushed stone and cobble can be selected according to their compactness and the filling material
4. Refer to this table for the ultimate frictional resistance of dug pile

where A_g is the total sectional area of tensile reinforcement, m^2; A is the sectional area of pile, m^2; $[R_g]$ is the design tensile stress of reinforcement, kPa; $[R]$ is the ultimate tensile strength of pile material, kPa. Other symbols are the same as in Formula 5.5.

It is worth noting that during the construction, the pile surface within the seasonal freezing depth range should be kept regular and smooth to avoid the increase of tangential frost heaving force caused by the unequal pile diameters and increased surface roughness. The consistency of pile diameter should be maintained near ground surface and remove the excess concrete to avoid the increasing of pile

Table 5.12 Ultimate friction resistance of pile and foundation soil f_j / kPa

Soil type	State	Precast concrete pile	Underwater bored pile	Pipe sinking cast-in-place pile	Drilled pile for dry operation
Fill soil		20 ~ 28	18 ~ 26	15 ~ 22	18 ~ 26
Mud		11 ~ 17	10 ~ 16	9 ~ 13	10 ~ 16
Mucky soil		20 ~ 28	18 ~ 26	15 ~ 22	18 ~ 26
Cohesive soil	$I_L > 1$	21 ~ 36	20 ~ 34	16 ~ 28	20 ~ 34
Red clay	$0.7 < \alpha_\omega \le 1$	13 ~ 32	12 ~ 30	10 ~ 25	12 ~ 30
Silt	e > 0.9	22 ~ 44	22 ~ 40	16 ~ 32	20 ~ 40
Silty fine sand	Slightly dense	22 ~ 42	22 ~ 40	16 ~ 32	20 ~ 40
Medium sand	Medium density	54 ~ 74	50 ~ 72	42 ~ 58	50 ~ 70
Coarse sand	Medium density	74 ~ 95	74 ~ 95	58 ~ 75	70 ~ 90
Gravelly sand	Medium dense and dense	116 ~ 138	116 ~ 135	92 ~ 110	110 ~ 130

Note: 1. For the fill soil that has not completed self-weight consolidation and miscellaneous fill soil mainly consisting of household waste, their side resistance is not considered in the calculation
2. $\alpha_\omega = W/W_L$; W is water content, and W_L is water content of liquid limit
3. For Prefabricated pile, according to soil layer depth (h), f_j multiply by the correction coefficient in Table 5.13

Table 5.13 Correction coefficient

Soil depth(m)	≤5	10	20	≥30
Correction coefficient	0.8	1.0	1.1	1.2

Note: Data in this table are taken from Soil Mechanics and Foundation Engineering, edited by Zhao Minghua, Wuhan University of Technology Press, 2000

diameter at the ground surface to form an enlarged pile top, which will generate normal frost heave force and increase the frost heave force (Fig. 5.30 and Fig. 4.6). In order to increase the anchoring and bearing capacity of the pile foundation, the surface roughness of the cast-in-place pile can be increased below the maximum freezing depth.

5.3.1.2 Checking Calculation of Anti-Frost Heaving Stability of Pile Foundation with Expanded Base

When the foundation soil is frozen, the volume expansion produces the same expansion stress around it. Under natural conditions, at ground surface, it shows free expansion, namely, frost heave amount. With the increase of freezing depth, the frozen expansion of the soil under the ground will produce upward expansion stress,

Fig. 5.30 Frost jacking caused by enlarged pile top of Zhang Wanguan Bridge in Jilin

and also produces the same stress on the underlying soil layer (in the freezing test of saturated coarse-grained soil, the water level in another connecting pipe rises). When the foundation is constrained, the foundation soil will produce tangential frost heaving force on the pile through the pile-soil interaction, generating an upward shear frost jacking force on the pile, and also produce an expansion stress in the opposite direction below, commonly referred to as downward frost heaving force. The Third Survey, Design and Research Institute of the Ministry of Railways observed the downward frost heaving force distribution test of the expanded foundation slab, which shows that there is frost heaving pressure on the expanded foundation slab during the freezing process of the foundation soil (Fig. 5.31), which is distributed in a "saddle" shape. With the increase of the area, the peak of downward frost heaving force moves from the pile side to the edge of the expanded foundation slab. The field test of anti-frost jacking of pile foundation with expanded slab base is carried out at Wanjia Permafrost Test Field of Heilongjiang Institute of Water

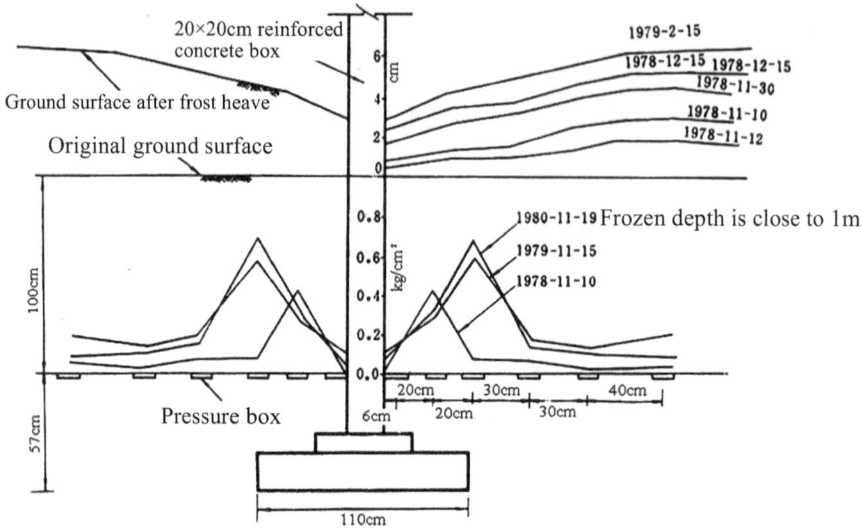

Fig. 5.31 Distribution of measured downward frost heaving force (Cui & Zhou, 1983)

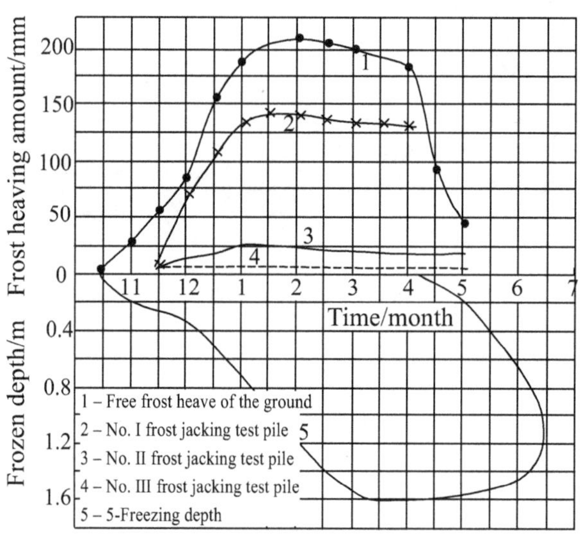

Fig. 5.32 Frozen depth, ground surface frost heave and frost jacking curve of test pile at Wanjia permafrost test site (Sui & Zuo, 1983)

Resources Science (Fig. 5.32). The burial depth of test pile is 1.6 m, and test pile I is a square pile with a cross-section of 0.2 m × 0.2 m, the frost jacking amount reaches 140 mm, accounting for 68% of the free frost heave amount of the ground surface. Test pile II has a cross-section of 0.2 m × 0.2 m and an expanded slab base of 0.5 m × 0.5 m, and its frost jacking amount was 20 mm, which is 86% less than that of test pile I. The cross-section of test pile III is 0.2 m × 0.2 m with an expanded slab base of 1.0 m × 1.0 m, and it is completely anchored without frost jacking. Laboratory model test results indicate that when the freezing depth reaches 0.4 m, the frost heave reaction force measured on the 0.5 m × 0.5 m expanded slab base of the 0.1 m × 0.1 m test pile is 50 kPa (pressure box) or 30 kPa (water bag pressure gauge), with an average of 38 kPa, which is similar to the tangential frost heave force measured on the straight pile (38 kPa).

If considering downward frost heaving force (Fig. 5.33), the size of foundation slab should meet the following requirements:

$$kuH_f\sigma_\tau < P + G + P_A + \rho HF_0 + f_j\left(F_1 + F_2\right) \tag{5.8}$$

where k is the safety factor, which is 1.3; u is the perimeter length of the pile in the frozen soil layer, m; H_f is the thickness of frozen soil layer, m; σ_τ is the tangential frost heaving force, kPa; P is the upper load, kN; G is the foundation weight (including pile column, coping and foundation plate), kN; ρ Is the density of the fill soil on the foundation plate, kN/m^3; H is the filling height on the foundation plate, m; F_0 is the area of foundation plate excluding pile column, m^2; P_A is downward frost heaving force, kN; f_j is the frictional resistance of non-frozen soil layer, kPa; F_1 is the total area around the pile in the non-frozen soil layer, m^2; F_2 is the total area around the base plate, m^2.

The maximum distribution range of downward frost heaving force is the limit range of ground frost heave deformation around the pile, roughly equivalent to the freezing depth. The downward frost heaving force acting on the foundation plate is a part. In Fig. 5.33, as for this kind of situation, the downward frost heaving force acting on a unit length can be checked as follows:

$$P_A = 2T_k\left[\frac{2l}{L}\left(2 - \frac{l}{L}\right) - 1\right] \tag{5.9}$$

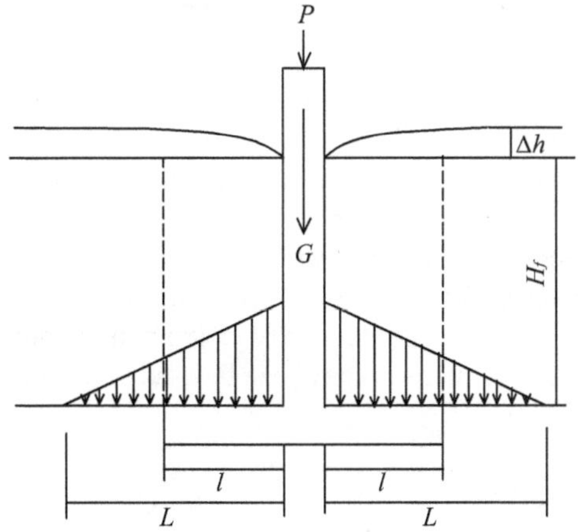

Fig. 5.33 Downward frost heaving force diagram

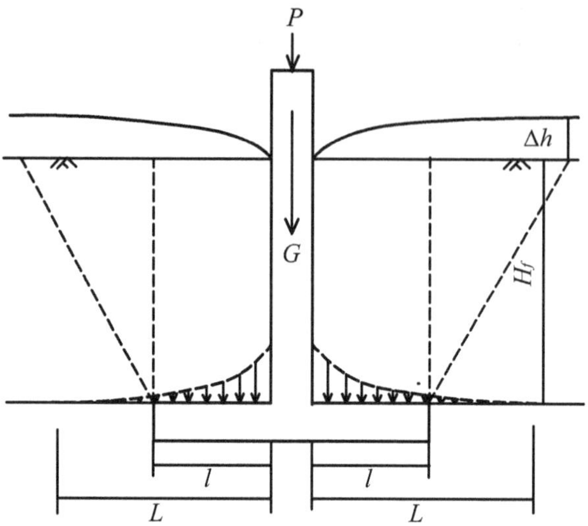

Fig. 5.34 Stress of pile foundation with expanded base

where T_k is the tangential frost heaving force per linear meter of the foundation side, kN/m; L is the action range of downward frost heaving force, m, $L = H_f$ can be supposed; l is the extended length of the expanded slab base of pile foundation, m;

In the process of freezing, the weight of the soil column above the foundation plate and the shear strength of frozen soil are important factors to prevent the foundation from being frost jacked. If the fill soil is tightly connected to the surrounding soil, the shear angle of jacking is about 30° (Fig. 5.34).

5.3.2 Anti-Frost Heaving Measures for Pile, Column and Pier Foundations

The frost damage of buildings with pile, column or pier foundation is the result of the tangential frost heaving force at the pile side. The adfreezing force between pile and soil depends on the degree of ice cementation (ice film). When the ice film thickness reaches the optimum, the adfreezing force will be the maximum. If the ice bonding force between the pile and soil is weakened or eliminated, and the pile is isolated from the foundation soil, the frost heave of the foundation soil at the pile side will not produce frost jacking on the pile, column or pier foundation. The measures mentioned above can be used to prevent frost heaving of pile, column and pier foundation.

5.3.2.1 Replacement Method

The non-frost heaving soil is used to replace the frost heaving soil around the pile, column and pier foundations. Coarse grained soil such as sand gravel is usually used, but the silt clay content in the backfill should not exceed 5 ~ 10%, and a drainage outlet during freezing is needed.

Regarding the horizontal range, non-frozen soil is generally replaced within 0.3 ~ 0.5 m around the pile.

The replacement depth is 2/3 of the design value of freezing depth in Class II frost heaving foundation soil. In Class III frost heaving foundation soil, it is necessary to replace the soil within the whole design freezing depth.

5.3.2.2 Physical Chemistry Method

As described in the slab foundation, hydrophobic materials should be used to fill around the pile side to isolate the frost heaving foundation soil from the pile or column foundation, or hydrophobic materials and sand should be mixed to form improved soil to fill around the pile (Fig. 5.35).

Salinization of foundation soil between salts is adopted. Low valent cationic salts such as NaCl, KCl, $CaCl_2$ are commonly used, but the mixing amount should be controlled (see Formula 5.3). When the salt content in the soil is less than 0.5%, the physical and mechanical properties of the foundation soil will not be affected. The effective duration of salt-based anti-frost heaving measures is relatively short, generally lasting 2 to 3 years, and salt content may diminish after that. Anti-frost

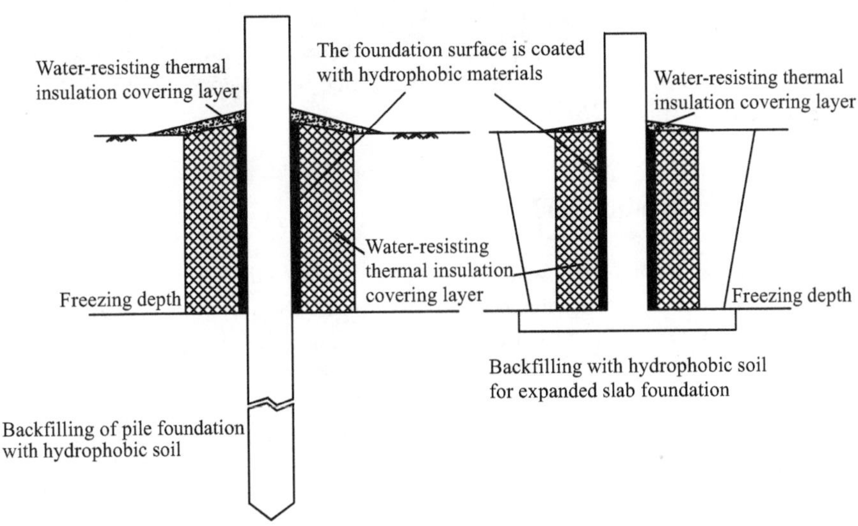

Fig. 5.35 Schematic diagram of replacement of hydrophobic soil for pile foundation and expanded slab foundation

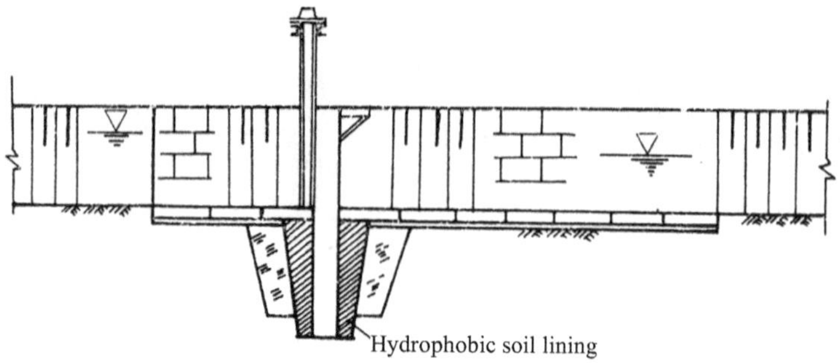

Fig. 5.36 Replacement and filling of the foundation of the I-shaped sluice with hydrophobic soil (both according to Tong & Guan, 1985)

heaving with crystalline salt or saturated solution should be implemented several months before the freezing period,

Hydrophobic materials include residual oil, asphalt (8% ~ 10% used oil should be mixed to make it plastic at low temperature), coal tar and other petroleum by-products. Chemical surfactants include wood sulfate with sodium dichromate, polyester, polyamide, polyolefin slurry, polyurethane foam plastic, silicon organic colloidal compound (such as sodium ethylsilanol solution, sodium methylsilanol solution, etc.), hydrophobic surfactant, etc.

Hydrophobic materials should be mixed with air-dried crushed loose soil, and hot mixing (temperature up to 120 ~ 150 °C) should be adopted for backfilling and compaction in layers. The backfilling width at pile side is 200 ~ 300 mm. The replacement depth should reach the maximum freezing depth. Grease and residual oil can be used as coating on the surface of pile foundation, and the coating thickness is typically 2 ~ 5 mm.

In hydraulic structures, hydrophobic soil can be backfilled on the front and rear foundation surfaces and around the I-shaped sluice body to prevent the tangential frost heave force (Fig. 5.36).

5.3.2.3 Thermal Insulation Method

The purpose of the thermal insulation method is to reduce the freezing depth. Industrial thermal insulation materials are used mostly as a thermal insulation layer under the slab foundation of hydraulic structures. The pile foundation uses thermal insulation materials to cover the surface around the pile. The surface should be paved with a protective layer to prevent the surface water from infiltrating or wetting, and reduce the thermal insulation performance of the thermal insulation materials.

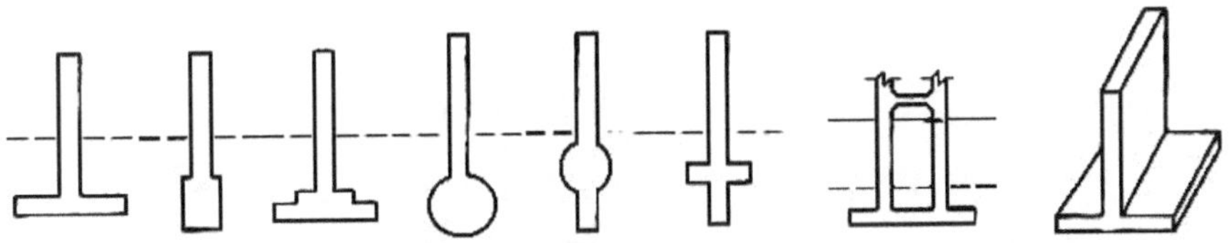

Fig. 5.37 Main types of expansion plate foundation. (**a**) Pile foundation with an expanded slab base; (**b**) Tapered pile foundation; (**c**) Pile foundation with a Stepped expanded slab base; (**d**) Pile foundation with an explosively expanded base; (**e**) Spherical expanded pile foundation; (**f**) Ring-shaped expanded pile foundation; (**g**) Bent pile foundation with an expanded slab base; (**h**) Wall-shaped pile foundation with an expanded slab base

5.3.2.4 Structural Method

The structural method is to change the form of the foundation structure to reduce the contact area between the foundation and the foundation soil, or anchor the foundation to achieve the purpose of frost heaving prevention, such as I-shaped sluice, pile foundation, deep pile foundation, expanded plate foundation, explosion expanded piling, pile- bent, etc. (Fig. 5.37)

1. **Deep foundation**

On the frost heaving foundation soil, when the building foundation is designed according to the principle that no deformation is allowed, in order to avoid the action of tangential frost heaving force and normal frost heaving force, deep foundation is often used. The embedded depth of pile foundation should not only meet the requirements of bearing capacity but also meet the stability requirements of anti-frost jacking. In permafrost regions, the stability requirement of anti-frost jacking is often a controlling condition. Under the action of tangential frost heaving force, the pile foundation should meet the following requirements:

$$P+G+F \geq kS\sigma_{\tau} \tag{5.10}$$

where S is the side surface area of pile foundation in the frozen soil layer, m^2, $S = H_f \cdot \pi D$; F is the total friction force, kN. $F = f_j \cdot (L - H_f) \cdot \pi D$; D is the diameter of pile foundation, m.

The required buried depth (L) for the stability of piles to resist frost jacking should be calculated and determined according to the following formula:

Circular pile

$$L = \frac{H_f\left(k\sigma_{\tau} + f_i\right)}{f_i} - \frac{P+G}{\pi D f_i} \tag{5.11}$$

Square pile

$$L = \frac{H_f\left(k\sigma_\tau + f_i\right)}{f_i} - \frac{P+G}{4af_i} \tag{5.12}$$

where a is the length of each side of the square pile, m.

Since the dead weight (G) of the pile is a function of the pile length, it is necessary to assume the pile length (or the buried depth) when calculating the buried depth (L), and obtain the reasonable burial depth of the pile through trial calculation. In addition, it is also necessary to check the anti-frost jacking strength of the pile section and the steel bar section under the maximum frost heaving force. The reinforcement amount of pile column at the maximum frozen depth section should meet the requirements of the following formula:

$$A_g = \frac{S\sigma_\tau - P - G_1}{\sigma_g} \tag{5.13}$$

where G_1 is the dead weight of the pile above the maximum frost depth section, kN; A_g is the sectional area of longitudinal load-bearing steel bar, m^2; σ_g is the allowable tensile stress of steel bar, kPa.

In order to ensure that the pile or column has sufficient anti-frost jacking strength, it is generally required that the longitudinal load-bearing steel bar should go deep into 3/4 buried depth of the pile or the column, and the strength of this section should be checked. In underwater construction of piles or columns, the requirement is often difficult to meet, and the concrete bearing tensile stress is generally not considered in the section where reinforcement is discontinued, so it is necessary to meet the requirements of the following formula:

$$P + G_2 + \pi D\left(\frac{3}{4}L - H_f\right)f_i - S\sigma_\tau > 0 \tag{5.14}$$

where G_2 is the dead weight of pile column over (3/4) L, kN.

If the above formula cannot meet the requirements, the longitudinal reinforcement should be set to the bottom of the pile.

2. **Expanded pile foundation**

The pile foundation with an expanded base forms an expanded base under the frozen soil layer (Fig. 5.41d and e), which can be expanded by blasting, drilling, reaming machine or water gun. This method is applicable to the cohesive soil foundation with low groundwater level, but not to the sand soil foundation as it is prone to collapse.

It is difficult to control the size of the expanded base by blasting reaming. Measures should be taken to strictly control the construction quality to meet the

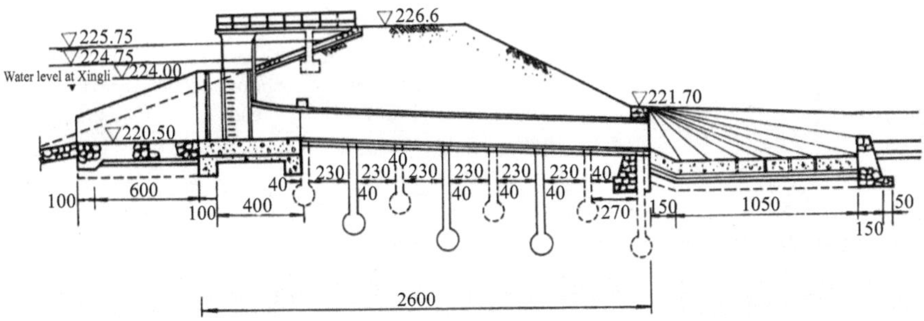

Fig. 5.38 Spillway tunnel using pile foundation with a blasting reamed base at Qingfeng Reservoir in Nehe County, Heilongjiang Province (Unit: cm) (Tong & Guan, 1985)

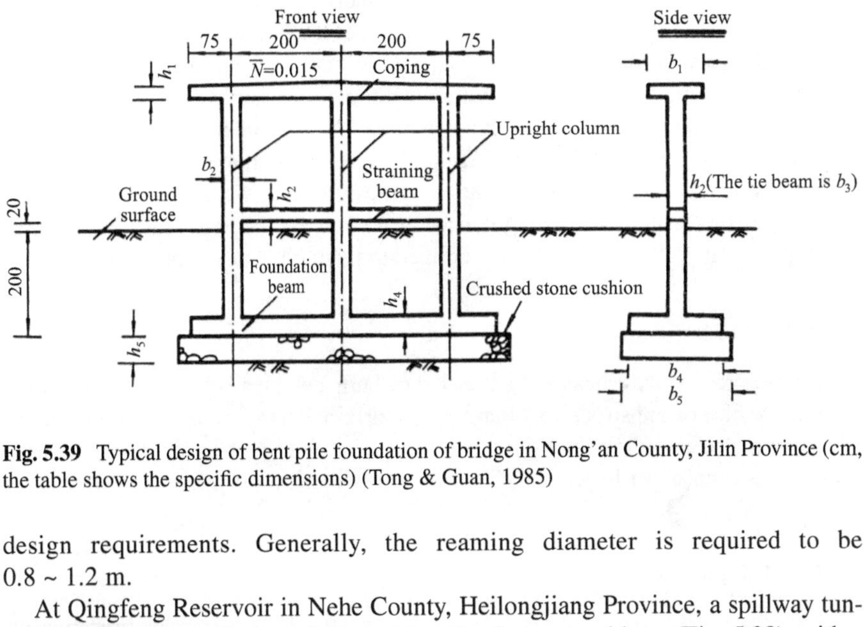

Fig. 5.39 Typical design of bent pile foundation of bridge in Nong'an County, Jilin Province (cm, the table shows the specific dimensions) (Tong & Guan, 1985)

design requirements. Generally, the reaming diameter is required to be 0.8 ~ 1.2 m.

At Qingfeng Reservoir in Nehe County, Heilongjiang Province, a spillway tunnel was built using pile foundation with a blasting reamed base (Fig. 5.38), with a good performance.

3. **Bent pile foundation**

Bent pile foundation is connected by an expanded slab base, which is made into dual-pile or triple-pile type as needed. The expanded slab base is used to increase the self anchoring capacity of the pile foundation to achieve the purpose of anti-frost jacking. In Fig. 5.39, it is a typical design sketch of a bent pile foundation of a bridge in Nong'an County, Jilin Province.

Pier height/m	Span/m	Span/m									
		Coping		Upright column		Straining beam		Foundation beam		Foundation gravel cushion	
		Width b_1	Hight h_1	Width b_2	Hight h_2	Width b_3	Hight h_3	Width b_4	Hight h_4	Width b_5	Hight h_5
≤2.5	1.5 ~ 4.0	0.45	0.3	0.3	0.3	0.3	0.25	0.8	0.35	1.0	0.3
	5.0 ~ 8.0	0.6	0.3	0.3	0.3	0.3	0.25	1.0	0.35	1.2	0.5
2.51 ~ 4.0	1.5 ~ 4.0	0.45	0.3	0.35	0.3	0.35	0.3	0.8	0.35	1.0	0.3
	5.0 ~ 8.0	0.6	0.3	0.35	0.3	0.35	0.3	1.0	0.35	1.2	0.5

4. **I-shaped sluice**

The I-shaped sluice (Fig. 5.40) is widely used for small and medium-sized sluices in cold regions. The sluice with a large sluice chamber area is replaced with a compressed I-shaped sluice, which reduces the contact area between the sluice foundation and the foundation soil, so as to reduce the normal frost heaving force at the foundation bottom. Meanwhile, the tangential frost heaving force at the lateral side of the foundation is eliminated by replacement or physical chemistry methods, so as to effectively prevent frost damage. The applications of I-shaped sluice in Xinjiang, Heilongjiang and other regions indicate that most of them work well. The I-shaped sluice takes the place of the octagonal wing wall which is prone to frost damage, and replaces it with a slope lining connection which can adapt to frost heave deformation.

5. **Method of built-on-piles**

In order to prevent frost heave, the method of built-on-piles is to elevate the whole building or part of the structure, changing the original slab foundation to pile foundation, and also changing the original strip foundation to pier or pile foundation, to avoid the normal frost heaving force. It is also widely used in water conservancy

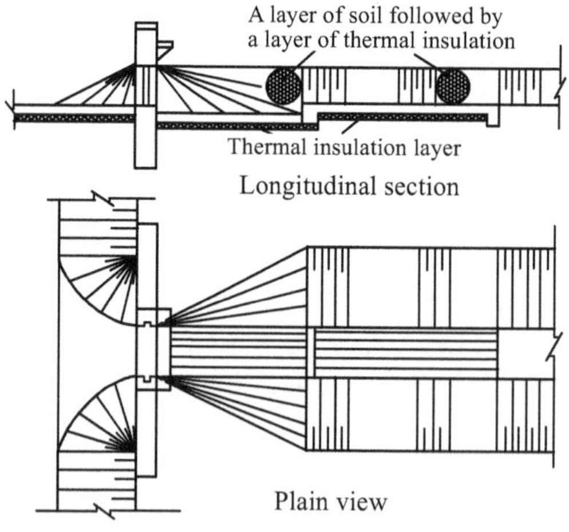

Fig. 5.40 I-shaped sluice in Xinjiang

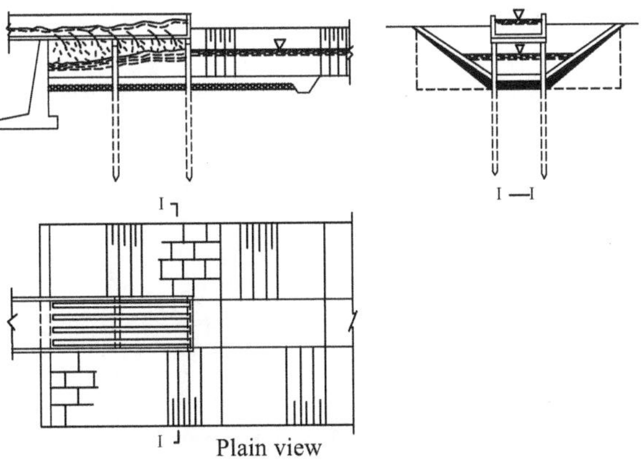

Fig. 5.41 Mesh screened drop (Tong & Guan, 1985)

Fig. 5.42 Entrance of collar type water conveyance tunnel

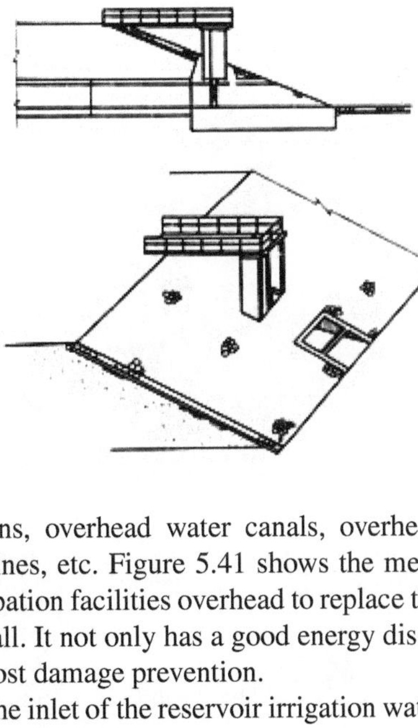

projects, such as overhead pumping stations, overhead water canals, overhead energy dissipation facilities, overhead pipelines, etc. Figure 5.41 shows the mesh screened drop, which makes the energy dissipation facilities overhead to replace the easily frozen stilling basin and outlet side wall. It not only has a good energy dissipation effect, but also has a good effect of frost damage prevention.

The frost damage of the retaining wall at the inlet of the reservoir irrigation water conveyance tunnel is common in cold regions. The retaining wall at the inlet could be replaced to a collar type inlet (Fig. 5.42), which gained a positive result. In order to prevent inlet scouring, a certain range of dam slope and inlet bottom should be protected by lining.

6. **Casing method**

The casing method, or isolation method, is to isolate the pile foundation from the surrounding soil by casing, so that the lateral surface of the pile foundation is not frozen together with the surrounding soil, by which the tangential frost heaving

Fig. 5.43 Oil wrapped pile

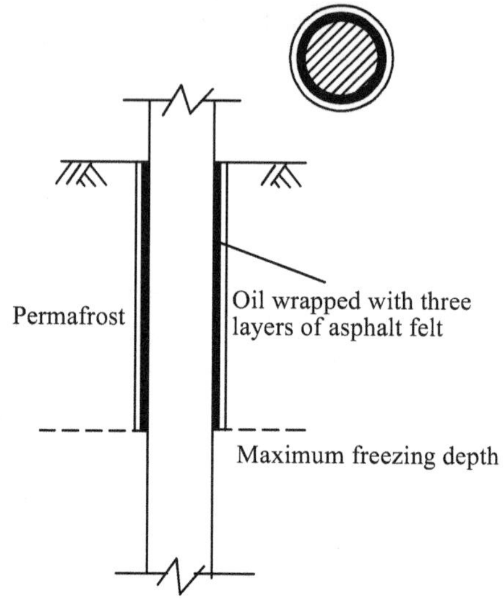

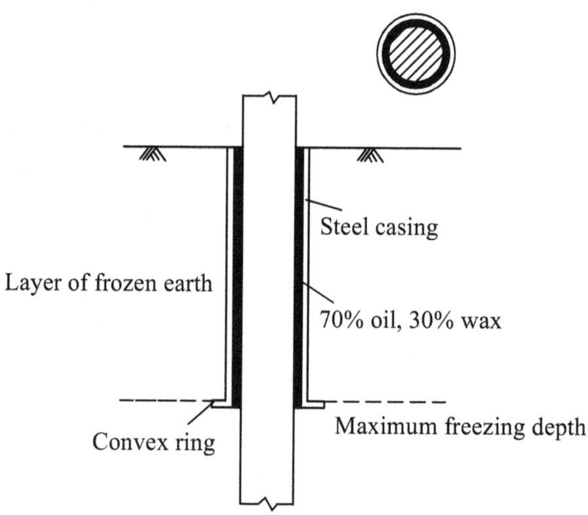

Fig. 5.44 Casing tube of anti-frost jacking (both according to Tong & Guan, 1985)

force is eliminated. The commonly used casing methods include oil wrapped pile (Fig. 5.43) and casing tube (filled with grease mixture, Fig. 5.44).

The coating material on the lateral foundation should be convergent and non-freezing hydrophobic coating, such as resin, residual oil, asphalt mastic, etc. The wrapping layer should be mostly asphalt felt, polyethylene film plastic (to be thick), etc.

The casing method is applied to pile foundation, that is, within the maximum freezing depth, the pile foundation is applied with a casing tube with a convex ring

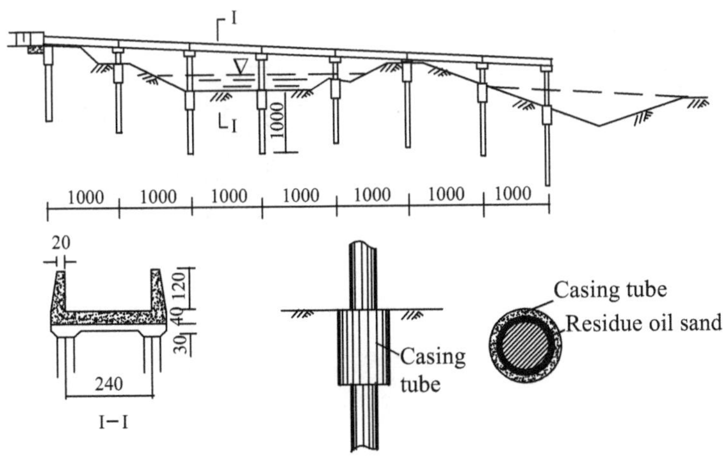

Fig. 5.45 Slope water aqueduct crossing over a canal (unit: cm)

at the foot. A gap of 20 ~ 50 mm is left between the casing tube and the pile foundation, which is filled with butter, asphalt and waste engine oil mixture, industrial vaseline oil, oil and wax mixture etc. The protruding convex ring at the bottom of the casing tube plays an anchoring role and is arranged to prevent the casing from being frost jacking. The loss of oil mixture filler should be replenished every few years, and the casing tube should be laid down again.

In this method, when the water level in the middle of slope of the irrigation canal system is higher than the canal water level, the slope water aqueduct scheme is adopted to make the water cross over a canal, and then the flow is discharged by the flip bucket. For example, for the aqueduct (Fig. 5.45) crossing the canal on the slope of the North Heilongjiang Nenjiang River Diversion Project, in order to prevent the pile foundation of the aqueduct from frost jacking, casing tube is added to the pile foundation to ensure the operation safety of the infrastructure.

7. **Comprehensive measures**

In the actual project, the anti-frost heave scheme should be determined in combination with the building grade, operation requirements, foundation conditions, climate environment, local materials, etc. Under the condition that the foundation conditions are simple, also with simple small buildings, and having large capacity to adapt to frost heaving deformation, a single anti-frost heaving measure can work very well. If the foundation condition is poor, with complex building structure, as well as with large area and high standard of the deformation requirements, comprehensive measures need to be taken to meet the design requirements of frost heaving prevention. In many cases, the combination of two or more measures can achieve more effective frost heave prevention with less effort.

The combination of drainage method and thermal insulation method has achieved better frost heave prevention than a single measure (Fig. 5.46). The former eliminates the moisture on the back of the wall and reduces frost damage, but it often fails

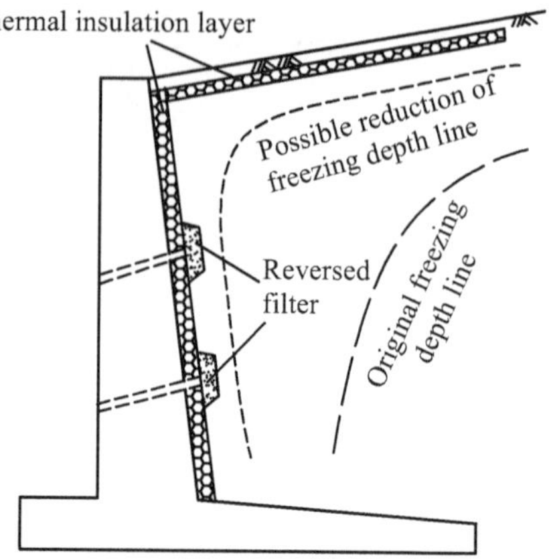

Fig. 5.46 Measures for insulation, drainage and frost protection of retaining wall

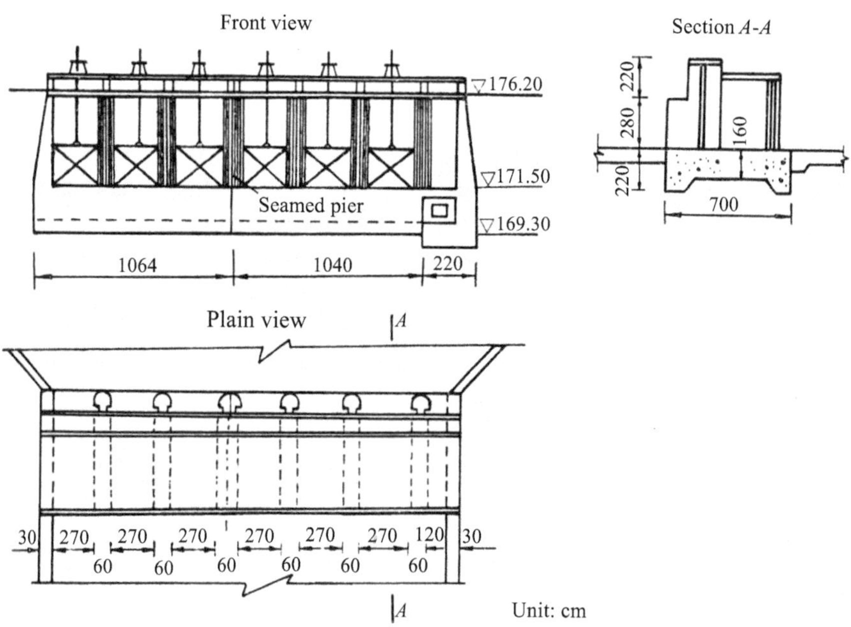

Fig. 5.47 General layout and structure of flood discharge sluice of Heilongjiang Jinhe River Reservoir (Tong & Guan, 1985)

to reduce the moisture in the soil layer on the back of the wall to reach the extent of weak frost heaving, so that the frost damage may still occur. If a thermal insulation layer is laid on the wall back and the soil surface to reduce the freezing depth of the wall back, it can achieve a satisfactory prevention of frost heave.

The example is the flood gate of Heilongjiang Jinhe River Reservoir (Fig. 5.47). The sluice has six holes, with a total length of 23.24 m and a width of 7 m. Every three holes are connected into a whole, and joint piers are set in the middle. The

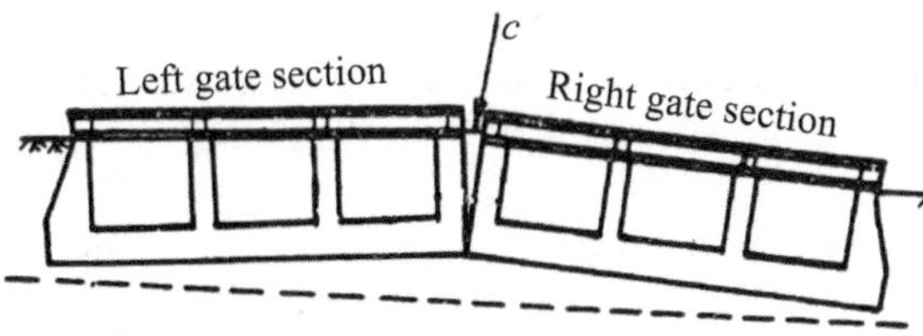

(a) Frost heaving and lifting of whole sluice

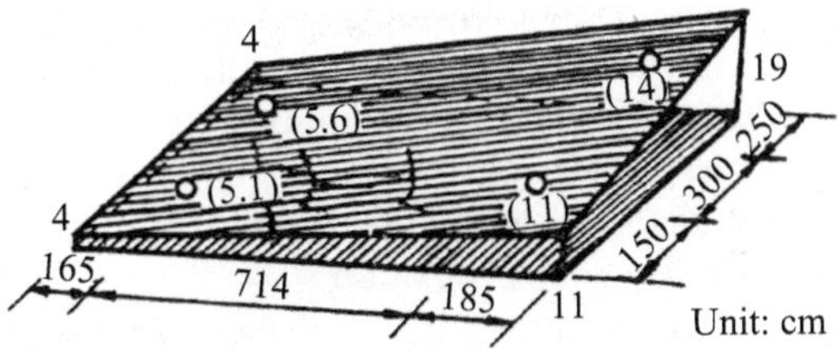

(b) Frost heaving and lifting of left sluice section

Fig. 5.48 Schematic diagram of sluice chamber frost heave lifting (Tong & Guan, 1985)

sluice foundation soil is heavy silty clay, with plastic limit of water content of 21%, pre freezing water content of 27.7%, and the maximum freezing depth of 2.1 m. The original design measures to prevent anti-frost heave are the following. ① The thickness of the bottom slab of the sluice chamber is 1.6 m, and the front and rear teeth are 2.2 m deep. ② Because the sluice chamber is too long (23.24 m), a joint pier is set in the middle to divide the sluice into 2 units with high rigidity.

According to field observation, the frost heaving started in December 1963 and reached the maximum at the end of March. The frost heaving uplift of the whole sluice body reached 180 mm, and the maximum opening width of the joint pier was 110 mm (Fig. 5.48). Thaw settlement began at the sluice foundation by the end of May next year. The residual width of the middle pier joint was 30 mm on July 10, which was not completely restored. Although there was no strength damage to the bottom slab of the sluice, the vertical water stops of the joint piers were all broken, which could not meet the anti-seepage requirements, and also, the frost heaving deformation of the sluice chamber exceeded the allowable deformation limit, which suggest that under the joint action of the normal frost heaving force under the sluice gate and the tangential frost heaving force of the front and rear teeth, the 973.06 tons weight sluice could still be lifted up by frost heave.

Therefore, a small sluice pier was added in front of the sluice pier, and a temporary small sluice was also set in winter to store water. Ice and water were used for thermal insulation to eliminate the freezing depth and normal frost heaving force under the sluice gate and ensure the stability of the chamber. When the reservoir water level was low, firewood should be used for thermal insulation. Since the renovation in 1964, it had been operating well.

This example shows that the frost damage cannot be eliminated only by increasing the rigid strength of the building. Insulation measures can be applied to reduce the freezing depth of the foundation, achieving the goal of reducing or eliminating frost damage.

Another example is the Dingjiayao Pumping Station in Inner Mongolia. The integral raft foundation (1.1 m thick, close to the freezing depth) was adopted in the design. After completion, dozens of cracks appeared in the factory workshop. For this reason, the following measures had been taken: ① The surrounding of the factory workshop, the bottom of the outlet pipe, and the space between the rear wall and the anchor block had been replaced with gravel, with a depth of 1.4 ~ 1.5 m and a width of 3 ~ 5 m. ② The factory workshop and anchor block were surrounded by concealed drainage pipes (pipe diameter 200 mm) to reduce the groundwater level by 1 ~ 1.2 m; ③ The expansion ring had been set in the pressure pipe between the rear wall and the anchor block to adapt to the frost heaving deformation and reduce the frost heaving uplift force on the rear wall.

After treatment, there was no vertical frost heave deformation in the factory workshop. It shows that single structure anti-frost heave measures fail to prevent frost damage. The replacement method plays an important role in weakening or eliminating the frost heaving force.

Still another example is No. 6 sluice gate of West Main Canal of Dahaizi, Xinjiang. The following are its anti-frost heaving design. ① The sluice chamber plate was 0.35 m thick, 0.8 m Gobi sand was used as a replacement under the plate, the bottom was connected with the vertical wall drain hole, and the water can be discharged to the water apron. The side wall of the sluice was connected with the bottom plate as a whole, and deep teeth were set at the front and rear of the sluice chamber. ② At the inlet and outlet, the partition wall was a trapezoidal section with a small upper section and a large lower section, and the surface was thickly coated with asphalt. Replacement and drainage measures under the concrete slabs of bottom and slope were adopted. The bottom lining and gate bottom clay blanket were deeply buried, and a thermal insulation layer was laid on it. The water outlet was U-shaped, whose bottom was filled with sand gravel.

5.4 Issues Related to Dykes and Dams in Frozen Soil Engineering

Permafrost regions are widely distributed in China. The water resources in Heilongjiang, Inner Mongolia, Xinjiang, Qinghai, and Xizang are abundant and await further development. However, with the development of the national economy, this is bound to be put on the agenda. Chinese experts have conducted geological studies on the engineering planning of comprehensive utilization of water resources in the Erguna River and the boundary part of Heilongjiang River (Xu & Wang, 1995). The survey focuses on studying the frozen soil conditions and engineering frozen soil zoning in this area, and predicting the changes of frozen soil conditions and their impacts on engineering geological conditions and buildings during the construction and operation of various water conservancy projects. The Be'erci River in the Mangui region of Daxing'anling Mountains in Inner Mongolia had also been planned.

5.4.1 Design Principles and Key Points

Permafrost regions are located in a cold climate zone, with mean annual air temperature below zero, as well as having great range of temperature differences both in daily and annual air temperature, and with a long winter. The repeated process of freezing in winter and thawing in summer occurs not only in the foundation soil, but also on the structures, especially on the hydraulic structures that do not have internal heating system, which often experience many engineering problems related to frost damage. The continuity of permafrost regions in China is not high, that is, permafrost in China is not always distributed in a continuous way. It takes continuous, sporadic and isolated patchy forms. In permafrost regions, major river basins often have penetrated thawed zones, while smaller rivers mostly have non-penetrated thawed zones. Discontinuous or isolated patches of permafrost is distributed on both sides, and various permafrost phenomena occur, with significant ice-related issues and the continued impact caused by the presence of ice loads. With dry winter and flooded summer, the sudden spring floods are commonplace. In mountains of the Qinghai Xizang Plateau and Xinjiang, valleys are deep, where the seasonal freezing and thawing often make it hard to construct. The above characteristics in cold regions should be considered in the design of hydraulic structures. As mentioned in Sect. 5.1 of this chapter, on the basis of complying with the design requirements of non-frozen soil areas, some special projects and researches should be added according to the environment of frozen soil areas and the characteristics of frozen soil foundation.

When carrying out engineering construction, it is necessary to strengthen the engineering geological survey of frozen soil, and on this basis, to carry out the engineering geological evaluation of frozen soil. ① Evaluation of engineering geological conditions of frozen soil. Classifying good or bad engineering geological sections will help to determine the site, construction plan, design principles, engineering treatment, etc. of hydraulic structures. ② Evaluation of the ground temperature of permafrost. The mean annual ground temperature of permafrost is the basic factor to determine the stability of permafrost. Determining whether the building is located in low-temperature or high-temperature permafrost directly impacts the determination of design principles and technical schemes of hydraulic structures. ③ Evaluation of ice content of frozen soil. It directly affects the settlement and deformation properties of frozen soil foundation. According to the distribution and quantity of ice content of frozen soil, the thawing settlement level of frozen soil foundation is determined according to the engineering classification of permafrost. The thawing settlement distribution map (horizontal and vertical directions) of frozen soil foundation is needed to work out, so as to decide the design and construction scheme.

The thermal stability of permafrost first depends on the mean annual ground temperature of the frozen soil. The lower is the ground temperature, the better is the thermal stability of the frozen soil. At the same time, it also depends on the thermal effects of the external environment, such as climate warming, artificial heat sources, water infiltration and thermal erosion.

Rockfill dams and reinforced concrete gravity dam should be constructed as the water retaining structures on the permafrost foundation. Overflow dam often brings greater thermal effect, causing greater thawing of the frozen soil foundation under the dam foundation. This can result in differential settlement in the foundation and increase seepage, accelerating permafrost thaw. Rockfill dams are often used in permafrost regions, while concrete gravity dams are often used in penetrated thawed zones in river basin. Overflow dams are not recommended on permafrost foundation.

Because the foundation soil has different physical and mechanical properties, strength indexes, deformation characteristics and thermal stability in the freezing and thawing states. When it transforms from the freezing state to the thawing state the strength decreases and the deformation increases. When permafrost is used as a foundation, the design principle should be determined according to the freezing or thawing state of the foundation soil.

Therefore, the foundation design of hydraulic buildings in permafrost regions should also follow the general design principles in permafrost regions. According to the engineering geological conditions of permafrost and the characteristics of buildings, the design principles for maintaining the freezing state of the foundation or the thawing state of the foundation soil should be determined. For low temperature hard frozen soil (the compression coefficient is not greater than 0.01 MPa^{-1}), it is more economical and reasonable to adopt the design principle (principle I) of maintaining the freezing state of the foundation. For high-temperature plastic frozen soil (the compression coefficient is greater than 0.01 MPa^{-1}), the design principle of thawing state (Principle II) is generally adopted. It is worth noting that the gradual thawing design principle is not suitable for dam structures.

Principle I. Permafrost (perennial frozen rock) in the foundation should always be frozen during construction and operation, and the waterproof measures of buildings, the impermeability, and anti-seepage stability of foundations, and the contact zone with the buildings should be guaranteed. When frozen soil is used as the foundation and bank slope of hydraulic structures according to Principle I, different types of refrigeration equipment should be equipped, and thermal insulation and waterproof measures should be taken to prevent the frozen soil foundation from thawing, improve its bearing capacity and reduce deformation.

Principle II. Manually thaw the permafrost to a specified depth before building construction. When the high-temperature frozen soil is used as the building foundation according to Principle II, the fissured rock and coarse frozen soil should be pre-thawed and grouted before the reservoir impoundment to prevent dam foundation from leakage and also reduce the deformation of the building and ensure the bearing capacity of the foundation.

For hydraulic structures, the depth affected by thermal erosion of water flow is often deep, exceeding the maximum thawing depth under dry and flood-free conditions. Therefore, the design principle of gradual thawing of frozen soil foundation is difficult to meet the deformation requirements of hydraulic structures during operation. Therefore, it is more reasonable to adopt two design principles: keeping frozen state (Principle I) or thawing state (Principle II).

Different design principles are allowed for different structures of a same water conservancy, but special measures should be taken to isolate the contact zones of structures designed according to different principles. Generally, freezing devices are used to avoid the thermal erosion of frozen foundations by thawing foundations.

Freezing and thawing processes take place on the surface layer of foundation in frozen soil regions every year. In the design of hydraulic structures in permafrost regions, in addition to selecting a reasonable foundation structure type according to the engineering geological characteristics of permafrost, and calculating the bearing capacity, deformation and stability, the anti-frost heave checking calculation should also be carried out.

Any engineering construction in permafrost regions should implement national environmental protection laws and regulations to protect natural resources, restore and improve the environment when the work is done. The design of environmental protection in permafrost areas should reflect the principle of giving priority to prevention and paying equal attention to construction and protection. Relevant environmental protection measures should be formulated according to the natural conditions of the site and the characteristics of the permafrost environment. A large number of practices have proved that protecting the frozen soil environment is also to protect the thermal stability of the frozen soil foundation. The damage to the environment will inevitably affect the stability of the building foundation, bringing serious consequences. For this reason, strict requirements should be imposed on soil taking and dumping, construction shortcuts, temporary camps, etc.

The ice layer containing soil within permafrost cannot be used as the bearing stratum of building foundation. When the ice content of the ice layer containing soil exceeds 50%, the water content of coarse-grained soil, fine sand and silty sand exceeds 44%, and the water yield reaches 10% ~ 20% after thawing. When the water content of cohesive soil is higher than 36% ~ 48%, which is its plastic limit water content, it is flowing after thawing. When the ground ice layer is subjected to external forces, it is easily subject to its rheological properties, and the long-term strength is significantly reduced.

After the impoundment of the reservoir, under the thermal erosion of the reservoir water and the impact of waves, the ground ice in the frozen soil on the reservoir bank is easy to thaw, causing the collapse, and constantly forming new reservoir banks. Therefore, during the design, not only the outline and section of the reservoir bank reconstruction should be predicted according to the frozen soil engineering geological conditions and thermal calculation results, but also the relevant protective measures should be considered in advance.

5.4.2 Types of Dykes and Dams

According to the experience of hydraulic structure construction in cold regions, more than 90% of the dams are built with local earth and rock materials, which take more consideration of the characteristics of frozen soil than concrete dams and reinforced concrete dams, and are more substantive.

5.4.2.1 Concrete Dam and Reinforced Concrete Dam

In permafrost regions, both concrete dams and reinforced concrete dams tend to be constructed as gravity dams, and usually the gravity buttress dams with baffle plates downstream would be selected.

During dam design, the physical and mechanical changes after thawing of frozen soil foundation during dam construction and operation, as well as the influence of temperature regime on the strength and stability of dam concrete should be considered. If necessary, lay thermal insulation layer on the exposed concrete surface, and do a good job of waterproof and thermal insulation treatment in the water level fluctuation area. The thermal insulation material can be PVC foam plastic with waterproof performance.

For concrete dams designed according to the melting principle (Principle II), measures should be taken to prevent the concrete on the downstream surface from freezing when the dam foundation and water transmission pipeline facilities are not working.

The thickness of thermal insulation layer should be determined according to thermal calculation. When determining the thickness of waterproof and thermal insulation layer within the variation range of water level, three states of dam surface temperature and external temperature should be considered.

1. Case 1: the thickness of the protected concrete surface in frozen state. When the water level drops to the optimal level, determine the thickness of the thermal insulation layer on the concrete surface under the condition of rapid freezing during the external negative temperature in autumn, winter and spring. This thickness should keep the concrete surface of the dam body from thawing during the maximum water level when it rises.
2. Case 2: reduce the number of freeze-thaw cycles to reach the thickness of concrete that will not be damaged. Under the conditions of operation, building structure and frost resistance grade selection, determine the optimal thickness of concrete surface without damage when reducing the number of times of freeze-thaw cycles.
3. Case 3: The thickness of dam concrete that is not allowed to freeze. Determine the thickness of dam concrete that will not freeze when subjected to temperature impact during operation.

The concrete is in the thawing state at the very beginning. When water level in the water level fluctuation area drops in each season, that is, under the impact of negative temperature, the thickness of the thermal insulation layer should be able to keep the concrete from freezing.

5.4.2.2 Earth and Rock Dam

The earth and rock dam is constructed with local earth and rock materials. There are four types of earth and rock dams according to materials, construction methods and waterproof components:

1. Rolled earth dam filled with sandy soil, cohesive soil or large gravelly soil (or gravel). Filling under dead-weight or manual compaction (layered compaction) or directly in water, such as river, creek, lake or artificial pond.
2. Filling earth dam with fine grained soil (sandy soil, cohesive soil, artificial tailings sand, etc.) and large gravelly soil (excluding coarse gravel) is a structure filled with hydraulic machinery according to the standard of hydraulic filling.
3. A large gravel earth dam filled with large gravelly soil and boulder soil. The soil waterproof body is composed of standard compacted weakly permeable cohesive soil or sand, or not less than 30–40% sand-clay filling (fine grained soil with particle size less than 2 mm).
4. A rockfill dam filled with large pieces of gravelly soil and boulder soil (rock block, most commonly rock block or rockfill). Non geotechnical materials (metal plate, cutoff wall, sheet pile, reinforced concrete wall, cutoff wall, asphalt concrete inclined wall, cutoff wall, geomembrane, etc.) are used to make the waterproof body.

Rolling construction method is widely used in earth and rock dam construction in China. The main dam types are homogeneous dam, zonal waterproof earth dam and zonal waterproof non-earth dam. After rolling, the waterproof soil should meet the following requirements: permeability coefficient, which should not be greater than 1×10^{-4} cm/s for homogeneous dam, and core wall and inclined wall should not be greater than 1×10^{-5} cm/s; water soluble salt content is not more than 3% (mass meter); organic matter content (mass meter) should be no more than 5% for homogeneous dam, and no more than 2% for core wall and inclined wall; good plasticity and permeability stability; little volume change takes place during immersion and dehydration.

Frozen soil area can be divided into freezing type dam, thawing type dam and freezing-thawing (or thawing-freezing) transition type according to the initial cryogenic fabric and temperature state during dam construction and operation.

Frozen dam is a permeable dam, and its waterproof stability and seepage strength are guaranteed by the frozen waterproof body and foundation. That is, the permafrost dam foundation should be kept in a natural freezing state during construction and operation, and the impervious core in the dam body should be gradually frozen before the reservoir with the highest water level, and then be frozen during the whole operation period.

Thawing dam refers to the dam body, impermeable core and dam foundation in thawing state or partially frozen state, allowing a limited amount of permeability to the dam body and dam foundation (or just the dam foundation). That is to say, before the construction, operation and impoundment of the dam body, the foundation will be melted to the design depth, and the soil waterproof facilities will be maintained. The earth dam can be in the thawing-freezing state, while in operation, it is mainly in the melting state.

For the transitional thawing-freezing dam, the water level drop section in front of the dam is allowed to be designed according to different principles. If there is a through talik under the river bed of a big river, and the floodplain is underlain by permafrost with large thickness and thaw compressibility, a concrete and earth hybrid dam can be built. Concrete dams can be built in the thawing zone in river bed according to the thawing principle, and earth dams can be built on the permafrost on both banks according to the freezing principle. Freezing and cooling devices should be set at the joint of the two to prevent the frozen soil from thawing. Thermosyphon can be the cooling device applied.

When there is a deep talik or penetrated talik at the bottom of the river, and there is a strong thaw compressibility permafrost layer (within 6 ~ 8 m in thickness) on the floodplain, but there is a soil layer with low thaw compressibility below it, the construction of a thawing earth dam should be considered. At this time, the dam foundation soil with strong compressibility should be removed and replaced with soil with low compressibility. The highly compressible soil can be treated by manual thawing and compaction.

The dam structure is determined according to the earth materials, construction methods, construction principles of earth dam and impermeable core, and the temperature state of dam body and dam foundation (Table 5.14).

The type and structure of dam body and geotextile of river section in cold region are determined comprehensively according to the following conditions: hydrological and climatic conditions in construction area; dam foundation and river bank topography; engineering geology and frozen soil conditions; ice content and thawing settlement of dam foundation frozen soil; the scale of the talik under the riverbed; regional seismic intensity; storage of local building materials; general layout of water conservancy projects; construction organization and process mode; construction period and operating conditions; economic demonstration; environmental protection measures; human working environment guarantee, etc.

The earth dam structure can be built on the rock foundation or non-rock foundation, which can be either in the thawing state or in the freezing state for many years. When the volumetric ice content of the foundation is more than 0.4, it is better to build a frozen dam. The dam body waterproof equipment and the soil of the dam foundation in the riverbed talik should be frozen, and the frozen state should be preserved during operation. Types of liquid-gas, liquid or gas seasonal refrigeration devices (thermosyphon) can be used to achieve the frozen state. The impermeability of the frozen dam should be guaranteed by keeping the dam body and part of the

Table 5.14 Basic types of earth dams

Dam type characteristics	Thawing dam	Freezing dam
Categorized by dam structure	Homogeneous earth dam Heterogeneous earth dam: Earth core dam, earth inclined wall dam Central impermeable prismatic dam Upstream waterproof prismatic dam Dams with heterogeneous materials: Diaphragm wall, grouting core wall, inclined wall	Homogeneous earth dam Heterogeneous earth dam: Core dam Central impervious prismatic dam Upstream anti-seepage prismatic dam
Categorized by dam foundation waterproof equipment structure	A dam with blanket Dam with key wall Dam with grouting (cement grouting) curtain Dam with diaphragm wall (wall in soil, asphalt concrete or earth rock concrete wall sheet pile)	Dam with ice soil wall Dam with key wall Layered filling, mechanical compaction, and then freezing of dam foundation and dam soil water layered filling with refrigeration device, and then freezing of dam foundation and dam soil with refrigeration device
Categorized by dam construction method	Filling soil in layers and mechanical compaction Filling in water	Fill the soil with small thickness layer by layer in winter, and mechanically compact layer by layer and freeze it by natural negative air temperature

dam foundation in a frozen state. During the operation of the dam, permeation through the dam body and dam foundation is not allowed, and the foundation soil is prevented from thawing when passing through the waterproof equipment. When the ice content ratio of the foundation is 0.03 ~ 0.4 for the warm permafrost containing ice, it is appropriate to build a thawing dam with a lower dam height. The waterproof equipment should be set in the dam body and dam foundation to ensure its impermeability, while the homogeneous dam should have sufficient length for underground profile.

The following are the priority conditions for s engineering geology and frozen soil when deciding the type between freezing dam and thawing dam.

Freezing dam. ① When the frozen soil thaws, it is categorized as strongly thaw-settlement soil, that is, it exceeds the thaw settlement of Level III or above. ② The flowing velocity at the bottom of the river bed in the talik should not faster than 3×10^{-5} m/s, and the talik depth is less than 5 m. When the infiltrating velocity and the thickness of the thawing zone are larger, the possibility of freezing should be determined through thermal calculation. It is recommended to conduct cement, clay grouting or chemical (such as silicification method) treatment in advance.

Thawing dam. ① When the frozen soil thaws, it is categorized as Level I and II hard or semi-hard non-thaw settlement or weak thaw settlement frozen soil. ② There are deep or penetrated talik in the riverbed, and the river valley edge and floodplain are non-thaw settlement or weak thaw settlement frozen soil.

The core wall or inclined wall waterproof body should adopt the key wall, and its depth should reach or be embedded in the soil layer with low compressibility. The core wall and inclined wall of embankment dam, and the connection part of asphalt concrete core wall of rockfill dam and thawing bank should be embedded in the bank. If the bank is situated on fissured rock or non-rock soil with strong water permeability, the waterproof curtain should be set to connect with the anti-seepage facilities of the dam.

When the rolling earth dam is designed for the low-pressure plastic and medium compressible non-stone foundation soil with the frozen soil thawing, the freezing dam with core wall is recommended, while the freezing dam with central waterproof prism (core wall) is recommended for the ice-rich frozen soil foundation.

The following are several kinds of soil forbidden to be used for building earth dams.

1. Saline soil with chloride salt content more than 5% (by mass) and chloride-sulfate salt content more than 10%;
2. Organic mineralized soil and similar soil, sludge, peat, peat soil, etc. containing more than 5% (by mass) of incompletely decomposed organics or organisms (plant roots and residues, etc.) or more than 8% (by mass) of fully decomposed amorphous or liquid organic compounds (such as humus, oil, etc.);
3. Ice-rich and ice-bearing frozen soil.

The most widely used waterproof elements in dam construction in cold regions mainly include core wall, inclined wall, prism and waterproof blanket. The soil waterproof facilities of the freezing dam [central and downstream waterproof prisms, core walls and waterproof blanket) should ensure that a freezing curtain (thermosyphon using working medium such as liquid ammonia or kerosene (Fig. 5.49)) is set up on the dam body, which should be closely combined with the permafrost of the dam foundation and dam shoulder to ensure that the dam body is watertight.

The dam body of earth and rock dam and rockfill dam should be filled with relatively solid, cold resistant, mild frost heaving or non-frost heaving large rubble or non-sorting rock (or rockfill), which has relatively high stability.

The thawing dam can be built on the stone or non-stone foundation soil with low compressibility when thawing. When the frozen soil thaws and turns into foundation soil with moderate to strong thaw settlement, it is recommended to build a freezing earth and rock dam.

The requirements for soil impermeable core (core wall, inclined wall), transition layer and inverted filter of earth and rock dam are the same as those of earth dam. It is needed to focus on the freezing of working layer during construction, and take measures to avoid affecting the compaction density. The gradient of dam slope is determined according to the stability of the whole dam body. The slope ratio of inclined wall earth dam is usually 1: (2.5 ~ 3.0), and that of core wall earth dam is 1: (2 ~ 2.5).

Fig. 5.49 Thermosyphon (kerosene working fluid) on Russian dam body (Photo by Tong Changjiang)

The joint parts of the dam body to the permafrost dam foundation and the dam shoulders are the most important parts of the dam body in cold regions, and are often the main parts where accidents occur. The waterproof facilities [core wall, inclined wall, prism, intercepting diaphragm (sheet pile), etc.] of the thawing earth dam and dam foundation should be grouted from the gallery through the concrete key wall to form a curtain in the dam foundation.

If the dam foundation is a non-stone foundation with weak compressibility and weak permeability, the core wall (inclined wall) joint part should be embedded into the dam foundation through the softening layer at the upper part. If there is riverbed alluvium, the key wall should be deep into the rock under the dam foundation.

The joint part between the frozen earth and rock dam and the dam foundation should also extend the freezing curtain below the permafrost table, and pre-freeze the talik under the riverbed before the reservoir impoundment.

Drainage facilities should be arranged at nonfreezing parts and with anti-freezing measures in order to avoid ice cone at dam body or bank connection.

5.4.2.3 Calculation of Temperature Regime of Earth Dam

When constructing earth dams in permafrost regions, in addition to the calculation of porous flow, seepage stability, skid resistant stability, stress and deformation, the calculation of temperature regime stability during the operation of the dam body is also important. The main thermal calculations are the following.

Thawing dam: predict the temperature of the dam body, dam foundation and the joint between the dam and the shoulders during construction and operation;

Freezing dam: calculate the temperature states of the dam body, dam foundation, the joint between the dam and the shoulders and dam foundation freezing system in the short term and long-term operation, and the thickness and joint status of the freezing curtain under the action of seasonal cooling devices (such as thermosyphon) or other freezing systems;

Reservoirs with thawing and freezing dams: predict the temperature state around the dam body, reservoir bank, upstream reservoir bottom and downstream riverbed during the construction and operation of the dam and reservoir.

The calculation should take into account: ① the temperature changes of dam body and dam foundation caused by the destruction of natural state by construction and the temperature field of frozen dam when reaching the final state; ② predict the temperature stability of earth dam at any time during construction and operation (after 1, 3, 5, 10 years, etc.).

It is a very complex building thermophysical problem to determine the space temperature field of heterogeneous earth dam on permafrost foundation. The usual method is to separate the dam body from the dam foundation, the dam and the reservoir, and the upstream and downstream reaches to predict their temperature states separately. In this way, a simpler thermodynamic model and a simplified heat conduction problem can be used to calculate the stable and unstable temperature regimes. However, for heterogeneous earth and rock dams and rockfill dams, in the first few years, there will be convective runoff between water (upstream prism) and pore gas (downstream prism) in the dam body which is built layer by layer by block crushed stones and rocks, and water seepage will gradually appear in the joint of soil impermeable core (inclined wall, core wall, blanket), transition layer (inverted filter), thawed (or thawing) foundation and dam. The convection of heat and mass transport is increased, and the complex and uneven temperature state is formed in the dam body, dam foundation and dam shoulders.

The recent calculation is based on the heat conduction of soil impervious core, the heat and mass transfer convection model of rockfill dam, methods such as natural convection of water in the upstream prism and the convection of internal confined air in the downstream prisms have been applied to calculate. However, the calculation results indicate that the core wall and dam foundation must be cooled (natural or artificial) by means of freezing device to keep the dam body frozen, otherwise, the dam foundation will inevitably thaw.

5.4.3 Water Discharge and Intake Structures

For water discharge and intake structures in permafrost regions, attention should be paid to the influence of water passing on the temperature regime of frozen soil foundation and buildings and the physical and mechanical properties of foundation, as well as the changes in the thawing process.

The design principle of water discharge and intake structures in the dam body should be the same as that of the dam body, and the structures on the bank should be determined according to the engineering geology, frozen soil conditions and economic comparison. When adopting the design principle of maintaining the frozen state of frozen soil, thermal insulation and waterproof measures should be taken.

The water discharge and intake structures of the freezing dam should be arranged on the river bank. In case of frozen soil foundation situated on soil with ice content higher than saturated frozen soil, it is generally removed and replaced with weak thaw-settlement soil, or the foundation type (such as pile foundation) is changed, and thermal insulation and waterproof prevention measures are taken to avoid thawing of frozen soil foundation.

The water discharge and intake structures of the thawing dam should generally be arranged on the river bank at the dam side. When they are arranged within the dam body, they should be arranged on the axis section with the least compressibility (river bottom talik).

When the water discharge and intake structures are arranged, the ice passing conditions should be considered. When the upstream velocity is less than 0.5 s/m, the buildings are not allowed to pass through ice.

5.4.4 Channels, Culvert Gates and Revetments

In addition to the frost heaving effect in seasonally frozen soil areas, the thaw settlement damage caused by the thawing of frozen soil foundation should also be considered for the damages of canals, culverts, gates and retaining walls in permafrost regions.

The hydraulic structures on the permafrost foundation of class I and II thaw settlement soil can be designed in the thawing state, and the waterproof and anti-freezing and thawing stability can be designed according to the requirements of the seasonally frozen soil area.

When the frozen soil foundation belongs to class III thaw settlement soil or low temperature frozen soil with high ice content, thermal insulation and waterproof measures should be taken at the bottom of hydraulic structures, and the design principle of maintaining the frozen state of frozen soil foundation should be adopted:

1. The channel should be constructed by means of fill. When excavation is adopted, thermal insulation and waterproof measures for canal bottom and slope must be strengthened to keep the permafrost table within the design depth. During the design of canal slope, the freezing and thawing interface or waterproof material interface should be used as the sliding surface to check the slope stability. The shear strength of freezing and thawing interface soil should be determined by test or according to similar engineering design data.
2. The sluice chamber should adopt pile, column foundation or plate pile foundation, the base should be provided with thermal insulation and impermeable layer, and the wall side should be provided with anti-frost heave measures. In general, pure sand gravel should not be used for replacement in frozen soil sections with high ice content.
3. The culvert should be built in concrete circular pipe culvert, rectangular culvert, box culvert, slab culvert and metal corrugated pipe culvert. The I-shaped inlet and outlet form should be adopted. The bottom should be provided with thermal

insulation and waterproof measures. When the foundation is frozen soil with high ice content and is thick, pile foundation should be used.

4. The revetment should be built in prefabricated light and flexible structure, or concrete cantilever retaining wall, instead of gravity mortar rubble retaining wall. The buried depth of foundation should be 1.3 ~ 1.5 times of the natural permafrost table. At the back side of the revetment, it should be provided with thermal insulation layer, draining hole and other anti-frost heave measures.

5.4.5 Tunnel and Concealed Pipe Hydraulic Structures

Reinforced concrete lining should be generally considered for tunnel water conveyance in permafrost rock. The pressure tunnel scheme is adopted in the frozen rock to ensure a more stable and uniform working condition. When the non-pressure tunnel scheme is adopted, the influence of temperature change on the physical and mechanical properties of frozen rock mass under the construction and operation conditions of buildings should be considered.

In the case of underground hydraulic engineering structures with shallow depth, the thawing circle may be close to the freezing-thawing alternating zone or the ground surface, so it should be noted that surface subsidence may be caused. Especially for the rock mass with large ice content and frozen expansion, the possibility of causing the settlement of underground structures after its thawing should be considered.

Ground surface damage should be avoided near the tunnel entrance in permafrost area, and measures should be taken to protect the frozen soil and environment at the entrance. When the frozen soil with high ice content is excavated, the gravelly soil should be replaced and compacted in layers.

Waterproof board and thermal insulation layer should be laid outside the tunnel lining. The thermal insulation layer should be fully enclosed. The thickness of the thermal insulation layer should be determined by thermal calculation. When the tunnel base is located in the frozen soil section with high ice content, the soil should be excavated before concrete filling. When there is frozen soil with high ice content around, it should also be excavated and sealed with concrete, and the thickness should not be less than 0.3 m.

5.4.6 Power Station and Pumping House

The foundation of power (pumping) station buildings and forebay ice discharge chamber in permafrost regions should be rock, or frozen soil with thaw settlement of class I and II, or river bottom talik, and the foundation design can be carried out according to the convention.

The depth of the ground power (pumping) station factory workshop foundation should be greater than the maximum seasonal thawing depth, the ground and underground drainage facilities should be well prepared, and anti-freezing measures should be taken. When there is frozen soil with high ice content in the foundation, the pile foundation should be used, and its depth should exceed the maximum depth of the talik or reach the weakly weathered zone of the bedrock. Anti-frost heave measures should be taken for the seasonally frozen depth range.

5.4.7 Bank Protection Works

The ice conditions should be considered for the bank protection (wharf) works. It is better to arrange the wharf outside the ice formation area. If there is isolated permafrost, the wharf should be arranged in the talik. When conducting protection works in the strongly thaw settlement frozen soil area, the foundation treatment should be carried out in advance (such as pre-thawing, compaction, local replacement, or use of bored concrete piles), or the structure suitable for the large deformation of the foundation should be adopted.

When the height of the bank protection (wharf) work is high (more than 15 m) a reasonable platform should be set at the dynamic water level. In order to avoid or reduce the damage (or scouring) effect of ice load on the bank protection work, hydrophobic materials can be used to paint the front wall (or surface).

The frost heaving force (horizontal, tangential and normal frost heaving force) of the soil layer on the back of the bank protection (wharf) work is an important reason for the damage of the bank protection works. The main measures to eliminate the disease are still soil improvement (replacement, chemical treatment, etc.), waterproofing and drainage, and thermal insulation to reduce or eliminate the frost heaving force of the soil layer on the back of the building. Or change the foundation form, such as sheet pile wall, comprehensive measures (Fig. 5.50), or Penn's cage.

The water outlet of the drainage facilities should be arranged below the freezing layer of the building at the lowest water level, and the reversed filter should be used if necessary.

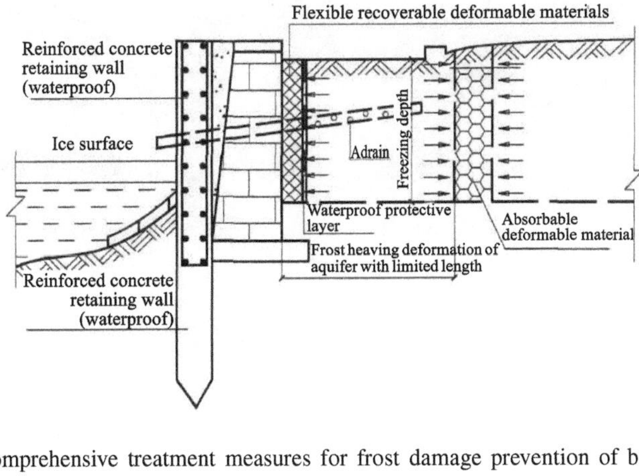

Fig. 5.50 Comprehensive treatment measures for frost damage prevention of bank protection works (Zhai, 2011)

5.5 Frozen Soil Engineering Problems in Reservoir Area

The engineering problems such as bank stability, reservoir leakage, reservoir siltation, induced earthquakes etc. of reservoirs in permafrost regions are similar to those in seasonally frozen soil regions. In addition to the fact that the induced earthquake depends on the geological structure, the problems of bank stability and reservoir siltation in permafrost regions are more prominent.

5.5.1 Reservoir Bank Reconstruction

There are generally ground ice layers of various thicknesses in the permafrost layer. Especially at the permafrost table, it has thick ground ice layers (right in Fig. 5.51), with the maximum thickness of 2 ~ 4 m. Under the thermal erosion and wave impact of lake water, reservoir water and river water, the bank is prone to thermal collapse and slide (left in Fig. 5.51), resulting in subsidence and siltation. Once the ground ice is exposed, it is often subject to retrogressive erosion. The slope toes and valleys of the mountain slope are rich in ground ice. Even the rock fractures also contain crack ice layer (Fig. 5.52), especially in the strongly weathered zone of bedrock, so that there are ground ice layers of different degrees in the weakly weathered zone of bedrock. The bank slope will still thaw and collapse under the thermal erosion of reservoir water.

Wave is another important force that causes the collapse of the reservoir bank in the frozen soil area. It is the main zone of wave impact between the normal water level and the lowest dead water level. It also includes the height of wave run-up at the normal water level and the depth of wave scouring under the dead water level,

Fig. 5.51 Lake bank collapse (left, photo by Lin Zhanju) and ground ice layer (right, photo by Sun Zhizhong) in permafrost region

Fig. 5.52 Bedrock crack ice

Fig. 5.53 Gravel Slope

which is the range of wave impact. The degree of collapse of the bank slope in the frozen soil area is much more severe than that in the non-frozen soil area, because the rock and soil layers have been subjected to the cryogenic weathering and freeze-thaw effect, the degree of fragmentation is more serious.

Within the bank slope with mild gradient, the soil layer contains ground ice layer. When it is a slope of more than 20° and the bedrock is strongly weathered or weakly weathered, it still contains crack ice (Fig. 5.52). The surface layer is subject to strong cryogenic weathering, and the rock is relatively broken. There are many rubble slopes and debris slopes (Fig. 5.53). Some valleys may exist relict moraines and rock glaciers, and the stability of the bank slope is poor.

Fig. 5.54 Thermal collapse of river bank (Photo by Tong Changjiang)

Fig. 5.55 Water flow scouring on slope (Photo by Tong Changjiang)

5.5.2 Reservoir Siltation

There are a lot of ground ice layers in the terraces and beaches of rivers and valley banks in permafrost regions. Under the impacts of water flow and lateral thermal erosion, the stabilities of these terraces and beaches are very poor, and it is very easy to produce thermal collapse (Fig. 5.54) and gelifluction, until the ground ice in the soil layer disappears. The surface layer of the soil hillside is subject to repeated freezing and thawing, the soil is loose, and the slope surface is strongly scoured by water flow (Fig. 5.55). In the heavy rain period, these soil hillsides may be the source of the materials for reservoir sedimentation.

Fig. 5.56 Land subsidence of the lake after the water recedes in the Yellow River source area (Photo by Zhang Senqi)

5.5.3 Ground Surface Settlement

When the reservoir water (lake water) is at a high level after submerged, the ground ice layer in the underwater frozen soil layer gradually thaws under the thermal erosion of the reservoir water (lake water), and the original ground surface sinks. When the reservoir water (lake water) is at a low level, the exposed ground shows a thaw settlement with a height difference of 1 ~ 2 m (Fig. 5.56).

5.5.4 Reservoir Leakage

Leakage in the reservoir area of permafrost regions often occurs in the joint zone between the dam body and the bank slope, which may lead to leakage around the dam.

The structural fault zone is often in the structural talik of the permafrost region, and the main channel of the subpermafrost water may become the leakage path of the reservoir.

Although the permafrost layer is impermeable, its impermeability is subject to the thickness of the permafrost layer. The permafrost layer in mountainous areas of China is distributed in a "shell" shape. The permafrost layer in the Western Area is determined by altitude. The higher the altitude is, the greater the thickness of the permafrost layer is. The permafrost layer in Northeast China is mainly determined by latitude, and then by altitude.

5.5.5 Frost Heaving and Ice Sheet

During the freezing period, the slope protection of the earth dam will swell and crack, with overall destruction. The masonry is out of position and the crack is expanded. Under the impact of stormy waves in spring, the masonry reversed filter will be hollowed out and collapsed.

The ice sheet of the reservoir will damage the dam body, sluice gate, sluice pier, slope protection, revetment, water tower, etc. The ice sheet climbs up along the dam slope, with a daily climbing height ranges from a dozen to tens of centimeters, and the whole winter climbing height can reach 2 ~ 3 m. In winter, the ice sheet pushes the dam slope masonry, and in spring melting period, the ice sheet gouges the dam slope. According to the indoor test data, the freezing strength between ice and concrete (Zhang et al., 1993) reaches 1.15 MPa when frozen at −7 °C and 0.9 MPa when frozen at −2 °C in an ambient temperature of −15 °C. The critical thickness of the ice sheet for buckling failure is 0.27 m (Sun et al., 2006a, b, c), which generates ice pressure on the dam slope. When the ice thickness is 1.0 m, the static ice pressure in horizontal direction is 245 kN/m.

Chapter 6
Pipeline Engineering in Frozen Soil Regions

Pipeline projects in permafrost regions include long-distance oil and natural gas pipelines, water supply and drainage pipelines in towns, stations and power plants, etc. This chapter mainly discusses the outdoor pipeline projects, especially the long-distance oil and gas pipelines crossing permafrost regions. With the development of Chinese national economy, long-distance water transmission pipeline project in permafrost regions may also be constructed.

Due to the special geographical environment in China, there are high latitude permafrost regions determined by latitude, plateau (high mountain) permafrost regions determined by altitude, high temperature (mean annual ground temperature ≥ -1.5 °C) permafrost regions and low temperature (mean annual ground temperature < -1.5 °C) permafrost regions. Under the control of a large pattern (latitude and altitude), affected by various factors (geology, topography, hydrology, vegetation, human activities, etc.), continuous permafrost, patchy permafrost, sporadic permafrost, and seasonally frozen soil are distributed, which makes the diseases (frost damage) of long line projects (roads, pipelines) occur frequently, and the foundation and their design and treatment are extremely complex.

6.1 Engineering Geological Route Selection in Frozen Soil Regions

The diseases encountered in linear engineering in permafrost regions are mainly thermal thaw settlement and frost heave. The former mainly occur in fine grained frozen soil rich in thick ground ice, while the latter mainly occur in fine grained soil experiencing repeated seasonal freezing and thawing with abundant occurrences of groundwater. Oil pipelines in permafrost regions cannot avoid these diseases. There are many reports in the former Soviet Union, the United States, Canada, etc., and

Q. Wu, C. Tong, *Frozen Soil Engineering in Cold Regions*,
https://doi.org/10.1007/978-981-95-2527-0_6

there have also been pipeline frost heave and thaw settlement diseases in China's Golmud-Lhasa Oil pipeline and China Russia oil pipeline (Figs. 6.1 and 6.2).

Thaw settlement of pipelines in permafrost regions is caused by the thermal impacts on the foundation soil layer containing ground ice from the external heat (the temperature of pipeline carrier and environmental changes). The Mohe-Daqing oil pipeline crosses about 512 km of permafrost region, and 119 km of high temperature and high ice content permafrost region. According to the oil temperature monitoring data of Mohe, Tahe and Gagedaqi stations (Fig. 6.3), the minimum and maximum oil temperatures are 0.42 °C and 16.2 °C respectively. The pipeline operates at a positive temperature throughout a whole year, heating the frozen soil foundation continuously, so that the frozen soil with high ice content gradually thaws, the foundation subsides, and the accumulated water in the pipe trench further accelerates the thawing of the frozen soil foundation, resulting in the settlement and deformation of the pipeline and the ground (Fig. 6.2, left). From October 2012 to June 2013, the ground penetrating radar detection results of the thawing depth and freeze-thaw interface under the pipeline near the milage of 40 km of the Mo-da Line (Fig. 6.4) showed that the thawing depth under the bottom of the pipeline reached 1.0 m in October, and 1.5 m under the pipeline in June of the following year. As the operating time increases, the thawing depth of the frozen soil foundation under the

Fig. 6.1 Frost heave of oil pipeline in permafrost region. (**a**) Canadian Roman Well buried pipeline (photo by Jin Huijun); (**b**) Golmud-Lhasa buried pipeline (photo by Liu Yongzhi); (**c**) Russian Yakutia Southern Overhead Pipeline (photo by Jin Huijun)

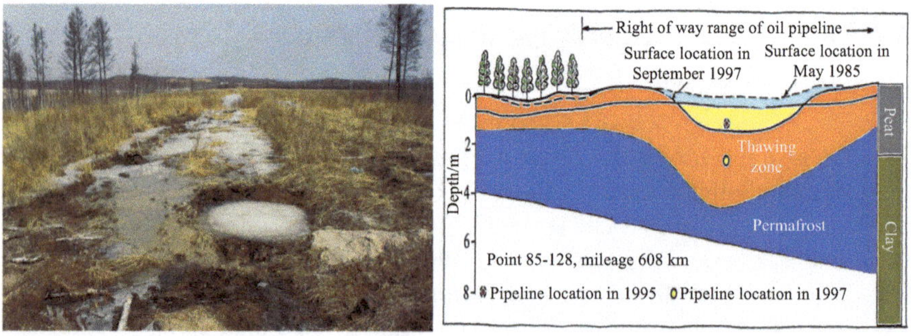

Fig. 6.2 Thaw settlement of oil pipeline in permafrost region. (**a**) Trough caused by thaw settlement on the China Russia oil pipeline (photo by Tong Changjiang); (**b**) Diagram of Thaw Settlement Measurement of Canadian Oil Pipeline (Jin et al., 2006)

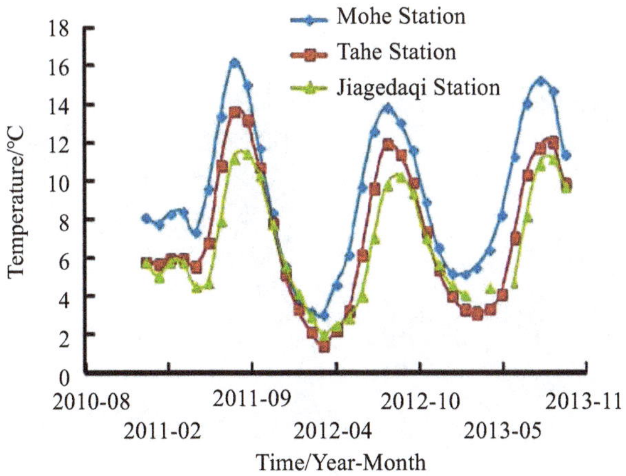

Fig. 6.3 Outflowing oil temperatures from 2011 to 2012 (Li et al., 2015)

pipeline will continue to increase. JM7 monitoring site (Tan et al.) is in frozen swamp, with a silty clay layer of ice-rich or ice-saturated. After the pipeline was put into operation for half a year (December 9, 2010-May 9, 2011), the pipeline had moved down by 0.4 m. 17-year observation of Roman Well oil pipeline in permafrost region of northwest Canada (right figure of Fig. 6.2) showed that the thawing depth increased to 3 ~ 5 m (in lacustrine deposition) or 5 ~ >7 m (in coarse grained soil), and the ground surface subsidence was 0.7 ~ 1 m.

When the pipeline is buried in swamp or wetland in permafrost region, frost heave of the foundation soil causes the pipeline to frost heave and warp, such as the frost heave of the Roman Well oil pipeline in Canada (Fig. 6.1a). The pingo causes the pipeline to freeze and warp, and the warping height of Golmud Lhasa oil pipeline due to freeze is 0.5 ~ 1.0 m (Fig. 6.1b). As the oil pipeline of the Mohe-Daqing Line is in positive temperature operation all year round, frost heave of the pipeline is generally not easy to occur, but it is found that the lower part of the pipeline is frozen in winter (Fig. 6.5). Ice cones and pingos appear in low-lying valleys in winter, which makes the pipeline frost heave, such as in the valley wetland around the milage of 276 km (near Xiufeng) on the Jiamo Highway, where there are abundant springs and water sources above the pipeline. In this place, ice cones and pingos are formed in winter, with the thickness of the ice layer reaching 1.5 m and the amount of frost heave nearly 3.0 m, and the upper part of the pipeline forms an ice channel (Fig. 6.6). When the pipe temperature decreases in winter (the lowest temperature is only 0.4 °C), the saturated soil layer at the pipe bottom may freeze and cause frost heave. For another example, the ice-covered pipeline (Fig. 6.7) in around the mileage of 115 km in valley wetland of the Jiagedaqi-Mohe Highway often has ice cones, which are nearly 1.0 m high. In order to prevent the pipeline from frost heave, it is often necessary to expose and drain the water. This phenomenon is found at the mileages of 33 km and 300 km of Jiagedaqi-Mohe Highway, Xiufeng, Bizhou, and 7 km of Mohe-Daqing Highway.

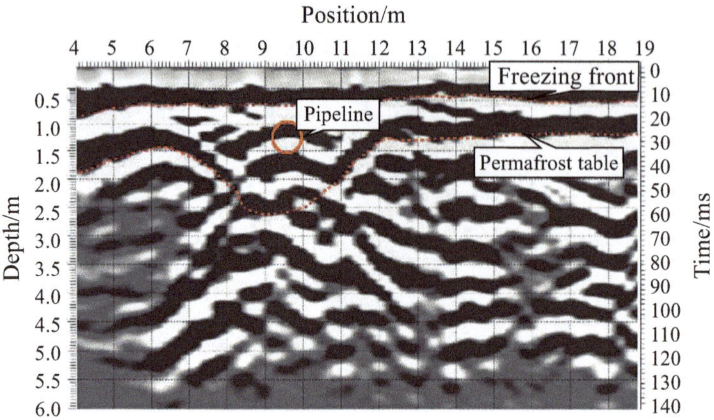

(a) October 14, 2012

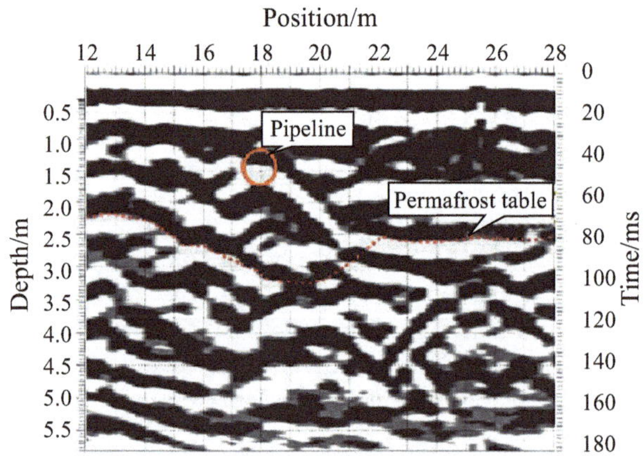

(b) June 17, 2013

Fig. 6.4 Ground penetrating radar profile under the ground surface near the mileage of 40 km of the pipeline of Mohe-Daqing (Li et al., 2015)

Fig. 6.5 Ice layer under the Pipeline of Mohe-Daqing Line at the mileage of 7 km

Fig. 6.6 Spring aufeis at the mileage of 276 km of the pipeline along the Jiagedaqi-Mohe Highway. The figure at the top right corner shows the outlet of spring water

Fig. 6.7 Pipeline covered by aufeis at the mileage of 115 km along the Jiagedaqi-Mohe Highway (Photos from Figs. 6.5, 6.6 and 6.7 by Tong Changjiang)

The frost heave of overhead pipelines is mostly caused by the frost heaving of the pile columns. For example, the pipeline deformation of Russia's Siberian oil pipeline is caused by the frost heaving of pile columns (Fig. 6.1), and the frost heaving of railways, highways and houses in Daxing'anling Mountains can be seen everywhere.

In addition, features related to frozen ground (or undesirable features related to frozen ground) in permafrost regions, such as thaw slumps, thaw subsidence, pingos, ice cones, permafrost swamps, thick ground ice, etc., may threaten the stability of pipelines.

For the route selection of long-distance pipeline in permafrost regions, in addition to conducting in-depth investigation and research, and collecting meteorological, hydrological, geological, engineering geological and hydrogeological data as in non-permafrost regions, it is more important to find out the characteristics of frozen soil along the pipeline in permafrost regions (distribution, ground temperature, engineering type of frozen soil, etc.), and the distribution range, type, size, relevant factors (ground ice, groundwater, landform, hydrogeology, etc.) of undesirable features related to frozen ground, the hazard potential, etc., optimize the route and propose bypass and passing plans and measures. On the basis of the basic principles of pipeline route selection, the following principles should be followed:

1. The regional classification of permafrost is determined according to the complexity of construction site and foundation (Table 6.1). The route should avoid or reduce the passing length at complex sites as much as possible.

Table 6.1 Classification of pipeline foundation in permafrost regions

Complexity of construction site	Engineering geological characteristics of frozen soil	Relative subsidence when permafrost thaws	Foundation complexity
Complex site	The terrain is complex, with a variety of frozen landforms, including frozen swamps. There is a diverse range of lithology, and the area is in continuous or isolated permafrost regions. It includes permafrost of saturated ground ice with thick ice layer, or ice layers containing soil, with a mean annual ground temperature higher than −1.0 °C. Groundwater is well developed, and there are strong manifestations of permafrost phenomena and adverse geological effects	>0.2	I
Comparatively complex site	The terrain and landforms are relatively complex, with well-developed groundwater and poor drainage. The area is in continuous or discontinuous permafrost regions, featuring well-developed ground ice in ice-rich permafrost or icy permafrost. The mean annual ground temperature ranges from −1.0 °C to −2.0 °C. The lithologies are mostly fine-grained soils containing blocky gravel. There are pronounced permafrost phenomena and adverse geological effects	0.03 ~ 0.2	II
Simple site	The terrain and landforms are simple, with a simple lithology, mostly consisting of coarse-grained soil, with good drainage. It is either in talik zone or discontinuous permafrost regions (permafrost table deeper than 8 m). The permafrost is ice-poor with little ground ice. There are no pronounced permafrost phenomena or adverse geological effects.	0.0 ~ 0.03	III

2. The route should avoid or minimize passage through frozen swampy or wetland. It is not suitable to be laid in low-lying wetland zones (such as the aufeis shown in Fig. 6.7). Instead, it is advisable to choose to pass along the edges. In wetland or low-lying sections, measures should be taken to prevent erosion and sub-erosion (Fig. 6.8) to avoid pipeline frost heave.
3. When the line passes through mountains and hills, it should pass through gentle slopes, but it should avoid the areas where thick ground ice is developed; otherwise, it should strengthen the engineering measures against thaw slumping and gelifluction.
4. When the route passes through the valley section, it is better to select the river talik zone or the section in ice-poor or icy permafrost on the high terrace of the sunny slope. It is better not to select the valley frost-thaw overlap zone, and the transition zone should pass through the shortest distance.
5. The line should avoid the sections where undesirable features related to frozen ground are formed. If it is necessary to pass, the following options should be selected.
   ① In the section with thick ground ice, the area with the thinnest and narrowest ground ice should be selected to pass through.
   ② In the section of thaw slump and gelifluction, the line should pass under the slumping mass, and the buried depth of the top of the pipeline should be deeper than the sliding floor at the bottom of the slumping mass.
   ③ It is better to pass through the spring outlet point from above. If the pipeline is buried when constructing, it should be kept away from the spring outlet point. When intercepting and waterproof measures are taken above the spring outlet point, the runoff should not be cut off or blocked to prevent pingos from forming and water from intruding into the pipe trench.

Fig. 6.8 Erosion and scouring of pipelines by surface water and groundwater (photo by Tong Changjiang)

Fig. 6.9 Pingo in the North of new forest area

Fig. 6.10 Pingo of South Songling along the Mohe-Daqing Line

④ Avoiding areas with developed groundwater and pingos, or areas prone to form such things [Figs. 6.9 and 6.10 (Survey Engineering Department of Daqing Oilfield Engineering Co., Ltd., 2008)]. It is advisable to pass beneath them at a certain distance. Avoid exposing or obstructing groundwater flow during construction to prevent the formation of aufeis.

⑤ In the thermokarst lake section, the line should avoid or pass a certain distance from the edge, and the lake bank slope should have protective measures to avoid collapse and bank reconstruction.

6. The mountain passing section should avoid thick ground ice and undesirable features related to frozen ground. There is often ground ice occurring on the slope and at the foot of the slope. Thermal insulation and protection measures should be taken to prevent the formation of thaw slump. Erosion prevention measures should be taken on the slope surface, mainly for drainage, and drainage ditches should be set on both sides (energy dissipation measures should be taken at the steep slope section).
7. It is better to cross the river at the section of the talik zone and with the shortest distance. The valley section usually belongs to non-penetrating thawing area, so it is better to choose the section of ice-poor or icy permafrost. Protect the bank slope to avoid thermal collapse. Measures that can prevent aufeis and pingos should be taken. If pile foundation is adopted, anti-frost heave measures should be taken.
8. The tunnel should not be located in the area where groundwater is developed and thick ground ice is formed. The entrances should not be located in the area where undesirable features related to frozen ground are formed. Protective measures should be strengthened for the entrance slope. Waterproof and drainage measures should be set in the tunnel.
9. The pumping station and oil storage tank should be located in the rock bed or coarse-grained soil section in icy permafrost, avoiding the section with ice-rich permafrost or soil with strong frost heave and undesirable features related to frozen ground. The foundation should be provided with anti-frost heave measures, and waterproof and drainage measures should be taken.

6.2 Pipeline Design Principles

Pipeline engineering in permafrost regions will inevitably cross permafrost regions, and the heat carried by oil pipeline is an artificial heat source that causes permafrost to thaw. Within the influential range of pipeline engineering, its impact is far greater than that of solar heat source, and it is much more intense. According to the mean annual ground temperature and ice content of the permafrost foundation, the design principle of keeping the frozen state of the frozen soil foundation or the design principle of thawing the frozen soil foundation should be selected.

The thawing design principle is suitable for isolated or sporadic permafrost areas in the southern limit of frozen soil. The mean annual ground temperature is relatively high, generally −0.5 °C ~ 0 °C, and the thickness of permafrost is 3 ~ 10 m. Under the thermal influences of natural and oil pipeline, it is very easy to thaw and difficult to prevent it from thawing. The design principle of pre-thawing or gradual thawing depends on the thickness and ice content of permafrost. In general, the isolated or sporadic high temperature and high ice content permafrost is mostly located in the swamp or wetland zone. When the thickness is less than 5 m, the appropriate time (preferably the end of winter and the beginning of spring) should be selected to design according to the principle of replacement of non-frost heave soil after removal, or pier column foundation should be used to pass through the frozen soil layer, and the end should be located at a certain depth of the non-frozen soil layer. The pier column foundation spacing depends on the pipeline strength. When the permafrost is with low ice content, it can be designed according to the principle of gradual thawing when the thaw settlement deformation meets the requirements of pipeline stability within the design thaw depth. Otherwise, engineering measures should be taken to maintain the stability of pipeline works.

The design principle of maintaining the frozen state of the frozen soil foundation is suitable for large continuous permafrost areas, where the mean annual ground temperature is generally lower than −2 °C, the thickness of permafrost is 60~ >100 m, and the permafrost is with high ice content. The cooling device and thermal insulation measures should be taken to keep the foundation frozen. For buried oil pipeline, except for wraping it with thermal insulation layer, the pipeline bottom should be laid with thermal insulation layer, or thermosyphons should be installed on both sides of the pipeline (the distance from the pipeline wall should not be more than 0.5 m), and they should be inserted obliquely through the thermal insulation layer at the pipeline bottom to the soil under the pipe. Or pier column pile foundation should be adopted, and the depth of the foundation end should meet the stability requirements of the pipeline. For overhead pipeline, the thermosyphon can be inserted into the reinforced concrete pile (or steel pipe concrete pile) to form a thermal pile that can cool the frozen soil foundation.

In the continuous and discontinuous permafrost areas, when the mean annual ground temperature is −0.5 °C ~ −2.0 °C, the thickness of frozen soil is generally less than 60 m, and the permafrost is with low ice content, if the settlement deformation within the controlled frozen soil thaw depth meets the design requirements, it is allowed to design according to the frozen soil thawing principle. However, when the permafrost is with high ice content, it is difficult to maintain the frozen state for the frozen soil foundation by means of only thermal insulation measures, so the thermosyphon should be added to cool the frozen soil foundation.

In addition, attention should be paid to the design of pipeline engineering in areas with undesirable features related to frozen ground, such as aufeis, pingos, thaw slumps, gelifluction etc. In addition to strengthening the engineering geological survey of frozen soil, in principle it is better to "avoid". If it is impossible to avoid, the following conditions should be considered in the design.

1. It should be selected on the Dry and Sunny Gentle Slope
2. The pipeline located in pingo or aufeis section should pass through the section at the edge beyond the lower influential distance, but should be keeping at an appropriate height to avoid selecting route in low-lying section. When the route is located above it, the influence of groundwater on pipeline stability should be considered.
3. The thaw slump and gelifluction sections should pass by its lower edge, and sufficient distance should be kept.
4. In frozen swamp and wetland, the pipeline should be laid on the ground or overhead, and anti-frost heave and anti-thaw settlement measures should be taken. For buried pipeline, special anti-freezing and anti-thaw measures should be taken.
5. Permafrost with high ice content is mostly distributed in the fine grain soil zone on the shady slope with a slope of 4° ~ 8°, and the thick ground ice is rarely distributed on the slope with a slope greater than 16°.
6. The foundation treatment length of transition section (frozen soil and thawing soil, high and low ice content permafrost, high and low temperature permafrost) is generally not less than 20 m.

Therefore, the following principles should be considered in pipeline design.

1. In the bedrock, talik and ice-poor permafrost sections, there is no risk of thaw settlement, due to the good foundation condition. Some thaw settlement deformation will occur in ice-poor permafrost section, which will not cause pipeline safety problems. The "buried + replacement" method can be used for design.
2. In the section with icy or ice-rich permafrost, when the pipeline is buried, the thermal effect of oil transportation causes a certain range of thawing ring around the pipeline. In warm permafrost area, under the influence of 10 °C oil temperature, the pipe bottom has thawed to a depth of 7 m in 50 years (Cold and Arid Regions Environmental and Engineering Research Institute, Chinese Academy of Sciences, et al., 2009) (Fig. 6.11). Estimated with a maximum thaw settlement

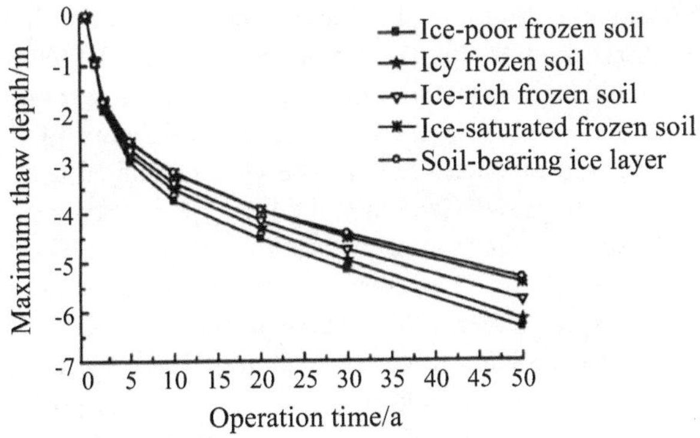

Fig. 6.11 Maximum thaw depth for the engineering types of frozen soil under the Pipeline Bottom (Fu et al., 2012)

coefficient of ice-rich permafrost, which is 10%, the settlement is approximately 0.7 m. When the design of "buried + replacement + geotextile + thermal insulation (i.e. pipeline thermal insulation)" is adopted, the thawing rate under the pipeline bottom is slowed down.

3. In the section with ice-saturated permafrost, based on the maximum thaw settlement coefficient of 25%, the settlement is approximately 1.8 m, and the anti-thaw settlement measures need to be strengthened. When the design of "burial + replacement + geotextile + thermal insulation (pipe bottom insulation board) + thermal insulation" is adopted, the thaw depth under the pipeline bottom could be reduced by about 30%.
4. The frozen swamp is often a permafrost area with high ice content. Its thaw settlement depends on the thickness of the ice layer. According to the thaw settlement of permafrost, its thawing settlement coefficient is higher than 25%, and its thaw settlement is more than 1.8 m. When the design of "buried + replacement + geotextile + thermal insulation (pipe bottom insulation board) + thermal insulation" is adopted, the thaw settlement is still large though the thaw depth under the pipeline bottom is reduced. If the thickness of the ice layer containing soil at the permafrost table is not large, the pipeline can be deeply buried after the ice layer containing soil is removed and replaced with non-frost heave soil, to achieve the effect of anti-thaw settlement. If the thickness of the ice layer containing soil is large, the cooling measures such as thermosyphon should be added or the pile pier should be taken with anti-freezing measures.

6.3 Pipeline Laying

Generally, there are three modes of pipeline laying: being buried, laying on ground and being overhead. The laying mode of the trunk pipeline in permafrost region depends on the characteristics of permafrost in the section or region and the annual average temperature of the material transported within the pipeline.

Based on the temperature of the pipeline transported material, it can be divided into different sections: high-temperature pipeline section—the temperature of the transported material is high throughout the year (sometimes up to 40 °C ~ 70 °C); warm pipeline section—only the annual average temperature of the transported material is positive; cold pipeline section—the average annual temperature of the transported material is negative. Generally, the first type occurs in oil pipelines and gas pipelines with short distance after pressurization, and the rest two types only appear in gas pipelines. There are two cases of cold pipelines. One is that the temperature of the transported material is negative all the year round (it may occur in Siberia, Russia, but it is unlikely to occur in China). The other is that the average annual temperature of the transported material is negative or positive, but it may have a positive value or close to 0 °C at a certain time in a year. According to the characteristics of permafrost and their ground temperature in different sections of the pipeline, their burial modes should also be different (Table 6.2).

Table 6.2 Trunk pipeline laying modes in permafrost regions

Type of pipeline section	Classification of foundation complexity		
	I	II	III
High temperature pipeline section	Being overhead	Laying on ground or being overhead	Being buried
Warm pipeline section	Laying on ground	Being buried or laying on ground	Being buried
Cold pipeline section	Being buried	Being buried	Laying on ground

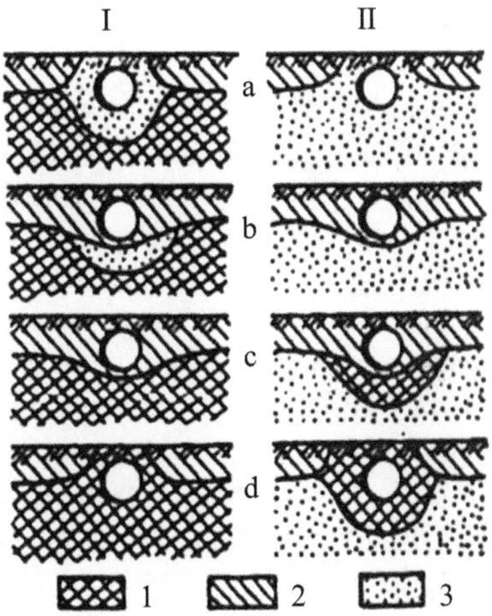

Fig. 6.12 Thermal interaction between buried pipeline and soil(Э.Д. Ye, 2010) (T_{pr}—Annual average temperature; T_{max}—Maximum temperature; T_{min}—Different combinations of minimum temperature) I—Permafrost region; II—Talik area. 1—Permafrost; 2—Seasonally frozen soil; 3—Thawed soil

In permafrost regions, pipelines with different temperatures of transported material have different thermal interactions with permafrost foundation (Fig. 6.12). When the annual average temperature of the transported material within pipeline is with positive or high temperature, the buried pipeline thaws the permafrost foundation, and large thaw rings form in thick ground ice sections (Fig. 6.12a), and both the foundation soil and pipeline will subside, causing deformation of the pipeline. Although the average annual temperature of the transported material is positive, when the minimum temperature in winter reaches a negative value, a seasonally frozen layer (Fig. 6.12b) may form around the pipeline, which is likely to lead to frost heave in the pipeline. When the annual average temperature of the transported material within the pipeline is negative, it cools the foundation soil and forms a permafrost layer. Even if the maximum temperature during the warm season is positive, a seasonally frozen layer may still form in winter (Fig. 6.12c), causing frost heave deformation of the pipeline. If the pipeline transported material and the maximum temperature are both negative, the permafrost layer will be formed and the ground temperature of the permafrost layer will be reduced (Fig. 6.12d), which is conducive to maintaining the frozen state of the permafrost foundation, but may cause frost heave.

The laying mode of being buried is widely adopted in the design of long distance oil and gas pipelines in China. The laying of being buried can better avoid pipeline fracture, reduce gas pipeline accidents, prevent forest fires, cause no obstacles to surface water runoff, and have no impact on wildlife migration, but it is subject to the impact of frozen soil conditions, which may cause pipeline freezing and thawing diseases. It shows in Fig. 6.12 that in warm permafrost regions with high ice content, especially in the thick ground ice sections, such as in complex sections (Class I foundation in Table 6.2), if the high-temperature pipeline is buried, a large thawing ring will form around and at the lower part of the pipeline, resulting in the forming of thawing caves. In this circumstance, super strong insulation and cooling measures should be taken (Fig. 6.13). In permafrost areas with high ice content, the high-temperature (heating) pipes are usually laid overhead rather than buried. Figure 6.14 shows the Alaska overhead oil pipeline with thermosyphons crossing the high temperature and high ice content permafrost zone.

In the 1970s, 112,000 thermosyphons (thermosyphons + steel pipe piles) had been used in the 1200 km oil pipeline project in Alaska, USA. The diameter of the thermosyphon was 50 ~ 750 mm, the length was 8 ~ 18 m, and the working fluid is ammonia. After the installation of the thermosyphons, the temperature of the pipeline support and the permafrost at 6 m deep dropped rapidly. By the beginning of January 1975, the ground temperature reached −24 °C, rose to −7.0 °C in April, and remained below 0 °C in summer. It has been in operation for more than 30 years.

In 1976, EPS thermal insulation layer was also used in the 120 km coarse grained soil cushion of the oil pipeline in Alaska, with a thickness of 100 mm, reducing the thaw depth and achieving satisfactory results.

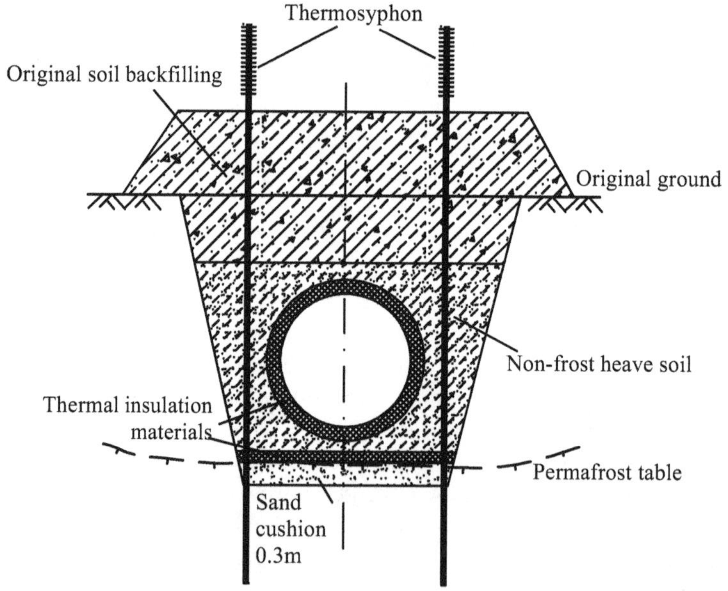

Fig. 6.13 Buried pipeline with cooling and insulation measures

Fig. 6.14 The Alaska overhead oil pipeline (the left picture shows thermosyphons) crosses the warm permafrost with high ice content in permafrost section (photo by Jin Huijun)

6.3.1 Buried Pipeline

When there is no need to consider the longitudinal deformation or building the compensation section of underground pipeline, the pipeline can be designed as buried laying. The buried depth of the pipeline should be determined according to the terrain, geology and landform, groundwater level, characteristics of permafrost, external load and pipeline stability. According to the specification, the thickness of soil above the top of the pipeline should not be less than 0.8 m.

The bottom width of the pipe trench should be determined according to the external diameter of the pipeline, the number of pipelines in a same trench, the excavation method and the working season, the assembly and welding process, the frozen soil engineering geological conditions and other factors.

1. When the trench depth is less than 5 m, the trench bottom width B is calculated as the following.

$$B = D_0 + b \tag{6.1}$$

where D_0 is the outer diameter of the steel pipeline structure, m. When multiple pipelines are laid in a same trench, D_0 is the sum of the outer diameter of each pipeline plus the net distance between pipelines; b is the width of the trench bottom and the margins, m. The value should be taken according to the references in Table 6.3 (the characteristics of permafrost should be considered when using in the frozen soil area).

2. When the trench depth is equal to or greater than 5 m, it should be determined according to the soil layer category and its physical and mechanical properties.

In permafrost regions, the stability of the permafrost layer is related to permafrost table depth, ice content, construction season, and groundwater conditions. In warm season, the air temperature, the solar radiation and the groundwater immersion make the trench wall unstable, collapsed or submerged; in cold season, the frozen soil layer and the frozen groundwater (Fig. 6.15) make the strength of the frozen soil

Table 6.3 Width of the trench bottom and the margins /m (GB 50253-2014)

Conditional factors		Welding in trench				Manual arc welding under trench				Welding elbow under the trench and pipeline trench at the joint
		Soil pipe trench		Rock blasting pipeline trench	Trenches at hot bends and cold bends	Soil pipeline trench		Rock blasting pipeline trench	Pipe trench for semi-automatic welding under trench	
		Trench with water	Trench without water			Trench with water	Trench without water			
b value	Trench depth within 3 m	0.7	0.5	0.9	1.5	1.0	0.8	0.9	1.6	2.0
	Trench depth 3 m ~ 5 m	0.9	0.7	1.1	1.5	1.2	1.0	1.1	1.6	2.0

Fig. 6.15 Ice formation in water flooded pipeline trench

Fig. 6.16 Pipeline trench excavation by drilling and loose blasting (Both photos by Tong Changjiang)

layer reaches dozens of times of that in thawing state, all of these circumstances make it difficult to excavate (Fig. 6.16).

Frozen clay and sandy soil usually contain different degrees of ground ice. In case of ice-poor permafrost, the pipeline should be well waterproofed (usually coated with asphalt). In case of icy and ice-rich permafrost, the pipeline should be wrapped with thermal insulation layer, and the waterproof layer should be painted after being wrapped with plastic outside. In case of ice-saturated permafrost and ice layer containing soil, a 0.2 m sand cushion should be laid at the bottom of the pipeline trench. After another layer of thermal insulation layer is laid, the pipelines with thermal insulation layer and waterproof treatment should be placed on it. Non-frost heave soil should be backfilled around the pipe. If necessary, thermosyphons should be added to cool the frozen soil layer around the pipeline and the permafrost under the pipeline bottom, and drainage measures should be taken to maintain the frozen state of the permafrost foundation (Fig. 6.13). The frozen clay with high ice content is suitable for pipelines in warm and cold sections, but not for pipelines in high temperature (heating) sections.

Massive gravel soils mostly belong to permafrost with low ice content. When laying the pipeline, in order to avoid the pipeline thermal insulation layer being punctured, 0.2 m sand cushion should be laid at the bottom of the pipeline, 0.3 m plain soil (excluding rubbles) or sand should be laid around the pipeline, and then the non-frost heave soil should be backfilled synchronously. 0.3 m of plain soil or sandy soil should be kept at the top of the pipeline before being backfilled with large pieces of gravelly soil or gravel soil (Fig. 6.17). The warm section, cold section and high temperature (heating) section of the pipeline are suitable for this treatment.

Buried pipelines are set in continuous permafrost area, suitable for transporting warm or cold materials. The positive temperature material must be transported in insulated pipeline or pipeline added with cooling device (thermosyphon), instead of bare pipe. The insulation layer and thermosyphon can reduce the temperature of the permafrost foundation, reduce the thawing depth of the permafrost, and meet the requirements for the allowable settlement of the pipeline. In permafrost area with high ice content, especially in fine-grained soil with thick ground ice, high temperature (heating) pipelines will not only cause the permafrost foundation to thaw, but also form slurry suspension after the thick layer of ground ice thaws, causing the pipeline to float and lose its stability. Of course, in discontinuous permafrost area, the rise of groundwater level may float the pipeline, especially in winter, the

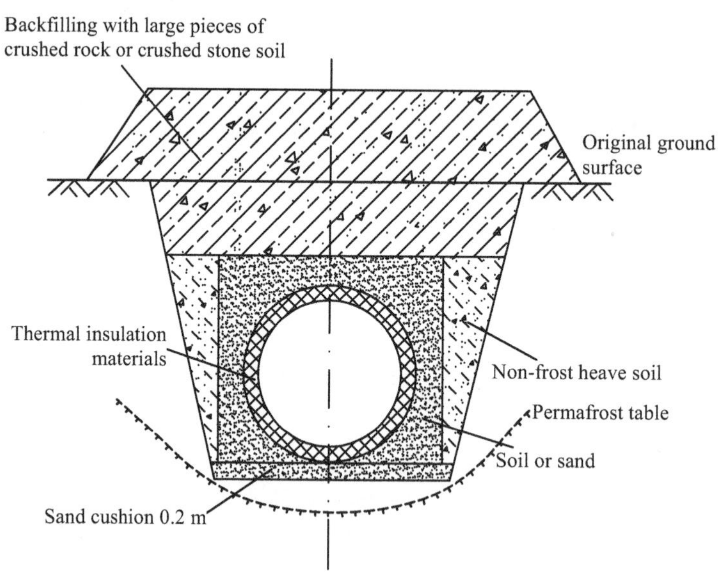

Fig. 6.17 Thermal insulation pipeline of large pieces of crushed stone soil

seasonal freezing depth gradually increases, and the groundwater may change from non-pressure water to confined water, forming a confined swelling water dome (swelling water dome is common in the Greater Hinggan Mountains), which will lift the pipeline. The cold pipeline with transported material with negative temperature is suitable for the permafrost foundation to keep the frozen state. However, with the cooling effect of the pipeline, a freezing ring is formed around the pipeline, and soil moisture or groundwater will migrate to the periphery of the cold pipeline and accumulate ice (Fig. 6.18), forming ice layers of different thicknesses, which will lead to frost heave effect on the pipeline.

In continuous permafrost, buried pipelines are used to transport high-temperature (warming up) materials. In addition to pipeline insulation and laying thermal insulation layer at the bottom of the pipeline, thermosyphons should also be added (Fig. 6.13). Or, the buried pile foundation (steel pile or reinforced concrete pile) pier and abutment can be used to fix the insulation pipe on the pier and abutment, and it is needed to keep the end of the pile foundation pier and abutment exceed the calculated maximum thaw depth (Fig. 6.19). In order to reduce the length of pier and abutment piles, thermosyphons can be inserted into reinforced concrete piles to form thermal piles and reduce the thaw depth of permafrost.

Bare pipelines are only suitable for laying in talik, bedrock areas or in the ice-poor permafrost area in soil with large pieces of gravel. They are not suitable for the permafrost with high ice content such as icy and ice-rich permafrost.

The advantages of the buried pipeline are briefly described, but it also has disadvantages. First, it is difficult to find and deal with accidents; secondly, it is difficult to predict the thermal effect of the pipeline on the permafrost foundation, because the oil temperature within the pipeline is not constant during transportation; third, the excavation workload is large, and the excavation in frozen soil and swamp often brings many difficulties.

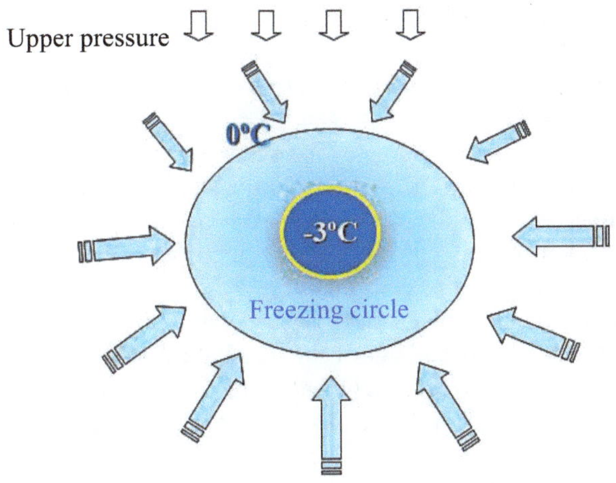

Fig. 6.18 Freezing and icing of cold pipeline (Jin Huijun)

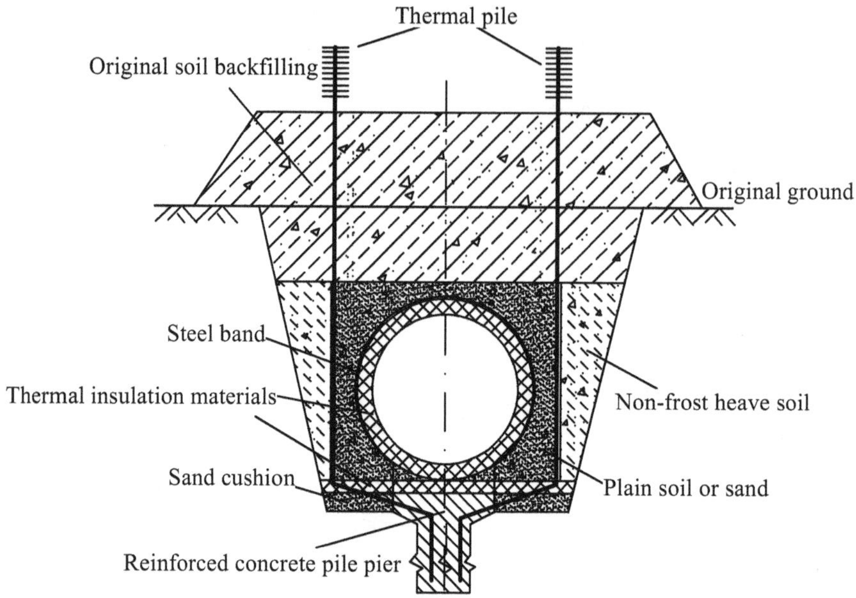

Fig. 6.19 Reinforced concrete pile abutment (or thermal pile abutment)

6.3.2 Pipeline on the Ground

Pipeline on the ground is usually used in areas with well-developed swamp or areas with intricate topographic changes. In permafrost areas, pipeline on the ground is mainly laid in frozen swamps or wetland with high temperature and high ice content. Pipeline on the ground can be classified into the embankment type and the non-embankment type. The former is buried the pipeline in the embankment, and the latter is laying the pipeline on the ground cushion (Fig. 6.20), and it can be made as the longitudinal deformation compensatory pipeline.

The embankment type pipeline, like the buried pipeline, has no deformation compensation, however, it can be built into a compensated section. The embankment type pipeline in permafrost region is similar to the road embankment project. In addition to increasing the heat absorption, an artificial heat source is added, which has a great impact on the surface heat exchange conditions of the permafrost foundation. It may increase the thawing effect on the permafrost foundation, form a deeper talik, and cause pipeline deformation.

In construction of the embankment type pipeline, the bottom of the pipeline should be filled with 0.1 ~ 0.2 m sand cushion, and the pipeline embankment should be filled with non-frost heave soil. The height of the pipeline embankment should be 0.5 m or higher than the top of the pipeline, the width of the top should be more than 1.5 m or no less than 1.5 times of the pipeline diameter, and the slope gradient should be no less than 1:1.5. In continuous permafrost, the sand cushion at the

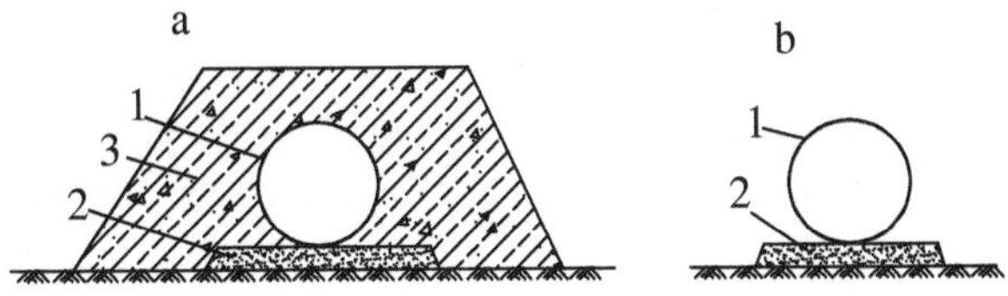

Fig. 6.20 Ground laying main pipeline. A—Embankment type; B—Non-embankment type; 1—Pipeline; 2—Sand cushion; 3—Local soil

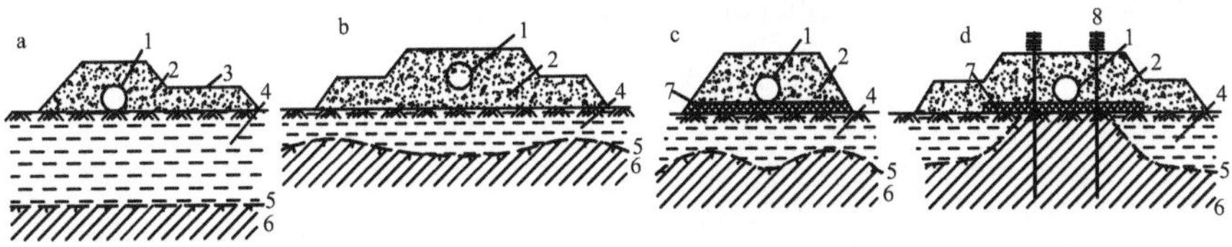

Fig. 6.21 Schematic diagram of ground laying of trunk pipeline (Yershov, 2016). a—Cold and warm pipeline (Class III site); b—Warm pipeline (Class II site); c—High temperature pipeline (Class II site); d-Warm pipeline (Class I and II sites). 1—Pipeline; 2—Filling; 3—Supplementary filling profile in stable state after settlement; 4—Thawed soil under the base; 5—Permafrost table; 6—Permafrost layer; 7—Insulation layer; 8— Thermosyphon

pipeline bottom is directly laid on the turf or peat layer, the slope can be covered with the turf layer, and the top of the cold pipeline can be covered with the peat layer. Warm and high-temperature (heating) pipelines can be laid on the ground in permafrost regions in four ways. ① In bedrock, talik, or coarse and ice-poor permafrost, they can be directly laid on the ground. Even if the soil layer at the bottom of the pipeline thaws, its thaw settlement can still meet the deformation requirements of the pipeline (Fig. 6.21a). ② In a simple site with ice-poor and icy permafrost, between the pipeline bottom and the original ground surface, it should be filled with a certain thickness of filler, and the pipeline embankment slope should be filled with berms to reduce the thermal impact of insulated pipelines (Fig. 6.21b). ③ For class II complex sites with ice rich permafrost, the insulation layer is laid on the natural ground surface and then the insulated pipeline is laid on it to reduce the thaw depth of the pipeline base (Fig. 6.21c). ④ In class I complex sections with ice-saturated permafrost and ice layer containing soil, in addition to laying insulation layer and building berm on the original ground surface, the pipeline should be wrapped with insulation layer of sufficient thickness (determined by calculation), and cooling devices (thermosyphons) should be set at the places no more than 0.5 m on both sides to reduce the ground temperature of the permafrost under the pipeline base and raise the permafrost table as much as possible (Fig. 6.21d).

It is worth noting that the surface water on both sides of the embankment type pipeline should be drained without ponding. Drainage culverts or pipelines should be set according to the natural conditions. It is not advisable to forcibly change the surface water runoff conditions. If necessary, a water retaining twist should be set above the pipeline, but the surface water must be led into the culvert or drainage pipe, and should not flow into the toe of the pipeline embankment. The engineering practice has proved that the accumulated water on both sides of the pipeline on the ground will have a thermal impact on the permafrost under the pipeline base, increase the thawing depth of the permafrost foundation and soften the foundation soil, resulting in thaw settlement and affecting the stability of the pipeline.

The main advantage of the pipeline on the ground is that it avoids heavy excavation, and it is easy to find the accident site. Its disadvantage is that the volume of earthwork is large, and it is easy to be impacted by external forces, and it also affects and blocks the surface water flow, hinders the migration of wild animals, and increases the additional thermal effect on the permafrost foundation.

6.3.3 Overhead Pipeline

There are low overhead, high overhead and swinging overhead laying methods for overhead pipelines. Excluding the flood plain with flowing ice, the overhead pipeline can be used in any environment, especially in the frozen swamp or wetland with high temperature and high ice content permafrost. Common pipeline support piles (piers) include reinforced concrete, metal, earth embankment, gabion, etc. Reinforced concrete piles and metal piles are used for high support, and thermal piles are used in the high temperature and high ice content permafrost zone, that is, thermosyphons are inserted into reinforced concrete piles and metal piles (Figs. 6.22 and 6.23). The piles mainly bear pipeline loads, and the thermosyphons are used to cool the permafrost foundation and keep it in a frozen state. Due to the high temperature and heating of the high support pipelines, there is still a large amount of heat transferred into the permafrost through the support piles, which raises the temperature and reduces the freezing force (or freezing strength) between the permafrost and the pile foundation, that is, reduces the stability of the high-support pile.

High support pipelines are mostly laid at the residential areas and intersections of the transport highways. High support pipelines are also required in the wilderness and wildlife migration areas. The height of high support is generally 4.5 ~ 5.5 m, which may be higher in special sections. Most pipeline works adopt double piles and cross beam structure (Fig. 6.23), which has good stability but high in cost. Single pile support is also used to increase the depth of pile foundation to maintain the stability of overhead pipeline. No matter with single pile or double piles, the anti-frost heave measures of pile foundation must be strengthened within the active layer, otherwise, under the action of tangential and normal frost heave forces, the pile foundation will be frost jacked (Figs. 6.1c and 6.24), affecting the stability of overhead pipelines. In the construction process, in addition to keeping the side surface of the pile column smooth and clean within the seasonally active layer, the side surface should adopt anti-frost heave measures, such as chemical treatment. It is absolutely not allowed to pile concrete on the side surface of the pile column foundation, or it will form an enlarged foot and generate normal frost heave force.

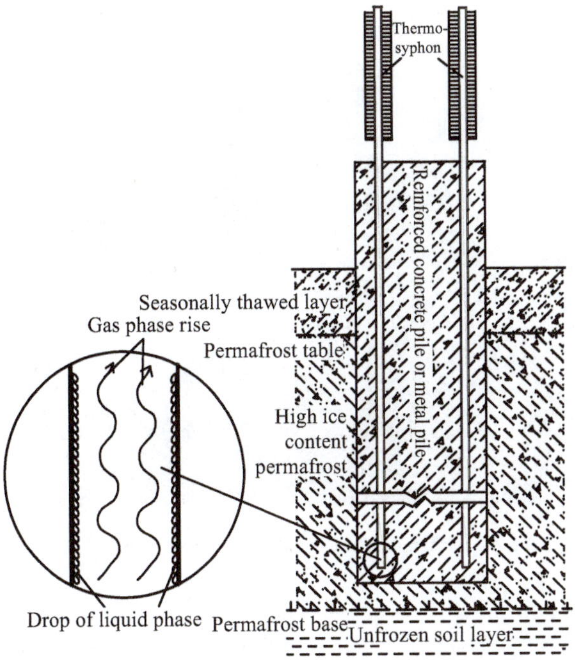

Fig. 6.22 High support thermosyphon pile

Fig. 6.23 Alaska high support oil pipeline (photo by Jin Huijun)

Fig. 6.24 Frost jacking of Russian overhead pipeline (photo by Wang Enliang)

The low support pipeline is mainly laid in the open area outside the industrial blocks and residential areas. The height of the low support is generally 0.5 ~ 1.5 m, for keeping the ventilation effect under the pipeline and prevent the heat of high temperature or heating pipeline from entering the permafrost foundation (Fig. 6.25). The low supports are usually built on earth embankments with coarse grained soil. The concrete slab foundation of the support is laid at the upper part of the embankment, and the insulation layer is laid below the slab, and then the pipeline supports are built. Support forms include single support and double supports (Fig. 6.26).

Movable supports have various types which are longitudinally movable (only along the pipeline axis), freely movable (both vertically and horizontally along the pipeline axis) and swinging. Movable supports are used to offset the thermal deformation of the pipeline, and movable plates, rollers, pulleys and suspension devices are used to ensure the movement of the support. The swinging support (Fig. 6.27) is generally applicable to the laying of pipelines with smaller diameters and pipelines in seismic zones. Figure 6.28 shows that in Alaska, a lateral sliding support was employed to make the pipeline passed through the seismic zone.

The overhead pipeline should be designed to compensate the thermal deformation of the pipeline. The pipeline generally are laid as "Z" shaped, "Γ" shaped or "V" shaped, in a straight line or in a folded line, with different support spacing.

The biggest advantage of overhead pipeline is that it can adapt to the high temperature and high ice content frozen swamp or wetland, and the pipeline has little impact on the permafrost foundation. It is convenient for maintenance and reduces the accident probability, costing less to eliminate accidents. It does not affect surface water runoff and animal migration. The disadvantage of it is that the cost is high, and it is relatively fragile under the impact of external force. When the

Fig. 6.25 Russian low overhead oil pipeline (photo by Jin Huijun)

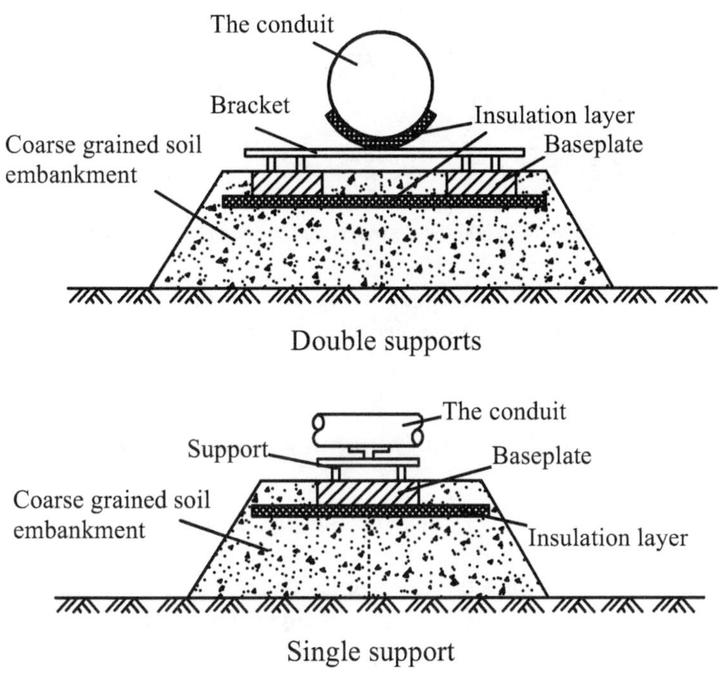

Fig. 6.26 Low overhead pipeline support

Fig. 6.27 Schematic diagram of swinging support

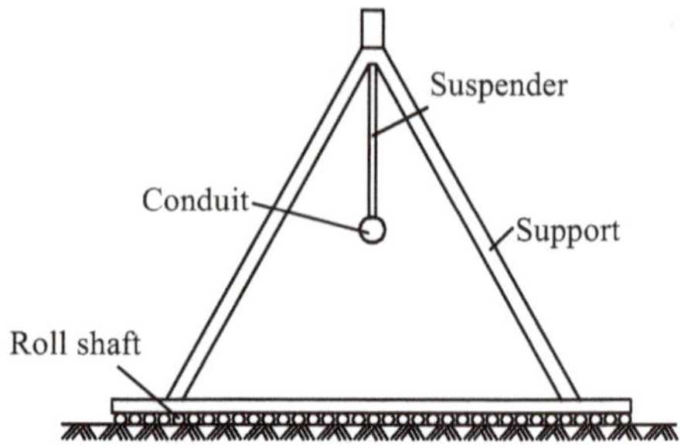

Fig. 6.28 Movable low overhead oil pipeline in Alaska seismic belt (photo by Jin Huijun)

anti-freezing measures of pile column are dysfunctional, it could be frost jacked, which affects the stability of the pipeline.

6.4 Freezing and Thawing Dynamics of Pipeline Foundation

During the operation of buried warm or heating pipelines, the surrounding soil temperature will rise, the permafrost foundation will be warmed and then thawed, which leads to the subsidence of the pipelines. According to the designed oil temperature condition (Fig. 6.29), there will be no frost heave problem in the cold season of the Mohe-Daqing oil pipeline, and the problems are mainly caused by thaw settlement. According to the prediction of the variations of the thawing depth under the condition of 3 °C ~ 10 °C of the inlet oil temperature (Table 6.4), the maximum thaw depth around the bare pipeline in 50 years is 9.1 m, and the maximum thawing depth under the pipeline bottom is 5.6 m, which is greater than the maximum thaw depth under the pipeline bottom with the initial design oil temperature. However,

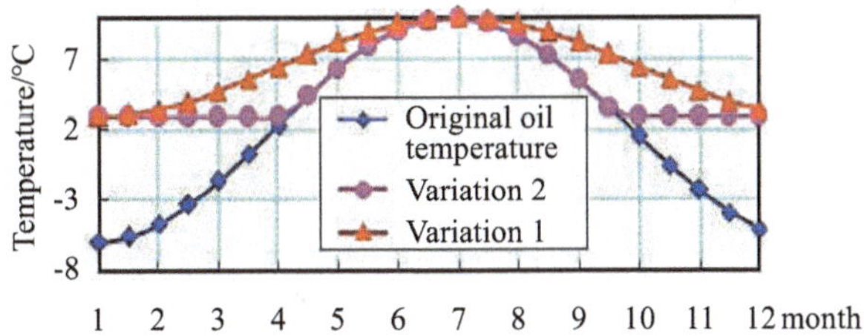

Fig. 6.29 Variations of inlet oil temperature regardless of phase transition (Cold and Arid Regions Environmental and Engineering Research Institute, 2009)

Table 6.4 Comparison of maximum thaw depth/m under the bottom of pipeline in high temperature permafrost with varying inlet oil temperatures of 0 km in 30 and 50 years (Cold and Arid Regions Environmental and Engineering Research Institute, Chinese Academy of Sciences, 2009)

Operating life	Condition of pipe bottom	Oil temperature conditions	Ice-poor frozen soil	Ice medium permafrost	Ice rich permafrost	Ice Saturated permafrost	Ice layer containing soil
Thirty years	Bare pipe	Original	3.5	2.7	2.2	1.7	1.3
		Variation 1	7.1	6.3	5.7	5.0	4.2
		Variation 2	6.4	5.7	5.0	4.3	3.9
	Insulated pipe	Original	2.2	1.6	1.0	0.4	0.2
		Variation 1	3.8	3.1	2.5	2.0	1.5
		Variation 2	3.5	2.8	2.1	1.5	1.1
Fifty years	Bare pipe	Original	4.7	3.7	2.5	2.0	1.4
		Variation 1	9.1	8.1	7.1	6.0	5.2
		Variation 2	8.2	7.2	6.1	5.2	4.7
	Insulated pipe	Original	4.0	2.9	2.0	1.2	0.5
		Variation 1	5.6	4.6	3.5	2.7	2.0
		Variation 2	5.2	4.2	3.3	2.3	1.6

Note: 1. The pipe diameter is 914 mm; The thickness of insulation layer is 80 mm; Transmission capacity 1500×10^4 t/a
2. The maximum oil temperature is 10 °C, and the minimum oil temperature is −6 °C
3. Variation 1: the maximum oil temperature is 10 °C, and the minimum oil temperature is 3 °C. The numerical calculation is carried out according to the change of sinusoidal law
4. Variation 2: the maximum oil temperature is 10 °C, the minimum oil temperature is −6 °C, but when it is lower than 3 °C, the numerical calculation should be conducted as per 3 °C

the actual deformation calculated according to 2 ~ 3 years of pipeline operation is greater than the predicted thaw depth under the pipeline bottom.

During the operation of cold buried pipeline, freezing ring and ice accumulation will be formed around the pipeline. When the inlet oil temperature is −6.41 °C ~ 3.65 °C, the freezing depth of bare pipe bottom is large (Table 6.5) regardless of the friction heat and pump temperature rise, which should be noted, but the use of insulated pipelines can effectively reduce the freezing depth. With the increase of kilometers away from the initial station, and taking account of the pump temperature rise and friction heat, the freezing depth is decreasing.

Table 6.5 Maximum freeze depth (m) under the pipeline bottom along the pipeline in the talik at the inlet oil temperature of −6.41 ~ 3.65 °C (Cold and Arid Regions Environmental and Engineering Research Institute, Chinese Academy of Sciences, 2009)

Pipe diameter 914 mm, regardless of pump temperature rise and friction heat							Pipe diameter 813 mm, taking account of pump temperature rise and friction heat						
	W = 20%		W = 25%		W = 35%			W = 20%		W = 25%		W = 35%	
Mileage/km	The second year	The 30th year	The second year	The 30th year	The second year	The 30th year	Mileage/km	The second year	The 30th year	The second year	The 30th year	The second year	The 30th year
0	2.18	1.67	1.58	1.33	1.38	1.16	0	1.62	0.95	1.25	0.80	1.21	0.75
172	1.95	1.52	1.38	1.18	1.23	1.10	80	1.34	0.92	1.00	0.74	0.88	0.68
313	1.76	1.39	1.25	1.08	1.10	0.96	172	1.26	0.86	0.95	0.69	0.84	0.54
423	1.56	1.25	1.10	0.98	0.98	0.87	240	1.26	0.83	0.90	0.69	0.85	0.53
							313	0.92	0.60	0.69	0.49	0.62	0.45
							403	0.80	0.41	0.61	0.36	0.55	0.34

Note: *W* represents the moisture content of the soil

At the initial operation stage of the Mohe-Daqing oil pipeline, the required maximum temperature of the inlet oil should not be higher than 3.65 °C in warm season, and the required minimum temperature should not be lower than −6.41 °C in cold season. The design of alternating conversion of positive and negative temperatures will help to maintain the stability of the frozen soil foundation. However, this is not the actual situation after operation. In fact, this pipeline is actually operated in positive temperature for both seasons. According to the measured data in 2011, the lowest temperature of Mohe Station was about 2 °C, with a highest of 13.25 °C (Zhang et al., 2014). In 2012, the temperature decreased slightly, but the average temperature still reached 5.30 °C (Fig. 6.30).

According to the measured data of the Mohe-Daqing oil pipeline, the pipeline hydraulic monitoring system is used in the river marsh section (1.5 km long) of ice-rich to ice-saturated permafrost with silty clay, gravel silty clay and gravel sand (Fig. 6.31). From May 2012 to the first 10 days of June 2013, the maximum vertical displacement of the pipeline reached 49.8 cm (Fig. 6.32). According to the

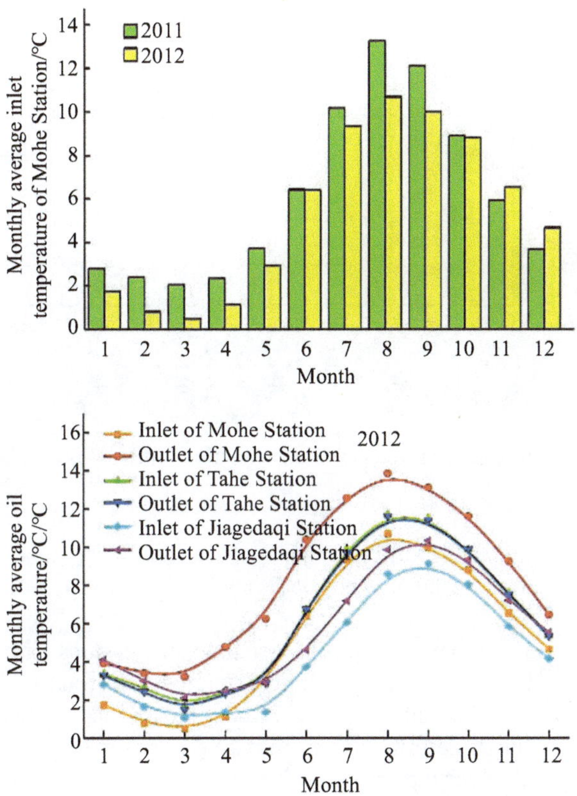

Fig. 6.30 Variations of monthly average inlet oil temperature of Mohe Station (left) and variations of monthly average inlet and outlet oil temperatures of Mohe-Jiagedaqi Station (right) (Feng et al., 2014)

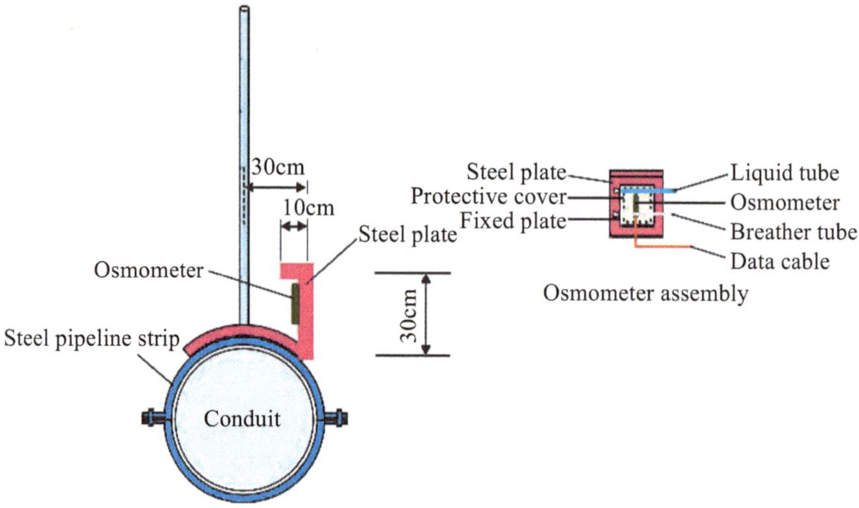

Fig. 6.31 Schematic diagram of pipeline hydraulic monitoring system

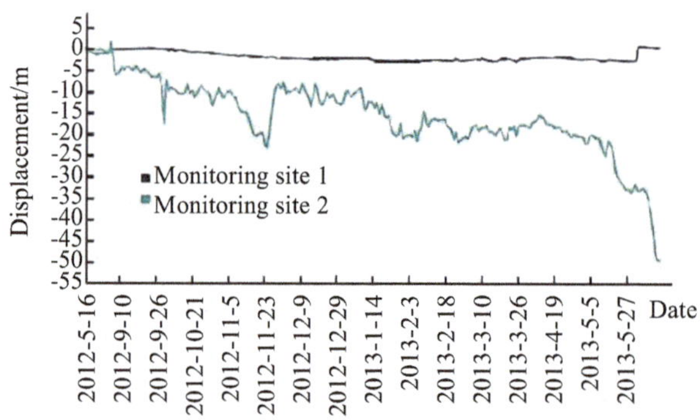

Fig. 6.32 Observation results of vertical displacement of pipeline (Both according to Feng et al., 2014)

maximum thaw settlement coefficient (25%) of ice-saturated permafrost, the thawing depth of permafrost at the pipe bottom is about 2 m after the pipeline started its service for 2–3 years.

The original design scheme of anti-thaw settlement measures for pipeline in frozen swamp or wetland in high-temperature permafrost area was "buried + replacement + staying cold + thermal insulation". Later, the thermal insulation layer was removed, causing large deformation in many sections, indicating that only the staying-cold layer used for pipeline is not enough to prevent thawing caused by oil temperature, and cooling measures must be taken to reduce the thaw depth under the pipeline bottom. The in-situ measurement of the exemplary section of K305 of Mohe-Daqing pipeline where thermosyphons are used to prevent the thaw settlement of pipeline shows that the temperature at the bottom of the pipeline is

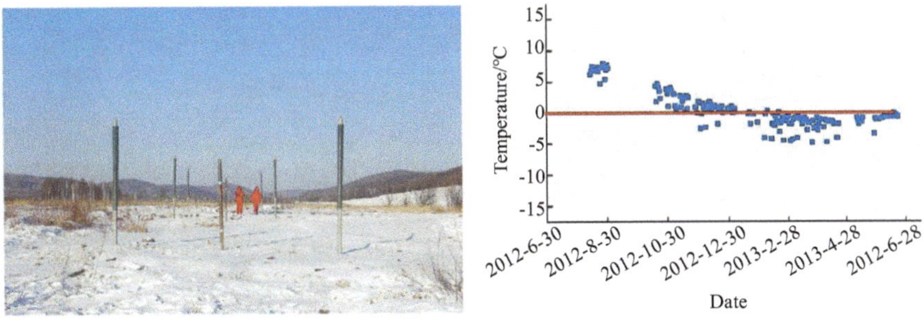

Fig. 6.33 Thaw settlement prevention test of thermosyphon of Mohe-Daqing oil pipeline (left figure) and variations of ground temperature at pipeline bottom with time (right figure) (Feng et al., 2014)

generally between 0 °C and 5 °C during the warm season (September to October), drops to 0 °C in November, and reaches −5 °C as the lowest in late June of the following year (Fig. 6.33), basically fluctuating around 0 °C. At a vertical distance of 2.5 m from the pipeline center, the soil temperature at a depth of 2.5 m in August is 8 °C ~ 10 °C, dropped to about 0 °C in November, and drops to −10 °C ~ −8 °C in March of the next year, and rises to 0 °C in May. If the pipeline bottom is laid on an insulating layer, the stability of the permafrost foundation under the pipeline bottom can be maintained.

According to the actual oil temperature in operation of the Mohe-Daqing pipeline, the thermal regime of the soil in 50 years after the completion of the pipeline is calculated by using the finite element analysis (Zhang, 2014).

1. When the buried depth of the pipeline top is 2.0 m and there is no thermal insulation layer, with an average oil temperature of 10 °C, the disturbance to the low-temperature permafrost area is small, and the permafrost has basically reached the equilibrium thermal state since the 10th year of operation. However, the high-temperature permafrost area is strongly disturbed (Fig. 6.34). The maximum thaw depth of high and low temperature permafrost in the 10th year is 5.12 m and 4.35 m, with a difference of 0.77 m; and the maximum thaw depth in the 50th year is 6.17 m and 4.71 m, with a difference of 1.46 m. Because the mean annual ground temperature of high temperature permafrost is high, it is very easy to lose its thermal stability when disturbed by external heat. At this time, thermal insulation materials should be laid or cooling measures (thermosyphon) should be taken depending on the circumstances to reduce the thaw depth of the permafrost under the bottom of the pipeline. If the permafrost foundation is of ice-poor or icy, the thawing design principle can be adopted.
2. Laying thermal insulation materials can effectively reduce the maximum thaw depth of the permafrost. The greater the thickness of the thermal insulation layer, the smaller the disturbance of the pipeline to the permafrost (Fig. 6.35). However, it doesn't mean the thicker the better, as the thaw depth decreases slightly when the thickness of the thermal insulation layer is greater than 8 cm (Fig. 6.36).

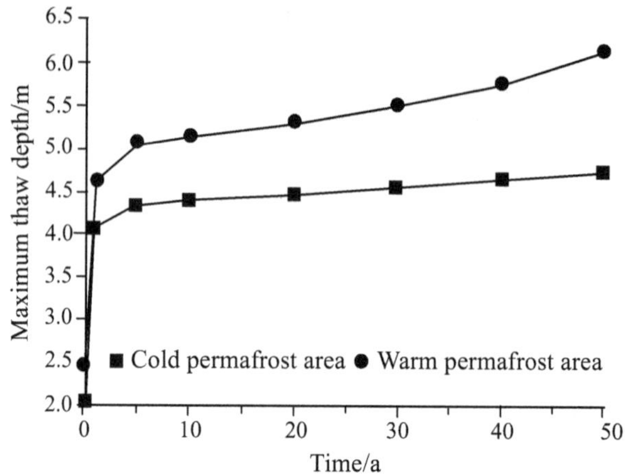

Fig. 6.34 Variations of maximum thaw depth of warm and cold permafrost under the pipeline bottom

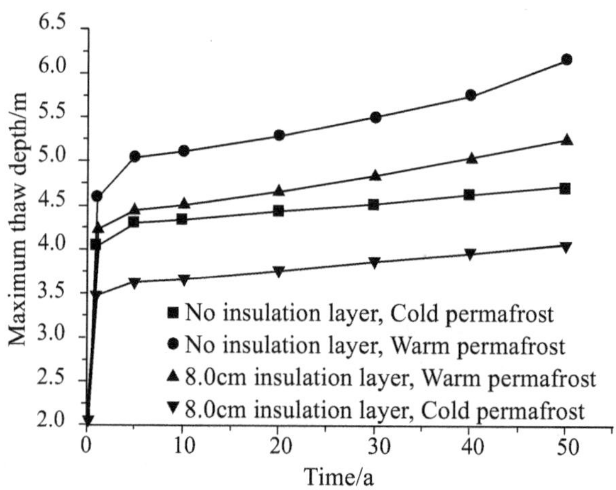

Fig. 6.35 Maximum thaw depth under the pipeline bottom with or without insulation layer (Both according to Zhang, 2014)

3. The average temperature of the transported oil has a great influence on the distribution of the thermal regime of permafrost. With the increase of the average oil temperature, the maximum thaw depth of permafrost gradually increases. When the annual average ground surface temperature is −3.5 °C and without thermal insulation layer, the oil temperature increases from 3 °C to 15 °C. The maximum thaw depth in the 50th year increases from 4.0 m to 4.73 m when the

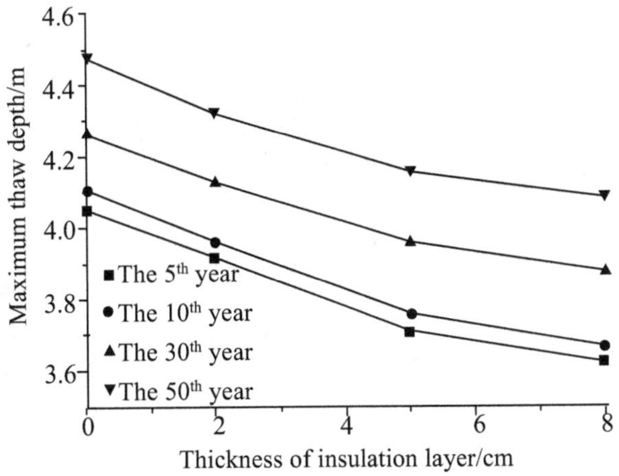

Fig. 6.36 Relationship between the maximum thaw depth under the pipeline bottom and the thickness of insulation layer

buried depth of the pipeline top is 1.5 m. The maximum thaw depth increases from 4.31 m to 5.36 m when the buried depth of the pipeline top is 2.5 m. When the oil temperature is 15 °C the maximum thaw depth increases by 3.41 m after the 50th year.

6.5 Pipeline Crossing Works

For the method of oil and gas pipeline crossing water areas, mountains and gullies, the buried pipeline is typically adopted, using methods such as trenching, horizontal directional drilling and tunnel (including drilling and blasting, shield tunneling, pipeline jacking, etc.).

The Mohe-Daqing oil pipeline employs directional drilling to cross rivers, gullies and highways, such as those crossing the Heilongjiang River and Huma River. In the Heilongjiang directional drilling crossing project, according to the drill core analysis and indoor rock physical and mechanical property test, the optimal depth of the Heilongjiang directional drilling is determined to be 38.0 m (Wu et al., 2011), with a length of 1100 m. The river bed of the Daxing'anling Mountains belongs to the permeating talik. The sunny slope of the river bank is basically exposed bedrock, and the floodplain and terrace on the shady slope may have permafrost. When crossing the highway, it should be selected in the rock foundation, talik or section with ice-poor permafrost. Some sections of permafrost with high ice content should be provided with thaw-prevention measures to avoid thaw settlement caused by pipeline temperature thawing the permafrost under the base.

Fig. 6.37 Wood chip insulation measures for the oil pipeline crossing the mountain slope of the Mohe-Daqing Line (left: Bizhou, right: 63 km of Jiagedaqi-Mo Highway) (photo by Tong Changjiang)

Trenching is adopted for crossing mountains along the Mohe-Daqing oil pipeline. As there is still permafrost on the mountain slope, in order to prevent the instability caused by the thawing of the permafrost on the slope, wood chip insulation measures (Fig. 6.37) are taken in the section where the ground ice in the permafrost layer at the foot of the slope is relatively developed. The wood chip has small thermal conductivity, large pores, and cold air seeps into the bottom to prevent the warm air from coming in, which plays a role in maintaining the relative stability of the permafrost. However, surface water erosion prevention measures should be strengthened. The 63 km slope of Jiagedaqi-Mohe Road is severely eroded by surface water.

6.6 Cathodic Protection

The buried metal pipeline should adopt the joint protective measures of anti-corrosion coating and cathodic protection, which has been proved an effective corrosion control technology. It is specified that the cathodic protection potential of the pipeline should be −850 mV (CSE) or more negative. The cathodic protection potential of the pipeline in the environment of 100 Ω·m to 1000 Ω·m soil electric resistivity should be negative than −750 mV (CSE). The cathodic protection potential of the pipeline in the environment where the soil electric resistivity is greater than 1000 Ω· m should be negative than −650 mV (CSE). It can be seen that the reasonable potential on the pipeline is an important parameter to determine the cathodic protection system, and is the minimum protection potential required to stop the corrosion reaction of the metal pipeline.

Soil electric resistivity is one of the important parameters affecting cathodic protection of buried pipelines. Soil type, structure, water content, temperature and pH are important factors affecting soil electric resistivity. The Mohe-Daqing oil pipeline passes through large continuous and patchy talik, continuous permafrost areas, isolated and sporadic permafrost areas, seasonally frozen soil areas and river taliks. The soil layers within the buried depth range repeatedly undergo the alternating process of freezing and thawing every year. The soil electric resistivity varies repeatedly with freezing and thawing, which must be considered in the effective implementation of pipeline cathodic protection.

Soil changes from unfrozen state to frozen state, transitioned from three-phase system to four-phase system, and the water-ice in frozen soil is in dynamic equilibrium along with the varying of negative temperature, and till −73 °C, there is still unfrozen water. In the process of soil cooling, different cooling rates and directions lead to the migration of water to the freezing front and the formation of different frozen soil structures due to the characteristics of ice accumulation, so that the electric resistivity of frozen soil changes greatly with its structural characteristics. The conductive channels of frozen soil include double layer conductive channels on the surface of soil particles, unfrozen water conductive channels, ice-water mixture conductive channels, ice conductive channels, soil particle-ice-water mixture conductive channels, etc.

Different lithology basically has the same change trend that the electric resistivity increases rapidly with the decrease of temperature. Figure 6.38 shows the test results of three typical soils in Northeast China, and Table 6.6 shows the test results of five kinds of soil in Northwest China. In the thawing state, when the temperature drops from 22 °C to 3 °C, the soil electric resistivity increases by one fold, and the electric resistivity ranges from tens to 6000 Ω· m. In the process of ±0 °C phase change of soil pore water, the duration is long, the liquid water gradually freezes into solid ice, and the soil electric resistivity changes by leaps and bounds, and the value increases sharply (Fig. 6.39 shows a more intuitive curve of soil resistivity from freezing to thawing). The electric resistivity increases with the soil temperature decreasing.

The Mohe-Daqing Line is a buried warm (heat increasing) pipeline. There is a thawing ring around the pipeline, which shrinks or expands in cold or warm seasons, and the permafrost outside the ring changes accordingly. The seasonal variation of frozen soil and thawed soil, sometimes hard to predict their distributions along the pipeline, leads to complex variations in soil electric resistivity around the pipeline, so attention should be paid to select the pipeline cathodic protection method.

Fig. 6.38 Variation curve of electric resistivity of different soils with temperature (Liu, 2015)

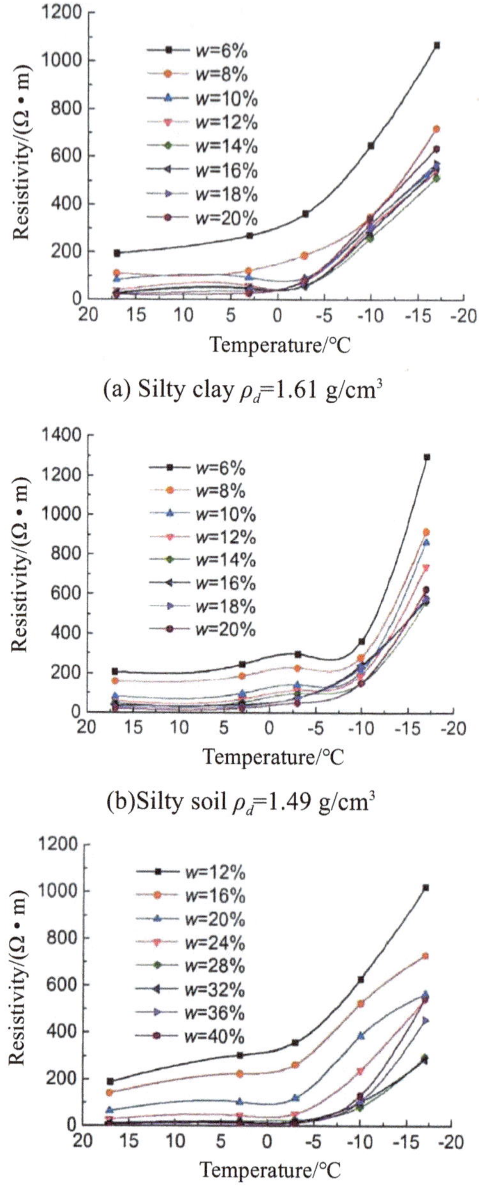

(a) Silty clay ρ_d=1.61 g/cm^3

(b)Silty soil ρ_d=1.49 g/cm^3

(c) Mudstone ρ_d=1.29 g/cm^3

Table 6.6 Soil electric resistivity at different temperatures /Ω· m (Cao et al. 2007)

Soil type	Water content/%	Soil temperature (°C)													
		22	18	13	9	3	1	0_+	0_-	-1	−3	−5	−9	−13	−18
Loess	19.6	67	75	81	102	129	138	145	8020	11,600	26,200	52,400	106,000	164,000	363,000
	10.6	689	991	1060	1200	1440	1550	1570	17,800	18,600	138,000	296,000	591,000	1,030,000	1,380,000
	4.75	2160	2960	3690	4180	5970	22,700	27,600	41,800	75,600	460,000	1,030,000	2,070,000	4,140,000	–
Sandy soil	16.2	52	66	83	131	156	113	120	1312	2162	5800	13,000	42,000	138,000	437,000
	9.36	229	299	355	376	539	543	582	6100	10,000	10,200	26,700	99,000	317,000	793,000
	4.48	751	824	936	1080	1280	2040	4830	7800	13,700	41,200	102,000	317,000	793,000	3,170,000
Gravel soil	17.3	83	111	150	185	283	317	340	1070	1660	4110	9720	37,600	97,300	223,000
	7.83	176	242	310	324	412	471	477	1020	1650	4820	11,300	40,700	99,200	317,000
	3.73	218	258	299	352	470	515	556	8080	9540	14,400	27,400	70,500	198,000	453,000
Sand	11.2	149	219	331	377	505	579	624	24,100	45,300	61,000	79,000	128,000	203,000	–
	9.1	215	267	330	417	593	663	714	9020	11,600	19,200	26,400	54,700	106,000	212,000
	3.52	351	437	518	641	815	881	933	6610	8310	14,600	25,200	55,700	93,300	212,000
Gravel	9.64	400	550	662	792	1000	1360	1490	18,100	35,400	57,000	81,000	132,000	224,000	–
	6.1	820	971	1030	1360	1770	1920	2030	9170	13,100	27,500	40,700	85,000	183,000	366,000
	1.25	895	1130	1590	1720	2180	2370	2480	4830	5930	14,400	27,800	62,700	119,000	297,000

Note: "-" in the table means out of measurement range

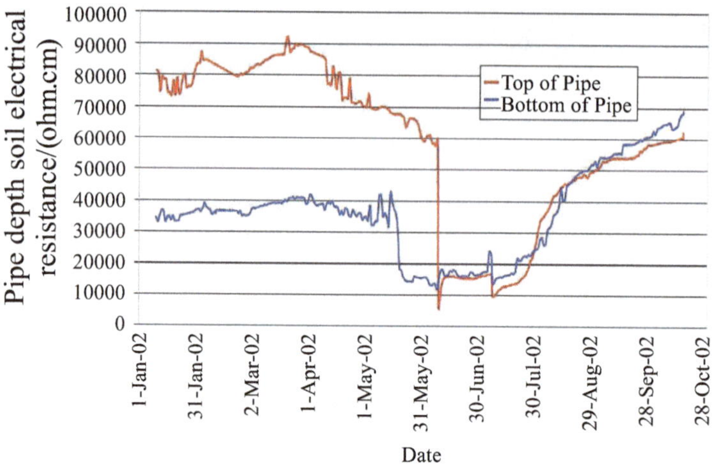

Fig. 6.39 Variation patterns of soil electric resistivity at the top and bottom of an oil pipeline in Canada within 1 year (Hao, 2012)

6.7 Station and Tank Foundation Treatment

6.7.1 Site Selection and Survey

The foundation conditions in permafrost regions are complex. The site selection of pipeline stations, valve chambers, storage tanks and other buildings and structures should avoid high ice content permafrost sections, as well as features related to frozen ground development sections and sections with direct hazards or potential threats to the project.

The valve chamber and other buildings of the pipeline have a relatively small area and lower bearing capacity requirements. The site selection, survey and design can be carried out according to the *"Geological Survey of Frozen Soil Engineering"* for geological survey of frozen soil engineering in construction engineering.

The site selection, survey and design of steel storage tank structures for pipelines should be carried out according to relevant regulations.

Contents of foundation requirements for geotechnical investigation can be specified as: site topography, geological structure, seismic effect, adverse geological phenomena, stratum stratification conditions, physical and mechanical properties of each rock and soil layer, site stability, homogeneity of rock and soil, characteristic value of bearing capacity, compression coefficient, compression modulus, corrosivity of groundwater, soil and water to building materials standard freezing depth of soil and engineering problems possibly caused by engineering construction.

Permafrost foundation is a special foundation. In addition to general foundation requirements, extra survey data should be added, such as the distribution, table and base, mean annual ground temperature, material composition, ice content, frozen soil structure, thickness and distribution of ground ice, physical and mechanical properties, thawing and subsiding properties, distribution, characteristics and causes of talik, cryogenic phenomena of frozen soil, frozen soil engineering types classified by frozen soil ice content, seasonally thawed layer thickness, material composition, water content, ice content, frost heaving, groundwater characteristics, etc.

Compared with the thawed soil foundation, the complexity and heterogeneity of the permafrost foundation are more prominent. In addition to the material composition, it is manifested in the heterogeneity of the ice content, the frozen soil structure, the table variation, and the frost heave and thaw settlement caused by the freezing and thawing process under the thermal impact.

Three characteristics of tank foundation are noted: large diameter, generally 20 ~ 40 m, and large tank diameter up to 80 ~ 100 m; the additional pressure is large, generally up to 150 ~ 300 kN/m^2; the resulting settlement is also large, with the settlement in general soil reaching tens of centimeters, and in soft soil foundation reaching 1~2 m. Due to the height of the tank, the requirements for uneven settlement are higher, especially for floating roof tanks.

The site for storage tank should be selected in the section with flat terrain, good foundation conditions, low groundwater level, homogenous soil and with small differential deformation. However, it is not easy to meet all the above requirements in permafrost regions. Because the distribution of ground ice in permafrost is often heterogeneous, the thaw settlement deformation will be differential after thawing. Considering the safety and stability of the storage tank, it is better to select the section with shallow bedrock burial depth, or the terrace of coarse grain and low ice content frozen soil. Even so, partial sections of terraces and completely or strongly weathered bedrock zones still have different degrees of ice content (horizontal and vertical). The engineering geological survey should know in detail regarding the distribution and characteristics of permafrost and talik under the tank foundation, the composition of permafrost, the type, structure, range and thickness of the frozen soil engineering, the distribution characteristics of ground ice layer, the physical and mechanical properties of frozen soil, the table and base distribution of permafrost and the conditions of groundwater.

Therefore, the engineering survey should be carried out for the storage tank construction site in permafrost area according to the requirement of "complex site". The number and spacing of survey points in the low ice content permafrost area and the section with thin thickness of the permafrost layer should meet the requirements in Table 6.7. For the permafrost section with high ice content and large thickness of permafrost layer, when the tank diameter is greater than 40 m, a special study should be carried out. The determination of the depth of the survey borehole, that is, The determination of the survey borehole depth is based on calculating the deformation values in a soil layer with a thickness of $\triangle Z$ upwards from the depth of deformation calculation. The ratio of the calculated deformation value in this layer to the total calculated deformation value in the depth range of the compression layer is less than or equal to 0.025, which serves as the criterion for controlling the calculation depth of foundation deformation. When there is no influence of adjacent loads, the simplified formula for calculating the deformation depth of the foundation center point is:

$$\Delta Z = b\left(2.5 - 0.4\ln b\right) \tag{6.2}$$

where, b is the width of tank foundation, m.

Table 6.7 Number and layout of exploratory points in detailed exploration stage of tank engineering (GB 50324-2014)

Inner diameter of tank bottom ring Dt (m)	Number of exploration points	Layout of exploration points
$D_t < 20$	1 ~ 3	It can be arranged in the center of the tank or along the periphery
$20 \leq D_t < 30$	3 ~ 5	One in the center of the tank, and the rest are distributed along the periphery
$30 \leq D_t < 40$	5 ~ 7	
$40 \leq D_t < 60$	7 ~ 9	
$60 \leq D_t < 80$	9 ~ 13	One in the center of the tank, and the rest are uniformly distributed in concentric circles
$80 \leq D_t < 100$	13 ~ 17	
$D_t \geq 100$	≥23	

Note: 1. When the geological conditions are complex, the number of exploration points should be increased by one level compared to the values listed in the table;
2. In the table, the number of exploration points with large tank diameter and small tank diameter at the same level should be taken as the large value and the small value respectively

Table 6.8 Depth of general exploratory boreholes/m (GB 50473-2008)

Nominal volume of storage tank (m^3)	General foundation	Soft soil foundation
≤5000	1.0 ~ 1.2 D_t	1.0 ~ 1.5 D_t
10,000	1.0 ~ 1.2 D_t	1.0 ~ 1.5 D_t
20,000 ~ 30,000	0.9 ~ 1.0 D_t	1.0 ~ 1.1 D_t
50,000	0.7 ~ 0.8 D_t	0.8 ~ 0.9 D_t
≥100,000	0.6 ~ 0.7 D_t	0.7 ~ 0.8 D_t

It is determined according to special soil requirements. The depth of control survey borehole should be increased by 10 m compared to the regular depth of general survey borehole, and in case that it is in rock foundation, the depth should be increased by 5 m (Table 6.8). When bedrock is underlying the permafrost within the estimated depth, it should penetrate the permafrost layer and enter the moderately weathered layer of bedrock for 2 ~ 3 m.

6.7.2 *Design Principles and Foundation Treatment*

1. **Design principles**

The design principle of tank foundation in permafrost area can be determined according to the application conditions of the design principle of building foundation. According to the permafrost conditions of the oil storage tank site, the design principle of maintaining the frozen state of the permafrost foundation or the design principle of pre-thawing can be adopted. The design principle of gradual thawing can be adopted for coarse grained soil and bedrock foundation of permafrost with large thickness of permafrost layer, high mean annual ground temperature of permafrost, and low ice content.

As the formation of ice content in permafrost is related to climate, landform, geology, hydrogeology and other conditions, its changes are complex, and the heterogeneity of permafrost foundation increases, affecting the diameter and capacity of single tank and the layout of tank groups. The smaller diameter of a single tank is easy to adapt to the heterogenous foundation conditions, but the tank group occupies a larger area, and the steel consumption per unit volume is also large. According to the data statistics (Xu et al., 1999), the larger the storage tank, the less the total investment, the less the steel consumption per unit volume, and also the smaller the floor area of the tank group general plan. The storage tank capacity of the reasonable general plan is 125,000 m^3. The total volume and number of tanks in a tank group, involving the volume of a single tank, should be determined according to the reasonable selection of the design principles of tank foundation in permafrost regions and the treatment scheme of permafrost foundation.

2. **Permafrost tank foundation and foundation treatment**

At present, the commonly used types of tank foundations are slope protection foundations, ring wall foundations, outer ring wall foundations and pile foundations.

The type selection of tank foundations should be comprehensively considered according to the type, volume, site geological conditions, foundation treatment methods, construction technical conditions and economic rationality of the tank. Tank foundation types applicable to site and geological conditions are the following.

1. When the characteristic value of the bearing capacity of the natural foundation is greater than or equal to the average pressure of the base, the allowable value of foundation deformation meets the provisions of Table 6.8, and the site is unrestricted, the slope protection type tank foundation should be used (Fig. 6.40). Ring wall foundation (Fig. 6.41) or outer ring wall foundation (Fig. 6.42) can also be used. In permafrost regions, the weathered zone in the frozen bedrock and the coarse-grained soil zone with ice-poor and icy permafrost have the characteristics of high bearing capacity and small deformation.
2. When the characteristic value of the bearing capacity of the natural foundation is less than the average pressure of the base, but the foundation deformation meets the allowable value, and can meet the requirements of the bearing capacity after the foundation treatment or water filling preloading, the ring wall type foundation (Fig. 6.41), the outer ring wall type foundation (Fig. 6.42), or the slope protection type tank foundation (Fig. 6.40) should be used. In the weathered zone of frozen bedrock in permafrost region and the crushed stone section of permafrost with ice-poor and icy in dry zone, the bearing capacity of the foundation can be greatly improved after strong vibration rolling, and the uneven deformation can be reduced to meet the design requirements of the foundation.
3. When the characteristic value of the bearing capacity of the natural foundation is less than the average pressure of the foundation base, and the foundation deformation cannot meet the allowable value, and the foundation has liquefied soil layer under the impact of earthquake, the ring wall foundation should be used when the bearing capacity requirements and allowable value requirements can be met after the foundation treatment or water filling preloading, or the degree of elimination of liquefied soil layer meets the relevant regulations (Fig. 6.41). When the foundation treatment is difficult or not treated, the pile foundation should be used (Fig. 6.43). The icy and ice rich permafrost in gravel silty clay layer in permafrost region may produce large differential deformation after

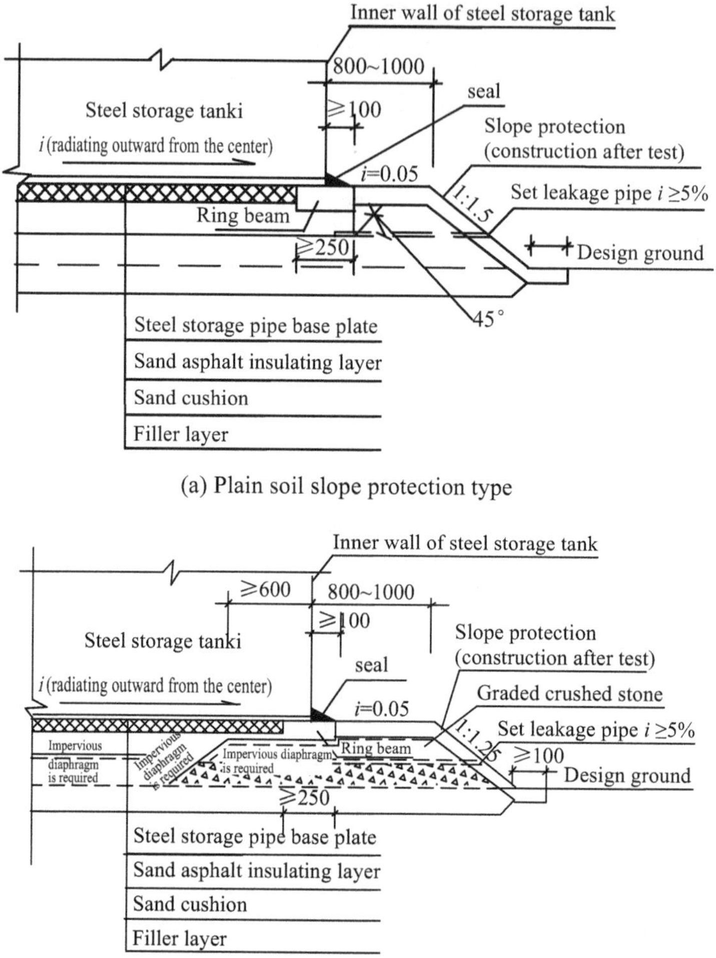

(a) Plain soil slope protection type

(b) Slope protection in gravel environment

Fig. 6.40 Slope protection tank foundation (unit: cm)

thawing. The gravel pile (or pile foundation) is used, and the surface gravel layer is rolled (compacted) with strong vibration to form a gravel cover + impermeable geotextile raft, forming a gravel pile (pile foundation) composite foundation.

4. When the construction site is limited and the tank equipment has special requirements, the ring wall foundation should be used (Fig. 6.41). Due to uneven distribution of ground ice in permafrost regions, uneven deformation of permafrost foundation increases, and storage tank construction site is greatly limited. Foundation treatment measures of pile foundation and concrete raft can be taken to expand the demand for storage tank construction site (Fig. 6.44).

When the site selection of the storage tank cannot avoid the permafrost section with high ice content, the engineering geological survey of permafrost must be strengthened. In addition to the lithological composition of the soil layer, the frozen soil

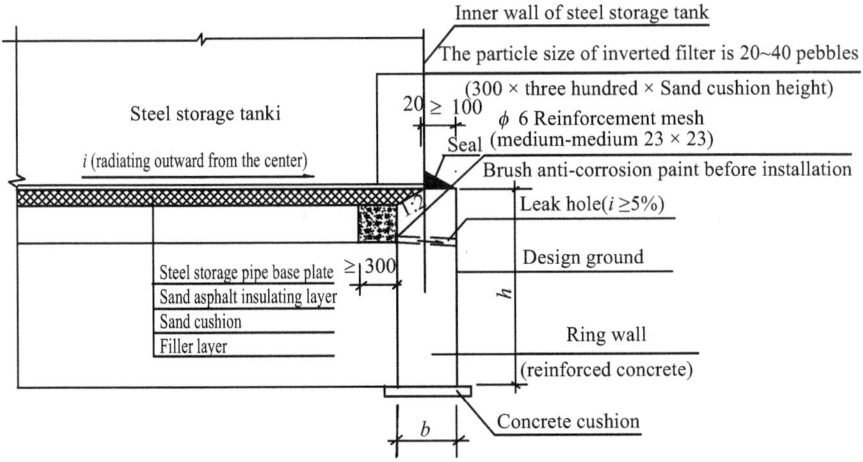

Fig. 6.41 Foundation of ring wall tank (unit: cm)

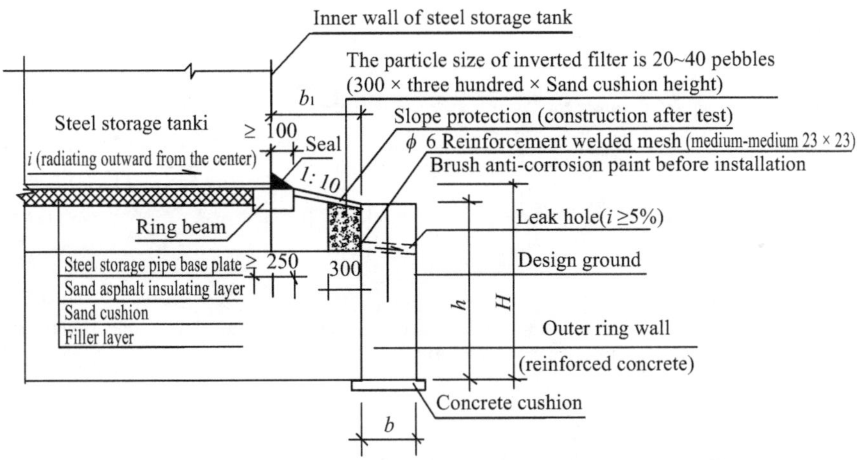

Fig. 6.42 Foundation of outer ring wall tank (unit: cm). b—ring wall thickness/m; h—ring wall height/m; b_1—distance from the inner side of outer ring wall to the inner side of tank wall/m; H—height from tank bottom to the outer ring wall bottom/m

characteristics of the site, the distribution, thickness and the depth of ground ice, and the physical and mechanical properties of the permafrost foundation (especially the plane and vertical distribution of the thawing settlement of the frozen soil foundation) must be known in detail. The foundation should be treated according to the characteristics of frozen soil foundation to form a composite foundation of pile foundation, rubble and ventilation tube (Fig. 6.44). When necessary, reinforced concrete bearing platform plate can be added above the pile foundation, or elevated ventilation pile foundation can be used (Fig. 6.45).

At present, the foundation treatment methods in permafrost regions include replacement method, replacement dynamic compaction, block gravel cushion,

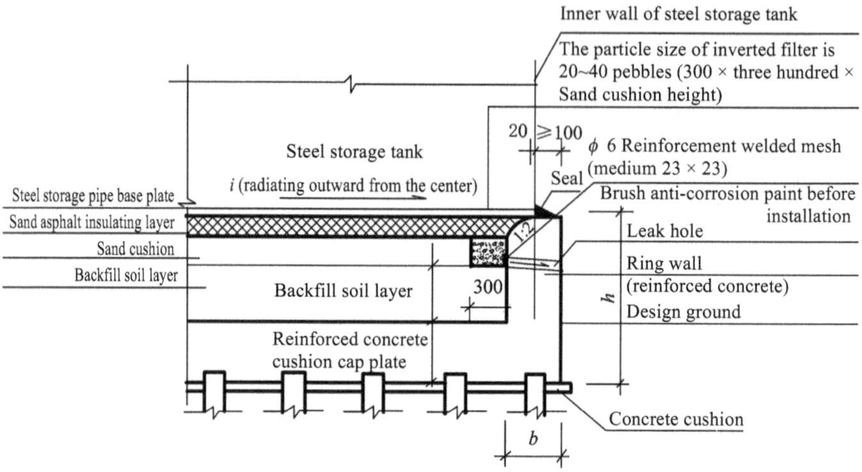

Fig. 6.43 Pile foundation

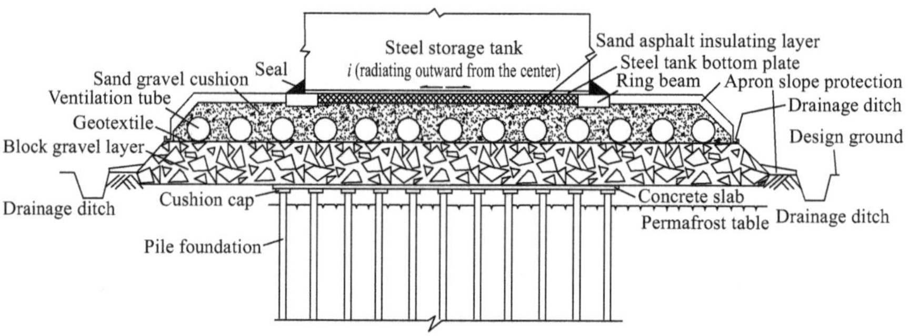

Fig. 6.44 Schematic diagram of composite foundation for ventilation tube pile foundation of broken stone bearing platform in permafrost regions

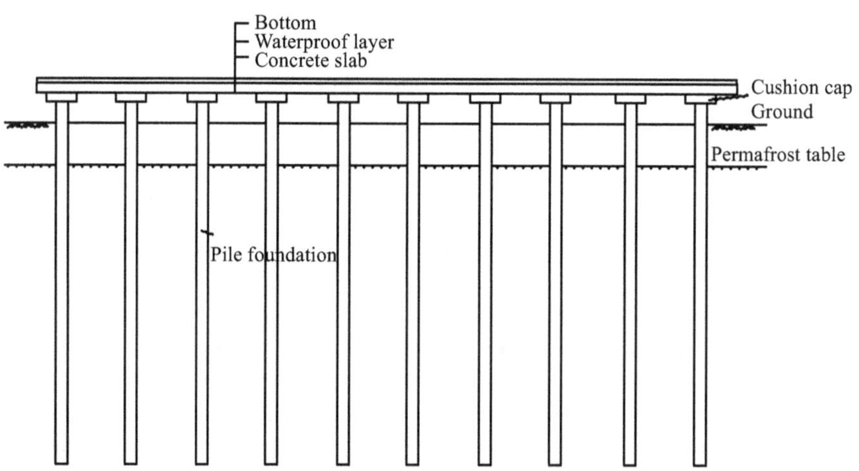

Fig. 6.45 Elevated ventilation pile foundation

ventilation tube, thermosyphon, insulation layer, pile foundation, cement fly ash lime pile, and composite measures such as ventilation tube + block gravel cushion, thermal insulation layer + thermal probe. If possible, the electric heating preconstruction thaw of foundation method should be adopted in high temperature and high ice content permafrost section. No matter what foundation treatment method is adopted, measures must be taken to prevent frost heaving of seasonally active layer, drainage around the construction site and protect the frozen soil environment, so as to avoid more thermal erosion caused by surface water thermal erosion and damage to the frozen soil environment (increase surface heat radiation, heat conduction, water heat and other erosion) and affect the thermal stability of the permafrost foundation.

The countries like the United States, Canada and Russia have accumulated some experiences in building oil storage tanks in permafrost regions. The basic methods are pile foundation, thermal insulation foundation and cooling permafrost foundation. Pile foundation (Belinov, 1996) and prefabricated square pile are adopted in Turmin Oil Fields of Russia. The length of the pile depends on the engineering geological conditions of the permafrost at the site, and the spacing is 3 ~ 5 times of the pile diameter. The pile cap is covered with concrete slabs, and the lower part is elevated ventilated. There are concrete piles, wood concrete mixed piles, thermal piles and steel pipe piles.

6.8 Water Supply and Drainage Works in Permafrost Regions

With the vigorous development of China's national economy and the prosperity of the country, the permafrost areas have also changed from sourcing to ecological construction, and the ecological environment and resources protection and construction of the "Natural Forest Protection Project" and "Three-river-source Area Ecological Protection" have been implemented. From the original small villages and small forest farms to the urbanization construction centered on the county and town (or forestry bureau), the water supply and drainage works have gradually developed to the requirements of urbanization.

The severe cold climate in permafrost regions, the widespread distribution of permafrost and the fragile ecological environment have brought many difficulties and engineering problems to the construction of water supply and drainage projects in permafrost regions.

6.8.1 Characteristics of Water Supply and Drainage Works in Permafrost Regions

Urban water supply and drainage works include water sources, pipelines, water storage structures, water supply and drainage facilities, etc. However, the difficulty and cost of water supply and drainage engineering construction are much higher than those in non-permafrost regions due to severe cold climate and permafrost layer.

1. **Water source**

The water sources in permafrost regions are mainly from river water, intrapermafrost water and subpermafrost water. Suprapermafrost water should not be used as the water source for urban water supply.

As water sources of rivers, large rivers in permafrost regions are situated in areas classified as penetrated talik or non-penetrated talik. Most of the rivers in the Daxing'anling Mountains are brown in color, and the closer to the forest area, the thinner the river and the darker the color. According to published data (Niu et al., 2010), Gan River in Jiagedaqi and Duobukuer River in Songling are basically of calcium carbonate, double calcium carbonate and double magnesium carbonate, with a total hardness of 0.01 ~ 0.15 g/L and a pH of 6.0 ~ 8.0, belonging to extremely soft neutral water. From 1975 to 1984, the average water quality for many years was Class I ~ II, and from 1986 to 2006, it was Class III ~ IV. All indexes were more inferior to before, and the water body was in a slightly polluted state. The amount of water flow changed seasonally.

Cl^- and mercury in the water source well of Tuotuo River along the Qinghai-Xizang Railway exceed the standard, and Cl^-, SO_4^{-2} and manganese in the water quality at the Yangtze River source grand Bridge and the guarded water source well exceed the standard (Xue, 2015), all of which need to be treated to meet the standard. The rising spring water in the structural talik in the south of Tuotuo River is the main domestic water supply source in the future (Tan et al., 2016), with a maximum water outflow of 9.26 L/s and a minimum of 5.04 L/s. The hydrochemical type is $HCO_3 \cdot SO_4$-Ca, with a salinity of 0.728 g/L, a chromaticity of less than 5, there is no other foreign colors, the turbidity is less than 1NTU, there is no smell, no taste, no visible matter, and the pH value is 8.08.

The suprapermafrost water, except for that in the river terrace talik, which has good water quality and quantity, is usually not used as the urban water supply source. In continuous permafrost area, it is frozen in winter.

The intrapermafrost water is usually developed in the structural fault zones, such as the artesian water of Erdaogou along the Qinghai-Xizang Highway. Its recharge source is subpermafrost water, with abundant amount of water flow, and the water quality meets the drinking water standard. In the Songling area of the Daxing'anling Mountains, there is also intrapermafrost water (Fig. 6.46) in the sand gravel layer of the river valley terrace, which is pressurized. The source of recharge is the Duobukuer River. The water quality is better than the river water, and the amount of water flow has seasonal changes. The water temperature is low.

The subpermafrost water is generally stored in the structural belt, such as the spring water from the Budongquan at the Qinghai-Xizang Highway, which is supplied by the groundwater from the frozen layer of the structural fault zone. The water quality is good and it flows continuously all year round. The structural fault zone with a large depth also exists in Daxing'anling Mountains area. For example, the water source of Mohe Waterworks passes through permafrost and andesite stratum to give water at a depth of about 150 m. For example, the water source of Dalai Valley Forestry Bureau in Daxing'anling Mountains of Inner Mongolia also comes from the fault zone, which is at the depth of about 130 m. The water temperature is usually low.

Fig. 6.46 Intrapermafrost water in the Daxing'anling Mountains (Photo by Tong Changjiang)

In general, the water temperature of water sources in permafrost regions is generally low, and the water sources are prone to freezing, so anti-freezing measures should be taken. From the recharge area to the drainage area, the chemical type of groundwater varied from simple to complex, the salinity varied from low to high, and the water quality varied from good to bad. Except for the subpermafrost water, the water volume is affected by seasons and changes greatly, and the water source in urban areas is vulnerable to the pollution of human activities.

2. **Pipe network**

Among the towns in permafrost regions, for the place where the regional special offices are located, the county scale is relatively large and has a population of about 80,000–150,000, with an area of about 20,000 km^2. Except that, the towns and forestry bureaus are equivalent to the northern villages and towns in non-permafrost regions, with a population of about 10,000–50,000 and an area of about 10,000 km^2. The length of water supply pipeline in these towns and forestry bureaus are relatively short. Except that small timber and food processing industries have 24-h water consumption, water consumption is basically stopped at night. In villages and towns such as forest farms, the intermittent water supply is adopted. Therefore, the pipeline is easy to freeze in winter, and the outdoor faucet freezes and heaves. The impacts of repeated freeze and thaw of the foundation make the pipeline deform and crack easily (Fig. 6.47).

During urbanization, the buildings are packed which causes the permafrost in cities and towns to degrade, or to form discontinuous permafrost. However, the buried depth of pipelines still needs to be set under the maximum seasonal thawing depth, which is generally 2.5 ~ 4.0 m. It is not easy to find the accident point when the pipeline is buried deeply, and it is very likely to cause high maintenance work, especially in winter. When the freezing depth is more than 0.7 m, the machinery can no longer be employed, and it is all manual excavation.

Fig. 6.47 Fracture of water supply pipeline (left) and freezing and frost heaving of faucet (right)

3. **Water storage buildings and ancillary facilities**

Larger towns with tall buildings often use roof water storage tanks, while other small towns use water tower for water storage. The thermal insulation of the vertical pipes and water storage tank in the water tower is the focus of antifreeze measures. At present, polystyrene, polyurethane and other thermal insulation materials are used to replace the previous using of fire stove for antifreeze. The water tank is mostly made of reinforced concrete, and the thermal insulation layer should be used around the tank at the bottom and at the top cover. If necessary, electric heating equipment should be provided to prevent the water from freezing when it stays in the water tower for a long time. Generally, 0.2 m thick ice layer is formed around the tank, which will not affect the regular use of the tank. The steel water tank is easy to rust, and rust removal and rust prevention are required every year. The vertical pipes for water supply are not allowed to freeze, and heating cables need to be laid on the pipe wall. In railway and other station areas, it is recommended to use the nearby high slope terrain to build water storage tanks, but it is necessary to take non-frost heave soil replacement around the tank and prevent frost heave and take thermal insulation measures at the bottom of the tank, and meanwhile, the heat insulation and increasing the buried depth of the water pipeline are also needed.

The outdoor public water hydrant is easy to freeze in winter (the right photo in Fig. 6.47), and indoor tap water is recommended.

The water supply inspection well often suffers from frost damage in winter, in which, the shaft wall is broken or displaced due to frost heaving, and water and ice are accumulated.

4. **Drainage facilities**

There are many problems with the drainage system in permafrost regions. In addition to the special drainage system facilities in large cities and towns, most small cities and towns lack drainage pipe networks, so that the on-the-spot drainage is regular. Existing problems are the following. ① Production and domestic sewage are not treated according to the current national drainage standards. ② The drainage pipe network is often frozen and blocked. ③ In winter, drainage seepage wells in permafrost regions often cannot work properly due to freezing or the presence of permafrost at the bottom, resulting in overflow and environmental pollution.

6.8.2 Water Supply Project Construction

The common problems of water supply and drainage engineering in permafrost regions are: freezing of water source wells, freezing, cracking, blocking, frost heaving and thawing of water supply and drainage pipelines, frost heaving damage of ancillary facilities, etc. In addition to the conventional design, the core of the design is supposed to focus on the anti-freezing and thermal insulation measures.

6.8.3 Design of Water Supply Well

In the past, the water supply source in permafrost regions was mainly the suprapermafrost water. Because the amount of the water was subjected to seasons, it was difficult to meet the requirements of the sanitary standards for drinking water. With the development of the country and the requirements of people's health, deep groundwater is gradually developed in permafrost regions, and the subpermafrost water is used as the source of drinking water. In the fault fracture zone of the permafrost region, there are many confined water sources with large water volume and good water quality. The water outflow of a single well is about 10 ~ 50 m^3/h, with a general depth of 40 ~ 50 m and the maximum depth of 90 ~ 130 m. The water quality in large areas of continuous permafrost on the Qinghai-Xizang Plateau is slightly inferior and needs to be treated. However, most of the water sources are surrounded by permafrost, so antifreeze measures must be taken when designing water source wells.

The subpermafrost water in the permafrost area pressurized, and the water temperature is low. The water source well pipe needs to pass through the permafrost layer of about 100 m (the mean annual ground temperature of the permafrost with a thickness up to 100 m is low, which is defined as low temperature permafrost). The pipe well needs thermal insulation and standby heating cables (Fig. 6.48) to avoid freezing of the water source of the pipe well. The thermal insulation layer is usually made of polyurethane foam, and the length of the thermal insulation layer should be 10 m below the base of the permafrost. The outside of the protective pipe wall should be filled with gravel to prevent frost jacking. In the aquifer in non-permafrost layer, it is designed according to the conventional water source well. During the construction, the filter pipe of wire wrapped filter and the well wall pipe should be put into the well hole at the same time.

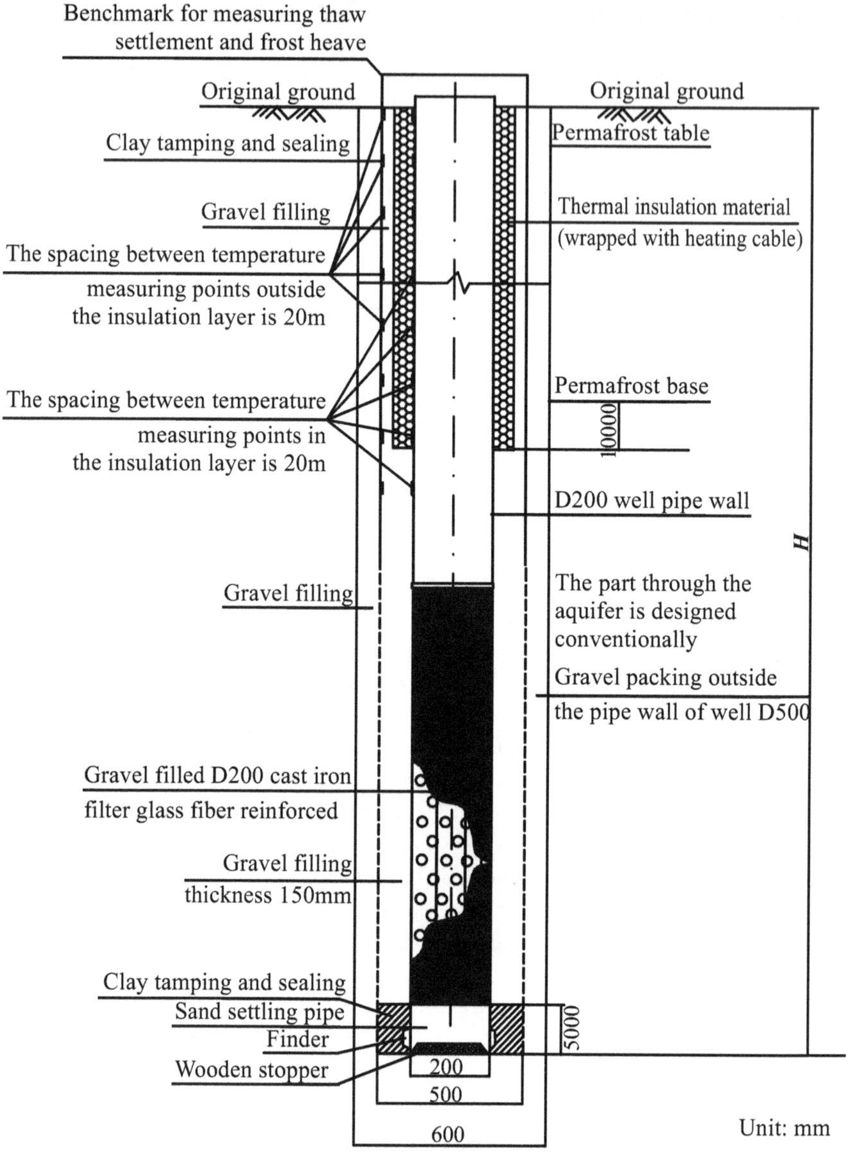

Fig. 6.48 Schematic Diagram of Water Source Well Pipe in Permafrost Regions (Su et al., 2002)

Temperature sensors are arranged in the thermal insulation layer of the inner wall of the protective tube and the outer wall of the well pipe to measure the variations of the water temperature of the well pipe. An automatic control system is set outside the well. When the temperature in the thermal insulation layer is ≤0.1 °C, the heating cable is started immediately to heat the water supply pipe. When the temperature in the thermal insulation layer is ≥1.0 °C, the heating is stopped to ensure that the water source well pipe is not frozen and the whole system is working properly. In the 1970s and 1980s, the freezing of water source well pipes often occurred in Daxing'anling Mountains area. Heating cables and temperature control systems were used to eliminate the freezing of water source well pipes. The pipe wall thermal insulation layer, heating cable, monitoring sensors and other equipment must be firmly fixed, and can only be put into the well bore after passing the strict inspection on the ground, and then the heating system and various electronic equipment should be tested item by item.

The drilling diameter of many water source wells is 600 mm. It is better to use casing to drill with pipe, which can avoid borehole collapse and shrinkage. When encountering bedrock, no casing is needed. If encountering soft rock, casing is still needed, but in the aquifer part, the pulping casing should be used.

In order to improve the water quality, especially when the water from the supra-permafrost water with low temperature, low turbidity and dark color is used as the water source, water purification facilities are usually built near the well for water treatment.

6.8.4 Water Supply Pipeline Network

In recent years, cities and towns in permafrost regions have changed greatly. In the Daxing'anling Mountains area, urban construction is gradually concentrated, buildings are gradually increased, bungalows are reduced, and many adopt central heating and water supply. Except for government agencies, in the Qinghai-Xizang Plateau, the bungalows are the relatively dominated form for residence.

The layout of the pipeline network should take the terrain and ground features as the control points. The main pipeline should be close to the water end users. The pipeline network should be used for both transmission and distribution. The pipeline should be straight to reduce corners. The branch pipeline should not be too long, and should be connected with the main pipeline as far as possible to form a circular pipeline network if conditions permit.

The laying forms of pipelines include overhead, buried (shallow or deep buried), trench and embankment, each with different advantages and disadvantages. The form that is suitable for local conditions should be selected. A large number of engineering practices show that thermal insulation pipelines can achieve good results for water supply and drainage pipelines in permafrost regions (including cold regions), and its heat output can be reduced by 80% compared to that of the deep buried pipes.

1. Overhead thermal insulated pipeline
   The overhead thermal insulation pipeline is suitable for the following sections: the sections where the pipeline passes through marshes, bogs, depressions, ditches, thick ground ice, strong frost heave soil, flood threatened areas and other areas with unfavorable geological conditions. Additionally, if the pipes are directly laid in the soil layer, there may be difficulties to carry out the pipe-laying

construction in the soil layer or it is not easy to maintain the pipes after they are buried.

① Pile foundation or column foundation

Reinforced concrete structure or steel structure (filled with fine aggregate concrete) should be adopted, and its buried depth should be determined according to the calculation of frost heaving force. Generally, it should not be less than twice of the maximum seasonal thawing depth. Anti-frost heave measures should be taken within the maximum seasonal thawing depth. The pile foundation should adopt borehole pile or explosive expanded pile, and the pile end should be equipped with expanded part, and the pile diameter should not be less than 200 mm. Single pile (Fig. 6.49) or double pile can be used. (Fig. 6.50)

② There are fixed and sliding pipeline supports. The cushion cap is connected with the pile (column) as a whole. The pipeline is fixed on the cushion cap by the support, and a movable cover is installed outside. The thermal insulation layer is made of polystyrene, polyphenolic aldehyde or polyurethane foam with small thermal conductivity and water absorption, and the thickness is not less than 60 mm or determined by thermal calculation.

③ The main and branch water distribution pipelines should be laid with one-way slope as far as possible, and the minimum slope should not be less than 5%. A vent pipe should be set at the lowest point of the main pipeline.

2. Buried pipeline
   The Buried pipeline includes the types of shallow buried (thermal insulation pipeline) and deep buried (without thermal insulation layer).

① Shallow buried thermal insulation pipeline is applicable to sections with low frost heave and thaw settlement foundation. The burial depth should be less than 2/3 of the maximum seasonal freezing depth (the main frost heaving zone), generally

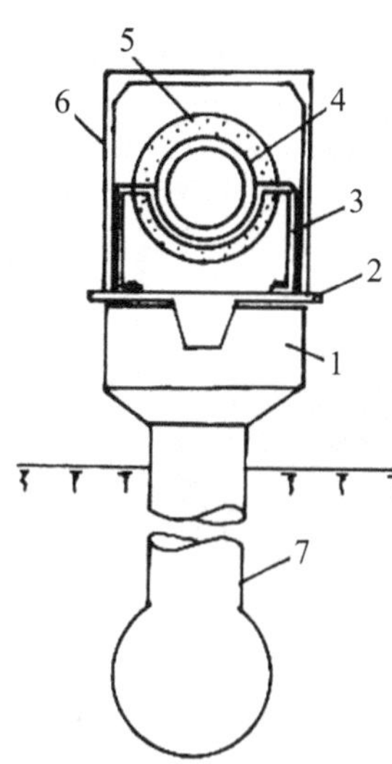

Fig. 6.49 Overhead thermal insulation pipeline structure (the Third Survey and Design Institute of the Ministry of Railways, 1994). 1—cushion cap; 2—"T" beam; 3—Support; 4—Water pipeline 5—Insulation layer; 6—Protective cover; 7—Explosion expanded pile

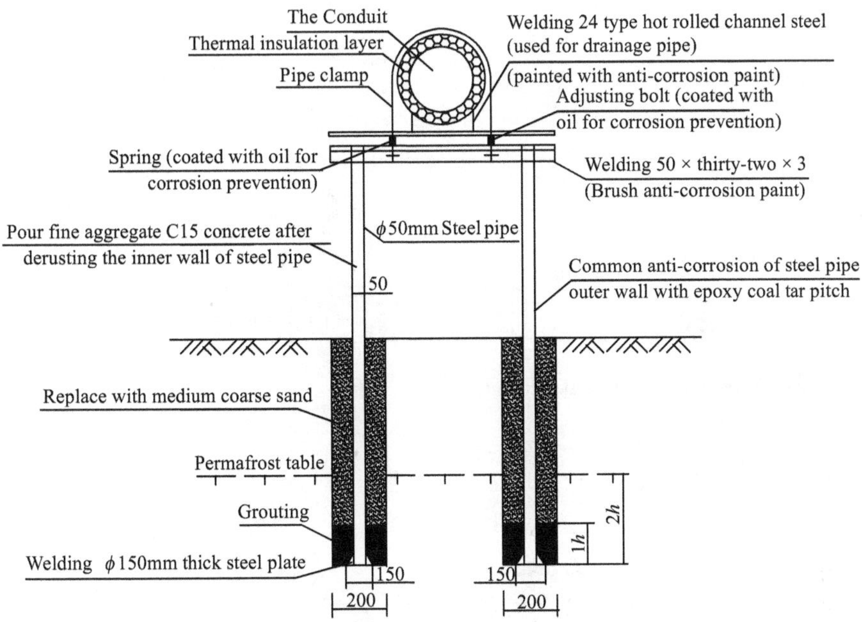

Fig. 6.50 Structure of overhead thermal insulation pipeline (Double Pile)/Unit: mm (Xue, 2015)

1.5 ~ 2.0 m below the ground. The base should be replaced with non-frost heaving coarse grained soil with thickness of not less than 0.5 m. The pipeline should be backfilled with coarse grained soil of a certain thickness before being backfilled with original soil. The thickness of the thermal insulation layer is generally not less than 60 mm, or determined by thermal calculation. When the groundwater level is high, a protective cover should be provided to prevent the thermal insulation layer from being wetted. Waterproof measures should be taken at the joints of the protective cover (Fig. 6.51), and a drainage system should be set inside. In areas with low ground temperature, heating cables (Fig. 6.52) should be added to ensure the normal operation of the pipeline. The heating cable with heating power of 30 W/m can be selected if the pipeline diameter is ≤200 mm. The starting temperature is 0.5 °C, the power off temperature is controlled at 3 °C, and the automatic control system is used for control.

② Deep buried pipeline is applicable in taliks in permafrost area or cold seasonally frozen soil area. The maximum seasonal freezing depth in these areas can reach 2.5 ~ 4 m. The water pipeline should be buried 0.2 ~ 0.3 m below the bottom of the maximum seasonally frozen layer. The simple management is the greatest advantage of this type, with low operation cost, and not affected by power supply, but it also has disadvantages which are with large earthwork volume and difficult maintenance, etc. Generally, after an operation test through a winter, under the influence of water temperature, it does not freeze even if the water supply is stopped for a long time.

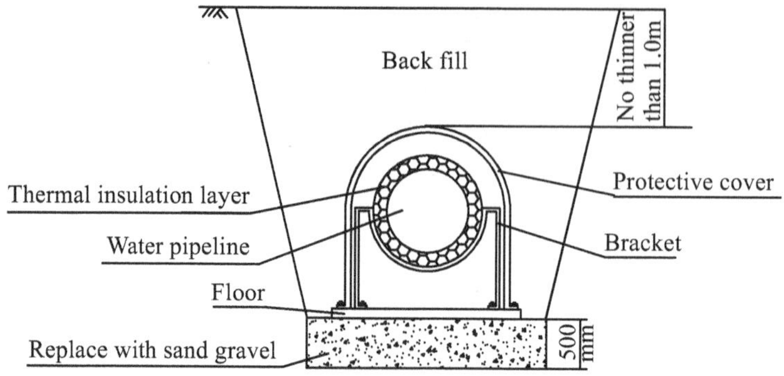

Fig. 6.51 Protection of shallow buried pipeline

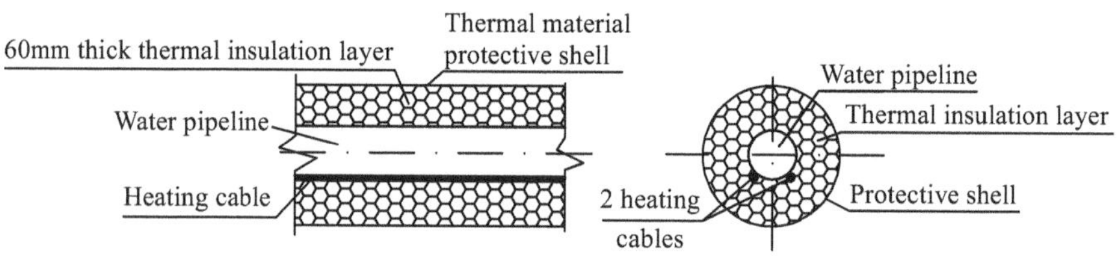

Fig. 6.52 Plan of heating cable for shallow buried pipeline

In the warm permafrost area, the ground temperature at the permafrost table is relatively high, and the pipeline can also be laid by deep burial. According to the ground temperature observation data at Lingding Station of Yakeshi-Linhai Line in Daxing'anling Mountains (the Third Survey and Design Institute of the Ministry of Railways, 1994), the heat loss of water pipeline at 1 m deep was 100%, at 2 m deep was 48.6%, and at 3 m deep was 31.2%. Therefore, deep buried pipeline has advantages to antifreeze. Before the design, the geological conditions of the section where the pipeline passes, the ground temperature variations within the range of 0 ~ 4 m, the soil water content and the thaw settlement data of the permafrost table should be obtained. The soil layer 0.5 ~ 1 m from the base of the pipeline foundation should be replaced with sandy gravel or non-thaw settlement soil. The water pipeline interface should be made of flexible filler, such as rubber pad interface, so that it can adapt to deformation. The construction season should be from April to June or from September to October. The construction quality of foundation treatment, water pipeline interface and backfill tamping density should be ensured. Frozen soil should not be used as backfill. The water temperature of water tower or mountain pool should be maintained at about 4 °C. The first water supply should avoid the severe cold season and minimize the water retention time in the pipeline.

③ Trench laying pipeline is a shallow buried pipeline, which should adopt the thermal insulation layer. With the development of urbanization, the areas where many cities and towns are situated have gradually changed from continuous permafrost regions to discontinuous permafrost regions, and the depth of permafrost table has decreased to 6 ~ 8 m, such as the Yakeshi New Area in Inner Mongolia and Mohe Airport, etc. In some warm permafrost regions with a population of more than 10,000, it has turned into seasonally frozen soil regions, such as Jiagedaki, Yakeshi and other cities. Trench pipes are widely used for centralized heating in cities and towns. Water supply and drainage pipelines can be laid in the trench together with heat supply pipelines, such as Hailar and other areas.

3. Water supply structure

   1. Water tower. Most of the cities and towns have water towers, which should be of reinforced concrete structure. Since there is no heating facility in the tower, the foundation should be designed according to the principle of keeping the foundation frozen. The vertical pipelines in the tower should be provided with an industrial thermal insulation layer, the thickness of which is determined by thermal calculation, generally 50 ~ 80 mm. If necessary, heating cables should be added. In addition to the bearing capacity of the foundation, the stability under the frost heaving force should also be considered in the design of the tower foundation, and pile foundation should be adopted.
   2. Water storage tank. For villages and water supply in transportation stations, underground or semi-underground reinforced concrete water storage tanks can be built when there are suitable terrain conditions. The foundation and surroundings of the water tank should be replaced with non-frost heaving soil, and the replacement depth of the foundation should not be less than 0.5 m. The industrial thermal insulation layer (about 100 mm thick) should be pasted around the outside of the water tank, and the residual oil protective layer should be brushed outside. In addition, medium coarse sand should be used as the protective layer, and then non-frost heaving soil should be replaced with a thickness not less than 0.5 m. Reinforced concrete cover plate is used for the tank, and the thermal insulation layer is laid on it, and then the wooden roof is made.
   3. Inspection wells. Inspection wells should be set at the branch of the pipeline network, and inspection wells on the straight pipeline section should be generally spaced from 200 m to 300 m. The prefabricated reinforced concrete integral structure (or in-situ casting) is adopted. The base is replaced with non-frost heaving soil, with a thickness of about 0.5 m. The external wall of the well is pasted with an industrial thermal insulation layer (100 mm thick), coated with a residual oil protective layer, and then backfilled with non-frost heaving soil using medium coarse sand as the protective layer. The well cover (special well cover is used on the road, and reinforced concrete or wooden well cover can be used in other sections) is provided with industrial thermal insulation layer or filled with other insulation materials (Fig. 6.53).
   4. Hydrants. The shared water hydrants should be installed as close to the main pipeline as possible in a dedicated water hydrant room, and thermal insulated pipelines should be used to avoid freezing due to small diameter of branch pipelines and water consumption at night.

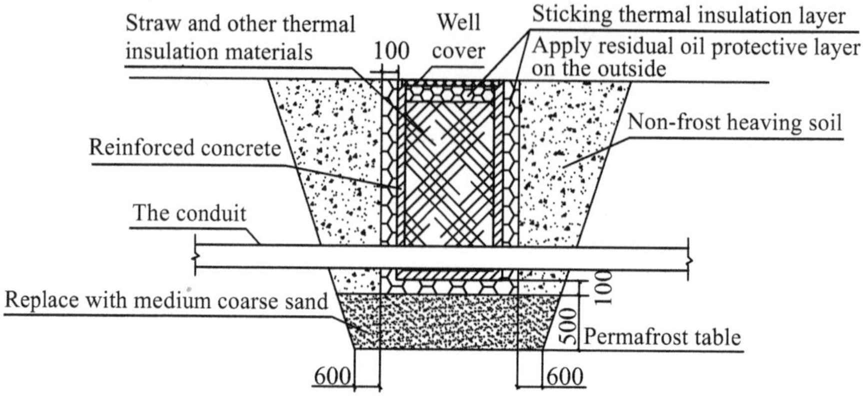

Fig. 6.53 Schematic diagram of inspection well (Unit: mm)

5. Fire Hydrant. Outdoor fire hydrants should be set in special thermal insulated wells or used with inspection wells.

6.8.5 *Drainage Works*

With the development of the national economy and the improvement of people's living standards, our country puts forward the higher requirements for environmental protection, changes the sanitary environment in villages and towns, ensures that the drinking water and the drainage of domestic sewage meets the national standards. Additionally, to keep the drainage and treatment systems work properly is an important step to protect the water environment.

The low temperature in the high latitude permafrost region and the low temperature, low pressure and low oxygen environment in the high-altitude permafrost region strictly restrict the anhydrous treatment process and its outcome.

1. **Drainage pipe**

   1. Laying depth of buried drainage pipelines. The low temperature environment in permafrost regions has caused great difficulties in the drainage of domestic sewage. Avoiding the freezing of drainage pipelines is the main factor to ensure the reliability of regular operation. Generally speaking, when the pipeline is deeply buried the ground temperature could be relatively high, so the possibility of freezing of the pipeline during idle time is less; but the project cost of deeply buried pipeline is higher, and the maintenance difficulty is increased. In the taliks or the isolated discontinuous permafrost area, the drainage pipeline should be buried below the maximum seasonal freezing

depth. In continuous permafrost areas, it is advisable to increase the thermal insulation layer of drainage pipelines and shorten the pipeline length. The burial depth of the drainage pipeline can be reduced. According to the survey and test data of Daxing'anling Mountains Railway Station and residential buildings, the burial depth of drainage pipelines should be 1.5 ~ 2.0 m below the ground surface (the Third Survey and Design Institute of the Ministry of Railways, 1994), or 2/3 of the maximum freezing depth.

2. Diameter, gradient and flow rate of drainage pipeline. The diameter of the pipeline is related to the flow, minimum gradient and flow rate. The minimum diameter of the drainage pipeline is 300 mm, and the corresponding minimum design gradient is 3‰. The designed flow rate of the pressure pipeline should be 0.7 ~ 2.0 m/s. The drainage pipelines in permafrost regions often suffer from blockage or frost damage. In addition to hydraulic conditions, thermal conditions should also be considered when selecting pipeline diameter. Too large or too small pipeline diameter is unfavorable. According to the survey data, the diameter of outgoing pipeline should not be less than 200 mm, and the gradient should not be less than 7‰. The diameter and gradient of main pipeline should also be appropriately increased. The pipeline section with flow rate less than 0.7 m/s should be equipped with flushing facilities. The survey data has testified that the proper increase of the diameter and slope of the drainage pipeline to make it maintain a certain flow is an important condition to ensure the normal operation of the drainage pipeline in permafrost area.
3. Treatment of drainage pipeline base. The thaw settlement of ice containing sections in permafrost regions is the main reason for pipeline deformation. In addition to thermal impact, drainage pipelines with small diameters and slopes are prone to local full flow, slow down or siltation of water flow, and the leakage of interfaces may lead to the thawing of the permafrost in pipeline foundation and then differential settlement. Therefore, the non-thaw settlement soil should be used for replacement in the sections where the drainage pipeline foundation may have a talik zone through thermal calculation. The drainage pipeline interface should be socketed or collared, and filled with flexible materials.
4. Anti-freezing of drainage pipelines. If the distance of drainage pipeline is too long, it could freeze due to the decrease of water temperature. The distance when the temperature drops to 0 °C during the flow in the drainage pipeline can be estimated according to the following formula (Third Survey and Design Institute of the Ministry of Railways, 1994):

$$l = -1163R \cdot q \cdot \ln \frac{-t_d}{t_1 - t_d} \tag{6.3}$$

where R is the thermal resistance of water pipeline, m·°C/W; q is the drainage volume, m^3/h; t_d is the ground temperature at the pipeline axis, °C; t_1 is the starting point temperature of the calculated pipeline section, °C.

When the critical distance is exceeded, the drainage pipeline may freeze. Generally, the critical distance is extended by raising the sewage temperature, increasing the sewage flow or the thermal resistance of the drainage pipeline. If these methods are difficult to conduct, only thermal insulation materials can be used for extra-long pipelines to prevent freezing of them.

2. **Ancillary structures**

1. Inspection wells. The whole structure of reinforced concrete is adopted. The outside of the well wall is backfilled with non-frost heave soil. The well foundation also needs to be replaced with coarse grained soil. The wooden shaft cover is covered with thermal insulation materials.
2. Water outlet. The outlet of the drainage pipeline is often frozen due to low air temperature, so the concealed outlet should be used. A sedimentation well (pool) is set at the outlet of the drainage ditch, the well wall is connected with the radial rubble dry masonry drainage ditch, and a reinforced concrete cover plate is set on the sedimentation well. When discharging, the whole outlet is covered with local peat layer and soil layer, with a thickness of 1 ~ 1.5 m, forming a half cone. The distance from the sedimentation well to the toe of the fill slope is about 3 m. The practice shows that (the Third Survey and Design Institute of the Ministry of Railways, 1994), the temperature in the sedimentation well is 1.3 °C, the water temperature is 3.2 °C, and the temperature of the sewage flowing out of the masonry is 2.7 °C under the environmental air temperature of 37 °C. However, the sedimentation well should be cleaned in summer.

3. **Sewage treatment structures**

Infiltration facilities using soil purification, such as seepage wells and underground filter fields, are not suitable for permafrost regions. Anti-freezing and anti-thawing measures should be taken for septic tank. The side wall and bottom of the septic tank should not only be laid with thermal insulation layer, but also the frozen soil foundation under the tank should be replaced with non-thaw settlement soil. The Third Survey and Design Institute of the Ministry of Railways has designed a "double-layer septic tank" in the Daxing'anling Mountains permafrost region to solve the freezing problem. The inner septic tank is the same as the general septic tank, and the outside and bottom are additionally provided with thermal insulation layer and heating equipment (equivalent to a jacket box). The observation data show that the anti-freezing effect of the "double-layer septic tank" is good. Even if no heating measures are used, the water temperature in the inner tank can be maintained at about 1 °C, and the annual average temperature is 5.2 °C. However, the duration in which the temperature exceeds 10 °C is only 2 months, which makes the feces in the tank be difficult to biodegrade.

6.9 Pipeline Calculation

Thermal calculation of pipelines in permafrost regions is very complex, which needs to be calculated under stable and unstable operation modes, respectively. Generally speaking, the transportation mode of oil pipeline is basically stable, and the situation along the pipeline is unstable. The water supply and drainage transportation mode is unstable. At present, the academia basically uses computational software for numerical simulation to obtain numerical solutions. However, the numerical simulation method requires more calculation parameters. Based on the design requirements, it is necessary to simplify the thermal calculation formula of relevant pipelines.

6.9.1 Temperature Variations of the Transported Material in the Pipeline

When the transported material in the buried pipeline flows or just in static state, its heat loses along the pipeline or over time, and the temperature gradually decreases. The formulas for calculating the decreasing of temperature of the transported material along the pipeline (Cold and Arid Regions Environmental and Engineering Research Institute, Chinese Academy of Sciences, 2009) are the following.

When the loss of heat caused by friction occurring in flowing of the transported material is considered:

$$t_L = (t_d + b) + \left[t_s - (t_d + b)\right]e^{-aL} \tag{6.4}$$

When the influence of the loss of heat caused by friction occurring in flowing on the temperature of the transported material is not considered:

$$\begin{gathered} t_L = t_d + (t_s - t_d)e^{-aL} \\ a = K\pi d / Gc, \\ b = gi / ca \end{gathered} \tag{6.5}$$

where G is the mass flow of the transported material; c is the specific heat capacity of the transported material; d is the outer diameter of the pipeline; L is the pipeline length; K is the total heat transfer coefficient of the pipeline; t_s is the starting point temperature of the pipeline; t_L is the temperature at pipeline at the length of L; t_d is the temperature of the surrounding medium; i is the hydraulic gradient; g is the acceleration of gravity.

Calculation formula for heat loss of stable thermal conductive pipeline (according to the Third Survey and Design Institute of the Ministry of Railways, 1994; He, 1983):

$$q = \frac{t_s - t_d}{R} = \frac{2\pi\lambda(t_s - t_d)}{\ln\left[\frac{2h}{d} + \sqrt{\left(\frac{2h}{d}\right)^2 - 1}\right]} \quad (6.6)$$

where t_s is the temperature of the transported materials (water, oil and gas) in the pipeline; t_d is the ground temperature at the axis; h is buried depth of pipeline; d is the pipeline diameter; λ Is the thermal conductivity of the soil layer.

Because $\left[\frac{2h}{d} + \sqrt{\left(\frac{2h}{d}\right)^2 - 1}\right]$ only related to the buried depth and diameter of the pipeline, it is called the shape factor.

When $(h/d) > 2$, then, the error caused by $R = \frac{1}{2\pi\lambda}\ln\frac{4h}{d}$ should not exceed by 1%.

The heat loss of the transported material during flowing in the pipeline is equal to the heat released by the temperature drop of the transported material. The heat balance established in the overall movement of each unit length of the pipeline is the integral of the auxiliary equation. The heat is measured in m³/h, so the calculation formula of the decreasing of the temperature of the transported material (water) during transportation is:

$$t_L = (t_s - t_d)e^{\frac{1}{1163Rq}} + t_d \quad (6.7)$$

where t_L is the end point temperature of pipeline with a length of L, °C; t_s is the temperature of the starting point of the pipeline, °C; t_d is the ground temperature at the depth aligning with the pipeline axis, °C; R is the thermal resistance of the pipeline, m·°C/W; q is the heat loss of the pipeline, J/(m·s).

6.9.2 *Calculation of Pipeline Thawing Ring*

Heat loss from the pipeline to the edge of thawing ring:

$$Q_1 = \frac{2\pi\lambda^+(t_s - 0)}{\ln(D/d)}$$

Heat loss from the edge of the thawing ring to the soil layer:

$$Q_2 = \frac{2\pi\lambda^-(0 - t_d)}{\ln(4h/D)}$$

From $Q_1 = Q_2$ it is obtained:

Diameter of the pipeline thawing ring:

$$D = e^{\frac{\lambda^{+} t_s \ln(4h) - \lambda^{-} t_d \ln D}{\lambda^{+} t_s - \lambda^{-} t_d}} \tag{6.8}$$

where λ^{+} and λ^{-} are the thermal conductivity of the soil layer when it is frozen and thawed, respectively, and D is the diameter of the thawing ring, h is the burial depth of the pipeline.

For simplifying the calculation, the medium thermal resistance of little influence can be ignored, such as the heat release resistance from the transported material to the wall of the pipeline, the thermal resistance of the directly buried pipeline without protective cover, etc. The heat loss of the thermal insulated pipeline is:

$$Q = \frac{(t_s - t_d)}{R} \mathrm{J}/(\mathrm{m{\cdot}s}) \tag{6.9}$$

where Q is the heat loss per hour per unit length of the pipeline, kJ/m· h or J/(m·s); R is the total thermal resistance of the pipeline, m·°C/W.

Total thermal resistance of the pipeline:

$$R = \frac{1}{2\pi\lambda_g} \ln \frac{d_{gw}}{d_{gn}} + \frac{1}{2\pi\lambda_B} \ln \frac{d_{Bw}}{d_{gw}} \tag{6.10}$$

where λ_g and λ_B are the thermal conductivities of the pipeline wall and thermal insulation layer, respectively, W/(m·°C); d_{gw} and d_{gn} are the outer diameter and inner diameter of the pipeline, respectively, m; d_{Bw} is the outer diameter of the thermal insulation layer, m; α is the heat release coefficient, W/(m^2·°C).

When the pipeline is buried near the table of the permafrost, initially, the surrounding area of the pipeline is roughly thawing in a circular shape. In warm season, the active layer thaws downwardly, which could overlap with the pipeline thawing process. Then the residual thawing index of the active layer will continue to thaw the permafrost below the pipeline, which is equivalent to that the active layer above the pipeline is thawed by the heat from ground surface and the heat of the pipeline thaws the permafrost below the maximum thaw depth. This interpretation of heat is reasonable, so the impact of the thaw depth of the natural ground on the thawing ring of the pipeline need to be considered. Under the impact of global warming, the thaw depth of permafrost will increase correspondingly and the permafrost table will decrease. Therefore, the depth of pipeline thawing ring in permafrost regions should be corrected as the following:

$$H_t = \mathrm{h_t} + \Delta \mathrm{h_1} + \Delta \mathrm{h_2} \tag{6.11}$$

where $\mathrm{h_t}$ is the thickness of the thawing ring of pipeline in permafrost (bare pipeline or thermal insulated pipeline); $\Delta \mathrm{h_1}$ is the correction value of the thawing of ground surface, which is calculated according to the Grabellon formula of multi-medium (Table 6.9); $\Delta \mathrm{h_2}$is the correction value of the decrease of permafrost table caused by climate change, which can be obtained from Stefan's formula. The square of the

Table 6.9 Correction coefficient of the thawing of ground surface (Xu et al. 2009)

	Bare pipeline	Thermal insulated pipeline
Ice-poor permafrost foundation	1.24	1.44
Icy permafrost foundation	1.18	1.39
Ice-rich permafrost foundation	1.14	1.31
Ice-saturated permafrost foundation	1.12	1.25
Value	1.20	1.35

depth of permafrost table is directly proportional to the air temperature thawing index. If the air temperature in the Daxing'anling Mountains in Northeast China rises by 2.4 °C in 50 years, it is 1.2 times of the thawing index of the original air temperature and 1.1 times of the original depth of permafrost table of natural ground. According to the engineering experience and safety considerations, 1.3 times of the original permafrost table is taken. That is:

Bare pipe: $H_t = 1.2h_t + 0.3h_o$
Thermal insulated pipeline: $H_t = 1.35h_t + 0.3h_o$

where *ho* is permafrost table.

6.9.3 *Heat Loss of Water Pipeline Working Intermittently*

In fact, most of the water pipelines work intermittently, especially in cities and towns without industrial water. During the day, the pipelines are transporting water, and at night, they generally stop working. The heat loss process of the pipelines is unstable, and the soil layer around the pipeline is in an alternating process of heat absorption and heat loss.

Assume that the total thermal resistance of the pipeline is *R*, the time when the water in the pipe starts to be static at $T = 0$, and the initial temperature of the water is t_0. After *T*, the temperature in the pipeline decreases to *t*, and the amount of heat loss of the pipeline is $Q = (t\text{-}t_d)/R$. The heat loss of the pipeline in *dT* should be *dQ*, then

$$T_1 = 913d^2R \cdot \ln\frac{t_0 - t_d}{t_1 - t_d} = 2102d^2R \cdot \log\frac{t_0 - t_d}{t_1 - t_d} \tag{6.12}$$

where T_1 is the time when the water temperature decreases from t_0 to t_1, h; *d* is the pipeline diameter, m; *R* is the total thermal resistance of the pipeline, m·°C/W; t_0 is the initial temperature when the water in the pipeline starts to be static, °C; t_1 is the water temperature in the pipeline after T_1, °C.

Chapter 7
Engineering of Overhead Transmission Line in Frozen Soil Regions

The pole and tower foundations are the important part of transmission lines. In addition to the upper structure, the special characteristics of frozen soil and its variations directly affect the foundation selection, buried depth, construction procedure and quality, treatment design and optimization for freeze-thaw diseases.

7.1 Engineering Geologic Route Selection in Frozen Soil Regions

Climate, human activities and ecological environment variations affect the hydrothermal dynamics of frozen soil, resulting in frost heave, thaw settlement and thaw slumping and other unfavorable cryogenic phenomena, which is different from the engineering geological problems of conventional foundation in non-frozen soil area. In addition, micro-topography, geological structure and hydrogeological environment, etc., also have a significant impact on the optimization of engineering design.

7.1.1 Frozen Soil Engineering Geological Conditions and the Main Frozen Soil Engineering Problems

The main frozen soil engineering geological conditions of transmission lines in permafrost regions are ice content in frozen soil foundation, mean annual ground temperature, and the depth of permafrost table, etc. The main frozen soil engineering geological problems are frost heave in permafrost foundation, thaw settlement,

Q. Wu, C. Tong, *Frozen Soil Engineering in Cold Regions*,
https://doi.org/10.1007/978-981-95-2527-0_7

rheological and creep properties of permafrost, unfavorable cryogenic phenomena, etc.

Permafrost ice content: for low-ice content permafrost (ice-poor frozen soil, icy frozen soil), the water content after thawing is 12–18% (the water content in silt is 21%), and the thaw settlement coefficient is 1–3%, the frost heave coefficient is 1–3.5%, and the lithology is mostly coarse-grained soil, and when the water content in fine-grained soil is less than$\omega_p + 4$, fine-grained soil also belongs to low-ice content permafrost; for high-ice content permafrost (ice-rich permafrost, ice-saturated permafrost and ice-bearing soil layer), the water content after thawing is 18–44%, the thaw settlement coefficient is 3–25%, the frost heave coefficient is 3.5–12%, and the lithology is mostly fine-grained soil, and when coarse-grained soils contain thick layers of ground ice lenses, it is also classified as this category. When the permafrost with high ice content is warmed and under pressure, large thaw settlement and creep occurs, which has a great impact on the stability of the tower foundation. And engineering measures should be taken to maintain the stability of the pole and tower foundations.

Mean annual ground temperature of permafrost: according to its mean annual ground temperature (T_{cp}), permafrost can be divided as extremely warm and unstable permafrost ($-0.5\ °C \leq T_{cp} \leq 0\ °C$), warm and unstable permafrost ($-1.0\ °C \leq T_{cp} \leq -0.5\ °C$), cold and basically stable permafrost ($-2.0\ °C \leq T_{cp} \leq -1.0\ °C$) and cold and stable permafrost ($T_{cp} < -2.0\ °C$). The warm permafrost has a poor capacity of cold storage, and is easily thawed under external thermal disturbance and then loses its stability. In plateau and alpine permafrost regions, it is mostly located in the southern and northern limits of permafrost, the fringe area of taliks, river valleys and the fault basin. There are also warm permafrost regions in mountain areas with low or medium heights in high-altitude plains and high-latitude permafrost regions. The cold permafrost has a large amount of cold storage, which is not easy to thaw under external thermal disturbance, and it has good stability. The cold permafrost is widely distributed in swamps and wetlands, plateaus and medium or high mountains; and sometimes the ground temperature in low mountain areas is also lower, where cold permafrost also occurs.

Permafrost table: the permafrost table is in the near-surface layer, which directly participates in heat exchange between atmosphere, the surface and the lithosphere. Most of the heat exchange between the permafrost and the atmosphere happens in the active layer, which is directly related to the thermal state of the lower permafrost layer and the characteristics of the cryogenic structures, and it is the main depths where the tower foundation is buried. In the active layer above the permafrost table, during the freezing period, the soil freezes, which has a adhesive frost heave effect on the building foundation; during the thawing period, the frozen soil thaws, the ground ice thaws, and the soil body subsides, which affects the stability of the building foundation. Generally, the permafrost table in flood plain, bare land with sparse vegetation and bedrock sections is relatively deep, which is deeper in coarse-grained soil than that in fine-grained soil. In relatively dry soil with low water content, the permafrost is deeper than that in soil with higher water content. Compared to the areas in continuous permafrost or cold permafrost, the permafrost table in isolated permafrost areas or in warm permafrost is deeper. A large number of investigation and survey data show that there are ground ice layers with different thicknesses at the permafrost table, which is also a sign of permafrost table, and within this depths around the permafrost table, it is an unstable foundation section. Thus, the buried depth of the tower foundation should exceed the permafrost table to maintain the stability of the building foundation.

Frost heave of foundation soil: due to the frost heave of the foundation soil, the tower foundation in the transmission line engineering is frost jacked and then reconstructed (Fig. 7.1), broken, inclined and collapsed (Fig. 7.2). Pingos (Fig. 7.3) and aufeis (Fig. 7.4) are generated due to the changes of the permafrost environmental conditions, which endangers the stability of the tower foundation.

Thaw settlement of permafrost foundations: the ground ice in permafrost foundations thaw under external thermal erosion, resulting in differential thaw settlement of the foundation. Global warming deepens the permafrost table, and this situation should be taken into account in the design for the tower foundation. And the thermal disturbances caused by the construction of the tower foundation and the heat conduction of the foundation materials will increase the ground temperature of the permafrost foundation, weaken or lose the thermal stability of the permafrost foundation, and cause thaw slumping and thaw settlement (Fig. 7.5). Meanwhile, the heat will also change the surrounding permafrost environment and the foundation pit could be immersed by the groundwater (Fig. 7.6), and it will bring in huge heat and negative consequences, resulting in thaw settlement of the permafrost foundation and also causing frost heave of the foundation, pingos or aufeis.

Creep and rheology of frozen soil: stress relaxation and creep occur under long-term external loads on permafrost, which leads to viscoplastic flow of ground ice. In the slope section, due to unbalanced force posed on the plane of the tower

Fig. 7.1 Frost heave of 220 kV transmission tower toundation in Haiya (Jiang & Liu, 2006)

Fig. 7.2 Poles toppling in Tanggula due to frost heave (Qian et al., 2009)

Fig. 7.3 Pingo at the foundation of the Qinghai-Xizang Direct-current Transmission Line (photo by Yu Qihao)

foundation, the foundation deformed under shear stress, and the steel tower slips (Fig. 7.7). The slope slipping of the steel tower may also be caused by the uneven horizontal frost-heave force or the normal frost-heave force on the foundation. Additionally, the human or natural factors may disturb the thermal state of the permafrost on the slope, causing thaw collapse.

Features related to unfavorable cryogenic phenomena: unfavorable cryogenic phenomena mainly include thermokarst lakes, gelifluctions, aufeis, pingos, etc., and some of them are caused by natural factors, and some are from the consequential cryogenic phenomena caused by engineering activities. For example, the pingos in

Fig. 7.4 Aufeis at the tower foundation of the Qinghai-Xizang Transmission Line

Fig. 7.5 Thaw slumping and thaw settlement caused by excavation of foundation pit

Fig. 7.3 and the aufeis in Fig. 7.4 are caused by the changes in the hydrothermal state of the permafrost foundation due to the excavation for tower foundation. The thick ground ice in the fine-grained soil of the permafrost foundation thaws due to thermal erosion, and further changes the hydrogeological environment, and then it becomes the main places of groundwater processes (Fig. 7.4).

Fig. 7.6 Water accumulation in the foundation pit (all photos by Yu Qihao)

Fig. 7.7 Slipping of the steel tower on the slope at the Kunlunshan Pass (Liu et al., 2009)

7.1.2 Selections of the Route and Location of Frozen Soil Engineering Geology

In addition to the general requirements, the route selection for the transmission line should fully consider the permafrost engineering geological conditions and the particularity of frozen soil engineering problems: in the permafrost region, the natural geography and geological and geomorphological environment are special, the development of transportation and economy are relatively lagging behind and the ecological environment is relatively fragile.

7.1.2.1 Route Selection of Frozen Soil Engineering Geological for Transmission Lines

1. The route is highly recommended to be laid along the "engineering corridor" of linear engineering and engineering constructions such as the existing highways and railways, etc., or along the river valley. In the permafrost region of Tianshan Mountains in Xinjiang, China, the route should be selected to pass through the areas with less glaciers and less snow cover, and keep the route with short length as long as possible. The "Engineering Corridor" is a permafrost region with relatively developed economy, and also with convenient transportations, better terrain and permafrost conditions. Additionally, for the engineering corridor, there has rich research materials and engineering construction experience, which is beneficial to the construction of power transmission projects.
2. Compared with non-frozen soil areas, the engineering geological survey and investigation in permafrost regions require more time, more contents, with more diverse methods, and more complicated prediction and optimization. The investigation includes not only the frozen soil survey, hydrogeology and other data in the construction area, but also the optimization of the route on the basis of studying and compiling with the frozen soil engineering geological map (including the characteristics of frozen soil, ground temperature state of frozen soil, distribution of ground ice, lithological conditions, unfavorable cryogenic phenomena and geological and geomorphological conditions, etc.). By these means, it is possible to predict and avoid the possible frozen soil engineering problems.
3. It is advisable to avoid or shorten the route in the section with thick ground ice. Thick ground ice mostly develops in fine-grained soil layers (mainly silty clay), on the terrain with the slopes of 4–8°, on shady slopes and swamps and wetlands, in the area with lush vegetation, in lower parts of gentle slopes and low-lying areas in low mountain and hilly areas.
4. The route should be selected in the medium or high mountains and on sunny slopes where the buried depth of bedrock is shallow, in the area with bare and dry surface, and in the section with good frozen soil engineering geological conditions such as talik, ice-poor permafrost, icy permafrost and in coarse-grained soil, etc.
5. The route should avoid the areas with aufeis, pingos, thaw slumps, gelifluctions, thermokarst lakes and frozen swamps, and it should be arranged at a certain distance from the edge of those areas when it is inevitable.

7.1.2.2 Selection of Tower Location

Transmission lines in permafrost regions may pass through seasonally frozen soil, isolated permafrost, continuous permafrost, talik and other types of frozen soil areas, so the selection of tower location should be determined according to the characteristics of frozen soil. Generally, the selection of tower location in seasonally frozen soil areas and taliks can be determined according to the principles for non-frozen soil areas; the tower location in isolated permafrost region should be selected in the thawing zone or the zone with deeper permafrost table; the tower location in permafrost region should avoid thick ground ice and unfavorable cryogenic phenomena, and should be selected in ice-poor permafrost or icy permafrost area, or cold permafrost area. In summary, the principle for tower location selection in permafrost regions (Qian et al., 2009; Liu et al., 2008a, b, c) include: selecting the higher location instead of the lower, selecting the sunny slope instead of the shady slope, selecting the dry location instead of the wet, selecting the thawing location instead of the frozen, selecting the bare ground instead of the covered, selecting the coarse-grained soil instead of the fine-grained, selecting the avoidance of the risky section instead of passing through, and selecting the straight line instead of the winding.

1. Selecting the higher location instead of the lower. Under regular circumstances, the soil layers in the valleys and low-lying areas are mostly fine-grained soils with high ice content, high frost heave and thaw settlement features. The soil layers on the mountains or on the slopes are mostly crushed stone soils with low ice content, low frost heave and thaw settlement features, but measures for the stability of slopes still need to be taken. If the tower has to be built in valley areas, the terraces or tableland in higher places should be selected for the tower location.
2. Selecting the sunny slope instead of the shady slope. The tower location on the hillside should be chosen on the sunny slopes. On the sunny slope, the permafrost temperature is higher, the ice content is lower, the depth of permafrost table is deeper, and the frost heave and thaw settlement in the permafrost foundation are relatively small. When in a similar environment, the conditions of permafrost on the shady slopes are just opposite.
3. Selecting the dry location instead of the wet. The tower location should be selected in the area where the ground surface is relatively dry, with fewer surface ponds or strong soil permeability, or the terrain is relatively high. Generally, if the ice content of permafrost is low, the ground temperature will be high, and most of them belong to low-ice content permafrost, so the permafrost foundation has a relatively good stability. For a zone with relatively wet surface, and a relatively low terrain position, and also with a ground surface that is easy to accumulate water, as well as with soil of low permeability, it is likely that this zone is located in ice-rich permafrost area.
4. Selecting the thawing location instead of the frozen. There are river taliks, lake taliks, and radiation taliks in permafrost regions, as well as taliks at the edge of isolated permafrost regions. The tower should be located in the radiation talik or the talik outside the isolated permafrost, or on the high terrace of the river talik. The location for the tower should keep a considerable distance from the edge of the permafrost area, and it should avoid to be situated in the transitioned zones between thawing and freezing areas (the principle that frozen soil and thawed soil are not allowed to exist simultaneously under the building foundation should be implemented).

5. Selecting the bare ground instead of the covered. In the areas with high vegetation coverage, the terrain is usually low-lying, the ground surface is likely with ponds, the ice content of the permafrost is high, the ground temperature is relatively low, and the soil layer is mostly fine-grained soil with weak water permeability. Generally, the permafrost in these areas is relatively developed, and thick ground ice is very likely to occur. On the contrary, when the ground surface is relatively bare, the ground surface is unlikely with ponds, the terrain is likely higher, the permafrost is less developed, and the depth of permafrost table is deeper.
6. Selecting the coarse-grained soil instead of the fine-grained. The tower location should be selected in coarse-grained soil layers such as bedrock, pebble soil, crushed stone soil, etc., and need to avoid fine-grained soil layers such as thicker humus soil, cohesive soil, and fine silty sand. The coarse-grained soils are mostly permafrost of low ice content, with weaker frost heave and thaw settlement; while in fine-grained soils, it usually contains more ground ice and is likely to be permafrost with high ice content.
7. Selecting the avoidance of the risky section instead of passing through. In areas with unfavorable cryogenic phenomena such as thick ground ice, pingos, aufeis, thaw slumps, etc., "avoidance" should be the main strategy. When it is inevitable to passing through the risky section, the tower location should be determined according to the following principles: the thick ground ice section should be passed within the shortest distance in the area with narrow and thin ice layer; the section with thaw slumps and gelifluction should be passed keeping a certain distance away from the gentle slope; it is advisable to avoid passing through thermokarst lake; the section with pingos and aufeis should be passed on the basis of the local hydrogeological conditions by keeping a certain distance away, and drainage measures should be taken; the design principles of permafrost protection and special engineering measures should be adopted in frozen swamp section.
8. Selecting the straight line instead of the winding. Under the impact of long-term lateral force, the deformation caused by the lateral creep of the high ice content permafrost may lead to the inclination of the tower foundation, especially in the slope area. This impact may act with frost-heave force, increasing the instability of tower foundation.

7.2 Foundation Design

In the determination for the tower foundation type of transmission line in frozen soil regions, in addition to considering the safety level and types of the towers, the particularity of the frozen soil area should also be considered, such as the characteristics of the frozen soil (the types, ice content, mean annual ground temperature, permafrost table, etc.), unfavorable cryogenic phenomena, frozen soil environment, traffic conditions, construction techniques, etc. According to the operations of medium-high voltage alternating current and direct current transmission lines in the seasonally frozen soil areas and permafrost regions of Northeast China, the Qinghai-Xizang Plateau, and the Tianshan Mountains of Xinjiang, China, and with reference to foreign experience, the foundation types can be basically classified as excavated foundation, dug foundation and drilled foundation.

7.2.1 Buried Depth of Foundation

Frozen soil areas include the seasonally frozen soil areas and the permafrost regions. In these areas, the problems and methods are different relating to the buried depth of foundation.

7.2.1.1 Seasonally Frozen Soil Regions

In addition to meeting the stability requirements, the buried depth of tower foundation of transmission line in seasonally frozen soil regions mainly depends on the seasonal freezing depth and frost heave of the foundation soil. The freezing depth can be obtained by local weather stations or determined according to relevant data. The frost heave of foundation soil can also be obtained from surveys and tests.

When the foundation soil is non-frost heave, the buried depth of the foundation should be determined according to the requirements of conventional transmission lines.

When the foundation soil is weak-frost heave, the buried depth of the foundation should not be less than the design freezing depth.

When the frost heave of foundation soil is frost heave, strong frost heave and extremely strong frost heave, the buried depth of base plate should be set at the place 0.5 m below the design depth line.

Judging from the factors and laws of frost heave of foundation soil, the conditions of the foundation soil that lead to weak-frost heave include: first, the fine-grained soil whose water content is about at the plastic limit of water content, or the coarse-grained soil with the lowest water-holding capacity; second, the depth of groundwater level is very deep, or the distance between the groundwater level before freezing and the design freezing depth is more than 1.5 m, and there is no water migration during the period of freezing. A large number of experiments and field observation data show that the frost heave amount of foundation soil is distributed along the freezing depth, mainly concentrated in the depth of 1/3 to 2/3 of the freezing depth below the surface, accounting for about 80% of the total frost heaving amount. On the basis of meeting the requirements for the maximum frost jacking force of the tower and economic comparison, after the foundation sides are painted with grease or residual oil or replaced with 0.2 m coarse-grained soil, and the foundation bases should be replaced with 0.5 m coarse-grained soil, the buried depth of the foundation can be considered to be placed within the range of design freezing depth.

7.2.1.2 Permafrost Regions

In addition to meeting the stability requirements, the buried depth of tower foundation of the transmission line in permafrost regions mainly considers the characteristics of the permafrost (types of permafrost, ice content, mean annual ground temperature), permafrost table, frost heave and thaw settlement of permafrost foundation, and then determines the design principle for the foundation, that is, to maintain the permafrost foundation in a frozen state, gradually thawing, pre-thawing design principle, and determine the types and buried depth of the foundation.

The design principle of maintaining the frozen state of permafrost foundation is suitable for the areas with low temperature, high ice content and continuous permafrost, and the section with thick permafrost layer, low mean annual ground temperature and good thermal stability of permafrost foundation, or the sections where the freezing strength and the bearing capacity of the permafrost within the buried depth of the foundation can still meet the stability requirements for the tower foundation if the permafrost temperature rises. The buried depth of foundation should be determined by frost jacking resistance, stability of frost-heave force and thermal calculation, but the minimum buried depth should not be less than 0.5 m below the maximum (or design) seasonal thawing depth. Generally, the following requirements can be roughly met: the minimum buried depth of excavated foundation and dug foundation is obtained based on the design thawing depth of (Z_d^m) plus 1 m; and the pile foundation is obtained based on the design thawing depth plus 2 m.

The design thawing depth should adopt the natural maximum seasonal thawing depth in the building section. If there is no measured data, the standard thawing depth can be obtained by using the calculation formula of permafrost thawing depth, and then corrected by factors such as soil quality, humidity, topography and surface conditions, etc.

The calculation formula for design thawing depth is:

$$Z_d^m = Z_0^m \cdot \psi_s^m \cdot \psi_\omega^m \cdot \psi_c^m \cdot \psi_{to}^m \tag{7.1}$$

where Z_0^m is the standard thawing depth, which should be determined according to the local measured data. If there is no measured data, it can be calculated according to the formula 7.2~7.6.

ψ_s^m, ψ_ω^m, ψ_c^m and ψ_{to}^m are the correction coefficients for the effects of soil, humidity (thaw settlement), surface cover and topographic factors on the thawing depth, respectively, and the values are taken according to Table 7.1.

In high-altitude permafrost regions (on the Qinghai-Xizang Plateau):

$$Z_0^m = 0.195\sqrt{\Sigma T_m} + 0.882 \tag{7.2}$$

In high-latitude permafrost regions (in Northeast China):

$$Z_0^m = 0.134\sqrt{\Sigma T_m} + 0.882 \tag{7.3}$$

Table 7.1 Correction coefficients of various factors affecting thawing depth

	ψ_s^m				ψ_ω^m					ψ_{to}^m			ψ_c^m	
Items	Effects of soil quality (lithology)				Effects of humidity (thaw settlement)					Effects of terrain			Effects of surface cover	
Content	Cohesive soil	Fine sand, silty sand and silty soil	Medium sand, coarse sand and gravelly sand	Crushed stone soil	Non-thaw settlement	Weak thaw settlement	Thaw settlement	Strong thaw settlement	Thaw collapse	Flat	Shady slope	Sunny slope	Bare ground surface	Turfy cover on ground surface
ψ	1.00	1.20	1.30	1.40	1.00	0.95	0.90	0.85	0.80	1.00	0.90	1.10	1.00	0.70

where $\sum T_m$ is the standard value of the thawing index of local measured air temperature (degree · month), and in the mountain area without measured data, the value can be calculated according to the following formula.

In Northeast area:

$$\Sigma T_m = (7532.8 - 90.96L - 93.57H) / 30 \tag{7.4}$$

In Qinghai region:

$$\Sigma T_m = (10722.7 - 141.25L - 114.00H) / 30 \tag{7.5}$$

In Xizang region:

$$\Sigma T_m = (9757.7 - 71.81L - 140.48H) / 30 \tag{7.6}$$

where L is the latitude of the tower foundation (°); H is the altitude of the tower foundation (100 m).

The design principle for gradually thawing state is suitable for bedrock foundation and coarse-grained soil foundation with ice-poor or icy permafrost, and the mean annual ground temperature of the permafrost belongs to the unstable ground temperature zone ($-1.0\ °C < T_{cp} < -0.5\ °C$). If the permafrost foundation is fine-grained soil with low ice content, and the bearing capacity and thaw settlement deformation of the foundation after the thawing can meet the requirements for the tower foundation, the principle for gradually thawing state design can be adopted. Otherwise, corresponding engineering measures should be taken. The buried depth of the tower foundation should be determined by the frost jacking resistance, the stability of frost-heave force and thermal calculation.

The design principle for pre-thawing state is suitable for the frozen soil with high ice content where the mean annual ground temperature of the permafrost is in an extremely unstable state ($T_{cp} > -0.5\ °C$), or the discontinuous permafrost region with the depth of permafrost table exceeding 4–7 m, or the isolated permafrost region with thinner permafrost, or sparse and sporadic permafrost regions with very thin permafrost, or taliks. If the bearing capacity and deformation of the permafrost foundation after thawing can meet the requirements for the tower foundation, it can be determined according to the relevant requirements of the seasonally frozen soil areas.

7.2.2 Foundation Types

The types of tower foundations in permafrost regions can be classified as excavated foundations, dug foundations and drilled foundations according to construction technology.

7.2.2.1 Excavated Foundation

Excavated foundations include tapered column foundation, straight (cylindrical) expanded foundation, assembled foundation, independent foundation, etc. In seasonally frozen soil areas and permafrost taliks, the main purpose is to prevent frost heave. Under the condition of meeting the requirements for the maximum frost jacking force of the steel tower, the buried depth of the foundation should usually be placed on the top surface of the foundation plate below the maximum seasonal freezing depth. In areas with good geological conditions, it can be buried below the depths of 2/3–4/5 of the maximum seasonal freezing depth. In the permafrost region, due to the great thermal disturbance caused by the excavation of the foundation (heating and groundwater erosion, see Figs. 7.6 and 7.8), when the ground temperature of the permafrost rises, and the depth of the permafrost table increases. The design depth of the foundation not only needs to consider these factors listed in Table 7.1, but also needs to consider the thermal effects brought by the heavy excavation. Combined with different topography and frozen soil conditions, considering the change of the influence depth of 1–2 m during freeze-thaw cycle, the design buried depth should be increased accordingly. At present, the buried depth of heavy-excavated foundation for transmission line of the Qinghai-Xizang direct-current power is generally 2–3 m below the permafrost table, and the maximum depth is up to 4 m (Yan et al., 2012).

1. **Tapered column foundation**

Tapered column foundation is suitable for seasonally frozen soil regions, permafrost with low ice content, coarse-grained ice-rich frozen soil areas with even ice content distribution, taliks and bedrocks. Methods, such as heavy excavation of the foundation pit, pouring the base plate, hanging the fiberglass reinforced plastic mold of the tapered column (for anti-freezing) and then pouring the concrete, are adopted

Fig. 7.8 Steel cylinder wall protection and pumping measures

(Figs. 7.9 and 7.10). However, it has great damage to the frozen soil environment, and has great thermal erosion to the permafrost, and also has a large disturbance to the hydrogeological conditions of the frozen soil. Besides, the laboring is highly in need, and the construction in bedrock area is difficult.

During the freezing the trapezoidal sectional foundation can make the frozen soil and the foundation slope cracked and separated, and the frost jacking of the tangential frost-heave force on the foundation can be reduced. In Russia, the reinforced concrete foundation is made into a trapezoid, with the slope angle of 1–3°. According to the test on the pile foundation of the buildings by Heilongjiang Institute of Building Sciences in cold regions, the minimum slope angle for weakening the tangential frost-heave force on the foundation side is 9°. If the skew angle of the arris of the tapered column foundation is considered as 9°, the optimization can be carried out by tests.

If the tapered column foundation is a sloping structure (Fig. 7.11), a precast glass fiber reinforced plastic mold, with the characteristics of smooth surface and weak hydrophilicity can be used, which can significantly reduce the tangential frost-heave force on the side of the foundation. The buried depth of the base plate all exceeds the maximum seasonal freezing depth or the permafrost table.

On the basis of giving full play to the frost heave resistance of the tapered column foundation structure, the upright column is optimized and improved (Fig. 7.12) to avoid the waste of materials and reinforcement due to the oversized diameter of the upright column bottom.

1. Tapered column and benched foundation (Fig. 7.12a): the deadweight is large. The frost jacking force is larger, the surrounding soil is light, the groundwater level is shallower, and the buried depth of foundation is shallower, which has

Fig. 7.9 Cast-in-place tapered column foundation

Fig. 7.10 Construction of assembled foundation (all photos by Yu Qihao)

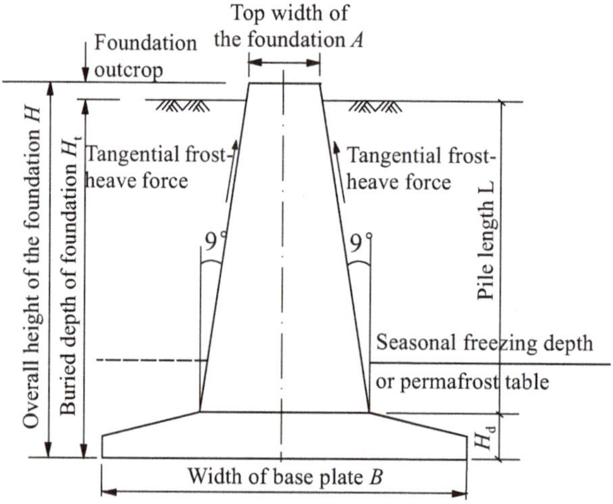

Fig. 7.11 Schematic diagram of conventional tapered column foundation

good applicability to the foundation tower position where the pit wall is easy to collapse.

2. Tapered column and straight column and mat foundation (Fig. 7.12b): it is suitable for the foundation with large buried depth, to eliminate frost-heave force to the greatest extent and save concrete materials, and the sectional diameter at the bottom of the tapered column is beneficial to the control of reinforcement and the spacing of main reinforcement.
3. Straight column and tapered column and mat foundation (Fig. 7.12c): it is suitable for excavated foundation of tower in gentle slope areas, which is used in permafrost regions with weak frost heave. It saves material and is conducive to the control of reinforcement arrangement and the spacing of the main reinforcement of the column.

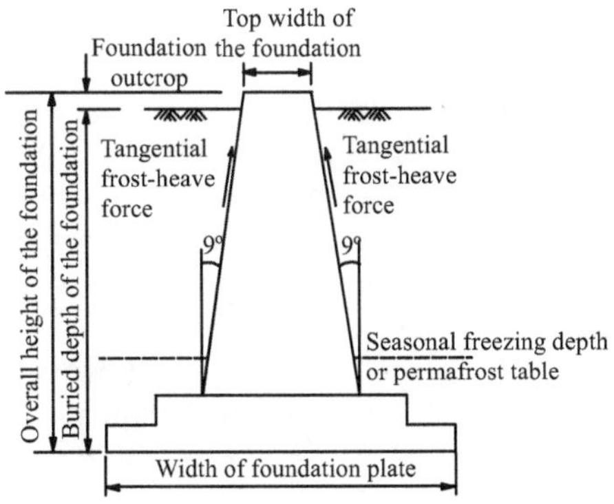

(a) Tapered column and benched foundation

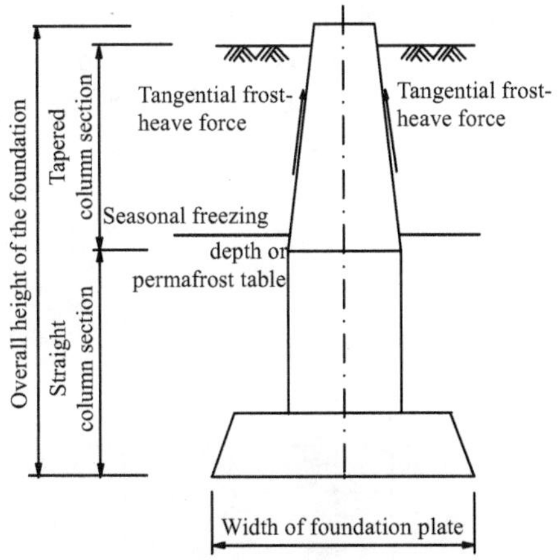

(b) Tape ring column and straight column and foundation base plate

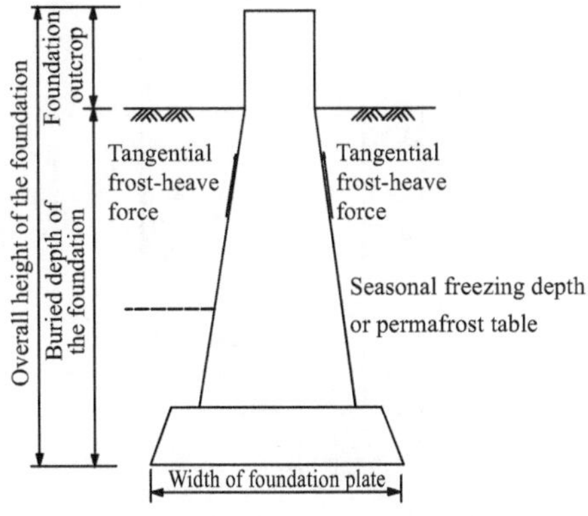

(c)Straight column and tapered column and foundation base plate

Fig. 7.12 Conventional and optimized tapered column foundations (Hu et al., 2012). (**a**) Tapered column and benched foundation; (**b**) Tape ring column and straight column and foundation base plate; (**c**) Straight column and tapered column and foundation base plate

2. **Assembled foundation**

Assembled foundations are suitable for non-frozen soil areas, taliks, bedrock in permafrost regions, areas of low ice content permafrost and ice-rich coarse-grained permafrost, and the foundations without frost heave or with weak frost heave. It has the advantages of prefabrication, easy quality assurance, on-site assembly, short construction period, good frost jacking resistance, compression and overturning resistances, etc., (Fig. 7.13). However, it is necessary to have convenient transportation and hoisting, and the use of heavy excavation process will have a large thermal erosion effect on the permafrost, and will cause great damage to the environment. Winter construction can be employed in the mountainous areas with deeper seasonal freezing depth and high plains in permafrost regions.

3. **Independent foundation**

The guyed V-type tower (Liu et al., 2008a, b, c) is a level-one tower, with only one independent foundation set for each tower. It is built with small excavation area, less thermal disturbance to the frozen soil and lower probability in occurrences of differential deformation, and it can be used in taliks and permafrost areas with low temperature and high ice content. If the large-diameter rotary drilling technology is used, it can also be applied to warm permafrost with high ice content. Due to the long refreezing time, it is easily affected by strong winds and icy cover, etc., so it should not be used in the area with strong winds, aeolian sand and easily covered with ice.

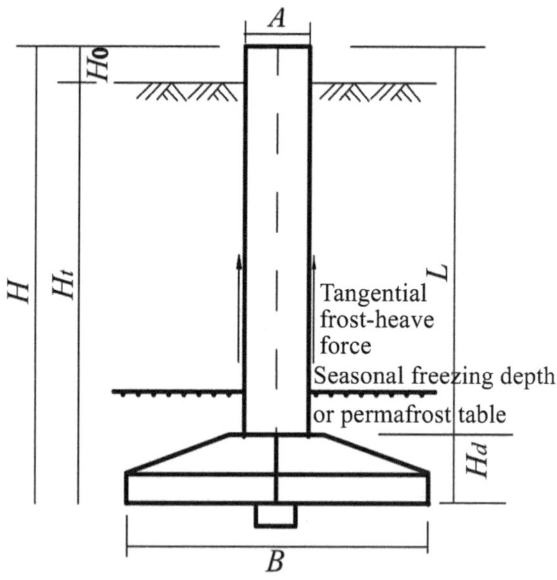

Fig. 7.13 Schematic diagram of assembled foundation

7.2.2.2 Dug Foundation

The dug foundation (Fig. 7.14) is suitable for non-permafrost regions, permafrost bedrock, taliks, permafrost areas with low ice content, and foundation soils without frost heave or with very weak frost heave. If the geological conditions are good (such as mudstone or sandstone), the groundwater level is deep, the soil is fine, the section is easy to be shaped, and the dug foundation is usually applied on slopes. The dug foundation can reduce the disturbance to the original soil of the foundation and give full play to the bearing capacity of the frozen soil foundation. In order to maintain the stability of the hole wall and avoid its collapse, the construction should be carried out in winter. And it is advisable to increase the buried depth of the foundation, make full use of the lateral soil pressure, reduce the overturning moment of the horizontal force test, reduce the bending moment of the foundation pile body and the eccentric stress at the bottom of the foundation, and improve the lateral stability and bearing capacity of the foundation.

The dug foundation includes the excavated foundation and the manual-dug foundation. The buried depth of the excavated foundation is shallow while the buried depth of manual-dug foundation is deep. According to engineering experience, the ratio of the diameter of the expanded base of the dug foundation to the diameter of the pile body should not be greater than 3.0, and the difference between the two should not be greater than 2.0 m.

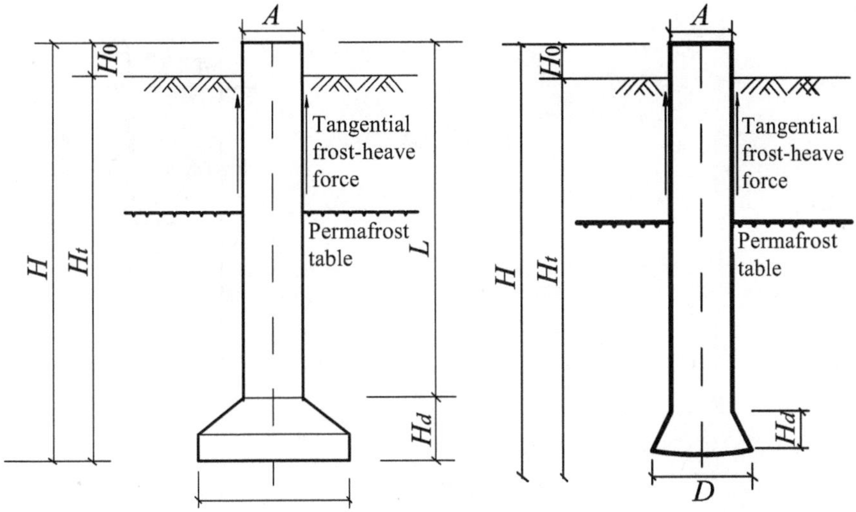

Fig. 7.14 Schematic diagram of dug foundation (on the Left) and hand-dug foundation (on the right)

7.2.2.3 Drilled Foundation

The drilled foundation includes cast-in-place pile foundation, shallow pile foundation, small pile cap foundation, expanded-base pile foundation, and explosion-expanded pile foundation. It can be used in seasonally frozen soil and permafrost regions where the tower location is not suitable for tapered column foundation. In permafrost regions, it is more used for the permafrost with high temperature and high ice content and the permafrost with low temperature and high ice content.

1. **Pile foundation**
The pile foundation (Fig. 7.15) is suitable for permafrost regions and seasonally frozen soil areas with all kinds of loose materials. The pile foundations currently used are mainly cast-in-place pile and insert pile in borehole. The Pore-forming methods include: rotary drilling, large-aperture drilling, augering, percussion drilling, under-reamed boring, explosion-expanded pile, etc. For the cast-in-place piles which poured the concrete on site, it disturbs the ground surface very badly, especially the thermal disturbance to the permafrost, which causes the longer refreezing time. The inserted pile uses piles that precast, which disturbs the ground surface less, but the traffic work is larger, which is difficult for applying in mountainous areas.

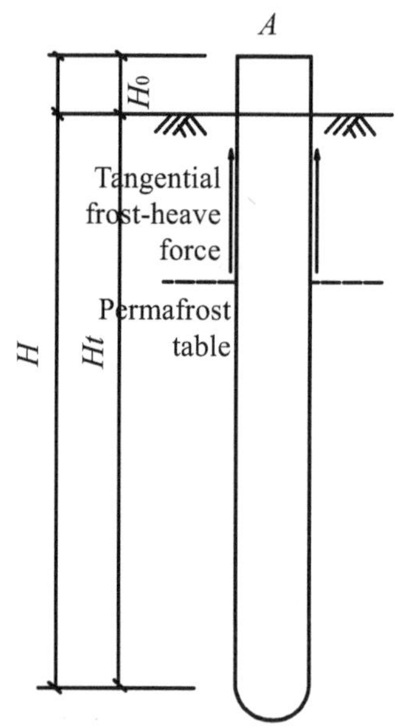

Fig. 7.15 Pile foundation

2. **Special-shaped pile foundation**

Due to the limitations of site and construction conditions, it can be designed as other types of pile foundations.

1. Small pile cap foundation (Fig. 7.16a): due to the limitations of drilling tools, site conditions or geological conditions, the small pile diameters, the multiple-pile foundations, and the bearing platform connecting mode can be used as tower foundations.
2. Ramming expanded pile foundation (Fig. 7.16b): in the fine-grained soil section with warm and extremely unstable permafrost zone, if the buried depth of the foundation is shallow, the ramming expanded pile construction can be used to expand the pile tip and increase the bearing capacity of the foundation and strengthen the anti-frost jacking performance to maintain the stability of tower foundation.
3. Explosion-expanded pile Foundation (Fig. 7.16c): in the warm permafrost region, if the buried depth of the foundation is shallow and it is not suitable for excavated foundation construction, the explosion-expanded pile construction can be used to expand the pile tip and increase the bearing capacity of the foundation and strengthen the anti-frost jacking performance to maintain the stability of the tower foundation.

The above mentioned three kinds of foundation types are basically the tower foundation types used for transmission lines in the seasonally frozen soil areas and the permafrost regions. For the fact that the route of the transmission line is in hilly and mountainous areas, especially in the seasonally frozen soil area, in the taliks in

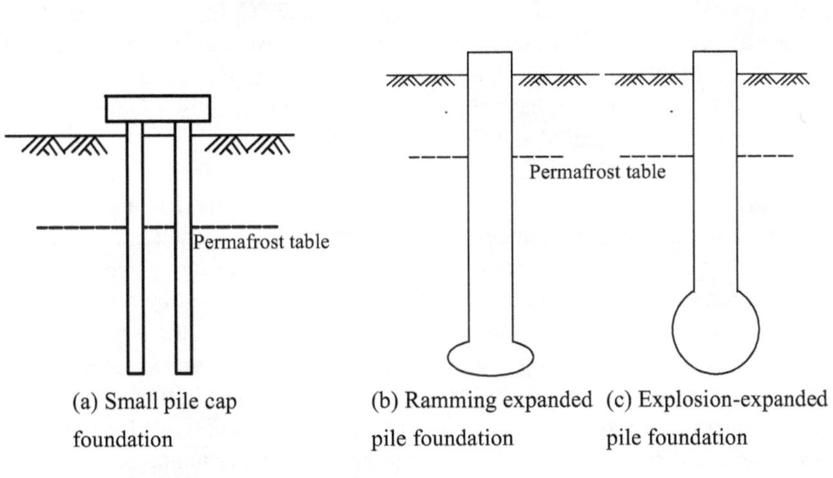

Fig. 7.16 Special-shaped pile foundation. (**a**) Small pile cap foundation; (**b**) Ramming expanded pile foundation; (**c**) Explosion-expanded pile foundation

permafrost region, and the coarse-grained frozen soil areas with high temperature and low ice content, the excavated foundation is widely used. The advantages, applicable conditions and construction requirements of various foundation types are summarized in Table 7.2.

7.2.3 Ground Temperature State of the Foundation

7.2.3.1 Ground Temperature of Tower Foundation During Freezing Period

The construction method of the excavated foundations is likely to cause great thermal disturbance to the permafrost foundation. Thus, the construction period is basically selected in winter, and the rapid construction of the foundation pit excavating, pouring and backfilling process should be achieved. After the completion of the project, the observation data of the overall temperature state of the tower foundation after the first freezing period show that (Yu et al., 2013), 10 monitoring tower foundations are all in the frozen state, and the temperature state above the permafrost table is basically the same as the natural temperature, and there are differences below the permafrost table; the cold permafrost areas or the permafrost areas with low ice content are basically the same, while the warm permafrost areas and the permafrost areas with high ice content are quite different, especially for the tower foundation constructed at the end of autumn when the ambient temperature is high, the differences are more prominent. For example, in the Xieshuihe site, the mean annual ground temperature of the permafrost is −0.07 °C, the tower foundation is greatly affected by the construction and there is a positive temperature zone higher than 0 °C within the 2 m depth range of the tower foundation. After 4 months of freezing process, although it has been frozen, the temperature is still about 0.25 °C higher than that in the natural site (Fig. 7.17).

During the thawing period, the ground temperature at the tower foundation increases progressively, which is shown as follows: ① after the thawing stage in the warm season, the permafrost under the base of the tower foundation is still in frozen state; ② the variations of ground temperature, the warming rate and the thawing depth around the tower foundation are mainly affected by the types of tower foundation and permafrost conditions; ③ the ground temperature at the bottom of the excavated foundation is slightly higher than that of the natural ground temperature, and the increase is less than 0.5 °C; ④ due to the construction in winter, the backfill cannot reach the density required by the design and there are a large number of pores and fissures; the water infiltration leads to the temperature rising of permafrost, increases the thawing rate and thawing depth of the tower foundation, and aggravates the frost heave and thaw settlement deformation of the tower foundation.

Table 7.2 List of evaluations of various foundation types

Foundation types	Code names	Names	Schematic diagrams	Applicable conditions	Advantages and disadvantages	
					Advantages	Disadvantages
Excavated foundation	I	Taper-faced rigid foundation	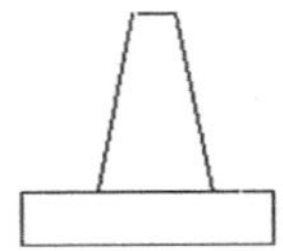	1. non-permafrost regions; 2. permafrost bedrocks and taliks; 3. permafrost regions with low ice content; 4. the coarse-grained soil section with ice-rich frozen soil where the underground ice is distributed uniformly; 5. the above-mentioned areas with frost heave.	The construction is easy; there are not terrain requirements; it has certain anti-frost heaving performance;	It has great damages to the environment; it has great thermal erosion effects on frozen soil;
	II	Straight-faced rigid foundations	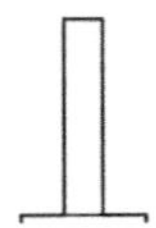	1. non-permafrost regions; 2. permafrost bedrocks and taliks; 3. permafrost regions with low ice content; 4. the coarse-grained soil section with ice-rich frozen soil where the underground ice is distributed uniformly; 5. the foundation soil without frost heave or with very weak frost heave.	The construction is easy; there are no terrain requirements	It has great damages to the environment; it has great thermal erosion effects on the frozen soil
	III	Columnar assembled foundations	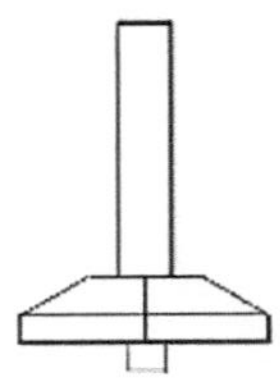	1. non-permafrost regions; 2. permafrost bedrocks and taliks; 3. permafrost regions with low ice content; 4. the coarse-grained soil section with ice-rich frozen soil where the underground ice is distributed uniformly; 5. the foundation soil without frost heave or with very weak frost heave.	The construction is easy; there are no terrain requirements.	It has great damages to the environment; It has great thermal erosion effects on frozen soil.
Dug foundations	IV	Manuel-dug foundation	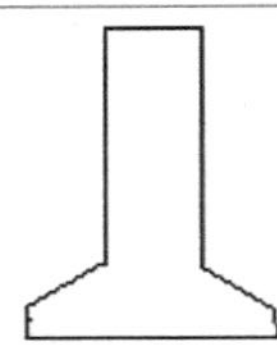	1. non-permafrost regions; 2. permafrost bedrocks and taliks; 3. permafrost regions with low ice content; 4. the foundation soil without frost heave or with very weak frost heave; 5. the section with fine-grained soil and deep groundwater level.	The construction is easy; there are no terrain requirements; quarry rubbish is less, which is beneficial to environmental protection	It has damages to the environment; it has great thermal erosion effects on frozen soil; Requirements for the molding of soil body are high.

(continued)

Table 7.2 (continued)

Foundation types	Code names	Names	Schematic diagrams	Applicable conditions	Advantages and disadvantages	
					Advantages	Disadvantages
Drilled foundations	V	Cast-in-place deep pile foundation	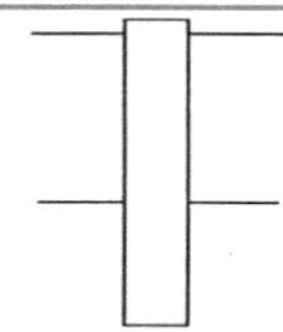	1. permafrost regions with flat terrain, high or low temperature with high ice content; 2. the sections where machinery and equipment can be easily transported.	It adapts to frozen soil protection	The relatively flat terrain is required; it takes longer time for permafrost to refreeze
	VI	Shallow pile foundations	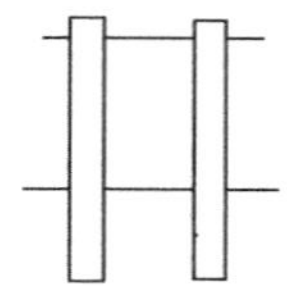	1. permafrost regions with flat terrain, high or low temperature and high ice content; 2. the sections where machinery and equipment can be transported.	It is conductive for frozen soil protection.	Drilling rig and hoisting equipment are required; there are certain requirements for the terrain; more transportation works are needed; It takes a long time for the frozen soil around the cast-in-place pile to refreeze.
	VII	Small pile cap foundation	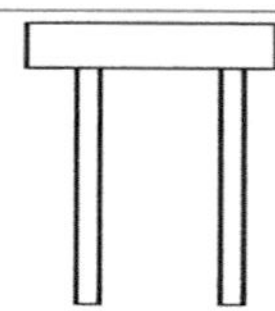	1. permafrost regions with high or low temperature and high ice content; 2. the sections where machinery and equipment can be transported.	It is conductive for frozen soil protection.	Drilling equipment is required
	VIII	Ramming expanded foundation	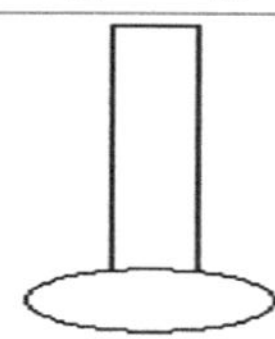	1. the section with finer and softer active layer; 2. permafrost sections with the permafrost table deeper than 2 meters; 3. the section with relatively flat terrain; 4. the sections where machinery and equipment can be easily transported to.	It adapts to frozen soil protection.	Special equipment is required.

Foundation types	Construction technology		Environmental evaluations	Existingproblems
	Technological requirements	Construction season		
Excavated foundation	1. The heavy excavation is needed; 2. sun-shading measures are required for the construction in permafrost regions; 3. pumping equipment is required to avoid water accumulation in the foundation pit; 4. backfilling should be distributed uniformly layer by layer to meet compaction requirements; 5. environmental protection measures should be strictly taken; 6. the operation should be carried out in strict accordance with the requirements of the thermosyphon construction technology.	In seasonally frozen soil area: from May to October; in permafrost regions: from June to October.	It has great damage to the environment, especially the frozen soil environment; the operation should be carried out in strict accordance with environmental protection requirements; areas that taken soil from or discarded soil to should meet environmental protection requirements.	How to prevent the thermal effects of construction on frozen soil; how to deal with the construction when the groundwater level is high
	1. The heavy excavation is needed; 2. sun-shading measures are required for the construction in permafrost regions; 3. pumping equipment is required to avoid water accumulation in the foundation pit; 4. backfilling should be distributed uniformly layer by layer to meet compaction requirements; 5. environmental protection measures should be strictly taken; 6. anti-frost heaving measures should be taken on the side of the foundation.			
Dug foundations	1. Construction of open caisson method is required; 2. Measures for preventing the collapse during digging and expanding should be well prepared; 3. Measures for the environment protection should be strictly taken.			Reliable measures for preventing the collapse during digging and expanding should be taken.

(continued)

Table 7.2 (continued)

Foundation types	Construction technology		Environmental evaluations	Existingproblems
	Technological requirements	Construction season		
Drilled foundations	1. Rotary drilling should be used for the construction; 2. the work should be carried out in strict accordance with the operating procedures; 3. measures for environment protection should be taken strictly; 4. the molding temperature of pouring concrete should be controlled at 5 to 10 °C.	The construction season is from October of the current year to June of the next year	Protective measures for the bringing of machineries to the site and discarding spoils, etc., should be taken in strict accordance with the requirements for environmental protection.	It has high requirements for construction equipment and staff
	1. Drilling machinery for construction should be used; 2. work should be carried out in strict accordance with the operating procedures; 3. taking thermal insulation measures can extend the construction period in winter; 4. measures for environment protection should be well prepared; 5. the molding temperature of pouring concrete should be controlled at 5 to 10 °C; 6. The operation should be carried out in strict accordance with the requirements for the construction technology of thermosyphon.	For precast driven piles: from July to November; for cast-in-place pile: from September to November or from February to July.		The construction is difficult.
	1. Drilling machinery for construction should be used; 2. work should be carried out in strict accordance with the operating procedures; 3. taking thermal insulation measures can extend the construction period in winter; 4. measures for environment protection should be well prepared; 5. the molding temperature of pouring concrete should be controlled at 5 to 10 °C.	Construction season: from September to November or from February to July.		
	1. Tamping machine for construction should be used; 2. works should be carried out in strict accordance with the operating procedures; 3. taking thermal insulation measures can extend the construction period in winter; 4. measures for environment protection should be well prepared.	Construction season: from August to November.		It has high requirements for the construction equipment and staff.

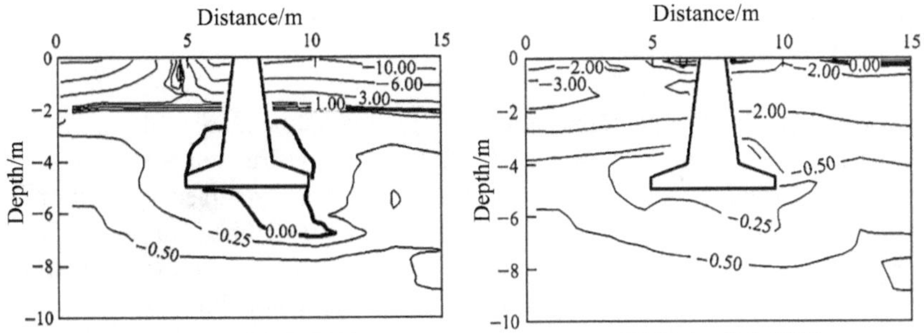

(a) In January 5, 2011 (b) In April 5, 2011

Fig. 7.17 Ground temperature isotherm of Xieshuihe tower foundation (°C) (Yu et al., 2013). (**a**) In January 5, 2011; (**b**) In April 5, 2011

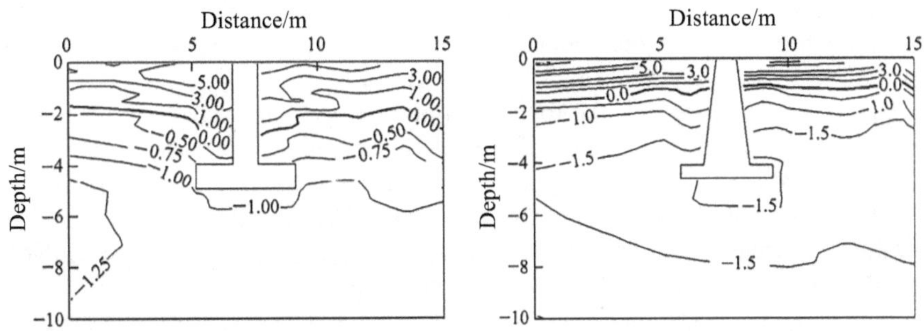

(a) Chumar river monitoring field in the permafrost area with low ice content (b) Beiluhe monitoring field in the permafrost area with high ice content

Fig. 7.18 Ground temperature isotherms of tower foundations in the permafrost areas with high ice content or low ice content (°C) (Yu et al., 2013). (**a**) Chumar river monitoring field in the permafrost area with low ice content; (**b**) Beiluhe monitoring field in the permafrost area with high ice content

7.2.3.2 Ground Temperature of Tower Foundation Under Different Permafrost Conditions

During the thawing period, in the warm permafrost area, although the cooling measure of thermosyphon is used, the effect is still poor. The thawing of ground ice below the permafrost table absorbs a lot of heat and delays the temperature rise of the permafrost. The thawing depth in the warmest month in the permafrost area with low ice content is 1.0 m deeper than that in the natural site, whereas it is basically the same in the permafrost area with high ice content (Fig. 7.18), and when the differences in thawing depth between the two is about 1.5 m, the thawing rate of the permafrost with low ice content will be 2–3 times higher than that of the permafrost with high ice content. The maximum thawing depth in cold permafrost area is 10–20 days earlier than that in warm permafrost. The thawing depth of excavated foundation is 1.0 m deeper than that of natural site, and some can reach 2 m.

7.2.3.3 Ground Temperature of Tower Foundation of Different Foundation Types

The influences of the foundation types on the ground temperature of the tower foundation are shown as follows. ① The influence from the construction process is that the construction techniques of tapered column foundation, assembled foundation and other foundations are proceeded by means of excavating and backfilling. Due to the large amount of excavation (the foundation pit of the tapered column foundation is 4–6 m, and the foundation pit of the assembled foundation is 4–5 m), the thermal state of permafrost foundation is strongly disturbed, the stratigraphic structure and hydrogeological conditions are destroyed, the hydrothermal dynamics and state of permafrost foundation are changed, and a large amount of heat is brought in so that the temperature of permafrost foundation is increased. Cast-in-place pile and dug foundation have less thermal disturbance to the permafrost foundation. ② The compactness of backfill for excavated foundations cannot meet the design requirements, especially in winter construction, there are many voids and cracks remaining in the backfill of frozen soil block, which becomes the channel for surface water and groundwater to seeping into the foundation pit during the thawing period and also delays the temperature refreezing of the tower foundation. However, there are no backfilling problems for cast-in-place piles and dug foundations. ③ The heat of hydration from concrete pouring of tower foundation has an influence on the ground temperature of permafrost, while there is no influence of hydration heat on the assembled foundation, for the tapered column foundation is larger than the pile foundation.

Summarizing the variations of ground temperature in the initial stage of the construction of tower foundation, it indicates that the thermal disturbance caused by excavation construction technology to the permafrost foundation appears to be more intense, and the influence gradually weakens along with the time. After 1.5 freeze-thaw cycles (from January, 2011 to April, 2012), the ground temperature monitoring from different foundation types (Fig. 7.19) shows that the ground temperature around the cast-in-place pile and dug foundation is basically the same as that of permafrost under natural surface, and the ground temperature of permafrost in the backfill for excavated foundations rises slightly, with the increase of less than 0.5 °C, but the thawing depth is larger than that of the natural site.

7.2.3.4 Ground Temperature of the Tower Foundation Under Engineering Measures

Under global warming, the ground temperature of permafrost has also increased, causing the thawing of ground ice and the deformation of the tower foundation. For this reason, in the permafrost region, the function of convective heat transfer of the thermosyphon is employed to increase the cold storage of the permafrost foundation, decrease the ground temperature, and protect the thermal stability of the permafrost under the tower foundation.

The monitoring data of the ground temperature of tower foundation after installing the thermosyphon at different sites show that the thermosyphon has a very

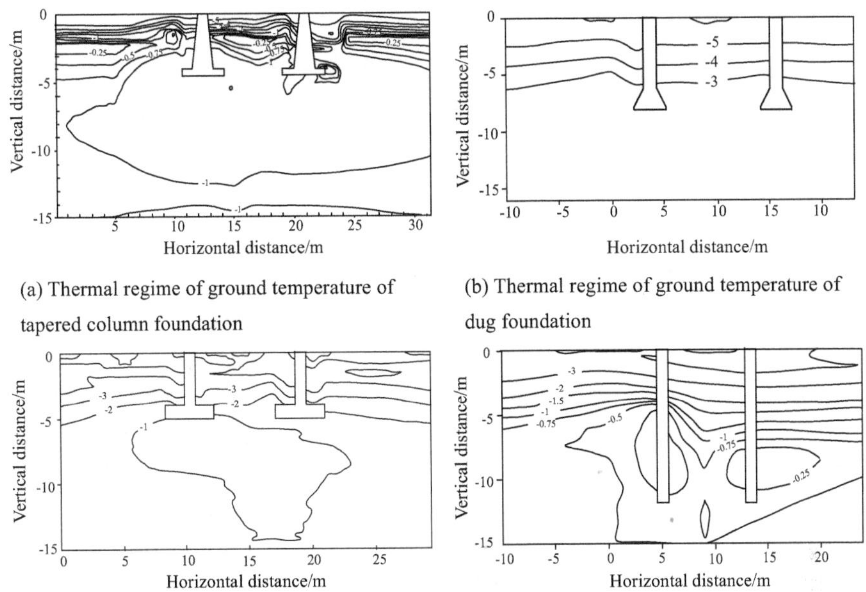

(a) Thermal regime of ground temperature of tapered column foundation

(b) Thermal regime of ground temperature of dug foundation

(c) Thermal regime of ground temperature of assembled foundation

(d) Thermal regime of ground temperature of cast-in-place pile foundation

Fig. 7.19 Influence of foundation types on the ground temperature of tower foundation (Huang et al., 2016). (**a**) Thermal regime of ground temperature of tapered column foundation; (**b**) Thermal regime of ground temperature of dug foundation; (**c**) Thermal regime of ground temperature of assembled foundation; (**d**) Thermal regime of ground temperature of cast-in-place pile foundation

obvious effect on reducing the ground temperature of permafrost under the tower foundation (Fig. 7.20). Within the depth range above the base of the tower foundation (at −4 m to −1 m), the ground temperature difference at the same depth of tower foundation between thermosyphon is 1.0–1.05 °C, while the ground temperature difference is 1.5–2.0 °C within the depth range below the bottom (at −10 m to −4 m). Under the cooling effect of thermosyphon, the cooling amplitude at the same depth of the foundation is 1.0–1.5 °C larger than that of the natural site, and the difference in the depth of the isotherm of ground temperature is 2 m, and the maximum is up to 3.2 m.

7.2.4 *Deformation of Tower Foundation*

From the monitoring data, the vertical deformation of the tower foundation mainly includes settlement deformation (compaction deformation and thaw settlement deformation) and frost-heaving deformation. After 1.5 freeze-thaw cycles, most of the vertical deformations tends to be stable (Fig. 7.21a). The vertical deformation of tower foundation mainly occurs in the initial stage of construction, under the impact

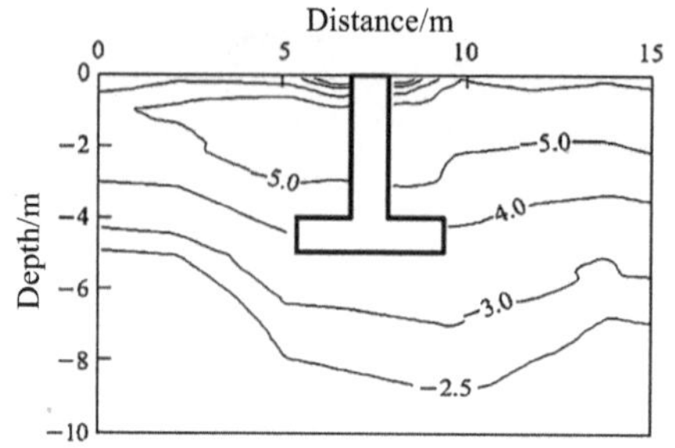

(a) Sites with thermosyphon

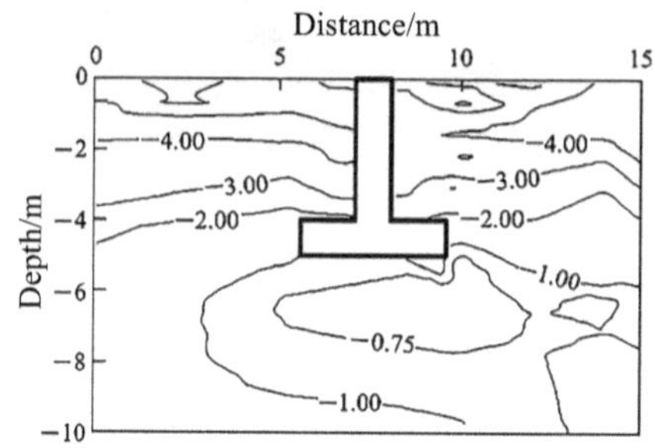

(b) Sites without thermosyphon

Fig. 7.20 Influence of the thermosyphon on the thermal regime of ground temperature at the tower foundation (Yu et al., 2013). (**a**) Sites with thermosyphon; (**b**) Sites without thermosyphon

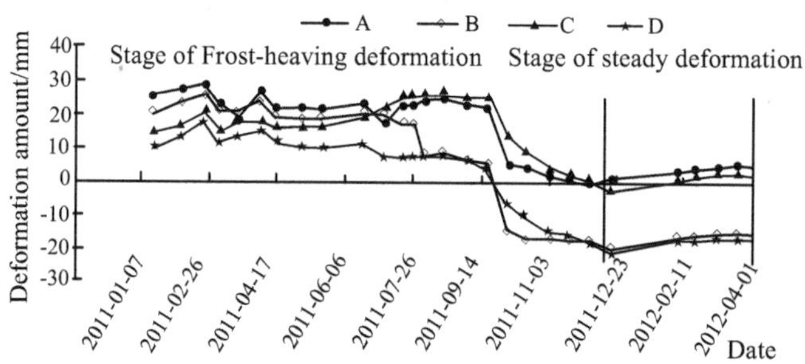

(a) Schematic diagram of vertical deformation of tower foundation

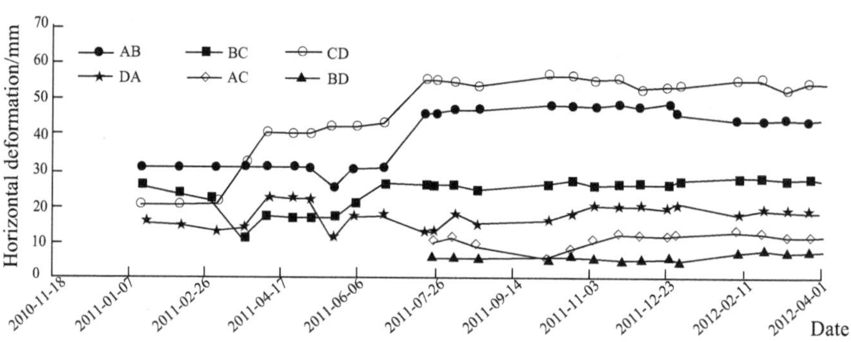

(b) Schematic diagram of the horizontal deformation of the tower foundation

Fig. 7.21 Schematic diagram of the tower foundation deformation (Huang et al., 2016). (**a**) Schematic diagram of vertical deformation of tower foundation; (**b**) Schematic diagram of the horizontal deformation of the tower foundation

of the self-weight of the tower foundation and the heat storage of construction (including water infiltration), the compaction of the backfill and the thaw settlement of the frozen soil (especially the warm permafrost with high ice content). After the completion of the construction in winter (in December, 2010), the tower foundation was frozen (Fig. 7.22), resulting in frost-heaving deformation; and then, thaw settlement deformation occurred in the warm season of the following year, and some ground surface subsided significantly (Fig. 7.23). Under the impacts of self-weight and tower load, the vertical deformation of the tower foundation gradually faded, and the tower foundation were in stable state.

The plane deformation of tower foundation is mainly reflected in the variations of the tower-leg spacing. In addition to the large change in the initial stage, under the impact of vertical pressure or tensile force, it remains basically stable in the subsequent process, with insignificant impact (Fig. 7.21b).

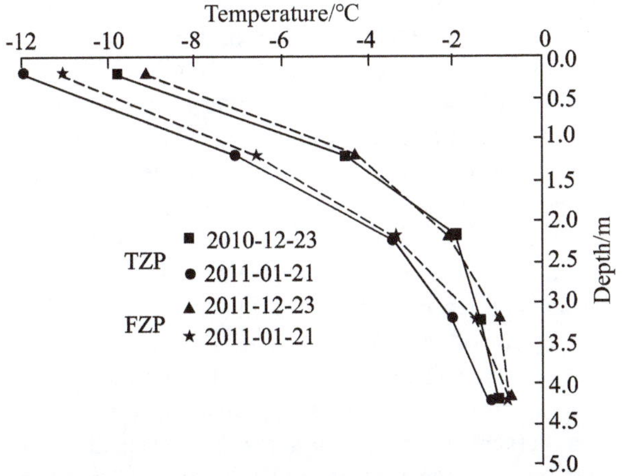

Fig. 7.22 Distribution of ground temperature along the depth of the tower foundation backfill (Cheng et al., 2012)

Fig. 7.23 Ground surface subsidence of tower foundation pit

7.3 Treatment for Frozen Soil Foundation

The difficulty related to the stability of engineering buildings in frozen soil areas lies in ensuring the stability of frozen soil foundations and the base under various hydrothermal environments. In addition to "detailed frozen soil engineering investigation", "elaborate frozen soil foundation design" and "reasonable construction of frozen soil foundation", it is still necessary to strengthen the structure design for frozen soil foundation and foundation treatment.

In the foundation design, according to the frost heave and thaw settlement characteristics of the permafrost foundation, the tapered column foundation is used to prevent the foundation from frost jacking damages, and the dug foundation, the expanded-base pile foundation and the deep pile foundation are also used to prevent the foundation from thaw settlement deformation. Even if the new form of foundation structure is adopted, the frozen soil foundation may still have the properties of frost heave and thaw settlement, and some auxiliary measures need to be added to prevent the tower foundation from frost heave and thaw settlement.

According to the reasons for the instability of the tower foundation, it is concluded that the prevention of the impact from hydrothermal dynamics, attached with the physiochemical measures and reasonable construction techniques, these comprehensive measures can be used to achieve a multiple effect. According to the existing engineering experience, the hydrothermal erosion brought by surface water and groundwater flow can often make all engineering measures fail, and strengthening the protection of frozen soil environment can enhance the hydrothermal stability of the frozen soil foundation. The positive and negative basic factors that cause frost heave and thaw settlement in frozen soil areas are soil quality (silt and silty clay with freezing and thawing sensitivity), moisture (liquid migratory moisture and solid accumulative moisture), and temperature (cooling negative temperature and thawing positive temperature). In order to prevent buildings from frost damages, it is necessary to start with the treatment for foundation soil and the judgment regarding the adaptability of foundation structure (rigid or flexible), and the main methods are as follows: ① waterproofing and drainage method; ② physiochemical method; ③ thermal insulation and cooling method; ④ structural method; ⑤ method of comprehensive measures. Methods for anti-frost heave and thaw settlement prevention for frozen soil foundation of transmission line are shown in Fig. 7.24.

7.3.1 *Waterproofing and Drainage Method*

Waterproofing and drainage methods, that is, preventing the intrusion of surface water and groundwater into the foundation (or foundation pit) during the construction and after the completion. The design and construction of waterproofing and drainage should fully consider the hydrological and hydrogeological conditions in the construction site. Before the construction of the foundation project, drainage measures for surface water should be taken efficiently, and the surface water should

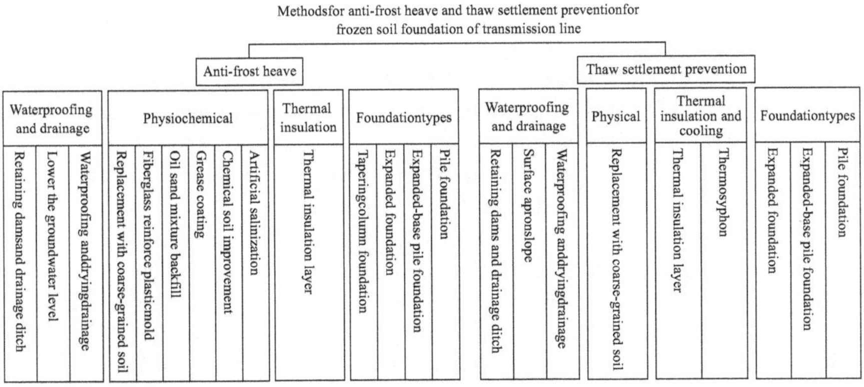

Fig. 7.24 Methods for anti-frost heave and thaw settlement preventionfor frozen soil foundation of transmission Line

be drained out a certain distance of the construction site as far as possible to avoid intrusion into the foundation pit (Fig. 7.6). Measures should be taken to lower the groundwater level if necessary. According to the frozen soil conditions, the construction time should be reasonably selected to prevent the thaw collapse of the foundation pit (Fig. 7.5).

Before the construction of the foundation pit, the water-stop ridge and drainage ditch should be built outside a certain distance of the foundation pit to discharge the surface water along the slope to ensure that there is no water-logged depression between the foundation pit and the water-stop ridge or drainage ditch. And a sump pit can be dug outside the thermal influence range of the foundation pit at a lower location, and a water pump should be used to drain the accumulated water in time to lower the groundwater level.

After the completion, a water-stop ridge or drainage ditch should be built outside a certain distance around the tower foundation (or the pre-construction facilities should be retained and reinforced), reasonable form of slope protection should be adopted and waterproofing and anti-scouring treatment measures should be taken. The accumulated water in the pre-construction sump pit should be drained, and then the sump pit should be backfilled and compacted to the natural state. And the surface apron slope on the foundation should be expanded in area in order to avoid the accumulation of surface water.

7.3.2 *Physiochemical Methods*

Physiochemical methods are methods of replacement, glass fiber reinforced plastic mold and artificial chemical treatment.

1. Replacement method: the foundation soil with frost heave (or thaw settlement) is replaced with coarse-grained soils without frost heave (or thaw settlement) such as sandy gravel, etc. Usually, the coarse-grained soil is replaced on the foundation side, and it should be replaced in the form of inverted filter, the replacement width is 400 mm on both sides of the width of the base plate, and the replacement thickness is 400 mm. Under the foundation base, coarse sand, gravelly sand or sandy gravel should be used to replace the foundation soil with frost heave and thaw settlement, thicker than 0.5 m. The replacement depth depends on the changes of frost heave in the foundation soil. In the foundation soil with weak frost heave, the replacement depth reaches 2/3 of the design freezing depth. In the foundation soil with frost heave, strong frost heave or extremely strong frost heave, the replacement depth should reach the design freezing depth.
2. Glass fiber reinforced plastic mold: tapered column foundation (Fig. 7.25) should be cast with fiberglass reinforced plastic mold formwork supporting or the fiber glass reinforced plastic mold formwork supporting should be fixed to the side wall of pile foundation within the maximum freezing depth.
3. Artificial chemical treatment method: use heavy oil, asphalt with 5% used engine oil, industrial vaseline, engine oil, chemical surfactant, etc. to coat on the side of the foundation, or fill the side surface of the foundation with non-hydrophilic materials such as grease and oiled sand etc., and the coating thickness of the foundation side surface is generally 2–5 mm (by multiple brushing), or use the oil sand made by hot-mixed residual oil and medium sand, and then fill it around the foundation side, with the thickness of 0.2–0.3 m to eliminate the freezing strength between the frost-heaving soil and the foundation. The depth of treatment should reach 0.5 m below the maximum freezing depth. Or make a solution with NN-22 octadecyl ethylenediamine surfactant (accounted for 0.1% of the soil weight) and water, and mix the solution with diesel fuel (miscella at −10 °C or 0 °C) (water and diesel accounted for 6% of the soil weight) and soil (heated to 120–150 °C) and the hot-mixed uniform water-repellent soil(hydrophobic soil) can be obtained, and then backfill it (with the thickness of 0.15–0.25 m) around the foundation. Or use polymerizing agents with high-valence cation Fe^{3+}, Al^{3+} etc., to mix with soil to form granular foundation soil (similar to coarse-grained soil), to weaken the water migration ability of soil.

The low-valence cation salts such as NaCl, KCl and $CaCl_2$ are used to properly salinize the foundation soil and reduce the freezing temperature of the foundation soil, but the admixture amount should be controlled.

7.3.3 Thermal Insulation and Cooling Method

Thermal insulation method: to block the thawing effects on the permafrost foundation by heat transfer from the surface or the structure, or to block the freezing effects of air temperature on the foundation soil. Generally, industrial thermal insulation

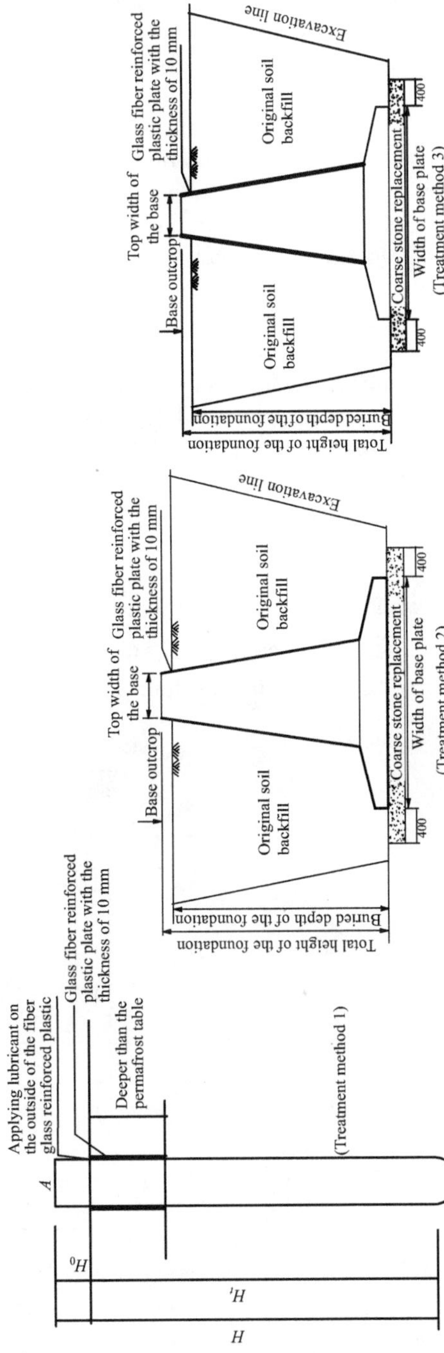

Fig. 7.25 Treatment method for the foundation pouring with the fiberglass reinforced plastic mold form work supporting

materials are used, and the stress-bearing part at the bottom of the foundation can use extruded polystyrene foam board (XPS) with greater strength, and the part with less stress on the surface or foundation side can use polystyrene foam board (EPS) or polyurethane board (PU), and the irregular parts can use in-situ foaming polyurethane (PU).

Cooling permafrost foundation method: at present, the method used for cooling permafrost foundation of the steel tower foundation is mainly thermosyphon. The thermosyphon has the function of actively cooling the foundation, which can raise permafrost table, thicken the permafrost layer, enhance the strength of the frozen soil and increase the cold storage of permafrost layer. According to the relevant experimental observations, after the thermosyphon experiences a freezing cycle, its effective freezing radius can reach 2 m, and the soil temperature can be reduced by 2–4 °C.

When the buried depth of the foundation is less than 7 m, the buried depth of the thermosyphon needs to be 7 m; when the buried depth of the foundation is greater than 7 m, the buried depth of the thermosyphon needs to be 9 m. For installation of thermosyphons at the tower position, comprehensive consideration should be given to the foundation types, foot distance, base plate width and permafrost conditions, etc., and the thermosyphon should be arranged as close to the foundation as possible (Fig. 7.26). The arrangements of thermosyphons for tower foundation of Qinghai-Xizang direct-current networking project are as follows.

Scheme A: built-in scheme (Fig. 7.27a). When the foundation column is wide enough, the thermosyphon can be set into the foundation column. The method of

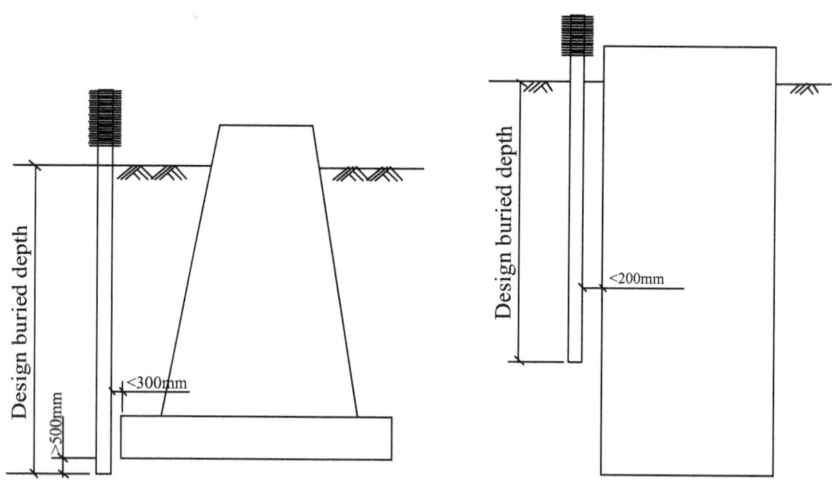

(a) Installation position of the thermosyphons for the tapered column foundation

(b) Schematic diagram of the longitudinal section of the installation position of the thermosyphons for the pile foundation

Fig. 7.26 Schematic diagram of the installation position of the thermosyphon

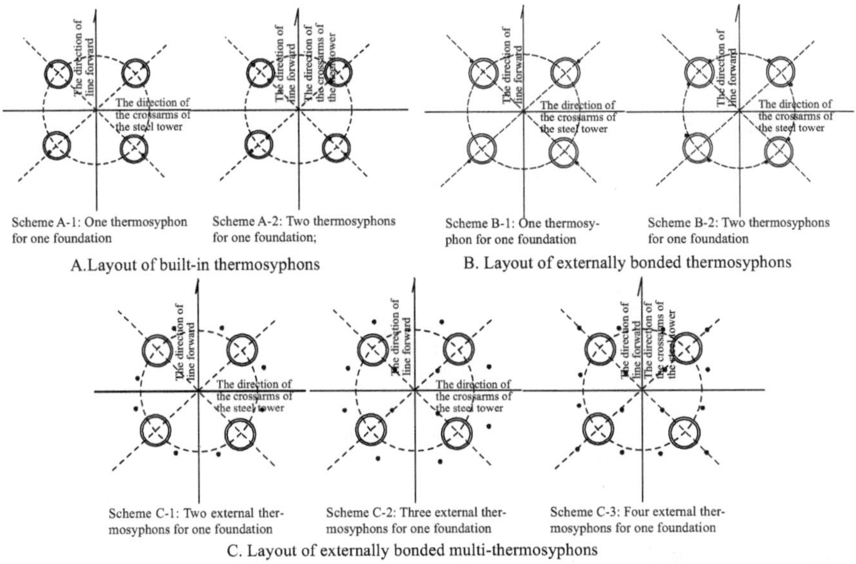

Fig. 7.27 Layout of thermosyphon for the steel tower foundation (Northwest Electric Power Design Institute of China Electric Power Engineering Consulting Group, 2008)

one thermosyphon for one foundation (A-1) or two thermosyphons for one foundation (A-2) can be adopted.

Scheme B: externally bonded scheme (Fig. 7.27b). When the width of the foundation cannot meet the built-in scheme, the scheme of bonding the thermosyphon to the edge of the foundation column can be adopted. The method of one thermosyphon for one foundation (B-1) or two thermosyphons for one foundation (B-2) can be adopted according to the foot distance and the geological conditions of the foundation.

Scheme C: external scheme (Fig. 7.27c). The thermosyphon should be attached to the edge of the base plate as much as possible. According to the foot distance and geological conditions of the foundation, the method of two external thermosyphons for one foundation (C-1), three external thermosyphons for one foundation (C-2) or four external thermosyphons for one foundation (C-3) can be adopted.

The application conditions of foundation treatment as an auxiliary measure to prevent frost heave and thaw settlement are listed in Table 7.3.

7.3.4 *Structural Method*

Structural method means to change the shape of the foundation section to adapt to the frost heave and thaw settlement deformation of the permafrost foundation (Table 7.2).

Table 7.3 Auxiliary measures for anti-frost heaving and thaw settlement prevention of the foundation

Auxiliary measures	Conditions of use and functions	Advantages and disadvantages		Existing problem	Suggestions
		Advantages	Disadvantages		
Thermosyphon	The use conditions and functions of thermal probes are as follows: Usage conditions and functions: 1. Cold areas with the freezing index greater than 500 °C•d;2. Warm permafrost regions with high ice content; 3. Accelerate the refreezing of permafrost foundation around cast-in-place concrete pile in the permafrost section; 4. Accelerate the freezing under the foundation plate and reduce the normal frost-heave force on the foundation; 5. Maintain the stability of the permafrost table and prevent the permafrost from thaw settlement.	It can accelerate the freezing of the foundation soil, decrease the ground temperature of the permafrost, stabilize the permafrost table, avoid the thawing of the permafrost, increase the adfreezing force of the permafrost and the foundation, and strengthen the anchoring strength of the permafrost foundation.	It has to pay attention to the protection of vulnerable parts; special transportation methods should be used; it is not easy to monitor its operating state; it needs to drill and then bury.	There are the existing problems, including: 1. Anti-corrosion treatment measures; 2. The methods for leak detection; 3. The display of working state indicator signs.	It should be applied in the warm permafrost areas with high ice content;it can be used for anti-thaw settlement treatments.

Oil sands	Usage conditions and functions: 1. The cold areas with frost-heave foundation soil; it can: 2. Prevent the frost jacking effects of the seasonally frozen layer on the foundation; 3. Weaken or eliminate the tangential frost-heave force on the foundation.	The industrial waste residual oil can be recycled; the production method is simple (medium sand hot-mixed with 15% asphalt).	The construction is a little more troublesome.		It should be applied to the foundation soil with frost heave. It can be used for anti-frost heave treatment.
Thermal insulation	Usage conditions and functions: 1. Decrease the seasonal freezing depth of foundation soil; 2.Weaken the external thermal disturbance to the permafrost foundation and the thawing of permafrost;3. Reduce the normal frost-heave force under the foundation plate in the seasonally frozen soil area.	It is simple and easy to operate and it has good thermal insulation performance.	It not only insulates but also prevents the permafrost foundations from freezing in winter.	There are the existing problems, including: 1. The index control of thermal insulation performance; 2. The damage rate control of transportation and construction.	It should be applied to the rigid foundation, the light-fill and high-fill rigid foundation. It can be used for anti-frost heave and anti-thaw settlement treatment.
Sandy gravel iInverted filter	Usage conditions and functions: 1. The frost-heave foundation with water permeable conditions; 2. Reduce the tangential frost-heave force and normal frost-heave force on the foundation.	It has the easy access to materials and the construction is simple.	It has to control the content of silt and clay in the sandy gravel, which is vulnerable to contamination of surrounding clay soil and leads to consequent failure, affecting environmental protection.	There are the existing problems, including: 1. The determination of the reasonable thickness of the inverted filter; 2. Long-term anti-frost heave effects.	It can be applied to anti-frost heave treatment if the economical and reasonable conditions meet requirements for the environmental protection.

The tapered Column Foundation, whose foundation section is trapezoidal and the slope angle is about 9°, can reduce the tangential frost-heave force of most foundation soil.

For the columnar rigid expanded foundation, the width of the foundation plate should be buried below the maximum seasonal freezing (or thawing) depth, to increase the anchoring force and bearing capacity of the foundation, and lubricants should be smeared on the side of the column pile or the side of the column pile should be replaced with sandy gravel, oil sand, and hydrophobic composite soil, etc., to reduce the tangential frost-heave force. The gravelly sand should be replaced under the foundation plate to prevent the normal frost-heave force, or the thermal insulation board should be laid to prevent the thawing of the permafrost foundation.

Assembled foundation, with the prefabricated components assembled in-situ, is suitable for areas with convenient transportation conditions. The foundation is treated in the same method as the columnar rigid expanded foundation.

Dug foundation and expanded-base pile foundation can use manual digging or drilling construction method to expand the area of the pile toe, which is used for the permafrost layer to increase the freezing strength of the foundation side and the bearing capacity of the pile tip. Anti-frost heave measures should be taken within the range of seasonally thawed layer.

The pile foundation adopts the drilling construction method. It is used in permafrost regions to reduce construction disturbance and increase the freezing strength, anchoring force and bearing capacity of the foundation side. Anti-frost heave measures should be taken within the range of seasonally thawed layer.

The pile cap foundation is used in permafrost regions to reduce the depth of the pile foundation and increase the freezing strength. Except that anti-frost heave measures should be taken within the range of seasonally thawed layer, the raise pile cap above grade by at least the expected ground frost-heave amplitude; provide taper and clearance to avoid ponding, the cavity should be retained, and the ground should be tapered to avoid water accumulation.

7.3.5 Comprehensive Measures Method

According to the experiences of preventing thaw settlement and frost heave in permafrost regions and in seasonally frozen soil areas, the effect of using a single measure to prevent thaw settlement and frost damages of engineering buildings is not ideal, and the results are less effective. By comprehensively considering frozen soil, hydrogeology, foundation types, construction conditions and foundation treatment materials etc., and combining the above foundation treatment measures with each other, it can often play a better anti-freezing effect. For example, the thermosyphons can be used to prevent thaw settlement, reduce the temperature of the permafrost under the foundation, increase the strength of the permafrost foundation, and achieve the design principle for maintaining the permafrost foundation in frozen state. If only the measure of thermosyphon is used, the range of cooling and freezing in cold season will rebound in warm season every year, leading to the increase of temperature or the decrease of the freezing area, and the cold accumulation will be limited. If the thermosyphon is combined with the thermal insulation layer to form a composite measure, the cold storage obtained in cold season will not cause too much loss due to the warming in warm season, the cold storage of the frozen soil foundation will increase year by year, and the stability of the frozen soil foundation can be reached in a short time.

7.4 Foundation Construction Technologies

The ground ice gives the permafrost foundation some special properties. During construction, the permafrost foundation is susceptible to collapse, water erosion and thermal erosion due to the influence of temperature and human activities, which not only causes great difficulties to construction, but also brings some troubles to the later operation of the tower foundation because of the thermal disturbances brought by construction. Therefore, the tower foundation construction in permafrost region should choose the appropriate construction time, shorten the excavation time, reduce the days of exposing the foundation pit under the sun, and prevent the permafrost in the foundation pit from thawing. And the operation from excavating to pouring should be continuously carried out as fast as possible, to ensure the quality and meet the requirements for rapid construction.

On the basis of the design principle for maintaining the frozen state of the permafrost foundation, it is basically in the area with thick ground ice, bogging surface and large runoff. The excavation time of foundation pit should be selected in cold season, and the optimal time is from September to March of the next year.

On the basis of the design principle of allowing thawing of the permafrost foundation, it's advisable to excavate in the cold season. When it is inevitable to construct in warm season, apart from sun-shading and rainproof measures, water pumping, drainage and protective measures for preventing the collapse of the foundation pit wall should also be well taken.

Preparatory work should be done before the excavation construction, such as the foundation reinforcement cage should be lashed in responding shapes outside the pit in advance, hoisting machinery and equipment and concrete batching should also be prepared in advance, and the following construction experiences of the Qinghai-Xizang direct-current networking project can be summarized and referred to (Ma et al., 2013).

7.4.1 Excavation of Foundation Pit

1. **Mechanical excavation**

If the water accumulated in the foundation pit is not much during excavation, the construction time can be adjusted according to the characteristics of the great temperature difference between the day and the night in the permafrost area. It can start at 2 a.m. in the morning if the temperature is low, and the excavation of the foundation pit should be completed at 8 a.m. or 9 a.m. in the morning. Since the permafrost layer is not easy to collapse during the excavation process, the steel bars hoisting and formwork board pouring can be carried out immediately and the foundation pouring is completed before the permafrost layer is thawed.

For the foundation pits with heavy excavation works, they are with large seepage and their walls are prone to collapse, the wall protection and collapse prevention measures must be taken during the excavation process.

The excavated foundation pit must be poured quickly to avoid the occurrence of "overnight pouring". If the construction cannot be carried out in time, when the excavation of the foundation pit is close to the design depth, the soil layer with the thickness of more than 500 mm should be reserved as the insulation layer, and the foundation pit should be tamped and then dug to the design depth before pouring. After the excavation of the foundation pit, sun-shading and cooling measures should be taken to avoid the thawing of the permafrost and the water accumulation in the foundation pit and the collapse of the pit wall.

In order to protect the environment and facilitate the timely recovery of the environment after construction, the turf on the ground surface should be transplanted to the vicinity of the tower position for storage and maintenance before excavation; and the turf should be restored to the original state after the foundation is backfilled.

2. **Blasting excavation technology**
Excavation by blasting is suitable for earthworks with thick permafrost layer and large area (Zhang et al., 2011). The blasting excavation technology needs blasting control according to the parameters of smooth blasting technology. At the same time, blasting equipment and materials should be reasonably selected.

3. **Digging**
The manual-dug (pile) foundation should be conducted in cold season. During the excavation process, the wall protection procedure should be strictly carried out once every 500 mm, and the steel cylinder wall protection should be used when the excavation reaches the part of expanded head.

4. **Pile foundation**
The cast-in-place pile foundation can be quickly drilled by the dry process of the rotary drilling rig. The key is to choose the drill bit according to the local conditions. The proper drill bit can greatly improve the drilling speed and the pore-forming quality. The rotary drilling rig can be used for borehole reaming.

7.4.2 Concrete Pouring

The hard concrete with lower water-cement ratio should be used. For the cast-in-place concrete pile foundation, the molding temperature of the concrete should be controlled within 5–10 °C.

7.4.3 Foundation Pit Backfill

The backfill density should not be less than 93%. The backfill soil should be compacted layer by layer, with the thickness of 0.3 m for each layer, and the backfill soil should be 500 mm higher than the ground surface as the anti-subsidence layer. The fill should be backfilled with the homogeneous soil as much as possible, and the water content should be controlled. When backfilled with borrow soil, it should be layered and filled regularly according to the category. The fill should start from the lowest point of the filling section and be layered and filled horizontally from bottom to top. The silt, debris, ice cubes and water-soaked mud in the fill should be removed. If there is groundwater or stagnant water, the drainage measures should be taken to remove them.

7.4.4 Thermosyphon Construction

The hoisting equipment, drilling machinery and backfill should be prepared well.

1. **Thermal probe thermosyphon installation**
    1. After the installation location of the thermosyphon is decided, the vertical borehole or the inclined borehole can be drilled. The diameter of the borehole should be 50–80 mm larger than the diameter of the pipe shell of the thermosyphon. The dry drilling should be used and the chilled drilling fluid can be added according to the stratigraphic conditions. Small circulation and small drilling footage should be used, and the borehole depth should be 100–200 mm deeper than the design depth.
    2. The hole diameter and hole depth should be checked, and the drilling slurry should be removed.
    3. After the borehole is prepared, the crane lifts the thermosyphon and moves it vertically to the borehole, manually pushes the thermosyphon, and slowly lowers down into the foundation trench in the borehole for positioning. After the vertical position is in accordance with the standard, the thermosyphon will be fixed.
    4. When backfilling, after filling the borehole with cold water, the medium-coarse sand should be slowly poured into the gap between the thermosyphon and the borehole wall, and the amount of sand poured should be consistent with the calculated amount. When backfilling, the oil contamination, sawdust, stones and other sundries should not be mixed in, to avoid gaps or hollowness;
    5. After the sand-filling is frozen, the support should be removed. After the installation of the thermosyphon, the testing instrument provided by the manufacturer should be used to inspect the thermosyphon at least twice and the data should be recorded in detail. If the thermosyphon is damaged, it should be replaced in time.

2. **Requirements for thermosyphon installation**

    1. The height of the top of the thermosyphon should be unified, the height error of the installation of the thermosyphon for the single-leg foundation should be within 30 mm, the elevation error of the installation of the thermosyphon for one foundation should be within 50 mm, the error of the foundation pit depth of thermosyphon should be within −20–50 mm, and the allowable deviation of the verticality of the thermosyphon body should be 1% of the length of the thermosyphon body.
    2. The relative error of the installation of two thermosyphons for the single-leg foundation should be within 50 mm, the relative error of the installation of 8 or 16 thermosyphons for one foundation should be within 100 mm, and the absolute error of the installation position of all the thermosyphons should be within 30 mm.

When the thermosyphon is designed with observation equipment, the observation equipment should be buried at the same time as the thermosyphon is installed. The working state of thermosyphon should be regularly checked.

Chapter 8
Airport Construction in Frozen Soil Regions

In permafrost and seasonally frozen soil Regions, the construction of airport runways is unlikely to avoid the problems of frozen soil engineering.

8.1 Main Problems of Frozen Soil Engineering in Airports in Frozen Soil Regions

No matter in permafrost or deep seasonally frozen soil Regions, the frost heave of foundation soil can cause differential deformation, cracks and slab staggering to airport runways, and mud squeezing in warm season. In an airport in Northeast China, the frost heaving of foundation soil can make the concrete pavements rise by 30 ~ 100 mm, bringing about problems such as the mud squeezing and other problems because of freezing and thawing, etc. (Shi et al., 2014). The concrete pavement runway, with a high thermal conductivity, can increase the heat transmitted to the underlying permafrost in warm seasons, leading to differential thawing and deformation of permafrost foundation. If the drainage facilities of the foundation trench are damaged or destroyed during the construction, the groundwater on both sides will intrude the foundation trench, which will cause frost heave and uplift for the foundation soil in cold season; and the frozen foundation will thaw and subside in warm season, which will not only cause deformations of runways, but also cause fractures and frost boil of the pavements in warm season.

In permafrost area the main frozen soil engineering problems of airport runways are differential deformations caused by thaw settlement of permafrost foundation and frost heave in active layer. Its manifestations are similar to the concrete pavement deformation in permafrost regions, including thaw settlement, longitudinal and transverse cracks, etc., and the reason is that the permafrost foundation contains ground ice. Airport runways in permafrost regions must be closed if there is thaw settlement deformation which is alike to the deformation in highways (Figs. 8.1, 8.2

Q. Wu, C. Tong, *Frozen Soil Engineering in Cold Regions*,
https://doi.org/10.1007/978-981-95-2527-0_8

Fig. 8.1 Thaw settlement of Mobei Expressway

Fig. 8.2 Longitudinal cracks of Jiamo Highway

and 8.3) and cracks on the pavements. As the running time of the project increases, the heat transmitted into the permafrost foundation accumulates year by year, and the thawing depth and thaw settlement increase year by year as well.

The natural seasonal freezing depth in the deep seasonally frozen soil areas in the cold zone is more than 2 m, and the maximum seasonal freezing depth in taliks in the permafrost regions can reach 4 ~ 5 m, or even deeper. The frost heaving of foundation soil depends on soil texture, water content and negative temperature (related to freezing depth). The silty clay, silt, and coarse-grained soils with a content of more than 15% of fine-grained soil (less than 0.075 mm) have frost heave susceptibility, which have 4% water content exceeding the plastic limit, and can cause differential frost heave and deformation to airfield pavements when there is water migration. When the soil quality and freezing depth are relatively constant, the depth of groundwater level becomes an important factor of frost heave.

Fig. 8.3 Traverse of Mobei Expressway

The thickness of the anti-freezing layer should be considered according to the extreme temperatures that may occur during the service life, and the seasonal freezing depth should be determined according to the maximum local freezing depth within 30 years. For the design of the thickness of the minimum antifreeze layer of the airfield pavements of Yichun Airport, it is about 70% of the maximum depth (Xing & Hao, 2010).

The frost damages of concrete pavements of airport runways in frozen soil areas are generally serious. There are poor durability, early alligator cracks, scaling, and chipped edges and corners. According to the investigation (Miao, 2002), the damages of airport pavements are mainly manifested as: the ruptures of pavements, the longitudinal, transverse and oblique cracks, fractures and the fragmentation in corners of slab, the fragmentation and detachment of the slabs near the joint, the loose spalling of the surface of slab, the bulge and slab staggering among adjacent slabs, etc. The main reason is that the poor frost resistance and impermeability of concrete cause the poor freeze or thaw resistance, impermeability and durability. In cold climate conditions, it is easy to cause freeze-thaw denudation of concrete pavements, and the frost action of foundation soil is very likely to damage airport pavements.

The erosion of foundation trench by groundwater in warm season can also cause freeze-thaw damage to airport runways in cold regions. The drainage measures on both sides of the runway foundation trench must be effectively protected. Once the drainage facilities are damaged in construction process, the suprapermafrost water on the side wall of the foundation trench will intrude into the foundation trench, resulting in serious freeze-thaw damage to the runways. First, the intrusion of groundwater will bring a large amount of heat to the frozen soil foundation, causing thaw settlement of the runways. Secondly, the frozen soil foundation is an impermeable layer, so the intrusion of groundwater will soften the foundation soil of the trench, and increase its water content. The water freezes in cold season, bringing the frost heave and the deformation to the runways, and then causing the thaw settlement in warm season.

8.2 Site Selection of Engineering Geology for Airports in Permafrost Regions

The site selection of civil airports is a complex systematic project with many influencing factors. To meet the clearance, the spatial domain, the meteorological conditions, the coordination with urban planning and development and adjacent airports, should be considered; the good engineering geology, the hydrogeology, the topography, and the environmental protection conditions should be chosen; the supplies of water, electricity, oil, gas, the transportation and communication conditions to maintain the operation of airports need to be guaranteed; the engineering quantities of airport construction, the occupied cultivated land and forest land and the workload of dismantling and moving, and engineering investment, etc. also should be taken into account.

The construction conditions of airports are one of the important parts of the site selection. It is basic requirement for engineering investigation to understand the stratigraphic distribution, lithological characteristics, groundwater conditions, structural characteristics, seismic effects and adverse geological process within the range of site, and this chapter only involves the special issues related to the site selection of airports in permafrost regions. The frost heave of foundation soil and thaw settlement of frozen soil foundation are the main frozen soil engineering problems of airport runways in frozen soil area. The site selection of airports in deep seasonally frozen soil areas focuses on knowing the impact of frost heave of foundation soil on airport construction. As for the site selections in permafrost regions, it is necessary to comprehensively know the effects of the frost heave of active layer and thaw settlement of permafrost layer on airport construction.

8.2.1 Characteristics of Frozen Soil Distribution

The deep seasonally frozen soil areas (with freezing depth ≥2 m) are mainly distributed in Heilongjiang, Inner Mongolia, Xinjiang, Qinghai, Xizang and other places in China. The regions with freezing depth ranging 1 ~ 2 m are Jilin, Liaoning, Ningxia, and parts of Gansu, Hebei and Shanxi provinces. When the maximum freezing depth is greater than 0.5 m, an anti-freezing layer is needed.

The permafrost is mainly distributed in Daxing'anling and Xiaoxing'anling Mountains in Northeast China, the Tianshan Mountains in Xinjiang and the Qinghai-Xizang Plateau. Daxing'anling and Xiaoxing'anling Mountains belong to the high-latitude permafrost regions, and the higher the latitude, the broader the areas of the permafrost distribution, and the thicker the permafrost is. From north to south, the permafrost is distributed in large continuous, isolated and sporadic forms and the permafrost is mainly distributed in the low-lying swamps and wetlands, as well as the lower parts of slope. The thickness of the permafrost layer in high mountains and ridges is relatively small, and the taliks are mostly distributed in flood plains and the first-grade terraces of large rivers. The Tianshan Mountains and the Qinghai-Xizang Plateau in the west belong to the alpine and plateau permafrost regions. The altitude of the lower limit of the permafrost distribution increases with the height of altitude, ranging from the Tianshan Mountains (with the altitude of 2200 m), the northern pass of the Kunlun Mountains on the Qinghai-Tibet Plateau (with the altitude of 4200 m), the Gangdise Mountains (with the altitude of 4700 m) to the north of the Himalayas (with the altitude of 5100 m). As the altitude increases, the distribution areas of the permafrost are expanding, the thickness of permafrost layer is enlarging. The permafrost is mainly distributed in swamps and wetlands, slopes and high plains, and the seasonally frozen soil areas are mainly in flood plains and the first-grade terraces of large rivers (with the altitude of 4272 m and the freezing depth of 2.4 m in Nagri Gunsa Airport) (Sun & Zhang, 2013). There are various types of taliks and seasonally frozen soil areas in permafrost regions, mainly distributed in river valleys, isolated and sporadic permafrost regions.

At present, the plateau airports with the altitude of more than 2800 m in China are located in the seasonally frozen soil areas. Jiagedaqi, Manzhouli, Arxan, Yichun, Baishan, Heihe and Genhe and other airports in Daxing'anling and Xiaoxing'anling Mountains are in the seasonally frozen soil layer in the permafrost regions, but only Mohe Airport is located in the permafrost layer.

8.2.2 Causes of Frost Heave of Foundation Soil

The frost heave resistance of foundation soil can be ranked according to the following sequence: silty soil, silty clay, clay, crushed and gravel soil (the content of silty clay particles with the particle size less than 0.075 mm should be more than 12%), coarse sand and sandy gravel (the content of silty clay particles with the particle size less than 0.075 mm should be less than 5%).

As for the initial water content of frost heave of soil, when the water content of clay soil is less than $0.8\omega_p$ (water content of plastic limit) and the water content of coarse-grained soil is less than 12%, the foundation soil can be regarded as non-frost heave. When there is external water migration, the amount of frost heaving depends on the amount of migration during freezing.

The Negative temperature and the freezing depth are directly proportional to the local freezing index (annual cumulative value of negative temperature, degree·day). The deeper the freezing depth is, the larger the frost heaving of foundation soil becomes.

The Negative temperature depends on the location of airport, and the location of airport usually cannot change the freezing index of the climate. For the site selection of airport, it is advisable to choose a good foundation soil dominated by coarse-grained soil and a site where groundwater is easy to drain. From the perspective of foundation, the coarse-grained soil foundation and the dry fine-grained soil foundation are usually the preferred sites.

8.2.3 Nature of Thaw Settlement of Permafrost Foundation

The nature of thaw settlement of permafrost foundation is that the permafrost contains different degrees of ground ice, which is formed in the natural, historical and geological environment. The heat brought into the permafrost foundation by human activities (pavement structure, the removal of vegetation and construction, etc.) can change the surrounding environment of permafrost. The combination of the two may cause the thaw settlement of permafrost foundation. The characteristics of permafrost with high ice content are as follows:

1. Lithology. The particle size of soil with the strongest ability to segregate ice is 0.02 ~ 0.005 mm. According to the ability of segregating ice, they are ranked by lithology as the silty soil, the silty clay, the clay, and the silty sand. There are always soil-bearing ice layers in peat soil layers. When the content with particle size less than 0.05 mm in the coarse-grained soil is more than 15%, the permafrost with high ice content can also be formed under the conditions of saturated water.
2. The formation of soil. The proportion of permafrost with high ice content is the largest in lacustrine and slope deposits (including mud flow deposit), followed by deluvium-diluvial and diluvial deposits, and the permafrost with high ice content in alluvial deposits is the smallest.
3. The ground temperate zone of permafrost. Most areas of continuously distributed permafrost have permafrost with high ice content, and the permafrost with low ice content is common in isolated permafrost regions (except swamps and wetlands with high ice content). The frozen soil with high ice content occupies a large proportion in frozen soil area with low temperature.

The permafrost with high ice content is mainly distributed in the fine-grained soil and peat soil. In permafrost regions, the airports should be built in the coarse-grained soil or the dry fine-grained soil, and the swamps and wetlands should be avoided.

8.2.4 Particularity of Hydrogeological Conditions in Frozen Soil Area

The hydrogeological conditions in seasonally frozen soil areas are the same as those in non-frozen soil areas. It is advisable to choose the areas with a deep groundwater level and non-erosive water to avoid flood disasters of natural water system. The deep depth of groundwater level is conducive to weakening or eliminating the effects of capillary water and reducing the frost heave of the foundation soil. The water intrusion will bring thaw settlement to permafrost.

There are three types of groundwater in permafrost regions, namely suprapermafrost water, intra-permafrost water and sub-permafrost water. The suprapermafrost water from atmospheric precipitation and the upstream (upper) groundwater directly affects the airport runways. Due to the impermeability (with small water permeability, the permafrost layer can be regarded as the impermeable layer) of the underlying permafrost layer, groundwater is usually buried at a shallow depth and often occurs in the pores (commonly known as mountain stream water) in the soil layer (or crushed stone soil, or gravel soil) near the surface. The suprapermafrost water is intermittent, liquid in warm season and solid (ice) in cold season. During cold season, as the freezing depth of the surface layer increases gradually, the suprapermafrost water changes from unpressurized to pressurized state, and tends to migrate to the zone with shallower freezing depth (higher ground temperature), forming pingos in the downstream overflow zone (Fig. 8.4), or forming aufeis exposed out of the surface, or overflowing from the excavated foundation trench (Fig. 8.5).

8.2.5 *Relationship Between Topography and Frozen Soil*

The development zone of permafrost with high ice content is closely related to topography and geomorphology, which in essence reflects the relationship between the material composition of the soil layer and hydrogeology. Generally, in the slope areas, the material composition in the upper part is relatively coarse, mostly crushed stone and block stone, with good drainage conditions. The loose deposits turn into fine materials in the middle and lower part, and water is easy to gather to form thick ground ice and soil-bearing ice layer at the toe of the slope.

Fig. 8.4 Pingos in valley plains (swamps and wetlands)

Fig. 8.5 Aufeis formed by the overflow of suprapermafrost water on Jiamo Highway

1. Topography. The Low-lying swamps and wetlands are usually located in fine-grained soil with sufficient water, and often with the permafrost of high ice content. The mean annual ground temperature is lower in low mountains and hilly areas in the western region, where there is sufficient precipitation, and developed mud flow deposits, a relatively large proportion of the permafrost with high ice content, especially with high proportion of the soil-bearing ice layer. Followed by the middle and high mountains, although there is lower temperature and higher water content, the deposits are thinner and coarser, with a smaller proportion of the permafrost with high ice content. The valley plains are thermally affected by the river, with a higher ground temperature and widely developed coarse-grained soil, and the least development of the permafrost with high ice content. The mountain deposits in Northeast China have coarser and thinner particles, and mostly belong to permafrost with low ice content due to the influence of the temperature inversion layer.
2. Low mountains and hilly areas. The shady slope receives less solar radiation, and the ground ice is fully developed and more than that on the sunny slope. There is thick fine-grained soil and sufficient water on the slope with gradient of 4° ~ 8°, which is most conducive to the growth of ground ice. The ground ice develops poorly on the slope of 10° ~ 16°. The thick ground ice cannot be found on the slope greater than 16°. There is generally only ice in cracks on the slope more than 25° where the denudation is dominant. The top of the hillside is mostly residual sediments, which is difficult to store water and is mainly permafrost with low ice content. The permafrost with high ice content gradually develops in the middle and lower parts, and the deposits in the lower part of ditches are relatively coarse, mainly the permafrost with low ice content. There are thick fine-

grained soil and sufficient water in intermountain basins and valleys, mainly containing the permafrost with high ice content.

3. Valley plains. The sediments of flood plains and the first-grade terraces are the coarse-grained soil, mostly the permafrost with low ice content. The rear edges of the second and third grade terraces are close to the piedmont gentle slope, where the fine-grained soil develops, the water accumulates, and there is permafrost with high ice content.
4. Mountain-front alluvial fans and inclined ice-water sediment plains. In the middle and upper parts, the particles of deposits are coarse, and most of them are the permafrost with low ice content. In the lower part and the depressions between fans there are the permafrost with high ice content, with a high content of fine-grained soil and sufficient water. The permafrost with high ice content can be found in wet, low-lying areas with lush vegetation, as well as in the depressions of lacustrine deposits.

The Valley plains are affected by the thermal erosion of river and belong to fluvial taliks. The airport runways should be chosen in the valley plains with coarse-grained soil.

8.2.6 Influences of Unfavorable Frozen Soil Phenomenon on Airport Construction

The gelifluction and thaw slump are mainly developed in the fine-grained soil with gentle slopes containing thick ground ice (Fig. 8.6) or in the intermountain depressions. The pingos mostly appear at the junction of the piedmont gentle slope and the

Fig. 8.6 Thaw slumps on the slope

Fig. 8.7 Pingos on the piedmont gentle slope and valley plains

Fig. 8.8 Stone rings in permafrost regions (photo by Li Shude)

valley plain (Fig. 8.7) or above the blocked groundwater flow. The aufeis mostly occurs in the groundwater overflow areas, often exposed in the form of spring water.

Not all the flood plains or the first-grade terraces in permafrost regions are taliks, and there are still permafrost layers. Although they are the coarse-gained soil layers, they have a high content of fine-grained soil and sufficient water content, and in the active layer after frequent and repeated freezing and thawing processes, the materials with different particle sizes are differentiated, sorted and recombined under the effect of differential frost heaving to form polygonal stone rings (Fig. 8.8) and ice wedges, indicating that they are formed in the permafrost with low mean annual ground temperature. The sand wedges and soil wedges are often found in the deep seasonally frozen soil areas, which indicates that there were areas with severe cold climates and unfavorable frozen soil phenomenon historically, and that the foundation soil belongs to the permafrost layer with differential distribution of ground ice.

The sections with sand wedges and soil wedges represent the heterogeneous property of the foundation soil.

8.3 Design of Airport Runways in Frozen Soil Regions

In the geotechnical engineering design of civil airport, the safety and regular use of the airport should be guaranteed first, and then the advanced technology, the economic rationality, the resource saving and environmental protection should be taken into consideration. The foundation deformation and slope stability of airport runways in frozen soil area should conform to the requirements of the basic control indexes of the regulations. According to the particularity of differential deformation and slope instability of the roadbed, except for the material composition and water content of the soil foundation in the non-frozen soil area, the repeated freeze-thaw processes caused by the alternation of positive and negative temperature (air temperature and ground temperature) leads to the fundamental changes in the soil foundation and materials, and there are usually coupling effects of water and heat and the effects of force and heat. In the geotechnical engineering design of airport runways in frozen soil area, the following issues should be considered.

8.3.1 Frost Heave

In cold season, the pavement and its roadbed below (the base course and parts of the soil foundation) in cold region freeze, and the water (the inherent water in soil and the water migrated in the freezing process) will freeze and turn into ice, and the volume will expand by 9% (the inherent water) and 1 + 9% (the migrated water), causing the displacement of soil particles (in roadbed) and the frost heave deformation of the pavements. In the design of airport runways in seasonally frozen soil or permafrost regions, more attention should be paid to this special phenomenon of frost heave and an anti-freezing layer with adequate thickness should be set, or other measures should be taken to eliminate or weaken the frost heave of the roadbed.

In the design of geotechnical engineering, it is necessary to judge whether the foundation soil has frost heave, and determine the level of frost heave according to the soil texture, the water content and the groundwater state. Second, the freezing depth can be determined by the meteorologic stations, the maps of standard freezing depth or the calculations. Third, the prevention and control measures should be put forward, such as the waterproof and drainage, the soil improvement, and the temperature control, etc.

Since the roads and airport runways belong to multi-layer structures, the freezing depth obtained by the formula for calculating the seasonal freezing depth of natural soil cannot be directly applied as the freezing depth of the roads. According to the freezing index, the seasonally frozen soil areas can be divided into four categories

(Table 8.1), and the three main factors regarding the material composition of the pavement structural layer, the state of subgrade filling and excavation, and the humidity condition of soil foundation can be simplified into three coefficients, and the maximum freezing depth (Z_{max}) under the road can be calculated by the following simplified formula:

$$Z_{\max} = abcZ_d \tag{8.1}$$

where Z_d is the ground standard depth, m; *a* is the thermophysical properties coefficient of the pavement and subgrade materials, as shown in Table 8.2; *b* is the humidity coefficient of the subgrade, as shown in Table 8.3; *c* is the section form coefficient of the subgrade, as shown in Table 8.4.

According to the freezing index, the calculation formula of freezing depth is proposed as follows:

$$h_f = abc\sqrt{\Sigma T_f} \tag{8.2}$$

Table 8.1 Divisions of frost regions (JTG D50-2006)

Divisions of frost regions	Heavy frost regions	Middle frost regions	Light frost regions	Non-frost regions
The freezing index (°C·d)	≥2000	2000 ~ 800	800 ~ 50	≤50

Table 8.2 Thermophysical properties coefficient *a* of the pavement and subgrade materials

Subgrade materials	Clay soil	Silty soil	Silty sand	Fine-grained soil gravel Clayed sand	Gravel (sand) containing fine-grained soil
Thermophysical properties coefficient	1.05	1.1	1.2	1.3	1.35
Pavement materials	Cement concrete	Asphalt concrete	Lime-ash soil and cement soil	Lime-fly ash macadam and cement macadam (gravel)	Graded crushed stone
Thermophysical properties coefficient	1.4	1.35	1.35	1.4	1.45

Note: The value of *a* is taken as the weighted average value of materials of each layer of subgrade and pavement within the geodetic freezing depth

Table 8.3 Humidity coefficient *b* of subgrade

Humidity	Dry	Middle wet	Wet	Over wet
Humidity coefficient	1.0	0.95	0.90	0.80

Table 8.4 Section form coefficient *c* of the subgrade

Types of fill and excavation	The height of subgrade fill: Without fill	2 m	4 m	6 m	Fill height more than 6 m	Depth of subgrade excavation: 2 m	4 m	6 m	Fill height more than 6 m
Section form coefficient	1.0	1.02	1.05	1.08	1.10	0.98	0.95	0.92	0.90

Table 8.5 Thermophysical properties coefficient *a* of materials in pavement structural layer (JTJ 014-1997)

Thickness of insulation material layer /m	0 ~ 0.10	0.10 ~ 0.20	0.20 ~ 0.30	0.30 ~ 0.40	>0.40
Northeast China	2.20	2.20 ~ 2.10	2.10 ~ 2.00	2.00 ~ 1.90	1.90 ~ 1.80
North China	2.15	2.15 ~ 2.05	2.05 ~ 1.95	1.95 ~ 1.85	1.85 ~ 1.75
Northwest China	2.10	2.10 ~ 2.00	2.00 ~ 1.90	1.90 ~ 1.80	1.80 ~ 1.70

Note: (1) The thickness of insulation material layer refers to the pavement structural layer composed of cinder, slag and fly ash and other industrial wastes
(2) The material with good thermal insulation performance takes the low value, and the material with poor thermal insulation performance takes the high value

Table 8.6 Coefficient of fill and excavation *b* of subgrade section

	Fill			Excavation		
Height of excavation/m	0 ~ 0.5	0.50 ~ 2.0	>2.0	0 ~ 0.5	0.50 ~ 2.0	>2.0
Northeast China	1.80 ~ 2.00	2.00 ~ 2.20	2.25	1.80 ~ 1.70	1.70 ~ 1.55	1.50
North China	1.85 ~ 2.05	2.05 ~ 2.25	2.30	1.85 ~ 1.75	1.75 ~ 1.60	1.55
Northwest China	1.90 ~ 2.10	2.10 ~ 2.30	2.35	1.90 ~ 1.80	1.80 ~ 1.65	1.60

Note: If the fill is high, the maximum value can be taken; if the excavation is deep, the minimum value can be taken

Table 8.7 Humidity coefficient *c* of subgrade

Humidity types of subgrade	Over wet	Wet	Middle wet
Northeast China	1.00 ~ 1.05	1.05 ~ 1.07	1.07 ~ 1.10
North China	1.01 ~ 1.06	1.06 ~ 1.08	1.08 ~ 1.10
Northwest China	1.02 ~ 1.07	1.07 ~ 1.09	1.09 ~ 1.11

Note: When the humidity coefficient of the subgrade is low, the maximum value can be taken

where h_f is the depth from the road surface to the depth of road frozen line, cm; a is the thermophysical properties coefficient of the pavement structural layer, as shown in Table 8.5; b is the section (fill and excavation) coefficient of pavement, as seen in Table 8.6; c is the humidity coefficient of the subgrade, as seen in Table 8.7; T_f is the average value of the freezing index in the last 10 years, that is, the accumulated value of negative temperature in winter (°C·d), which is calculated and determined according to the observation data of meteorological department.

Table 8.8 Humidity of soil foundation (MHT 5004-2010)

Humidity of soil foundation	Sandy soil	Clayed soil	Silty soil
Dry	$B_m \geq 1.20$	$B_m \geq 1.10$	$B_m \geq 1.05$
Middle wet	$1.20 > B_m \geq 1.00$	$1.10 > B_m \geq 0.95$	$1.05 > B_m \geq 0.90$
Wet	$1.00 > B_m \geq 0.85$	$0.95 > B_m \geq 0.80$	$0.90 > B_m \geq 0.75$
Over wet	$B_m < 0.85$	$B_m < 0.80$	$B_m < 0.75$

Table 8.9 Humidity of subgrade (JTGD 50-2006)

Humidity of subgrade	B_m average consistency B_m	General characteristics
Dry	>1.00	The subgrade is dry and stable, and the strength of the upper soil layer of the subgrade is not affected by the groundwater and the standing water on the surface, $H > H_1$
Middle wet	0.75 ~ 1.00	The upper soil layer of the subgrade is in the transition zone affected by the groundwater or the standing water on the surface, $H_2 < H \leq H_1$
Wet	0.50 ~ 0.75	The upper soil layer of subgrade is in the capillary influence zone of the groundwater or the standing water on the surface, $H_3 < H \leq H_2$
Over wet	<0.50	The subgrade is extremely unstable, and the frozen areas thaw and occurs the mud squeezing in spring, and the non-frozen areas are soft in rainy seasons, $H \leq H_3$

Note: H is the height of the lowest point of the road trough from the groundwater or the standing water on the surface in unfavorable season, m; H_1, H_2 and H_3 are the critical heights of water levels in dry, middle wet and wet states, respectively, m

The category of the humidity of the soil foundation is determined according to the measured average consistency B_m of soil at a depth of 0.8 m below the top surface of the soil foundation in unfavorable season, as shown in Table 8.8, or is determined according to the natural division, soil types, drainage conditions and the height between the top surface of the soil foundation and the groundwater level or the height of standing water on the ground surface (Table 8.9).

The average consistency B_m of soil can be calculated according to the following formula:

$$B_m = (\omega_L - \omega_m) / (\omega_L - \omega_p) \quad (8.3)$$

where ω_L is the liquid limit water content of soil (measured by the combined liquid-plastic limit tester), %; ω_m is the plastic limit water content of soil (measured by the combined liquid-plastic limit tester), %; ω_m is the average water content within the range of the ballast bed, %.

The critical height of water level in dry, middle wet and wet states of the soil foundation can be determined based on the local data and experiences. In addition to the original water content (or in a wet state), the frost heave of the soil foundation in frozen soil area also depends on the minimum distance between the maximum depth of freezing depth line (or the design freezing depth) and the groundwater level (which can be replaced by the groundwater level before freezing), which directly affects the value of water migration in the freezing process.

The formula for calculating the freezing depth (h_f) of soil foundation under pavements based on the heat conduction theory (Bing, 1989, Han et al., 2001) is as follows:

$$h_f = \frac{T_0 - T_s}{B \cdot i_t} \frac{\lambda_f}{\lambda_u} - H \frac{\lambda_f}{\lambda_p} \tag{8.4}$$

where λ_f is the thermal conductivity of soil foundation in frozen state, kcal/(m·h·°C); λ_u is the thermal conductivity of soil foundation in unfrozen state, kcal/(m·h·°C); λ_p is the weighted average thermal conductivity of the pavement structure, kcal/(m·h·°C); T_0 is the temperature at the freezing line of the soil foundation, °C (approximately 0 °C); T_s is the surface temperature of pavements, taken by the local mean negative temperature of the coldest month, °C; i_t is the temperature gradient of soil foundation, °C/m, as seen in Table 8.10; H is the total thickness of the pavement structure, m; B is the correction coefficient, generally taking 1 ~ 1.25.

With the same water content and soil texture, the deeper the freezing depth of the soil foundation is, the greater the frost heave of the pavement becomes. When the pavement begins to frost heave, the summation of the freezing depth that does not cause frost heave and deformation due to the relative elongation of the pavement and the freezing depth at which the pavement structural strength can resist the frost heave and deformation bending moment is less than or equal to the maximum freezing depth of the soil foundation when the pavement has frost heaving and deforming, the pavement will not experience differential frost heave and deformation, and this freezing depth is the allowable freezing depth of the roadbed (Bing, 1989; Han et al., 2001):

$$h_a = 0.84 \times \sqrt[4]{\frac{\rho H}{E \cdot \eta}} L + 0.95 \times \sqrt{\frac{\varepsilon}{\eta}} L \tag{8.5}$$

where ε is the ultimate relative elongation of the surface layer, which is generally taken as 0.0003 for regular pavements and 0.0006 for asphalt pavements; η is the frost heave ratio of soil, %; ρ is the average bulk density of pavement structure, kN/m^3; E is the freezing modulus of pavement structure, MPa; L is the width of pavement, m; H is the total thickness of pavement structure determined according to strength, m.

Table 8.10 Standard values of the maximum temperature gradient (JTG D40–2011)

Natural division of highways	II V	III	IV VI	VII
Maximum temperature gradient /(°C/m)	83 ~ 88	90 ~ 95	86 ~ 92	93 ~ 98

Note: When the altitude is high, a high value can be taken; when the humidity is high, a low value can be taken

Obviously, the soil texture and water content of roadbed are the basic conditions for determining the frost heave ratio. In addition to harnessing the mechanical properties of the pavement material, changing the thermophysical properties of the pavement structural layer to reduce the freezing depth can also prevent frost heave. According to the combined thermophysical parameters of the pavement structural layers, the freezing depth of runways (h_r) can be calculated, and the allowable freezing depth (h_a) of soil foundation can be calculated by the thickness (H) of pavement determined by the strength. And the minimum anti-freezing layer thickness of the pavement is the difference between the maximum freezing depth of the runway and the allowable freezing depth of the soil foundation under pavements, namely, $h_{anti} = h_r—h_a$. The thickness of the anti-freezing layer of the runway is the summation between the pavement thickness calculated by its strength and the thickness of the anti-freezing cushion.

The minimum thickness of anti-freezing layer given by the civil airport specification (Table 8.11) is larger than that of the highway specification (Table 8.12) because the stress on airport runways is greater than that on the highways. However, according to the calculation of the minimum anti-freezing layer thickness of the pavement structure of Yichun Airport, the frost heave can be eliminated only when the thickness of the pavement structure is 68% of the local maximum freezing depth (Xing & Hao, 2010), and the minimum anti-freezing layer thickness given by the specification cannot meet the anti-freezing requirements. The statistical data show that the effective freezing depth that can increase the value of frost heave is about 70% of the maximum depth. From the perspective of safety, under harsh conditions, it is advisable to take the maximum freezing depth as the basis for calculating the minimum anti-freezing layer thickness of the pavement.

Table 8.11 Minimum thickness of anti-freezing layer for civil airport runways/m (MHT 5004-2010)

Humidity of subgrade	Soil texture of subgrade	Local maximum freezing depth within the design life /m			
		0.50 ~ 1.00	1.00 ~ 1.50	1.50 ~ 2.00	>2.00
Middle wet sections	Low, medium and high liquid limit clay	0.30 ~ 0.50	0.40 ~ 0.60	0.50 ~ 0.70	0.60 ~ 0.95
	Silt, low and medium silty clay	0.40 ~ 0.65	0.50 ~ 0.80	0.60 ~ 0.95	0.70 ~ 1.20
Wet sections	Low, medium and high liquid limit clay	0.40 ~ 0.65	0.50 ~ 0.80	0.60 ~ 1.10	0.75 ~ 1.30
	Silt, low and medium silty clay	0.50 ~ 0.80	0.60 ~ 0.90	0.80 ~ 1.20	0.90 ~ 1.50

Note: ① High values should be used in sections with large freezing depth or excavation and high groundwater level, or the foundation and cushion with poor high-temperature performance; low values can be used in sections with shallow freezing depth or fill, or the foundation and cushion with good high-temperature performance

② In wet sections in frozen regions, lime soil should not be used as the foundation (cushion) layer

③ In areas where the frost depth is less than 0.5 meters, the anti-freezing layers may not be necessary

Table 8.12 Minimum thickness of anti-freezing layer for pavement structural layer of highway cement concrete (JTGD 40-2011)

Humidity of subgrade	Soil texture of subgrade	Local maximum freezing depth within the design life /m			
		0.50 ~ 1.00	1.00 ~ 1.50	1.50 ~ 2.00	>2.00
Middle wet sections	Frost heave soil	0.30 ~ 0.50	0.40 ~ 0.60	0.50 ~ 0.70	0.60 ~ 0.95
	Extreme frost heave soil	0.40 ~ 0.60	0.50 ~ 0.70	0.60 ~ 0.85	0.70 ~ 1.10
Wet sections	Frost heave soil	0.40 ~ 0.60	0.50 ~ 0.70	0.60 ~ 0.90	0.75 ~ 1.20
	Extreme frost heave soil	0.45 ~ 0.70	0.55 ~ 0.80	0.70 ~ 1.00	0.80 ~ 1.30

Note: ① the frost heave soil refers to the fine-grained soil gravel (GM, GC), the fine-grained soil sand (SM, SC) except the extremely fine-grained silty sand, and the clayed soil with plasticity index less than 12 (CL, CH);
② The extreme frost heave soil refers to the silty soil (ML, MH), the extremely fine-grained silty sand (SM), and the clayed soil (CL) with plasticity index between 12 and 22
③ The Low values can be used in sections with shallow freezing depth or fill, or the foundation and the cushion layer with good high-temperature performance. The high values can be used in sections with deep freezing depth or excavation and high groundwater level, or the foundation and the cushion layer with poor high-temperature performance
④ The anti-freezing layer thickness of the structural layer may not be considered in the areas where the freezing depth is less than 50 cm

The frost heave soil, freezing depth and the distance from groundwater level are the factors affecting the frost heave of the roadbed (subgrade). When the content of particles smaller than 0.075 mm in foundation soil is less than 10%, it still has a frost heave ratio of 1%. In extremely cold regions where the maximum freezing depth exceeds 2.5 m, the standard of less than 10% content of particles smaller than 0.075 mm to classify soil as non-frost-heave is not sufficient for airport runways and high-speed railways (roads).

When the groundwater level is 1.5 ~ 3 m away from the bottom of the ballast bed, the frost heave soil or extreme frost heave soil will experience frost heave due to the sufficient groundwater recharge.

Therefore, the sand, sandy gravel and other granular materials, or the low-dose inorganic binders can be used in the anti-freezing cushion to stabilize the granular materials, and the content of fine-grained particles smaller than 0.075 mm should not exceed 5%. However, it should be noted that if the clay (smaller than 0.005 mm) accounts for the majority, it can be considered that there is no or very little water migration in the sandy soil when its content is no more than 1% (Wang, 1980).

8.3.2 *Thaw Settlement*

In the design of runways for civil airports in permafrost regions, the frost heaving of the active layer, as well as the thaw settlement of permafrost layer below the permafrost table should be considered together. The causes of thaw settlement in permafrost foundation are as the follows:

1. The internal factors. The permafrost foundation contains ground ice, and usually there is thick ground ice distributed at the permafrost table, with the ice content of 40% ~ 70% and a thickness of generally 0.5 ~ 2.0 m, with a maximum thickness of 3 ~ 5 m. The permafrost table is 1.0 ~ 2.5 m in the cold permafrost, and 1.0–3.5 m in the warm permafrost. The maximum depth in the isolated permafrost region is about 7 m. In the fine-grained soil areas, the distribution depth of high ice content can reach 10 ~ 15 m, and 5 ~ 8 m in the coarse-grained soil areas.
2. The external factors. The thermal influences are, first, climate warming; second, the change of the natural ground surface; third, human activities (including engineering construction). For example, the depth of the artificial permafrost table of the black asphalt pavement reaches 4 ~ 7 m, about double of the natural permafrost table depth. The artificial permafrost table depth of the gravel pavement is about 1.6 times of the natural permafrost table. When the height of the embankment is less than 0.5 m, the permafrost table has reduced in 83% of the sections. The reflectivity of asphalt pavements is 0.06 and that of concrete pavements is 0.20. The temperature of the turf-covered surface is about twice as high as the air temperature. The temperature of asphalt concrete pavements is about five times of the air temperature. The temperature of concrete pavements is about 3.5 times of the air temperature. The thermal conductivity of the pavement structural layer (in regular conditions) increases, and that of the cement concrete pavements is about 0.7 ~ 0.9 times larger than that of the fine-grained soil. The heat during construction, and the heat storage brought in after completion, as well as the intrusion of surface water and groundwater into the foundation trenches will increase the temperature of permafrost foundation and thaw the permafrost.

In design, it is necessary to master the permafrost conditions at the runway base, such as the lithology, the mean annual ground temperature, permafrost table, the engineering types of frozen soil, the ground ice content (or expressed by water content) and its distribution (horizontally and vertically), the thaw settlement (the horizontal and vertical coefficient of thaw settlement), the physical and mechanical properties and the hydrogeological conditions, etc.

8.3.3 Bearing Capacity of Frozen Soil Foundation

The bearing capacity of permafrost foundation is affected by factors such as the material composition, density, water content (ice content), temperature of the foundation soil, among which the temperature is the key factor affecting the variation of bearing capacity of permafrost foundation. With the ground temperature increasing, the content of unfrozen water in permafrost also increases but the strength decreases, and vice versa. When calculating the bearing capacity of the permafrost foundation, the value of bearing capacity provided during the survey should be considered. In addition, there are more factors needed to be considered during the operation period of the airport runways, including the environment of permafrost (the removal of surface vegetation, etc.), the property of pavement structural layer, the upper load (the static load and the dynamic load), the foundation treatment, the construction methods and the processes and global climate change and so on, any change of which will cause variations in thermal regime of the permafrost foundation, and make the permafrost temperature increase or the frozen foundation thaw. Since the width of the cross-section of the airport runway is greater than 45 m, the variations (horizontal and vertical direction) in the thermal regime of permafrost foundation caused by pavements is much greater than that of the highway and railway subgrades, so it is necessary to consider the engineering safety and fully harnessing the strength properties of permafrost foundation in calculation.

When the mean annual ground temperature of permafrost foundation is low (cold permafrost), the design principle of protecting the frozen state is usually adopted. Because the unfrozen water content in permafrost is low but the bearing capacity is high, the bearing capacity of permafrost foundation can meet the requirements of airport construction if proper measures are taken to maintain the foundation in frozen state. When the mean annual ground temperature of permafrost foundation is high, it belongs to the warm plastic permafrost with a large compression modulus, and the compression and settlement deformation under load cannot be ignored.

The landing of the airplane will have an impact on the pavement structural layer, which will be transmitted to the permafrost foundation through pavement concrete slabs. Its effect is similar to the vibration effect of the train load (its continuity and frequency of vibrations are far less than those of train load, which is with intermittency), and it differs from the impact loads (such as blasting, earthquakes and mechanical shock) directly acting on the permafrost. The former is often studied with a low-temperature dynamic triaxial material testing machine, and the latter is usually completed with a Hopkinson pressure bar test. In general, the dynamic strength of permafrost is less than its static strength, and the strength of permafrost foundation appears to be weakened under dynamic load. The experimental research has shown that (Gao et al., 2009; Zhu et al., 2008), the compressibility was large under low vibration frequencies. With same water content, vibration frequency, confining pressure and temperature, the larger is the dynamic stress amplitude, the faster the residual strain increases. Under different temperatures and confining pressures, the generalized hyperbolic model can be used to describe the dynamic stress and dynamic strain relationship of the backbone curve of the warm permafrost with high ice contents. Under the same confining pressure, the dynamic modulus increases with the decrease of soil temperature. The damping ratio of frozen silty clay with high temperature increases with the increase of strain amplitude.

8.3.4 Freeze-Thaw Durability of Concrete Pavement

The concrete pavements in cold regions are often subjected to freeze-thaw cycles, leading to deterioration of durability such as cracking, spalling, peeling, outcrop sand and even outcrop stone on concrete pavements. This deterioration can result in cracked slabs, fractured slab corners and fragmented edges near the joints etc. According to the test data of the airport pavement concrete performance and mix ratio design requirements, the loss of flexural strength is the greatest in freeze-thaw cycles. After the specimen has been frozen and thawed for 100 times according to the benchmark mix ratio, the compressive strength is reduced by 9.30%, 42.27% for the flexural strength, and 76.3% for the relative dynamic elastic modulus. The relative dynamic elastic modulus of the concrete specimen mixed with 0.8% air-entraining and water-reducing agent is 94.7% after freezing and thawing for 150 times. The dynamic elastic modulus decreases with the decresing of the concrete strength, which indicates that the deterioration of the durability appears gradually. From the effect of freeze-thaw cycles on the mass loss of pavement concrete (Fig. 8.9) and the spalling of surface, the mass loss of the concrete specimen with 0.4% admixture after freeze-thaw cycles for 150 times is 0.58%, and more than 50% of the surface is peeled off. The mass loss of concrete specimen with 0.8% admixture after freeze-thaw cycles for 500 times is 1.91%, and the surface is almost completely peeled off. The addition of air-entraining and water-reducing admixture can improve the frost resistance of concrete and reduce the mass loss and the spalling of surface caused by freeze-thaw cycles. According to the freeze-thaw test of concrete core columns (Zhao & Ba, 2002), the edges of core columns are falling after freeze-thaw cycles for 250 times in water, and the strength is decreased by 32.9%. The surface and edges of core column have a large number of cracks after freeze-thaw cycles for 270 times in deicing fluid, and the strength is decreased by 20.6%, because the freezing point of deicing fluid is low.

In the design of pavement in cold regions, it is necessary to consider the improvement of the frost resistance and durability of pavement concrete. For a specific area, among the factors that cause the freeze-thaw damages to concrete, the times of annual natural freeze-thaw cycles have the greatest impact on frost resistance, accounting for 90%, followed by the daily minimum temperature, accounting for 8%, and finally the maximum cooling rate, accounting for 2%. The highest requirements for frost resistance should be met in Qinghai-Xizang Plateau, Western

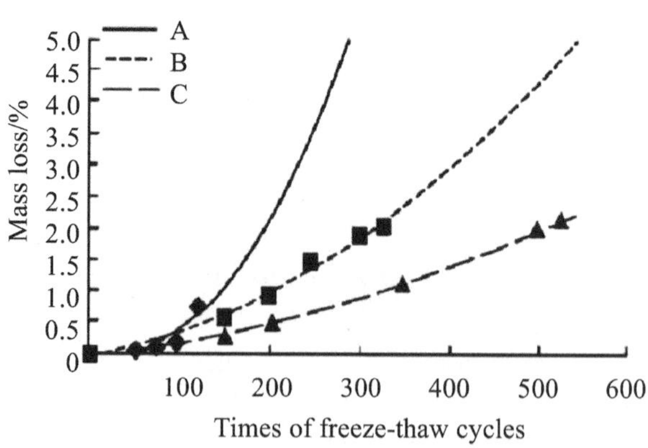

Fig. 8.9 Relationship between the times of freeze-thaw cycles and the trend of mass loss (Li et al., 2011). A—Benchmark mix ratio; B—With a mixture of 0.4% air-entraining and water-reducing admixture; C—With a mixture of 0.8% air-entraining and water-reducing admixture

Sichuan Plateau, Western Yunnan Plateau, Daxing'anling Mountains in Northeast China and other areas, and special designs should be made to deal with frost, followed by the border of Heilongjiang and Inner Mongolia, and most areas of Northwest and Northeast, and general frost resistance need to be designed in the south of Northeast China and North China.

8.3.5 Drainage Works

In general, the drainage systems are set up in different zones according to topographic and hydrogeological conditions. In the pavement areas, the vertical and horizontal thermal insulation underground ditches for drainage can be arranged at the bottom of the foundation trench of runways, and the vertical underground ditches for percolation (Fig. 8.10) can be arranged on the side of the foundation trench (one side or both sides). The slope ratio of the bottom of the foundation trench is not less than 1%, and the groundwater can be drained into the underground ditches for percolation on the side, and then the groundwater in the foundation trench can be discharged into the deep drainage system of the pavement areas together. In the area with a soil surface, the thermal insulation underground ditches for percolation (Fig. 8.11) can be arranged to drain the surface water and shallow groundwater on both sides of the pavement areas and around the soil surface areas into the shallow drainage system. If the groundwater level is high, it is advisable to change the thermal insulation underground ditches for drainage into thermal

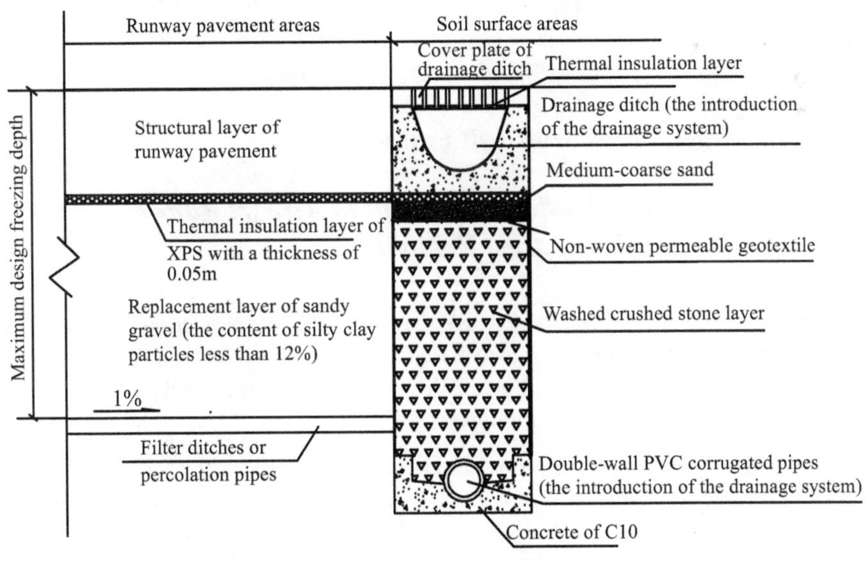

Fig. 8.10 Schematic diagram of lateral filter ditches for percolation of runways

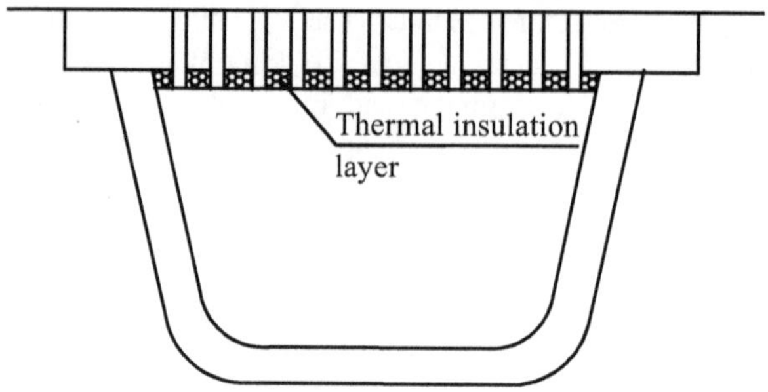

Fig. 8.11 Thermal insulation drainage in soil surface areas

insulation underground ditches for percolation (such as a lateral underground ditches for percolation on the pavement) to reduce the groundwater level. The thermal insulation underground ditches for percolation should be arranged in the soil surface area at a certain distance from the pavement to reduce the groundwater level if necessary.

The width of lateral underground ditches for percolation (Fig. 8.10) of the foundation trench is 1.0 ~ 1.5 m, filled with washed crushed stone or cobble and gravel, and lined with a layer of 450 g/m^2 non-woven permeable geotextile outside, and laid with 0.1 m thick medium-coarse sand above the geotextile. The drainage pipes at the bottom of the underground ditches for percolation adopt Φ200 mm double-wall PVC corrugated pipes with a wall thickness of 50 mm. The holes with a diameter of 8 mm are drilled within 240° above the pipe wall in a staggered layout, and the vertical spacing is 150 mm, and a layer of 450 g/m^2 non-woven permeable geotextile can be warped outside, with the overlapping length between the cloths more than 0.2 m.

8.4 Treatment of Frozen Soil Foundation

8.4.1 Types of Foundation Treatment of Airport Runways in Foreign Permafrost Regions

According to the report (Liu et al., 2015), the main types of foundation treatment for airport runways in foreign permafrost regions include the bedrock runways, the runways with thermal insulation layer, the runways with thermosyphons, and the runways excavated and replaced with sandy gravel.

1. The bedrock runway is the most ideal runway to avoid engineering problems caused by frozen soil, such as Courcheville International Airport in the French Alps, limited by the topography, which is a small airport with a runway length of about 500 m. A large number of airport runways along the coast of Greenland are built directly on bedrock. Limited by topographic and geological conditions,

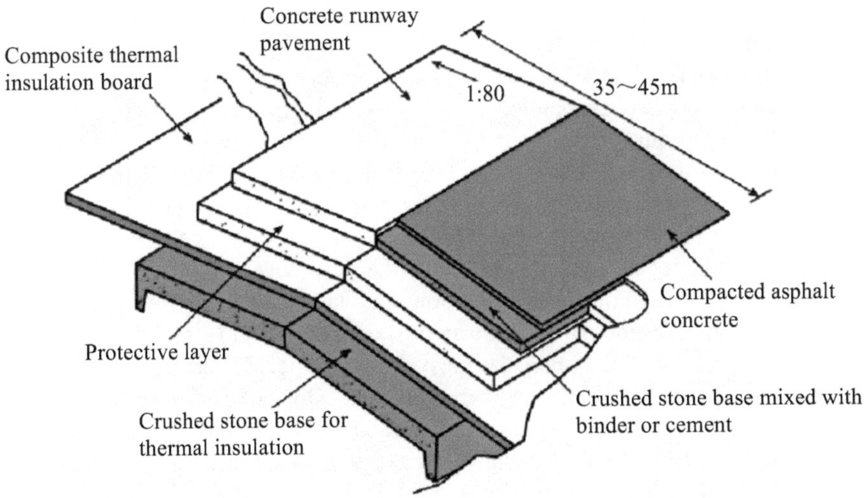

Fig. 8.12 Structural diagram of runways with thermal insulation layer (Liu et al., 2015)

bedrock runways make it difficult to meet the requirements for runway length and the safe takeoff and landing of large aircrafts.

2. The runways with thermal insulation layer can prevent the external heat from entering the permafrost foundation through the pavement, causing changes in the thermal state of the permafrost foundation, ensuring that the thawing depth of the permafrost foundation on the airport runways is less than or equal to the design thawing depth, and maintaining the thermal stability of the permafrost foundation. Without doubt, the heat of the permafrost foundation is also blocked off from dissipating outward in winter. In other words, the external cold energy cannot be transmitted into the permafrost foundation, which is not conducive to the cooling of the permafrost foundation. In 1969, the first runway with thermal insulation layer was built in Kotzebue, Western Alaska, adopting expansionary extruded polystyrene foam boards (XPS). According to the on-site monitoring, test, calculation and evaluation of the performance of the runway and the different thermal insulation materials, the polyurethane foam was negated because of its large water absorption and low strength, while the expansionary extruded polystyrene foam boards had a better performance, and still had an excellent thermal insulation in service to the 15th year, and the design life could be more than 20 years. Figure 8.12 is a typical structural diagram of an airport runway in permafrost areas with thermal insulation layer. The runway has an inclination of 1:80 from the center to both sides, with a protective layer and a thermal insulation layer at the lower part, and crushed stone at the bottom. In the case of asphalt pavements, a layer of crushed stone base mixed with binder or cement will be added on the lower part.

It should be noted that the permafrost in Alaska belongs to the cold permafrost, with the mean annual ground temperature lower than −4 ~ −5 °C in extremely cold climate. The mean annual ground temperature at sites suitable for construction in permafrost regions of China is –0.5 ~ –2 °C, and the solar radiation is also higher than that of Alaska. Therefore, the subgrade with thermal insulation layer is not promoted for use on the Qinghai-Xizang Railway.

3. The runways with thermosyphons (thermal probes) installed. In order to solve the problem of thaw settlement and frost heave of the runways, the thermosyphons were installed in an airport in Alaska in 1982. The State Department of Transportation proposed that the top elevation of the cooling fin of the thermosyphon could not exceed the elevation of the runway centerline, and the runway could not be closed during construction. Two layers of the thermosyphon were arranged on both sides of the runway in staggered layout (Fig. 8.13). The horizontal slope of the upper thermosyphon was 1:15 and the length was 24.38–42.68 m (80–140 ft), while the horizontal slope of the lower thermosyphons was 1:10 and the length was 24.38 m (80 ft). It had been proved by monitoring that the arrangement of thermosyphons had a good effect on maintaining the stability of the runway foundation.

   The airport runways with thermosyphon are categorized as the "embankment type" runways. There are "embankment type" airport runways (Fig. 8.14) in permafrost regions of Canada, which are suitable for small and medium-sized aircrafts. After compaction, the ground is paved with block stone (similar to the crushed rock embankment on the Qinghai-Xizang Railway), with crushed stone soil laid on the upper part, and the sand stone pavement is adopted.

   The "embankment type" runways are not allowed to use in the airports in China, so it is difficult to adopt thermosyphons to reduce the temperature of the permafrost foundation to maintain the thermal stability of it.

4. The runways constructed by excavation and then replacement with sandy gravel. The permafrost is excavated and replaces with crushed stone or gravel to build runways. For the airport runways in the coastal plains of the Arctic Ocean in Alaska, runways are generally built with gravel replacement (with the thickness of the gravel layer of 1.5 m). Considering that the permafrost thaws in summer, it is often necessary to multiply the engineering safety factor by 1.25 to elevate

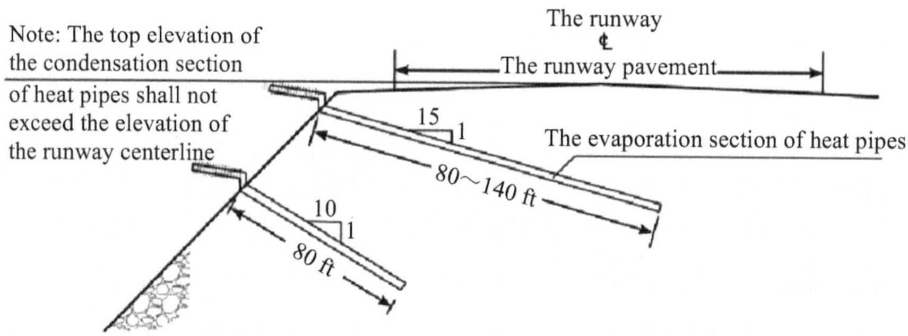

Fig. 8.13 Structural diagram of airport runway with thermosyphon in Alaska (Liu et al., 2015)

Fig. 8.14 "Embankment type" runways of U Miujaq Airport in Canada (photo by Yu Wenbing)

the permafrost table into the backfilled gravel layer. In the Brooks Mountains of northern Alaska, the replacement thickness of gravel layer needs to be increased by 0.3 m. The washed gravel is required in the pavement 1.2 m lower to prevent the capillary effect from the surface water, and the fine sand needs to be added in the top 0.3 m to enhance cohesion and achieve good compaction.

For the permafrost layer of the original wetland on the runway of Mohe Airport in China, the original ice-rich permafrost layer of frozen silty clay is replaced with sandy gravel. The replacement depth was decided when the water content of the permafrost layer was slightly greater than the plastic limit water content, and the depth was about 13 m (with the pavement structure layer and the foundation trench of the sandy gravel cushion layer of 4.8 m). The vertical and horizontal drainage pipes were set at the bottom of the foundation trench. After nearly 10 years of operation, the runway was still in good performance.

8.4.2 *Foundation Treatment of Airport Runways in Permafrost Regions*

The major problems in frozen soil engineering of airport runways in permafrost regions include frost heave and thaw settlement of the permafrost foundation, as well as the high fill and the water supply and drainage projects.

1. Anti-frost Heave Measures for Pavement Foundation

The Anti-frost heave measures for airport runways include the following aspects:

1. Improvement of Soil Texture of Roadbed Foundation to Prevent Frost Heave

The soil improvement of roadbed foundation mainly focuses on two aspects: the material composition and the compactness.

The foundation soil with frost heave susceptibility is the soil whose particle size is less than 0.1 mm. Under the same condition, when the content of particles smaller than 0.075 mm in the soil exceeds 50%, it exhibits the maximum frost heave. Therefore, soils belonging to the fine-grained soil can be classified as the soil with frost heaving susceptibility. Among them, the silty soil, silty clay, low liquid limit clay, and the coarse-grained soil with more than 15% of the particles not greater than 0.075 mm have strong frost heave. The coarse-grained soil, such as the coarse sand, gravel sand, gravel and crushed stone, with less than 15% of particles smaller than 0.075 mm (it should be controlled within 5% for the airport runways and the high-speed railways), belongs to the soil with weak frost heave susceptibility. The Fine-grained soil, with strong water holding capacity, thick water film, and high rising height of capillary water, will result in great frost heave due to strong water migration during freezing. The coarse-grained soil does not have such a capacity, and it also has "water repulsive effect" during freezing process. Using the coarse-grained soil to replace the frost heaving susceptibility soil can effectively block the elevation of capillary water, reduce the water migration to the freezing front, and then reduce the frost heave of foundation soil.

The increase of compaction density of the foundation soil can inhibit its frost heave. A large number of test data show that when the dry density of cohesive soil reaches 1.7 g/cm^3, the increase in density can make the soil into a two-phase system, and the excess water will be squeezed out, resulting in the decrease of frost heave (Fig. 5.4). In unsaturated fine-grained soil, the maximum value of frost heave is within the range of 80%–90% of the compaction degree. When the compaction degree reaches 90% the relative value of frost heave decreases proportionally (Liu, 2009). With the compaction density increases, the weakly bound water dominates the pores of silty clay, blocking the elevation of capillary water and the infiltration of surface water. According to *Specifications for Design of Cement Concrete Pavement of Civil Airports* and the highway department, the compaction degree of the soil foundation within the range of freezing depth should be more than 95%. The sandy gravel foundation soil generally can meet the requirements of compaction density (Table 8.13).

The coarse-grained soil such as the sandy gravel needs to meet three conditions to prevent frost heave:

Firstly, the purity of the sandy gravel backfills: the content of silty clay particles smaller than 0.075 mm in sandy gravel replacement layer of airport runway

Table 8.13 Test results of ground vibration rolling in aaline soil wetlands of Ali Airport in Xizang (Zhang and Zhang, 2012)

Rolling times	Before vibration rolling	4 times	6 times	8 times	10 times
Dry density /g·cm^3	2.00	2.05	2.11	2.14	2.16
Compaction degree /%	90.8	93.0	95.8	97.1	98.1

Note: It is composed of the round gravel and the sandy pebbles, and the surface layer in the wet saline soil is the miscellaneous fill, plain fill and silty fine sand

should be controlled within 5%, and the impurities such as grass roots cannot be included.
Secondly, the replacement depth: the depth of frost-heave soil replaced by the sandy gravel is closely related to the groundwater level and the distance between the groundwater level, and the maximum design freezing depth should be greater than 1.0 m. Normally, it is advisable for the replacement depth to be greater than 70 ~ 80% of the maximum design freezing depth. In Hokkaido, Japan, the average replacement rate is 80% of the maximum freezing depth. In Sweden, it takes the freezing depth of 0.9 m as the boundary. Within this boundary, full replacement is conducted, while for areas exceeding this boundary, the replacement rate is 80%. The replacement rate of flexible airport runways in the United States Army Corps of Engineers is 80% of the total freezing depth after replacement. In permafrost regions, the two-way freezing occurs in the active layer, that is, the surface is frozen from top to bottom due to the atmosphere, and the permafrost table is frozen from bottom to top, and the two-way frost-heaving is caused in the soil, forming a "K" curve. The amount of frost heave depends on the enrichment of groundwater, and the replacement rate should be greater than that in seasonally frozen soil areas, or up to 100%.
Thirdly, the water in the replacement layer can be drained out. During the freezing process, the pore water in the sandy gravel layer freezes to form a positive pore water pressure. Under the impact of the pressure gradient, the excess water is discharged to reduce frost heave. The drainage channels (open system) should be provided or set up under the sandy gravel replacement layer to ensure the anti-frost heave effect of the sandy gravel replacement.

2. Thermal Insulation

The thermal insulation method, or the thermophysical method, can reduce the freezing depth of the foundation soil by adjusting its ground temperature state, so as to reduce the frost heave and deformation of the pavement. According to the thermal insulation effect, the closer the thermal insulation layer to the surface, the better the thermal insulation effect is. A protective layer should be set between the pavement and the thermal insulation layer. The cement-stabilized sandy gravel base under the pavement plays a dual role, which not only ensures the strength, but also protects the underlying thermal insulation layer (Fig. 8.15). For example, the anti-frost heave measures for the test section of the Harbin-Dalian Passenger Line: the XPS insulation boards of 0.1 m and 0.06 m were laid below the subgrade bed (with a strain at 3%, the compressive strength is more than or equal to 700 kPa, the absorptivity is less than 1%, and the thermal conductivity is less than or equal to 0.03 W/m • K); the bottom layer was replaced with an anti-freezing layer of A and B filler with the thickness of 1.0 m, and the subgrade bed was sealed by composite geomembrane. The observation data show that the ratio (average) of the temperature under and above the thermal insulation board is 29.01% and 31.61%, respectively (Kong, 2012). The difference of thermal insulation effect of the two thicknesses is only increased by 2.5%, which can reduce the maximum frost heave on the top surface of the subgrade by 50.5% and 35.3% respectively, indicating that the thermal

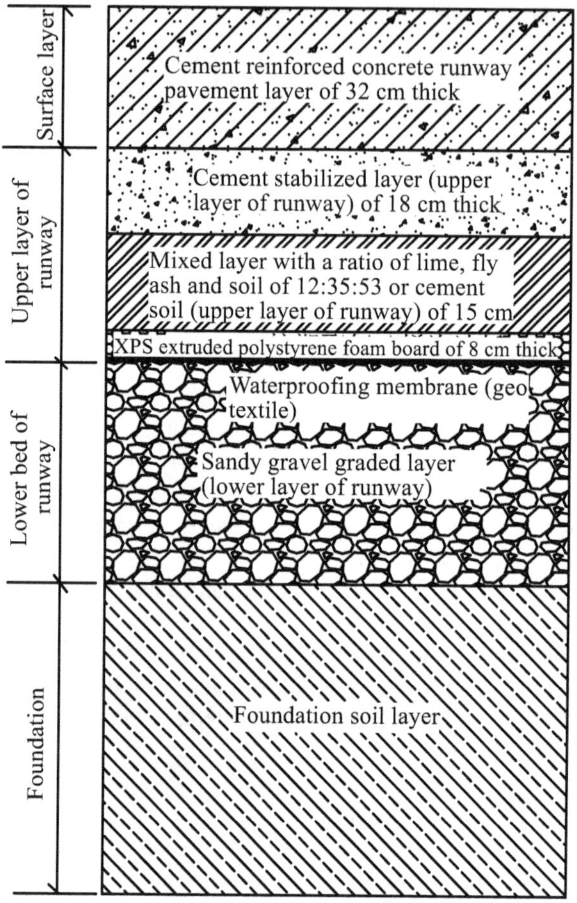

Fig. 8.15 Structural diagram of airport runways

insulation board has a good thermal insulation effect. However, the thickness has increased by 66.66%, the improvement in the thermal insulation effect is not large, which indicates that the thickness of the thermal insulation board is not proportional to its thermal insulation effect.

Experience shows that, in general the thermal insulation board (EPS) of 10 mm can reduce the freezing depth by 0.1 ~ 0.13 m. According to the observation data of the main channel in Minqin County, Gansu Province, the thermal insulation board of 10 mm can reduce the freezing depth by 0.3 m (Chen et al., 2006), which may be related to the factors such as the soil layer properties, the water content and the performance of the thermal insulation boards.

Comparing the physical properties of the XPS and EPS insulation materials, the XPS board is suitable for highways, railways and airport runways, with higher compressive strength, lower absorptivity and thermal conductivity. The thickness of the thermal insulation board under airport runways shall not exceed 100 mm.

Based on the experience of the Qinghai-Xizang Highway and Railway, the buried depth of the thermal insulation board should be set at 0.3 ~ 0.5 m below the design elevation of the subgrade. When the subgrade height exceeds 3.0 m, the influence of lateral heat transfer of the subgrade should be considered, and the buried depth of thermal insulation board should be reduced accordingly. Although the pavement area of airports is level with the soil surface area, the soil surface area on both sides of the airport runway is 1.5 times larger than the area of the pavement, which has the freezing depth approximate to that of the natural freezing depth and may affect the freezing depth of the pavement area, so the buried depth of the thermal insulation board should be appropriately increased, and it should be buried 0.3 ~ 0.5 m below the bottom surface of the pavement layer (concrete slab).

3. Waterproof and Drainage

It is an important measure to drain off the water in foundation soil to prevent frost heave. The drainage facilities in the pavement area and the soil surface area should form a system, and the on-site drainage system should be combined organically with the off-site drainage system. A geotextile membrane can be used at the bottom to cut off the capillary water.

The Longitudinal underground ditches for percolation with thermal insulation should be set on the side of the foundation trench in the pavement area (Fig. 8.11), and the transverse underground ditches for drainage should be set on the bottom of the foundation trench to connect with the lateral underground ditches for percolation with thermal insulation, so that the groundwater of the foundation trench can be discharged from the pavement area in appropriate topographical sections. The underground ditches or pipes for percolation can be used at the bottom of the foundation trench (Fig. 8.16).

In the sections with high groundwater levels in seasonally frozen soil areas, the groundwater level should be controlled at 0.5–1.0 m below the maximum freezing

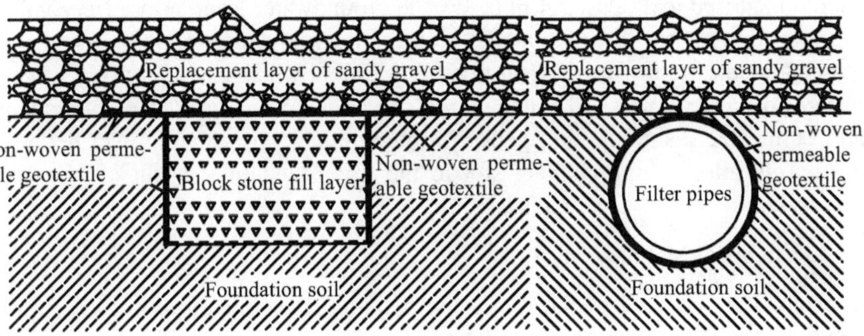

Fig. 8.16 Schematic diagram of cross-section of transverse underground ditches or filter pipes at the bottom of runway foundation trench

depth line. There are several methods to be taken: firstly, the thermal insulation layer can be used to reduce the seasonal freezing depth; secondly, the depth of lateral underground ditches for percolation of the foundation trench can be increased and the groundwater level can be reduced; third, the permeable water-resisting layers of sandy gravel and crushed stone, the impermeable geotextile water-resisting layers, or the water-resisting layers mixed residual oil with soil can be laid at the bottom of the pavement structural layer or the anti-freezing layer to block the groundwater; fourth, the embankment-type fill foundation can be used to raise the pavement elevation, such as in Kunsha Airport in Ngari Prefecture, Xizang. The first three methods should be adopted in combination to achieve twice the result with half the effort (Fig. 8.17).

The buried depth of drainage and percolation ditches can be calculated according to the following formula:

$$h_s = Z + d + h_0 + h_j \tag{8.6}$$

where h_s is the buried depth of the percolation ditch, m; Z is the freezing depth (the difference between the natural freezing depth and the equivalent influence depth of thermal insulation layer), m; h_0 is the water depth in the percolation ditch, m; h_j is the safe height (the controlled depth of groundwater level), m; d is the maximum vector moment of the precipitation curve of the percolation ditch, m, namely: $d = I_0 \cdot L$; I_0 is the average slope value of precipitation curve; L is the distance from the edge of the percolation ditch to the centerline of the pavement, m.

The drainage of foundation trench in pavements in permafrost regions can be referred to Fig. 8.10 and the maximum design freezing depth should be changed to the maximum design thawing depth.

An airport in Xizang adopts the method of elevating the channel slots (Sun et al., 2012). The seasonal freezing depth of the site is 2.4 m. Considering the shallow buried depth of the groundwater in the airport and with large variation, and a part of the area lies in wetlands, so the rising height of capillary water in sandy gravel fill is 0.4 m. By comprehensive analysis of the critical freezing depth, the rising height of the capillary water and the safe height, the design elevation of the channel slot should be raised to at least 2.4 m below the groundwater level before freezing, and the elevation of the entire channel slot of the foundation should be raised to 2.6–3.8 m considering the 50-year flood prevention, which can effectively solve the hazards of saline frozen soil and the threat of floods. At the same time, impermeable and permeable geotextile and geotextile membrane can be laid 1.0 m below the design elevation of the channel slot in the soil foundation to cut off the capillary water.

4. Chemical Methods

The physical-chemical method is to use artificial material to inhibit the forming conditions to weaken or eliminate frost heave.

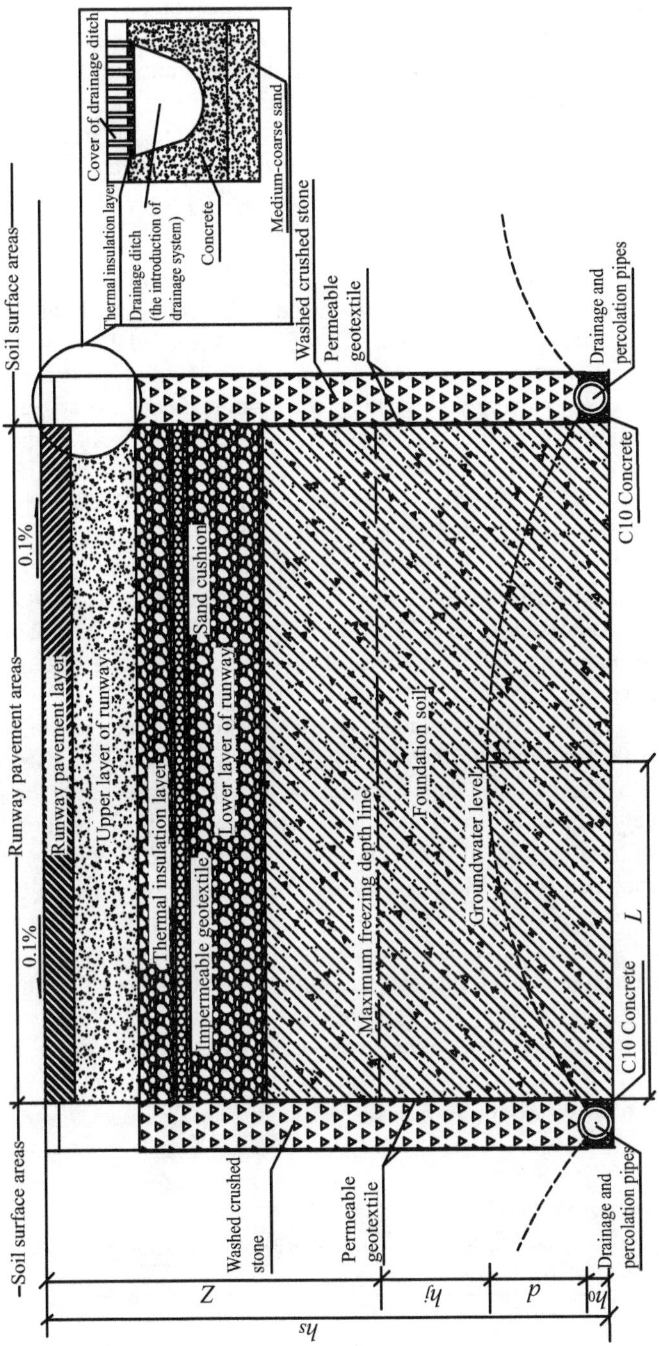

Fig. 8.17 Precipitation methods of percolation ditches in the sections with high groundwater level

① The foundation soil is improved by soil salinization, and then the freezing point of the foundation soil is lowered to inhibit frost heaving. The usage of soluble inorganic salts, such as $CaCl_2$, KCl, NaCl is common, and among them, NaCl is the most common one. After the soil salinization, the concentration of inorganic salt ions in pore water solution increases. The cations in the soil are replaced, which causes the coalescence of soil particles and reduces the freezing temperature of the soil (Table 8.14), so that the soil will not be frozen in a certain temperature range and have no frost heaving. The method is easy to apply and there are extensive materials sources.

However, it should be noted that the resistance to deformation and destruction of soil after salinization are reduced significantly. The cohesion of soil is reduced by 2/3 ~ 5/6 under slight salinization (Z = 0.5% ~ 1.0%), and the bearing capacity is also reduced correspondingly, but the amount of deformation is much higher. In the process of freezing and thawing, the desalinization will appear, and the period of validity preventing frost heave is 2 ~ 5 years. The main method to prolong the validity of salinization is to tamp the artificial salinized soil in layers to make the dry density reach 0.9 ~ 1.03 times of the optimal dry density, so as to improve the strength of salinized soil, reduce the freezing point of pore water solution, make the water not migrate or migrate slightly during freezing. For example, when the dry density of silty clay reaches 2.0 g/cm^3 and the concentration of pore water solution is 0.15 g/cm^3, its frost heave resistance performance can still be maintained after 16 freeze-thaw cycles in an open system, and the stable relative frost heaving is 1% ~ 2%.

② The hydrophobic materials, including petroleum products such as the heavy oil, liquid petroleum asphalt, surfactants, etc., mixed with soil can be used to make the surface of soil particles have hydrophobicity, which can change the percolation channel of the soil layer, reduce the water content of the foundation soil, and weaken the frost heaving of soil. The surfactants are the polyurethanes containing 12-carbon bonds and 18-carbon bonds. For example, NN′- dioctadecyl ethylenediamine can reduce the frost heave by 90%, with 60%~80% decrease of the elevation of groundwater and 98% decrease of infiltration rate of surface water.

③ Adding the coagulant to the silty soil and sandy clay can increase the particle size of soil and reduce the rising height of the capillary water. The main polymerizing agents are: maleic polymer, sodium poly-acrylate, polyvinyl alcohol and high-valence Fe^{3+}, Al^{3+} and other polymerizing agents. The frost heave can be reduced by 90% with the dosage of 0.5% ~ 1.0%.

Table 8.14 Initial freezing temperature of soil with different salinity

	Freezing temperature of soil with different salinity Z (%)/°C			
Soil texture	0.4	0.55	0.85	1.2
Silty clay	−1.2	−1.5	−2.6	−3.0
Silty soil	−1.3	−1.7	−2.5	−3.4
Fine sand	−1.3	−1.6	−2.5	−3.5

On the contrary, adding the sodium phosphate dispersant to the fine-grained soil, such as silty clay, can disperse the natural soil micelles, increase the content of fine particles and the compactness, resulting in weakening the frost heave of the soil. The dispersants mainly include the sodium tetra metaphosphate, sodium hexametaphosphate and sodium tripolyphosphate. The dosage is 0.5% ~ 1.0% of the soil weight.

④ Weaken the water permeability of the isolation layer of the soil: adding the water-soluble salt materials to the soil can increase the viscosity of the soil pore water, or forming insoluble compounds on the surface of soil particles can improve the impermeability of the soil. There are physical and chemical methods.

(a) Chemical materials to reinforce the isolation layer:

The lignosulfonates include the sulfite spent pulping liquor +$K_2Cr_2O_7$, + $CaCl_2$, and + $Ca(OH)_2$ and so on, which can completely eliminate the frost heave with the increase of the dosage.

The silicone liquids include the sodium methyl silanolate, ethyl-silanol natrium, polyvinyl silanol, etc. Adding 10% ~ 20% silicone liquids to sandy clay and fine sand, the foundation soil will not disintegrate after soaking in water for half a year, and the frost heave of soil is significantly decreased.

The carboxymethyl cellulose (CMC) is used to increase the viscosity of pore water or external water sources. Adding 0.5% CMC in the aqueous solution containing 1% CMC and the viscosity of 160×10^{-3} Pa·s (the viscosity of pure water at 20 °C of 1×10^{-3} Pa·s), the viscosity can reach 30×10^{-3} Pa·s, and the frost heave is greatly reduced.

(b) Physical materials to reinforce the isolation layer:

The cement-soil is made of the cement, soil and water in a certain proportion and the common water-cement ratio is (0.7:1.0) ~ (1.4:1.0), and the cement-soil ratio is 1:6 ~ 1:8. It is suitable for the sandy soil, gravel sand, gravel and the soil mixed with industrial wastes. The sandy clay should be mixed with lime, and the soil containing organic matter should be mixed with $CaCl_2$. The heavy clay and soil with large amounts of organic matter are not suitable. The thickness of the cement-soil asan isolation layer is 0.15 ~ 0.2 m.

The lime content in the lime soil (3% ~ 20%): the content of active CaO in lime is larger than or equal to 52%, and there are more than 90% with the particle size smaller than 0.35 mm. it will exert better effect and impermeability after adding a small amount of liquid asphalt.

For the asphalt soil, using the emulsified asphalt to treat the fine-grained soil (silt, clay), and then wrapping it with the high-strength, anti-aging polyethylene plastic film, can replace the coarse-grained soil as the subgrade fill, and has a good effect of thermal insulation and water resistance. Nowadays, if the asphalt and cement are mixed in use, the frost, deformation and corrosion resistances will be improved.

5. Dynamic Consolidation

The dynamic consolidation refers to using the shock wave generated by the free fall of rammers to densify the foundation, and the gas is removed and the soil particles close to each other are rearranged, from a natural disordered state to a stable state, and the pore volume is greatly reduced. In the saturated soil, a large tensile stress produced by dynamic consolidation brings a series of vertical cracks in the soil, and the pore water is discharged from the cracks to accelerate the consolidation of the saturated soil. In seasonally frozen soil areas, the dynamic consolidation makes the foundation soil compacted, and decreases the water content (Table 8.15) as well as the frost heave (Table 8.16).

It proved that the dynamic consolidation has a good effect on foundation soils, including consolidating the crushed rock soil, sand soil, silt soil, unsaturated cohesive soil, collapsible loess and artificial fill, but has a little and even counterproductive effect on the soft and flow plastic cohesive soil, as well as the saturated silt and silty soil, because the water is hard to discharge. At this time, the dynamic consolidation can be used to strengthen the foundation effectively.

The principle of dynamic consolidation replacement is to form the replacement through the strong impact and the filler, so that the replacement body and the original foundation soil to form a composite foundation to bear the load together (Fig. 8.18). The principle is that when the cylindrical heavy hammer falls from high, the hammer will penetrate the soil and release a large amount of energy, causing three influences on the reinforced soil: firstly, the soil under the hammer is compressed and moved downward rapidly due to the huge impact force from the hammer bottom, and the bottom of rammed pit is compacted (Fig. 8.18a), and the density is greatly increased; secondly, the soil on the side of the hammer is also subjected to the huge punching force from the edges of the hammer, and vertical shear failure occurs, forming a nearly vertical cylindrical deep pit (Fig. 8.18b); thirdly, the soil will vibrate violently at the same time under the effect of stamping

Table 8.15 Changes in physical properties and strength of foundation soil before and after dynamic consolidation (Tang, 1992)

Load status	Soil texture	Sampling depth /m	Dry density /g·cm^3	Water content /%	Bearing capacity of the foundation /kPa
Before the dynamic consolidation	Silty clay	1.0	1.58	24	150
After the dynamic consolidation	Silty clay	1.0	1.72	18.5	250

Table 8.16 Pavement frost heave before and after dynamic consolidation (Tang, 1992)

Sections	Amount of frost heave before the treatment/mm	Amount of frost heave after the treatment/mm
Longwo highway	100	30
Zhongqi road	60	20

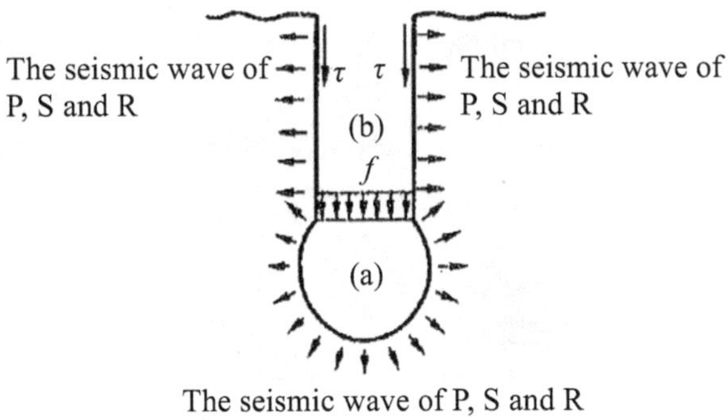

Fig. 8.18 Schematic diagram of dynamic consolidation replacement (Gong, 2008)

and punching shear from the hammer, and spread into the deep soil in three kinds of seismic waves (P wave, S wave and R wave). Besides, under the joint action of vibrating liquefaction, drainage consolidation and vibration compaction, the soil around the replacement body is also reinforced.

The shape and size of the replacement body of the dynamic consolidation replacement are related to the construction technology:

① The diameter and depth of the crown-shaped cylindrical replacement body are related to the diameter of the compaction energy (Fig. 8.19). If the replacement depth is deep, the compaction energy must be increased by effectively increase the replacement depth and times.
② The replacement body is required to have a small diameter and a large depth-diameter ratio. In addition to using a small-diameter hammer, it is also necessary to increase the single-blow compaction energy to effectively increase the penetration depth of each hammer. The diameter of the replacement body is generally 1.5 ~ 1.8 times the hammer diameter, which is related to the size of the single-blow compaction energy, the replaced soil and the properties of the replaced materials.
③ The properties of the replacement materials have a great influence on the shape of the replacement body, and the coarse-grained materials such as the crushed rock have a good replacement effect and a clear outline of the replacement body. The sand, gravel and mountain soil should not be used as replacement materials in saturated soft soil.
④ The larger the uplift of the replacement ground, the worse the degree of the replacement soil, and the closer it is to a simple extrusion replacement process. The wide and thick crown-shaped compaction areas indicate that the replacement has a good compaction and reinforcement effect on the original soil. When the replacement soil is difficult to compact, it has reached a non-compactable state. Otherwise, the ground will uplift.

In the isolated permafrost sections of the first-grade highway of Hailar Provincial Highway 203, Inner Mongolia, the thickness of the degraded permafrost reaches 1.5 ~ 3.5 m, and the lithology is the muddy silty clay and soft plastic-flow plastic

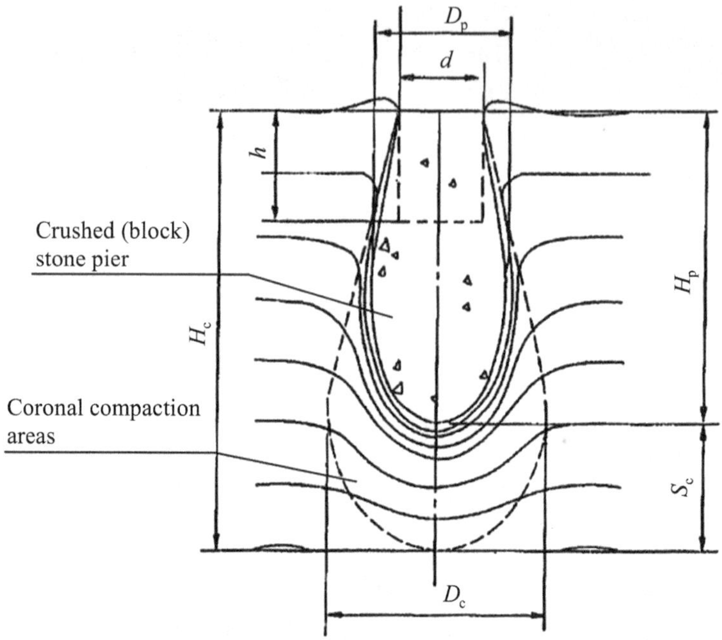

Fig. 8.19 Typical section of dynamic consolidation replacement body (Gong, 2008). *d*—diameter of the rammed pit; *h*—depth of the rammed pit per replacement; D_p—diameter of the replacement crushed stone pier; H_p—depth of the replacement body; D_c, H_c and S_c—diameter, depth and bottom thickness of the coronal compaction areas under the replacement body, respectively

cohesive soil deposited by swamps and lakes. When the dynamic consolidation is used to treat the permafrost foundation, the particle size of the flake stone pier is no more than 150 mm and the treatment depth is 7.0 m (Fig. 8.20). The spot ramming hammer has a diameter of 1.2 m, with a hammer weight of 17.5 t, 3000 kN·m for the compaction energy, and 10 ~ 13 for the total tamping times; while the full ramming hammer has a diameter of 2.0 m, with hammer weight of 12 t, 1000 kN·m for the compaction energy, and 3 for the total tamping times. The treatment requirements are as follows: the flake stone pier must penetrate the permafrost and soft soil layers below the foundation, and the average ramming settlement of the last two blows is less than 50 mm. The ground around the rammed pit will not have large uplifts, and the overlap of full rammed impression is not less than 1/4 of the hammer diameter, and the particle size of cushion material is no more than 50 mm and the mud content is not more than 5%. This method has a lot of advantages in the treatment of isolated permafrost foundation, including the high strength, low frost heave, rapid consolidation of foundation soil, small range of lateral deformation, simple construction machinery, fast speed, easy quality control, and simple detection methods and so on.

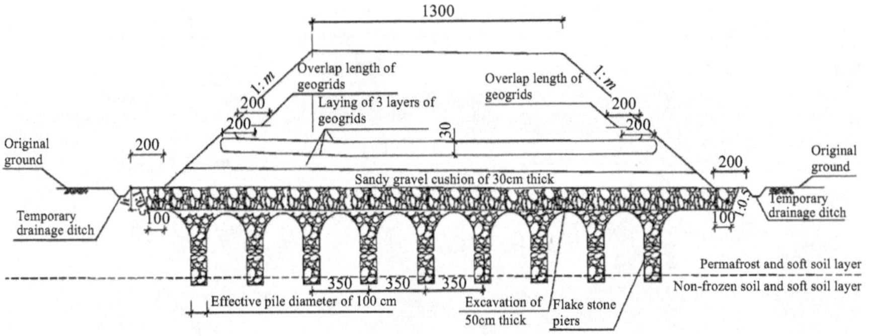

Fig. 8.20 Cross-section diagram of island permafrost foundation treated by dynamic consolidation replacement method on Hailar Provincial Highway 2013, Inner Mongolia (Unit: cm) (Zhao & Zhao, 2014)

2. Measures Preventing Thaw Settlement for Frozen Soil Foundation

The buried (excavation) methods against thaw settlement applicable to airport runways mainly include the replacement method, the thermal insulation method, the method of pile cap foundation, the method of dynamic consolidation replacement, the pre-thawing method (hydraulic thawing, steam thawing, electro-thermal thawing, thermochemical thawing), etc. When the airport runway is constructed by the fill, the thermosyphon (thermal probes) method and the thermal insulation can constitute a composite measure. No matter what is adopted, the underground and surface drainage measures must be strengthened and implemented.

When using the pre-thawing method, the thawed soil should be compacted and reinforced after pre-thawing, and the main methods include the mechanical method, physical method, chemical method, and electrochemical method, etc.

1. Replacement Method

The coarse-grained materials, such as the crushed (block) rock and sandy gravel can be used for replacement in the warm permafrost with high ice content and thin thickness. The replacement method should meet the following requirements: the content of silty clay particles (smaller than 0.075 mm) in fillers should not exceed 5% and the drainage measures should be taken to ensure the draining when freezing. Besides, the standing water should be avoided to soak the underlying frozen soil layer.

Mohe Airport in the Daxing'anling Mountains is the first airport built on permafrost in China. The site is located in low mountains and hills, with 0 ~ −0.1 °C mean annual ground temperature of permafrost (Fig. 8.21) and more than 15 m thickness (the maximum thickness of permafrost in hydrogeological investigation was 18 ~ 22 m). The engineering types and grades of thaw settlement (Fig. 8.22) are as follows: the ice-poor permafrost (I), the ice medium permafrost (II), the ice-rich permafrost (III) and the ice saturated permafrost (IV). The lithology included the

Fig. 8.21 Ground temperature curve of permafrost in Mohe Airport Runways in 2008

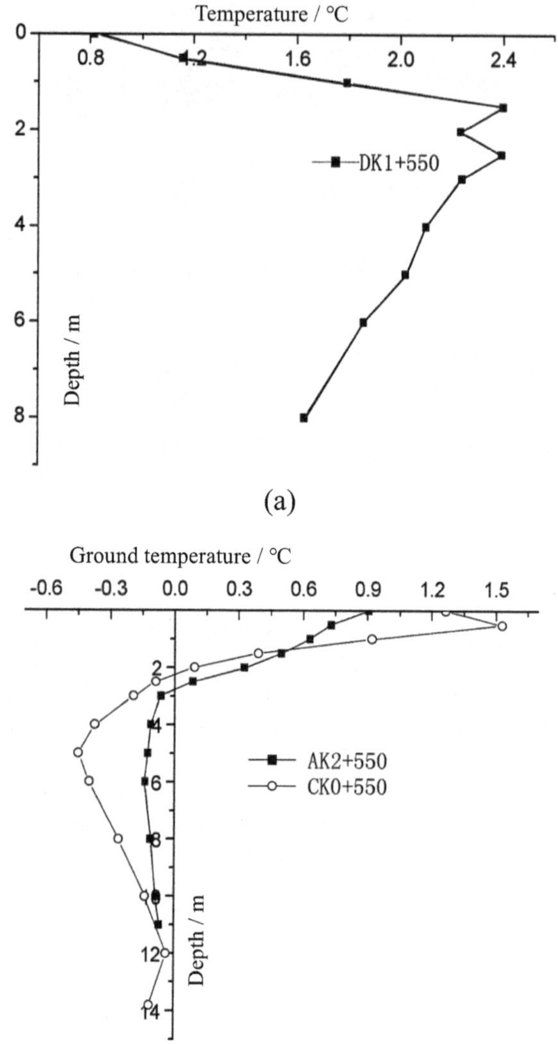

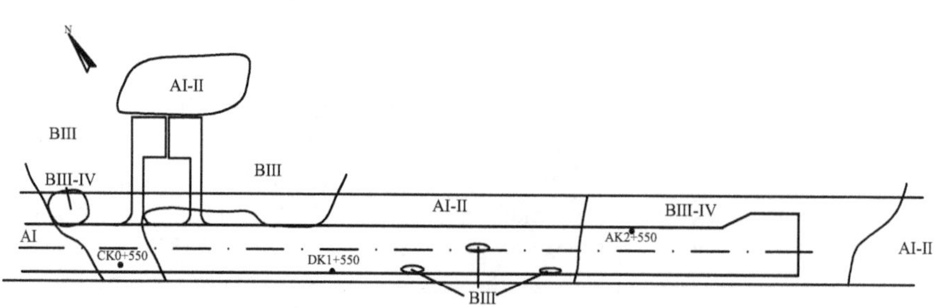

Fig. 8.22 Schematic diagram of classification of thaw settlement of permafrost in Mohe Airport Runway

breccia, crushed rock, gravel silty clay, silty clay and clay. The maximum seasonal freezing depth is 3.5 m. At the southeast end of the runway, there are the ice-rich and saturated permafrost (F-B, III-IV), the silty clay interbedded with gravelly sand and clay. The thickness of ice layers at the depth of 3 ~ 5 m is 20 ~ 50 mm, and a large number of ice crystals could be found at the depth of 6 ~ 7 m. The total water content above this depth reaches 17.2% ~ 57.8%, and the coefficient of thaw settlement is 10% ~ 25%, which belongs to the soil with strong thaw settlement. The total water content below is 17% ~ 26.7%, which belongs to the soil with weak thaw settlement (Table 8.17).

The runway area belongs to the warm permafrost with high ice content, so it is not suitable to adopt the design principle of protecting permafrost, but to adopt the design principle of pre-thawing. After comparison, the crushed rock and sandy gravel can be used to replace the frozen silty clay and clay, which can avoid thaw settlement of the permafrost and meet the design requirements of strength with economical rationality. The replacement depth shall be determined by the time when the water content of permafrost reaches the plastic limit, namely, 8 m, and the drainage measures shall be well implemented.

The maximum seasonal freezing depth in the site is 3.5 m. The thermal conductivity of 1.5 2 m pavement structural layer (the thickness of the concrete slab is 0.32 m, the thickness of the cement-stabilized sandy gravel is 0.4 m, and the thickness of the graded gravel is 0.8 m) is 25%–40% larger than that of the silty clay, so the seasonal freezing depth might increase by about 1.0 m after the construction of the pavement. Because the fine-grained soil within the maximum seasonal freezing depth of the pavement areas has high water content, and the increase of the freezing depth may cause frost heave, so the silty clay within the maximum freezing depth could be replaced with sandy gravel.

The operational performance over the past decade indicates that the methods employed for handling the foundation of warm permafrost with high ice content at Mohe Airport are reasonable, safe, and cost-effective.

Table 8.17 Classification of permafrost thaw settlement

Level	Levels of thaw settlement	Characteristics	Coefficient of thaw settlement/%;	Types of permafrost
A	Non-thaw settlement I–Weak thaw settlement II	I: Breccia, crushed and block rock, $W < 20\%$ II: Gravel silty clay, $W < 20\%$	1 ~ 3	Ice-poor permafrost Icy medium permafrost
B	Thaw settlement III - Strong thaw settlement IV	III: Silty clay, clay, etc., for example, ice crystals, $W = 17\% \sim 39.3\%$ IV: Silty clay, clay, etc., for example, the ice layer of 20 mm, $W = 30.6\% \sim 45\%$	3 ~ 10 10 ~ 25	Ice-rich permafrost Ice saturated permafrost

2. Thermal Insulation Method

The following conditions should be considered when adopting the thermal insulation method to prevent permafrost foundation from thaw settlement. ① The freezing index should be 3 times greater than the thawing index because of its extremely cold climate in permafrost regions. ② The mean annual ground temperature of the permafrost should be lower than −2.0 ~ −3.0 °C. ③ The thermal insulation material has low thermal conductivity, high strength and low water absorption. ④ The suprapermafrost water is easy to drain.

In China, the states and characteristics of permafrost are suitable for the frozen soil engineering construction. Except the alpine permafrost regions, the mean annual ground temperature for most permafrost regions is higher than −2.0 °C and the air temperature freezing index is just 1 ~ 2 times higher than the thawing index. The thermal insulation layer can only isolate the heat transferred by the atmosphere, reduce the seasonal thawing depth to a certain degree, and also isolate the transfer of atmospheric cold energy to the permafrost foundation in winter, reducing the accumulation of cold energy in permafrost foundation, which will lead to the increase of the thaw depth of permafrost foundation year by year.

For example, the mean annual freezing index is −3306 °C·d and the thawing index is 1056 °C·d in Kotzebue Airport in permafrost regions of Alaska. An EPS insulation board of 0.1 m is laid in the runway fill, with a buried depth of 0.23 m. A concrete base of 0.15 m is directly laid on the thermal insulation layer, and an asphalt concrete pavement of 0.08 m is added on the base layer. Observations show that the thawing depth could be reduced by 1 ~ 1.35 m with the thermal insulation layer of 0.1 m (the thawing depth of the section without thermal insulation layer is 3.0 m, and the thawing depth of the section with thermal insulation layer is 1.65 ~ 2.0 m).

The permafrost regions in Inuvik, northwestern Canada, has a mean annual temperature of −8.7 °C, a freezing index of −4690 °C·d and a thawing index of 1210 °C·d. On the Mackenzie Highway, the embankment height is 1.5 m, and the thickness of EPS thermal insulation layer, laid on the three sections of the subgrade, is 50 mm, 90 mm and 115 mm, respectively, and the buried depth is 0.7 m. Observations show that the center of the road with the thermal insulation layer of 50 mm subsides by 0.3 m in the first year, and very little settlement in other road sections. The observation in the 6th year shows that only EPS thermal insulation layer of 90 mm and 115 mm could effectively prevent permafrost thawing.

3. The Method of Pile Cap Foundation

In the warm permafrost regions with high ice content and large thickness, the temperature of permafrost foundation will increase due to the construction of pavement structural layers and changes in ground conditions (the removal of vegetation and humus layers, the bare ground, and the artificial materials coverage, etc.), and the thickness of active layer and the settlement deformation increase. Under global warming, the reinforced concrete pavements and the exposed soil surfaces, the aircraft operations and various human activities will make the underlying permafrost layer warm up and the permafrost table deepens and degrades year by year, and the deformation caused by the thawing of permafrost will also increase gradually. When it is difficult to maintain the permafrost foundation in a frozen state by adopting protection measures, the design principle of gradual thawing can be adopted. During the service life of the airport runway, the deformation caused by the thawing of permafrost can meet the design requirements of the pavement.

The method of pile cap foundation (Fig. 8.23) is used to treat the permafrost foundation for two purposes during the operation period of the airport. Firstly, the buried depth (h_f) of the pile foundation in the permafrost is still placed in the permafrost layer, and the bearing capacity of the pile foundation can still meet the design requirements of bearing capacity. Secondly, the consolidation settlement and deformation generated when the permafrost foundation thaws can be adjusted through the pile cap to meet the design requirements of deformation.

The vertical bearing capacity of the pile foundation can be calculated as the following formula:

$$R_a = q_{fpa} \cdot A_p + U_p \left[\sum_{i=1}^{n} f_{cia} l_i + \sum_{j=1}^{m} q_{sja} l_l \right] \tag{8.7}$$

where R_a is the characteristic value of the vertical bearing capacity of a single pile, kN; q_{fpa} is the characteristic value of the end resistance of the permafrost layer at the pile tip, kPa; f_{cia} is the characteristic value of the freezing strength around the pile of the ith layer of permafrost, kPa; q_{sja} is the characteristic value of lateral resistance of the soil around the pile at the jth layer, kPa. When the soil in the active layer is with strong frost heave or extremely strong frost heave, it will produce negative friction on the pile foundation during thawing. If the value cannot be determined, 10 kPa can be taken and used as a negative value. l_i and l_i are the lengths of the pile in each section divided by soil layer, m; A_p is the cross-sectional area of the pile bottom, m^2; U_p is the perimeter of the pile, m; n is the number of the permafrost layers; m is the number of the layers of thawed soil.

The value of the pile tip resistance and the freezing strength around the pile in the permafrost layer should be determined according to the temperature.

In different seasons, the soil temperature in the upper part of the permafrost layer (usually in the depth range of 0 ~ 8 m) will be affected by the changes of air temperature. In spring and summer, the active layer thaws. In autumn, the soil temperature in the upper part of the permafrost layer reaches the highest, and the freezing strength of the permafrost is the weakest. In winter and spring, the active layer freezes, resulting in frost heave. In late spring, the soil temperature in the upper part of the permafrost layer reaches the lowest and the freezing strength is the highest.

Therefore, when calculating the bearing capacity of the pile foundations, the soil temperature during the period with the highest ground temperature in the permafrost should be selected as the basis for the values of the freezing strength and the pile tip resistance. During the construction of the pile cap (raft), a sand cushion layer of 0.2 m (not fully filled, not compacted) should be laid between the bottom

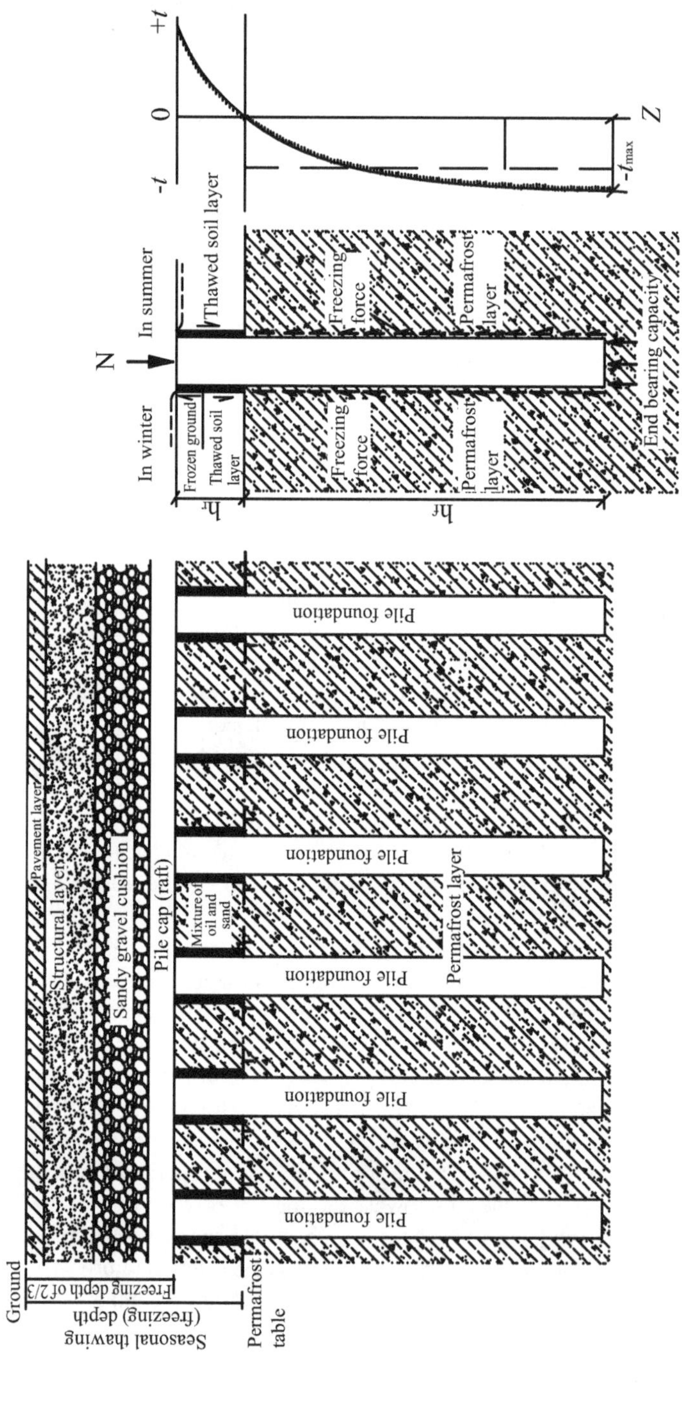

Fig. 8.23 Pile cap foundation

and the top of the pile to prevent the heave force caused by the frost heaving of the active layer on pile cap.

4. Thermosyphon (Thermal Siphon) with Thermal Insulation Layer to Cool the Foundation

In basically stable permafrost regions with large thickness of permafrost layer and high ice content, the method of "prefering filling instead of excavating" can be used to protect the permafrost foundation. However, because of the airport construction and environmental changes, the ground temperature and the active layer thickness will increase, so it cannot maintain the thermal stability of the permafrost foundation only by a thermal insulation layer. And it is necessary to take measures to cool the foundation to increase the cold storage capacity of the permafrost foundation. The combination of the thermosyphon and the thermal insulation layer can be used for the airport runways in permafrost regions to build a composite cooling runway of "thermosyphons with thermal insulation layer" (Fig. 8.24).

The thermosyphons used on airport runways and their power should be specially designed. The "horizontal" thermosyphons should be adopted. The length of the evaporation section should reach 1/2 of the pavement width, and the buried depth should be below the pavement structural layer (the combination of the pavement layer, the structural layer and the thermal insulation layer), with the horizontal slope gradient of thermosyphon of 1:10. The condensation section is set above the ground in the soil surface, and its height cannot be higher than the central elevation of the pavement (or the height of the border light of pavements). In order to increase the cooling area of the condensation section, it can be designed according to the environmental conditions, or it can be designed as "Y" shape. The thermosyphons on both sides are arranged in a staggered manner, and the center spacing between each two neighboring thermosyphons on one side is not more than 4 m. The thermosyphons can be buried after excavating slots, and be filled with coarse sand of 0.2 m around them (with no crushed stone in the backfill).

5. Pre-thawing Method

In the degraded permafrost regions with extremely high temperature, in the areas in large crushed rock soil, gravel soil and fine-grained soil with low ice content (ice-poor permafrost), the design principle of allowing the permafrost foundation to

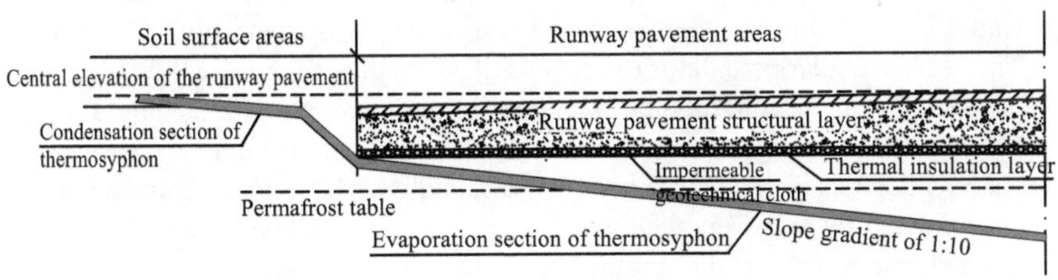

Fig. 8.24 Structural schematic diagram of the airport runways with the combination of thermosyphons and thermal insulation

thaw can be adopted when the calculated settlement of the permafrost foundation after thawing should not exceed the allowable limit settlement during the construction and operation of the runway. It is worth noting that many conditions should be considered, including the physical properties of permafrost foundation (water content, unfrozen water content, dry density, etc.), the thermal parameters (the thermal conductivity of frozen soil in frozen and thawed states, etc.), the strength properties (cohesion, internal friction angle and the bearing capacity in thawed state, etc.), the deformation properties (thaw coefficient of frozen soil and the relative compression coefficient during thawing) and the basic information of buildings (structures) (geometric dimensions, pressure transmitted to the frozen soil foundation, etc.). The calculation procedure should be as the follows: ① to determine the maximum thawing depth of permafrost foundation under the pavement at different times and stable periods; ② to determine the design strength (bearing capacity) of the thawing foundation soil; ③ to determine the frost heave of the pavement.

In part of the deep sections of permafrost regions with extremely high temperature, if there is permafrost with ice content that could lead to thaw settlement, the permafrost foundation will thaw under the thermal impacts of climate warming and human engineering construction during the operation of the airport runways. When the calculated thaw settlement of permafrost foundation exceeds the limit settlement, the pre-thawing method needs to be taken to make the settlement of the permafrost foundation reduce to the design requirements before the construction of runways.

The pre-thawing methods include the using of natural heat or artificial heat to thaw the permafrost.

The method of using the natural heat to thaw permafrost requires sufficient time. In order to accelerate the thawing process, the surface cover layer and vegetation can be eradicated, or the ground surface can be blackened, or transparent plastic film can be used to cover the ground to increase the ground temperature of the building areas. Generally, the ground temperature of the exposed ground is 3–5 °C higher than the air temperature, which can thaw the soil layer to a depth of 3 ~ 4 m within a summer. However, the anti-freezing measures should be taken to prevent refreezing in winter, or to reduce the freezing depth.

There are many methods to thaw the permafrost by using the artificial heat, which are selected according to the conditions of permafrost foundation and funds:

① Hydraulic thawing method (Fig. 8.25a). Permafrost is thawed with cold or hot water. It is suitable for coarse-grained warm permafrost with high ice content. The needle tubular percolator can be inserted into the drill hole and water can be pressed into the needle tubular percolator to thaw the permafrost. The distance between the two neighboring needle tubes is 3–4 m, and the depth is 7-8 m. The coarse-grained soil can be drained by itself, and it usually can be compacted and consolidated by self-weight in 2 ~ 3 weeks. If this method is applied in fine-grained frozen soil, the fine-grained soil will be washed away by water.

② Steam thawing method (Fig. 8.25b). The steam is used to thaw permafrost (open and closed types). The steam needle tube is inserted into the fine-grained perma-

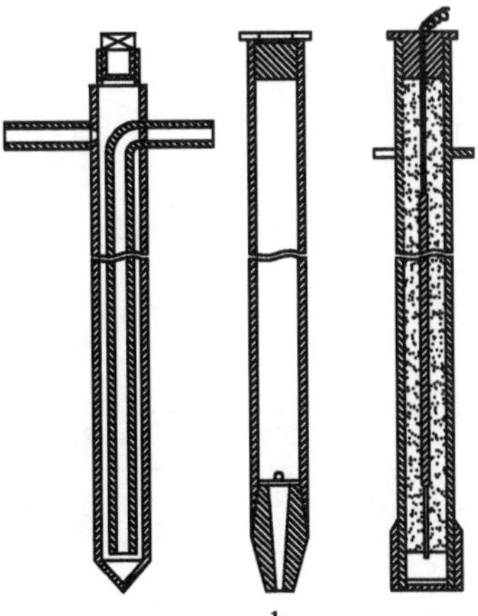

Fig. 8.25 Structural schematic diagram of metal needle tubes for thawing permafrost. a—For hydraulic thawing; b—For steam thawing; c—For electro-thermal thawing

frost, and the heat can thaw the permafrost. The open type means that the steam can directly enter the permafrost through the end of the needle tube to thaw it, but it will increase the water content of the thawed soil. The closed type means that the steam circulates in needle tubes and thaws the permafrost through the heat of the tube wall. The steam needle tube gradually sinks into the fine-grained permafrost under its self-weight and with slightly hammered, and the thawing depth is generally about 10 m.

③ Electro-thermal thawing method (Fig. 8.25c). Permafrost is thawed with alternating currents (according to the design depth). The electro-thermal thawing method generally does not increase the water content of the thawed soil, and may even make it partially dry. Figure 8.26 shows the use of electro-thermal thawing method to thaw permafrost in the former Soviet Union. The voltage is 120 ~ 400 V or higher, and the operation time is 1 ~ 2 months; the electric power required every square meter of foundation is 0.1 ~ 0.3 kW and the electric energy consumed per cubic meter of foundation soil by electro-osmosis is 60 ~ 80 kW h. The ice lens and the ice interlayer will thaw in electric heating, and the mineral interlayers are arranged closely, causing the thawing soil to be compacted. Each electrode is powered by a power source and is also functioning as a drainage pipe. For safety, the electro-thermal site must be surrounded by fences, and warning signs must be set up, and guards on duty can be set if necessary.

④ Thermochemical thawing. High-heat cement, quick lime and other materials are used to heat and thaw permafrost during the process of condensation, hardening and ripening.

Quick lime will produce some physical effects in the ripening process, such as water absorption, expansion and heat generation, and will cause gelling reaction among the admixtures. Several main functions of lime pile composite foundation are as follows:

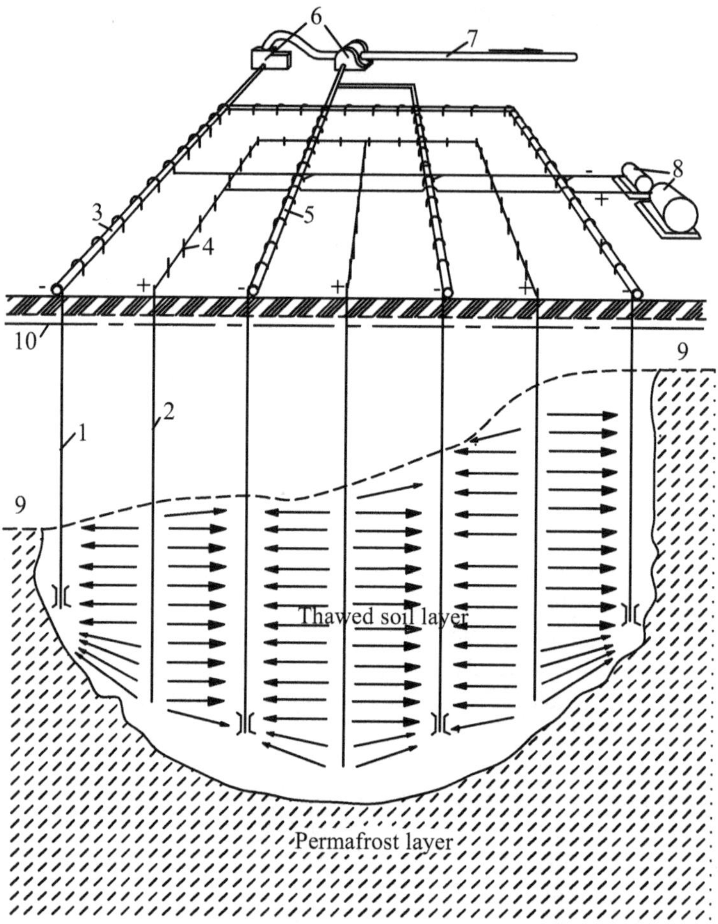

Fig. 8.26 Schematic diagram of soil reinforcement by electro-osmotic pre-thawing (H.A. Tsytovich, 1985). 1—Cathodic needle filter tube; 2—Anodic needle filter tube; 3—Connecting hose; 4—Electrode; 5—Catch-pipe to lower the water level; 6—Water pumps; 7—Drainage pipes; 8—Direct current power supply; 9—Permafrost table; 10—Groundwater level

(a) The high temperature effect increases the soil temperature among piles. In theory, 1 kg of quick lime can release 950 kJ of heat after ripening. The highest temperature of pure quick lime pile measured in Japan is 400 °C, and the highest temperature of lime pile with admixtures in China is 200 ~ 300 °C. With the normal replacement ratio, the temperature of the soil between piles can reach 40 ~ 50 °C. High temperature effect promotes the dehydration of soil around the pile, and the heat generated by the ripening of quick lime can be used to thaw the frozen soil around the pile.

Table 8.18 Coefficients of volume expansion of pile materials under different pressures (Gong, 2008)

	Materials				
		The mix proportion of fly ash to lime		The mix proportion of volcanic ash to lime	
Pressures /MPa	Quick lime	2:8	3:7	2:8	3:7
50	1.49	1.40.	1.34	1.35	1.26
100	1.37	1.33	1.28	1.28	1.19
150	1.29	1.26	1.22	1.21	1.12

Table 8.19 Permeability coefficient test of lime piles (Gong, 2008)

Specimen numbering	Sampling locations	The permeability coefficient/cm·s^{-1}	Specimen numbering	Sampling locations	The permeability coefficient/cm·s^{-1}
3c1	The center of lime piles	6.13×10^{-5}	12c1	The center of lime piles	7.58×10^{-5}
9c1	The center of lime piles	4.07×10^{-3}	13c1	The center of lime piles	6.53×10^{-5}
6c1	The center of lime piles	1.29×10^{-3}	天1	The center of lime piles	7.30×10^{-3}

(b) Quick lime absorbs water and expands to compact the soil among piles. When quick lime is ripened, the solid can be disintegrated and the void volume, the specific surface area of the particles, adsorbate, and the volume of solid particles are increased. Due to the difference in quality of quick lime, its volume increases to 1.5 ~ 3.5 times after ripening in the natural state, that is, the coefficient of volume expansion, and the coefficient of volume expansion of the first-grade quick lime is about 3 ~ 3.5 times. In general, the increase coefficient of pile diameter with admixtures is 1.1 ~ 1.2, which is equivalent to 1.2 ~ 1.5 times the coefficient of volume expansion (Table 8.18), resulting in obvious compaction on the surrounding soil. According to indoor test data, the axial expansion force of quick lime can reach 11.3 MPa in a closed thick steel cylinder with a diameter of 100 mm. When the pile diameter is 150 mm, the maximum measured pressure at 225 mm (1.5 d) and 300 mm (2 d) away from the pile center is 16 ~ 70 kPa. When the pile diameter is 300 mm, the lateral tension at different depths at 1.5 d and 2 d away from the pile center is 60 ~ 300 kPa. But the lateral tension cannot be stabilized at a certain value. The expansion of piles also causes uplift at the top of pile.

(c) Drainage consolidation. When admixtures (fly ash and soil) with different mix proportion are used, their permeability coefficients range from 7.30×10^{-3} to 6.13×10^{-5} (Table 8.19), which is equivalent to the permeability of silty sand and fine sand, and 10–100 times larger than that of general cohesive soil. Lime piles have a large void ratio of 1.3 ~ 1.7 and large grain compositions. The observation data show that, 1 h after the pile is formed, the pore water pressure of the soil among piles rapidly drops to the level before the pile is driven, and continues to drop. When coal cinder, slag and steel slag and other coarse particles are used as pile admixtures, the drainage consolidation can exert more obvious effects.

(d) Chemical consolidation effect. The reaction of active admixture and quick lime under specific conditions is very complicated. In general, $Ca(OH)_2$ reacts with SiO_2 and Al_2O_3 in active admixtures to form calcium silicate and calcium aluminate hydrate, which is insoluble in water and can be hardened in soil with a high water content. The quick-acting effects caused by the ripening of quick lime, such as water absorption, expansion and heat generation, generally tend to be stable in a short period of time (4 weeks). However, the ionization (water absorption) with the soil among piles will continue, and the calcium ion Ca^{2+} generated by the ionization of $Ca(OH)_2$ is exchanged with the cations on the surface of the clay particles and adsorbed on the surface of the particles, changing the electrification state of the clay particles and thinning the weak bound water on the surface. Soil particles coagulate together, increasing the agglomerate and shear strength and decreasing the plasticity. The consolidation reaction of the calcium ion Ca^{2+} and SiO_2 and Al_2O_3 in clay will continue. The carbonation reaction of lime will be slow as the water content decreases, but will continue to increase in strength.

A section of the Mohe Airport Expansion Project is fully weathered tuff rock (silty clay), which belongs to a degraded warm permafrost, with a mean annual ground temperature of −0.1 ~ 0.0 °C; some parts of the section (with the buried depth of 8 ~ 15 m) with rich ice is thaw settlement permafrost (level III), with a total water content of 17% ~ 35%, and a final estimated thaw settlement amount of 150 ~ 250 mm, and the unit settlement under 200 kPa pressure is 18.8%. It is not suitable to adopt the design principle of keeping the permafrost in frozen state, but the design principle of gradually thawing and pre-thawing can be adopted. According to the engineering environment conditions of airport runways, three methods are proposed to treat the permafrost foundation: the replacement foundation, the piled raft foundation and the artificial thawing of permafrost.

The lime pile test was conducted from November the 4th, 2016, with the pile diameter (d) of 400 mm, the pile spacing of 2.5 d and 3 d. And the piles were arranged in equilateral triangle, with the pile depth of 18 m. There were16 test piles and 18 monitoring holes.

The mix proportion of the quick lime to soil (silty sand, $W = 11.6\%$) and cement is 50%~45% and 5%.

Purpose of the test: the thawing state of permafrost; the compaction degree of soil among piles; the bearing capacity, the deformation modulus and the construction technology.

Construction: using the twisted drill to drill boreholes and mixing 6 barrels (180 kg) of material at a same time, and filling in the boreholes in 3 times; mixing the soil material to maintain the water content of 20%, and add 2.311 kg water each time. The filling height was 0.4 m and the rammer weight was 75 kg, with the distance of fall of 0.4 ~ 0.5 m; after hammering for 15 times, the residual height was 0.3 m. According to the laboratory test, the expansion rate of lime pile is 1.3 (the lime pile with the diameter of 400 mm expands to 530 mm).

Monitoring: all the temperature measuring holes in the center of the piles were in positive temperatures on November 11, with highest temperatures of 18 ~ 32 °C. A week later, the cumulative swelling amount of the vertical deformation of piles was 26 ~ 31 mm, and the cumulative swelling amount of soil among piles was 18 ~ 30 mm. After 5 months (April in 2017), the average temperature of the thawed soil (with the depth of 8–16 m) in each temperature measuring hole was more than 1 °C (Fig. 8.27), and the cumulative deformation of the piles and the soil among piles basically remained unchanged.

The test results showed that the high temperature generated by the ripening of quick lime can thaw the extremely unstable permafrost (−0.1 ~ 0.0 °C) around the pile (2.5 ~ 3.0 *d*), and the ground temperature was still higher than 1 °C after a cold winter. The ripening of quick lime resulted in volume expansion and brought vertical expansion of 30 mm to soils among the piles, indicating that expansion and compaction existed in the radial direction. 15 days after the tests, the standards implementation was used in soils among piles and the dynamic detection was used in piles. The test results showed that the bearing capacities of the piles and the soil among piles below 4 m were 266 kPa and 308 kPa respectively, meeting the design requirements.

The commonly used lime pile borehole has a diameter of 300–600 mm, arranged in triangle, and the center-to-center distance between pile boreholes can be 2 ~ 3 times the diameter of the pile borehole.

3. High Fill and Slope Stability

There are often high fill projects in permafrost regions, and the stability requirements of the fill projects are consistent with those in non-frozen soil areas. The core difference is that there is ice in the permafrost layer of the slope. With the overlying high fill soil layer, the underlying permafrost layer will produce some changes that need to be considered in the design.

1. Changes in Permafrost Table

According to the experimental research on the Qinghai-Xizang Highway and Railway, in the unstable permafrost regions on the Qinghai-Xizang Plateau, when the subgrade height of fill is 2.5 ~ 5 m, the permafrost table below the subgrade may elevate. As for the excessive high embankment fill, the increase of slope area on both sides will result in the increase in heat absorption and the decrease in the permafrost table below the subgrade, causing thaw settlement and deformation of the subgrade. The condition is roughly similar to that of Daxing'anling Mountains in Northeast China. According to the survey (The Third Railway Survey and Design Institute, 1994), the embankment height is less than 0.5 m and the decrease of the artificial permafrost table accounts for 83%, 54% for the embankment height of 0.5 ~ 1.0 m, 44% for 1.0 ~ 1.5 m and 21% for 1.5 ~ 2.0 m. Of course, the lower the air temperature is, the lower the mean annual ground temperature of the permafrost

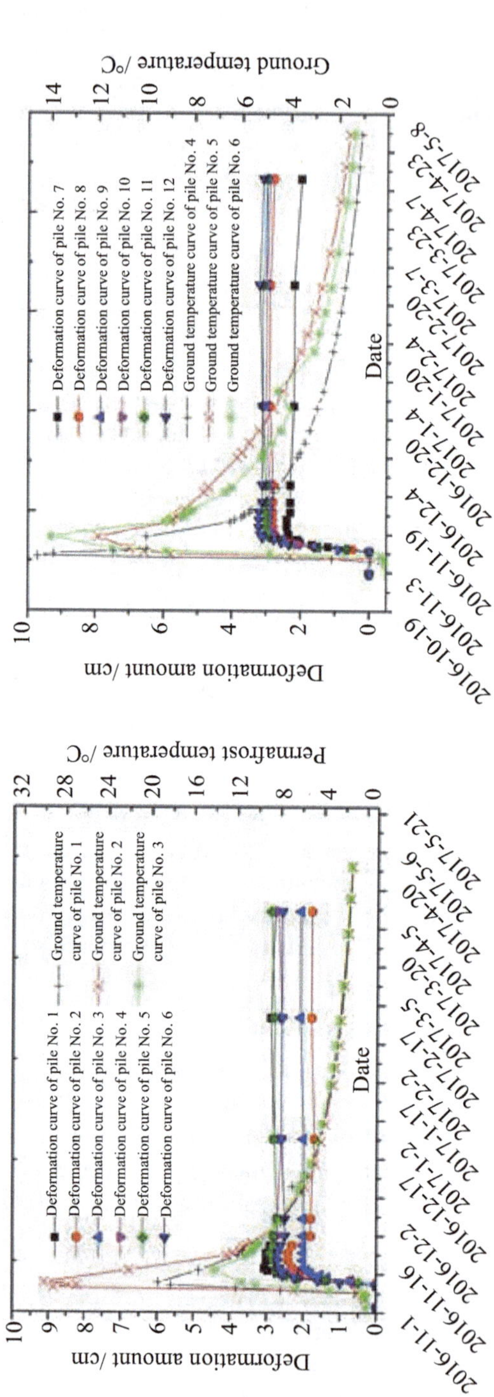

Fig. 8.27 Cumulative deformation and ground temperature curves of lime piles No. 1-6 (the Left) and No. 7-12 (the Right) in Mohe Airport (Positive Means upward) (Cheng Jia)

is, and the stronger the thermal stability of the permafrost is, and the filling height can be increased correspondingly.

The maximum height of the high fill in the northwest of Mohe Airport is about 40 m. The annual air temperature in Mohe area is −5.5 °C, with the lowest air temperature of −52.3 °C and the maximum seasonal freezing depth of 3.5 m. In the past 50 years (1958–2008), the variations of the average air temperature shows a rising trend (0.33 °C/10a), especially with a significant increase in winter (0.67 °C/a) (Gui et al., 2009). The airport construction and climate warming have gradually degraded the permafrost in Mohe Airport, where the mean annual ground temperature is 0 ~ −0.1 °C, and the artificial permafrost table is declined to about 7 m, but there are still ice-rich permafrost in some regions.

During the survey period of Mohe Airport in 2005, the permafrost table on the northwest slope was 1.2 m on average, and the permafrost table on the runway was about 1.8 m. The filling width was about 700 m and the minimum and maximum filling height was 6–8 m and about 43 m, respectively. As the cold energy brought in by cold seasons was less than the heat brought in during the construction period, the higher the filling height, the greater the heat storage of the construction is. Ten years later, the survey of high fill sections in soil surface areas of the airport runway showed that the permafrost table in the fill soil at the toe of slope (with the filling height less than 7 m) was slightly elevated (still in loose sediments), the artificial permafrost tables in the rest of the boreholes were decreased, from the original layer of loose silty clay sandwiched with crushed rock to the original bedrock surface (Fig. 8.28). According to the ground temperature observation data of the railway subgrade larger than 5 m in Daxing'anling Mountains, it takes about 5 ~ 6 years to re-stabilize the permafrost table after it drops to the artificial permafrost table because of construction. The ground temperature observation data of the borehole (with a filling height of about 7 m) in autumn 2015 showed that the mean annual ground temperature (with a depth of 17 m) was −0.12 °C, and the ground temperature at a depth of 20 m was −0.09 °C.

The decrease of the permafrost table may cause soil settlement, especially in sections with high ground temperature and high ice content, because the decrease of the permafrost table will lead to settlement and deformation of the filled soil. The amount of settlement deformation depends on the ice content, density and superimposed load of thepermafrost layer.

2. Compaction and Deformation of Permafrost

Permafrost is not inherently incompressible, especially for the warm permafrost with high ice content, which has a great compressibility under load. The compaction process of permafrost is a very complex physical and mechanical process, including the decrease of crystallization temperature of permafrost and ice (an increase in thawing temperature) under pressure, the migration (drainage) and infiltration of unfrozen water (formed before and during the pressure), and the structural reconstruction (at sections with lower pressure) after the thawing of the corners of ice crystals and their aggregate viscoelastic fluid.

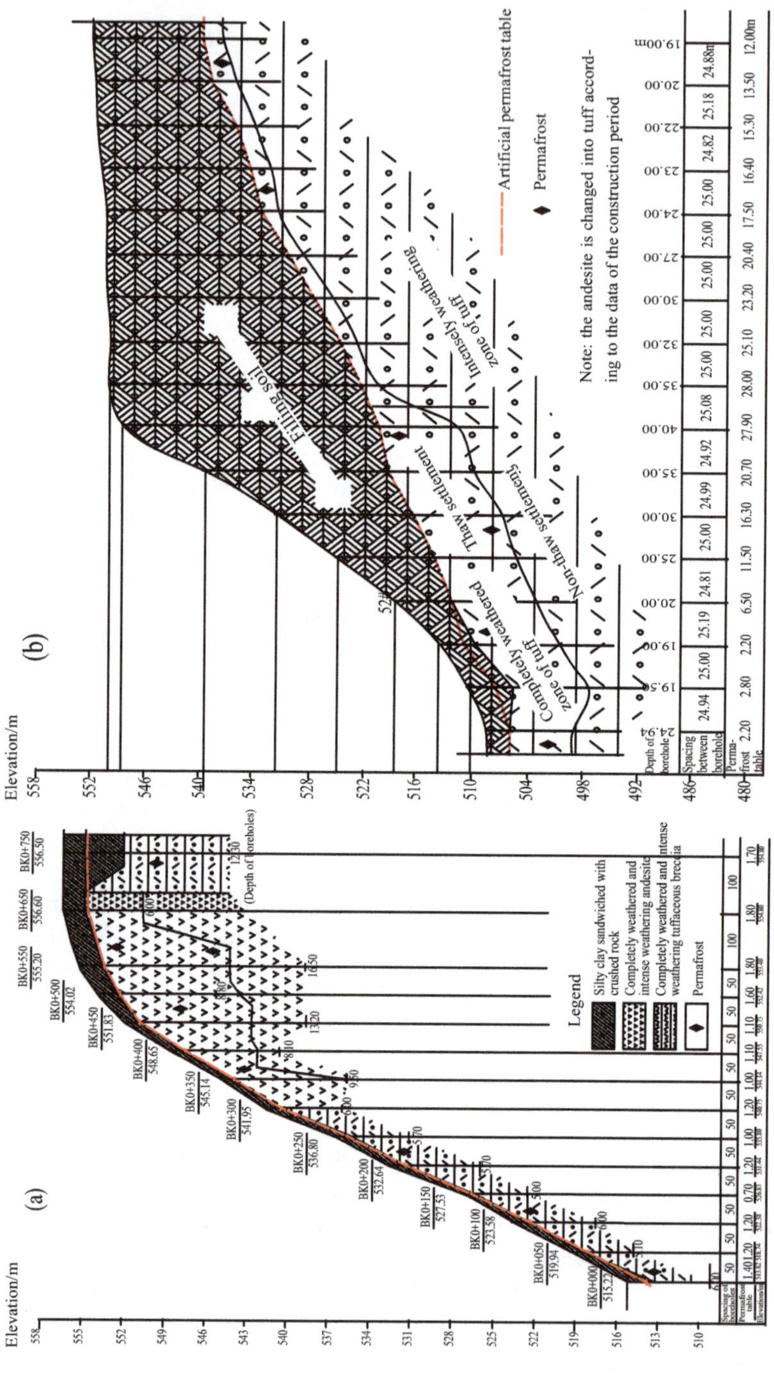

Fig. 8.28 Engineering geological profile of the original permafrost in Northwest of Mohe Airport runway [(**a**) construction period, (**b**) expansion period]

A large number of experimental data show that, even under the impact of low pressure, parts of ground ice in warm permafrost with high ice content thaws into unfrozen water, and then migrates, infiltrates, and recrystallizes; at the same time, the solid ice and mineral grains will also be compacted and displaced, which will bring about the changes in volume and structure to the permafrost during the loading process. During the compaction process of permafrost, the mass of soil particles will not decrease, and the unfrozen water will migrate, discharge or sublimate, and the total pore volume of soil will decrease with the increase of pressure. That is to say, under the pressure of 100 kPa, the void ratio of permafrost due to the extrusion of unfrozen water caused by the thawing of ice at the contact point of soil particles is as follows:

$$\Delta e = G_S \cdot \Delta\omega_u = 2.78 \times 0.0007 \approx 0.00195$$

where Δe is the variation of void ratio; G_S is the specific gravity of soil particles; $\Delta\omega_u$ is the variation of unfrozen water, and the increment of unfrozen water caused by external pressure of 1 kPa is set as 0.0007.

Under the impact of external load, the compressibility of permafrost is usually expressed by the volume compression coefficient (m_v). The water content and soil temperature of permafrost are the main factors affecting its compressibility. The lower the soil temperature, the smaller the compressibility, and it increases rapidly with the increase of soil temperature. When the soil temperature is −0.5 °C and −0.3 °C, the amount of compression deformation accounts for more than 70% of the total amount of deformation. For frozen silty clay, when the water content is near the plastic limit (ω = 13–17%), the compression coefficient is small and increases with the increase of water content. When the water content is near the saturated condition (ω = about 40%), the compression coefficient reaches the maximum value. After that, the compression coefficient will decrease with the increase of water content and finally tends to the compression coefficient of ice.

The compression coefficient of the warm permafrost with high ice content has obvious compressibility as soil temperature increases (Fig. 8.29). In the regions with high temperature, when the soil temperature rises from −1.5 °C to −0.3 °C, the compression coefficient increases by 7 times from 0.04 MPa^{-1} to 0.29 MPa^{-1}. In the process of compression, there will be unfrozen water discharged to cause infiltration and deformation.

3. Shear Strength of Frozen soil

The shear strength of soil is usually classified into cohesive strength and frictional strength, and it is difficult to separate them completely for the shear strength of cohesive soil. When the shear strength of frozen cohesive soil and sandy soil is 9.8 MPa, it can still be expressed as:

$$\tau = C + \sigma \cdot \tan\varphi$$

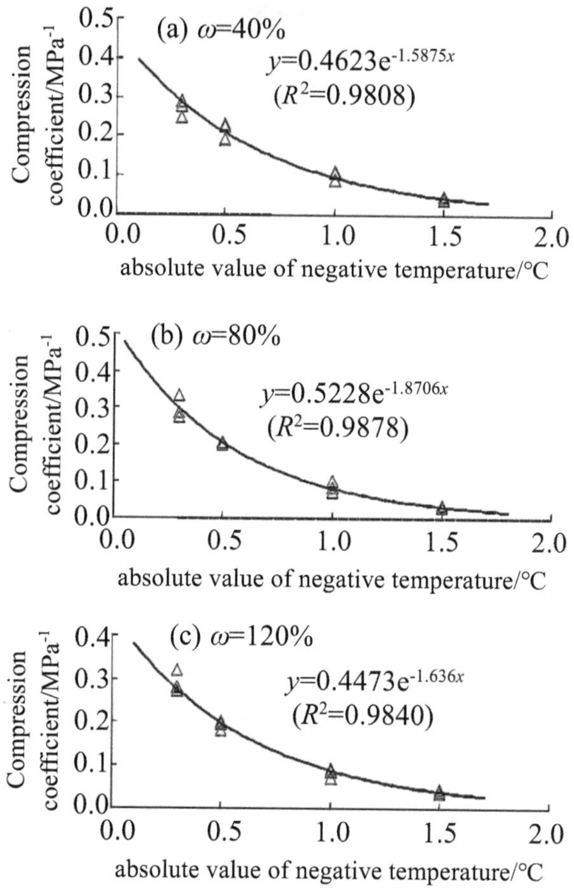

Fig. 8.29 Variation of compression coefficient of permafrost with different water content along with soil temperature (Zheng et al., 2009)

The main factors affecting the shear strength of permafrost are the soil lithological composition, the soil temperature and the duration of action.

After freezing or freeze-thaw cycles, the structure and composition of soil will change a lot, from a three-phase system to a four-phase system, and the pores and cracks will also enlarge.

The two parameters (cohesion and friction angle) of the shear strength of the ephemeral frozen soil are affected by soil temperature (Wu & Ma, 1994):

$$C = c_0 |t|^{\alpha} \quad (t < -0.2\ ^\circ\mathrm{C}) \tag{8.8}$$

where c_0 and α are the test parameters. Figure 8.30 shows the logarithmic relationship curve between the long-term cohesion of cohesive soil and the negative temperature of the soil layer. Under a same negative temperature, it decreases with the increase of water content of frozen soil layer (Table 8.20). As to the medium sand in

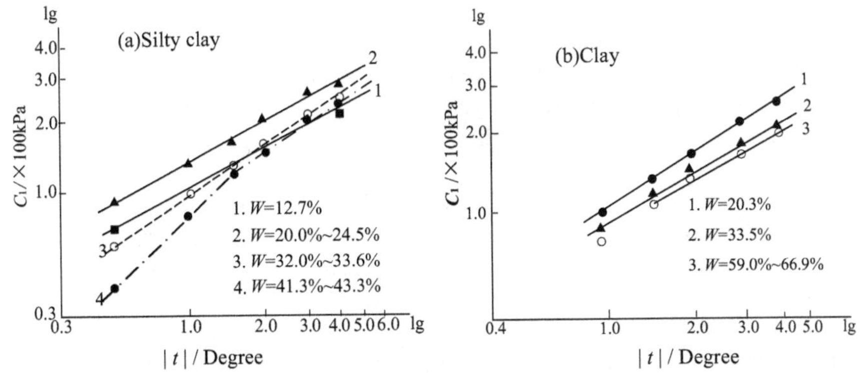

Fig. 8.30 Logarithmic relationship between frozen silty clay and clay Cl and the negative temperature |t| (Ma, 1983)

Table 8.20 Parameters c_0 and α of soil with different water content

Soil texture	Water content/%	C_0/×100 kPa·°C	α
Silty clay	12.7	1.02	0.55
	20.0 ~ 24.8	1.32	0.55
	32.0 ~ 33.6	0.98	0.69
	41.3 ~ 43.3	0.93	0.67
Clay	20.3	0.91	.070
	33.5	0.83	0.60
	59.0 ~ 66.9	0.78	0.60

Huainan, C_0 is 4.6 and α is 0.355. The long-term cohesion is small, generally (1/3) ~ (1/5) or even 1/10 of the instantaneous cohesion.

$$\varphi = a + k|t| \quad (t < -0.2\ °\mathrm{C}) \tag{8.9}$$

where a and k are the test parameters, which are determined by the test. As for the medium sand in Huainan, a is 24.4 and k is 0.5.

The artificial permafrost table is the interface between the thawed soil and the permafrost, and contains a thick layer of ice. There is a frozen layer of certain thickness below the thawing surface, and the soil temperature is close to the freezing temperature of the soil, which belongs to the transition zone between thawed soil and permafrost, and is also the weak layer of permafrost foundation on the slope of high fill (high embankment). Compared with the thawed soil layer above, the soil temperature in the transition zone is lower, and the water in the soil has a greater viscosity, which increases the cohesion of the soil. Compared with the frozen soil layer below, it has a higher temperature, a larger amount of unfrozen water, a large viscoplasticity and strong rheological properties, and its shear strength lies between thawed soil and frozen soil. Under the impact of the overlying load, the compression and thawing of the ground ice layer on the transition zone makes it contain more unfrozen water, and the unfrozen water and the standing water in the underlying permafrost layer form excess pore water pressure together, which may make this interface become a good lubricant surface.

The freeze-thaw transition zone has the characteristics of frozen soil and has a long-term strength.

According to the experimental observation of high embankments, the slope gradient of the foundation is also an important factor affecting the slope stability. When the slope gradient is greater than 1:6, the horizontal displacement will increase rapidly.

The theoretical basis for the calculation and evaluation of slope stability is Mohr-Coulomb strength criterion, which is based on the limit equilibrium method. The underlying soil layer of the thawed soil at the freeze-thaw interface is in a frozen state, which prevents the infiltration and drainage of pore water, resulting in the forming of excess pore water pressure. When suprapermafrost water flows along the slope, the safety factor can be derived by the effective stress principle:

$$
\begin{aligned}
F &= \frac{C' + \left\{\left[(1+m)\gamma + m(\gamma_{sat} - \gamma_w)\right]\right\} z\cos^2\alpha \cdot \tan\varphi'}{\left[(1-m)\gamma + m\gamma_{sat}\right] z\sin\alpha \cdot \cos\alpha} \\
&= \frac{C' + (\gamma_{sat} - \gamma_w) z\cos^2\alpha \tan\varphi'}{\gamma_{sat} z\sin\alpha\cos\alpha} \\
&= \frac{C'}{\gamma_{sat} z\sin\alpha\cos\alpha} + \frac{(\gamma_{sat} - \gamma_w)\tan\varphi'}{\gamma_{sat}\tan\alpha}
\end{aligned}
\tag{8.10}
$$

where C' is the effective cohesion; φ' is the effective internal friction angle or the underwater internal friction angle; z is the depth of the sliding surface; m is the height of the water head above the sliding surface, with the value between 0 and 1; α is the slope angle; γ_{sat} is the varying density of the slope; γ_w is the water density; γ is the natural density of the slope soil.

According to the geological profile data of the depth of permafrost table in high fill at the northwest end of Mohe Airport (Fig. 8.21), the permafrost table is basically parallel to the original ground slope, with an average slope angle (α) of 7.09°. The effective cohesion C' of the gravelly sand is about 0; its effective internal friction angle φ' is 24°; its varying density of slope γ_{sat} is 1.74 g/ cm^3, and its depth of the sliding surface z is 19.2 m. As for the silty clay, the values of the four parameters are 8 kPa, 15°, 1.75 g/cm^3 and 16.1 m respectively. After checking the slope stability, it can be derived that the safe factor F is 1.522 for the gravel sand and 3.241 for the silty clay. The project has been in operation for nearly 10 years, and the high fill project is stable, except for the rainfall erosion on the surface (Table 8.21).

Table 8.21 Physical and mechanical properties of permafrost table in the high fill at the Northwest of Mohe Airport

Borehole number	Permafrost table/m	Water content/%	Density/ g·cm^{-3}	Plastic limit water content/%	Thaw settlement coefficient/%	Lithology
49	2.2	22.5	1.82	18.3	3.15	Silty clay
51	2.2	20.6	1.72	15.1		Silty clay
53	11.5	31.2	1.65	26.9	3.32	Silty clay (completely weathered tuff)
55	20.7	22.7	1.82	18.3	3.36	Silty clay (completely weathered tuff)
57	28	17.9	(1.73)	(1.65)	5.12	Gravel sand (completely weathered tuff)
59	23.2	16.7	(1.81)	(16.8)	4.12	Gravel sand (completely weathered tuff)
61	17.5	17.1	(1.68)	(16.5)	4.59	Breccia (completely weathered tuff)
63	15.3	19.2	1.73	17.8	5.96	Gravel sand (completely weathered tuff)
65	12.0	17.3	(1.75)	(16.4)	4.62	Gravel sand (completely weathered tuff)

References

AASHTO. (2007). *AASHTO LRFD bridge design specification*. American Association of State Highway and Transportation Officials.

B.E. Vinetev Water Engineering Research Institute of the Soviet Union and S.Ya Zhukov Design Institute Leningrad Branch of the Ministry of Power and Electrification. (1992). *Design code for hydraulic structures in permafrost distribution areas (CH30-83)*. Translated by Research Institute of Songliao Water Resources Commission of the Ministry of Water Resources.

Bai, Q. G., Qui, W. G., & Zhang, J. R. (2007). Study on the classification standards for frost heaving forces in frozen soil tunnels. *Tunnel Construction*, 104–108.

Belinov, M. B. (1996). *Foundation Design of Large Vertical Steel oil Tanks in cold Regions*. Translated by Ma, C. S. Foreign oilfield engineering, 43-44.

Bing, W. S. (1989). *Theoretical basis and application of road surface frost damage prevention and control*. Harbin Institute of Technology Press.

Cai, H. C., et al. (2016). Variation of temperature and permafrost along the Qinghai-Tibet Railway. *Chinese Journal of Rock Mechanics and Engineering, 35*(7), 1434–1444.

Cao, X. B., et al. (2007). Study of the temperature impact on soil resistivity. *Transactions of China Electrotechnical Society, 22*(9), 1–6.

CCCC First Highway Survey and Design Institute Co., Ltd., et al. (2006). *General research report on subgrade stability technology in permafrost regions*. CCCC First Highway Survey and Design Institute.

Chang, X. L., et al. (2013). Review of permafrost monitoring in the Northern Da Hinggan Mountains, Northeast China. *Journal of Glaciology and Geocryology, 35*(1), 93–100.

Chen, T. (2011). *The study on stability of typical structure embankment of the Qinghai-Tibet Railroad at the permafrost regions under train dynamic load.* Thesis for Master's Degree of Lanzhou Institute of Seismology, China Earthquake Administration.

Chen, J. X., & Luo, Y. B. (2007). Calculation method of anti-freezing layer thickness in cold region tunnel. *Journal of Traffic and Transportation Engineering, 7*(2), 76–79.

Chen, X. B., et al. (1979). On anti-heave measure by replacing clayey soil with gravels. *Chinese Science Bulletin, 20*, 935–939.

Chen, X. B., et al. (2006). *Frost action of soil and foundation engineering*. Science Press.

Chen, Y. M., et al. (2008a). An integrated experimental study of harmful rising height of capillary water of highway subgrade in seasonally frozen ground regions. *Journal of Glaciology and Geocryology, 30*(4), 641–645.

Chen, J. B., et al. (2008b). Formation and mechanism of high subgrade diseases of Qinghai-Tibet Highway. *Journal of Chang'an University (Natural Science Edition), 28*(6), 30–35.

Chen, J., et al. (2011). Observation and study of cooling radius of heat pipes in swampy permafrost regions along Qaidar-Muli Railway. *Journal of Glaciology and Geocryology, 33*(4), 897–901.

Q. Wu, C. Tong, *Frozen Soil Engineering in Cold Regions*,
https://doi.org/10.1007/978-981-95-2527-0

Cheng, G. D. (2003). The influence of local factors on the permafrost occurrence and their implication for Qinghai-Tibet Railway Design. *Science in China (Series D), 33*(6), 602–607.

Cheng, G. D., & Wang, S. L. (1982). On the zonation of high-altitude permafrost in China. *Journal of Glaciology and Geocryology, 4*(2), 1–16.

Cheng, G. D., Tong, B. L., & Luo, X. B. (1983). The impact of embankment filling on the permafrost table. In *Professional papers on the permafrost studies of Qinghai-Xizang Plateau* (pp. 195–203). Science Press.

Cheng, Y. F., et al. (2012). Monitoring and analysis of coarse-grained frozen soil temperature in Qinghai-Tibet DC Transmission Line Engineering. *Chinese Journal of Rock Mechanics and Engineering, 31*(11), 2363–2371.

China Railway First Survey & Design Institute Group Co., Ltd. (2006). *Study on key technologies of bridges, culverts and tunnels in the Qinghai-Tibet Railway test project*. China Railway First Survey & Design Institute Group, Ltd..

China Railway First Survey & Design Institute Group Co., Ltd. (2007). *Research report on engineering technology of permafrost tunnels in the Qinghai-Tibet Railway*. China Railway First Survey & Design Institute Group, Ltd..

China Railway First Survey and Design Institute Group Co., Ltd, et al. (2007). *Research achievement report on permafrost subgrade engineering technology of the Qinghai - Tibet Railway*. Xi'an.

China Railway Northwest Science Research Institute. (2000). *Summary of main research achievements and application condition analysis of embankment engineering in plateau frozen soil zones*. China Railway Northwest Science Research Institute.

China Railway Northwest Science Research Institute Co., Ltd., et al. (2006). *Long-term comprehensive observation and engineering experiment research on permafrost at Fenghuoshan of the Qinghai-Tibet Railway*. China Railway Northwest Science Research Institute Co., Ltd..

Chu, S. H., Wang, P., & Zhang, T. Q. (2017). Degradation of permafrost and the preliminary evaluation of its change tendency. *Inner Mongolia Forestry Investigation and Design, 40*(2), 89–92.

Cold and Arid Regions Environmental and Engineering Research Institute, Chinese Academy of Sciences, et al. (2009). *Heat and stress analysis and numerical simulation of buried pipelines in permafrost regions*. Daqing Oilfield Engineering Co., Ltd.

CSA Group. (2010). *CAN/CSA-S6–06: Canadian Highway bridge design code*. CSA International.

Cui, G. Y. (2013). *The state investigation and the maintenance analysis of culverts in the permafrost regions*. Degree of Master for Chang'an University.

Cui, G. X., & Li, Y. (1994). The study on freezing point of wet sand under loads. *Journal of Glaciology and Geocryology, 16*(4), 320–326.

Cui, C. H., & Zhou, K. J. (1983). Experimental study on frost heaving reaction. In *Proceedings of the 2nd National Conference on permafrost* (pp. 257–264). Lanzhou:Gansu People's Publishing House.

Dai, J. B. (1983). Variation regularity of artificial upper table for railway embankments in permafrost regions in Northeast China. In *Proceedings of the 2nd National Conference on permafrost* (pp. 379–384). Lanzhou:Gansu People's Publishing House.

Dai, H. S., Sheng, Y., & Chen, J. (2006). Longitudinal roadbed cracks in Qinghai-Tibet Highway and its forming regularity. *Highway*, (1), 85–88.

Ding, Y. H. (2002). Projection for the future environmental in Western China (Volume 2). In Q. Dahe (Ed.), *Assessment of environmental evolution in Western China*. Science Press.

Ding, J. K., & He, G. S. (2000). The critical value of annual average temperature-an important factor for designing the critical height of embankments in permafrost regions of the Tibet Plateau. *Journal of Glaciology and Geocryology, 22*(4), 333–339.

Ding, J. K., et al. (2011). *Permafrost and railway engineering*. China Railway Publishing House.

Ding, S. J., et al. (2012). Loading test of the assembly foundation in frozen soil of the Qinghai-Tibet AC and DC interconnected project. *Electric Power, 45*(12), 42–47.

Dong, Y. H. (2014). *Research of construction technology about high speed railway tunnel located high altitude cold regions*. Master Degree thesis of Southwest Jiaotong University.

Dou, M. J., et al. (2002). Distribution regularities of subgrade diseases in permafrost section of Qinghai-Tibet Highway. *Journal of Glaciology and Geocryology, 26*(6), 780–784.

Dou, B. S., Bao, W. M., & Wang, C. B. (2009). Discussion on application of geosynthetic material to anti-seepage of canal. *Water Resources and Hydropower Engineering, 40*(4), 27–29.

Duan, D. M. (2010). *Research on subgrade engineering in permafrost wetlands*. China Railway Publishing House.

Editorial Committee of "Qinghai-Tibet Railway". (2016). *Qinghai-Tibet Railway: Science and technology volume · permafrost chapter*. China Railway Publishing House.

Feng, S. G., et al. (2014). Thaw settlement prevention and temperature monitoring of Mohe-Daqing crude oil pipeline in permafrost march zone. *Oil & Gas Storage and Transportation, 33*(5), 488–492.

Fu, J., Zhang, J. Z., & Liu, G. (2008). Design of metal corrugated pipe culverts on the China - Nepal Highway. *Highway*, (9), 318–320.

Fu, Z. G., et al. (2012). Thermal effect of buried oil pipelines in permafrost regions. *Journal of Engineering Thermophysics, 33*(12), 2163–2166.

Fu, J., Xie, Q. B., & Qi, H. Y. (2014). A review of culvert engineering in permafrost regions of the Qinghai-Tibet Plateau. *Highway*, (3), 28–32.

Gao, Z. H., et al. (2009). Experimental study of the dynamic strength characteristics and residual strain of ice-rich Frozen Soil. *Journal of Glaciology and Geocryology, 31*(6), 1143–1149.

Goering, D. J., & Kumar, P. (1996). Winter-time convection in open-graded embankments. *Cold Regions Science and Technology, 24*, 57–74.

Gong, X. N. (2008). *Foundation treatment manual* (3rd ed.). China Architecture & Building Press.

Gui, H. L., et al. (2009). Analysis of climate change trend in Mohe County in the past 50 years. *Modern Agriculture*, (6), 24–26.

Guo, P. F. (1983). Permafrost in Qilian Shan Area. In *Proceedings of the 2nd National Conference on permafrost* (pp. 30–35). Lanzhou:Gansu People's Publishing House.

Guo, H. X., Yuan, S. C., & Zhang, L. X. (2009). Fundamental energy condition of the application of low-temperature heat pipes in Qinghai-Tibet Railway. *Journal of Southeast University (Natural Science Edition), 39*(5), 967–972.

Han, H., Hou, Y., & Cai, L. (2001). Analysis of roadbed bank stability and determination of road surface thickness for preventing frostbite in higher cold region. *Journal of Inner Mongolia University of Technology, 20*(4), 304–309.

Hao, H. N. (2012). *Study on cathodic protection potential of underground pipelines*. China University of Petroleum Doctoral Dissertation.

Hao, Z. C., et al. (2013). Estimation method for maximum frozen depth of seasonal frozen soil in source region of the Yellow River. *Water Resources and Power, 31*(5), 73–76.

He, J. (1983). Laying of water supply pipelines and thermal calculation in permafrost regions. *Railway Standard Design*, 22–28.

He, G. S., Ding, J. K., & Li, Y. Q. (2006). Thermal transmission characteristics and application of dumped gravel layers. In *Long-term comprehensive observation and experimental research on permafrost at Fenghuoshan of the Qinghai-Tibet railway, attachment III* (pp. 546–552). China Railway Northwest Research Institute, Ltd.

Heilongjiang Transportation Science Research Institute. (2006). *Technical research on bridge and culvert engineering in permafrost regions*. Heilongjiang Transportation Science Research Institute.

Hu, T. F., & Du, S. T. (2015). Analysis of defect characteristics and protective measures for cutting slopes in permafrost regions along Qinghai-Tibet Railway. *Railway Standard Design, 59*(11), 26–31.

Hu, M. J., et al. (2004). An experimental study on cooling effect of the perforated ventilation pipes on Qinghai-Tibet Railway roadbed. *Chinese Journal of Rock Mechanics and Engineering, 23*(24), 4195–4199.

Hu, Z. Y., et al. (2012). Type-selection optimization of cone cylindrical foundation in permafrost areas at high altitudes. *Electric Power Construction, 33*(6), 34–36.

Huang, X. M. (1983). Determination of critical height of railway embankments in permafrost regions on Qinghai-Tibet Plateau. In *Proceedings of the 2nd National Conference on permafrost* (pp. 391–397). Lanzhou:Gansu People's Publishing House.

Huang, X. M. (2001). Design principles of railway subgrades for plateau frost soil zone and its application. *China Railway Science, 22*(1), 23–31.

Huang, X. M. (2006). Empirical formula for determining the thickness of insulation layer in cuttings in permafrost regions of the Qinghai-Tibet Plateau. In *Long-term comprehensive observation and experimental research on permafrost at Fenghuoshan of the Qinghai-Tibet Railway, attachment III* (pp. 145–165). China Railway Northwest Research Institute Co., Ltd..

Huang, X. M., & Shu, D. D. (1983). Design and construction of cuttings in massive ground ice areas. In *Proceedings of the 2nd Chinese National Conference on Geocryology* (pp. 385–390). Gansu People's Publishing House.

Huang, M. K., Yu, C. L., & Wang, R. (2004). Analysis to the application of ventilating embankment in permafrost regions engineering. *Soils Engineering and Foundations, 18*(6), 38–40.

Huang, Y. S., et al. (2016). Ground temperature and deformation monitoring of tower foundations for Qinghai-Tibet DC Transmission Engineering in permafrost regions. *Electric Power, 49*(2), 84–89.

Industry Recommended Standards of the People's Republic of China. (2004). *General code for design of highway bridges and culverts (JTG D60-2004)*. People's Communications Press.

Industry Recommended Standards of the People's Republic of China. (2011). *Code for design of soil and foundation of building in frozen soil region (JGJ 118-2011)*. China Architecture & Building Press.

Institute of Metals & Chemistry, China Academy of Railway Science, et al. (1978). Study on the application of surfactants in chemical anti-frost heave of railway subgrade. In *Railway science and technology: Track engineering* (Vol. 4). Railway Publishing House.

Ji, C. (2015). *Temperature field analysis of bored piles in island permafrost regions*. Degree of Master for Northeast Forestry University.

Jia, Q. (2000). *Experimental study on normal frost heave force of slab foundation of hydraulic structures in seasonal frozen soil regions*. Master Degree thesis of Dalian University of Technology.

Jia, X. Y., Li, W. J., & Zhu, Y. Q. (2003). Analysis on hydration influence of grouting pile concrete in permafrost regions. *Journal of Shijiazhuang Railway Institute, 16*(4), 88–90.

Jiang, H. J., & Cheng, E. Y. (1989). Calculation of minimum burial depth of building foundations considering soil frost heave. In *Proceedings of the 3rd National Conference on permafrost* (pp. 266–271). Science Press.

Jiang, H. P., & Liu, Z. R. (2006). ±500kV direct current transmission lines ground and foundation design in frozen earth area. *Inner Mongolia Electric Power Technology, 24*(4), 1–4.

Jiang, F. Q., & Yang, Y. P. (2008). Research on the heat transfer law of ventilation pipeline embankment in permafrost regions of the Qinghai-Tibet Plateau. *Journal of Railway Engineering Society*, (11), 27–48.

Jin, D. W. (2004). *Study on slope stability in permafrost regions of the Qinghai-Tibet Plateau*. Chang'an University Doctoral Dissertation.

Jin, H. J., et al. (1998). Study on CH_4 fluxes from alpine wetlands at Huashixia Permafrost Station, Tibetan Plateau. *Journal of Glaciology and Geocryology, 20*(2), 172–174.

Jin, H. J., et al. (2006). Degradation of permafrost in the Da and Xiao Hingan Mountains, Northeast China, and Preliminary assessment of its trend. *Journal of Glaciology and Geocryology, 28*(4), 467–476.

Khrustalev, L. N. (2005). *Fundamentals of geotechnics in the cryolithozone*. Moscow State University Press.

Khrustalev, L. N., & Pustovoit, G. P. (1988). *Probabilistic-statistical analysis of foundation calculations in permafrost regions*. Nauka, Siberian Branch. (see: Permafrost translation collection on transportation construction and environmental engineering in cold regions, Vol. 14, by the Cold and Arid Regions Environmental and Engineering Research Institute, Chinese Academy of Sciences).

Kong, X. Y. (1999). *Advanced mechanics of fluids in porous media*. University of Science and Technology of China Press.

Kong, J. J. (2012). *The anti-freeze expansion of the insulation board and the dynamic response analysis in seasonal frozen soil subgrade*. Central South University.

Kudryavtsev, B. A.. (1992). *Principles of geocryology prediction in engineering geological research*. Translated by Guo Dongxin, et al. Lanzhou University Press.

Lai, J. X. (2008). *Study on temperature field feature and StructureCharacteristics of high altitude localities road Tunnelwith complex surrounding rock*. Chang'an University.

Lai, Y. M., et al. (2006). Laboratory study of particle size for optimal cooling effect of closed crushed-rock layers. *Journal of Glaciology and Geocryology, 28*(5), 755–759.

Lai, Y. M., et al. (2009). *Theory and application of cold region engineering*. Science Press.

Li, Z. L. (2006). *Study on design and construction of corrugated steel pipe culverts*. Degree of Doctor of Philosophy for Chang'an University.

Li, Y. Q. (2008). *Study on state of subgrade engineering in permafrost regions during operation period of Qinghai-Tibet Railway*. Degree of Doctor of Philosophy for Lanzhou University.

Li, Y. (2014). *Study on distribution of temperature field and cold insulation technology on tunnel in high altitude and cold zones*. Degree of Master for Chongqing Jiaotong University.

Li, G. L., et al. (1980). Investigation of pile foundations in permafrost region. *Journal of the China Railway Society, 2*(1), 83–94.

Li, Y., Cui, G. X., & Lu, H. L. (1996). A study on freezing point of wet clay under loading. *Journal of Glaciology and Geocryology, 18*(1), 43–46.

Li, D. Q., et al. (1999). Modeling analysis on the heat stability of embankment for the continuous permafrost district with rich moisture in Hua Shixia Valley. *Chinese Journal of Geotechnical Engineering, 21*(1), 17–20.

Li, N., Quan, X. J., & Li, G. Y. (2005a). Approach on temperature field analysis and design principle of bed for ventilating pipeline in frozen soil. *China Civil Engineering Journal, 38*(2), 81–86.

Li, D. L., et al. (2005b). Analyses on changes of surface temperature over the Qinghai-Tibet Plateau. *Plateau Meteorology, 24*(3), 291–298.

Li, N., Wei, Q. Z., & Ge, J. J. (2006). Structure type and work state study on heat pipe subgrade of Qinghai-Tibet Railway. *Journal of Beijing Jiaotong University, 30*(4), 22–25.

Li, W. L., Wu, Y. G., & Qiu, L. J. (2011). Study on freeze-thaw test of airport pavement concrete in cold area. *Subgrade Engineering*, (2), 123–125.

Li, P., et al. (2014). Status Quo and trend of soil stabilizers development. *Subgrade Engineering*, (3), 1–8.

Li, G. Y., et al. (2015). Frost hazards and mitigative measures following operation of Mohe-Daqing line of China-Russia crude oil pipeline. *Rock Mechanics, 36*(10), 2963–2973.

Liu, T. L. (1983). A summarization of formulas of calculating frozen or melted depth abroad. *Journal of Glaciology and Geocryology, 5*(2), 85–95.

Liu, Y. (2005). *Research on the load bearing property of bored single pile in permafrost regions*. Degree of Master for Southeast University.

Liu, H. J. (2007). *Study on consolidation settlement and stability of wetland soft soil foundation in cold region*. Institute of Engineering Mechanics Doctor Dissertation, China Earthquake Administration.

Liu, Y. P. (2009). *The research on frost heaving characteristics of roadbed at XiangPi Mountain section in 109th national highway*. Beijing Jiaotong University.

Liu, R. Q. (2014). *Research on setting technology for winter protection drainage hole of highway tunnels in high altitude and cold regions*. Degree of Master for Chongqing Jiaotong University.

Liu, Y. (2015). *Establishment and experimental study of frozen soil resistivity*. Northeast Forestry University.

Liu, Y. Z., et al. (2002). Deformation of highway roadbed in permafrost regions of the Tibetan Plateau. *Journal of Glaciology and Geocryology, 24*(1), 10–15.

Liu, J. K., et al. (2004). Observation and analysis of a new type embankment-bridge transition section in permafrost regions. *Journal of Glaciology and Geocryology, 26*(6), 800–805.

Liu, G., et al. (2008a). Changes of permafrost under the engineering conditions of Qinghai-Tibet Highway. *Journal of China & Foreign Highway, 28*(1), 25–27.

Liu, J. K., et al. (2008b). Study on transition structure from a cut-section to a fill-section of roadbed on permafrost regions. *Journal of Glaciology and Geocryology, 30*(1), 153–156.

Liu, H. J., et al. (2008c). The route, site and foundation's select in frozen ground region of Qinghai-Tibet direct current transmission line. *Electric Power Survey & Design*, (2), 12–16.

Liu, H. J., et al. (2009). Study on permafrost engineering problems and engineering countermeasure of transmission line. *Geotechnical Investigation & Surveying*, (4), 32–36.

Liu, W. B., et al. (2015). Techniques of airport runway construction in permafrost regions: A review. *Journal of Glaciology and Geocryology, 37*(6), 1599–1610.

Luo, Y. B. (2010). *Study on frost damage grades and its prevention and control technologies in cold region tunnel*. Beijing Jiaotong University Doctoral Dissertation.

Luo, D. L., & Jin, H. J. (2014). Variations of air temperature and precipitation from 1953 to 2012 in the Madoi station in the sources areas of the Yellow River. *Journal of Arid Land Resources and Environment, 28*(11), 185–192.

Luo, D. L., et al. (2012). Degradation of permafrost and cold-environmental on the interior and eastern Qinghai Plateau. *Journal of Glaciology and Geocryology, 34*(3), 538–546.

Ma, S. M. (1983). Experimental study on shear strength of frozen soil. In *Professional papers on the permafrost studies of Qinghai-Xizang Plateau* (pp. 106–111). Science Press.

Ma, W. Y. (1990). Prediction and prevention of water inflow in pits (tunnels). *Journal of Railway Engineering Society*, (4), 90–95.

Ma, H. (2006). *Study on the cooling mechanism and adaptivity of crushed-rock berm structure in stabilizing the Qinghai-Tibet Railway on permafrost*. China Academy of Railway Sciences.

Ma, H. J. (2015). *Zonation and frost resistance parameter Design for Durability of airport pavement concrete in cold region*. Nanjing University of Aeronautics and Astronautics.

Ma, H. T., Song, X. J., & Wang, Q. L. (2013). Foundation constructing technique of steel tower cast-in-situ concrete in permafrost regions. *Qinghai Electric Power, 32*(1), 25–29.

Miao, J. (2002). Discussion on causes and countermeasures of airport runway diseases in Northeast China. In *Proceedings of the 4th international conference on road and airport pavement technology* (pp. 588–592). China Communications Press.

Ming, Y. (2011). Design content of highway steel corrugated pipe culverts. *Heilongjiang Transportation Science and Technology*, (1), 101–102.

Mu, W. K., He, B. Q., & Dong, D. H. (2006). Application of heat pipe technology in Yichun island permafrost subgrade. *Heilongjiang Transportation Science and Technology*, (11), 7–8.

Mu, Y. H., et al. (2010). Comparative analysis of cooling effect of crushed rock embankment along the Qinghai-Tibet Railway. *Rock and Soil Mechanics, 31*(Supplement 1), 284–292.

Na, W. J., & Yuan, A. l. (2007). Prevention and anti-frozen technical mehtods of the channels in Alpine Region. *Geotechnical Engineering Technology, 21*(4), 204–206.

Na, W. j., Zhao, H. X., & Zhao, B. Y. (1997). The valuation for designing of horizontal frost heaving force in Retaining Wall in seasonally frozen soil. *Journal of Heilongjiang Hydraulic Engineering College*, (3), 25–28.

National Standards of the People's Republic of China. (2014). *Code for engineering geological investigation of frozen ground (GB 50324-2014)*. China Planning Press.

Niu, Y. F., Li, C., & Niu, Y. j. (2010). Analysis on water quality change trend in Greater Khingan Range. *Heilongjiang Science and Technology of Water Conservancy, 38*(5), 113–114.

Niu, F. J., et al. (2011). Study of the influencing factors of roadbed settlement in embankment-bridge transition section along Qinghai-Tibet Railway. *Rock and Soil Mechanics, 32*(Suppl.2), 372–377.

Northwest Electric Power Design Institute Co., Ltd. of China Power Engineering Consulting Group. (2008). *Research on foundation selection and design in high - altitude permafrost regions*. Northwest Electric Power Design Institute Co., Ltd.

Orlando B. A., & Branko, L. (2011). *Frozen ground engineering*. Translated by Yang, R. Q., & Li, Y. Z., China Architecture & Building Press.

Orlov, B. Q., et al. (1980). *Frost heave of soil and its effect on buildings*. Translated by Li Yongsheng. Changchun: Jilin Provincial Water Conservancy Science and Technology Information Center.

Pei, Z. Q., & Lu, J. Q. (1982). Research on physical and chemical methods for preventing soil frost heave at home and abroad. *Journal of Glaciology and Geocryology, 4*(1), 71–78.

Permafrost Subgrade Research Group of Qinghai-Tibet Highway. (1988). *Roadbed engineering in permafrost region*. Lanzhou University Press.

Qian, J., et al. (2009). Permafrost engineering geological characteristics and discussion of route selection in Qinghai-Tibet Plateau. *Journal of Engineering Geology, 17*(4), 508–515.

Qinghai Provincial Local Railway Administration, Lanzhou University. (2010). *Research report on key Technologies for Application of low-temperature Thermosyphons in Qaidam local railway of Qinghai Province*. Qinghai Provincial Local Railway Administration.

Qinghai-Tibet Highway Research Group. (1983). Determining of the height of subgrade with bitumen pavement in permafrost regions on Qinghai-Tibet Highway. In *Proceedings of the 2nd National Conference on permafrost* (pp. 364–371). Lanzhou:Gansu People's Publishing House.

Qiqihar Railway Research Institute of Harbin Railway Bureau, et al. (2007). *Research report on test of subgrade disease treatment in existing permafrost regions of northeast railway*. Qiqihar Railway Research Institute of Harbin Railway Bureau.

Qiu, W. G., & Sun, B. (2010). Model test study of frost heaving pressures in tunnels excavated in fractured rock mass in cold regions. *Journal of Glaciology and Geocryology, 32*(3), 557–560.

Qu, X. M., & Zhang, H. Y. (2006). *Introduction to engineering permafrost*. Harbin Engineering University Press.

Qu, X. M., & Zhang, B. (2008). *Anti-frost technology for hydraulic structures in seasonal frozen soil regions*. China Water & Power Press.

. (1983). Distribution regulaities of high ice-content permafrost along the Qinghai-Tibet Highway. In *Research Group of Qinghai-Tibet Highway, Proceedings of the 2nd National Conference on permafrost* (pp. 43–51). Lanzhou:Gansu People's Publishing House.

Shi, B. H., et al. (2014). Prevention of frost heave damage to airport pavement in cold region. *Road Machinery & Construction Mechanization*, 62–65.

State Construction Committee of the Soviet Union. (1990). *Building regulations and norms foundations and foundations on permafrost* (СНиП 2.02.04–88) [S]. Translated by Ding Jingkang, Xu Xueyan. Moscow.

Su, J., Wang, M. Y., & Ji, H. (2002). Design of Water source for Qinghai-Tibet Railway in permafrost zone. *Railway Design Standards*, (10), 30–33.

Sui, X. Z., & Zuo, L. (1983). The experimental research of dilatational counter-force. *Journal of Glaciology and Geocryology, 5*(2), 19–29.

Sun, B. (2012). Study on freezing damage and prevention classification of tunnel in cold region. *Railway Standard Design*, (4), 88–92.

Sun, J., & Zhang, X. (2013). Research on engineering geological properties and transformation measurements of salted frozen soil of the Ali Airport. *Hydrogeology and Engineering Geology, 40*(1), 93–99.

Sun, B. X., et al. (2004a). Impact of ballast grain sizes on natural convection cooling effect of embankment in permafrost regions. *Chinese Journal of Geotechnical Engineering, 26*(6), 809–814.

Sun, Z. Z., Ma, W., & Li, D. Q. (2004b). Experimental study on cooling effect of air convection embankment with crushed rock slope protection in permafrost regions. *Journal of Glaciology and Geocryology, 26*(4), 435–439.

Sun, B. X., et al. (2005). Study on thickness of Qinghai-Tibet railway fractured-rock revetment layer on the basis of convection cooling effect. *Journal of the China Railway Society, 27*(4), 96–103.

Sun, B. X., et al. (2006a). Ballast layer height of railway embankment in Qinghai-Tibet plateau permafrost region. *Engineering Mechanics, 23*(6), 127–134.

Sun, F. H., et al. (2006b). The contrast analysis on the average and extremum temperature trend in Northeast China. *Journal of the Meteorological Sciences, 26*(2), 157–163.

Sun, X. M., Hong, X. H., & Kong, D. (2006c). Analysis on static ice pressure of slop protection construction. *Journal of Heilongjiang Hydraulic Engineering College, 33*(2), 16–17. (Harbin).
Sun, G. Y., Jin, H. J., & Yu, S. P. (2008). The symbiosis models of marshes and permafrost—A case study in Daxing'an and Xiaoxing'an mountain range. *Wetland Science, 6*(4), 479–485.
Sun, J. B., Xu, C. H., & Xu, X. Y. (2011). Test research on negative friction of engineering pile in permafrost regions. *Construction Technology, 40*(344), 47–50.
Sun, J., et al. (2012). Study on engineering geological characteristics and prevention measures of salinized frozen soil at Ali Airport in Tibet. *Journal of China & Foreign Highway, 32*(5), 47–53.
Survey and Design Department of Daqing Oilfield Engineering Co., Ltd. (2008). *Investigation report on frost-related phenomena (ice cones, frost heave mounds) in permafrost of the Mohe-Daqing section of China - Russia crude oil pipeline project*. Survey and Design Department of Daqing Oilfield Engineering, Ltd.
Tan, Q. H., et al. (2012a). Application of hot rod Technology in Qinghai-Tibet DC project. *North China Electric Power*, (2), 16–19.
Tan, D. J., et al. (2012b). Displacement monitoring of Mohe-Daqing Oil Pipeline in the permafrost area. *Oil & Gas Storage and Transportation, 31*(10), 737–739.
Tan, X. J., et al. (2013). Study of temperature field of tunnel surrounding rock in cold regions considering effect of ventilation and length design of insulation material. *Chinese Journal of Rock Mechanics and Engineering, 32*(7), 1400–1409.
Tan, L. W., et al. (2016). Study on groundwater characteristics and development in permafrost region of Tuotuo River. *Yellow River, 38*(5), 62–67.
Tang, S. C. (1992). Application of reinforcing foundations by intensive tamping method in Daqing District. *Oilfield Surface Engineering, 11*(2), 58–60.
Tang, G. L., & Ren, G. Y. (2005). Reanalysis of surface air temperature change of the last 100 years over China. *Climatic and Environmental Research, 10*(4), 791–798.
The Third Railway Survey and Design Institute. (1994). *Frozen Soil Engineering*. China Railway Publishing House.
Tong, C. J., & Guan, F. N. (1985). *Frost heave of soil and prevention of frost damage to buildings*. Water Resources and Electric Power Press.
Tong, C. J., Wang, G. S., & Wu, Q. B. (1996). Current situation and tasks of environmental engineering geology research in cold regions of China. In *Proceedings of the 5th National Conference on Glaciology and Geocryology (Part 2)* (pp. 863–876). Gansu Culture Publishing House.
Tsytovich, N. A. (1985). *Mechanics of frozen soils*. Translated by Zhang, C. Q., et al. Science Press.
Wang, Z. Q. (1980). Influence of grain-size constituent on frost heave in fine sand. *Journal of Glaciology and Geocryology, 2*(3), 24–28.
Wang, X. Y. (1982). Frost heave and its distribution in different layers influenced by shallow groundwater. *Journal of Glaciology and Geocryology, 4*(2), 55–62.
Wang, Z. Q. (1986). Frost heaving classifications of the coarse grained soil. *Journal of Glaciology and Geocryology, 8*(3), 195–200.
Wang, Y. Y. (2003). Construction technique of soil removal and thermal insulation on frozen soil cutting at Kunlun Mountain pass in Qinghai-Xizang Railway. *Subgrade Engineering*, (3), 52–57.
Wang, H. N. (2004). *Research on the subgrade temperature field in permafrost regions on the Qinghai-Tibet Plateau*. Degree of Master for Chang'an University.
Wang, Z. Y. (2006). Research on the application of geosynthetic materials in canal seepage control in frost heave regions. *China Rural Water and Hydropower*, (3), 86–88.
Wang, F. (2011). *Research on treatment technology for subgrade settlement in segregated permafrost region of Boketu-Yakeshi expressway*. Degree of Doctor for Chang'an University.
Wang, X. (2013a). Experimental study of bearing characteristics of large-diameter cast-in-place bored pile under non-refreezing condition in low-temperature permafrost ground. *Chinese Journal of Rock Mechanics and Engineering, 32*(9), 1807–1812.
Wang, Y. L. (2013b). *Study on the stability of the slope and excavation technology of the permafrost tunnel portal in high altitude and cold regions*. Degree of Master for Chongqing Jiaotong University.

Wang, H. C. (2014). *Sensitivity analysis of frost-heave for tunnels in frozen zone and study on the technique of preventing frost damage with insulation*. Degree of Master for Chongqing Jiaotong University.

Wang, D. F. (2016). Design and calculation of cross-sectional dimensions of U-shaped channels. *Construction of water conservancy and hydropower*, (7), 151–152.

Wang, J. Y., & Hu, Y. F. (2004). A discussion on frost-heaving force on tunnel lining. *Journal of Glaciology and Geocryology, 26*(1), 112–119.

Wang, T., & Lu, L. G. (2007). The present situations of branch canal structure and lining damage causes in Ningxia. *China Rural Water and Hydropower*, (8), 34–38.

Wang, X., et al. (2004a). An experimental study on refreezing characteristics of large-diameter bored pile in different permafrost areas of the Qinghai-Tibet Plateau. *Chinese Journal of Rock Mechanics and Engineering, 23*(24), 4206–4211.

Wang, J. L., Huang, B., & Niu, F. J. (2004b). Experimental study of ventilation duct embankment. *Qinghai Transportation Science and Technology*, (2), 37–38.

Wang, S. J., Huo, Y., & Zhou, W. J. (2004c). Subgrade failure of Qinghai-Tebit Highway in permafrost area. *Highway*, (5), 22–26.

Wang, X., et al. (2005a). Study on refreezing processes of large diameter bored pile in permafrost area of the Qinghai-Tibet Plateau. *Journal of the China Railway Society, 27*(2), 102–107.

Wang, X., et al. (2005b). Experimental study on bearing features of bored pile under non-refreezing condition in permafrost region. *Chinese Journal of Geotechnical Engineering, 27*(1), 81–84.

Wang, A. G., Ma, W., & Wu, Z. J. (2005c). Study on influence of sand-and-gravel layer thickness upblock-stone railway embankment on cooling effect of frozen-soil foundation. *Chinese Journal of Rock Mechanics and Engineering, 24*(13), 2334–2341.

Wang, T., et al. (2006a). *1:4,000,000 map of the glaciers, frozen ground and deserts in China*. China Cartographic Publishing House.

Wang, A. G., Ma, W., & Wang, D. Y. (2006b). Comparison study on cooling effect of frozen-soil foundation on block-stone railway embankment with different thickness of block-stone. *Chinese Journal of Rock Mechanics and Engineering, 25*(Supplement 1), 3283–3288.

Wang, X. H., et al. (2006c). Identification of frozen and thawing zones for tunnel leakage treatment in Kunlun Mountain during curtain grouting. *Journal of Engineering Geology, 14*(5), 588–591.

Wang, D., et al. (2009). Research progress of the structure form for seepage prevention with resistance to frost heaving. *Journal of Water Resources and Water Engineering, 20*(1), 62–64.

Wang, A. G., Ma, W., & Wu, Z. J. (2011). Key factors influencing cooling effect of block-stone embankment on frozen soil foundation. *Chinese Journal of Geotechnical Engineering, 33*(Supplement 1), 211–215.

Wen, Z., et al. (2005). Evaluation of application of the insulation to embankment in Qinghai-Tibetan Railway. *Journal of Glaciology and Geocryology, 27*(5), 694–700.

Wu, Z. W. (1979). Engineering classification of permafrost. *Journal of Glaciology and Geocryology, 2*(2), 52–60.

Wu, Z. W. (1982). Classification of frozen soils in engineering constructions. *Journal of Glaciology and Geocryology, 4*(4), 43–46.

Wu, S. H. (2003). Design and construction of bridges and culverts of the Qinghai-Tibet Railway in permafrost regions. *Journal of Glaciology and Geocryology, 25*(Supplement 1), 59–68.

Wu, J. (2004). *Study on principle of tunnel freeze injure and calculation method of frost force*. Southwest Jiaotong University.

Wu, X. M. (2005). *Research on construction technique of bridges foundation of the Qinghai-Tibet railway in high-temperature & instable area permafrost regions*. Southwest Jiaotong University.

Wu, Z. W., & Liu, Y. Z. (2005). *Frozen soil foundation and engineering construction*. Ocean Press.

Wu, Z. W., & Ma, W. (1994). *Strength and creep of frozen soil*. Lanzhou University Press.

Wu, Z. W., & Zhang, J. Y. (1983). Preliminary experiment on thawing pressure of frozen soil. In *Professional papers on the permafrost studies of Qinghai-Xizang Plateau* (pp. 131–133). Science Press.

Wu, Q. B., Li, X., & Li, W. J. (2001). The response model of permafrost along the Qinghai-Tibetan Highway under climate change. *Journal of Glaciology and Geocryology, 23*(1), 1–6.

Wu, Q. B., Zhu, Y. L., & Liu, Y. Z. (2002). Assessment model of permafrost thermal stability under engineering activity. *Journal of Glaciology and Geocryology, 24*(2), 129–133.

Wu, Z. W., et al. (2003a). *Tunnel engineering in cold regions*. Ocean Press.

Wu, Q. B., Shi, Y., & Liu, Y. Z. (2003b). Interaction study of permafrost and highway along the Qinghai-Tibet Highway. *Science in China (Series D), 46*(2), 97–105.

Wu, G. Z., et al. (2003c). *Heat transfer calculation of buried pipelines*. Harbin Institute of Technology Press.

Wu, Y. P., et al. (2004). Influences of refreezing process of ground on bearing capacity of single pile and bridge construction in permafrost. *Chinese Journal of Rock Mechanics and Engineering, 23*(24), 4229–4233.

Wu, Q. B., Lu, Z. J., & Liu, Y. Z. (2005a). Permafrost monitoring and its recent changes in Qinghai-Tibet Plateau. *Advance in Climate Change Research, 1*(1), 26–28.

Wu, Q. B., et al. (2005b). Monitoring and analysis of cooling effect of block-stone embankment for Qinghai-Tibet Railway. *Chinese Journal of Geotechnical Engineering, 27*(12), 1386–1390.

Wu, J. M., Wang, S. J., & Zhang, J. J. (2005c). *The Highway engineering on permafrost regions*. China Communications Press.

Wu, Q. B., Dong, X. F., & Jiang, G. L. (2006). Cooling effect differences of soil beneath open and closed block-stone embankments. *Chinese Journal of Rock Mechanics and Engineering, 25*(12), 2565–2571.

Wu, Q. B., Cui, W., & Liu, Y. Z. (2010). The cooling effect of U-type crushed rock embankment on permafrost. *Journal of Glaciology and Geocryology, 32*(3), 532–537.

Wu, F. B., et al. (2011). Optimal cover depth of the directional drilling for the China-Russia Crude Oil Pipeline crossing Heilongjiang River. *Petroleum Planning & Engineering, 22*(5), 7–11.

Xie, Y. Q., & Wang, J. G. (1992). Technical measures for preventing frost damage of hydraulic structures in high altitude and cold regions. *Water Resources and Hydropower Engineering*, (3), 22–29.

Xing, Y. Z., & Hao, Y. F. (2010). Discussion on design of anti-freezing thickness for airport rigid pavement structure. *East China Highway*, (4), 52–53.

Xiong, S. M. (2012). *Force deformation characteristic analysed and design and calculated method for corrugated steel pipe culvert of highway subgrade*. Chang'an University.

Xiong, T., et al. (2009). Temperature distribution along piles in permafrost regions. *Rock and Soil Mechanics, 30*(6), 1658–1664.

Xiong, Z. W., et al. (2010). Disaster of cutting slope in permafrost region and treatment principle. *Journal of Railway Engineering Society*, (7), 6–10.

Xiu, L. (2015). *Analysis on factors causing frost heave damage to canal revetments in seasonal frozen regions*. Degree of Master for Heilongjiang University.

Xu, X. Z. (1994). The state of the art on description and classification of frozen soils and some suggestions. *Journal of Glaciology and Geocryology, 16*(3), 193–201.

Xu, L. J. (2006). *Ground temperature characteristic analysis on rubble slope embankment and gravel pavement embankment of Qinghai-Tibet railway permafrost region*. Southwest Jiaotong University.

Xu, X. T. (2013). *Study on frostance, anti-freezing and drainage technology of the Elashan Tunnel*. Chongqing Jiaotong University.

Xu, A. H. (2014). Analysis of the sensitivity of highway diseases in permafrost regions to ground temperatures and ice contents. *Journal of Glaciology and Geocryology, 36*(3), 622–625.

Xu, B. M., & Gu, Y. W. (1993). Frost damage of hydraulic structures and its prevention and control in Jilin Province. *Water Resources and Hydropower Engineering*, (12), 16–19.

Xu, B. M., & Lu, X. L. (1986). Influence of moisture condition on frost heave in clay. *Journal of Glaciology and Geocryology, 8*(3), 217–222.

Xu, B. M., & Wang, Z. R. (1995). Frozen soil engineering problems in the hydraulic construction of the Heilongjiang and Erguna Rivers. *Water Resources & Hydropower of Northeast China*, (1), 21–25.

Xu, Z. J., Xu, Q. Q., & Shen, M. J. (1999). *The design and treatment of foundation in large storage tank*. China Petrochemical Press.

Xu, X. Z., Wang, J. C., & Zhang, L. X. (2001). *Physics of frozen soil*. Science Press.

Xu, B. C., et al. (2009). Research on calculation method of buried pipe foundation melting laps in permafrost of Sino-Russian oil pipeline. In *Proceedings of the eighth international symposium on permafrost engineering* (pp. 208–214).

Xu, X. M., Wu, J. B., & Zhang, Z. Q. (2017). Responses of active layer thickness on the Qinghai-Tibet Plateau to climate change. *Journal of Glaciology and Geocryology, 39*(1), 1–8.

Xue, Z. (2015). Design of water supply and drainage systems in permafrost regions. *Railway Standard Design, 59*(4), 152–155.

Yan, J. F., Li, P., & Dong, X. L. (2012). Principal problems and solutions of the foundation engineering in the high-attitude permafrost region. *Electric Power, 45*(12), 34–41.

Yang, R. H. (2010). *Study of analysis and evaluation to the stability of slope embankment in permafrost regions of Qinghai-Tibet in operating period*. China Academy of Railway Sciences.

Yang, L. J., et al. (2011). Cooling capability of duct-ventilated highway embankment enhanced by self-windward vent. *Journal of Civil, Architectural and Environmental Engineering, 33*(1), 87–92.

Ye, E. H. (2010). *Research on technique of preventing frost damage in tunnel at cold area*. Chang'an University.

Yershov E. D. (2016). *Engineering geocryology*. Translated by Zhang, C. Q. Lanzhou University Press.

Yu, W. B., et al. (2003). Experimental study on the cooling effect of block stone and crushed stone. *Journal of Glaciology and Geocryology, 25*(6), 638–643.

Yu, B. H., Cheng, G. G., & Niu, F. J. (2004). Effect analysis of auto-temperature-controlled ventilation roadbed. *Chinese Journal of Rock Mechanics and Engineering, 23*(24), 4221–4228.

Yu, B. H., et al. (2013). Analysis of heat transfer characteristics of frozen soil foundation in Qinghai-Tibet DC transmission line project. *Chinese Journal of Rock Mechanics and Engineering, 32*(Supplement 2), 4025–4031.

Zang, E. M., & Wu, Z. W. (1999). *The degradation of permafrost and Highway engineering*. Lanzhou University Press.

Zhai, L. (2011). *Study on mechanism and prevention of frost damage of lakeshore structures in seasonal frozen region*. Degree of Master for Jilin University.

Zhang, J. Z. (2000). Discussion on the prevention of culvert engineering diseases in permafrost regions of the Qinghai-Tibet Highway. *Qinghai Transportation Science and Technology*, (2), 29–31.

Zhang, J. M. (2004). *Study on roadbed stability in permafrost regions on Qinghai-Tibetan Plateau and classification of permafrost in Highway engineering*. Institute of Cold and Arid Regions Environment and Engineering, Chinese Academy of Sciences.

Zhang, J. Z. (2006). Evolution of subgrade design principles and design heights in permafrost regions of the Qinghai - Tibet Highway. *Highway*, (10), 54–58.

Zhang, Y. L. (2014). *Numerical simulation study on effects of the buried oil pipeline on Permafrost temperature*. Beijing Jiaotong University.

Zhang, Y., & Lu, X. L. (2011). Research on controlled blasting excavation technology for power transmission line in permafrost soils. *Engineering Journal of Wuhan University, 44*(Supplement), 222–225.

Zhang, Z. D., & Wang, L. (2004). Discussion on the design of tunnels in high elevation and bitter cold region. *Modern Tunnelling Technology, 41*(3), 1–6.

Zhang, Z. Q., & Wu, Q. B. (2012). Predicting changes of active layer thickness on the Qinghai-Tibet Plateau as climate warming. *Journal of Glaciology and Geocryology, 34*(3), 505–511.

Zhang, Q. F., & Zhang, W. B. (2012). Experimental research on foundation treatment in the area of Ngari Kunsha Airport in Tibet. *Gansu Water Resources and Hydropower Technology, 48*(12), 24–27.

Zhang, X. P., Li, H. J., & Li, G. W. (1993). Laboratory simulation for freezing strength between ice and cement dike-dam. *Journal of Dalian University of Technology, 33*(4), 385–389.

Zhang, L. X., et al. (1998). Experimental study of the relationship between the unfrozen water content of frozen soil and pressure. *Journal of Glaciology and Geocryology, 20*(2), 124–127.

Zhang, K., Li, D. Q., & Tong, G. Q. (2011). Numerical analysis of the cooling effect of an embankment with ventilated ducts and closed block stones. *Journal of China University of Mining and Technology, 40*(1), 35–42.

Zhang, S. J., et al. (2014). Monitoring of frost heaving and thaw settlement of Mohe-Daqing Crude Oil Pipeline. *Oil & Gas Storage and Transportation, 33*(5), 488–492.

Zhao, W. J. (2007). *Stability study and engineering treatment analysis on embankment upon the slope at Ando section in the permafrost of Qinghai-Tibet Railway*. Beijing Jiaotong University.

Zhao, L. P. (2009). *Study of application of XPS Board in frozen subgrade works*. Degree of Master for Chang'an University.

Zhao, X. L., & Ba, Y. J. (2002). Damage mechanism of runway concrete in cold region. *Journal of Harbin University of Civil Engineering and Architecture, 35*(5), 81–83.

Zhao, F. N., & Zhao, Y. J. (2014). Application of dynamic compaction replacement method to treat special subgrade in island-shaped frozen soil. *Railway Engineering*, (10), 85–87.

Zheng, B., et al. (2007). Experimental study on transitional zone between embankment and cut over permafrost along Qinghai—Tibet Railway. *Chinese Railway Science, 28*(4), 12–18.

Zheng, B., et al. (2009). Study on compression deformation of warm and ice-enriched frozen soil. *Chinese Journal of Rock Mechanics and Engineering, 28*(Supplement 1), 3063–3069.

Zhou, J. W., et al. (2000). *Ceocryology in China*. Science Press.

Zhou, N. F., et al. (2005). Analyese on regional characteristics of temperature changes over Qinghai-Xizang plateau in recent 50 years. *Plateau Meteorology, 24*(3), 344–349.

Zhu, Y. L., & Zhang, J. M. (1982). Elastic and compressive deformation of frozen soils. *Journal of Glaciology and Geocryology, 4*(3), 29–38.

Zhu, Y. L., Zhang, J. M., & Wu, Z. W. (1983). Calculation of thaw settlement of frozen soil foundations. In *Professional papers on the permafrost studies of Qinghai-Xizang Plateau* (pp. 134–138). Science Press.

Zhu, Q., Shi, S. N., & Wu, F. X. (1988). Subsoil replacement using sand-gravel for preventing frost heave damage of canal lining. *Journal of Glaciology and Geocryology, 10*(4), 400–408.

Zhu, G. Y., et al. (2008). Study on deformation characters of frozen soil under long-term dynamic loading. *Low Temperature Architecture Technology*, (2), 112–114.

Index

Q. Wu, C. Tong, *Frozen Soil Engineering in Cold Regions*,
https://doi.org/10.1007/978-981-95-2527-0

D

E

F

U

V

W

Z

FSC
www.fsc.org
MIX
Papier | Fördert
gute Waldnutzung
FSC® C083411

Kolibri 360 GmbH
Ferdinand-Jühlke-Straße 7
99095 Erfurt, Deutschland
produktsicherheit@kolibri360.de